ENGINEERING DRAWING AND GRAPHIC TECHNOLOGY

ENGINEERING DRAWING AND GRAPHIC TECHNOLOGY

Thirteenth Edition

THOMAS E. FRENCH
Late Professor of Engineering Drawing, The Ohio State University

CHARLES J. VIERCK
Late Professor of Engineering Graphics, University of Florida, Gainesville

ROBERT J. FOSTER
Associate Professor of Engineering Graphics, The Pennsylvania State University

McGRAW-HILL BOOK COMPANY

New York   St. Louis   San Francisco   Auckland   Bogotá   Hamburg
Johannesburg   London   Madrid   Mexico   Montreal   New Delhi   Panama
Paris   São Paulo   Singapore   Sydney   Tokyo   Toronto

ENGINEERING DRAWING AND GRAPHIC TECHNOLOGY

Copyright © 1986, 1978, 1972 by McGraw-Hill, Inc. All rights reserved. A revision of *A Manual of Engineering Drawing for Students and Draftsmen*, copyright © 1966, 1960 by McGraw-Hill, Inc. All rights reserved. Copyright 1953, 1947, 1941, 1935, 1929, 1924, 1918, 1911 by McGraw-Hill, Inc. All rights reserved. Copyright renewed 1952, 1957, 1963, 1969 by Janet French Houston. Copyright renewed 1975 by Charles J. Vierck. Copyright renewed 1981 by Esther F. Vierck. Printed in the United States of America. Except as permitted under the United States Copyright Act of 1976, no part of this publication may be reproduced or distributed in any form or by any means, or stored in a data base or retrieval system, without the prior written permission of the publisher.

1234567890   HALHAL   89876

ISBN 0-07-022161-8

This book was set in Helvetica by York Graphic Services, Inc. The editors were Kiran Verma and J. W. Maisel; the designer was Elliot Epstein; the production supervisor was Phil Galea. New drawings were done by Felix Cooper. Cover photograph was taken by Peter Epstein.

Halliday Lithograph Corporation was printer and binder.

Library of Congress Cataloging-in-Publication Data

French, Thomas Ewing, 1871–1944.
  Engineering drawing and graphic technology.

  Includes index.
  1. Engineering graphics. I. Vierck, Charles J.
II. Foster, Robert J. (Robert Jay), 1934–
III. Title.
T353.F85   1986   604.2′4   85-16605
ISBN 0-07-022161-8

ABOUT THE AUTHORS

THOMAS E. FRENCH

Thomas E. French worked his way through college at Ohio State University as a patent draftsman and part-time assistant in the Department of Engineering Drawing, graduating in 1895. In 1906, he was appointed Head of the department, and subsequently earned for the university a national reputation in engineering and teaching methods.

In 1943, Professor French was awarded the Lamme Medal for Achievement in Engineering Education, presented by the Society for the Promotion of Engineering Education:

> To Thomas Ewing French for his genius in teaching the graphical language of engineering; for his fertility in producing successful textbooks of engineering drawing based on sound principles of teaching and upon a broad knowledge of the requirements of practical construction.

CHARLES J. VIERCK

Following undergraduate work at the University of Iowa, Charles Vierck joined the faculty of Ohio State University in 1929, where he studied under Thomas French. The two men collaborated closely in the years to come, and it was Vierck whom French chose as his successor from a group of Ohio State colleagues who had contributed to the text known in the engineering discipline as "French's Engineering Drawing."

Following Thomas French's death in 1945, Charles Vierck assumed full responsibility for revision through six editions, beginning with the Seventh Edition published in 1947. In 1954, after twenty five years of distinguished service at Ohio State, Professor Vierck retired, serving as Visiting Professor at the University of Florida until his death in 1980.

ROBERT J. FOSTER

Robert Foster is Associate Professor of Engineering Graphics at the University Park Campus of The Pennsylvania State University. He has taught engineering at the college level for twenty five years and is currently Program Chairman for Engineering Graphics. He has also worked in industry as a design engineer with the Bendix Corporation, the United States Navy, and the Erie Technological Corporation.

Professor Foster is the author of *Field Sketching* and *Graphic Science and Design* (with French and Vierck). He has been active in the American Society for Engineering Education (ASEE) and served as Chairman of its Engineering Design Graphics Division in the 1985–86 term.

CONTENTS

FOREWORD

It gives me special pleasure to introduce the Thirteenth Edition of *Engineering Drawing and Graphic Technology*, by Thomas E. French, Charles J. Vierck, and Robert J. Foster. Seventy-five years ago, McGraw-Hill launched the First Edition, with words that are as appropriate to this new edition as they were to the original text in August, 1911:

> This is a distinctive work, and it is worth a thorough investigation. It is a full, well rounded treatise, designed to give a sound training, and it is a useful book to every draftsman in later professional work. It is straightforward, clean cut, and logical in its presentation, beautifully illustrated, and practical.

The inception of this text started with a visit by Martin Foss (subsequently President of the McGraw-Hill Book Company) at the end of the day on February 9, 1910, to Professor Thomas E. French, at Ohio State University. Immediately intrigued by Professor French's idea that drawing is the language of engineering, Mr. Foss urged French to develop an innovative book, and the rest is history—a remarkable text that has been used by more than two million students.

Charles Vierck, French's student and coauthor through six editions, built upon French's legacy. Robert Foster has now revised this classic text, ensuring that it will continue to enjoy a special place among the important contributions to engineering education.

Harold W. McGraw, Jr.
Chairman of the Board
McGraw-Hill, Inc.

PREFACE

This textbook represents many years of study, not only in teaching, engineering experience, and writing, but also in painstaking attention to book design, principles, and usage.

These words of Charles J. Vierck, from the Preface to the Twelfth Edition are even more apt for this new revision of the text, originally published seventy-five years ago. Even though the thirteenth edition is a major revision, the integrity and thoroughness of this classic text, launched in 1911, have been maintained with thorough explanations, ample examples, and superior artwork.

The text is designed for students and teachers in high schools, community colleges, technical institutes, and universities. The book covers a wide range of topics in the fundamentals of graphics, while offering an up-to-date treatment, current standards, and clear organization, in a streamlined, modern format.

The new edition includes greater emphasis on recent developments, such as computer aided drafting (CAD), and a smoother integration of metric units. Much of the topic coverage, such as that of drafting equipment, tolerancing, welding, and mapping, has been revised to reflect new methodology and modern drafting implements, incorporating the most recent American National Standards Institute (ANSI) standards. The entire text has been carefully revised to enhance readability and to ensure accuracy.

The text is now organized into four major sections: Basic Graphics, Elements of Space Geometry, Applied Graphics and Design, and Special Topics. Each section is highly illustrated and includes a wide variety of problems, with two thirds of them based on inch-base units, and the remainder on metric units.

Section A, Basic Graphics (Chapters 1 through 8), includes introductory material and coverage of the use of instruments, lettering, and graphic geometry. Material on orthographic projection, dimensioning, and tolerancing has been thoroughly revised for greater breadth, clarity, and use of ANSI standards.

Section B, Elements of Space Geometry (Chapters 9 through 11), offers a concise section, with increased attention to spatial manipulation of lines and planes.

Section C, Applied Graphics and Design, reflects thorough revision. New to the text is a comprehensive chapter on CAD, written by Professor Richard F. Devon of The Pennsylvania State University, with provision for hands-on experience for students with access to an Apple computer. The chapters in Section C (12 through 15), also provide a solid overview of the design process and the development of charts and graphs.

The last section, D, Special Topics (Chapters 16 through 22), covers areas of particular interest to teachers and students. Topics covered include assembly elements, gears and cams, and welding. Also covered are electrical drawings, piping, structures, and mapping. Instructors are provided with maximum flexibility to select topics and problems to fit the needs of their individual courses.

The appendixes utilize the latest ANSI standards and now include tables on metric screw threads and metric cylindrical limits and fits. More rigorous and challenging than the text proper, these appendixes should prove to be especially useful to serious students.

A workbook, including a solutions manual, has been developed by Professor Hugh F. Rogers, of The Pennsylvania State University, to accompany the text. The workbook has been meticulously designed to provide students with a pedagogically effective learning tool of the highest quality.

I would like to acknowledge the generous assistance provided throughout the preparation of this edition. In addition to their substantial contributions noted above, Professors Richard F. Devon and Hugh F. Rogers have both provided useful commentary and advice. Additionally, the following reviewers of the manuscript have made constructive suggestions: Edwin T. Boyer, Lawrence C. Drake, Paul D. Mercado, Gary L. Waisner.

Comments from instructors and students on the approach, content, and usefulness of this new edition are most welcome. The task of revising a classic is a genuine challenge. The continued contribution of new ideas, however, will help to ensure that the merits of clarity, teachability, and state-of-the-art coverage that characterized the First Edition will be maintained in the future.

Robert J. Foster

ENGINEERING DRAWING AND GRAPHIC TECHNOLOGY

PART A

BASIC GRAPHICS

1. THE USEFULNESS OF ENGINEERING DRAWING AND GRAPHICS

Drawing is a language and a language has great usefulness. We are familiar with the languages of the written word, mathematics, and music. They are used throughout the world so that people can communicate with one another. Whether a story is to be read, a song is to be sung, or an equation is to express a physical truth, language is essential to understanding among peoples.

Drawing is a language that can be learned and used like any other. There is, of course, drawing as related to art, where creative expression evolves a form unique to the particular artist. The results can be of great inspiration and pleasure to those who view the finished artwork. You may agree that Fig. 1 is an example of good art. The scholarly-looking gentleman has on hand a book of floral sketches.

The language of engineering drawing developed within this book is of use also. However, it is used to express ideas related to the making of products that can serve society. If one wishes to make a new type of chair, a plan or drawing needs to be made so that workers can understand what the designer has in mind for the chair.

We will study many types of engineering drawings and how each has its own value and use. Traditionally, drawings are done by hand, as in Fig. 2, but drawings can also be done by computer, as seen in Fig. 3. We will be not overly concerned about the *process* of generating the drawing. We will concentrate on *knowledge* needed to ensure that any drawing will be clear and understandable to a user.

Our object is to study the language of engineering graphics so that we can write it clearly for those familiar with it and read it readily when written by another. To do this, we must know its basic theory and be familiar with its accepted conventions and abbreviations. Since its principles are essentially the same throughout the world, a person who has been trained in the practices of one nation can readily adapt to the practices of another.

This language is entirely graphic and written, and is interpreted by acquiring a visual knowledge of the object represented. A student's success with it will be

FIG. 1 Painting by Francois Clouet that shows careful attention to detail. *(Louvre Museum, Paris.)*

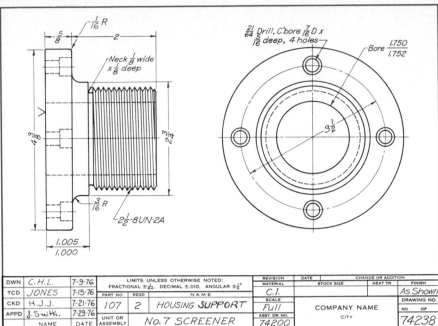

FIG. 2 Example of a two-view drawing drawn by hand-held instruments. The lettering is freehand.

indicated not only by skill in execution, but also by ability to interpret lines and symbols and to visualize clearly in space.

As a background for study, we will introduce in this chapter various aspects of graphics that will be discussed at length later. It is hoped that this preview will serve as a broad perspective against which you will see each topic, as it is studied, in relation to the whole. Since our subject is a graphic language, illustrations are helpful in presenting even this introductory material. Figures are thus used to clarify the text.

The book is divided into four major parts: basic graphics, elements of space geometry, applied graphics and design, and special topics. We will now take a quick look at each of these four sections.

2. BASIC GRAPHICS

A. Essentials of Graphics: Lines and Lettering Drawings are made up of lines that represent the surfaces, edges, and contours of objects. Symbols, dimensional sizes, and word notes are added to these lines, collectively making a complete description.

Lines are connected according to the geometry of the object represented, making it necessary to know the geometry of plane and solid figures and to understand how to combine circles, straight lines, and curves to represent separate views of many geometric combinations.

We will study the use of modern instruments for mechanical drawing. We will learn how to make basic geometric constructions, using graphic geometry, as in Fig. 4.

B. Methods of Expression There are three methods of writing the graphic language: freehand, with hand-held instruments, and by computer.

Freehand drawing is done by sketching the lines with no instruments other than pencils and erasers, as seen in Fig. 5. It is an excellent method to use during the learning process because of its speed and because at this stage the study of projection is more important than exactness. Freehand drawings are much used commercially for preliminary designing.

MECHANICAL DRAWING

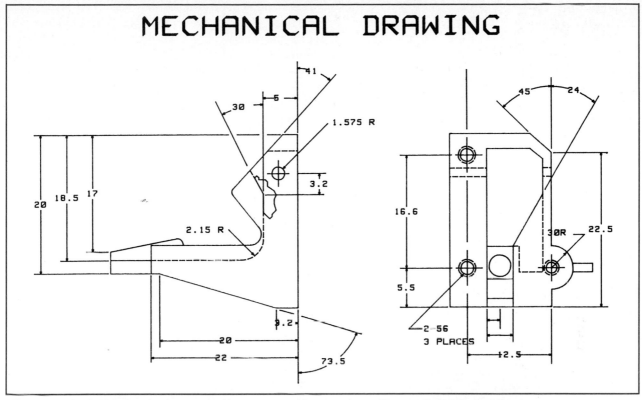

FIG. 3 Example of a two-view drawing generated by a computer-driven pen plotter. *(Courtesy Hewlett-Packard Corp.)*

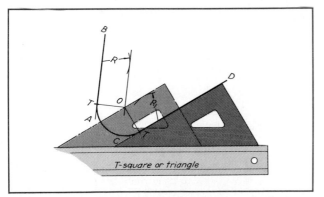

FIG. 4 Graphic geometry. Here one is making an arc tangent to two straight lines.

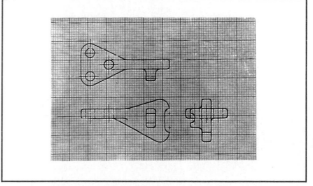

FIG. 5 Freehand drawings. The freehand method is fine for early study because it provides training in technique, form, and proportion. It is used commercially for economy.

Since most drawings are made to scale, instruments are used to draw straight lines, circles, and curves concisely and accurately. Figure 6 is an example of an instrument drawing.

The computer drawing seen in Fig. 3 represents a rapidly advancing aspect of engineering drawing. Computers can drive plotters which produce the actual drawing. Computer-generated drawings are at an advantage

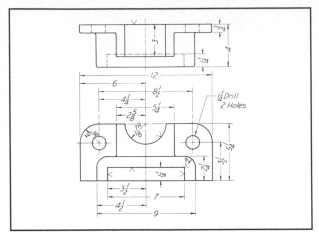

FIG. 6 Instrument Drawings. Because of the necessity of drawing "to scale," most drawings are made with instruments.

over hand-done instrument drawings when many drawings are needed but when each has a slight modification over the others, as, say, for designing drive shafts for a common model truck having only different wheelbase options. Also, computer drawings are valuable and almost necessary when complicated shapes, such as jet turbine blades, are to be drawn.

A person using the graphic language may choose between freehand lettering and lettering applied by use of templates or press-on transfers. Commercially available dry transfers, as illustrated in Fig. 7, can be of considerable help in creating professional-looking work in a minimum amount of time.

C. Methods of Shape Description Showing the *shape* of a part is a primary purpose of graphic communication.

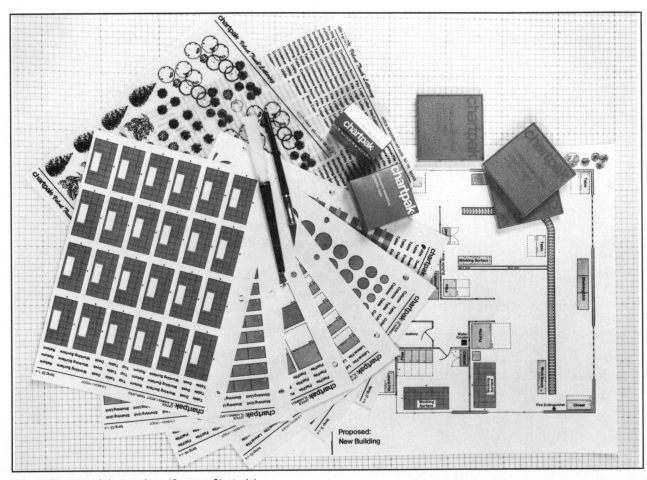

FIG. 7 Examples of dry transfers. *(Courtesy Chartpak.)*

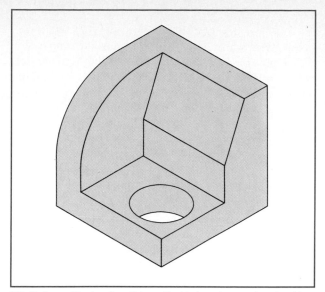

FIG. 8 Isometric drawing. This method is based on turning the object so that three mutually perpendicular edges are equally foreshortened.

The designer must select the best method available to show shape.

Two basic techniques are used to indicate shape: two-dimensional and three-dimensional. Figure 2 uses a two-dimensional technique known as multiview projection. Figure 8 uses a three-dimensional technique—in particular, the technique of isometric projection. The method of oblique drawing is seen in Fig. 9, while Fig. 10 shows the effects of using a perspective drawing.

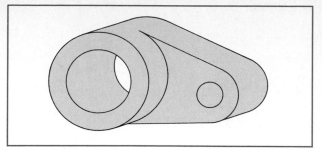

FIG. 9 Oblique drawing. This pictorial method is useful for portraying cylindrical parts. Projectors are oblique to the picture plane.

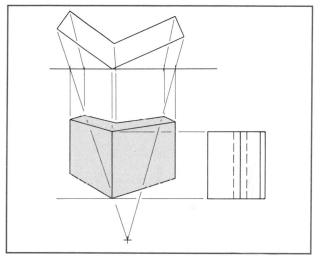

FIG. 10 Perspective drawing. This system is the most realistic of all but is also the slowest to construct.

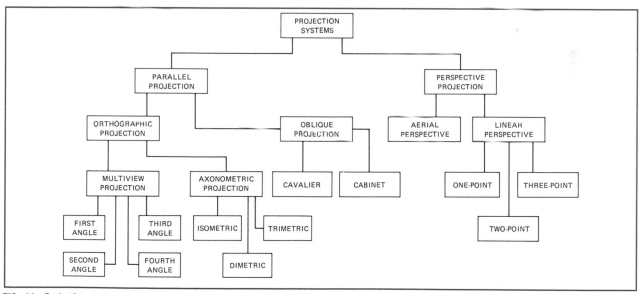

FIG. 11 Projection systems.

All available projection systems are shown in Fig. 11. Perspective projection, oblique projection, and axonometric projection are three-dimensional techniques of "pictorial representation." All others are two-dimensional. With some two-dimensional drawings one view is sufficient to describe a shape, as in Fig. 12. Some objects need two views to show shape, as in Fig. 13. Still other objects may need three views to fully describe shape.

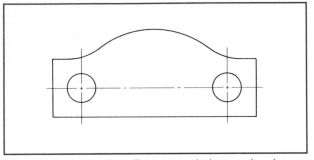

FIG. 12   One-view drawings. These are used whenever views in more than one direction are unnecessary, for example, for parts made of thin material.

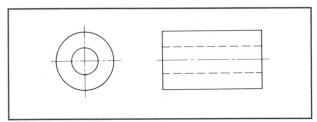

FIG. 13   Two-view drawings. Parts such as cylinders require only two views. More would duplicate the two already drawn.

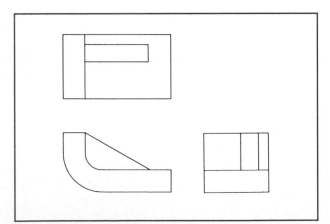

FIG. 14   Three-view drawings. Most objects are made up of combined geometric solids. Three views are required to represent their shape.

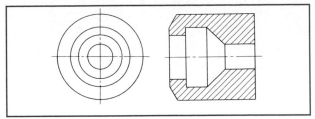

FIG. 15   Sectional views. These are used to clarify the representation of objects with complicated internal detail.

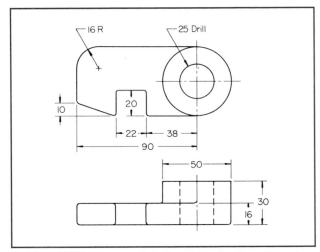

FIG. 16   METRIC. Dimensioning orthographic drawings. Dimensions showing the magnitude and relative position of each portion of the object are placed on the view where each dimension is most meaningful.

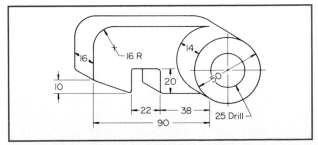

FIG. 17   METRIC. Dimensioning pictorial drawings. The descriptions of magnitudes and positions are shown on the pictorial by dimensions placed so as to be easily readable.

Figure 14 indicates such an object. Note, however, that all three figures (Figs. 12 to 14) use a two-dimensional projection system. We will learn in Chap. 5 how two-dimensional projection systems are developed and used. Chapter 7 deals with three-dimensional pictorial drawings (perspective, oblique, and axonometric projections).

We also will study sectional views. Such views are used to clarify the internal features of "hollow" parts by cutting into the object to show what is inside. Figure 15 shows two views of an object, with one view "cut in half" to let us see visibly what would otherwise be hidden. Chapter 6 is devoted to sectional views.

D. Methods of Size Description In addition to describing shape, one must give the *size* of features of an object. Size is given by "dimensions," which state linear distances, locations, diameters, and other necessary items.

A part drawn in multiview projection is shown dimensioned in Fig. 16. The same part is shown dimensioned in Fig. 17, using the oblique-projection form of three-dimensional drawing. It is also possible to use special techniques for dimensioning, such as in the photo drawing of Fig. 18.

As part of dimensioning, we will study tolerancing, which involves allowing a dimension to vary between a permitted maximum and minimum value. Figure 19 shows a part that is to fit into a slot. The part is always smaller in size than the slot. Therefore the part will fit into the slot with what is known as a "clearance fit."

Another type of tolerancing is geometric tolerancing, which controls the form of a part in regard to its flatness, perpendicularity, or roundness, for example. The boxed-in symbols in Fig. 20 are controls for various geometric tolerances. This is a rather involved figure, but don't be intimidated by its magnitude.

E. Methods of Measurement: English vs. Metric There is much more to the description of size than learning the standards and rules for the placement of dimensions on drawings.

First of all, the system of measurement and evaluation must be known and understood. History records many different evaluation systems. These are well known and documented. Of all recorded methods, two systems have withstood the test of time—the English system and the metric system. The metric system has been

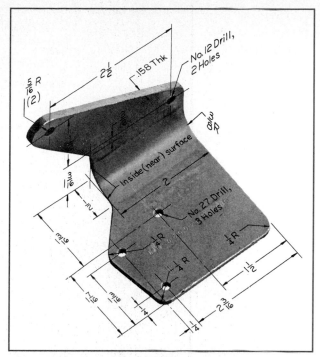

FIG. 18 Photo drawing with dimensioning. This method can save much time.

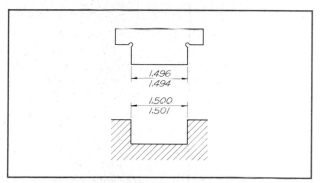

FIG. 19 An example of limit dimensioning. Tolerance on tongue, 0.002 in.; tolerance on groove, 0.001 in.; min. clearance, 0.004 in.

adopted and used by almost all countries except the United States.

It is not probable the United States will ever become a fully metric nation, but many U.S. industries are largely metric. Companies having extensive overseas sales, such as aerospace, construction equipment, and automotive, have made major commitments to the use of metric units. The international metric system is known as the *SI*

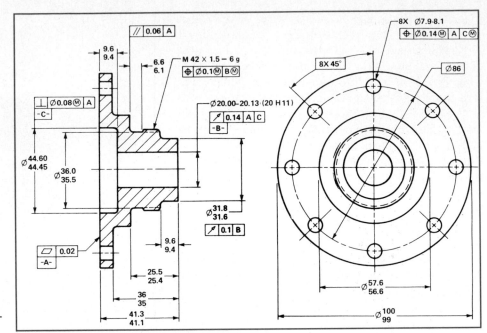

FIG. 20 Use of geometric tolerancing. *(ANSI Y14.5M—1982.)*

system and is approved by the National Bureau of Standards. Drafters must be reasonably familiar with the metric system. While designers must know well all units of metric (length, substance, current, temperature, light, etc.), a drafter needs to know mainly the unit of length. The metric unit of length is the meter; its subdivisions are the centimeter and millimeter. Common SI unit prefixes and symbols are given in the appendix.

3. SPACE GEOMETRY

Space geometry is the application of the theory of projection to the solution of space problems. Machines and structures are made up of geometric elements combined in various ways. The many points, lines, and planes are combined to form parts of a design. The determination of position, clearance, and movement of parts is a vital consideration, and may involve parallelism, perpendicularity, or angularity of the various lines and planes. Analyzing the relationships among points, lines, planes, and solids requires space geometry. An example of space geometry is seen in Fig. 21, where the point view of a line *AB* is constructed.

Auxiliary views can also help to clarify shape. Auxiliary views can be used to show the so-called normal view of a slanted surface. In Fig. 22 an auxiliary view is used to show the true shape of the semicircular portion of the part.

The accuracy obtained in manually solving space problems is often sufficient. For greater accuracy, mathematical or computer analyses are possible. In any case, a graphic representation is helpful for a complete understanding of the space problem.

Space geometry is a powerful tool for research, design, and development. Design usually begins with the graphic stage, and there will be a correlated activity involving engineering drawing, engineering geometry, and mathematics as they relate to the scientific aspects of the design.

4. APPLIED GRAPHICS AND DESIGN

The end point of all graphical knowledge is its use in design. Engineering and engineering technology are by definition design-oriented. Design can take various forms, but in general there are three basic forms of design: system design, hardware design, and software design.

System design involves complex networks of interlinking variables, all requiring research and study. An example of a system design is a subway system for a major city. The problem is complex. There are sociological considerations, such as the effect of building construction on neighborhoods. There are vast economic considerations. There are numerous engineering problems, involving many branches of engineering: civil, mechanical, electrical. To handle system design within a single book or even a single course is difficult and rarely undertaken. One usually contributes to system design by being part of a team working on the actual design project.

Hardware design involves finite bits and pieces brought together to form a working product, which could be anything from an electric iron to landing gear for an airplane to a microcomputer. Hardware design is considered to be a subset of system design and is therefore more manageable for development within a textbook or course.

Hardware design normally involves design sketches and eventually design drawings. The drawings may be done by hand or by computer. The result is the same:

information is made available to permit the construction of the product. Design sketches precede design drawings. The value of design sketches have been apparent for many years. Some 500 years ago Leonardo da Vinci sketched ideas for the mechanism seen in Fig. 23. Despite their antiquity, these sketches clearly show the role of shafts, wheels, and support members.

Design sketches evolve toward design drawings. Figure 24 shows a design sketch for a nutcracker. The subsequent design drawing is in Fig. 25. Further refinement would yield drawings of each and every individual part that needed to be created. Any existing parts (certain screws, pins, etc.), however, would simply be purchased.

Software design concerns final "products" that are, in essence, paper (i.e., "soft"). Computer programs fall naturally into this area of design. An engineer or technician puts creativity into producing a product that has real applications in engineering. A computer program can be used to instruct a computer to control automated tooling to produce parts, or to control flight paths of orbiting satellites, or to issue paychecks for employees. An example of a short computer program is seen in Fig. 26. This particular program instructs a microcomputer to provide solutions to the equation $V = IR$, given any two

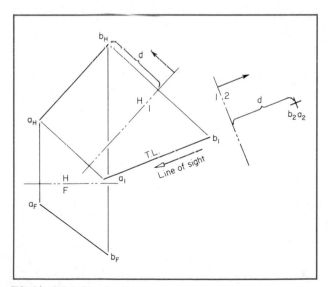

FIG. 21 Point view of a line.

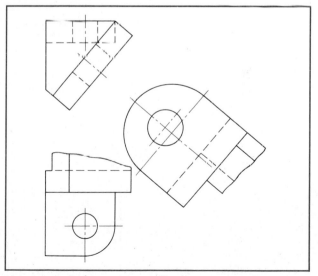

FIG. 22 Auxiliaries are used to show the *normal view* (true size and shape) of an inclined surface (at an angle to two of the planes of projection).

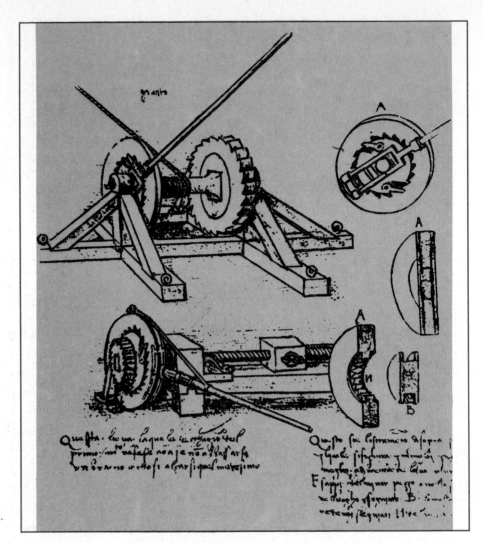

FIG. 23 An historical design drawing. Leonardo da Vinci (about 1500).

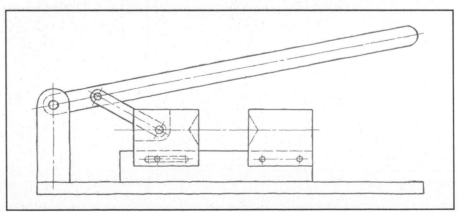

FIG. 24 Design sketch of a nutcracker.

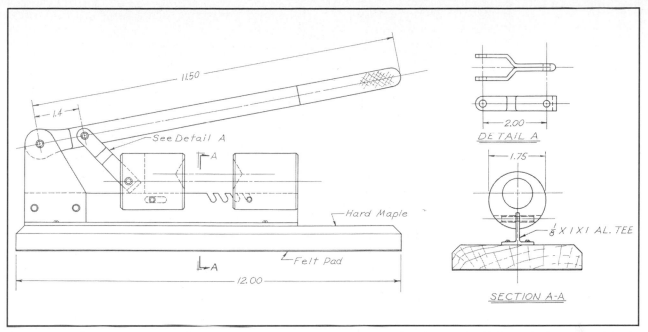

FIG. 25 Design drawing of nutcracker. This represents the final design, which is the culmination of progressive thinking and refinement in solving the problem.

```
]LIST

80   PRINT "OHM'S LAW"
85   PRINT "I=E/R"
86   PRINT "E=I*R"
87   PRINT "R=E/I"
90   PRINT "TYPE X FOR ANY UNKNOWN VALUE"
100  INPUT "VOLTAGE =?";E$
105  IF E$ = "X" THEN EE = 1
110  INPUT "RESISTANCE=?";R$
115  IF R$ = "X" THEN RR = 1
120  INPUT "CURRENT =?";I$
125  IF I$ = "X" THEN II = 1
130  IF EE = 1 THEN I =  VAL (I$):R =  VAL (R$):E = I * R: PRINT "VOLTAGE
     =";E
140  IF RR = 1 THEN E =  VAL (E$):I =  VAL (I$):R = E / I: PRINT "RESISTAN
     CE =";R
150  IF II = 1 THEN E =  VAL (E$):R =  VAL (R$):I = E / R: PRINT "CURRENT
     =";I
155  PRINT
160  END
```

FIG. 26 Software design in the form of a computer program. *(Courtesy R. Dunkle, Pennsylvania State University.)*

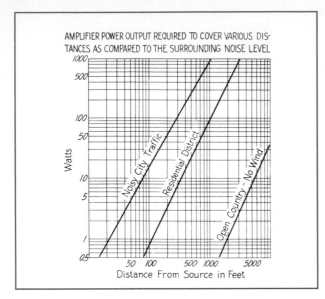

FIG. 27 A graph that relates amplifier power required to distance from the sound source.

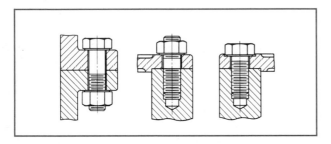

FIG. 28 Fasteners. Bolts and screws in a wide variety of forms are used for fastenings.

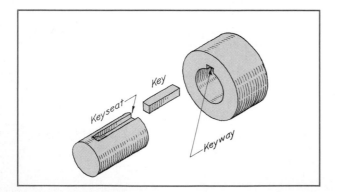

FIG. 29 Keys. Keys are used to prevent cylindrical elements such as gears and pulleys from rotation on their shafts.

of the three variables of voltage (V), current (I), or resistance (R).

Software design is a rapidly developing area of engineering, taking the full-time efforts of many competent persons. Such design, like others, is a specialty that can be learned through a combination of book instruction and hands-on experience.

Chapter 15 deals with software and its role in computer graphics. This area should prove both interesting and worthwhile.

Also of interest will be Chap. 14, which is devoted to charts and graphs. Figure 27 is an example of a graph. Charts, graphs, and diagrams fall roughly into two classes: (1) those used for purely technical purposes and (2) those used in advertising or in presenting information in a way that will have popular appeal. Engineers are concerned mainly with those of the first class, but they should also be acquainted with the preparation of those of the second and understand their potential influence. The aim here is to give a short study of the types of charts, graphs, and diagrams with which engineers and those in allied professions should be familiar.

5. SPECIAL TOPICS

This section of the book deals with specialized topics that, while not used continually by all drafters, are still of worth when needed. *Assembly elements*, for example, include items such as threaded fasteners (Fig. 28), keys (Fig. 29), and springs (Fig. 30). These elements allow parts to be fastened together, yet they allow for disassembly when needed. Such elements have a long history of use in design. They are discussed in Chap. 16 and are documented in the appendix.

Gears and cams (Figs. 31 and 32) are well-established devices to transmit motion and power through an angle or direction in a particular way. Bevel gears, for example, are often used to transmit power through a 90-degree angle. Cams allow for a carefully prescribed motion, such as the opening and closing of valves in an automotive engine. Chapter 17 covers gears and cams.

Welding and riveting are often considered to be permanent methods of assembly. Each is a highly specialized and standardized method, meaning that a drafter should consult standards carefully before using symbols and dimensioning. Chapter 18 deals with these methods

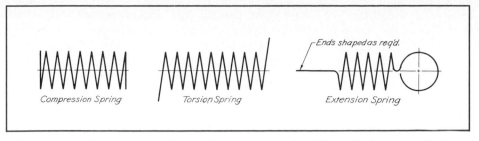

FIG. 30   Springs. A spring is an elastic body that stores energy when deflected. Basic forms are illustrated.

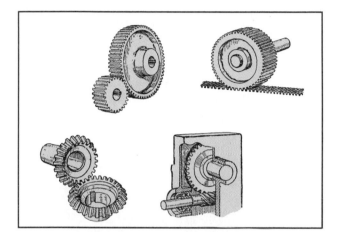

FIG. 31   Gears. Special practices include calculation of detailed sizes and their tabulation on drawings.

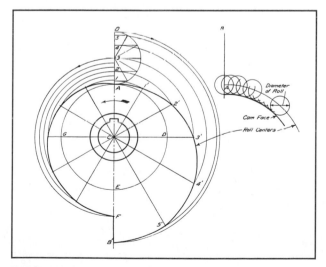

FIG. 32   Cams. Special practices include full-size layouts and also calculation and dimensioning of points on cam surfaces.

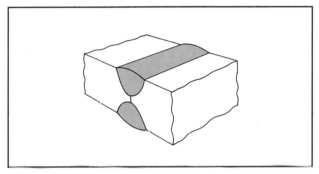

FIG. 33   Welds. A 3-D illustration that shows both sides of a U-groove weld.

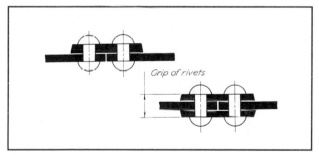

FIG. 34   Rivets. Rivets are permanent fasteners. They are cylinders of metal with a head on one end. When placed in position, the opposite head is formed by impact.

of fastening, both of which are well accepted and very reliable. Figure 33 shows one type of weld, while Fig. 34 indicates how rivets hold material together.

The drawing of *electrical components and circuits* is explained in Chap. 19. This area of drafting is highly technical and is often done by specialists in the field of electrical/electronic design. Figure 35 shows a schematic diagram of a design involving vacuum tubes. Other designs include transistors and integrated circuits. Manually done drafting is usually restricted to small, fairly simple designs. Templates are used to speed the drafting of common symbols, such as resistors and capacitors. Complicated circuits are commonly done by computer, with multileveled integrated circuits being a good example.

Piping is a topic going back in history many centuries. As a result, many well-refined standards have evolved to assist designers and drafters in making clear, easily understood symbols to depict the various pipes, elbows, tees, valves, etc., that make up a piping system. High-pressure systems, such as for a steam-powered electrical plant, have particularly stringent standards. A simple one-line system to show various components is seen in Fig. 36. Chapter 20 discusses the topic of piping.

Structures are dealt with in Chap. 22 from the viewpoint of steel beams, columns, and assemblies. The use of wood, concrete, and stone is treated more lightly. Again, we have in this area a set of highly specialized topics which a drafter must learn well before being allowed to develop plans. However, here we supply a brief survey. The general overview of a structure having many materials is seen in Fig. 37.

The last chapter looks at *mapping and topography* in an abbreviated manner, mainly so that the potential drafter can recognize the major elements that make up this field. It is likely that a person desiring to become a specialist in mapping and topography would do apprentice work in a firm having such work as a major activity.

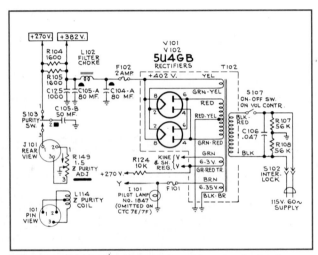

FIG. 35 Drawings of electric systems. Most electric drawings are made symbolically, using ANSI standards.

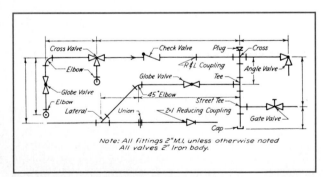

FIG. 36 Piping. Piping drawings are made to scale or symbolically. Symbols for pipe, valves, fittings, and other details are standardized by ANSI.

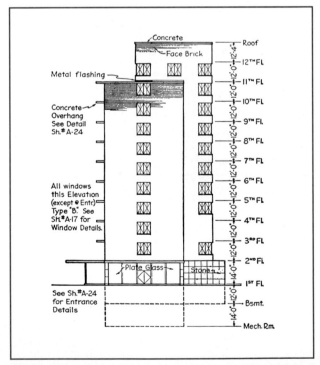

FIG. 37 Structures. Architectural standards of the American Institute of Architects (AIA) and structural standards of the American Institute of Steel Construction (AISC) prevail in the building industry.

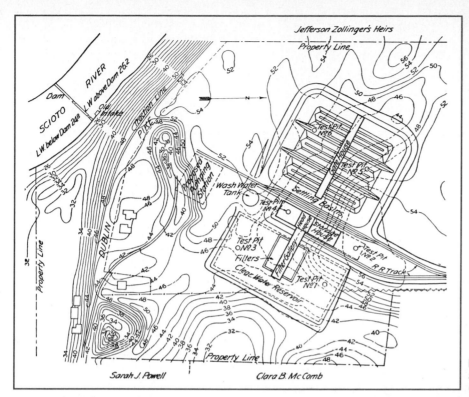

FIG. 38 Maps and topography. Maps and topographic drawings are made up chiefly of symbols.

This field has its own particular symbols and procedures which, like any others, can be learned and used. A municipal water facility is shown on the map given in Fig. 38.

Appendix: You will find in this section a diverse and thorough coverage of standards and symbols to support your study of the special topics of Chaps. 16 through 22.

In each case the standard given is the latest available. Some standards have recent updates, of 1982 and later; others date back into the 1960s without change.

The authors hope you will find your adventure in learning to be both enjoyable and useful. The text should serve as a valuable reference book for years to come.

2
GRAPHIC INSTRUMENTS AND THEIR USE

1. AN OVERVIEW

Graphic instruments are used to create designs in the world of technology. Some designs are drawn manually using basic instruments that have been developed and refined over many, many years. Other designs are drawn using automatic drafting equipment, often controlled by a computer. The end result is the same—a design to be produced and used to fulfill a need in society.

This chapter will describe the most common drafting instruments in use today and discuss how they are used. As one becomes more comfortable in the use of instruments over an extended period of practice, drawings will be made with greater speed, neatness, and accuracy. While only a minority of students become professional drafters, it is very satisfying to be able to draw at least well enough to convey sound information.

BASIC INSTRUMENTS

2. DRAWING BOARDS

We will study a number of basic graphic instruments. Before instruments can be used however, one must attach paper (or film) to a drawing surface. Sometimes one tapes paper directly on a drafting table. In other cases a separate drawing board may be used. Figure 1 shows a board of a type commonly used by students. Usually made of basewood or pine, the board sometimes includes an inserted steel edge for better wear. Another popular type of drawing board is one with a parallel straightedge (see Fig. 2). The straightedge serves the purpose of a T square.

3. DRAFTING MEDIA

There are several choices of materials on which you may draw. Basically there are four types of media: paper, vellum, cloth, and film.

A. Paper   There are two classes of paper. One, known as detail paper, is primarily for pencil work that is not to be reproduced by ozalid or other processes that require a degree of transparency of the paper. Detail paper is opaque. Any reproduction from it must be done by a photographic process. Detail paper is made in a variety of qualities and may be had in sheets or rolls. White drawing papers that will not turn yellow with age or exposure are used for finished drawings, maps, charts, and drawings for photographic reproduction. For pencil layouts and working drawings, cream or buff detail papers are preferred because they are easier on the eyes and do not show soil so quickly as white papers. In general, paper should have sufficient grain, or "tooth," to take the pencil, be agreeable to the eye, and have a hard surface not easily grooved by the pencil, with good erasing qualities.

The other class of paper is translucent paper, which is designed so that it can be used in common reproduction processes. Also known as tracing paper, it can be used for both pencil and ink work. Tracing paper is usually not coated. Only limited erasing may be done on it before damage occurs to the paper's surface. Lead lines erase well, but inked lines do not. Tracing paper is often 100 percent rag content.

B. Vellum   This medium for line work is 100 percent rag content that has been coated on one side with a synthetic resin. Vellum is somewhat less transparent than tracing paper, but its transparency is adequate for all reproduction methods. The resin coating serves a valuable purpose. Not only may lead be readily erased many times, but ink may also be erased, at least once. The coating serves as a protective shield that keeps erasers from abrading into the paper base of the sheet. However, repeated erasures of ink will remove the coating.

C. Cloth   Finely woven cloth coated with a special starch or plastic is used for making drawings in pencil or ink. The standard tracing cloth is used for inked tracings and specially made pencil cloth for pencil drawings or tracings. Cloth is more permanent than paper and will withstand more handling. However, cloth is considerably more expensive than paper or vellum. It is not widely used.

D. Film   The use of film is growing rapidly and is largely replacing use of cloth. Film is made of polyester plastic with a mat surface. Some films are designed primarily for pencil, but recently the films that accept ink well have become popular.

Film is very durable—resisting aging, cracking, and tearing. It is dimensionally stable and is little affected by moisture. Ink work is easily erased, a feature appreciated by the drafter. The higher cost of film over papers is offset by the quality and permanence over and beyond papers.

E. Standard Sizes   We are all familiar with the common-sized sheet, 8½ by 11 in. Drawing sheets have been sized in multiples of 8½ by 11 in for convenience in filing. Properly folded, standard sizes can be placed in folders and drawers that accept the 8½- by 11-in size.

FIG. 1 A basic drawing board. *(Courtesy Gramercy Corp.)*

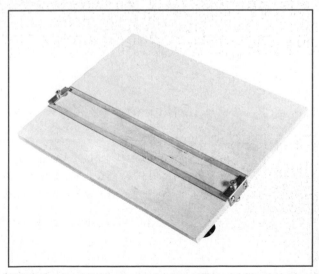

FIG. 2 Parallel straight-edge drawing board. This model features a handle (at top) and collapsible legs (not seen). *(Courtesy Gramercy Corp.)*

Figure 3 shows both U.S. drawing sizes and the international equivalents. These sizes conform to ANSI Standard Y14.1—1975.

In addition to these standard sizes, papers and films may be purchased in pads of various sizes, such as 9 by 12 in, 12 by 18 in, etc. Rolls are also available in 30-, 36-, and 42-in widths.

4. PENCILS AND LEADS

The proper choice of pencil and lead can do much to ensure a quality drawing or sketch. Neither pencils nor leads are expensive, but if properly selected they can do more than any other single instrument to assist you in obtaining clear, sharp, dark drawings.

A. Pencils The basic pencil contains graphite lead, made in various hardnesses. Figure 4 shows five varieties

USA size, in	Closest international size, mm
A (8.5 × 11.0)	A4 (210 × 297)
B (11.0 × 17.0)	A3 (297 × 420)
C (17.0 × 22.0)	A2 (420 × 594)
D (22.0 × 34.0)	A1 (594 × 841)
E (34.0 × 44.0)	A0 (841 × 1189)

FIG. 3 Comparison of paper sizes.

of pencils. Types A and B are ordinary pencils with the lead set into wood (type A is an American product and type B non-American). Both work well, but any wooden pencil must have the wood cut away to expose the lead. This takes time, and makes the pencil shorter.

Mechanical pencils, types C and D, are seen Fig. 4. Each has a chuck that can be opened by depressing a plunger on the end of the pencil so a desired lead can be inserted. Type C has a plastic handle and is lower in cost than type D, which has a metal handle. Both these mechanical pencils require sharpening.

A recently popular mechanical pencil is seen in Fig. 5. Sometimes called a thin-lead pencil, it features a 0.5-mm drawing lead that does not need sharpening because of its small diameter. A 0.3-mm lead is also available for extra-fine lines. Various lead hardnesses can be obtained for whatever size lead is selected.

Generally speaking, the initial higher cost of a mechanical pencil over a wooden pencil is rapidly offset by the lower cost of replaceable leads and the greater convenience of a fixed-length pencil. Also, it takes less time to sharpen the lead of a mechanical pencil than that of a wooden pencil.

B. Leads It is important to select the proper hardness for a particular job. Figure 6 lists the various grades of

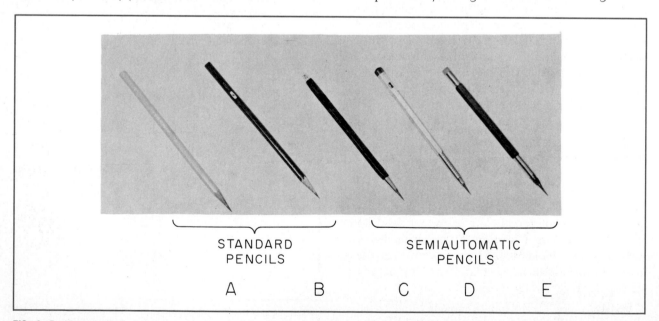

STANDARD PENCILS   SEMIAUTOMATIC PENCILS

A   B   C   D   E

FIG. 4 Drafting pencils.

FIG. 5 Thin-lead mechanical pencil. *(Courtesy J. S. Staedler, Inc.)*

lead hardness. Leads are graded by numbers and letters from 9H, an extremely hard lead, to 7B, a very soft lead. The grade of pencil must be selected carefully, with reference to the surface of the paper as well as to the line quality desired. For a pencil layout on detail paper of good texture, a pencil as hard as 5H or 6H may be used, while for finished pencil drawings on the same paper, 2H, 3H, or 4H pencils give the blacker line needed. For finished pencil drawings or tracings on vellum, softer pencils, H to 3H, are employed to get reproducible lines. The F pencil is much used for technical sketching, and the H is popular for lettering. In every case the pencil must be hard enough not to blur or smudge but not so hard as to cut grooves in the paper under reasonable pressure.

5. PENCIL POINTERS

For the money invested, a mechanical lead pointer such as the one shown in Fig. 7 is a good value. Such lead pointers sharpen the lead conveniently and rapidly, leaving the removed graphite within the container for later disposal. Simple, small, fixed-blade sharpeners are also available at low cost, but many drafters find them less satisfactory than the type seen in Fig. 7.

6. PENS AND INKS

Some people in the engineering field may say that inking is a lost art, seldom needed and seldom used. While it is true that inking seemed to decline in use for a period in the 1960s, it is now emerging as an increasingly useful tool. The permanence and quality of inked drawings have never been denied. The skill needed to create ink drawings, however, was in short supply. Now, in the 1980s, more and more mechanical aids are available to make inking easier. Pens have been improved, as well as inks and erasers. The increased use of polyester films and acetates for the drawing media, as opposed to conventional papers, has also made inking more attractive.

We will begin our study of inking with a look at the

Hard grades		Medium grades		Soft grades	
9H	Hardest	3H	Hardest	2B	Hardest
8H		2H		3B	
7H		H		4B	
6H		F		5B	
5H		HB		6B	
4H	Softest	B	Softest	7B	Softest

FIG. 6 Chart of lead hardness.

FIG. 7 Lead pointer. *(Courtesy Koh-I-Noor Rapidograph Inc.)*

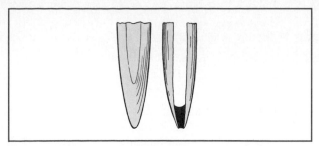

FIG. 8 Correct shape of pen nibs. An elliptical shape is best for ruling pens.

FIG. 9 A technical pen. *(Courtesy Koh-I-Noor Rapidograph Inc.)*

FIG. 10 Scriber set with lettering guides. *(Courtesy Koh-I-Noor Rapidograph Inc.)*

more traditional instruments and will then move on to more modern equipment.

A. Ruling Pens The ruling pen is for inking straight lines and noncircular curves. Several types are available. The important feature is the shape of the blades: they should have a well-designed ink space between them, and their points should be rounded (actually elliptical in form) equally, as in Fig. 8.

B. Technical Pens Ruling pens are traditional instruments that have been used for many decades. They do their particular jobs very well, but they are not easy to use until one becomes familiar with them.

More flexible and "forgiving" instruments have been developed to do much of the work formerly done only by ruling pens and compasses. The most widely used newer pen is the technical pen. One example is seen in Fig. 9, where the pen is hand-held. Also available are scriber sets (Fig. 10), which allow one to do lettering mechanically with the use of special guides. While not inexpensive, such scriber sets increase the professional quality of lettering for persons doing a lot of drafting.

Points for technical pens are available in stainless steel for use on paper, jewel tip for use on films, and tungsten carbide for use in automated drafting machines driven by a computer. Point sizes vary according to need. Figure 11 shows one system of point sizes with dual designation. The gray-toned squares are metric point sizes in the geometric progression of the square root of 2 (1.414), so that each line is 1.414 times wider than the preceding line.

A step beyond the scriber set are microprocessor-based systems which will letter a drawing in a variety of styles and sizes by merely typing in the desired lettering on a keyboard. Figure 12 shows one such machine. This system has three components, a plotter (1), which connects to a controller (2), which accepts preprogrammed modules (3) for particular symbols.

C. Inks Drawing inks are of various formulations depending on the particular paper or film being used. There are inks for paper and cloth, with drying rates from average to slow. Other inks are designed for polyester film, with fast to slow drying rates. One can also get inks for acetates, used often for overhead color projection. These inks may be wiped off with water or alcohol. Inks are available in numerous colors, including white.

6x0 .13	4x0 .18	3x0 .25	00 .30	0 .35	1 .50	2 .60	2½ .70	3 .80	3½ 1.00	4 1.20	6 1.40	7 2.00
.005 in.	.007 in.	.010 in.	.012 in.	.014 in.	.020 in.	.024 in.	.028 in.	.031 in.	.039 in.	.047 in.	.055 in.	.079 in.
.13 mm	.18 mm	.25 mm	.30 mm	.35 mm	.50 mm	.60 mm	.70 mm	.80 mm	1.00 mm	1.20 mm	1.40 mm	2.00 mm

FIG. 11 Dual-designation of point sizes for technical pens. *(Courtesy Koh-I-Noor Rapidograph Inc.)*

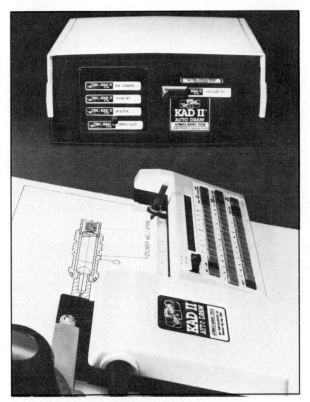

FIG. 12 An automated lettering system controlled by a microprocessor. *(Courtesy Koh-I-Noor Rapidograph Inc.)*

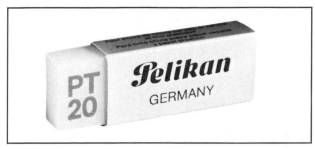

FIG. 13 Plastic eraser for drafting films. *(Courtesy Koh-I-Noor Rapidograph Inc.)*

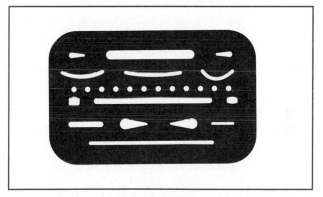

FIG. 14 An erasing shield. *(Courtesy Pickett Industries.)*

7. ERASERS

The choice of eraser to remove lines depends on whether the lines are in pencil or ink, and whether the drawing medium is paper or polyester film. The long-used artgum eraser is good for a gentle treatment of graphite lines, but plastic types have largely replaced the artgum type. You may get plastic types specifically made for graphite lines and for ink lines on paper. Also available, as shown in Fig. 13, are plastic erasers for use on drafting films. Consult a catalog of any major supplier of engineering drafting equipment for details on erasers.

A major concern in erasing is to minimize damage to the paper or film on which the lines are drawn. Overly

FIG. 15 An erasing machine. *(Courtesy Keuffel & Esser Co.)*

FIG. 16 An erasing brush. *(Courtesy Keuffel & Esser Co.)*

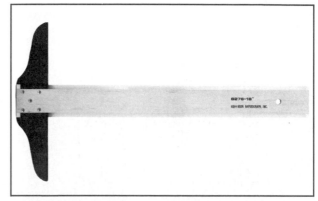

FIG. 17 Fixed-head T square. *(Courtesy Koh-I-Noor Rapidograph, Inc.)*

FIG. 18 Drafting machine. This model is attached to a drawing table. *(Courtesy Gramercy Corp.)*

vigorous erasing or use of too abrasive an eraser should be avoided. The paper or film should be on a sufficiently hard backing to aid thorough erasing. A triangle slipped under the drawing can give a firm surface.

An erasing shield (Fig. 14) is a helpful accessory to localize the erasing to where you wish it done. It is discouraging to erase an area larger than desired, spoiling lines you wish to keep. The professional drafter may also make use of an erasing machine, as shown in Fig. 15. These are efficient and save time. A brush (Fig. 16) is handy to whisk away the leftover particles of eraser material.

8. T SQUARES

The T square, Fig. 17, is a basic low-cost instrument that has been used for many decades. Its practical use still has merit, despite the more recent development of drawing machines of one type or another. Figure 18 illustrates a drafting machine which works very well but is impractical for the drafting student because of cost. Professional drafting offices of corporations provide such machines for their employees, and skills learned by the student using a T square and triangles can be transferred readily to drafting machines.

The fixed-head T square (Fig. 17), used for all ordinary work, is usually made of wood, with transparent plastic insert edges. Metal T squares are also available.

9. TRIANGLES

Triangles of the type seen in Fig. 19 are made of clear plastic. They come in various colors, but many drafters prefer the uncolored versions. Triangles should be kept flat to prevent warping. For most work a 6- or 8-in 45° and a 10-inch 30–60° are useful sizes.

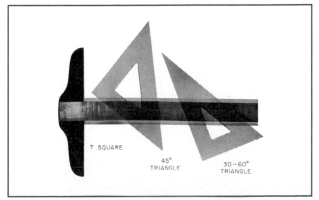

FIG. 19 T square and triangles. These are standard equipment for drawing straight lines.

10. SCALES

We are all familiar with everyday scales such as yardsticks or meter sticks. The foot ruler was used in grade school. Early exposure to some kind of scale is inevitable. In nearly all instances, however, full-size scales were used, where 1 in or 1 m on the paper represented the same value on the actual object.

In technical work it is common to show parts at reduced scale or enlarged scale, depending on the parts' size and the paper's size. An auto wheel might be drawn at half size; that is, the wheel as seen on the paper is one-half the actual size. A tiny part, such as an electronic component, might be drawn at 10-times size to increase the readability of the drawing. A large structure, such as a house, would be much reduced—to, say, one-forty-eighth size.

In every case the drawing must contain in a note the scale to which it is drawn. The title block often contains the scale note. The scale is given as

$$X = Y$$

where X is the unit on the drawing and Y is the representation on the actual object. For example, if 1 in = 5 ft, 1 in on the drawing represents 5 ft on the object. Such a scale would be a reduction scale of 1 to 60 (1 in = 60 in).

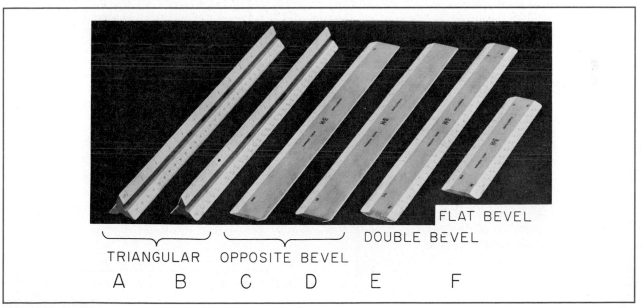

FIG. 20 Scale types.

Scales are made with various cross-sectional shapes, as shown in Fig. 20. The triangular form, *A* and *B*, has long been favored because it carries six scales as a unit and is very stiff. However, many drafters prefer the flat types because they are easier to hold flat to a board and have a particular working scale more readily available. The "opposite-bevel" scale, *C* and *D*, is easier to pick up than the "flat-bevel" scale, *F*; moreover, it shows only one graduation at a time. The "double-bevel" scale, *E*, in the shorter lengths is convenient as a pocket scale.

Scales are available in wood, plastic, and metal. Most students use plastic scales because of their low cost, yet they can be accurate and highly readable. Wood scales are seen less often, but some drafters choose them as a personal preference. Metal scales of aluminum or magnesium are more costly but are very durable and are dimensionally predictable with changes in temperature and humidity.

A. Types of Scales   There are basically two types of scales. One type is a decimal-base type, which includes both the inch and the metric (millimeter, centimeter, meter) subtypes. The other type is based on a foot-inch system.

Figure 21 lists the various categories of scales. It is helpful to describe each scale and its use, beginning with the civil engineer's scale.

B. Civil Engineer's Scale   The civil engineer's scale was originally designed for map work, but its convenience soon made it popular for any work involving decimal divisions of the inch. One can use a scale of 1 in = 5 lb as easily as 1 in = 5 ft. The use of the civil engineer's scale is so general that it is often called simply the engineer's scale.

C. Metric System   The industrial world has been increasingly using the metric system of units. The United States is the only major country that has been hesitant to make a full commitment to metric units. The metric system got a major boost following the French Revolution when France established the standard meter in the

---

SCALES

DECIMAL BASE

Civil engineer's
    10, 20, 30, 40, 50, or 60 divisions to the inch, representing any desired unit, such as ft, 10 ft, 100 ft, mi

Metric
    10, 20, 30, 40, 50, or 60 divisions to the centimeter, representing any desired unit, such as millimeter, centimeter, meter; other scales for special use are available, such as 1:33-$\frac{1}{3}$, 1:75, 1:250

FOOT-INCH BASE

Mechanical engineer's

$1'' = 1''$ (full size)	$\frac{1}{2}'' = 1''$ ($\frac{1}{2}$ size)
$\frac{1}{4}'' = 1''$ ($\frac{1}{4}$ size)	$\frac{1}{8}'' = 1''$ ($\frac{1}{8}$ size)

Architect's or mechanical engineer's

$12'' = 1'\text{-}0''$ (full size)	$1'' = 1'\text{-}0''$ ($\frac{1}{12}$ size)	$\frac{1}{4}'' = 1'\text{-}0''$ ($\frac{1}{48}$ size)
$6'' = 1'\text{-}0''$ ($\frac{1}{2}$ size)	$\frac{3}{4}'' = 1'\text{-}0''$ ($\frac{1}{16}$ size)	$\frac{3}{16}'' = 1'\text{-}0''$ ($\frac{1}{64}$ size)
$3'' = 1'\text{-}0''$ ($\frac{1}{4}$ size)	$\frac{1}{2}'' = 1'\text{-}0''$ ($\frac{1}{24}$ size)	$\frac{1}{8}'' = 1'\text{-}0''$ ($\frac{1}{96}$ size)
$1\frac{1}{2}'' = 1'\text{-}0''$ ($\frac{1}{8}$ size)	$\frac{3}{8}'' = 1'\text{-}0''$ ($\frac{1}{32}$ size)	$\frac{3}{32}'' = 1'\text{-}0''$ ($\frac{1}{128}$ size)

FIG. 21   Standard scales. Special scales are available.

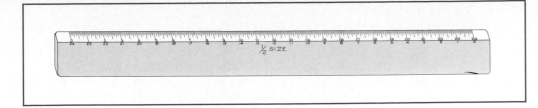

FIG. 22 A mechanical engineer's full- and half-size scale. Divisions are in inches to sixteenths of an inch. (The half-size scale is on the back of the full-size scale.)

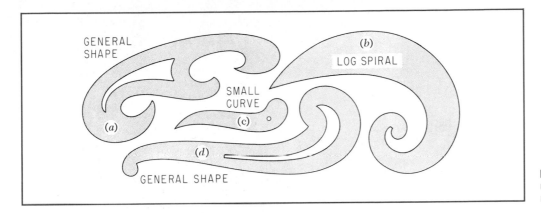

FIG. 23 Irregular curves. These are used for drawing curves where the radius of curvature is not constant.

1790s. The meter was defined as one ten-millionth of the straight-line quadrant connecting the earth's equator with the north pole.

The most recent form of the metric system is known as the SI system, named for the French *Système International d'Unités*. Since 1960, the meter has been defined in relation to the wavelength of the orange-red light of krypton 86. Metric standards for all SI units (length, mass, time, etc.) are administered by the International Standards Organization (ISO). The basic SI units and their abbreviations are given in Appendix 4.

Although the United States as a whole is not using the metric system, much of the machine-tool industry is. This is particularly true of companies heavily involved with export sales, such as agricultural, automotive, and aeronautical companies. The basic SI unit used is the millimeter (mm) for machine parts. The centimeter (cm), decimeter (dm), and meter (m) are not often used in the manufacturing trades.

D. Metric Scales Since a metric scale is on a decimal base just as is the civil engineer's scale, the use of the metric scale is almost the same as that for the civil engineer's scale. A desired scale is selected, say, 1 to 3. For the civil engineer's scale, 1 to 3 would be expressed as 1 = 3 (where 1 in = 3 in, ft, etc.). For the metric scale, 1 to 3 would be expressed as 1:3, using a colon instead of the equals sign.

The notation 1:3 means that one metric unit on the paper represents three metric units on the actual object. The metric unit is assumed to be millimeters for engineering drawings, unless otherwise stated. The same scale that gives a 1:3 could also be used for 1:30, 1:300, etc., by merely relocating the decimal point as the scale is read.

E. Mechanical Engineer's Scale Figure 21 shows that the mechanical engineer's scale is based on the foot-inch system. Four scales are normally used: full size, half size, quarter size, eighth size. The scale is laid off in inches and fractions thereof. Figure 22 shows a half-size scale. By using this scale the object is shrunk to one-half its actual size, but the dimensions remain the true dimensions, as they should. Mechanical engineer's scales are helpful in certain applications but are not used as widely as decimal scales for engineering work.

F. Architect's Scale The architect's scale finds considerable use in the drawing of buildings, cabinet work, and plumbing designs. All scales are in the foot-inch system, as seen in Fig. 21. All scales are reduction scales, except for the one marked "16" at the end of the scale. This scale is the full-size scale. The notation 16 means that the inch is subdivided into sixteenths.

For all reduction scales, some value in inches represents 1 ft. For example, if the sale is given as $1\frac{1}{2}''$ = $1'\text{-}0''$, each $1\frac{1}{2}$ in on the paper equals 1 ft (12 in) on the actual object. This scale would give eighth size, as previously noted in Fig. 21.

11. CURVES

Curved rulers, called "irregular curves" or "French curves," are used for curved lines other than circle arcs. The patterns for these curves are laid out in parts of ellipses and spirals or other mathematical curves in various combinations. For the student, one ellipse curve of the general shape of Fig. 23*a* or *d* and one spiral, either a logarithmic spiral (*b*) or one similar to the one used in Fig. 72, is sufficient. Figure 23*c* shows a useful small curve.

12. THE COMPASS

The compass has existed for centuries in one form or another, but modern compasses are excellent values compared with the relatively costly ones of earlier years. There are compasses for inking, lead, and a combination of ink and lead work.

Figure 24 shows a typical combination bow compass, as it is called. A 6-in overall length is common. The center-wheel adjustment is convenient for setting the particular arc to be drawn, and most students prefer the center-wheel adjustment to the friction-held adjuster shown in Fig. 27.

The standard 6-in compass will open to about 5 in. Lengthening bars can extend the maximum radius to be drawn. Such an arrangement is seen in Fig. 25. For very large radii, up to, say, 16 in, special beam compasses are recommended. Figure 26 shows one such beam compass.

The compass shown in Fig. 27 is an example of a combination lead and inking compass. An interchangeable steel inking point is shown instead of a possible lead insert piece.

13. DIVIDERS

Dividers are used for transferring measurements and for dividing lines into any number of equal parts. Dividers

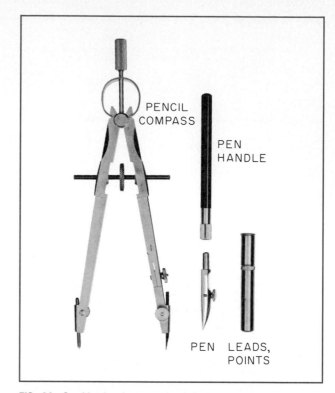

FIG. 24 Combination bow compass. Large bow. It performs triple service as dividers, pen, and pencil compass. Pen leg in handle makes ruling pen.

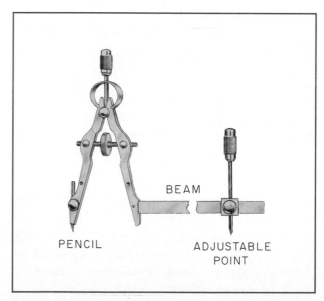

FIG. 25 Use of extender beam. Extender for large bow. This type extends the range of its center point, with the pen or pencil on the bow.

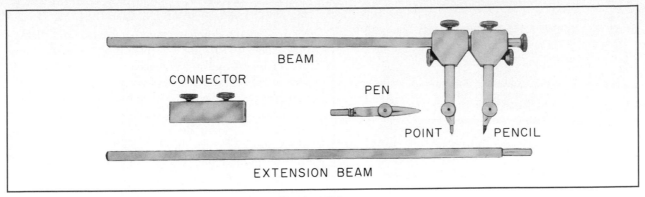

FIG. 26 A beam compass. This compass is adjustable for radii of 1 to 16 in.

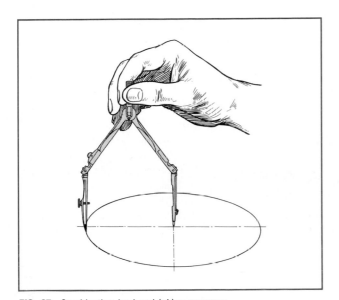

FIG. 27 Combination lead and inking compass.

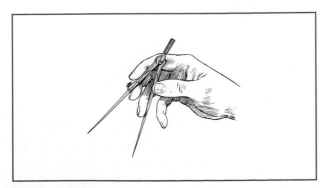

FIG. 28 Handling the dividers.

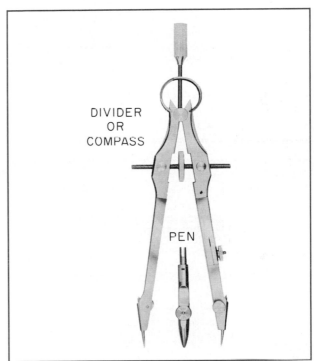

FIG. 29 Divider or compass.

are indispensable to the drafter, enabling transfers to be made quickly and accurately anywhere within a drawing. Figure 28 shows a popular style of divider which has a set position held by friction within the pivot joint. Another style is seen in Fig. 29; this is actually a combination compass and divider, depending upon whether a lead insert or needle point is used. The divider form is illustrated.

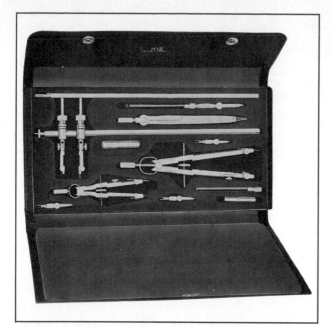

FIG. 30 A large-bow set, in traditional case containing many accessories.

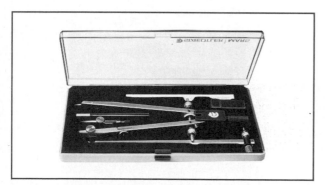

FIG. 31 Basic set in plastic case representing a sound value for the beginning student. *(Courtesy J. S. Staedtler, Inc.)*

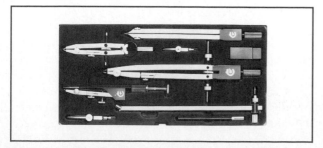

FIG. 32 Basic set with additional pieces. *(Courtesy J. S. Staedtler, Inc.)*

14. PURCHASING COMPASS AND DIVIDERS

Many students purchase their compasses and dividers individually to save the cost of full sets in cases. Some like the luxury and convenience of "case" instruments as seen in Figs. 30 to 32. The packaging of instruments is a matter of personal choice. A word of general advice is to purchase equipment of decent, solid quality, without getting into specialty items until one's work justifies the need. Specialty items, beyond the basic instruments we have been discussing, can be purchased as needed.

15. TEMPLATES

A wide range of templates are available to aid the drafter in making a variety of geometric shapes and symbols. Circle templates and ellipse templates are used often to speed drafting. Figures 33 and 34 show samples of such templates.

Specialized templates exist for architectural plumbing, electronic, and piping needs. Manufacturers of templates are glad to provide catalogs of the many template types on the market. Figures 35 to 37 give a sampling of template types.

16. PROTRACTORS

A protractor is an essential instrument when you need to measure or lay out a particular angle that is not a multiple of 15°. Some students make the mistake of purchasing a protractor that is too small, making measurements awkward and inaccurate. A protractor should be 5 or 6 in in length along the flat edge. Markings should be engraved, not just painted on. A common type of protractor that can serve well is shown in Fig. 38.

17. LETTERING DEVICES

The Braddock-Rowe triangle and the Ames lettering instrument are convenient devices used in drawing guide lines for lettering. See the discussion in the chapter on lettering (Chap. 3).

18. CHECKLIST OF INSTRUMENTS AND MATERIALS

Our discussion of basic instruments concludes with a list of the most-needed tools for the student and the beginning professional. Additions to this list can be made as desired, based on personal preference and need.

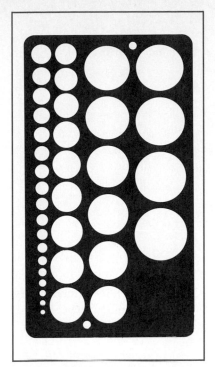

FIG. 33 Metric circle template. *(Courtesy Koh-I-Noor Rapidograph Inc.)*

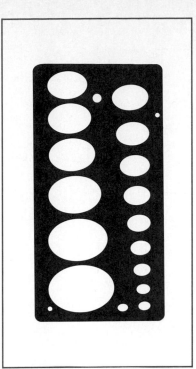

FIG. 34 45° ellipse template. *(Courtesy Koh-I-Noor Rapidograph Inc.)*

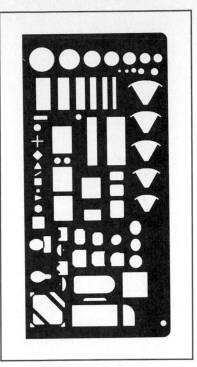

FIG. 35 Template for architectural symbols. *(Courtesy Koh-I-Noor Rapidograph Inc.)*

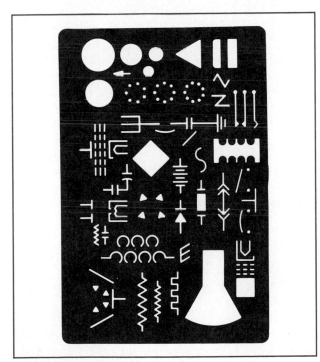

FIG. 36 Template for electronic symbols. *(Courtesy Koh-I-Noor Rapidograph Inc.)*

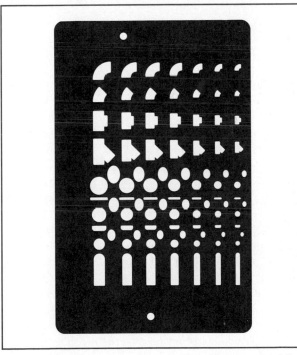

FIG. 37 Template for pipe fitting symbols. *(Courtesy Koh-I-Noor Rapidograph Inc.)*

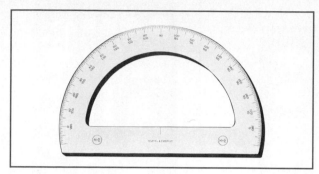

FIG. 38 A protractor of the semicircular type. *(Courtesy Koh-I-Noor Rapidograph Inc.)*

1. Bow compass
2. Dividers
3. T square
4. 45° triangle
5. 30–60° triangle
6. Civil engineer's scale
7. Metric scale
8. Architect's scale
9. Mechanical drafting pencil
10. 2H, H, and F leads
11. Irregular (french) curve
12. Circle template
13. Protractor
14. Eraser, plastic
15. Erasing shield
16. Drafting tape
17. Sandpaper pad or lead pointer

Specific sizes and brands are left to the student's or instructor's discretion. Inking equipment has not been listed, but could represent another investment when needed.

USE OF INSTRUMENTS

19. PREPARATION FOR DRAWING

In beginning to use drawing instruments, it is important to learn to handle them correctly. Carefully read the instructions and observe strictly all details of technique.

Facility will come with continued practice, but it is essential to adhere to good form from the outset.

It is best to make a few drawings solely to become familiar with the handling and feel of the instruments so that later, in working a drawing problem, you will not lose time because of faulty manipulation. Practice accurate penciling first, and do not attempt inking until you have become really proficient in penciling. With practice, the correct, skillful use of drawing instruments will become a subconscious habit.

For competence in drawing, *accuracy* and *speed* are essential, and in commercial work neither is worth much without the other. It is well to learn as a beginner that a *good* drawing can be made as quickly as a *poor* one.

The drawing table should be set so that the light comes from the left, and it should be adjusted to a con-venient height, that is, 36 to 40 in, for use while sitting on a standard drafting stool or while standing. There is more freedom in drawing standing, especially when working on large drawings. The board, for use in this manner, should be inclined at a slope of about 1 to 8. Since it is more tiring to draw standing, many modern drafting rooms use tables so made that the board can be used in an almost vertical position and can be raised or lowered so that the drafter can use a lower stool with swivel seat and backrest, thus working with comfort and even greater freedom than when an almost horizontal board is used.

The instruments should be placed within easy reach, on the table or on a special tray or stand that is located beside the table. The table, the board, and the instruments should be wiped with a dustcloth before starting to draw.

Select a paper type of your choice and attach the paper to the board with drafting tape, using a short strip across each of the four corners of the paper. Some drafters use special thumbtacks, but these are not so popular as in earlier years since advances have been made in easy-to-remove drafting tape. Attach the paper in a position comfortable for you, the drafter.

20. KEEPING A DRAWING CLEAN

A clean drawing is essential to producing clean reproductions. Smudges and random dirt spots on a drawing degrade the work of the drafter. More important, dirt can sometimes cause confusion in reading vital information contained in the drawing.

Producing a clean drawing involves common sense. First, one should have clean, dry hands. Second, one should remove excess graphite particles from a freshly sharpened lead. A simple wipe with a cloth or tissue will

do. Third, drafting instruments must be kept clean. A clean cloth kept readily available is helpful. Also an occasional washing with a mild liquid detergent will remove stubborn dirt. Dry promptly any wooden portions of equipment, such as a T square. Rubbing drafting instruments with erasers is not recommended since the surfaces will become slightly abraded, causing dirt to collect even faster.

21. ALPHABET OF LINES

We discuss the styles of lines used in drawings before discussing the use of instruments in greater detail. The lines seen in Fig. 39 cover virtually every situation that can occur in drawing. Specifics of conventions with re-

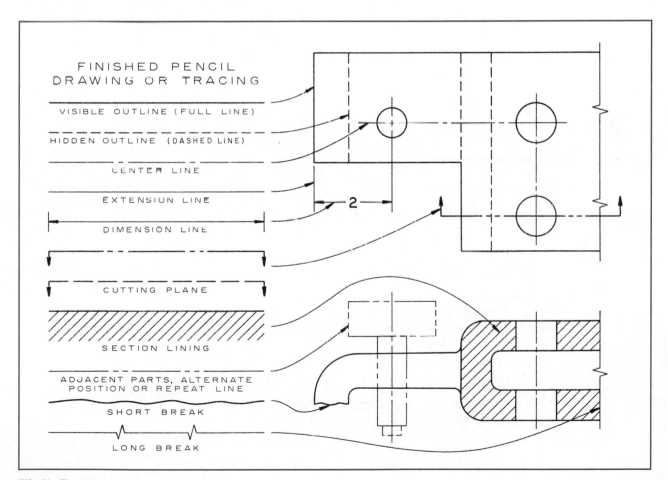

FIG. 39 The alphabet of lines for pencil drawings.

gard to lines may be found in ANSI Y14.2M—1979 of the American National Standards Institute.

The ANSI standard recommends two widths of lines for manually drawn work. The thin-line width should be about 0.016 in [0.35 mm], while the thick-line width should be about 0.032 in [0.7 mm]. A line gage is given in Fig. 40 for your information.

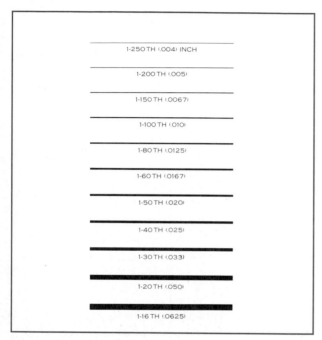

1-250 TH (.004) INCH

1-200 TH (.005)

1-150 TH (.0067)

1-100 TH (.010)

1-80 TH (.0125)

1-60 TH (.0167)

1-50 TH (.020)

1-40 TH (.025)

1-30 TH (.033)

1-20 TH (.050)

1-16 TH (.0625)

FIG. 40 Line gage. Draw a line on the paper to be used and apply it here to determine the width.

One sees in Fig. 39 that thin lines apply to the following lines: hidden, center, extension, dimension, section-lining, alternate-position (phantom), and long-break lines. Thick lines apply only to visible, cutting-plane, and short-break lines. Thin lines are often drawn with H or 2H lead grades, while thick lines can be done well with softer leads, such as F or H.

The line widths for ink work are virtually the same as for pencil work. The one exception is that widths for visible and hidden lines are slightly thicker.

If drawings are done mechanically, as for a computer-driven plotter, one width for all lines is often satisfactory. However, plotter pens are available in various widths (and colors). The width selected depends on the size of the drawing (large drawings are sometimes drawn with wider lines). Use of a drawing is also a factor in line width. If a drawing is to be reduced in size, a wider line can be chosen.

22. USE OF THE PENCIL

The grade of pencil must be selected carefully, with reference to the surface of the paper as well as to the line quality desired. For a pencil layout on detail paper of good texture, a pencil as hard as 5H or 6H may be used; for finished pencil drawings on the same paper, 2H, 3H, or 4H pencils give the blacker line needed. For finished pencil drawings or tracings on vellum, softer pencils, H to 3H, are employed to get printable lines. The F pencil is used for technical sketching, and the H is popular for lettering. In each case the pencil must be hard enough

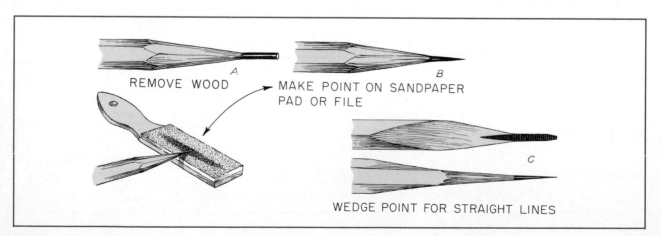

REMOVE WOOD

MAKE POINT ON SANDPAPER PAD OR FILE

WEDGE POINT FOR STRAIGHT LINES

FIG. 41 Sharpening the pencil.

not to blur or smudge but not so hard as to cut grooves in the paper under reasonable pressure. (Refer back to Fig. 6.)

A lead is not useful until it is sharpened. If you are using a wooden pencil, cut away the wood with a knife or preferably with a drafter's pencil sharpener, which removes the wood but doesn't sharpen the lead. As seen in Fig. 41, sharpen the lead by twirling the pencil as the lead is rubbed with long, even strikes against the sandpaper pad or file or placed in a special lead sharpener.

A flat or wedge point will not wear away in use so fast as a conic point, and thus some prefer it for straight-line work. The long wedge point illustrated at C is made by first sharpening, as at A, then making the two long cuts on opposite sides, as shown, then flattening the lead on the sandpaper pad or file, and finishing by touching the corners to make the wedge point narrower than the diameter of the lead.

Have the sandpaper pad within easy reach, and *keep the pencils sharp*. Most poor quality in pencil work is caused by dull leads, a simple fact that is often ignored by beginning students. Wipe the excess graphite from a freshly sharpened lead to keep loose granules from falling on the paper, later to be smudged by drafting instruments.

Not only must pencil lines be clean and sharp, but it is absolutely necessary that all the lines of each kind be uniform, firm, and opaque. This means a careful choice of pencils and the proper use of them. The attempt to make a dark line with too hard a pencil results in cutting deep grooves in the paper. Hold the pencil firmly, yet with as much ease and freedom as possible.

Keep an even, constant pressure on the pencil, and when using a conic point, rotate the pencil as the line is drawn so as to keep both the line and pencil sharp. Use a drafter's brush or soft cloth occasionally to dust off excess graphite from the drawing.

Too much emphasis cannot be given to the importance of clean, careful, accurate penciling. Never entertain the thought that poor penciling can be corrected in tracing.

Mechanical pencils may also be sharpened by use of a sandpaper pad. This is an effective and inexpensive means of sharpening, but one must be careful to prevent the graphite on the pad from falling onto the working surface.

The lead pointer, Fig. 7, is also a useful device.

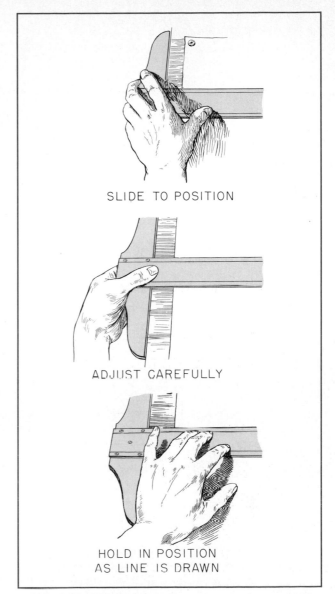

SLIDE TO POSITION

ADJUST CAREFULLY

HOLD IN POSITION
AS LINE IS DRAWN

FIG. 42 Manipulating the T square. The head must be firmly against the straight left edge of the board.

23. USE OF THE T SQUARE

The T square and the triangles have straight edges and are used for drawing straight lines. Horizontal lines are drawn with the T square, which is used with its head against the left edge of the drawing board and manipulated as follows: Holding the head of the tool, as shown in Fig. 42 (top), slide it along the edge of the board to a spot very near the position desired. Then, for closer ad-

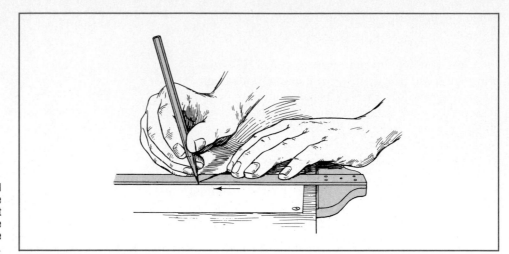

FIG. 43 Drawing a horizontal line. Hold the T square with the left hand; draw the line from left to right; incline the pencil in the direction of stroke, so that the pencil "slides" over the paper.

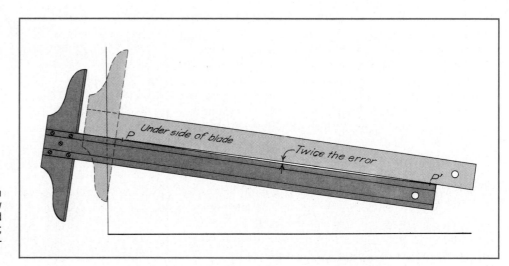

Under side of blade

Twice the error

FIG. 44 To test a T square, turn the T square upside down; draw the line; turn it right side up and align it with the original line; draw the second line and compare it with the first.

justment, change your hold either to that shown, in which the thumb remains on top of the T-square head and the other fingers press against the underside of the board, or, as is more usual, to that shown at the bottom of Fig. 42, in which the fingers remain on the T square and the thumb is placed on the board.

Figure 43 shows the position of the hand and pencil for drawing horizontal lines. Note that the pencil is inclined in the direction the line is drawn, that is, toward the right, and also slightly away from the body so that the pencil point is as close as possible to the T-square blade.

In drawing lines, take great care to keep them accurately parallel to the guiding edge of the T square. The pencil should be held lightly, but close against the edge, and the angle should not vary during the progress of the line. Horizontal lines should always be drawn from left to right. A T-square blade can be tested for straightness by drawing a sharp line through two points and then turning the square over and with the same edge drawing another line through the points, as shown in Fig. 44.

24. USE OF THE TRIANGLES

Vertical lines are drawn with the triangle, which is set against the T square with the perpendicular edge nearest the head of the square and thus toward the light (Fig. 45). These lines are always drawn upward, from bottom to top.

In drawing vertical lines, the T square is held in position against the left edge of the board by the thumb and little finger of the left hand while the other fingers of this hand adjust and hold the triangle. As the line is drawn, pressure of all the fingers against the board will hold the T square and triangle firmly in position.

As with the T square, care must be taken to keep the line accurately parallel to the guiding edge. Note the position of the pencil in Fig. 45.

The triangles must always be used in contact with a guiding straightedge. To ensure accuracy, never work to the extreme corner of a triangle; to avoid having to do so, keep the T square below the lower end of the line to be drawn.

With the T square against the edge of the board, lines at 45° are drawn with the standard 45° triangle, and lines at 30° and 60° with the 30-60° triangle (Fig. 46). With vertical and horizontal lines included, lines at increments of 45° are drawn with the 45° triangle as at *B*, and lines at 30° increments with the 30-60° triangle as at *A*. Both triangles are used for angles of 15, 75, 105°, etc. (Fig. 47). Thus any multiple of 15° is drawn directly; and a circle is divided with the 45° triangle into 8 parts, with

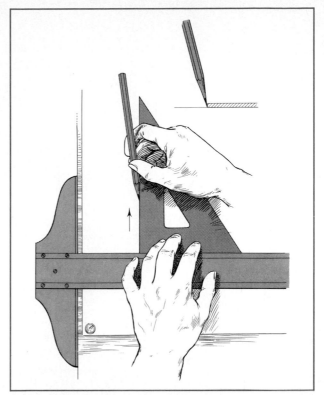

FIG. 45 Drawing a vertical line. With the T square and triangle in position, draw the line from bottom to top—always away from the body.

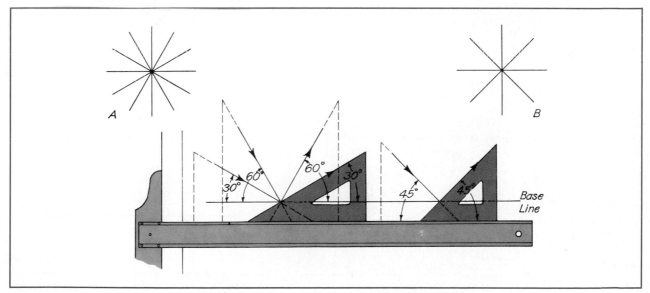

FIG. 46 To draw angles of 30°, 45°, 60°, multiples of 30° are drawn with the 30–60° triangle, multiples of 45° with the 45° triangle.

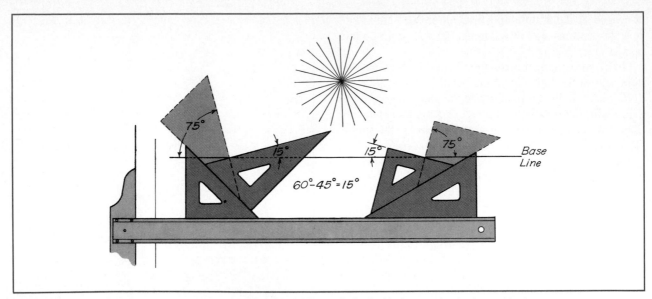

FIG. 47 To draw angles of 15° and 75°, angles in increments of 15° are obtained with the two triangles in combination.

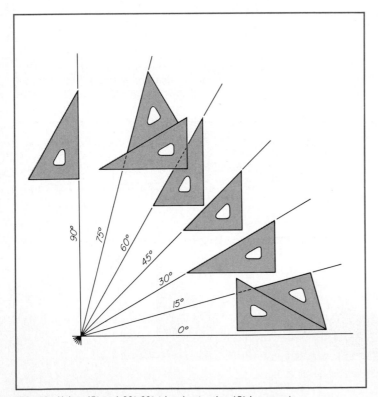

FIG. 48 Using 45° and 30°-60° triangles to give 15° increments.

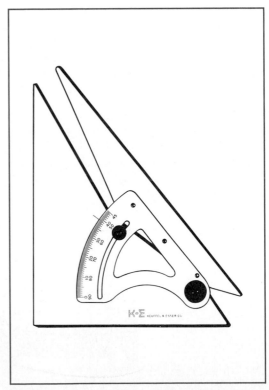

FIG. 49 An adjustable triangle. (*Courtesy Keuffel & Esser Co.*)

the 30-60° triangle into 12 parts, and with both into 24 parts, as seen in Fig. 48.

Figure 49 shows how the combination of a 45° and a 30-60° triangle can be manipulated to give any multiple of a 15° angle.

Some drafters add an adjustable triangle to their basic sets. Figure 50 shows one type of adjustable triangle. The versatility of such a triangle offsets its additional cost.

■ TO DRAW ONE LINE PARALLEL TO ANOTHER (Fig. 51) Adjust to the given line a triangle held against a straightedge, hold the guiding edge in position, and slide the triangle on it to the required position.

■ TO DRAW A PERPENDICULAR TO ANY LINE (Fig. 52) Place a triangle with one edge against the T square (or another triangle), and move the two until the hypotenuse of the triangle is coincident with the line, as at A; hold the T square in position and turn the triangle, as shown, until its other side is against the T square; the hypotenuse will then be perpendicular to the original line. Move the triangle to the required position. A quicker method is to set the triangle with its hypotenuse against the guiding edge, fit one side to the line, slide the triangle to the required point, and draw the perpendicular, as shown at B.

Never attempt to draw a perpendicular to a line with only one triangle by placing one leg of the triangle along the line.

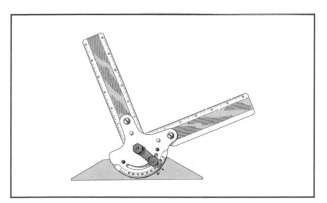

FIG. 50 Combined triangle, scale, and protractor. Several different types are made.

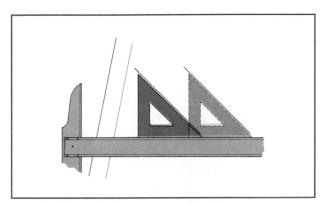

FIG. 51 To draw parallel lines. With the T square as a base, the triangle is aligned and then moved to the required position.

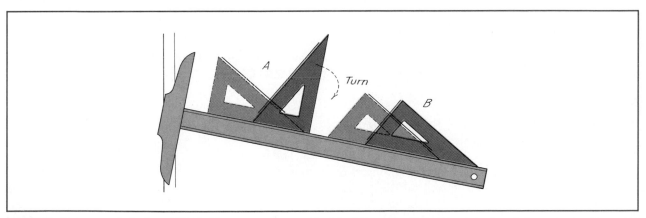

FIG. 52 To draw perpendicular lines. With the T square as a base and the triangle in position *A*, the triangle is aligned, then rotated and moved to perpendicular position; for position *B*, only the triangle is moved.

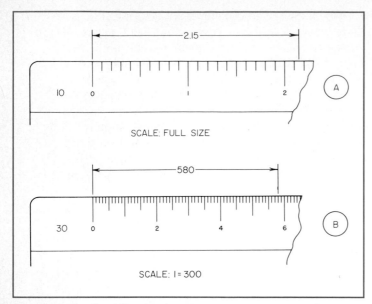

FIG. 53 Reading the civil engineer's scale.

FIG. 54 Making a measurement. Place the scale in position; the distance is marked on paper by short, light lines.

25. THE LEFT-HANDED DRAFTER

If you are left-handed, reverse the T square and triangles left for right as compared with the regular right-handed position. Use the head of the T square along the right edge of the board, and draw horizontal lines from right to left. Place the triangle with its vertical edge to the right, and draw vertical lines from bottom to top. The drawing table should be placed with the light coming from the right.

26. USE OF SCALES

A. Civil Engineer's Scale Use of the scale is not difficult. Refer to Fig. 53. The number of graduations per actual inch is shown at the end of the scale. In Fig. 53*a* one sees the full-size scale. The value illustrated is 2.15 in. The value could be assigned other units, such as dollars, if one were constructing a graph of some kind.

Note that the 10 given at the end of the scale tells only the number of divisions per inch. Remember that this is a full-size scale. When you look at the other scale given in Fig. 53*b*, the 30 at the end of the scale therefore tells you that there are 30 divisions to the inch. This scale automatically multiplies by 3 anything on the 10 scale.

The scaling on the 10 scale is therefore 1 (on paper) = 1 (actual), while the scaling on the 30 scale is 1 (on paper) = 3 (actual) *unless otherwise stated*. For the case given, 1 = 300, making it necessary to multiply by 100 any reading on the 30 scale. The reading of 5.8 is therefore read as 580. The units could be inches, feet, pounds, feet per second, etc. Using the scale as 1 = 300 means that the object on the paper is 1/300 size compared with the actual object.

To make a measurement, place the scale on the drawing where the distance is to be laid off, align the scale in the direction of the measurement, and make a *light* short dash with a sharp pencil at the proper graduation mark (Fig. 54).

Measurements should not be made on a drawing by taking distances off the scale with dividers, because this method is time-consuming and no more accurate than the regular methods. To avoid cumulative errors, successive measurements on the same line should if possible be made without shifting the scale.

B. Metric Scale Figure 55 gives examples of the use of a metric scale. The upper scale is 1:1, meaning 1 mm = 1 mm. Therefore, the value 43 is read as 43 mm. Do not be confused by the 10 at the left-hand side of the scale. This 10 merely means that 1 centimeter is divided into 10 parts.

The lower scale is to be read to a scale of 1:30, even though the scale itself multiplies by only 3 the values of the upper scale (e.g., 50 on the upper scale becomes 150 on the lower scale). Therefore, to read 1:30, not 1:3, the value of 155 mm must be multiplied by 10 to read 1550 mm.

C. Architect's Scale   The architect's scales, being "open divided" (that is, with units shown along entire length) and to stated reductions, such as 3″ = 1′-0″, may require some study by the beginner in order to prevent confusion and mistakes. As an example, consider the scale of 3″ = 1′-0″ (see Fig. 56). This is the first reduction scale of the usual triangular scale; on it the distance of 3 in is divided into 12 equal parts, and each of these is subdivided into eighths. This distance should be thought of not as 3 in but as a foot divided into inches and eighths of an inch. Note that the divisions start with the *zero* on the inside, the inches of the divided foot running to the *left* and the open divisions of feet to the *right*, so that dimensions given in feet and in inches may be read directly, as 1′-0½″.

On the other end will be found the scale of 1½″ = 1′-0″, or eighth size, with the distance of 1½ in divided on the right of the zero into 12 parts and subdivided into quarter inches, with the foot divisions to the left of the zero coinciding with the marks of the 3-in scale. Note again that in reading a distance in feet and inches, for example, the 2′-7⅛″ distance in Fig. 56, feet are determined to the left of the zero and inches to the right of it. The other scales, such as ¾″ = 1′-0″ and ¼″ = 1′-0″, are divided in a similar way, the only difference being in the value of the smallest graduations. The scale of 3/32″ = 1′-0″, for example, can be read only to the nearest 2 in.

27. USE OF THE DIVIDERS

A divider can be used to divide lines into any number of equal parts. For example, to bisect a line, the dividers are opened at a guess to roughly half the length. This distance is stepped off on the line, the drafter holding the instrument by the handle with the thumb and forefinger. If the division is short, the leg should be thrown out to half the remainder (estimated by eye), without removing the other leg from the paper, and the line spaced again with this new setting (Fig. 57). If the result does not come out exactly, the operation can be repeated.

Similarly, a line can be divided into any number of equal parts, say, five, by estimating the first division, stepping this lightly along the line, with the dividers held

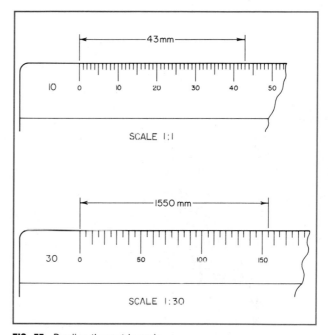

FIG. 55 Reading the metric scale.

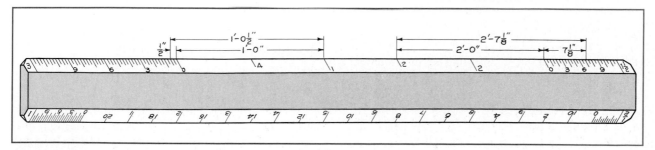

FIG. 56 Reading an open-divided scale. Feet are given on one side of the zero, inches and fractions on the other.

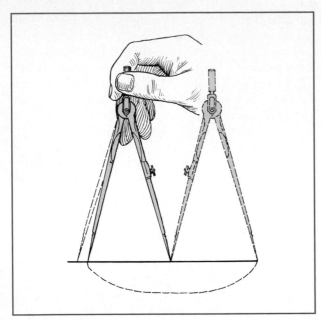

FIG. 57 Bisecting a line. Half is estimated; then the dividers are readjusted by estimating half the original error.

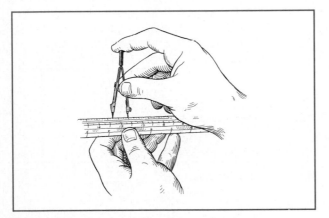

FIG. 58 Setting the compass to radius size. Speed and accuracy are obtained by adjusting directly on the scale.

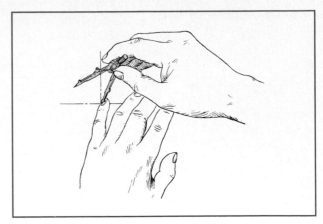

FIG. 59 Guiding the needle point. For accuracy of placement, guide with the little finger.

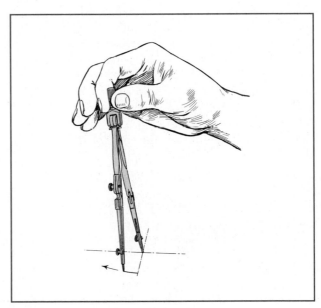

FIG. 60 Starting a circle. The compass is inclined in the direction of the stroke.

vertically by the handle, and turning the instrument first in one direction and then in the other. If the last division falls short, one-fifth of the remainder should be added by opening the dividers, keeping one point on the paper. If the last division is over, one-fifth of the excess should be taken off and the line respaced.

28. USE OF THE COMPASS

To use the compass for drawing a circle, set the compass on the scale, as shown in Fig. 58, and adjust it to the radius needed; then place the needle point at the center on the drawing, guiding it with the left hand (Fig. 59). Raise the fingers to the handle and draw the circle in one sweep, rolling the handle with the thumb and forefinger, inclining the compass slightly in the direction of the line (Fig. 60).

The position of the fingers after the rotation is shown in Fig. 61. The pencil line can be brightened, if necessary, by making additional turns. Circles up to perhaps 3 in in diameter can be drawn with the legs of the compass straight, but for larger sizes, both the needle-point leg and the pencil or pen leg should be bent at the knuckle joints so as to be perpendicular to the paper. The 6-in compass may be used in this way for circles up to perhaps 10 in in diameter; larger circles are made by using the extender beam, as illustrated in Fig. 25, or the beam compass. In drawing concentric circles, the *smallest* should always be drawn first, before the center hold has become worn.

The bow instruments are used for small circles, particularly when a number are to be made of the same diameter.

It should be noted that most compasses are supplied by the manufacturers with lead that is too hard. The result is light, gray arcs. It is suggested that lead be used that is one grade softer than is used for straight-line work.

If you use an H lead for straight lines, an F lead for your compass should give arcs just as dark as your straight lines. The softer F lead is recommended because it is not usually possible to press as hard on a compass to make arcs as it is on a pencil to make lines.

29. THE RULING PEN

The technical pens, described in Sec. 6, are used more widely than ruling pens. However, the ruling pen is still very useful on rough matted surfaces, such as poster board.

The important feature is the shape of the blades; they should have a well-designed ink space between them, and their points should be rounded (actually elliptical in form) equally, as in Fig. 62. If pointed, as in Fig. 63, the ink will arch up as shown and will be hard to start. If rounded to a blunt point, as in Fig. 64, the ink will flow too freely, forming blobs and overruns at the ends of the lines. Pens in constant use become dull and worn, as illustrated in Fig. 65.

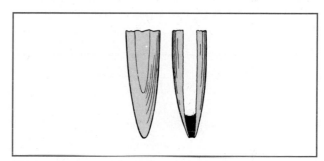

FIG. 62 Correct shape of pen nibs. A nicely uniform elliptical shape is best.

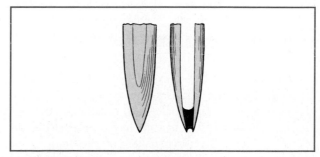

FIG. 63 Incorrect shape of pen nibs. This point is much too sharp. Ink will not flow well.

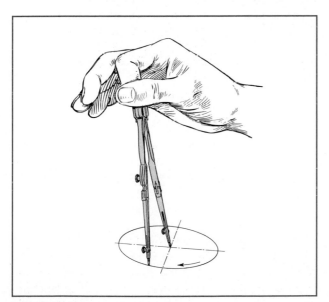

FIG. 61 Completing a circle. The stroke is completed by twisting the knurled handle in the fingers.

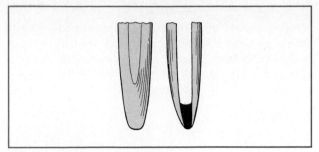

FIG. 64 Incorrect shape of pen nibs. This point is too flat. Ink will blob at the beginning and end of the line.

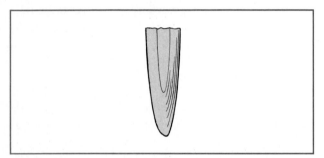

FIG. 65 Shape of worn pen nibs. This point needs sharpening to the shape of Fig. 62.

FIG. 66 Sharpening a pen. After bringing the point to the shape of Fig. 62 work down the sides by a rocking motion to conform to blade contour.

A. To Sharpen a Pen   Several materials may be used for sharpening: hard Arkansas knife stone, silicon carbide cloth or paper, or crocus cloth for finishing.

The nibs must first be brought to the correct shape, as in Fig. 62. Screw the nibs together until they touch and, holding the pen as in drawing a line, draw it back and forth on the stone, starting the stroke with the handle at 30° or less with the stone and swinging it up past the perpendicular as the line across the stone progresses. This will bring the nibs to exactly the same shape and length, leaving them very dull. Then open them slightly, and sharpen each blade in turn, on the outside only, until the bright spot on the end has just disappeared. Hold the pen, as in Fig. 66, at a small angle with the stone and rub it back and forth with a slight oscillating or rocking motion to conform to the shape of the blade. A stone 3 or 4 in long held in the left hand with the thumb and fingers gives better control than one laid on the table.

The blades should not be sharp enough to cut the paper when tested by drawing a line across it without ink. If oversharpened, the blades should again be brought to touch and a line swung very lightly across the stone as in the first operation. When tested with ink, the pen should be capable of drawing clean, sharp lines down to the finest hairline. If these finest lines are ragged or broken, the pen is not perfectly sharpened.

B. Use of the Ruling Pen   The ruling pen is always used in connection with a guiding edge—T square, triangle, or curve. The T square and triangle should be held in the same positions as for penciling.

To fill the pen, take it to the bottle and touch the quill filler between the nibs. Be careful not to get any ink on the outside of the blades. If a plastic squeeze bottle is used, place the small spout against the sides of the nibs and carefully squeeze a drop of ink *between* the nibs. Not more than 3/16 to 1/4 in of ink should be put in; otherwise the weight of the ink will cause it to drop out in a blot. The pen should be held in the fingertips, as illustrated in Fig. 67, with the thumb and second finger against the sides of the nibs and the handle resting on the forefinger. Observe this hold carefully, as the tendency will be to bend the second finger to the position used when a pencil or writing pen is held. The position illustrated aids in keeping the pen at the proper angle and the nibs aligned with the ruling edge.

The pen should be held against the straightedge or guide with the blades parallel to it, the screw on the outside and the handle inclined slightly to the right and always kept in a plane passing through the line and perpendicular to the paper. The pen is thus directed by the upper edge of the guide, as illustrated in actual size in Fig. 68. If the pen point is thrown out from the perpendicular, it will run on one blade and make a line that is ragged on one side. If the pen is turned in from the perpendicular, the ink is likely to run under the edge of the guide and cause a blot.

A line is drawn with a steady, even arm movement, the tips of the third and fourth fingers resting on and sliding along the straightedge, keeping the angle of inclination constant. Just before the end of the line is reached, the two guiding fingers on the straightedge should be stopped and, without stopping the motion of the pen, the line finished with a finger movement. Short lines are drawn with this finger movement alone. When the end of the line is reached, the pen is lifted quickly and the straightedge moved away from the line. The pressure on the paper should be light but sufficient to give a clean-cut line, and it will vary with the kind of paper and the sharpness of the pen. The pressure against the 'T' square, however, should be only enough to guide the direction.

If the ink refuses to flow, it may be because it has dried in the extreme point of the pen. If pinching the blades slightly or touching the pen on the finger does not start it, the pen should immediately be wiped out and fresh ink supplied. Pens must be wiped clean after using.

In inking on either paper or cloth, the full lines will be much wider than the pencil lines. You must be careful to have the center of the ink line cover the pencil line, as illustrated in Fig. 69.

Instructions in regard to the ruling pen apply also to the compass. The compass should be slightly inclined in

the direction of the line and both nibs of the pen kept on the paper; bend the knuckle joints, if necessary, to effect this.

It is a universal rule in inking that *circles and circle arcs must be inked first*. It is much easier to connect a straight line to a curve than a curve to a straight line.

C. Tangents   Note particularly that two lines are tangent to each other when the center lines of the lines are tangent and not simply when the lines touch each other; thus at the point of tangency, the width will be equal to the width of a single line (Fig. 70). Before inking tangent

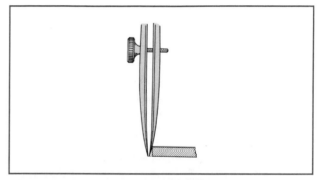

FIG. 68 Correct pen position. Even though the pen is inclined in the direction of the stroke, both nibs must touch the paper equally.

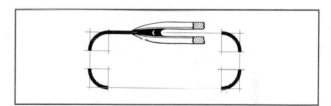

FIG. 69 Inking over pencil line. *Center* ink line over original layout line.

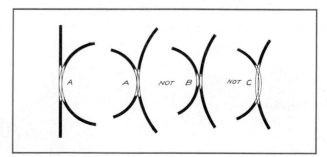

FIG. 70 Correct and incorrect tangents. Lines must be the width of *one* line at tangent point.

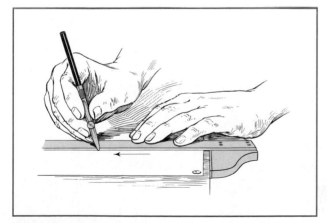

FIG. 67 Correct position of ruling pen. The pen is inclined in the direction of the stroke but held perpendicular to the paper, as in Fig. 68.

lines, the point of tangency should be marked in pencil. For an arc tangent to a straight line, this point will be on a line through the center of the arc and perpendicular to

the straight line, and for two circle arcs it will be on the line joining their centers.

D. Line Practice Using pencil, take a blank sheet of paper and practice making straight lines and circles in all the forms—full, dashed, etc.—shown in Fig. 39. Include starting and stopping lines, with special attention to tangents and corners.

In ink, proceed as for pencil practice and pay particular attention to the weight of lines and to the spacing of dashed lines and center lines.

Common problems in the use of a pen can lead to faulty lines as seen in Fig. 71. It is usual to experience some of these problems when you begin inking work for the first time. Don't become discouraged, but keep practicing so that your skill can improve.

30. USE OF THE FRENCH CURVE

The french curve is used as a pencil or pen guide for noncircular curves. Select a sufficient number of points to determine the curve desired. Then apply the french curve to the points, selecting a portion of the french curve that will fit a section of the desired curve most

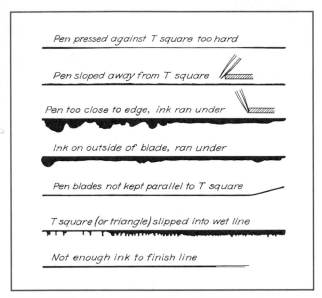

Pen pressed against T square too hard

Pen sloped away from T square

Pen too close to edge, ink ran under

Ink on outside of blade, ran under

Pen blades not kept parallel to T square

T square (or triangle) slipped into wet line

Not enough ink to finish line

FIG. 71 Faulty ink lines. The difficulty is indicated in each case.

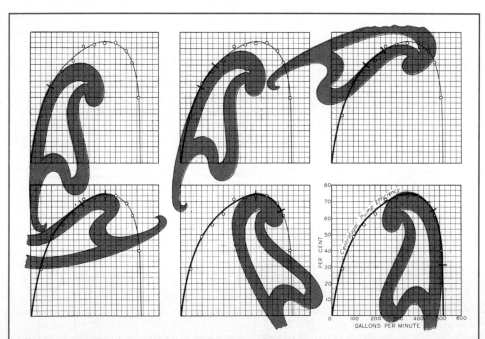

FIG. 72 Use of the french curve. The changing curvature of line and curve must match.

nearly. See that the french curve is placed so that the direction in which its curvature increases is the direction in which the curvature of the line increases (Fig. 72). In drawing the part of the line matched by the curve, *always* stop a little short of the distance in which the guide and the line seem to coincide. After drawing this portion, shift the curve to find another place that will coincide with the continuation of the line. In shifting the curve, take care to preserve smoothness and continuity and to avoid breaks or cusps. Do this by seeing that in its successive positions the curve is always adjusted so that it coincides for a short distance with the part of the line already drawn. Thus at each junction the tangents will coincide.

If the curved line is symmetrical about an axis, marks locating this axis, after it has been matched accurately on one side, may be made in pencil on the curve and the curve then reversed. In such a case take exceptional care to avoid a "hump" at the joint.

31. CAUTIONS IN THE USE OF INSTRUMENTS

To complete this discussion of instruments, here are a few points to remember:

Never use the scale as a ruler for drawing lines.

Never draw horizontal lines with the lower edge of the T square.

Never use the lower edge of the T square as a horizontal base for the triangles.

Never cut paper with a knife and the edge of the T square as a guide.

Never work with a dull pencil.

Never sharpen a pencil over the drawing board.

Never jab the dividers into the drawing board.

Never oil the joints of compasses.

Never use the dividers as reamers, pincers, or picks.

Never use a blotter on inked lines.

Never screw the pen adjustment past the contact point of the nibs.

Never leave the ink bottle uncorked.

Never hold the pen over the drawing while filling.

Never scrub a drawing all over with an eraser after finishing. It takes the life out of the lines.

Never begin work without wiping off the table and instruments.

Never put instruments away without cleaning them. This applies particularly to pens.

Never put bow instruments away without opening to relieve the spring.

Never work on a table cluttered with unneeded instruments or equipment.

Never fold a drawing or tracing.

SPECIALIZED INSTRUMENTS AND MACHINES

32. SPECIALIZED INSTRUMENTS

There are occasions when the drafter needs a special instrument or tool to accomplish a particular task. A number of special tools have been developed over the years, tools that still find use today. A selection of some of these instruments is shown in Figs. 73 through 79. By studying the figure captions one can sense the purpose of each tool. A professional drafter will discover additional specialized instruments in industrial practice.

33. DRAFTING MACHINES

Mention should be made of equipment beyond the basic instruments described to this point. More sophisticated equipment has value to the professional designer and engineer who may spend many hours per week in design work. An often-seen device is the drafting machine, mounted on the drafting boards in design rooms of corporations. Figure 18 is an example of a drafting machine illustrated previously. The rotary head can be indexed to

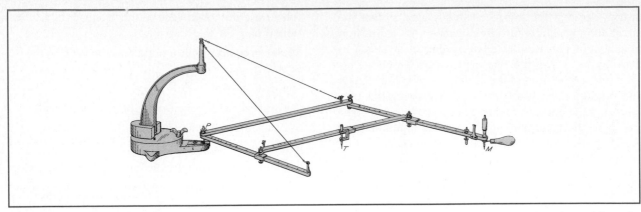

FIG. 73 A suspended pantograph. The illustration shows the instrument set up for enlarging: tracer point *(T)* following the points and lines of a drawing gives enlargement at marking point *(M)*. For reduction, *(T)* and *(M)* are reversed in position.

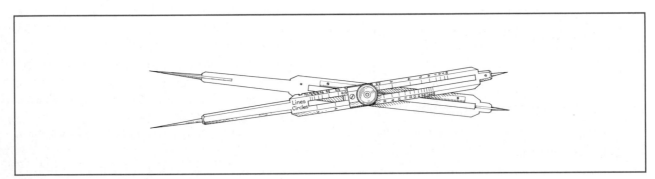

FIG. 74 Proportional dividers. The pivot is adjustable for different values of reduction or enlargement.

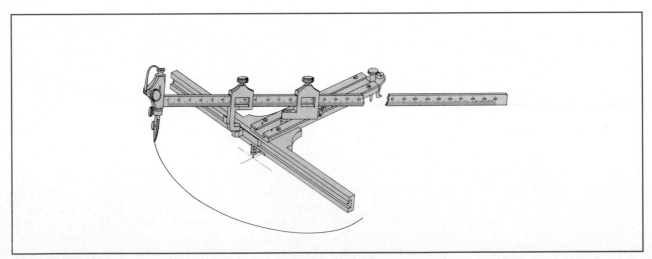

FIG. 75 An ellipsograph. Several styles are manufactured. This one is well made and accurate.

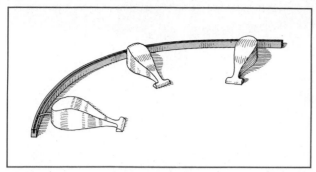

FIG. 76 A spline and "ducks." This flexible-curve ruler is used principally for drawing large curves, especially in lofting work.

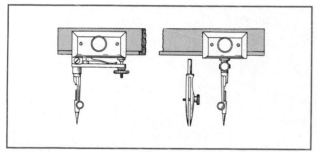

FIG. 77 A beam compass. This one employs a wooden beam, obtainable in various lengths up to 6 feet.

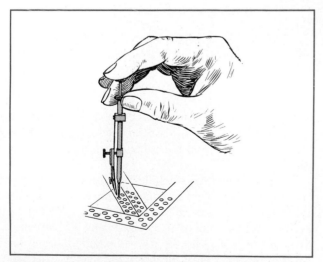

FIG. 78 A drop pen. This type of compass is particularly efficient for drawing a number of small circles of the same diameter. These are much used in structural drawing.

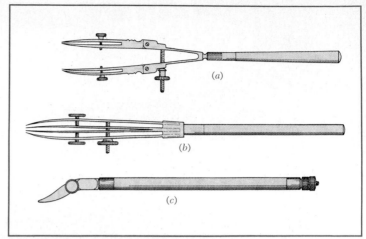

FIG. 79 Special pens. *(a)* "Railroad" pen for drawing double lines; *(b)* border pen for wide lines; *(c)* contour pen used in map drawing.

any angle. The horizontal and vertical scales rotate with the rotary head. In addition, horizontal and vertical movement of a rotary head–scales assembly is possible. This enables the drafter to move the scales to any position on the board. Such a machine includes the functions of T square, triangles, and protractor in one instrument, an obvious advantage to facilitate efficient drafting.

34. AUTOMATED DRAFTING

The ultimate drafting machine is one that drafts without the aid of the human hand. Automated drafting machines often take the form of an XY plotter, shown in Figs. 80 and 81. The plotter has a lead pointer, pen, or stylus which is controlled by input data coming from a computer. Large and intricate drawings are possible on plotters and can be done in a fraction of the time needed for hand drawings. Work is usually done in ink, using tungsten carbide tips.

The computer receives its instructions from a specific program written by an engineer or technician. The program can direct the computer to generate information for the XY plotter that will enable the plotter to draw any geometric shape that can be expressed mathematically. Circles, parabolas, and ellipses are shapes commonly used, as well as straight lines in any direction to an X or

FIG. 80 Computer-driven *XY* plotter.

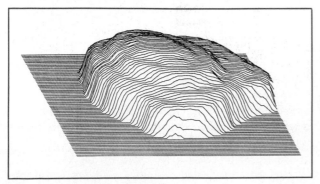

FIG. 81 Output from an *XY* plotter.

Y axis. The drafting stylus may be programmed to move in discrete XY steps, or in a polar mode, depending on the type of plotter and the particular instructions from the computer.

Automated drafting machines are part of a total machine system involving computer graphics. This rapidly evolving field is of special interest to companies having a need for many variations on a basic design. The automotive industry, for example, uses automatic drafting for creating varying windshield layouts. The aerospace industry makes much use of automated drafting for variations on airfoils.

35. EXERCISES IN THE USE OF INSTRUMENTS

The following problems can be used as progressive exercises for practice in using the instruments. Do them as finished pencil drawings or in pencil layout to be inked. Line work should conform to that given in the alphabet of lines.

PROBLEMS

GROUP 1. STRAIGHT LINES

1. An exercise for the T square, triangle, and scale. Through the center of the space draw a horizontal and a vertical line. Measuring on these lines as diameters, lay off a 4-in square. Along the lower side and upper half of the left side measure ½-in spaces with the scale. Draw all horizontal lines with the T square and all vertical lines with the T square and triangle.

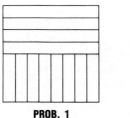 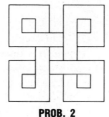

PROB. 1   PROB. 2

2. An interlacement. For T square, triangle, and dividers. Draw a 100-mm square. Divide the left side and lower side into seven equal parts with dividers. Draw horizontal and vertical lines across the square through these points. Erase the parts not needed.

3. A street-paving intersection. For 45° triangle and scale. An exercise in starting and stopping short lines. Draw a 4-in square. Draw its diagonals with 45° triangle. With the scale, lay off ½-in spaces along the diagonals from their intersection. With 45° triangle, complete the figure, finishing one quarter at a time.

 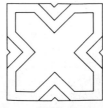

PROB. 3   PROB. 4

4. A square pattern. For 45° triangle, dividers, and scale. Draw a 4-in square and divide its sides into three equal parts with dividers. With 45° triangle, draw diagonal lines connecting these points. Measure ⅜ in on each side of these lines, and finish the pattern as shown in the drawing at the bottom of the page.

5. Five cards. Visible and hidden lines. Five cards 45 by 75 mm are arranged with the bottom card in the center, the other four overlapping each other and placed so that their outside edges form a 100-mm square. Hidden lines indicate edges covered.

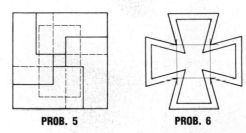

PROB. 5   PROB. 6

6. A Maltese cross. For T square, spacers, and 45° and 30-60° triangles. Draw a 100-mm square and a 35-mm square. From the corners of the inner square, draw lines to the outer square at 15° and 75°, with the two triangles in combination. Mark points with spacers 6 mm inside each line of this outside cross, and complete the figure with triangles in combination.

GROUP 2. STRAIGHT LINES AND CIRCLES

7. Insignia. For T square, triangles, scale, and compasses. Draw the 45° diagonals and the vertical and horizontal center lines of a 4-in square. With compass, draw a ¾-in-diameter construction circle, a 2¾-in circle, and a 3¼-in circle. Complete the design by adding a square and pointed star as shown.

PROB. 7

PROB. 8

8. A six-point star. For compass and 30-60° triangle. Draw a 100-mm construction circle and inscribe the six-point star with the T square and 30-60° triangle. Accomplish this with four successive changes of position of the triangle.

9. A stamping. For T square, 30-60° triangle, and compasses. In a 100-mm circle draw six diameters 30° apart. Draw a 75-mm construction circle to locate the centers of 8-mm-radius circle arcs. Complete the stamping with perpendiculars to the six diameters as shown.

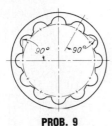

PROB. 9

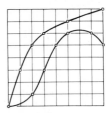

PROB. 10

10. A 24-point star. For T square and triangles in combination. In a 4-in circle draw 12 diameters 15° apart, using T square and triangles singly and in combination. With same combinations, finish the figure as shown.

GROUP 3. CIRCLES AND TANGENTS

11. Concentric circles. For compass (legs straight) and scale. Draw a horizontal line through the center of a space. On it mark off radii for eight concentric circles 6 mm apart. In drawing concentric circles, always draw the smallest first.

PROB. 11

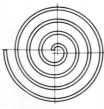

PROB. 12

12. A four-centered spiral. For accurate tangents. Draw a 3-mm square and extend its sides as shown. With the upper right corner as center, draw quadrants with 3- and 6-mm radii. Continue with quadrants from each corner in order until four turns have been drawn.

13. A loop ornament. For bow compass. Draw a 2-in square, about center of space. Divide AE into four ¼-in spaces with scale. With bow pencil and centers A, B, C, and D, draw four semicircles with ¼-in radius, and so on. Complete the figure by drawing the horizontal and vertical tangents as shown.

PROB. 13

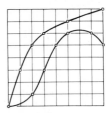

PROB. 14

14. A rectilinear chart. For french curve. Draw a 100-mm field with 12.5-mm coordinate divisions. Plot points at the intersections shown, and through them sketch a smooth curve very lightly in pencil. Finish by marking each point with a 2-mm circle and drawing a smooth line with the french curve.

GROUP 4. SCALES

15. Scale practice.

(a) Measure lines A to G to the following scales: A, full size; B, ½ size; C, 3″ = 1′-0″; D, 1″ = 1′-0″; E, ¾″ = 1′-0″; F, ¼″ = 1′-0″; G, ³⁄₁₆″ = 1′-0″.

(b) Lay off distances on lines H to N as follows: H, 3³⁄₁₆″, full size; I, 7″, ½ size; J, 2′-6″, 1½″ = 1′-0″; K, 7′-5½″, ½″ = 1′-0″; L, 10′-11″, ⅜″ = 1′-0″; M, 28′-4″, ⅛″ = 1′-0″; N, 40′-10″, ³⁄₃₂″ = 1′-0″.

(c) For engineer's scale. Lay off distances on lines H to N as follows: H, 3.2″, full size; I, 27′-0″, 1″ = 10′-0″; J, 66′-0″, 1″ = 20′-0″; K, 105′-0″, 1″ = 30′-0″; L, 156′-0″, 1″ = 40′-0″; M, 183′-0″, 1″ = 50′-0″; N, 214′-0″, 1″ = 60′-0″.

(d) For metric scale. Lay off distances on lines H to N as follows: H, 95 mm, full size; I, 700 mm, 1 mm = 10 mm; J, 1250 mm, 1 mm = 20 mm; K, 2200 mm, 1 mm = 30 mm; L, 3600 mm, 1 mm = 40 mm; M, 5100 mm, 1 mm = 50 mm; N, 4650 mm, 1 mm = 60 mm.

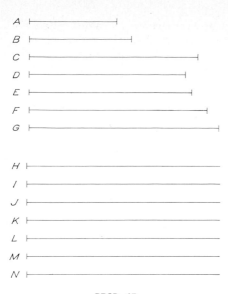

PROB. 15

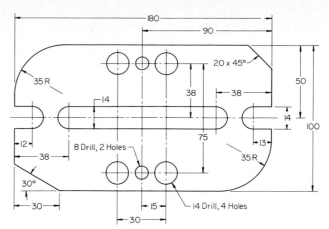

PROB. 18 METRIC. Fixture base.

GROUP 5. COMBINATIONS

16. A film-reel stamping. Draw to scale of 6″ = 1′-0″.

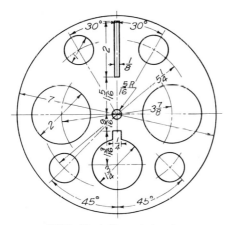

PROB. 16 A film-reel stamping.

17. Blank for wheel (not illustrated). Make a one-view drawing for stamping 125 mm OD; center hole 12 mm in diameter; eight spokes 10 mm wide connecting 40-mm-diameter center portion with 12-mm rim. Eight 6-mm-diameter holes with centers at intersection of center lines of spokes and 115-mm circle; 3-mm fillets throughout to break sharp corners.

18. Drawing of fixture base. Full size. Drill sizes specify the diameter.

19. Drawing of gage plate. Scale, twice size. Drill sizes specify the diameter.

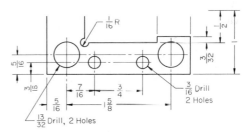

PROB. 19 Gage plate.

20. Drawing of milling fixture plate. Scale, twice size.

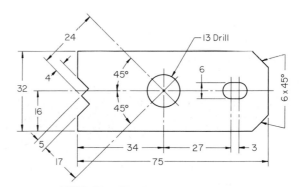

PROB. 20 METRIC. Milling fixture plate.

21. Drawing of dial shaft. Use decimal scale and draw 10 times size.

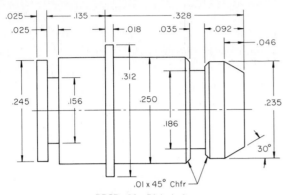

PROB. 21 Dial shaft.

22. Drawing of inner toggle for temperature control. Use decimal scale and draw 10 times size.

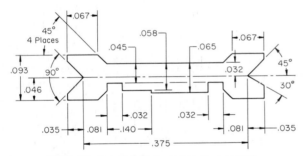

PROB. 22 Inner toggle for temperature control.

23. Drawing of mounting surface—O control. Scale, twice size.

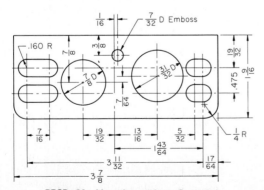

PROB. 23 Mounting surface: O control.

24. Drawing mounting leg—O control. Scale, full size.

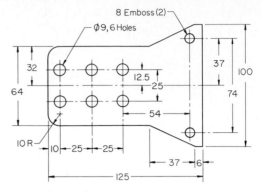

PROB. 24 **METRIC.** Mounting leg: O control.

25. Drawing of cooling fin and tube support. Full size.

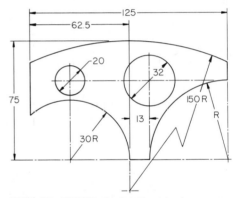

PROB. 25 **METRIC.** Cooling fin and tube support.

26. Drawing of cone-and-ball check. Scale, four times size.

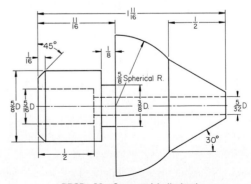

PROB. 26 Cone and ball check.

27. Drawing of bell crank. Stamped steel. Scale, full size.

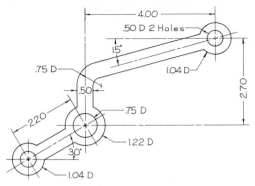

PROB. 27 Bell crank.

28. Drawing of control plate (aircraft hydraulic system). Stamped aluminum. Scale, full size.

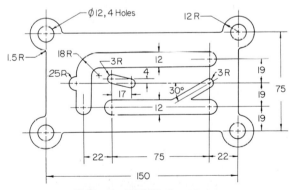

PROB. 28 METRIC. Control plate.

29. Drawing of torque disk (aircraft brake). Stamped steel. Scale, full size.

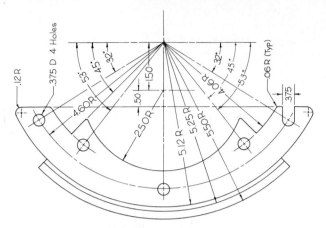

PROB. 29 Torque disk.

30. Drawing of brake shoe. Stamped steel with molded asbestos composition wear surface. Scale, half size.

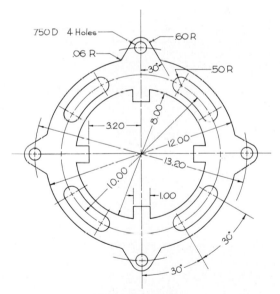

PROB. 30 Brake shoe.

3
CONSTRUCTIONAL GEOMETRY

1. INTRODUCTION

In order to create designs the drafter or designer needs to know how to construct various common geometric patterns. For example, how does one draw one line parallel to another, or how is a hexagon made? It is not necessary to memorize these manipulations, but a drafter must be able to refer to a reference text, such as this one, and follow the instructions with comfort and confidence.

It is true that the computer can be programmed to draw many shapes: circles, ellipses, parabolas. However, the computer needs instructions, and a person competent in geometric constructions is needed if the computer is to be properly programmed. If one needs a construction done only a few times, it is often faster to do the work manually than to put it on a computer. For a high volume of drawings of similar constructions, consider a computer-plotter system.

This chapter is divided into three sections for ease of grouping: line relationships, geometry of straight-line figures, and geometry of curved lines. Many constructions, with the exception of curved surfaces, are covered.

LINE RELATIONSHIPS

2. DRAWING STRAIGHT LINES

A. Single Line Straight lines are drawn by using the straight edge of the T square or one of the triangles. For short lines a triangle is more convenient. Observe the directions for technique in paragraphs 2B and C.

To draw a straight line through two points (Fig. 1), align the triangle or T square with points P and Q, and draw the line.

B. Parallel Lines

▪ (1) Using Triangles Parallel lines may be required in any position. Parallel horizontals or verticals are most common. Horizontals are drawn with T square alone, verticals with T square and triangle. The general case (odd angles) is shown in Fig. 2.

To draw a straight line through a point, parallel to another line (Fig. 2), adjust a triangle to the given line *AB*, with a second triangle as a base. Slide the aligned triangle to its position at point *P* and draw the required line.

■ **(2) Line through a Point and Parallel to a Given Line** (See Fig. 3) With the given point *P* as center and a radius of sufficient length, draw an arc *CE* intersecting the given line *AB* at *C*. With *C* as center and the same radius, draw the arc *PD*. With *C* as center and radius *DP*, draw an arc intersecting *CE* at *E*. Then *EP* is the required line.

■ **(3) Line Parallel to Another at a Given Distance**

■ FOR STRAIGHT LINES (Fig. 4) With the given distance as radius and two points on the given line as centers (as far apart as convenient), draw two arcs. A line tangent to these arcs will be the required line.

■ FOR CURVED LINES (Fig. 5) Draw a series of arcs with centers along the line. Draw tangents to these arcs with a french curve

C. Perpendicular Lines

■ **(1) Using Triangles** Perpendiculars occur frequently as horizontal-to-vertical (Fig. 6) but also often in other positions (Fig. 7). Note that the construction of a perpendicular utilizes the 90° angle of a triangle.

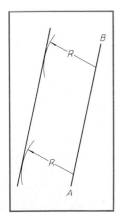
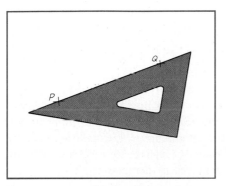
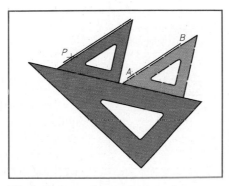
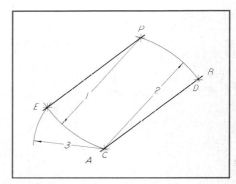

FIG. 1 To draw a line through two points.

FIG. 2 To draw a line parallel to another.

FIG. 3 Parallel lines. Two points determine the parallel.

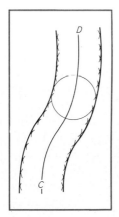
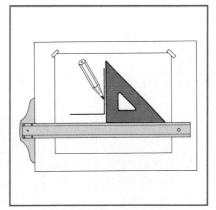

FIG. 4 Parallel lines. The parallel is drawn tangent to a pair of arcs.

FIG. 5 Curved parallel lines. The curves are drawn tangent to circle arcs.

FIG. 6 To draw a line perpendicular to another (when the given line is horizontal).

FIG. 7 To draw a line perpendicular to another (general position).

To erect a perpendicular to a given straight line (when the given line is *horizontal*, Fig. 6), place a triangle on the T square as shown and draw the required perpendicular.

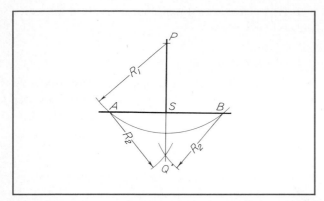

FIG. 8 A perpendicular from a point outside. Equal radii determine the perpendicular.

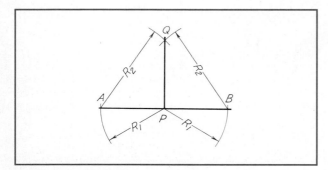

FIG. 9 (Left) A perpendicular from a point on a line. Equal radii determine a point on the perpendicular.

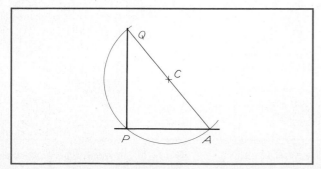

FIG. 10 (Right) A perpendicular from a point on a line. Geometrically, perpendicular chords of a circle produce this solution.

To erect a perpendicular to a given straight line (*general position*, Fig. 7), set a triangle with its hypotenuse against a guiding edge and adjust one side to the given line. Then slide the triangle so that the second side is in the position of the perpendicular and draw the required line.

▪ **(2) Perpendicular from a Point *to* a Given Line** (See Fig. 8) With point P as center and any convenient radius R_1, draw a circle arc intersecting the given line at A and B. With any convenient radius R_2 and with centers at A and B, draw intersecting arcs locating Q. The required perpendicular is PQ, with S the intersection of the perpendicular and the given line.

▪ **(3) Perpendicular from a Point *on* a Given Line**

▪ FIRST METHOD (Fig. 9) With point P on the line as center and any convenient radius R_1, draw circle arcs to locate points A and B equidistant from P. With any convenient radius R_2 longer than R_1 and with centers at A and B, draw intersecting arcs locating Q. PQ is the required perpendicular.

▪ SECOND METHOD (Fig. 10) With any convenient center C and radius CP, draw somewhat more than a semicircle from the intersection of the circle arc with the given line at A. Draw AC extended to meet the circle arc at Q. PQ is the required perpendicular.

3. MAKING TANGENTS

A *tangent* to a curve is a line that passes through two points on the curve infinitely close together. One of the most frequent geometric operations in drafting is the drawing of tangents to circle arcs and the drawing of circle arcs tangent to straight lines or other circles.

The points of tangency should be located by short cross marks. The method of finding these points is indicated in the following constructions. Note that the location of tangent points is based on one of these geometric facts: (1) The tangent point of a straight line and circle will lie at the intersection of a perpendicular to the straight line that passes through the circle center, and (2) the tangent point of two circles will lie on the circumferences of both circles and on a straight line connecting the circle centers.

A. Tangent Points See Fig. 11. To find the point of tangency for line AB and a circle with center D, draw

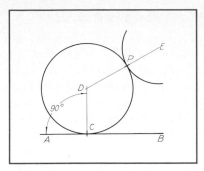

FIG. 11 Tangent points.

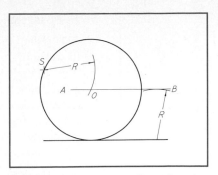

FIG. 12 A circle tangent to a line and passing through a point. The circle center must be equidistant from the line and the point.

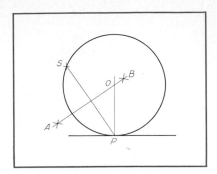

FIG. 13 A circle tangent to a line at a point and passing through a second point.

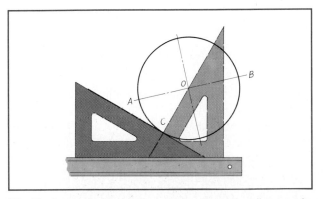

FIG. 14 A tangent at a point on a circle. The tangent line must be perpendicular to the line from the point to the center of the circle.

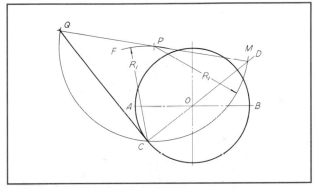

FIG. 15 To draw a tangent at a point on a circle. This solution sets up two right-angle chords of a circle. One chord passes through the center of the given circle.

DC perpendicular to line AB. Point C is the tangent point.

To find the tangent point for two circles (Fig. 11) with centers at D and E, draw DE, joining the centers. Point P is the tangent point.

B. A Circle of Given Size Tangent to a Line and Passing through a Point See Fig. 12. Draw a line AB, the given radius distance R away from and parallel to the given line. Using the given point S as center, cut line AB at O with the given radius. O is the center of the circle. Note that there are two possible positions for the circle.

C. A Circle Tangent to a Line at a Point and Passing through a Second Point In Fig. 13 connect the two points P and S and draw the perpendicular bisector AB (see paragraph 13). Draw a perpendicular to the given line at P. The point where this perpendicular intersects the line AB is the center O of the required circle.

D. A Tangent to a Circle at a Point on the Circle

▪ **(1) Using Triangles** (See Fig. 14) Given the arc ABC, draw a tangent at the point C. Arrange a triangle in combination with the T square (or another triangle) so that its hypotenuse passes through center O and point C. Holding the T square firmly in place, turn the triangle about its square corner and move it until the hypotenuse passes through C. The required tangent then lies along the hypotenuse.

▪ **(2) Using Construction** (See Fig. 15) Given the arc ABC and to draw a tangent at point C, draw the extended diameter CD and locate point M. Then with any convenient radius R_1, locate point P equidistant from C and M. With P as center and the same radius R_1, draw somewhat more than a semicircle and draw the line MPQ. The line QC is a tangent to the circle at point C.

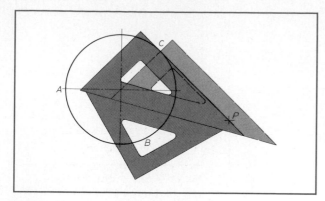

FIG. 16 A tangent to a circle from a point outside. The line is drawn by alignment through the point and tangent to the circle; the tangent point is then located on the perpendicular from the circle center.

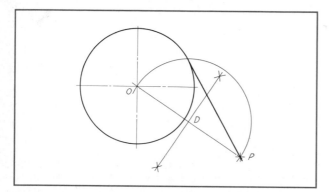

FIG. 17 To draw a tangent from an outside point. Here, as in Fig. 15, perpendicular chords determine the solution.

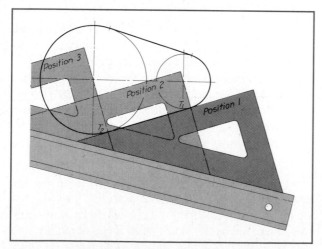

FIG. 18 Tangents to two circles (open belt). Tangent lines are drawn by alignment with both circles. Tangent points lie on perpendicular lines from circle centers.

E. A Tangent to a Circle from a Point Outside

▪ **(1) Using Triangles** (See Fig. 16) Given the arc *ACB* and point *P*, arrange a triangle in combination with another triangle (or T square) so that one side passes through point *P* and is tangent to the circle arc. Then slide the triangle until the right-angle side passes through the center of the circle and mark lightly the tangent point *C*. Bring the triangle back to its original position and draw the tangent line.

▪ **(2) Using Construction** (See Fig. 17) Connect the point *P* with the center of the circle *O*. Then draw the perpendicular bisector of *OP*, and with the intersection at *D* as center, draw a semicircle. Its intersection with the given circle is the point of tangency. Draw the tangent line from *P*.

F. A Tangent to Two Circles

▪ **(1) Open Belt** (See Fig. 18) Arrange a triangle in combination with a T square or triangle so that one side is in the tangent position. Move to positions 2 and 3, marking lightly the tangent points T_1 and T_2. Return to the original position and draw the tangent line. Repeat for the other side.

▪ **(2) Crossed Belt** (See Fig. 19) Arrange a triangle in combination with a T square or triangle so that one side is in the tangent position. Move to positions 2 and 3, marking lightly the tangent points T_1 and T_2. Return to the original position and draw the tangent line. Repeat for the other side.

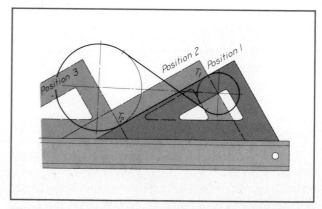

FIG. 19 Tangents to two circles (crossed belt). Tangent lines are drawn by alignment with both circles. Tangent points lie on perpendicular lines from circle centers.

4. TANGENT ARCS

A. An Arc of Given Radius Tangent to Two Lines See Fig. 20. Given the lines AB and CD, set the compass to radius R, and at any convenient point on the given lines draw the arcs R and R_1. Draw parallels to the given lines through the limits of the arcs. These parallels are the loci of the centers of all circles of radius R tangent to lines AB and CD, and their intersection at point O will be the center of the required arc. Find the tangent points by erecting perpendiculars, as in Fig. 16, to the given lines through the center O.

B. An Arc of Given Radius to Two Perpendicular Lines This may be done just as described in Sec. 4A. However, there is an alternative technique which you may wish to use. Using Fig. 21, draw an arc of radius R, with center at corner A, cutting the lines AB and AC at T and T_1. Then with T and T_1 as centers and with the same radius R, draw arcs intersecting at O, the center of the required arc.

C. An Arc of Given Radius Tangent to a Given Circle and Line In Fig. 22 let AB be the given line and R_1 the radius of the given circle. Draw a line CD parallel to AB at a distance R from it. With O as center and radius $R + R_1$, swing an arc intersecting CD at X, the desired center. The tangent point for AB will be on a perpendicular to AB from X; the tangent point for the two circles will be on a line joining their centers X and O.

D. An Arc of Given Radius to Two Given Circles

- **First Case** (Fig. 23)   The centers of the given circles are outside the required circle. Let R_1 and R_2 be the radii of the given circles and O and P their centers. With O as center and radius $R + R_1$, describe an arc. With P as center and radius $R + R_2$, swing another arc intersecting the first arc at Q, which is the center sought. Mark the tangent points in line with OQ and QP.

- **Second Case** (Fig. 23)   The centers of the given circles are inside the required circle. With O and P as centers and radii $R - R_1$ and $R - R_2$, describe arcs intersecting at the required center Q.

5. REVERSE CURVES

A. To Draw a Reverse, or Ogee, Curve Refer to Fig. 24. Given two parallel lines AB and CD, join B and C by

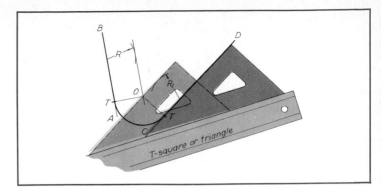

FIG. 20   An arc tangent to two straight lines.

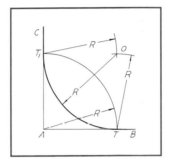

FIG. 21   An arc tangent at right-angle corner.

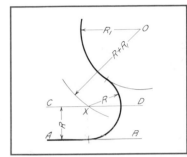

FIG. 22   An arc tangent to a straight line and a circle.

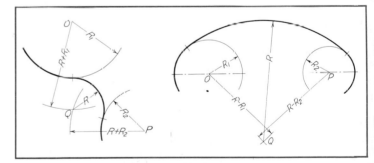

FIG. 23   An arc tangent to two circles.

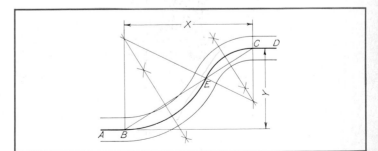

FIG. 24   An ogee curve.

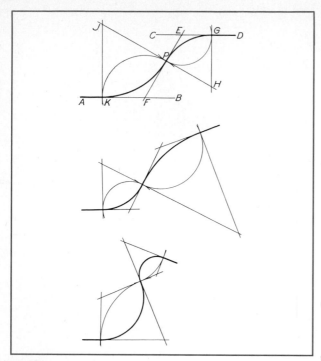

FIG. 25 Reverse curve tangent to three lines. Note that the arc centers must lie on a line perpendicular to the tangent line at the point of tangency.

a straight line. Erect perpendiculars at *B* and *C*. Any arcs tangent to lines *AB* and *CD* at *B* and *C* must have their centers on these perpendiculars. On the line *BC* assume point *E*, the point through which it is desired that the curve shall pass. Bisect *BE* and *EC* by perpendiculars. Any arc to pass through *B* and *E* must have its center somewhere on the perpendicular from the middle point. The intersection, therefore, of these perpendicular bisectors with the first two perpendiculars will be the centers for arcs *BE* and *EC*. This line might be the center line for a curved road or pipe. The construction may be checked by drawing the line of centers, which *must* pass through *E*.

B. To Draw a Reverse Curve Tangent to Two Lines and to a Third Secant Line at a Given Point Refer to Fig. 25. Given two lines *AB* and *CD* cut by the line *EF* at points *E* and *F*, draw a perpendicular *JH* to *EF* through a given point *P* on *EF*. With *E* as center and radius *EP*, intersect *CD* at *G*. Draw a perpendicular from *G* intersecting *JH* at *H*. With *F* as center and radius *FP*, intersect *AB* at *K*. Draw a perpendicular to *AB* from *K* intersecting *JH* at *J*. *H* and *J* will be the centers for arcs tangent to the three lines.

GEOMETRY OF STRAIGHT-LINE FIGURES

6. DIVIDING A LINE

A. To Bisect a Line See Fig. 26. From the two ends of the line, swing arcs of the same radius, greater than one-half the length of the line, and draw a line through the arc intersections. This line bisects the given line and is also the *perpendicular* bisector. Many geometric problems depend upon this construction.

B. To Trisect a Line. See Fig. 27. Using a 30-60° triangle and a T square, draw lines at 30° from *A* and *B* to intersect at *C*. Then at 60° to *AB* draw lines from *C* to give *U* and *V*, the desired points on *AB*.

C. To Divide a Line into any Number of Parts

■ First Method (Fig. 28) To divide a line *AB* into, say, five equal parts, draw any line *BC* of indefinite length. On it measure, or step off, five divisions of convenient length. Connect the last point with *A*, and using two triangles as shown in Fig. 2, draw lines through the points parallel to *CA* intersecting *AB*.

■ Second Method (Fig. 29) Draw a perpendicular *AC* from *A*. Then place a scale so that five convenient equal divisions are included between *B* and the perpendicular, as in Fig. 29. With a triangle and T square draw perpendiculars through the points marked, dividing the line *AB* as required.

7. ANGLES

A. To Lay Out a Given Angle

■ (1) Tangent Method (Fig. 30) The trigonometric tangent of an angle of a triangle is the ratio of the length of the side opposite the angle to the length of the adjacent side. Thus, tan *A* = *Y/X*, or *X* tan *A* = *Y*. To lay out a given angle, obtain the value of the tangent, as-

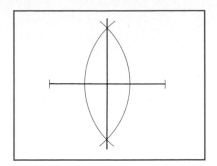

FIG. 26 To bisect a line (with compass).

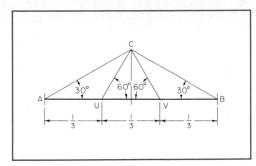

FIG. 27 To trisect a line (with T square and triangle). Third points are located by 30° and 60° angles.

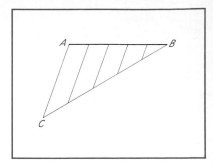

FIG. 28 To divide a line. Equal divisions on any line *BC* are transferred to *AB* by parallels to *AC*.

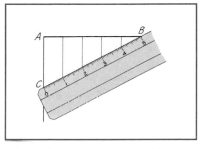

FIG. 29 To divide a line. Scale divisions are transferred to given line *AB*.

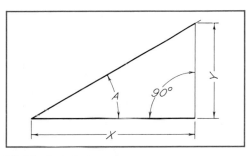

FIG. 30 Angle by tangent.

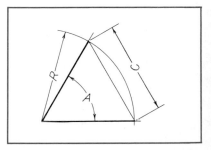

FIG. 31 Angle by chord. The proportion of *C* to *R* is obtained from a table of chords.

sume any convenient distance X, and multiply X by the tangent to get distance Y. Note that the angle between the sides X and Y must be a right angle.

■ **(2) Chordal Method** (Fig. 31) If the length of a chord is known for an arc of given radius and included angle, the angle can be accurately laid out. Given an angle in degrees, to lay out the angle, obtain the chord length for a 1-in circle arc from the table in the appendix. Select any convenient arc length R and multiply the chord length for a 1-in arc by this distance, thus obtaining the chord length C for the radius distance selected. Lay out the chord length on the arc with compass or dividers and complete the sides of the angle.

The chord length for an angle can be had from a sine table by taking the sine of one-half the given angle and multiplying by two.

B. To Bisect an Angle See Fig. 32. Given angle *SOR*, using any convenient radius shorter than *OR* or *OS*, swing an arc with *O* as center, locating *A* and *B*. Then with the same radius or another radius longer than one-half the distance from *A* to *B*, swing arcs with *A* and

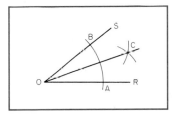

FIG. 32 To bisect an angle.

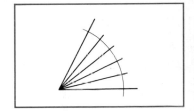

FIG. 33 To divide an angle into equal parts. Equal divisions (chords) are stepped off on an arc of the angle.

B as centers, locating *C*. The bisector is *CO*. With repeated bisection, the angle can be divided into 4, 8, 16, 32, 64, etc., parts.

C. To Divide an Angle into any Number of Parts
Sometimes an angle must be divided into some number of parts, say, 3, 5, 6, 7, or 9, that cannot be obtained by bisection.

To divide an angle into, for example, 5 parts (Fig. 33), draw any arc across the angle as shown, and use the bow dividers to divide the arc into equal parts. Lines through these points to the apex are the required divisions.

8. TRIANGLES

A. To Construct a Triangle Having Three Given Sides See Fig. 34. Given the lengths A, B, and C, draw one side A in the desired position. With its ends as centers and radii B and C, draw two intersecting arcs as shown. This construction is used extensively in developments by triangulation.

B. To Locate the Geometric Center of an Equilateral Triangle Refer to Fig. 35. With A and B as centers and radius AB, draw arcs as shown. These will intersect at C, the third corner. With center C and radius AB cut the original arcs at D and E. Then AD and BE intersect at O, the geometric center.

9. TO TRANSFER A POLYGON TO A NEW BASE

A. By Triangulation (Fig. 36)   Given polygon ABCDEF and a new position of base A'B', consider each point as the vertex of a triangle whose base is AB. With centers A' and B' and radii AC and BC, describe intersecting arcs locating the point C'. Similarly, with radii AD and BD locate D'. Connect B'C' and C'D' and continue the operation always using A and B as centers.

B. Box or Offset Method (Fig. 37)   Enclose the polygon in a rectangular "box." Draw the box on the new base and locate the points ABCEF on this box. Then set point D by rectangular coordinates as shown.

10. USES OF THE DIAGONAL

The diagonal is used in many ways to simplify construction and save drafting time. Figure 38 illustrates the diagonal used at A for locating the center of a rectangle, at B for enlarging or reducing a geometric shape, at C for producing similar figures having the same base, and at D for drawing inscribed or circumscribed figures.

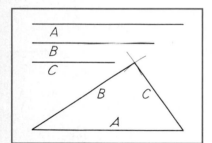

FIG. 34 To construct a triangle. The legs are laid off with a compass.

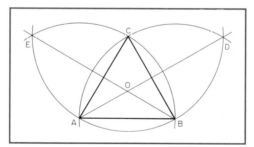

FIG. 35 To locate the geometric center of an equilateral triangle. Perpendicular bisectors of the sides intersect at the center.

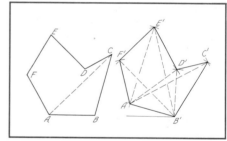

FIG. 36 To transfer a polygon: by triangulation. All corners are located by triangles having a common base.

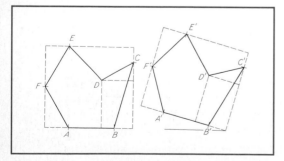

FIG. 37 To transfer a polygon: by "boxing." Each corner is located by its position on a rectangle.

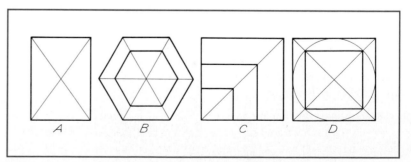

FIG. 38 Uses of the diagonal. The diagonals will locate the center of any regular or symmetrical geometric shape that has an even number of sides.

11. CONSTRUCTING REGULAR POLYGONS

A. To Construct a Regular Hexagon

■ **(1) Given the Distance across Corners**

■ FIRST METHOD (Fig. 39) Draw a circle with *AB* as a diameter. With the same radius and *A* and *B* as centers, draw arcs intersecting the circle and connect the points.

■ SECOND METHOD (WITHOUT COMPASS) Draw lines with the 30-60° triangle in the order shown in Fig. 40.

■ **(2) Given the Distance across Flats** The distance across flats is the diameter of the inscribed circle. Draw this circle, and with the 30-60° triangle draw tangents to it as in Fig. 41.

B. To Inscribe a Regular Pentagon in a Circle Using Fig. 42, draw a diameter *AB* and a radius *OC* perpendicular to it. Bisect *OB*. With this point *D* as center and radius *DC*, draw arc *CE*. With center *C* and radius *CE*, draw arc *EF*. *CF* is a side of the pentagon. Step off this distance around the circle with dividers.

C. To Inscribe a Regular Octagon in a Square Using Fig. 43 draw the diagonals of the square. With the corners of the square as centers and a radius of half the diagonal, draw arcs intersecting the sides and connect these points.

D. To Construct Any Regular Polygon, Given One Side Refer to Fig. 44. Let the polygon have seven sides. With the side *AB* as radius and *A* as center, draw a semicircle and divide it into seven equal parts with dividers. Through the second division from the left draw radial line *A-2*. Through points *3, 4, 5,* and *6* extend radial lines as shown. With *AB* as radius and *B* as center, cut line *A-6* at *C*. With *C* as center and the same radius, cut *A-5* at *D*, and so on at *E* and *F*. Connect the points.

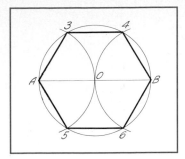

FIG. 39 A hexagon. Arcs equal to the radius locate all points.

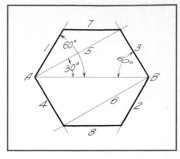

FIG. 40 A hexagon. The 30-60° triangle gives construction for all points.

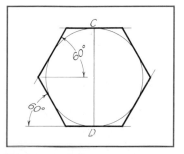

FIG. 41 A hexagon. Tangents drawn with the 30-60° triangle locate all points.

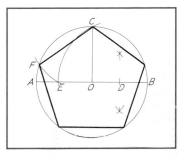

FIG. 42 To inscribe a regular pentagon in a circle.

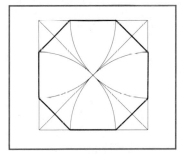

FIG. 43 To inscribe a regular octagon in a square.

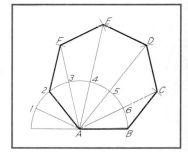

FIG. 44 To construct a regular polygon.

GEOMETRY OF CURVED LINES

12. USEFUL DEFINITIONS

A *curved line* is generated by a point moving in a constantly changing direction, according to some mathematical or graphic law. Curved lines are classified as single-curved or double-curved.

A *single-curved line* is a curved line having all points of the line in a plane. Single-curved lines are often called "plane curves."

A *double-curved line* is a curved line having no four consecutive points in the same plane. Double-curved lines are also known as "space curves."

A *normal* is a line (or plane) perpendicular to a tan-

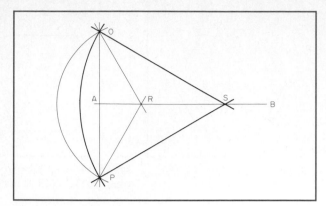

FIG. 45 Arc centers.

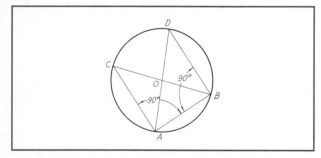

FIG. 46 Circle through three points.

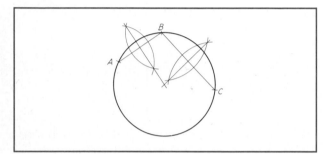

FIG. 47 To locate the center of a circle. Perpendicular chords to chord *AB* produce diameters *BC* and *AD*.

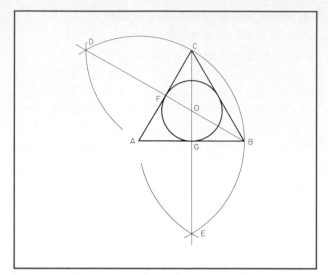

FIG. 48 To inscribe a circle in an equilateral triangle.

through two points *O* and *P* an infinite number of arcs may be drawn, but all will have centers such as *R* and *S* which lie on the perpendicular bisector *AB* of chord *OP*.

Using the above principle, only one circle can be drawn through three points, as illustrated in Fig. 46. The center must lie on the perpendicular bisectors of both chords, *AB* and *BC*. Incidentally, the center will also lie on the perpendicular bisector of chord *AC*. This construction can be used as a check on accuracy.

Any diameter of a circle and a third point on the circumference, connected to form a triangle, will produce two chords of the circle that are perpendicular to each other. Therefore, the center of a circle may be found, as in Fig. 47, by selecting any two points such as *A* and *B*, drawing the perpendicular chords *AC* and *BD*, and then drawing the two diameters *CB* and *DA*, which cross at center *O*.

14. CIRCLES WITH POLYGONS: TRIANGLES, SQUARES, ANY REGULAR POLYGON

A. To Inscribe a Circle within an Equilateral Triangle
See Fig. 48. With center *A* and radius *AB*, draw an arc as shown. With *B* and *C* as centers and radius *AB*, cut this arc at *D* and *E*. Then *EC* and *DB* intersect at *O*, the geometric center, and the radius of the subscribing circle is *OG* and/or *OF*.

gent line of a curve at the point of tangency. A normal to a single-curved line will be a line in the plane of the curve and perpendicular to a straight line connecting two consecutive points on the curve.

13. ARC AND CIRCLE CENTERS

The center of any arc must lie on the perpendicular bisector of any and all chords of the arc. Thus, in Fig. 45,

B. To Inscribe an Equilateral Triangle within a Circle
See Fig. 49. Using the radius of the circle and starting at
A, the known point of orientation, step off distances AB,
BC, CD, and DE. The equilateral triangle is ACE. Note
that six chords equal in length to the radius will give six
equally spaced points on the circumference.

C. To Inscribe a Circle in a Square See Fig. 50.
Given the circle ABCD, draw diagonals AC and BD,
which intersect at point O, the center. OP is the radius
of the circle.

D. To Inscribe a Square in a Circle See Fig. 51. From
A, the known point of orientation, draw diameter AC
and then erect BD, the perpendicular bisector of AC.
The square is ABCD.

E. To Inscribe a Circle in or on Any Regular Polygon
Using Fig. 52, draw perpendicular bisectors of any two
sides, for example PO and SO of AB and CD, giving O,

the center. The inner circle radius is OF or OG and the
external circle radius is OA, OB, etc.

15. TO LAY OFF ON A STRAIGHT LINE THE APPROXIMATE LENGTH OF A CIRCLE ARC

Refer to Fig. 53. Given the arc AB, at A draw the tangent AD and the chord produced, BA. Lay off AC equal
to half the chord AB. With center C and radius CB,
draw an arc intersecting AD at D; then AD will be equal
in length to arc AB (very nearly). If the given arc is between 45° and 90°, a closer approximation will result by
making AC equal to the chord of half the arc instead of
half the chord of the arc.

The usual way of rectifying an arc is to set the dividers
to a space small enough to be practically equal in length
to a corresponding part of the arc. Starting at B, step
along the arc to the point nearest A and without lifting
the dividers step off the same number of spaces on the
tangent, as shown in Fig. 54.

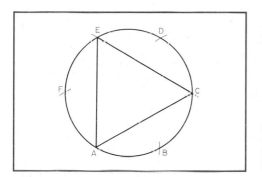

FIG. 49 To inscribe an equilateral triangle in a
circle.

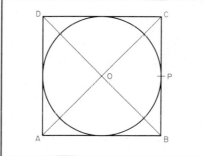

FIG. 50 To inscribe a circle in a square.
Diagonals locate the circle center.

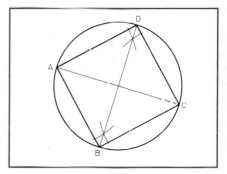

FIG. 51 To inscribe a square in a circle. The
perpendicular bisector of one diameter
locates the two corners to be found.

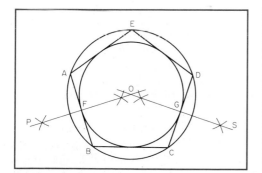

FIG. 52 To draw a circle on a regular polygon.

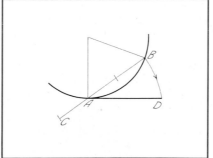

FIG. 53 To approximate the length of an arc.

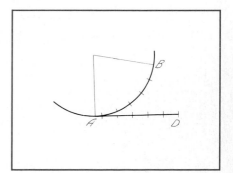

FIG. 54 To approximate the length of an arc.
The sum of short chords closely approaches
the arc length.

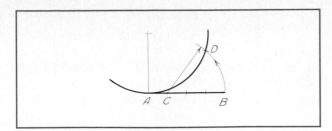

FIG. 55 To lay off, on an arc, a specified distance.

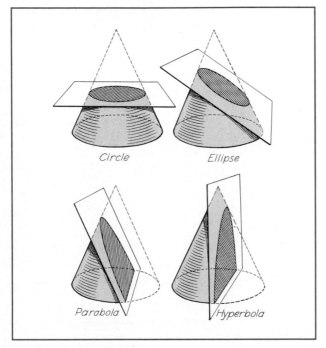

FIG. 56 The conic sections.

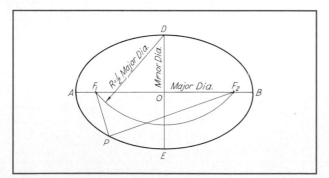

FIG. 57 The ellipse: major and minor diameters.

16. TO LAY OFF ON A GIVEN CIRCLE THE APPROXIMATE LENGTH OF A STRAIGHT ARC

Refer to Fig. 55. Given the line *AB* tangent to the circle at *A*, lay off *AC* equal to one-fourth *AB*. With *C* as center and radius *CB*, draw an arc intersecting the circle *D*. The line *AD* is equal in length to *AB* (very nearly). If arc *AD* is greater than 60°, solve for one-half *AB*.

17. OVERVIEW OF PLANE CURVES: THE CONIC SECTIONS

In cutting a right-circular cone (a cone of revolution) by planes at different angles, we obtain four curves called "conic sections" (Fig. 56). These are the *circle*, cut by a plane perpendicular to the axis; the *ellipse*, cut by a plane making a greater angle with the axis than do the elements; the *parabola*, cut by a plane making the same angle with the axis as do the elements; and the *hyperbola*, cut by a plane making a smaller angle than do the elements.

18. THE ELLIPSE

The ellipse is a commonly used conic section. It is worthwhile to review several ways in which the ellipse may be constructed. Refer to Fig. 57. An ellipse is the plane curve generated by a point moving so that the sum of its distances from two fixed points (F_1 and F_2), called "focuses," is a constant equal to the *major diameter AB*.

The *minor diameter DE* is the line through the center perpendicular to the major diameter. The focuses may be determined by cutting the major diameter with an arc having its center at an end of the minor diameter and a radius equal to one-half the major diameter.

The mathematical equation of an ellipse with the origin of rectilinear coordinates at the center of the ellipse is $x^2/a^2 + y^2/b^2 = 1$, where a is the intercept on the X axis and b is the intercept on the Y axis.

Any line through the center of an ellipse may serve as one of a pair of conjugate diameters. Two diameters are said to be *conjugate* when each diameter is parallel to the tangents at the ends of the other diameter. For example, in Fig. 58, *AB* and *CD* are a pair of conjugate diameters. *AB* is parallel to the tangents *MN* and *PQ*, and *CD* is parallel to the tangents *MP* and *NQ*. Also, each of a pair of conjugate diameters bisects all the chords parallel to the other. A given ellipse may have an unlimited number of pairs of conjugate diameters.

■ To determine the major and minor diameters, given the ellipse and a pair of conjugate diameters (Fig. 59)   The conjugate diameters are CN and JG. With center O and radius OJ, draw a semicircle intersecting the ellipse at P. The major and minor diameters will be parallel to the chords GP and JP, respectively.

A. Construction: Pin-and-String Method

This well-known method is often used for large work and is based on the definition of the ellipse. Drive pins at the points D, F_1, and F_2 (Fig. 57), and tie an inelastic thread or cord tightly around the three pins. If pin D is removed and a marking point is moved in the loop, keeping the cord taut, the point will describe a true ellipse.

B. Construction: Trammel Method for Major and Minor Diameters

■ (1) First Method (Fig. 60)   On the straight edge of a strip of paper, thin cardboard, or sheet of celluloid, mark the distance ao equal to one-half the major diameter and do equal to one-half the minor diameter. If the strip is moved, keeping a on the minor diameter and d on the major diameter, o will give points on the ellipse. This method is convenient as no construction is required, but for accurate results great care must be taken to keep the points a and d exactly on the major and minor diameters.

■ (2) Second Method (Fig. 61)   On a strip—as used in the first method—mark the distance do equal to one-half

the minor diameter and oa equal to one-half the major diameter. If this strip is moved, keeping a on the minor diameter and d on the major diameter, o will give points on the ellipse. This arrangement is preferred where the ratio between the major and minor diameters is small.

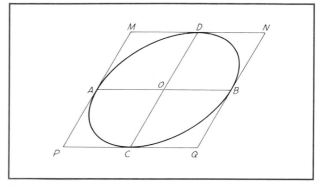

FIG. 58   The ellipse: conjugate diameters. Each is parallel to the tangent at the end of the other.

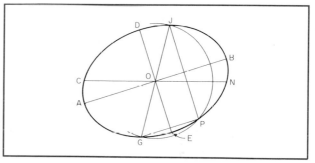

FIG. 59   Determination of major and minor diameters from conjugate diameters.

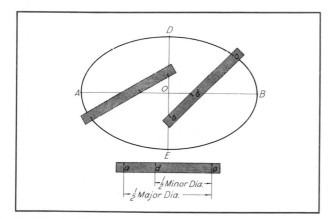

FIG. 60   An ellipse by trammel (first method).

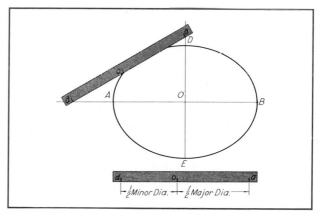

FIG. 61   An ellipse by trammel (second method).

C. Construction: Concentric-Circle Method for Major and Minor Diameters   See Fig. 62. This is perhaps the most accurate method for determining points on the curve. On the two principal diameters, which intersect at O, describe circles. From a number of points on the outer circle, as P and Q, draw radii OP, OQ, etc. intersecting the inner circle at P′, Q′, etc. From P and Q draw lines parallel to OD, and from P′ and Q′ draw lines parallel to OB. The intersection of the lines through P and P′ gives one point on the ellipse, the intersection of the lines through Q and Q′ another point, and so on. For accuracy, the points should be taken closer together toward the major diameter. The process may be repeated in each of the four quadrants and the curve sketched in lightly, freehand; or one quadrant only may be constructed and repeated in the remaining three by marking the french curve.

D. Construction: Circle Method for Conjugate Diameters   Refer to Fig. 63. The conjugate diameters AB and DE are given. On the conjugate diameter AB, describe a circle; then from a number of points, as P, Q, and S, draw perpendiculars as PP′, QO, and SS′ to the diameter AB. From S and P, etc., draw lines parallel to QD, and from S′ and P′ draw lines parallel to OD. The intersection of the lines through P and P′ gives one point on the ellipse, the intersection of the lines through S and S′ another point, and so on.

E. Construction: Parallelogram Method   See Figs. 64 and 65. This method can be used either with the major and minor diameters or with any pair of conjugate diameters. On the given diameters construct a parallelogram. Divide AO into any number of equal parts and AG into the same number of equal parts, numbering points from

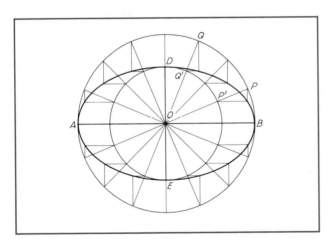

FIG. 62   An ellipse by concentric-circle method.

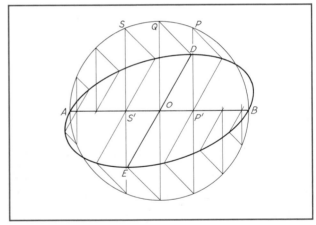

FIG. 63   An ellipse by circle method (conjugate diameters). After first construction, parallels plot points on the curve.

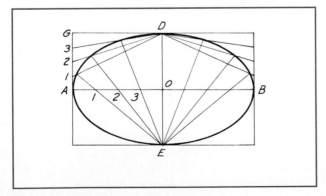

FIG. 64   An ellipse by parallelogram method.

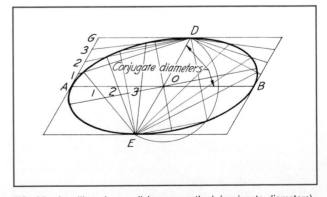

FIG. 65   An ellipse by parallelogram method (conjugate diameters).

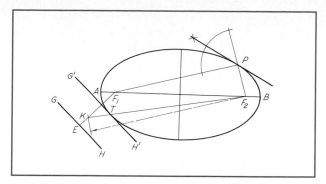

FIG. 66 Tangents to an ellipse at a point and parallel to a given line.

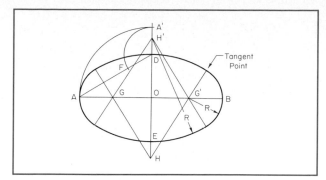

FIG. 68 A four-centered approximate ellipse.

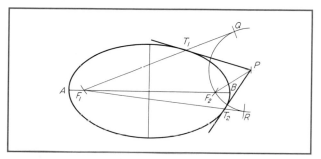

FIG. 67 Tangents to an ellipse from an outside point.

A. Through these points draw lines from *D* and *E*, as shown. Their intersections will be points on the curve.

F. Tangent to an Ellipse

▪ **(1) At a Given Point on the Curve** (Fig. 66) Draw lines from the given point *P* to the focuses. The line bisecting the exterior angle of these focal radii is the required tangent.

▪ **(2) Parallel to a Given Line** (Fig. 66) Draw F_1E perpendicular to the given line *GH*. With F_2 as center and radius *AB*, draw an arc cutting F_1E at *K*. The line F_1K cuts the ellipse at the required point of tangency *T*, and the required tangent passes through *T* parallel to *GH*.

▪ **(3) From a Point Outside** (Fig. 67) Find the focuses F_1 and F_2. With the given point *P* and radius PF_2, draw the arc RF_2Q. With F_1 as center and radius *AB*, strike an arc cutting this arc at *Q* and *R*. Connect QF_1 and RF_1. The intersections of these lines with the ellipse at T_1 and T_2 will be the tangent points of tangents to the ellipse from *P*.

G. To Make an Approximate Ellipse: Four-Centered Method
Using Fig. 68, join *A* and *D*. Lay off *DF* equal

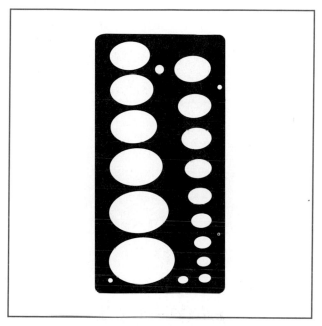

FIG. 69 Ellipse template for a 45° ellipse angle. *(Courtesy Koh-I-Noor Rapidograph Inc.)*

to *AO* − *DO*. This is done graphically as indicated on the figure by swinging *A* around to *A'* with *O* as center where now *DO* from *OA'* is *DA'*, the required distance. With *D* as center, an arc from *A'* to the diagonal *AD* locates *F*. Bisect *AF* by a perpendicular crossing *AO* at *G* and intersecting *DE* produced (if necessary) at *H*. Make *OG'* equal to *OG*, and *OH'* equal to *OH*. Then *G*, *G'*, *H*, and *H'* will be centers for four tangent circle arcs forming a curve *approximating* the shape of an ellipse.

H. Using an Ellipse Template
An ellipse template, such as the one shown in Fig. 69, can be a great time

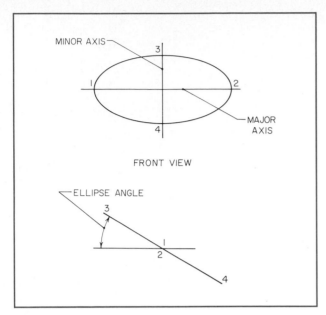

FIG. 70 The ellipse angle.

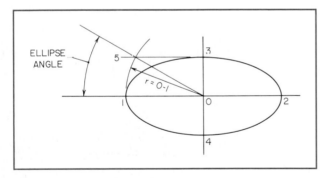

FIG. 71 Finding the ellipse angle.

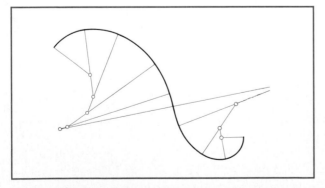

FIG. 72 A curve constructed with circle arcs. Note that lines through pairs of centers locate the tangent points of arcs.

saver. Ellipse templates are commercially available in a wide assortment of sizes and angle designations. The ellipse angle is that angle between the inclined edge of an ellipse and a horizontal line (see Fig. 70). If the ellipse angle is known, simply use an ellipse template of sufficient size for that angle.

If the ellipse angle is not known, a construction is available to determine the angle. Refer to Fig. 71. Using 0-1 as a radius, swing an arc from center 0 until the arc intersects line 3-5, drawn parallel to the major axis 1-2. Through the intersection point and center 0, draw a line. The angle this line makes with horizontal axis 1-2 is the ellipse angle. Select an ellipse template whose angle is the closest to the ellipse angle just found.

19. TO DRAW ANY NONCIRCULAR CURVE

See Fig. 72. This may be approximated by drawing tangent circle arcs: Select a center by trial, draw as much of an arc as will practically coincide with the curve, and then, changing the center and radius, draw the next portion, remembering always that *if arcs are to be tangent, their centers must lie on the common normal at the point of tangency.*

20. THE PARABOLA

The parabola is a plane curve generated by a point so moving that its distance from a fixed point, called the "focus," is always equal to its distance from a straight line, called the "directrix." Among its practical applications are searchlights, parabolic reflectors, road sections, and certain bridge arches.

The mathematical equation for a parabola with the origin of rectilinear coordinates at the intercept of the curve with the X axis and the focus on the axis is $y^2 = 2px$, where p is twice the distance from the origin to the focus.

To draw a parabola when the focus F and the directrix AB are given (Fig. 73), draw the axis through F perpendicular to AB. Through any point D on the axis, draw a line parallel to AB. With the distance DO as radius and F as center, draw an arc intersecting the line, thus locating a point P on the curve. Repeat the operation as many times as needed.

A. Construction: Parallelogram Method Usually when a parabola is required, the dimensions of the enclosing

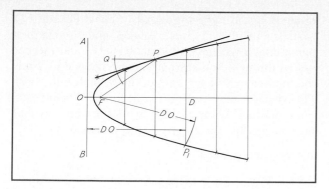

FIG. 73 A parabola. Points on the curve are equidistant from the focus *F* and directrix *AB*.

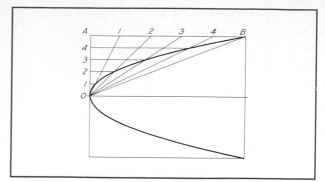

FIG. 74 A parabola by parallelogram method.

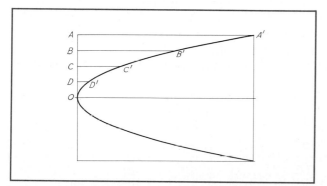

FIG. 75 A parabola by offset method. Offsets are proportionate to squares of divisions of *OA*.

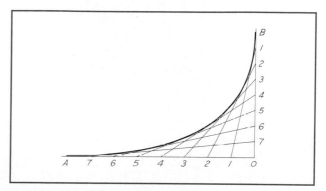

FIG. 76 A parabola by envelope method.

rectangle, that is, the width and depth of the parabola (or span and rise), are given, as in Fig. 74. Divide *OA* and *AB* into the same number of equal parts. From the divisions on *AB*, draw lines converging at *O*. From the divisions on *OA*, draw lines parallel to the axis. The intersections of these with the lines from the corresponding divisions on *AB* will be points on the curve.

B. Construction: Offset Method   See Fig. 75. Given the enclosing rectangle, the parabola can be plotted by computing the offsets from the line *OA*. These offsets vary in length as the square of their distances from *O*. Thus if *OA* is divided into four parts, *DD'* will be ¹⁄₁₆ of *AA'*; *CC'*, since it is twice as far from *O* as *DD'*, will be ⁴⁄₁₆ of *AA'*; and *BB'*, ⁹⁄₁₆. If *OA* had been divided into five parts, the relations would be ¹⁄₂₅, ⁴⁄₂₅, ⁹⁄₂₅, and ¹⁶⁄₂₅, the denominator in each case being the square of the number of divisions. This method is the one generally used by civil engineers in drawing parabolic arches.

C. Construction: Parabolic Envelope   See Fig 76. This method of drawing a pleasing curve is often used in machine design. Divide *OA* and *OB* into the same number of equal parts. Number the divisions from *O* and *B* and connect the corresponding numbers. The tangent curve will be a portion of a parabola—but a parabola whose axis is not parallel to either coordinate.

21. THE HYPERBOLA

The hyperbola is a plane curve generated by a point moving so that the difference of its distances from two fixed points, called the "focuses," is a constant. (Compare this definition with that of the ellipse.)

The mathematical equation for a hyperbola with the center at the origin of rectilinear coordinates and the focuses on the X axis is $x^2/a^2 - y^2/b^2 = 1$, where *a* is the distance from the center to the X intercept, *b* is the corresponding Y value of the asymptotes, lines that the tangents to the curve meet at infinity.

A. Construction: Given Focuses F_1 and F_2 and the Transverse Axis AB (Constant Difference) See Fig. 77. With F_1 and F_2 as centers and any radius greater than F_1B, as F_1P, draw arcs. With the same centers and any radius $F_1P - AB$, strike arcs intersecting these arcs, giving points on the curve.

B. Construction: To Draw a Tangent at Any Point P This construction is easily done by bisecting angle F_1PF_2 of Fig. 77.

C. Construction: Equilateral Hyperbola The case of the hyperbola of commonest practical interest to the engineer is the equilateral, or rectangular, hyperbola referred to as its asymptotes. With it, the law $PV = c$, connecting the varying pressure and volume of a portion of steam or gas, can be graphically presented.

Refer to Fig. 78. Let OA and OB be the asymptotes of the curve and P any point on it (this might be the point of cutoff on an indicator diagram). Draw PC and PD. Mark any points 1, 2, 3, etc., on PC, and through these points draw a system of lines parallel to OA and a second system through the same points converging at O. From the intersections of the lines of the second system with PD extended, draw perpendiculars to OA. The intersections of these perpendiculars with the corresponding lines of the first system give points on the curve.

22. CYCLOID CURVES

A cycloid is the curve generated by the motion of a point on the circumference of a circle rolled in a plane along a straight line (see Fig. 79). If the circle is rolled on the outside of another circle, the curve generated is called an "epicycloid"; if rolled on the inside, it is called a "hypocycloid." These curves are used in drawing the cycloid system of gear teeth.

The mathematical equation for a cycloid (parametric

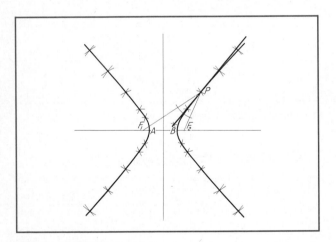

FIG. 77   A hyperbola.

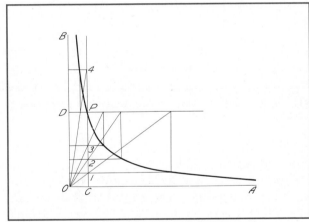

FIG. 78   An equilateral hyperbola. The asymptotes are perpendicular to each other.

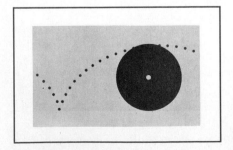

FIG. 79   Formation of a cycloid.

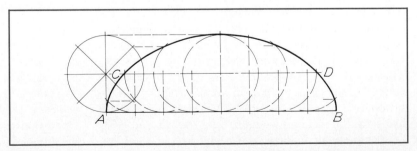

FIG. 80   A cycloid. A point on the circumference of a rolling wheel describes this curve.

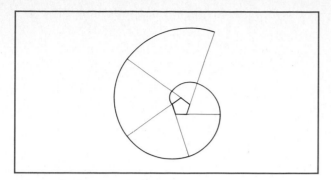

FIG. 81 An involute of a pentagon. A point on a cord unwound from the pentagon describes this curve.

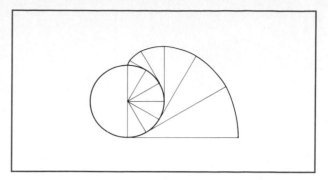

FIG. 82 An involute of a circle. As a taut cord is unwound from the circle, it forms a series of tangents to the circle.

form) is $x = r\theta - r\sin\theta$, $y = r - r\cos\theta$, where r is the radius of a moving point and θ is the turned angle, in radians, about the center from zero position.

▪ **To Draw a Cycloid** (Fig. 80) Divide the rolling circle into a convenient number of parts (say, eight), and, using these divisions, lay off on the tangent AB the rectified length of the circumference. Draw through C the line of centers CD, and project the division points up to this line by perpendiculars to AB. Using these points as centers, draw circles representing different positions of the rolling circle, and project, in order, the division points of the original circle across to these circles. The intersections thus determined will be points on the curve.

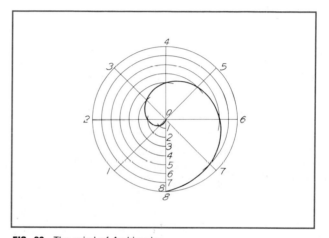

FIG. 83 The spiral of Archimedes.

23. THE INVOLUTE

An involute is the spiral curve traced by a point on a taut cord unwinding from around a polygon or circle. Thus the involute of any polygon can be drawn by extending its sides, as in Fig. 81, and, with the corners of the polygon as successive centers, drawing arcs terminating on the extended sides.

The equation of the involute of a circle (parametric form) is $x = r(\sin\theta - \theta\cos\theta)$, $y = r(\cos\theta + \theta\sin\theta)$, where r is the radius of the circle and θ is the turned angle for the tangent point.

In drawing a spiral in design, as, for example, of bent ironwork, the easiest way is to draw it as the involute of a square.

A circle may be conceived of as a polygon of an infinite number of sides. Thus to draw the involute of a

circle (Fig. 82), divide it into a convenient number of parts, draw tangents at these points, lay off on these tangents the rectified lengths of the arcs from the point of tangency to the starting point, and connect the points by a smooth curve. The involute of the circle is the basis for the involute system of spur gearing.

24. THE SPIRAL OF ARCHIMEDES

The spiral of Archimedes is the plane curve generated by a point moving uniformly along a straight line while the line revolves about a fixed point with uniform angular velocity.

▪ **To Draw a Spiral of Archimedes That Makes One Turn in a Given Circle** (Fig. 83) Divide the circle into a number

FIG. 84 Application of the helix.

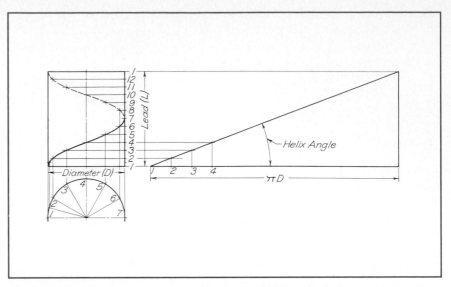

FIG. 85 The cylindrical helix and its development. Any point moves at a constant rate both around and along the axis.

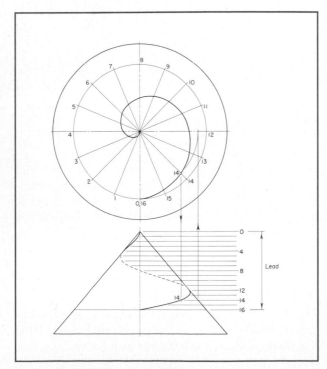

FIG. 86 The conic helix. A point traverses the surface of the cone while moving at a constant rate both around and along the axis.

of equal parts, drawing the radii and numbering them. Divide the radius 0-8 into the same number of equal parts, numbering from the center. With 0 as center, draw concentric arcs intersecting the radii of corresponding numbers, and draw a smooth curve through these intersections. The Archimedean spiral is the curve of the heart cam used for converting uniform rotary motion into uniform reciprocal motion.

The mathematical equation is $p = a\theta$ (polar form).

25. DOUBLE-CURVED LINES: THE HELIX EXAMPLE

The scarcity of geometric double-curved lines, by comparison with the numerous single-curved lines, is surprising. There are only two double-curved lines, the cylindrical and the conic helix, much used in engineering work. Double-curved lines will, however, often occur as the lines of intersection between two curved solids or surfaces. These lines are not geometric but are double-curved lines of general form.

The helix is a space curve generated by a point moving uniformly along a straight line while the line revolves uniformly about another line as an axis. If the moving line is parallel to the axis, it will generate a cylinder. Figure 84 is an example of a cylindrical helix. The word "helix" alone always means a cylindrical helix. If the moving line intersects the axis at an angle less than 90°, it will generate a cone, and the curve made by the point moving on it will be a "conic helix." The distance parallel to the axis through which the point advances in one revolution is called the "lead." When the angle becomes 90°, the helix degenerates into the Archimedean spiral.

A. Construction: A Cylindrical Helix Using Fig. 85, draw the two views of the cylinder (see Chap. 5). Then measure the lead along one of the contour elements. Divide this lead into a number of equal parts (say, 12) and the circle of the front view into the same number. Number the divisions on the top view starting at point 1 and the divisions on the front view starting at the front view of point 1. When the generating point has moved one-twelfth of the distance around the cylinder, it has also advanced one-twelfth of the lead; when halfway around the cylinder, it will have advanced one-half the lead. Thus points on the top view of the helix can be found by projecting the front views of the elements, which are points on the circular front view of the helix, to intersect lines drawn across from the corresponding divisions of the lead. If the cylinder is developed, the helix will appear on the development as a straight line inclined to the base at an angle, called the "helix angle," whose tangent is $L/\pi D$, where L is the lead and D the diameter.

B. Construction: A Conic Helix As in Fig. 86, make two views of the right-circular cone (see Chap. 5) on which the helix will be generated. Then lay out uniform angular divisions in the view showing the end view of the axis (in Fig. 86, the top view) and divide the lead into the same number of parts. Points can now be plotted on the curve. Each plotted point will lie on a circle cut from the cone by a plane dividing the lead and also on the angular-division line. Thus, for example, to plot point 14, draw the circle diameter obtained from the front view, as shown in the top view. Point 14 in the top view then lies at the intersection of this circle and radial-division line 14. Then locate the front view by projection from the top view to plane 14 in the front view.

26. GEOMETRIC SHAPES

A good summary of a wide variety of geometric shapes is shown in Fig. 87. A number of these shapes will be discussed throughout the course of the book. Others, such as the double-curved surfaces, are treated in advanced texts.

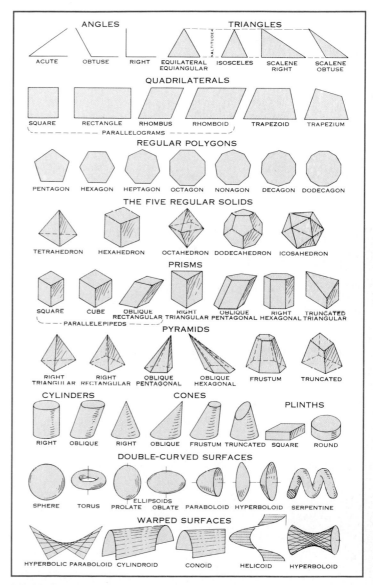

FIG. 87 Geometric shapes. These plane figures, solids, and surfaces are shown for reference.

PROBLEMS

GROUP 1. LINES AND PLANE FIGURES

1. Near the center of the working space, draw a horizontal line 4½ in long. Divide it into seven equal parts by the method of Fig. 28.

2. Draw a vertical line 30 mm from the left edge of the space and 90 mm long. Divide it into parts proportional to 1, 3, 5, and 7.

3. Construct a polygon as shown in the problem illustration, drawing the horizontal line AK (of indefinite length) ⅜ in above the bottom of the space. From A draw and measure AB. Proceed in the same way for the remaining sides. The angles can be obtained by proper combinations of the two triangles (see Chap. 2).

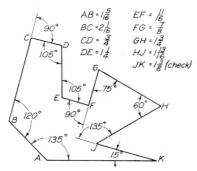

$$AB = 1\tfrac{5}{8} \qquad EF = \tfrac{11}{16}$$
$$BC = 2\tfrac{1}{16} \qquad FG = \tfrac{7}{8}$$
$$CD = \tfrac{3}{4} \qquad GH = 1\tfrac{3}{4}$$
$$DE = 1\tfrac{1}{4} \qquad HJ = 1\tfrac{13}{16}$$
$$JK = 1\tfrac{3}{8} \ (check)$$

PROB. 3   Irregular polygon.

4. Draw a line AK making an angle of 15° with the horizontal. With this line as base, transfer the polygon of Prob. 3.

5. Draw a regular hexagon having a distance across corners of 100 mm.

6. Draw a regular hexagon, distance across flats 80 mm.

GROUP 2. PROBLEMS IN ACCURATE JOINING OF TANGENT LINES

7. Draw the offset swivel plate in full size.

8. Draw two lines AB and AC making an included angle of 30°. Locate point P, 4 in from A and ½ in from line AB. Draw two circle arcs centered at point P, one tangent to line AB, the other to AC. Then draw two lines tangent to the opposite sides of the arcs and passing through point A. Locate all tangent points by construction.

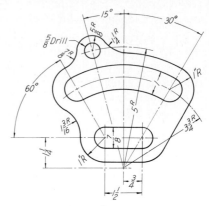

PROB. 7   Offset swivel plate.

9. Construct an ogee curve joining two parallel lines AB and CD as in Fig. 24, making $X = 4$ in, $Y = 2½$ in and $BE = 3$ in. Consider this as the center line for a rod 1¼ in in diameter, and draw the rod.

10. Make contour view of the bracket. In the upper ogee curve, the radii R_1 and R_2 are equal. In the lower one, R_3 is twice R_4. Draw full size.

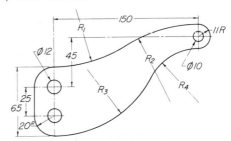

PROB. 10   **METRIC.** Bracket.

11. Draw an arc of a circle having a radius of 3¹³⁄₁₆ in, with its center ½ in from the top of the space and 1¼ in from the left edge. Find the length of an arc of 60° by construction; compute the length arithmetically, and check the result.

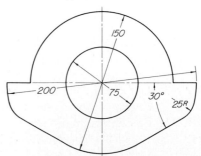

PROB. 12   **METRIC.** Washer.

12. Front view of washer. Draw half size.

13. Front view of shim.

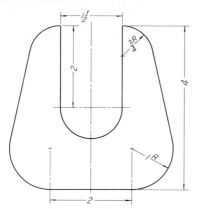

PROB. 13 METRIC. Shim.

14. Front view of rod guide.

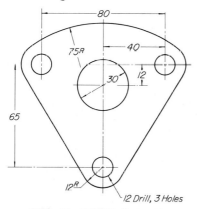

PROB. 14 METRIC. Rod guide.

15. Front view of a star knob. Radius of circumscribing circle, 2⅜ in. Diameter of hub, 2½ in. Diameter of hole, ¾ in. Radius at points, ⅜ in. Radius of fillets, ⅜ in. Mark tangent points in pencil. Draw full size.

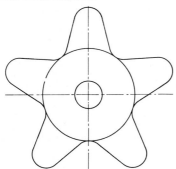

PROB. 15 Star knob.

16. Front view of a fan.

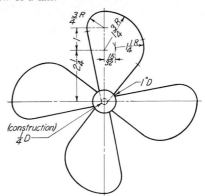

PROB. 16 Fan.

17. Front view of a level plate.

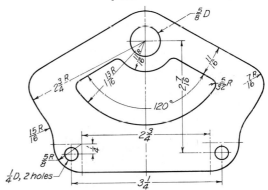

PROB. 17 Level plate.

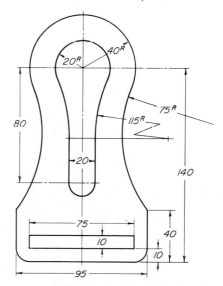

PROB. 18 METRIC. Eyelet.

18. Front view of an eyelet.

19. Front view of a stamping.

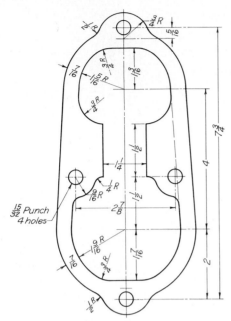

PROB. 19   Stamping.

20. Front view of spline lock.

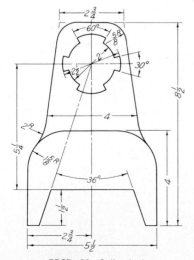

PROB. 20   Spline lock.

21. Front view of gage cover plate.

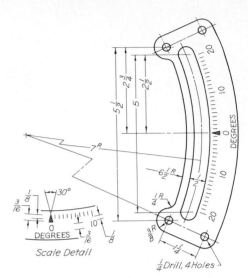

PROB. 21   Gage cover plate.

22. Drawing of heater tube.

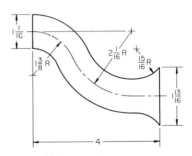

PROB. 22   Heater tube.

23. Drawing of toggle spring for leaf switch.

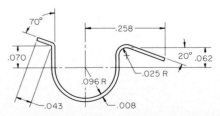

PROB. 23   Toggle spring for leaf switch.

24. Drawing of exhaust-port contour.

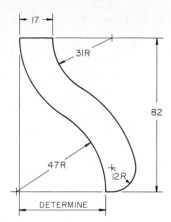

PROB. 24 METRIC. Exhaust-port contour.

25. Drawing of pulley shaft.

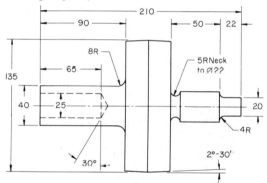

PROB. 25 METRIC. Pulley shaft.

26. Drawing of cam for type A control. Plot cam contour from the polar coordinates given.

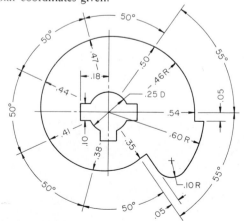

PROB. 26 Cam for type A control.

GROUP 3. PLANE CURVES

27. The conjugate diameters of an ellipse measure 3 and 4 in, the angle between them being 60°. Construct the major and minor diameters of this ellipse and draw the ellipse.

28. Using the pin-and-string method, draw an ellipse having a major diameter of 6 in and a minor diameter of 4¼ in.

29. Using the trammel method, draw an ellipse having a major diameter of 115 mm and a minor diameter of 75 mm.

30. Using the trammel method, draw an ellipse having a major diameter of 4½ in and a minor diameter of 4 in.

31. Using the concentric-circle method, draw an ellipse having a major diameter of 120 mm and a minor diameter of 40 mm.

32. Draw an ellipse on a major diameter of 4 in. One point on the ellipse is 1½ in to the left of the minor diameter and ⅞ in above the major diameter.

33. Draw an ellipse having a minor diameter of 2³⁄₁₆ in and a distance of 3¼ in between focuses. Draw a tangent at a point 1⅜ in to the right of the minor diameter.

34. Draw a five-centered arch with a span of 5 in and a rise of 2 in.

35. Draw an ellipse having conjugate diameters of 4¾ in and 2¾ in, making an angle of 75° with each other. Determine the major and minor diameters.

36. Draw the major and minor diameters for an ellipse having a pair of conjugate diameters 60° apart, one horizontal and 6¼ in long, the other 3¼ in long.

37. Draw a parabola, axis vertical, in a rectangle 4 by 2 in.

38. Draw a parabolic arch, with a 150 mm span and a 65 mm rise, by the offset method, dividing the half span into eight equal parts.

39. Draw an equilateral hyperbola passing through a point *P*, 15 mm from *OB* and 60 mm from *OA*. (Reference letters correspond to Fig. 78.)

40. Draw two turns of the involute of a pentagon whose circumscribed circle is ½ in in diameter.

41. Draw one-half turn of the involute of a circle 3¼ in in diameter whose center is 1 in from the left edge of the space. Compute the length of the last tangent and compare with the measured length.

42. Draw a spiral of Archimedes making one turn in a circle 4 in in diameter.

43. Draw the cycloid formed by a rolling circle 50 mm in diameter. Use 12 divisions.

44. Front view of cam.

45. Front view of fan base.

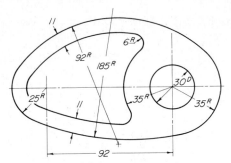

PROB. 44 METRIC. Front view of cam.

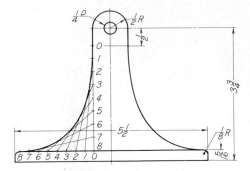

PROB. 45 Front view of fan base.

46. Front view of trip lever. Draw half size.

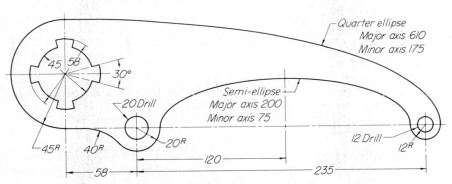

PROB. 46 METRIC. Front view of trip lever.

1. NEED FOR LETTERING

We have studied the various instruments used in drafting (Chap. 2) and also numerous geometric constructions that can be made with instruments (Chap. 3); It is now time to look at one more aspect of instrument use: lettering.

Lettering is needed for a host of items required on drawings, including dimensions of parts, notes of all kinds relating to the manufacture of parts, and descriptive titles. All lettering needs to be highly legible, uniform, and capable of being done rapidly.

There are several ways in which lettering can be done. These include both hand lettering and mechanical lettering. Mechanical lettering may be by typewriter, by use of templates and transfers, or by computer output. Mechanical lettering has seen a large increase in use in recent years and it is a very useful, efficient method of lettering. Typewriter, template, and transfer means of mechanical lettering are relatively inexpensive. Lettering a drawing by computer belongs in the realm of useful but more sophisticated and expensive techniques. However, computers are used more and more by firms that generate a large volume of drawings for costly projects.

In this chapter we will concentrate on hand lettering. Its use is universal and always of value. Our study will be devoted entirely to the commercial Gothic letter because this style predominates in all types of commercial drawings. However, other letter styles are used extensively on architectural drawings and on maps, in design and for display, and for decorative letters on commercial products.

2. SINGLE-STROKE LETTERING

By far the greatest amount of lettering on drawings is done in a rapid single-stroke letter, either vertical or inclined, and every engineer should have absolute command of these styles. The ability to letter well can be acquired only by continued and careful practice. It is not a matter of artistic talent or even of dexterity in handwriting. Many persons who write poorly letter very well.

The term "single-stroke," or "onestroke," does not mean that the entire letter is made without lifting the pencil or pen but that the width of the stroke of the pencil or pen is the width of the stem of the letter.

3. CONSIDERATIONS FOR GOOD LETTERING

The appearance of lettering can be enhanced by awareness of three areas: general proportions, the rule of stabil-

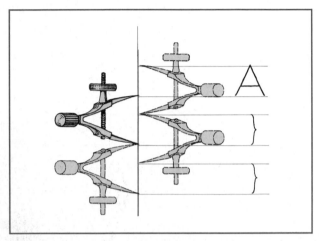

FIG. 1 To space guide lines. Bow dividers are spaced for the distance between base lines; this distance stepped off from capital and base lines locates successive lines of lettering.

ity, and use of guide lines. We will deal briefly with each in turn.

A. General Proportions There is no standard for the proportions of letters, but there are certain fundamental points in design and certain characteristics of individual letters that must be learned by study and observation before composition into words and sentences should be attempted. Not only do the widths of letters in any alphabet vary, from *I*, the narrowest, to *W*, the widest, but different alphabets vary as a whole. Styles narrow in their proportion of width to height are called "COMPRESSED," or "CONDENSED," and are used when space is limited. Styles wider than the normal are called "EXTENDED."

The proportion of the thickness of stem to the height varies widely, ranging from $\frac{1}{3}$ to $\frac{1}{20}$. Letters with heavy stems are called **"BOLDFACE,"** or **"BLACKFACE,"** those with thin stems "LIGHTFACE."

The vertical, single-stroke, commercial Gothic letter is a standard for titles, reference letters, etc. As for the proportion of width to height, the general rule is that the smaller the letters are, the more extended they should be in width. A low extended letter is more legible than a high compressed one and at the same time has a more pleasing appearance.

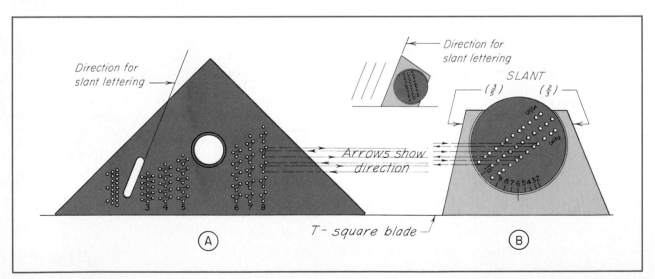

FIG. 2 (A) Braddock-Rowe triangle. Numbered sets of holes locate capital, waist, and base lines of successive lines of lettering. (B) Ames lettering instrument. A center disk, adjusted to letter-height number, gives equal spaces or 2:3 and 3:5 waist-line proportion for successive lines of lettering.

B. The Rule of Stability In the construction of letters, the well-known optical illusion in which a horizontal line drawn across the middle of a rectangle appears to be below the middle must be provided for. In order to give the appearance of stability, such letters as B, E, K, S, X, and Z and the figures 3 and 8 must be drawn smaller at the top than at the bottom. To see the effect of this illusion, turn a printed page upside down and notice the appearance of the letters mentioned.

C. Guide Lines Always draw light guide lines for both tops and bottoms of letters, using a sharp pencil. Figure 1 shows a method of laying off a number of equally spaced lines. Draw the first base line; then set the bow spacers to the distance wanted between base lines and step off the required number of base lines. Above the last line mark the desired height of the letters. With the same setting, step down from this upper point, thus obtaining points for the top of each line of letters.

The Braddock-Rowe triangle (Fig. 2A) and the Ames lettering instrument (Fig. 2B) are convenient devices for spacing lines of letters. In using these instruments, a sharp pencil is inserted in the proper row of countersunk holes, and the instrument, guided by a T-square blade, is drawn back and forth by the pencil, as indicated by Fig. 3. The holes are grouped for capitals and lowercase, the numbers indicating the height of capitals in thirty-seconds of an inch; thus, no. 6 spacing of the instrument means that the capitals will be $\frac{6}{32}$, or $\frac{3}{16}$, in high.

The Ames lettering instrument is also available with metric spacings, so one can use spacing in millimeters rather than inches, if desired.

4. LETTERING IN PENCIL

Good technique in lettering is as essential as it is in the drawing of lines and arcs. The lead selected for lettering should be chosen so as to make letters as dark and sharp as drawn lines. Many drafters use the same grade of lead for lettering as they do for dark lines. However, some persons prefer a lighter touch for lettering compared with drawing of lines. In this case, a lead one grade softer may be ideal. For example, if an H lead was used for drawn lines, an F lead might be selected for lettering. Sharpen the selected lead as would be done for line work, that is, with a long conic point.

The first requirement in lettering is to hold the pen or

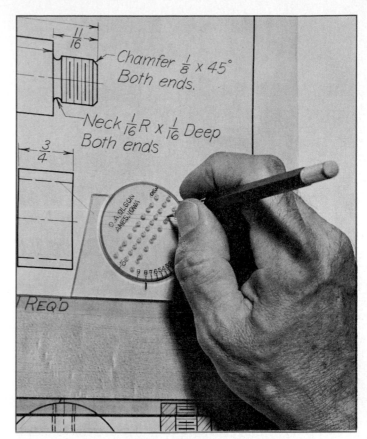

FIG. 3 Using the Ames lettering instrument. Lines are drawn as the pencil moves the instrument along the T square (or triangle).

pencil correctly. Figure 4 shows the pencil held comfortably with the thumb, forefinger, and second finger on alternate flat sides and the third and fourth fingers on the paper. Vertical, slanting, and curved strokes are drawn with a steady, even, *finger* movement; horizontal strokes are made similarly but with some pivoting of the hand at the wrist (Fig. 5). Exert pressure that is firm and uniform but not so heavy as to cut grooves in the paper. To keep the point symmetrical, form the habit of rotating the pencil after every few strokes.

5. VERTICAL LETTERING

Single-stroke engineering letters are derived from Gothic lettering, which includes the fancier styles of Old English Gothic and German Gothic. We will be concerned with only the single-stroke letters.

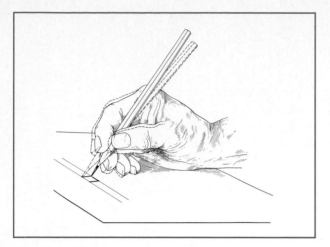

FIG. 4 Vertical strokes. These are made entirely by finger movement.

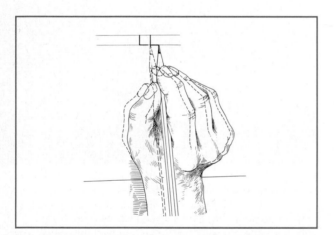

FIG. 5 Horizontal strokes. These are made by pivoting the whole hand at the wrist; fingers move slightly to keep the stroke perfectly horizontal.

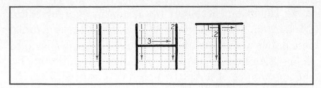

FIG. 6 The I-H-T group. Note the direction of fundamental horizontal and vertical strokes.

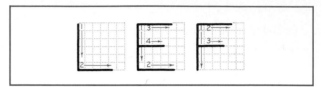

FIG. 7 The L-E-F group. Note the successive order of strokes.

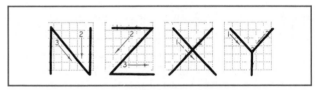

FIG. 8 The N-Z-X-Y group. Note that *Z* and *X* are smaller at the top than at the bottom, in accordance with the rule of stability.

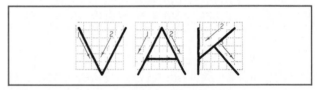

FIG. 9 The V-A-K group. The horizontal of A is one-third from the bottom; the second and third strokes of K are perpendicular to each other.

A. Vertical Capital Letters   In the following figures an alphabet of slightly extended vertical capitals has been arranged in family groups. Study the shape of each letter, with the order and direction of the strokes forming it, and practice it until its form and construction are familiar.

To bring out the proportions of widths to heights and the subtleties in the shapes of the letters, they are shown against a square background with its sides divided into sixths. Several of the letters in this alphabet, such as A and T, fill the square; that is, they are as wide as they are high. Others, such as H and D, are approximately five spaces wide, or their width is five-sixths of their height.

▪ **The I-H-T Group** (Fig. 6)   The letter I is the foundation stroke. You may find it difficult to keep the stems vertical. If so, draw direction lines lightly an inch or so apart to aid the eye. The H is nearly square (five-sixths wide), and in accordance with the rule of stability, the crossbar is just above the center. The top of the T is drawn first to the full width of the square, and the stem is started accurately as its middle point.

▪ **The L-E-F Group** (Fig. 7)   The L is made in two strokes. The first two strokes of the E are the same as for the L; the third, or upper, stroke is slightly shorter than the lower; and the last stroke is two-thirds as long as the lower and just above the middle. F has the same proportions as E.

■ **The N-Z-X-Y Group** (Fig. 8)   The parallel sides of N are generally drawn first, but some prefer to make the strokes in consecutive order. Z and X are both started inside the width of the square on top and run to full width at the bottom. This throws the crossing point of the X slightly above the center. The junction of the Y strokes is at the center.

■ **The V-A-K Group** (Fig. 9)   V is the same width as A, the full breadth of the square. The A bridge is one-third up from the bottom. The second stroke of K strikes the stem one-third up from the bottom; the third stroke branches from it in a direction starting from the top of the stem.

■ **The M-W Group** (Fig. 10)   These are the widest letters. M may be made in consecutive strokes or by drawing the two vertical strokes first, as with the N. W is formed of two narrow V's, each two-thirds of the square in width. Note that with all the pointed letters the width at the point is the width of the stroke.

■ **The O-Q-C-G Group** (Fig. 11)   In this extended alphabet the letters of the O family are made as full circles. The O is made in two strokes, the left side a longer arc than the right, as the right side is harder to draw. Make the kern of the Q straight. A large-size C and G can be made more accurately with an extra stroke at the top, whereas in smaller letters the curve is made in one stroke. Note that the bar on the G is halfway up and does not extend past the vertical stroke.

■ **The D-U-J Group** (Fig. 12)   The top and bottom strokes of D must be horizontal. Failure to observe this is a common fault with beginners. In large letters U is formed by two parallel strokes, to which the bottom stroke is added; in smaller letters, it may be made in two strokes curved to meet at the bottom. J has the same construction as U, with the first stroke omitted.

■ **The P-R-B Group** (Fig. 13)   With P, R, and B, the number of strokes depends upon the size of the letter. For large letters the horizontal lines are started and the curves added, but for smaller letters only one stroke for

each lobe is needed. The middle lines of P and R are on the center line; that of B observes the rule of stability.

■ **The S-8-3 Group** (Fig. 14)   The S, 8, and 3 are closely related in form, and the rule of stability must be observed carefully. For a large S, three strokes are used; for a smaller one, two strokes; and for a very small size,

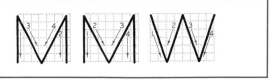

FIG. 10   The M-W group. M is one-twelfth wider than it is high; W is one-third wider than it is high.

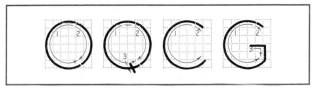

FIG. 11   The O-Q-C-G group. All are based on the circle.

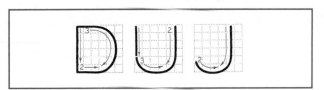

FIG. 12   The D-U-J group. These are made with combinations of straight and curved strokes.

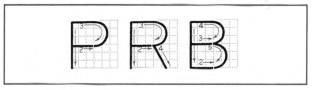

FIG. 13   The P-R-B group. Note the rule of stability with regard to R and B.

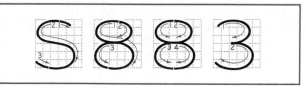

FIG. 14   The S-8-3 group. A perfect S and 3 can be completed to a perfect 8.

one stroke only is best. The 8 may be made on the S construction in three strokes, or in "head and body" in four strokes. A perfect 3 can be finished into an 8.

■ **The O-6-9 Group** (Fig. 15) The cipher is an ellipse five-sixths the width of the letter O. The backbones of the 6 and 9 have the same curve as the cipher, and the lobes are slightly less than two-thirds the height of the figure.

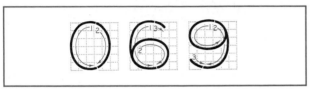

FIG. 15 The O-6-9 group. The width is five-sixth of the height.

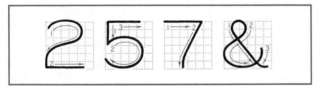

FIG. 16 The 2-5-7-& group. Note the rule of stability. The width is five-sixth of the height.

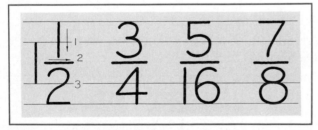

FIG. 17 Fractions. The total height of a fraction is twice that of the integer.

FIG. 18 Basic forms for lowercase letters. For standard letters, the waist-line height is two-thirds of capital height; capital line and drop line are therefore one-third above and one-third below the body of the letter.

■ **The 2-5-7-& Group** (Fig. 16) The secret in making the 2 lies in getting the reverse curve to cross the center of the space. The bottom of 2 and the tops of 5 and 7 should be horizontal straight lines. The second stroke of 7 terminates directly below the middle of the top stroke. Its stiffness is relieved by curving it slightly at the lower end. The ampersand (&) is made in three strokes for large letters and two for smaller ones and must be carefully balanced.

■ **The Fraction Group** (Fig. 17) Fractions are always made with a horizontal bar. Integers are the same height as capitals. The total fraction height is best made twice the height of the integer. The numerator and denominator will be about three-fourths the height of the integer. Be careful to leave a clear space above and below the horizontal bar. Guide lines for fractions are easily obtained with lettering instruments by using the set of uniformly spaced holes or by drawing the integer height above and below the center, the position of the horizontal bar.

B. Vertical Lowercase Letters The single-stroke, vertical lowercase letter is not commonly used on machine drawings but is used extensively in map drawing. The bodies are made two-thirds the height of the capitals with the ascenders extending to the capital line and the descenders dropping the same distance below. The basic form of the letter is the combination of a circle and a straight line (Fig. 18). The alphabet, with some alternative shapes, is shown in Fig. 19, which also gives the capitals in alphabetic order.

6. INCLINED LETTERING

A. Inclined Capital Letters Many drafters use the inclined, or slant, letter in preference to the upright. The order and direction of strokes are the same as in the vertical form.

After ruling the guide lines, draw slanting "direction lines" across the lettering area to aid the eye in keeping the slope uniform. These lines may be drawn with a special lettering triangle of about $67\frac{1}{2}°$, or the slope of 2 to 5 may be fixed on the paper by marking two units on a horizontal line and five on a vertical line and using T square and triangle as shown in Fig. 20. The Braddock-Rowe triangle and the Ames instrument both provide for

FIG. 19 Single-stroke vertical capitals and lowercase. Note the alternative strokes for small-size capitals and the alternative shapes for lowercase *a*, *g*, and *y*.

the drawing of slope lines. The form taken by the rounded letters when inclined is illustrated in Fig. 21, which shows that curves are sharp in all upper right-hand and lower left-hand corners and flattened in the other two corners. The alphabet is given in Fig. 22. Study the shape of each letter carefully.

Professional appearance in lettering is due to three things: (1) keeping to a uniform slope, (2) making the letters full and well shaped, and (3) keeping the letters close together. The beginner invariably cramps the individual letters and spaces them too far apart.

B. Inclined Lowercase Letters The inclined lowercase letters (Fig. 22) have bodies two-thirds the height of the capitals with the ascenders extending to the capital line and the descenders dropping the same distance below the base line. The lowercase letter is suitable for notes and statements on drawings.

All the letters are based on two elements, the straight line and the ellipse, and have no unnecessary hooks or appendages. They may be divided into four groups, as

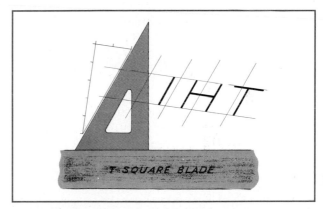

FIG. 20 Slope guide lines. The standard slope angle of 67½° can be approximated by a slope of 2 to 5.

FIG. 21 Form of curved-stroke inclined capitals. The basic shape is elliptical.

FIG. 22 Single-stroke inclined capitals and lowercase. Note the alternative strokes for small-size capitals and the alternative shapes for lowercase *a, g,* and *y.*

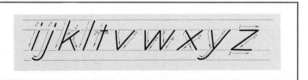

FIG. 23 The straight-line inclined lowercase letters. Note that the center lines of the letters follow the slope angle.

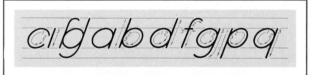

FIG. 24 The loop letters. Note the graceful combination of elliptical body, ascenders, and descenders.

FIG. 25 The ellipse letters. Their formation is basically elliptical.

FIG. 26 The hook letters. They are combinations of ellipses and straight lines.

shown in Figs. 23 to 26. The dots of *i* and *j* and the top of the *t* are on the "*t* line," halfway between the waist line and the capital line. The loop letters are made with an ellipse whose long axis is inclined about 45° in combination with a straight line.

The *c, e,* and *o* are based on an ellipse of the shape of the capitals but not inclined quite so much as the loop-letter ellipse. In rapid small work, the *o* is often made in one stroke, as are the *e, v,* and *w.* The *s* is similar to the capital but, except in letters more than ⅛ in high, is made in one stroke. In the hook-letter group, note particularly the shape of the hook.

The single-stroke letter may, if necessary, be much compressed and still be clear and legible (Fig. 27). It is also used sometimes in extended form.

7. FOR LEFT-HANDERS ONLY

The order and direction of strokes in the preceding alphabets have been designed for right-handed persons. The principal reason that left-handers sometimes find lettering difficult is that whereas the right-hander progresses away from the body, the left-hander progresses toward the body; consequently the pencil and hand partially hide the work done, making it harder to join strokes and to preserve uniformity.

For the natural left-hander, whose writing position is the same as a right-hander except reversed left for right, a change in the sequence of strokes of some of the letters will obviate part of the difficulty caused by interference with the line of sight. Figure 28 gives an analyzed alphabet with an alternative for some letters. In E the top bar

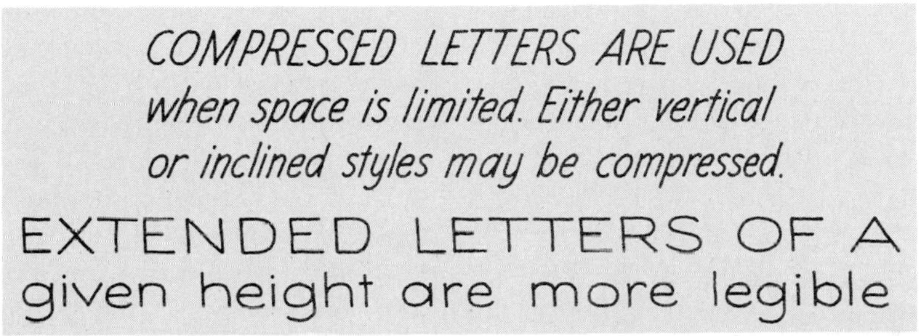

COMPRESSED LETTERS ARE USED
when space is limited. Either vertical
or inclined styles may be compressed.

EXTENDED LETTERS OF A
given height are more legible

FIG. 27 Compressed and extended letters. Normal width has *O* the same in width as in height and the other letters in proportion. Wider than high for *O* is extended, narrower compressed.

FIG. 28 Strokes for left-handers. Several alternative strokes are given. Choose the one that is most effective for you.

is made before the bottom bar, and M is drawn from left to right to avoid having strokes hidden by the pencil or pen. Horizontal portions of curves are easier to make from right to left; hence the starting points for O, Q, C, G, and U differ from the standard right-hand stroking. S is the perfect letter for the left-hander and is best made in a single smooth stroke. The figures 6 and 9 are difficult and require extra practice. In the lowercase letters a, d, g, and q, it is better to draw the straight line before the curve even though it makes spacing a little harder.

8. COMPOSITION

Composition in lettering has to do with the selection, arrangement, and spacing of appropriate styles and sizes of letters. On engineering drawings the selection of the style is practically limited to vertical or inclined lettering, so composition here means arrangement into pleasing and legible form.

Letters in words are not spaced at a uniform distance from each other but are arranged so that the areas of white space (the irregular backgrounds between the letters) are approximately equal, making the spacing *appear* approximately uniform. Figure 29 illustrates these background shapes. Each letter is spaced with reference to its shape and the shape of the letter preceding it. Thus adjacent letters with straight sides would be spaced farther apart than those with curved sides. Definite rules for spacing are not successful; it is a matter for the drafter's judgment and sense of design.

Figure 30 illustrates word composition. The sizes of letters to use in any particular case can be determined better by sketching them lightly than by judging from the guide lines alone. A finished line of letters always looks larger than the guide lines indicate. When capitals and small capitals are used, the height of the small capitals should be about four-fifths that of the capitals.

In spacing words, a good principle is to leave the space that would be taken by an assumed letter I connecting the two words into one, as in Fig. 31. The space would never be more than the height of the letters.

The clear distance between lines may vary from ½ to 1½ times the height of the letter but for the sake of appearance should not be exactly the same as the letter height.

FIG. 29 Background areas. Equal areas between letters produce spacing that is visually uniform.

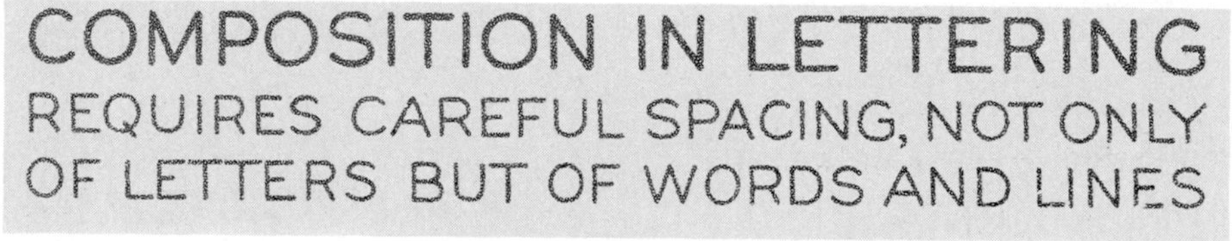

COMPOSITION IN LETTERING
REQUIRES CAREFUL SPACING, NOT ONLY
OF LETTERS BUT OF WORDS AND LINES

FIG. 30 Word composition. Careful spacing of letters and words and the proper emphasis of size and weight are important.

WORDSISPACEDIBYISKETCHINGIANIIIBETWEEN
WORDS SPACED BY SKETCHING AN I BETWEEN

FIG. 31 Word spacing. Space words so that they read naturally and do not run together (too close) or appear as separate units (too far apart).

9. TITLES

A. Composition The most important problem in lettering composition is the design of titles. Every drawing has a descriptive title. It gives necessary information concerning the drawing, and the information that is needed will vary with the different kinds of drawings.

The usual form of lettered title is the *symmetrical title*, which is balanced or "justified" on a vertical center line and designed with an elliptical or oval outline. Sometimes the wording necessitates a pyramid or inverted-pyramid ("bag") form. Figure 32 illustrates several shapes in which titles can be composed.

The lower right-hand corner of the sheet is the usual location for the title. The space allowed depends on the size and purpose of the drawing. On the United States size B (11 by 17 in), equivalent to the international size A3 (297 by 420 mm), a title is typically about 3 in [75 mm] long.

B. To Draw a Title When the wording has been determined, write out the arrangement on a separate piece of paper as in Fig. 33 (or, better, typewrite it). Count the letters, including the word spaces, and make a mark across the middle letter or space of each line. The lines must be displayed for prominence according to their relative importance as judged from the point of view of the persons who will use the drawing. Titles are usually made in all capitals. Draw the base line for the most important line of the title and mark on it the approximate length desired. To get the letter height, divide this length by the number of letters in the line, and draw the capital line. Start at the center line, and sketch lightly the last half of the line, drawing only enough of the letters to show the space each will occupy. Lay off the length of this right half on the other side, and sketch that half, working forward or backward. When this line is satisfactory in size and spacing, draw the remainder in the same way. Study the effect, shift letters or lines if necessary, and complete in pencil. Use punctuation marks only for abbreviations.

10. LETTERING STYLE FOR NOTES AND TITLES

By far the principal use of lettering on engineering drawings occurs on working drawings, where the dimensions,

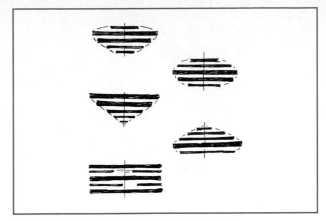

FIG. 32 Shapes in symmetrical composition. Design for clarity and emphasis.

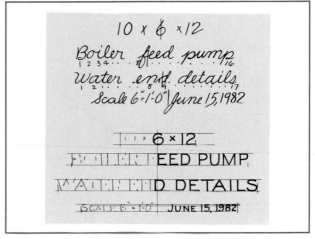

FIG. 33 Title composition. Sketch lightly; then when satisfactory, complete.

explanatory notes, record of drawing changes, and title supplement the graphic description of the object.

Styles vary, but the trend is to use capital letters. Most drafters are able to letter in capital letters with greater readability and fewer mistakes than in lowercase.

On any drawing the accuracy of the information given is critical. It is important to make the letter shapes correct, without personal flourishes. For example, a poorly made 3 may read as an 8. A poorly made 5 could be misread as a 6. Strive for a smooth, professional appearance with good readability.

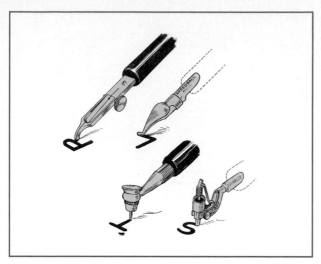

FIG. 34 Barch-Payzant, Speedball, Edco, and Leroy pens. These are used principally for displays, large titles, and number blocks.

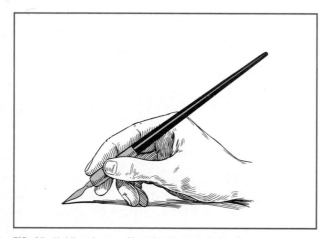

FIG. 35 Holding the pen. Note that the thumb, forefinger, and second finger make a three-point support.

FIG. 36 Lettering template. *(Courtesy Gramercy Corp.)*

11. HAND-HELD LETTERING PENS

To a great extent, hand-held lettering pens have been replaced by mechanical lettering devices discussed in the following section (12). There is still a place, however, for special pens made in sets of graded sizes. Figure 34 shows some of these special pens which can be useful, especially for large lettering.

Always wet a new pen and dry thoroughly so any oil film can be removed. A pen well broken in letters better than a new one, and should be given good care. A pen that has been dipped into writing ink should not be put into drawing ink. A pen being used should be wiped clean often, with a cloth pen wiper.

Use a penholder with cork grip (the small size) and set the pen in it firmly. Many prefer to ink a pen with the quill filler, touching the quill to the underside of the pen point, rather than dipping the pen in the ink bottle. If the pen is dipped, the surplus ink should be shaken back into the bottle or the pen touched against the neck of the bottle as it is withdrawn.

In lettering, the penholder is held in the fingers firmly but without pinching, in the position shown in Fig. 35. The strokes of the letters are made with a steady, even motion and a slight, uniform pressure on the paper that will not spread the nibs of the pen.

12. MECHANICAL LETTERING

Mechanical lettering, as a whole, refers to lettering done with the aid of devices that guide the hand-held pencil or pen. Into this category fall the various templates that can be purchased from numerous companies. Templates are obtainable for both vertical and slant lettering and for letter sizes covering a wide range. Figure 36 shows a typical template.

Templates are used by inserting a pencil or technical pen into the cut hole for a desired letter and following the contour of that letter. The template must be repositioned to begin another letter.

Quality of mechanical lettering is enhanced if instead of templates one uses controlled mechanical lettering systems beyond hand-held pencils and pens. Various systems are on the market, such as the Leroy lettering instrument of the Keuffel & Esser Company and the scriber sets of Koh-I-Noor Rapidograph, Inc.

Such systems, as seen in use in Fig. 37, rely on a pin to track the desired letter in a supplied template while a corresponding pen creates the actual letter on the paper. Inclined letters can be made from a vertical-letter template by adjusting the arm of the scriber. The ink reservoir seen in Fig. 37 may be interchangeable with that of the technical pen, discussed earlier.

Another type of mechanical lettering available is the dry-transfer sheet. The letters come in a very wide variety of styles. One company offers close to 500 styles. Each style has a choice of letter sizes. A sample style is seen in Fig. 38. Letters come on sheets that are nearly transparent. One merely places the letter selected over the drawing where the letter is desired. Rubbing the letter transfers it to the drawing. Another letter is selected, positioned, and transferred. In this way, words and sentences are composed. The drafter should, of course, use guide lines on the drawing to control vertical positioning. If certain words or sentences are needed over and over, custom-made transfer sheets can be used which contain the needed words.

Finally, mechanical lettering may be done totally by machine. The typewriter is a well-known device that has been adapted to produce lettering for drawings. Special typewriters with long carriages can accept a medium-sized drawing. One can then type directly on the drawing, as would be done for any paper inserted into a typewriter. There are portable typewriters, also, which can be laid directly on the drawing without the drawing being inserted into a carriage. Microprocessors are being used in some lettering machines, such as the one in Fig. 12, Chap. 2. This machine offers the flexibility of programmable characters, sizes, and spacing.

You can see that the drafter can choose from a wide range of instruments for aid in lettering. The choice may be limited by need and by money available to purchase such aids. Decisions are often based on balancing increased productivity in drafting with the cost of the device that can increase the productivity. Simple templates are relatively inexpensive; most drafters have one or more. Lettering machines of one sort or another may be limited to one per drafting office. In any event, the basic skills of lettering technique, style, and composition will serve the drafter well in whatever mode of lettering is selected.

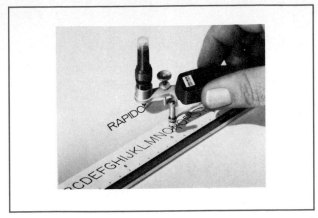

FIG. 37 A scriber set in use. *(Courtesy Koh-I-Noor Rapidograph Inc.)*

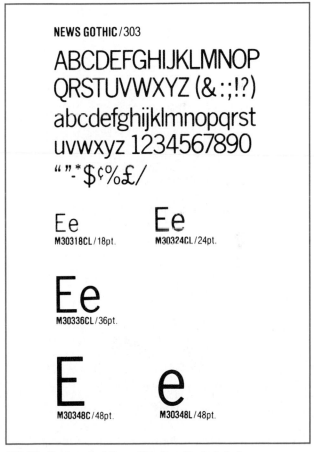

FIG. 38 Dry transfer letters. *(Courtesy Chartpak Co.)*

PROBLEMS

GROUP 1. LETTERING: VERTICAL CAPITALS

1. Large letters in pencil for careful study of the shapes of the individual letters. Starting ⁹⁄₁₆ in from the top border, draw guide lines for five lines of ⅜-in letters. Draw each of the straight-line letters *I, H, T, L, E, F, N, Z, Y, V, A, M, W,* and *X,* four times in pencil.

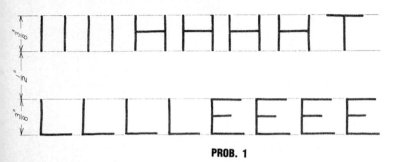

PROB. 1

2. Same as Prob. 1 for the curved-line letters *O, Q, C, G, D, U, J, B, P, R,* and *S.*

3. Same as Prob. 1 for the figures 3, 8, 6, 9, 2, 5, ½, ¾, ⅝, ⁷⁄₁₆, and ⁹⁄₃₂.

4. Composition. Same layout as for Prob. 1. Read Sec. 8 on composition; then letter the following five lines in pencil: (*a*) WORD COMPOSITION, (*b*) TOPOGRAPHIC SURVEY, (*c*) TOOLS AND EQUIPMENT, (*d*) BRONZE BUSHING, (*e*) JACK-RAFTER DETAIL.

5. Six-millimeter vertical letters in pencil and ink. Starting 6 mm from the top, draw guide lines for nine lines of 6-mm letters. In the group order given, draw each letter four times in pencil, as shown in the illustration.

PROB. 5

6. Composition. Make a three-line design of the quotation from Benjamin Lamme on the Lamme medals: "THE ENGI-

NEER VIEWS HOPEFULLY THE HITHERTO UNATTAINABLE."

7. Three-millimeter vertical letters. Starting 6 mm from the top, draw guide lines for 12 lines of 3-mm letters. Make each letter and numeral six times.

8. Composition. Letter the following definition: "Engineering is the art and science of directing and controlling the forces and utilizing the materials of nature for the benefit of all. Engineering involves the organization of human effort to attain these ends. It also involves an appraisal of the social and economic benefits of these activities."

GROUP 2. LETTERING: INCLINED CAPITALS

9 to 16. Same spacing and specifications as for Probs. 1 to 8, but for inclined letters.

GROUP 3. LETTERING: INCLINED LOWERCASE

17. Large letters in pencil for use with ⅜-in capitals. The bodies are ¼ in, the ascenders ⅛ in above, and the descenders ⅛ in below. Starting ⅜ in from the top, draw guide lines for seven lines of letters. This can be done quickly by spacing ⅛ in uniformly down the sheet and bracketing capital and base lines. Make each letter of the alphabet four times in pencil.

18. Lowercase for ³⁄₁₆-in capitals. Starting ½ in from the top, draw capital, waist, and base lines for 10 lines of letters (Braddock or Ames no. 6 spacing). Make each letter four times in pencil.

19. Composition. Same spacing as Prob. 18. Letter the opening paragraph of this chapter.

GROUP 4: LETTERING: TITLES

20. Design a title for the assembly drawing of a rear axle, drawn to the scale of 1 mm = 2 mm, as made by the Chevrolet Motor Co., Detroit. The number of the drawing is C82746. Space allowed is 75 by 125 mm.

21. Design a title for the front elevation of a powerhouse, drawn to ¼-in scale by Burton Grant, Architect, for the Citizens Power and Light Company of Punxsutawney, Pennsylvania.

GROUP 5. LETTERING: NOTES

22 to 24. Vertical capitals. Copy each note in no. 6 (Ames or Braddock) spacing; then in no. 4 spacing.

PAINT WITH METALLIC
SEALER AND TWO COATS
LACQUER AS PER CLIENT
COLOR ORDER.

PROB. 22

THIS PRINT IS AMERICAN
THIRD-ANGLE PROJECTION

PROB. 23

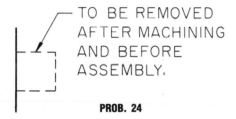

TO BE REMOVED
AFTER MACHINING
AND BEFORE
ASSEMBLY.

PROB. 24

25 to 27. Inclined capitals. Copy in no. 5 or no. 4 spacing.

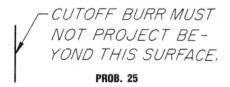

CUTOFF BURR MUST
NOT PROJECT BE-
YOND THIS SURFACE.

PROB. 25

ALTERNATE MATERIAL:
1ST. RED BRASS 85% CU.
2ND. COMM. BRASS 90% CU.
3RD. COMM. BRASS 95% CU.

PROB. 26

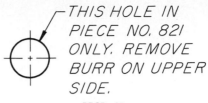

THIS HOLE IN
PIECE NO. 821
ONLY. REMOVE
BURR ON UPPER
SIDE.

PROB. 27

28 to 30. Vertical lowercase. Copy in no. 5 spacing.

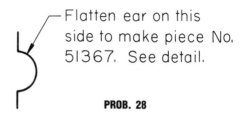

Flatten ear on this
side to make piece No.
51367. See detail.

PROB. 28

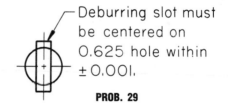

Deburring slot must
be centered on
0.625 hole within
±0.001.

PROB. 29

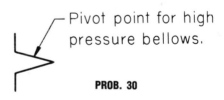

Pivot point for high
pressure bellows.

PROB. 30

31. to 33. Inclined lowercase. Copy in no. 4 spacing.

Material: #19 Ga (0.024)
C R Steel. Temper to Rockwell B-40 to 65.

PROB. 31

This length varies from
10 mm to 25 mm
See table Ⓐ below

PROB. 32

Extrude to 0.082
± .002 Dia.
3-56 Class 2 tap.

PROB. 33

34. For extra practice, select any of Probs. 23 to 34 and letter in a chosen style and size.

THEORY OF ORTHOGRAPHIC PROJECTION

1. BACKGROUND

The previous chapters have given you some introduction in the use of instruments, constructions, and lettering. The supporting information in this chapter makes it possible to move forward into describing real objects such that they could actually be understood and made.

A primary goal of engineers and others in technical work is to design machines and devices. Drawings must be prepared to show the shape and size of each part of the complete product. Drawing is a fundamental means of communication, whether it is done manually or by computer.

In this chapter we are concerned with methods of describing *shape*. The chapter on dimensioning and tolerancing (Chap. 8) will cover methods of describing *size*.

Shape is described by projection, that is, by the process of causing an image to be formed by rays of sight taken in a particular direction from an object to a picture plane. Methods of projection vary according to the direction in which the rays of sight are taken to the plane. When the rays are perpendicular to the plane, the projective method is *orthographic*. If the rays are at an angle to the plane, the projective method is called *oblique*. Rays taken to a particular station point result in *perspective projection*. By the methods of perspective the object is represented as it would appear to the eye.

Projective theory is the basis of background information necessary for shape representation. In graphics, two fundamental methods of *shape representation* are used:

1. *Orthographic views*, consisting of a set of two or more separate views of an object taken from different directions, generally at right angles to each other and arranged relative to each other in a definite way. Each of the views shows the shape of the object for a particular view direction and collectively the views describe the object completely. Orthographic projection *only* is used.

2. *Pictorial views*, in which the object is oriented behind and projected upon a single plane. Either orthographic, oblique, or perspective projection is used.

Since orthographic views provide a means of describing the *exact shape* of any material object, they are used for the great bulk of engineering work.

2. THEORY OF ORTHOGRAPHIC PROJECTION

Let us suppose that a transparent plane has been set up between an object and the station point of an observer's eye (Fig. 1). The intersection of this plane with the rays formed by lines of sight from the eye to all points of the object would give a picture that is practically the same as the image formed in the eye of the observer. This is perspective projection.

If the observer would then walk backward from the station point until reaching a theoretically *infinite* dis-

tance, the rays formed by lines of sight from the eye to the object would grow longer and finally become infinite in length, parallel to each other, and perpendicular to the picture plane. The image so formed on the picture plane is what is known as "orthographic projection." See Fig. 2.

Basically, orthographic projection could be defined as any single projection made by dropping perpendiculars to a plane. However, it has been accepted through long usage to mean the combination of two or more such views—hence the following definition: Orthographic projection is the method of representing the shape of an object by projecting perpendiculars from two or more sides of the object to projection planes.

3. ORTHOGRAPHIC VIEWS

The rays from the picture plane to infinity may be discarded and the picture, or "view," thought of as being found by extending perpendiculars to the plane from all

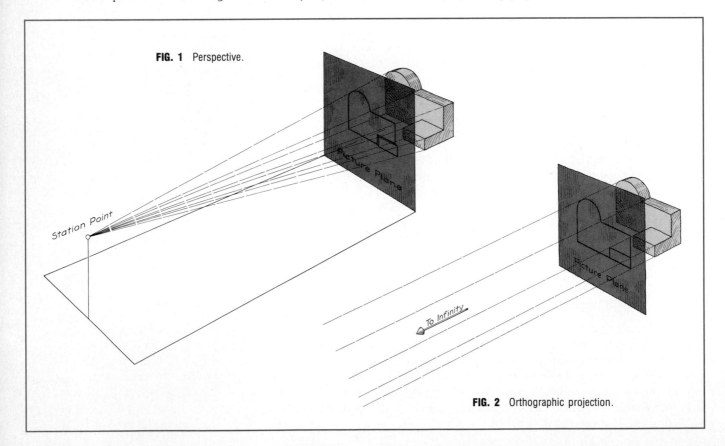

FIG. 1 Perspective.

FIG. 2 Orthographic projection.

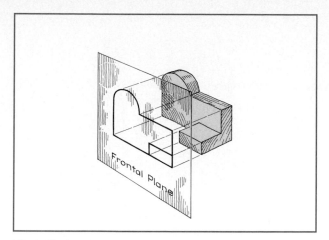

FIG. 3 The frontal plane of projection. This produces the front view of the object.

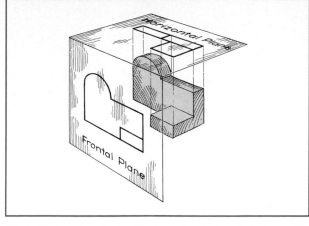

FIG. 4 The frontal and horizontal planes of projection. Projection on the horizontal plane produces the top view of the object. Frontal and horizontal planes are perpendicular to each other.

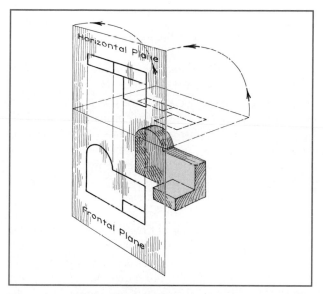

FIG. 5 The horizontal plane rotated into the same plane as the frontal plane. This makes it possible to draw two views of the object on a plane, the drawing paper.

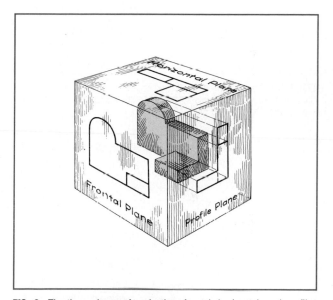

FIG. 6 The three planes of projection: frontal, horizontal, and profile. Each is perpendicular to the other two.

points of the object, as in Fig. 3. This picture, or projection on a frontal plane, shows the shape of the object when viewed from the front, but it does not tell the shape or distance from front to rear. Accordingly, more than one projection is required to describe the object.

In addition to the frontal plane, imagine another transparent plane placed horizontally above the object, as in Fig. 4. The projection on this plane, found by

extending perpendiculars to it from the object, will give the appearance of the object as if viewed from directly above and will show the distance from front to rear. If this horizontal plane is now rotated into coincidence with the frontal plane, as in Fig. 5, the two views of the object will be in the same plane, as if on a sheet of paper. Now imagine a third plane, perpendicular to the first two (Fig. 6). This plane is called a "profile plane," and a

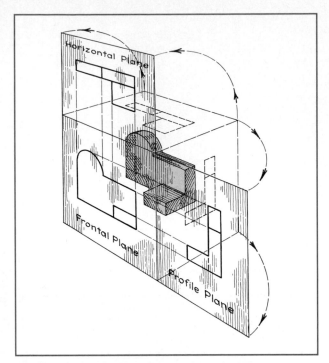

FIG. 7 The horizontal and profile planes rotated into the same plane as the frontal plane. This makes it possible to draw three views of the object on a plane, the drawing paper.

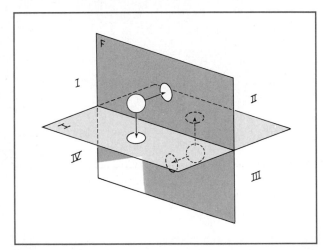

FIG. 8 Projection quadrants.

third view can be projected on it. This view shows the shape of the object when viewed from the side and the distance from bottom to top and front to rear. The horizontal and profile planes are shown rotated into the same

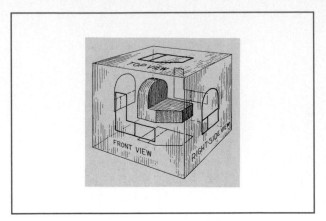

FIG. 9 "The transparent box." This encloses the object with another frontal plane behind, another horizontal plane below, and another profile plane to the left of the object.

plane as the frontal plane (again thought of as the plane of the drawing paper) in Fig. 7. Thus related in the same plane, they give correctly the three-dimensional shape of the object.

In orthographic projection the picture planes are called "planes of projection"; and the perpendiculars, "projecting lines" or "projectors."

In looking at these theoretical projections, or views, do not think of the views as flat surfaces on the transparent planes, but try to imagine that you are looking *through* the transparent planes at the object itself.

It should be noted that the projection box shown in Fig. 6 is not the only one available for use. Figure 8 gives a three-dimensional illustration that indicates four quadrants (I, II, III, IV). Each of these quadrants is available theoretically for use in projecting an object to picture planes. In Fig. 8, a ball (sphere) has been projected in the third quadrant to both the horizontal (H) plane and the frontal (F) plane. This practice conforms identically to that shown already in Fig. 6.

However, it is perfectly legitimate to use quadrant I for projection purposes instead of quadrant III. If this were done, the front view of the object would occur *above* the horizontal view, not below it as is the case for quadrant III. The use of quadrant I is called "first-angle projection," while the use of quadrant III is called "third-angle projection." North America uses third-angle projection. Europe and much of Asia use first-angle projection.

We will use third-angle projection exclusively. However, in today's extensive international business world, you will no doubt see plans drawn in the first-angle system. Being aware of the system should help you read such plans intelligently.

4. THE SIX PRINCIPAL VIEWS

Considering the matter further, we find that the object can be entirely surrounded by a set of six planes, each at right angles to the four adjacent to it, as in Fig. 9. On these planes, views can be obtained of the object as it is seen from the top, front, right side, left side, bottom, and rear.

Think now of the six sides, or planes, of the box as being opened up, as in Fig. 10, into one plane, the plane of the paper. The front is already in the plane of the paper, and the other sides are, as it were, hinged and rotated into position as shown. The projection on the frontal plane is the *front view*, *vertical projection*, or *front elevation*; that on the horizontal plane, the *top view*, *horizontal projection*, or *plan*; that on the side, or "profile," plane, the *side view*, *profile projection*, *side elevation*, or sometimes *end view* or *end elevation*.

By reversing the direction of sight, a *bottom view* is obtained instead of a *top view*, or a *rear view* instead of a *front view*. In comparatively rare cases a bottom view or rear view or both may be required to show some detail of shape or construction. Figure 11 shows the relative position of the six views. In actual work there is rarely an

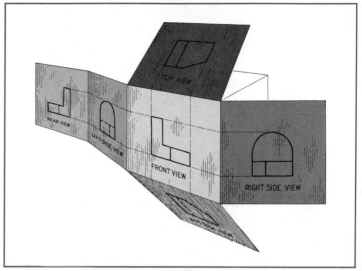

FIG. 10 The transparent box as it opens and all planes rotate to the plane of the frontal plane. Note that horizontal and profile planes are hinged to the frontal plane and that the "rear-view" plane is hinged to the left profile plane.

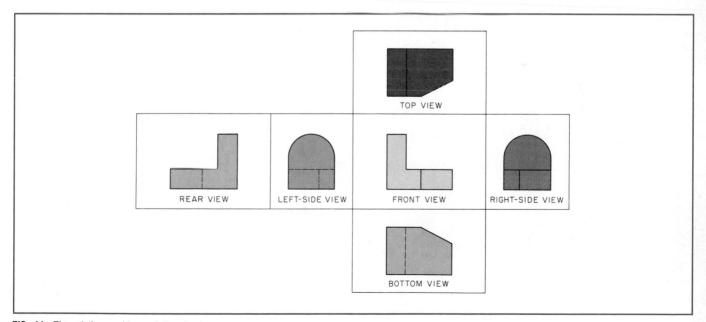

FIG. 11 The relative positions of the six views.

occasion when all six principal views are needed on one drawing, but no matter how many are required, their positions relative to one another are given in Fig. 11 (except as noted in Sec. 6). All these views are principal views. Each of the six views shows two of the three dimensions of height, width, and depth.

5. COMBINATION OF VIEWS

The most usual combination selected from the six possible views consists of the *top*, *front*, and *right-side* views, as shown in Fig. 12, which, in this case, best describes the shape of the given block. Sometimes the left-side view helps to describe an object more clearly than does the right-side view. Figure 13 shows the arrangement of *top*, *front*, and *left-side* views for the same block. In this case the right-side view would be preferred, because it shows no hidden edges.

6. "ALTERNATE-POSITION" VIEWS

The top of the enclosing transparent box may be thought of as in a fixed position with the front, rear, and sides hinged, as in Fig. 14, thus bringing the sides in line with

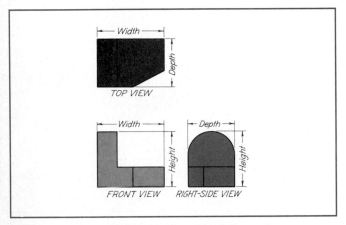

FIG. 12 Top, front, and right-side views. This is the most common combination.

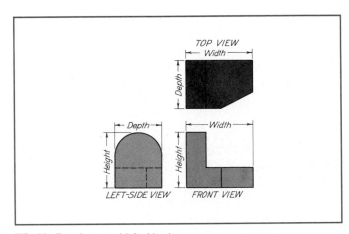

FIG. 13 Top, front, and left-side views.

FIG. 14 The transparent box opening for alternate-position views. Note that frontal and profile planes are hinged to the horizontal plane.

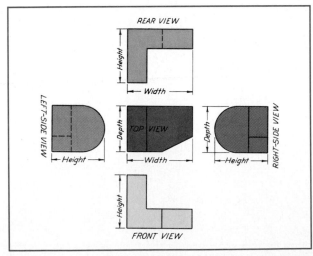

FIG. 15 Alternative position views. Study this figure carefully in conjunction with Fig. 14.

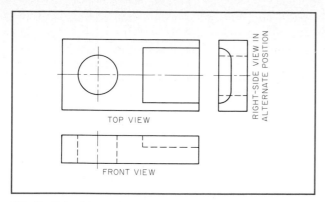

FIG. 16 Right-side view in alternative position. Note the saving in paper area (compared with regular position) for this broad, flat object.

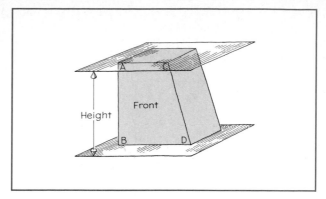

FIG. 17 Definition of height. This is the difference in elevation between two points.

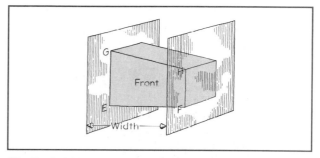

FIG. 18 Definition of width. This is the difference from left to right between two points.

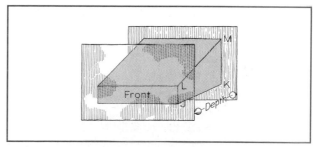

FIG. 19 Definition of depth. This is the difference from front to rear between two points.

the top view and the rear view above the top view, Fig. 15. This alternate-position arrangement is of occasional use to save space on the paper in drawing a broad, flat object (Fig. 16).

7. THE THREE SPACE DIMENSIONS

All objects have distinct limits that can be described in terms of height, width, and depth. It is desirable to define these three dimensions and to fix their directions.

Height is the difference in elevation between any two points, measured as the perpendicular distance between a pair of horizontal planes that contain the points, as shown in Fig. 17. Edges of the object may or may not correspond with the height dimensions. Edge *AB* corresponds with the height dimension, while edge *CD* does not, but the space heights of *A* and *C* are the same, as are *B* and *D*. Height is always measured in a vertical direction and has no relationship whatever to the shape of the object.

Width is the positional distance left to right between any two points measured as the perpendicular distance between a pair of profile planes containing the points. In Fig. 18 the relative width between points *E* and *G* on the left and *H* and *F* on the right of an object is shown by the dimension marked "width." The object edge *EF* is parallel to the width direction and corresponds with the width dimension, but edge *GH* slopes downward from *G* to *H*, so this actual edge of the object is longer than the width separating points *G* and *H*.

Depth is the positional distance front to rear between any two points measured as the perpendicular distance

between two frontal planes containing the points. Figure 19 shows two frontal planes, one at the front of the object containing points *J* and *L*, the other at the rear containing points *K* and *M*. The relative depth separating the front and rear of the object is the perpendicular distance between the planes as shown.

The three space dimensions—height, width, and

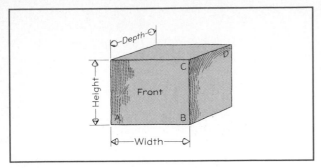

FIG. 20   Location of points in space. Height, width, and depth must be designated.

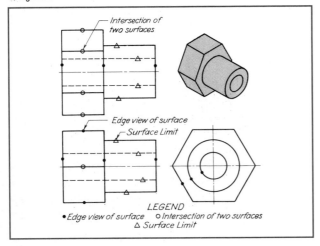

FIG. 21   What a line indicates.

depth—and the planes of projection are unchangeably oriented and connected with each other and with the view directions (Fig. 20). Each of the planes of projection is perpendicular, respectively, to its own view direction. Thus the frontal plane is perpendicular to the front-view direction, the horizontal plane is perpendicular to the top-view direction, and the profile plane is perpendicular to the side-view direction.

Therefore we can conclude that the front view will always show width and height, the top view width and depth, and the side view depth and height. We will see that any point in space can be located by giving its height, width, and depth relative to known picture planes.

8. WHAT LINES MAY INDICATE

Although uniform in appearance, the lines on a drawing may indicate three different types of directional change on the object. An *edge* view is a line showing the edge of a receding surface that is perpendicular to the plane of projection. An *intersection* is a line formed by the meeting of two surfaces when either one surface is parallel and one is at an angle or both are at an angle to the plane of projection. A *surface limit* is a line that indicates the reversal of direction of a curved surface (or the series of points of reversal on a warped surface). Figure 21 illustrates the different line meanings.

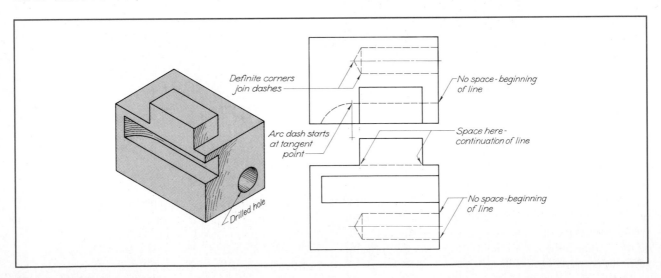

FIG. 22   Dashed-line technique.

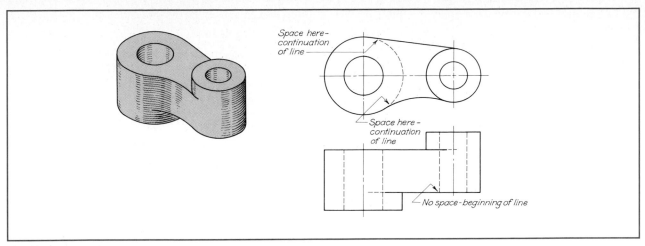

Space here- continuation of line

Space here- continuation of line

No space- beginning of line

FIG. 23 Dashed lines and arcs.

9. HIDDEN FEATURES

To describe an object completely, a drawing should contain lines representing all the edges, intersections, and surface limits of the object. In any view there will be some parts of the object that cannot be seen from the observer's position because they will be covered by portions of the object closer to the observer's eye. The edges, intersections, and surface limits of these hidden parts are indicated by a discontinuous line called a *dashed line*. In Fig. 22 the drilled hole that is visible in the right-side view is hidden in the top and front views, and therefore it is indicated in these views by a dashed line showing the hole and the shape as left by the drill point. The slot is visible in the front and side views but is hidden in the top view.

Dashed lines are drawn lighter than full lines, of short dashes uniform in length with the space between them very short, about one-fourth the length of the dash. It is important that they start and stop correctly. A dashed line always starts with a dash except when the dash would form a continuation of a full line; in that case a space is left, as shown in Fig. 23. Dashes always meet at corners. An arc must start with a dash at the tangent point except when the dash would form a continuation of a straight or curved full line. The number of dashes used in a tangent arc should be carefully judged to maintain a uniform appearance. Study carefully all dashed lines in Figs. 22 and 23.

10. CENTER LINES

In general, the first lines drawn in the layout of an engineering drawing are the center lines, which are the axes of symmetry for all symmetrical views or portions of views: (1) Every part with an axis, such as a cylinder or a cone, will have the axis drawn as a center line before the part is drawn. (2) Every circle will have its center at the intersection of two mutually perpendicular center lines.

The standard symbol for center lines on finished drawings is a fine line made up of alternate long and short dashes, as shown in the alphabet of lines (Chap. 2, Fig. 39). Center lines are always extended slightly beyond the outline of the view or portion of the view to which they apply. They form the skeleton construction of the drawing; the important measurements are made and dimensions are given to and from these lines.

11. PRECEDENCE OF LINES

In any view there is likely to be a coincidence of lines. Hidden portions of the object may project to coincide with visible portions. Center lines may occur where there is a visible or hidden outline of some part of the object.

Since the physical features of the object must be represented, full and dashed lines take precedence over all other lines. Since the visible outline is more prominent by space position, full lines take precedence over dashed lines. A full line could cover a dashed line, but a dashed line could not cover a full line. It is evident also that a dashed line could not occur as one of the boundary lines of a view.

When a center line and cutting-plane (explained in Chap. 6) line coincide, the one that is more important for the readability of the drawing takes precedence over the other.

Break lines (explained in Chap. 6) should be placed so that they do not spoil the readability of the overall view.

Dimension and extension lines must always be placed so as not to coincide with other lines of the drawing.

The following list gives the order of precedence of lines:

1. Full line
2. Dashed line
3. Center line or cutting-plane line
4. Break lines
5. Dimension and extension lines
6. Crosshatch lines

Note the coincident lines in Fig. 24.

12. WHAT AREAS MAY MEAN

The term "area" as used here means the contour limits of a surface or combination of tangent surfaces as seen in the different orthographic views. To illustrate, an area of a view as shown in Fig. 25 may represent (1) a surface in true shape as at A, (2) a foreshortened surface as at B, (3) a curved surface as at C, or (4) a combination of tangent surfaces as at D.

When a surface is in an oblique position, as is surface E of Fig. 26, it will appear as an area of all principal views of the surface. A study of the surfaces in Figs. 25, 26, and others will establish that a plane surface, whether it is positioned in a horizontal, frontal, profile, or inclined or skew position, will always appear in a principal orthographic view as a line or an area. Principal views that show an oblique surface as an area in more than one view will always show it in like shape.

No two adjacent areas can lie in the same plane. It is simple logic that if two adjacent areas *did* lie in the same plane, there would be no boundary between the areas and the two adjacent areas would therefore orthographically not exist. As an illustration, note that in Fig. 27 areas A, B, C, and D are shown in the front and side views to lie in different planes.

Hidden areas may sometimes be confusing to read because the areas may overlap or even coincide with each other. For example, areas A, B, and C in Fig. 28 are not separate areas because they are all formed by the slot on the rear of the object. The apparent separation

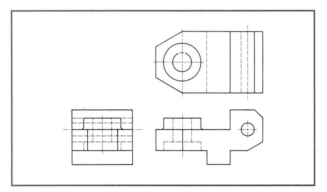

FIG. 24 Coincident-line study. Coincident lines are caused by the existence of features of identical size or position, one behind the other.

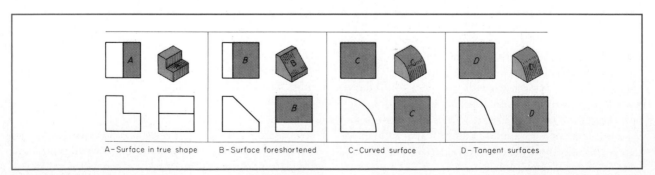

A–Surface in true shape   B–Surface foreshortened   C–Curved surface   D–Tangent surfaces

FIG. 25 The meaning of areas.

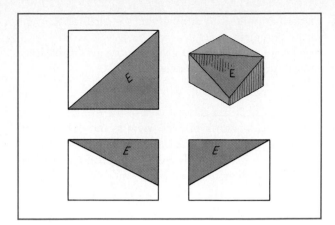

FIG. 26 Oblique surface.

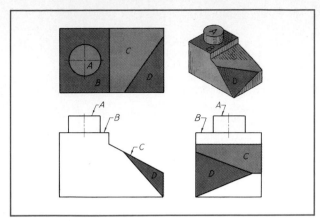

FIG. 27 Adjacent areas. *A, B, C,* and *D* all lie in different planes.

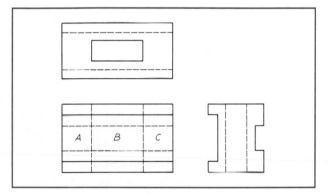

FIG. 28 Reading hidden areas.

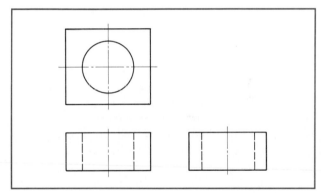

FIG. 29 Views to be read. Compare this drawing with Fig. 30.

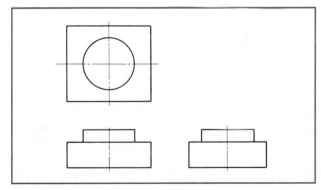

FIG. 30 Views to be read. Compare this drawing with Fig. 29.

into separate areas is caused by the dashed lines from the rectangular hole, which is not connected with the slot in any way.

13. READING LINES AND AREAS

A drawing is read by visualizing details one at a time and then mentally orienting and combining these details to interpret the whole object. Reading is primarily a reversal of the process of making drawings.

Consider the object in Fig. 29. A visible circle is seen in the top view. Memory of previous projection experience indicates that this must be a hole or the end of a cylinder. The eyes rapidly shift back and forth from the top view to the front view, aligning features of the same size ("in projection"), with the mind assuming the several possibilities and finally accepting the fact that, because of the dashed lines and their extent in the front view, the circle represents a hole that extends through the prism. Following a similar pattern of analysis, the reader will find that Fig. 30 represents a rectangular prism surmounted by a cylinder.

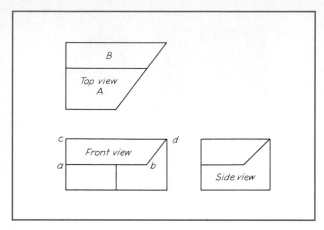

FIG. 31 Reading lines and areas.

The foregoing principles regarding the meaning of lines and areas must be used to analyze any given set of views by correlating a surface appearing in one view as a line or an area with its representation in the other views, in which it may appear as a line or an area. Study, for example, Fig. 31, first orienting yourself with the given views. From their arrangement, the views are evidently top, front, and right-side. An overall inspection of the views does not reveal a familiar geometric shape, such as a hole or boss, so an analysis of the surfaces is necessary. Beginning with the trapezoidal area A in the top view and then moving to the front view, note that a similar-shaped area of the same width is not shown; therefore, the front view of area A must appear as an edge, the line *ab*.

Next, consider area B in the top view. It is shown as a trapezoidal area (four sides) the full width of the view. Again, consider the front view for a mating area or line, where the area *abcd* is of similar shape and has the same number of sides with the corners in projection. Area *abcd* therefore satisfies the requirements of orthographic projection and is the front view of area B. The side view should be checked along with the other views to see if it agrees. Proceed in this way with additional areas, corre-

lating them one with another and visualizing the shape of the complete object.

The following outline gives a summary of the basic procedure for reading drawings.

First, orient yourself with the views given.

Second, obtain a general idea of the overall shape of the object. Think of each view as the object itself, visualizing yourself in front, above, and at the side, as is done in making the views. Study the dominant features and their relation to one another.

Third, start reading the simpler individual features, beginning with the most dominant and progressing to the subordinate. Look for familiar shapes. Read all views of these familiar features to note the extent of holes, thickness of ribs and lugs, etc.

Fourth, read the unfamiliar or complicated features. Remember that every point, line, surface, and solid appears in every view and that you must find the projection of every detail in the given views to learn the shape.

Fifth, as the reading proceeds, note the relationship between the various portions or elements of the object—such items as the number and spacing of holes, placement of ribs, tangency of surfaces, etc.

Sixth, reread any detail or relationship not clear at the first reading.

14. EXERCISES IN PROJECTION

The principal task in learning orthographic projection is to become thoroughly familiar with the theory and then to practice this theory by translating from a picture of the object to the orthographic views. Figures 32 and 33 contain a variety of objects shown by a pictorial sketch and translated into orthographic views. Study the objects and note (1) how the object is oriented in space, (2) why the orthographic views given were chosen, (3) the projection of visible features, (4) the projection of hidden features, and (5) center lines.

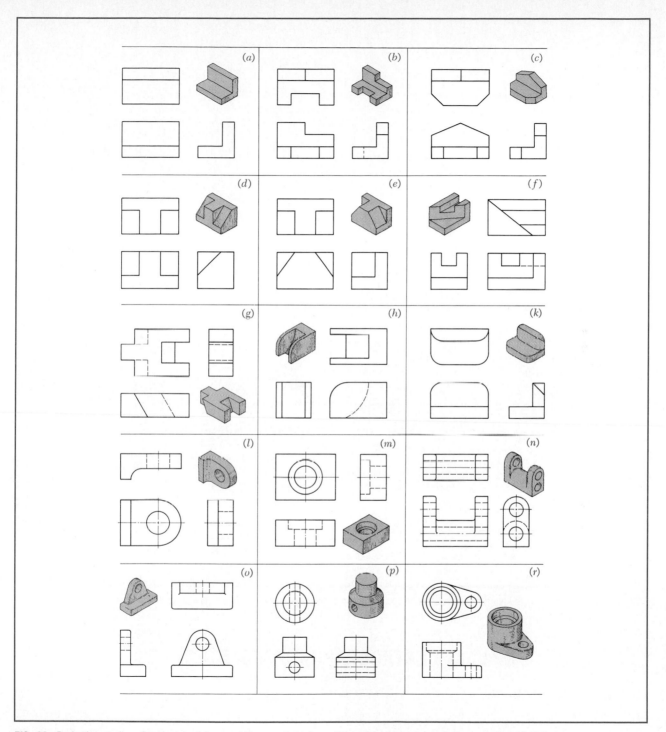

FIG. 32 Projection studies. Study each picture and the accompanying orthographic views and note the projection of all features.

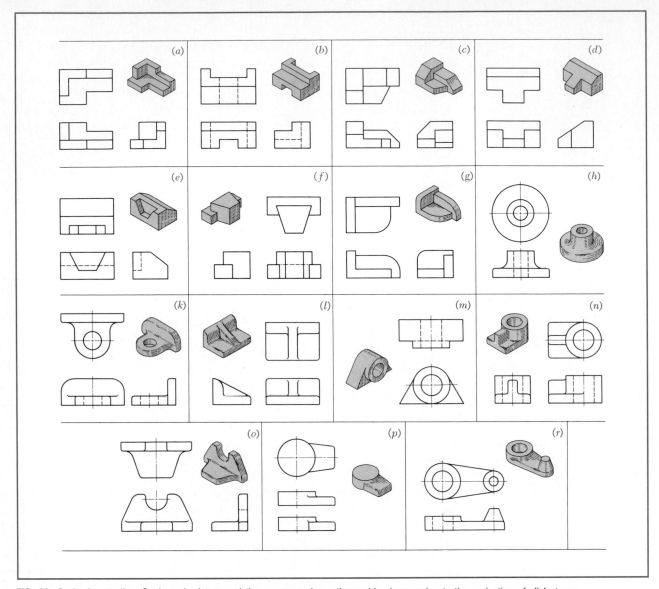

FIG. 33 Projection studies. Study each picture and the accompanying orthographic views and note the projection of all features.

USING ORTHOGRAPHIC PROJECTION

15. OBJECT ORIENTATION

An object can, of course, be drawn in any of several possible positions. *The simplest position should be used*, with the object oriented so that the principal faces are perpendicular to the sight directions for the views and parallel to the planes of projection, as shown in Fig. 34. Any other position of the object, with its faces at some angle to the planes of projection, would complicate the drawing, foreshorten the object faces, and make the drawing difficult to make and to read.

16. SELECTION OF VIEWS

In practical work it is important to choose the combination of views that will describe the shape of the object in the best and most economical way. Often only two views are necessary. For example, a cylindrical shape, if on a vertical axis, would require only a front and top view; if on a horizontal axis, only a front and side view. Conic and pyramidal shapes can also be described in two views. Figure 35 illustrates two-view drawings. Some shapes will need more than the three regular views for adequate description.

Objects can be thought of as being made up of combinations of simple geometric solids; the views necessary to describe any object can be determined by the directions from which it would have to be viewed for one to see the characteristic contour shapes of these parts. In the majority of cases the three regular views—top, front, and side—are sufficient to describe shape.

Sometimes two views are proposed as sufficient for an object on the assumption that the contour in the third direction is of the shape that would naturally be expected. In Fig. 36, for example, the figure at A would be assumed to have a uniform cross section and be a square prism. But the two views *might* be the top and front views of a wedge, as shown in three views at B. Two views of an object, as drawn at C, do not describe the piece at all.

With the object preferably in its functioning position

and *with its principal surfaces parallel to the planes of projection*, visualize the object, mentally picturing the orthographic views one at a time to decide on the best combination. In Fig. 37, the arrows show the direction

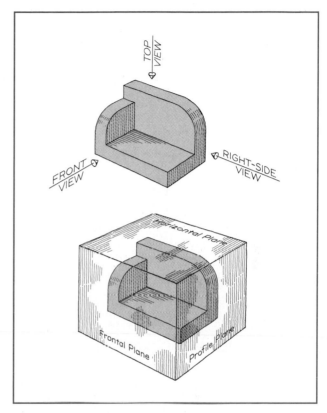

FIG. 34 Object orientation. Use the simplest position. It will give the clearest possible representation and be the easiest to draw.

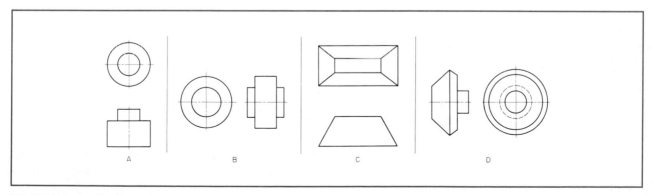

FIG. 35 Two-view drawings.

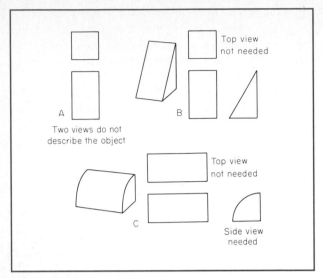

FIG. 36 A study of views.

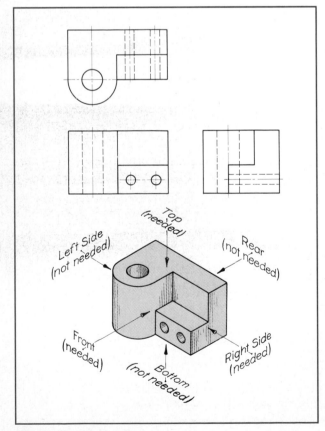

FIG. 37 Selection of views.

of observation for the six principal views of an object and indicate the mental process of the person making the selection. The person notes that the front view would show the two horizontal holes as well as the width and height of the piece, that a top view is needed to show the contour of the vertical cylinder, and that the cutout corner calls for a side view to show its shape. He or she notes further that the right-side view would show this cut in full lines, while the left-side view would give it in dashed lines, and observes also that neither a bottom view nor a rear view would be of any value in describing this object. The person has thus correctly chosen the front, top, and right-side views as the best combination for describing this piece. As a rule, the side view containing the fewer dashed lines is preferred.

17. SPACING THE SELECTED VIEWS

View spacing is necessary so that the drawing will be balanced within the space provided. A little preliminary measuring is necessary to locate the views. The following example describes the procedure: Suppose the piece illustrated in Fig. 38 is to be drawn full size on an 11- by 17-in sheet. With an end-title strip, the working space inside the border will be 10½ by 15 in. The front view will require 7¹¹⁄₁₆ in, and the side view 2¼ in. This leaves 5¹⁄₁₆ in to be distributed between the views and at the ends.

This preliminary planning need not be to exact dimensions; that is, small fractional values, such as ¹⁵⁄₆₄ in or ³¹⁄₃₂ in, can be adjusted to ¼ and 1 in, respectively, to speed up the planning. In this case the 7¹¹⁄₁₆-in dimension can be adjusted to 7¾ in.

Locate the views graphically and quickly by measuring with your scale along the bottom border line. Starting at the lower right corner, lay off first 2¼ in and then 7¾ in. The distance between views can now be decided upon. It is chosen by eye to separate the views without crowding, yet placing them sufficiently close together so that the drawing will read easily (in this case 1½ in). Measure the distance; half the remaining distance to the left corner is the starting point of the front view. For the vertical location: The front view is 4 in high, and the top view 2¼ in deep. Starting at the upper-left corner, lay off first 2¼ in and then 4 in; judge the distance between views (in this case 1 in), and lay it off; then a point

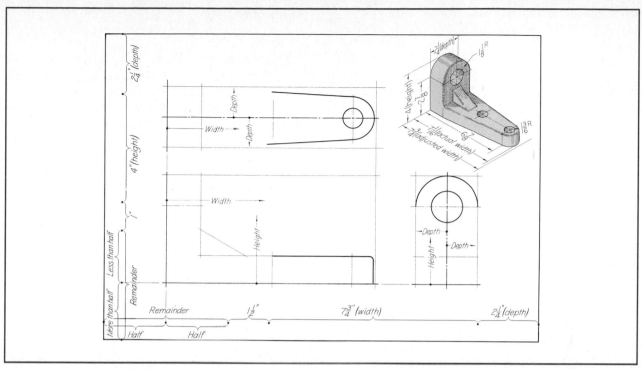

FIG. 38 Spacing the views on the paper. This is done graphically. Study the text carefully while referring to this figure and go through the steps by laying out the given object on a standard 11- × 17-in sheet.

marked at less than half the remaining space will locate the front view, allowing more space at the bottom than at the top for appearance.

Block out lightly the spaces for the views, and study the overall arrangement, because changes can easily be made at this stage. If it is satisfactory, select reference lines in each view from which the space measurements of height, width, and depth that appear in the view can be measured. The reference line may be an edge or a center line through some dominant feature, as indicated on Fig. 38 by the center lines in the top and side views and the medium-weight lines in all the views. The directions for height, width, and depth measurements for the views are also shown.

18. PROJECTING THE VIEWS

After laying out the views locate and draw the various features of the object. In doing this, carry the views *along together*, that is, *do not* attempt to complete one view

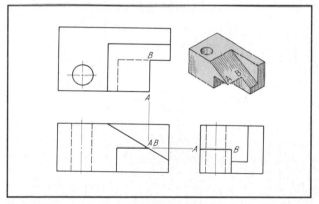

FIG. 39 Projection of lines. Carry all views along together. The greatest mistake possible is to try to complete one view before starting another.

before proceeding to another. Draw first the most characteristic view of a feature and then project it and draw it in the other views before going on to a second feature. As an example, the vertical hole of Fig. 39 should be drawn first in the top view, and then the dashed lines representing the limiting elements or portions should be projected and drawn in the front and side views.

In some cases, one view cannot be completed before a feature has been located and drawn in another view. Study the pictorial drawing in Fig. 39, and note from the orthographic views that the horizontal slot must be drawn on the front view before the edge AB on the slanting surface can be found in the top view.

Projections (horizontal) between the front and side views are made by employing the T square to draw the required horizontal line (or to locate a required point), as in Fig. 40.

Projections (vertical) between the front and top views are made by using the T square and a triangle as in Fig. 41.

Projections between the top and side views cannot be projected directly but must be measured and transferred or found by special construction. In carrying the top and side views along together, it is usual to transfer the depth measurement from one to the other with dividers, as in Fig. 42a, or with a scale, as at (b). Another method, used for an irregular figure, is to "miter" the points around, using a 45° line drawn through the point of intersection of the top and side views of the front face, extended as shown in Fig. 43. The method of Fig. 43, however,

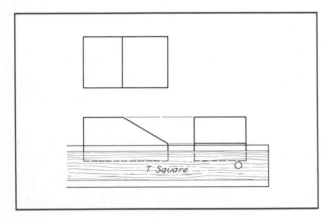

FIG. 40 Making a horizontal projection. This is the simplest operation in drawing. The T square provides *all* horizontal lines.

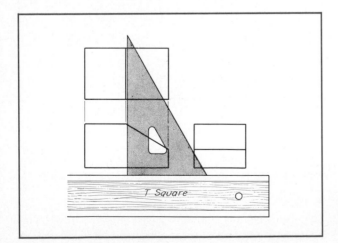

FIG. 41 Making a vertical projection. The 90° angle of a triangle with one leg on the horizontal T square produces the vertical.

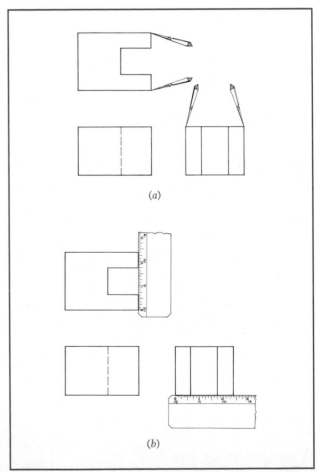

(a)

(b)

FIG. 42 Transferring depth measurements. Depth cannot be projected. Transfer the necessary distances with dividers, as at (a), or with scale, as at (b).

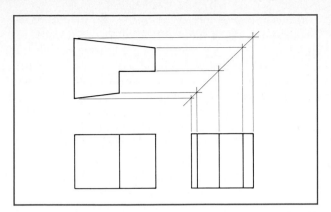

FIG. 43 Projecting depth measurements. A ''miter line'' at 45°, with horizontal and vertical projectors, transfers the depth from top to side view (or vice versa).

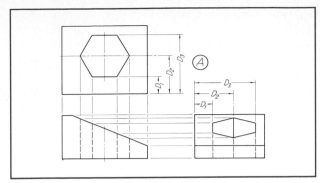

FIG. 44 Projections of surfaces bounded by linear edges.

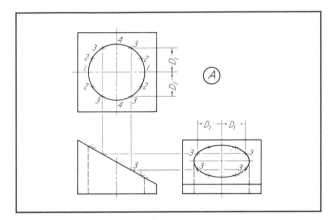

FIG. 45 Projection of an elliptical boundary.

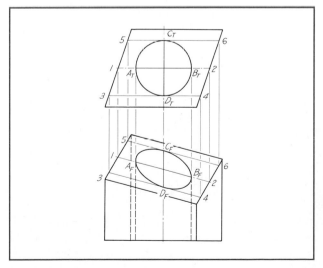

FIG. 46 Projection of an elliptical boundary by employing conjugate diameters.

requires more time and care than the methods of Fig. 42 and is therefore not recommended.

19. PROJECTION OF INCLINED SURFACES

In drawing projections of inclined surfaces, in some cases the corners of the bounding edges may be used, and in other cases the bounding edges themselves may be projected. In illustration of these methods Fig. 44 shows a vertical hexagonal hole that is laid out from specifications in the top view. Then the front view is drawn by projecting from the six corners of the hexagon and drawing the four dashed lines to complete the front view. To get the side view, a horizontal projection is made from each corner on the front view to the side view, thus locating the height of the points needed on the side view. Then measurements D_1, D_2, and D_3, taken from the top view and transferred to the side view, locate all six corners in the side view. The view is completed by connecting these corners and drawing the three vertical dashed lines.

20. PROJECTION OF AN ELLIPTICAL BOUNDARY

The intersection of a cylindrical hole (or cylinder) with a slanting surface, as shown in Fig. 45, will be an ellipse, and some projections of this elliptical edge will appear as another ellipse. The projection can be made as shown in Fig. 45 by assuming a number of points on the circular view and projecting them to the edge view (front) and then to an adjacent view (side). Thus points 1 to 4 are located in the top view and projected to the front view, and the projectors are then drawn to the side view. Measurements of depth taken from the top view (as D_1) will locate the points in the side view. Draw a smooth curve through the points, using a french curve.

If the surface intersected by the cylinder is oblique, as shown in Fig. 46, a pair of perpendicular diameters located in the circular view will give a pair of conjugate diameters in an adjacent view. Therefore, $A_T B_T$ and $C_T D_T$ projected to the front view will give conjugate diameters that can be used to draw the required ellipse.

In projecting the axes, extend them to the straight-line boundary of the oblique surface. Thus the line 1-2 located in the front view and intersected by projection of $A_T B_T$ from the top view locates $A_F B_F$. Similarly, lines 3-4 and 5-6 at the ends of the axis CD locate $C_F D_F$.

21. PROJECTION OF A CURVED BOUNDARY

Any nongeometric curve must be projected by locating points on the curve. If the surface is in an inclined position, as in Fig. 47A, points may be assumed on the curve laid out from data (assumed in this case to be the top view) and projected first to the edge view (front) and then to an adjacent view (side). Measurements, such as 1, 2, etc., from the top view transferred to the side view complete the projection. A smooth curve is then drawn through the points.

If the surface is oblique, as in Fig. 47B, elements of the oblique surface, such as 1'-1, 2'-2, etc., located in an adjacent view (by drawing the elements parallel to some known line of the oblique surface, such as AB), make it possible to project points on the curve 1, 2, etc., to the adjacent view, as shown.

22. PROJECTION BY IDENTIFIED CORNERS

In projecting orthographic views or in comparing the views with a picture, it is helpful in some cases to letter (or number) the corners of the object and, with these identifying marks, to letter the corresponding points on each of the views, as in Fig. 48. Hidden points directly behind visible points are lettered to the right of the letter of the visible point, and in this figure, they have been further differentiated by the use of "phantom," or dotted, letters.

23. ORDER OF DRAWING

The order of working is important, as speed and accuracy depend largely upon the methods used in laying down lines. Avoid duplications of the same measurement and kccp to a minimum changing from onc instrument to another. Naturally, *all* measurements cannot be made with the scale at one time or *all* circles and arcs drawn without laying down the compass, but as much work as possible should be done with one instrument before shifting to another. An orderly placement of working

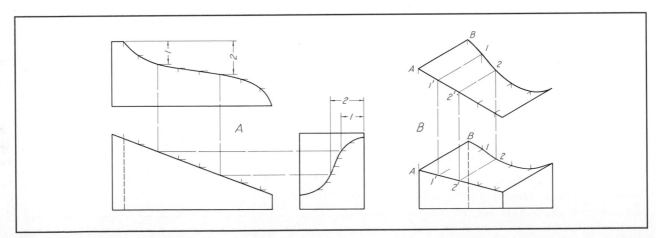

FIG. 47 Projection of a curved boundary.

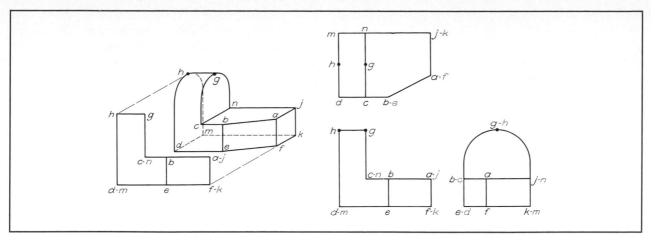

FIG. 48 Identified corners. Each corner is lettered (or numbered) as an aid in making projections.

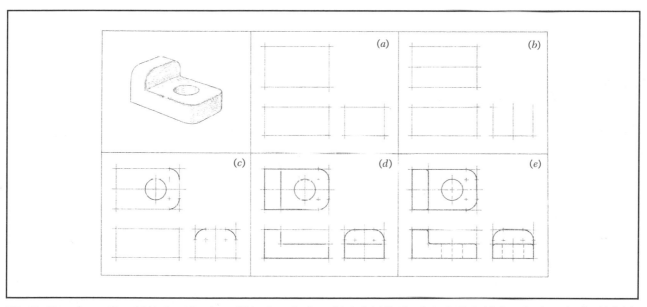

FIG. 49 Stages in penciling. (a) block out the views; (b) locate center lines; (c) start details, drawing arcs first; (d) draw dominant details; (e) finish. See text for explanation.

tools on the drawing table will save time when changing instruments. The usual order of working is shown in Fig. 49.

1. Decide what combination of views will best describe the object. A freehand sketch will aid in choosing the views and in planning the general arrangement of the sheet.

2. Decide what scale to use, and by calculation or measurement find a suitable standard sheet size; or pick one of the standard drawing-sheet sizes and find a suitable scale.

3. Space the views on the sheet, as described in Sec. 17.

4. Lay off the principal dimensions, and then block

in the views with light, sharp, accurate outline and center lines. Draw center lines for the axes of all symmetrical views or parts of views. Every cylindrical part will have a center line—the projection of the axis of the piece. Every circle will have two center lines intersecting at its center.

5. Draw in the details of the part, beginning with the dominant characteristic shape and progressing to the minor details, such as fillets and rounds. Carry the different views along together, projecting a characteristic shape, as shown in one view to the other views, instead of finishing one view before starting another. Use a minimum of construction and draw the lines to finished weight, if possible, as the views are carried along. *Do not make the drawing lightly and then "heavy" the lines later.*

6. Lay out and letter the title.

7. Check the drawing carefully.

ORTHOGRAPHIC SKETCHING

24. REASON FOR ORTHOGRAPHIC SKETCHING

Facility in making freehand orthographic drawings is an essential part of the equipment of every engineer, and since ability in sketching presupposes some mastery of other skills, practice should be started early.

Sketching is an excellent method for learning the fundamentals of orthographic projection and can be used by beginners even before they have had much practice with instruments. In training, as in professional work, time can be saved by working freehand instead of with instruments, because with this method more problems can be solved in an allotted amount of time.

Although some experienced teachers advocate making freehand sketches before practice in the use of instru-

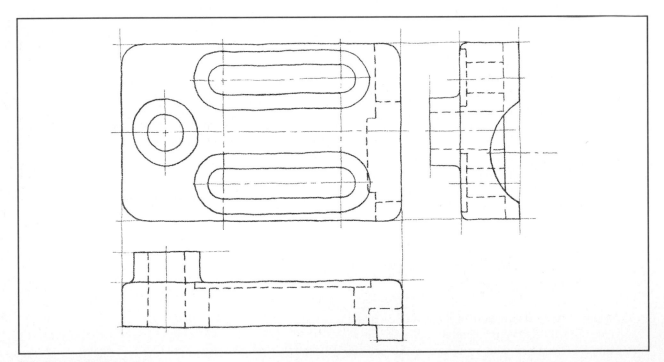

FIG. 50 A freehand drawing.

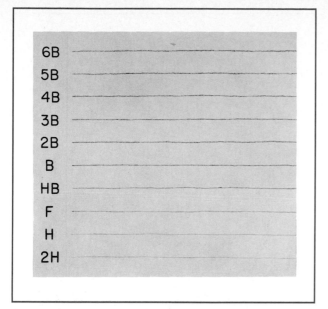

FIG. 51 Different pencil grades, using medium pressure on paper of medium texture.

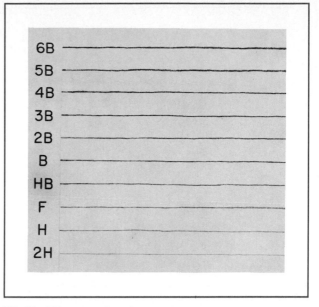

FIG. 52 Different pencil grades, using firm pressure on paper of medium texture.

ments, some knowledge of the use of instruments and especially of applied geometry is a great help because the essentials of line tangents, connections, and intersections as well as the basic geometry of the part should be well defined on a freehand drawing. Drawing freehand is, of course, an excellent exercise in accuracy of observation. Figure 50 is an example of a good freehand drawing.

25. PENCIL PRESSURE AND PAPER TEXTURE

The pencil grade, pressure, and paper texture all have an effect on the final result. Figure 51 shows the various pencil grades with medium pressure on paper of medium texture. To mark the difference in line quality obtainable by increasing the pressure, Fig. 52 shows the same paper but with firm pressure on the pencil. Note that the lines in Fig. 52 are much blacker than those in Fig. 51. The line quality of Fig. 52 is about right for most engineering sketches. The rather firm, opaque lines are preferred to the type in Fig. 51, especially if reproductions, by either photographic or transparency process, are to be made from the sketch.

To illustrate the difference in line quality produced by paper texture, Fig. 53 shows the same firm pressure used in Fig. 52, but this time on smooth paper. In Fig. 54 the same pressure has been used on rough paper.

26. PLAIN VS. COORDINATE PAPER

Sketches are made for many purposes and under a variety of circumstances, and as a result on a number of different paper types and surfaces. A field engineer in reporting information to the central office may include a sketch made on notebook paper or a standard letterhead. On the other hand, a sketch made in the home office may be as important as any instrument drawing and for this reason may be made on good quality drawing or tracing paper and filed and preserved with other drawings in a set.

The principal difficulty in using plain paper is that proportions and projections must be estimated by eye. A good sketch on plain paper requires better-than-average ability and experience. Use of some variety of coordinate paper is a great aid in producing good results. There are many kinds of paper and coordinate divisions available,

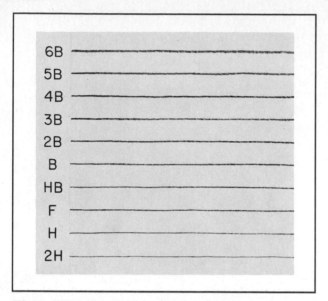

FIG. 53 Different pencil grades, using firm pressure on smooth paper.

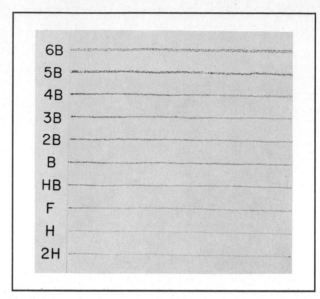

FIG. 54 Different pencil grades, using firm pressure on rough paper.

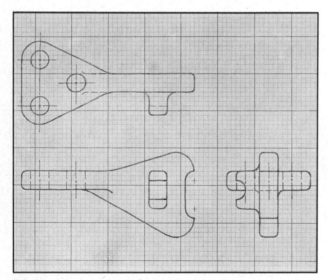

FIG. 55 A freehand drawing on coordinate tracing paper.

which the coordinate divisions do not show. Figure 55 is an example of a sketch on tracing paper with the coordinate divisions printed on the back of the paper in faint purplish-blue ink. Since the divisions are on the back, erasures and corrections can be made without erasing the coordinate divisions. Normally the divisions will not reproduce, so prints give the appearance of a sketch made on plain paper.

27. SKETCHING TECHNIQUES

The pencil is held with freedom and not close to the point. Vertical lines are drawn downward with a finger movement in a series of overlapping strokes, the hand somewhat in the position of Fig. 56. Horizontal lines are drawn with the hand shifted to the position of Fig. 57, using a wrist motion for short lines and a forearm motion for longer ones. In drawing any straight line between two points, keep your eyes on the point to which the line is to go rather than on the point of the pencil. Do not try to draw the whole length of the line in a single stroke. It may be helpful to draw a very light line first, as in Fig. 58A, and then to sketch the finished line, correcting the direction of the light line and bringing the line to final width and blackness by using strokes of convenient length, as at *B*. The finished line is shown at *C*. Do not

from smooth to medium texture and coordinate divisions of ⅛ or ¹⁄₁₀ to ½ in, printed on tracing paper or various weights of drawing paper.

The paper type, tracing or regular, is another factor to be considered. Reproduction by any of the transparency methods demands the use of tracing paper. If coordinate paper is used, it may be desirable to obtain prints on

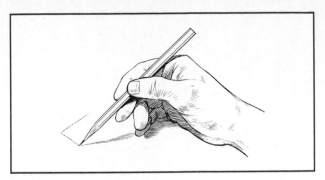

FIG. 56 Sketching a vertical line.

be disturbed by any nervous waviness. Accuracy of direction is more important than smoothness of line.

A. Straight Lines Horizontal lines are drawn from left to right as in Fig. 58; vertical lines from top to bottom as in Fig. 59.

Inclined lines running downward from right to left (Fig. 60) are drawn with approximately the same movement as vertical lines, but the paper may be turned and the line drawn as a vertical.

Inclined lines running downward from left to right (Fig. 61) are the hardest to draw because the hand is in a

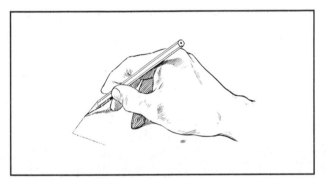

FIG. 57 Sketching a horizontal line.

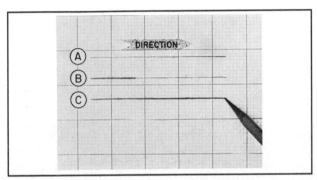

FIG. 58 Technique of sketching lines. (A) Set direction with a *light* construction line; (B) first stroke; (C) complete line with a series of overlapping strokes.

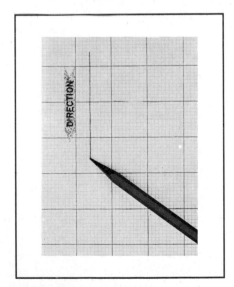

FIG. 59 Sketching a vertical line.

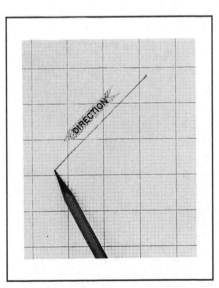

FIG. 60 Sketching an inclined line sloping downward from right to left.

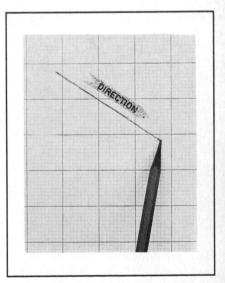

FIG. 61 Sketching an inclined line sloping downward from left to right.

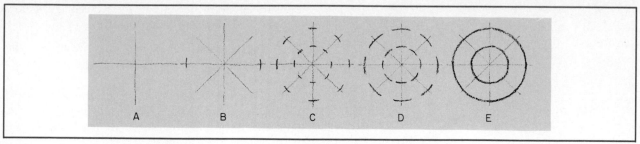

FIG. 62 Method of drawing freehand circles. (A) Draw center lines; (B) draw diagonals; (C) space points on the circle with *light*, short lines (by eye); (D) correct and begin filling in; (E) finish.

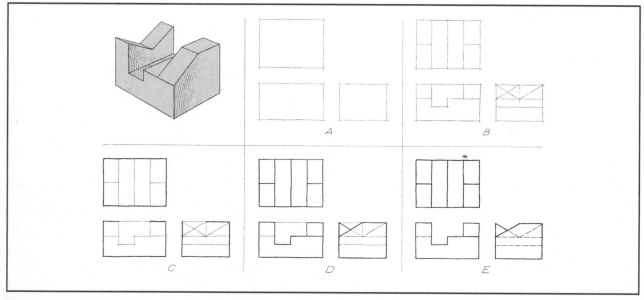

FIG. 63 Stages in making an orthographic freehand drawing.

somewhat awkward position; for this reason, the paper should be turned and the line drawn as a horizontal.

The sketch paper can easily be turned in any direction to facilitate drawing the lines, because there is no necessity to fasten the paper to a drawing-table top.

B. Circles Circles can be drawn by marking the radius on each side of the center lines. A more accurate method is to draw two diagonals in addition to the center lines and mark points equidistant from the center of the eight radii; at these points, draw short arcs perpendicular to the radii, and then complete the circle, as shown in Fig. 62. A modification is to use a slip of paper as a trammel. Another way of drawing a circle is to sketch it in its circumscribing square.

C. Projection In making an orthographic sketch, remember to apply the principles of projection and applied geometry. Sketches are *not* made to scale but are made to show fair proportions of objects sketched. It is legitimate, however, when coordinate paper is used, to count the spaces or rulings as a means of proportioning the views and as an aid in making projections. Take particular care to have the various details of the view in good projection from view to view.

28. THREE VIEWS BY SKETCHING

Practice in orthographic freehand drawing should be started by drawing the three views of a number of simple pieces, developing the technique and the ability to "write" the orthographic language, while exercising the constructive imagination in visualizing the object by looking at the three projections. Observe the following order of working:

1. Study the pictorial sketch and decide what combination of views will best describe the shape of the piece.

2. Block in the view, as in Fig. 63A, using a very light stroke of a soft pencil (2B, B, HB, or F) and spacing the views so as to give a well-balanced appearance to the drawing.

3. Build up the detail in each view, carrying the three views along together as at B.

4. Brighten the outline of each view with bold strokes as at C.

5. Brighten the detail with bold strokes, thus completing the full lines of the sketch as at D.

6. Sketch in all dashed lines, using a stroke of medium weight and making them lighter than the full lines, as at E, thus completing the shape description of the object.

7. Check the drawing carefully. Then cover the pictorial sketch and visualize the object from the three views.

OTHER WAYS TO SHOW SHAPES

29. USING PICTORIAL SKETCHING

A drawing is interpreted by mentally understanding the shape of the object represented. You can prove that you have read and understood a drawing by making the object in wood or metal, modeling it in clay, or making a pictorial sketch of it. Sketching is the usual method. Before attempting to make a pictorial sketch, make a preliminary study of the method of procedure. Pictorial sketching may be based on a skeleton of three axes, one

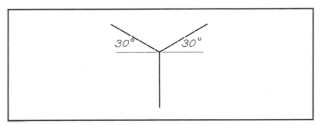

FIG. 64 Pictorial axes (isometric). This is the "framework" for sketching in isometric. See Fig. 65.

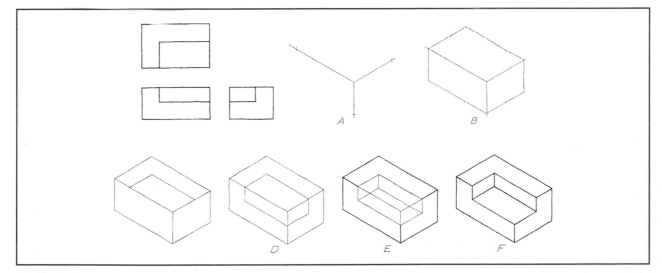

FIG. 65 Stages in making a pictorial sketch. *(A)* Draw axes; *(B)* block in the enclosing shape; *(C)* and *(D)* draw outline of detail on top, front, and side; *(E)* and *(F)* finish by completing surfaces represented on the orthographic drawing.

vertical and two at 30°, representing three mutually perpendicular lines (Fig. 64). On these axes are marked the proportionate width, depth, and height of any rectangular figure. Circles are drawn in their circumscribing squares.

Study the views given in Fig. 65. Then with a soft pencil (F) and notebook paper make a *very light* pictorial construction sketch of the object, estimating its height,

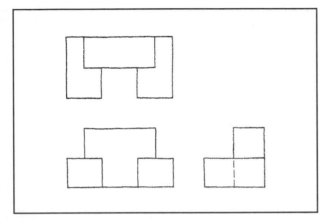

FIG. 66 A drawing to be read. This is the object modeled in Fig. 67.

width, and depth and laying the distances off on the axes as at *A*; sketch the rectangular box that would enclose the piece, or the block from which it could be cut, as at *B*. On the top face of this box sketch lightly the lines that occur on the top view of the orthographic drawing, as at *C*. Note that some of the lines in top views may not be in the top plane. Next sketch lightly the lines of the front view on the front face of the box or block, and if a side view is given, outline it similarly as at *D*. Now begin to cut the figure from the block, strengthening the visible edges and adding the lines of intersection where faces of the object meet, as at *E*. Omit edges that do not appear as visible lines unless they are necessary to describe the piece. Finish the sketch, checking back to the three-view drawing. The construction lines need not be erased unless they confuse the sketch.

30. USE OF MODELING

Modeling the object in clay or modeling wax is another interesting and effective aid in learning to read a drawing. It is done in much the same way as reading by pictorial sketching. Some shapes are easily modeled by

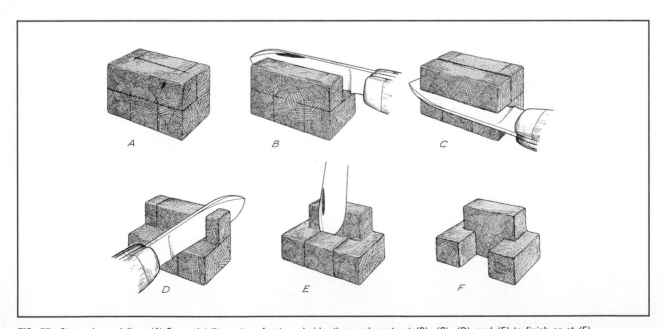

FIG. 67 Stages in modeling. *(A)* Score details on top, front, and side; then make cuts at *(B)*, *(C)*, *(D)*, and *(E)* to finish as at *(F)*.

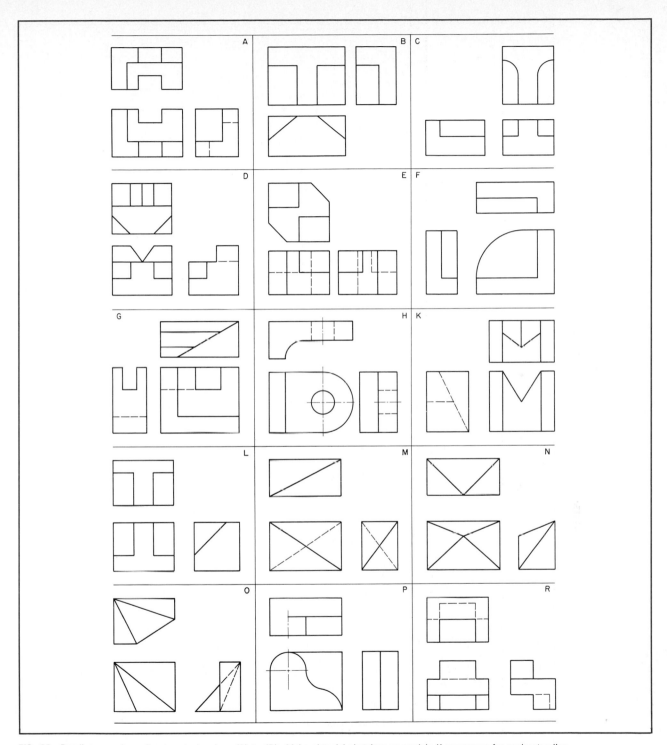

FIG. 68 Reading exercises. Read each drawing, (A) to (R). Make pictorial sketches or models if necessary for understanding.

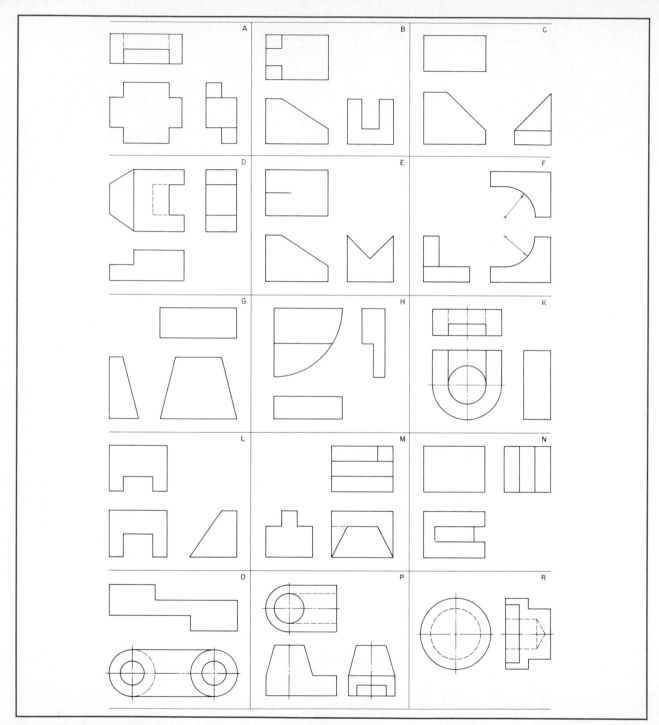

FIG. 69 Missing-line exercises. Read each drawing, A to K, and sketch the lines missing on the views. Check carefully. Use a model or sketch, if necessary, as an aid in locating all lines.

cutting out from the enclosing block; others, by first analyzing and dividing the object into its basic geometric shapes and then combining these shapes.

Starting with a rectangular block of clay, perhaps 30 mm square and 50 mm long, read Fig. 66 by cutting the figure from the solid. With the point of the knife or a scriber, scribe lightly the lines of the three views on the three corresponding faces of the block (Fig. 67A). The first cut could be as shown at *B* and the second as at *C*. Successive cuts are indicated at *D* and *E*, and the finished model is shown at *F*.

31. EXERCISES IN READING SHAPES

Figure 68 contains a number of three-view drawings of block shapes made for exercises in reading orthographic projections and translating into pictorial sketches or models. Proceed as described in the previous paragraphs, making sketches not less than 100 mm overall. Check each sketch to be sure that all intersections are shown and that the original three-view drawing could be made from the sketch. In each drawing in Fig. 69 some lines have been intentionally omitted. Read the drawings and supply the missing lines.

PROBLEMS

For practice in orthographic freehand drawing, select problems from the following group.

GROUP 1. FREEHAND PROJECTIONS FROM PICTORIAL VIEWS

The figures for Probs. 1 to 13 contain a number of pictorial sketches of pieces of various shapes which are to be translated into three-view orthographic freehand drawings. Make the drawings of fairly large size, the front view, say, 2 to 2½ in in length, and estimate the proportions of the different parts by eye or from the proportionate marks shown but without measuring. The problems are graduated in difficulty for selection depending on ability and experience.

Problem 13 gives a series that can be used for advanced work in freehand drawing or that can be used later on, by adding dimensions, as dimensioning studies or freehand working-drawing problems.

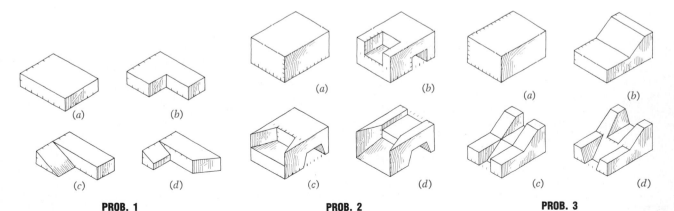

(a)   (b)   (c)   (d)

PROB. 1   **PROB. 2**   **PROB. 3**

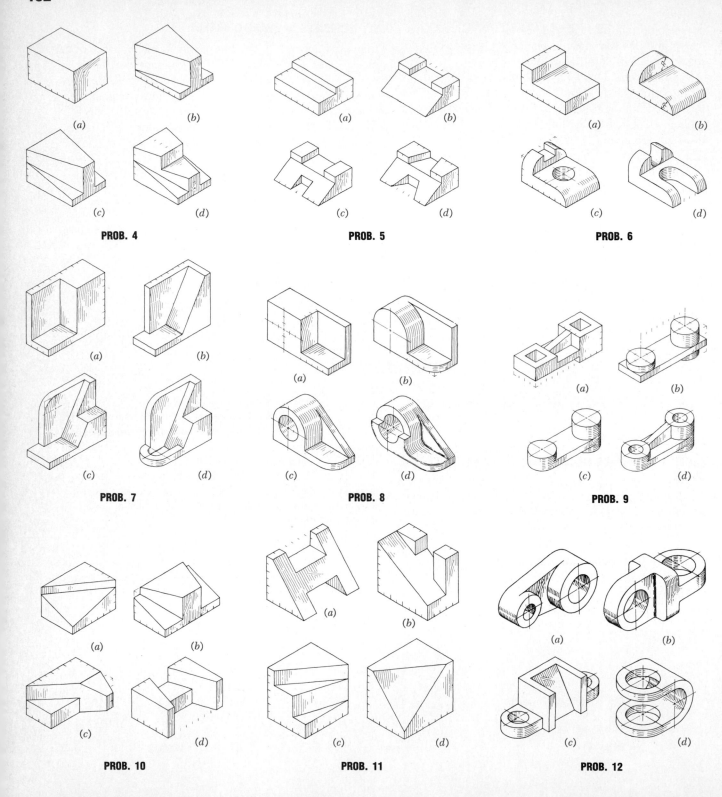

(a)   (b)

(c)   (d)

PROB. 4

(a)   (b)

(c)   (d)

PROB. 5

(a)   (b)

(c)   (d)

PROB. 6

(a)   (b)

(c)   (d)

PROB. 7

(a)   (b)

(c)   (d)

PROB. 8

(a)   (b)

(c)   (d)

PROB. 9

(a)   (b)

(c)   (d)

PROB. 10

(a)   (b)

(c)   (d)

PROB. 11

(a)   (b)

(c)   (d)

PROB. 12

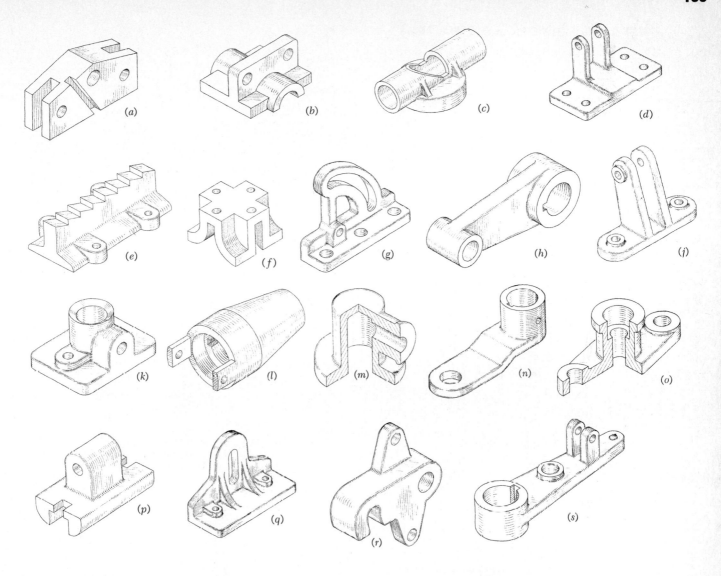

PROB. 13   Pieces to be drawn freehand in orthographic projection.

Select problems from the following groups for practice in projection drawing. Most of the problems are intended to be drawn with instruments but will give valuable training done freehand, on plain or coordinate paper.

The groups are as follows:

2. Projections from pictorial views

3. Use of decimal inch and metric scales

4. Views to be supplied, freehand

5. Views to be supplied

6. Views to be changed

7. Drawing from memory

GROUP 2. PROJECTIONS FROM PICTORIAL VIEWS

14. Draw the top, front, and right-side views of the beam support.

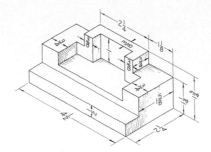

PROB. 14 Beam support.

15. Draw the top, front, and right-side views of the vee rest.

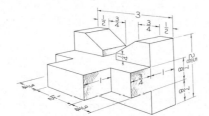

PROB. 15 Vee rest.

16. Draw three views of the saddle bracket.

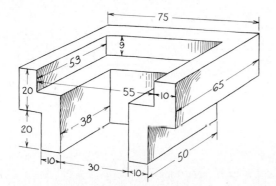

PROB. 16 METRIC. Saddle bracket.

17. Draw three views of the pivot block.

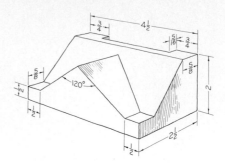

PROB. 17 Pivot block.

18. Draw three views of the inclined support.

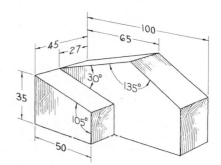

PROB. 18 METRIC. Inclined support.

19. Draw three views of the switch base.

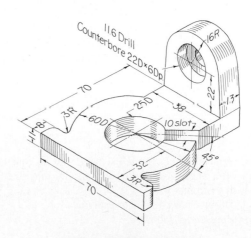

PROB. 19 METRIC. Switch base.

20. Draw three views of the adjusting bracket.

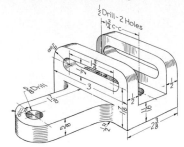

PROB. 20   Adjusting bracket.

21. Draw three views of the bearing rest.

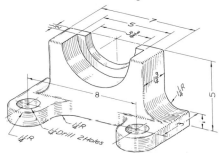

PROB. 21   Bearing rest.

22. Draw three views of the swivel yoke.

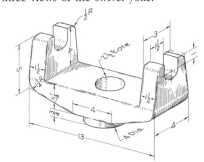

PROB. 22   Swivel yoke.

23. Draw three views of the truss bearing.

24. Draw three views of the sliding-pin hanger.

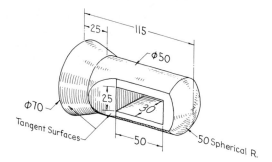

PROB. 24   METRIC. Sliding-pin hanger.

25. Draw two views of the wire thimble.

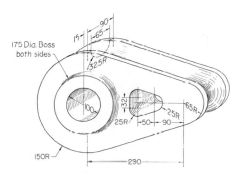

PROB. 25   METRIC. Wire thimble.

26. Draw three views of the hanger jaw.

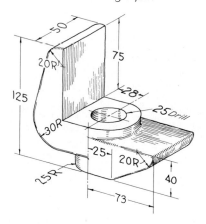

PROB. 26   METRIC. Hanger jaw.

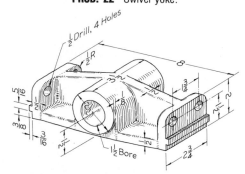

PROB. 23   Truss bearing.

27. Draw three views of the adjustable jaw.

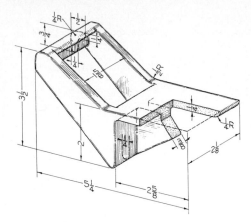

PROB. 27 Adjustable jaw.

28. Draw two views of the shifter fork.

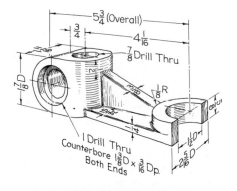

PROB. 28 Shifter fork.

29. Draw three views of the mounting bracket.

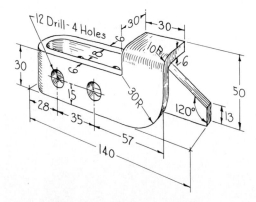

PROB. 29 METRIC. Mounting bracket.

30. Draw three views of the bedplate stop.

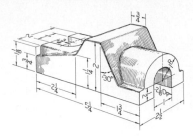

PROB. 30 Bedplate stop.

31. Draw three views of the clamp bracket.

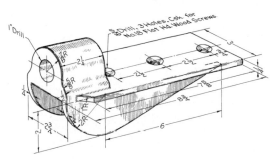

PROB. 31 Clamp bracket.

32. Draw three views of the tube hanger.

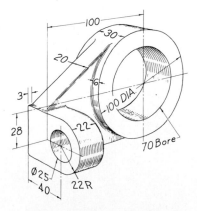

PROB. 32 METRIC. Tube hanger.

33. Draw three views of the shaft guide.

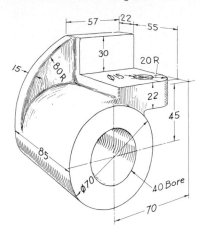

PROB. 33 METRIC. Shaft guide.

34. Draw three views of the clamp block.

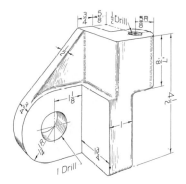

PROB. 34 Clamp block.

35. Draw three views of the angle connector.

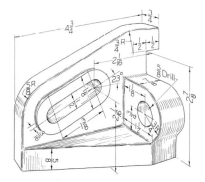

PROB. 35 Angle connector.

36. Draw two views of the end plate.

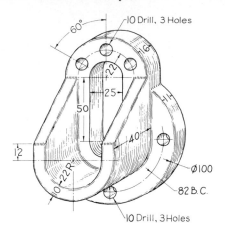

PROB. 36 METRIC. End plate.

37. Draw three views of the plastic switch base.

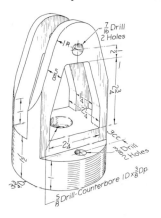

PROB. 37 Switch base.

38. Draw three views of the step-pulley frame.

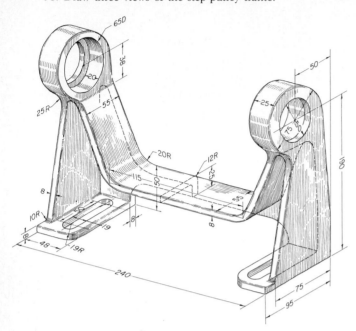

PROB. 38 METRIC. Step-pulley frame.

GROUP 3: USE OF DECIMAL INCH AND METRIC SCALES

39. Draw top, front, and right-side views.

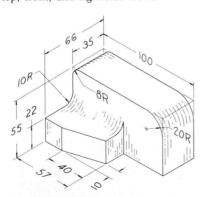

PROB. 39 METRIC. Motor mount.

40. Draw top, front, and right-side views.

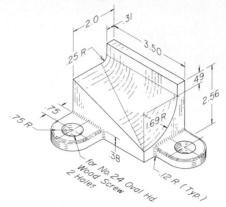

PROB. 40 Assembly-jig base.

41. Draw top, front, and right-side views.

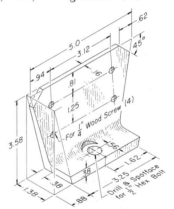

PROB. 41 Cargo-hoist tie-down.

42. Draw top, front, and left-side views.

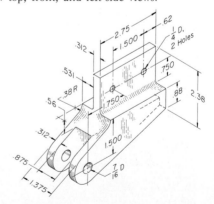

PROB. 42 Aileron tab-rod servo fitting.

43. Draw top, front, and right-side views. Bend radii and set-backs are 0.10 in.

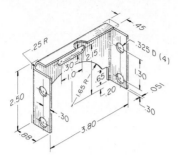

PROB. 43 Cover bracket.

44. Draw top, front, left-side, and right-side views. Show only *necessary* hidden detail.

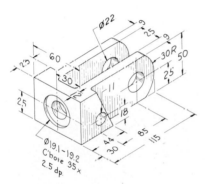

PROB. 44 **METRIC.** Latch bracket.

45. Draw top and front views.

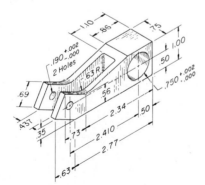

PROB. 45 Control crank.

46. Draw top, front, and left-side views. Bolts should have clearance. Undimensioned radii are ¼R. Convert fractions to decimals.

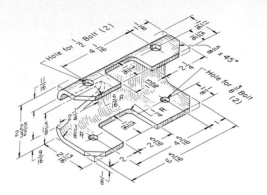

PROB. 46 Transformer mounting.

47. Draw top and front views.

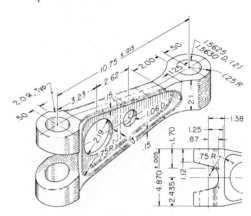

PROB. 47 Stabilizer link.

48. Make an orthographic drawing of the jet-engine bracket.

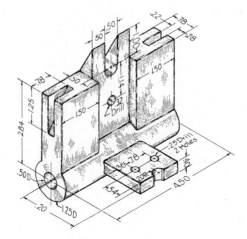

PROB. 48 Jet-engine bracket.

49. Make an orthogrpahic drawing of the missile gyro support.

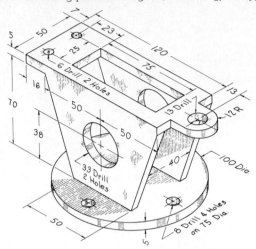

PROB. 49 METRIC. Missile gyro support.

50. Make an orthographic drawing of the reclining-seat ratchet plate.

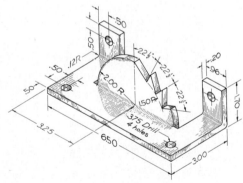

PROB. 50 Reclining-seat ratchet plate.

51. Make an orthographic drawing of the rigging yoke.

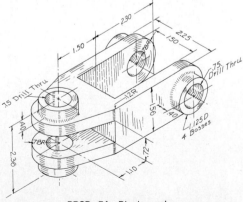

PROB. 51 Rigging yoke.

52. Make an orthographic drawing of the missile release pawl.

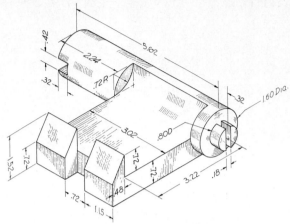

PROB. 52 Missile release pawl.

53. Make an orthographic drawing of the jet-engine inner-strut bracket.

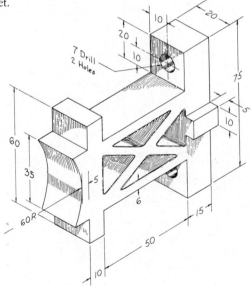

PROB. 53 METRIC. Jet-engine inner-strut bracket.

54. Make an orthographic drawing of the truss bearing.

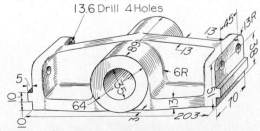

PROB. 54 METRIC. Truss bearing.

GROUP 4. VIEWS TO BE SUPPLIED FREEHAND

These problems (55 to 57) will give valuable training in reading orthographic views, as well as further practice in applying the principles of orthographic projection.

Study the meaning of lines, areas, and adjacent areas. Corners or edges of the object may be numbered or lettered to aid in the reading or to aid later in the projection.

A pictorial sketch may be used, if desired, as an aid in reading the views. This sketch may be made before the views are drawn and completed or at any time during the making of the drawing. For some of the simpler objects, a clay model may be of assistance.

After the views given have been read and drawn, project the third view or complete the views as specified in each case.

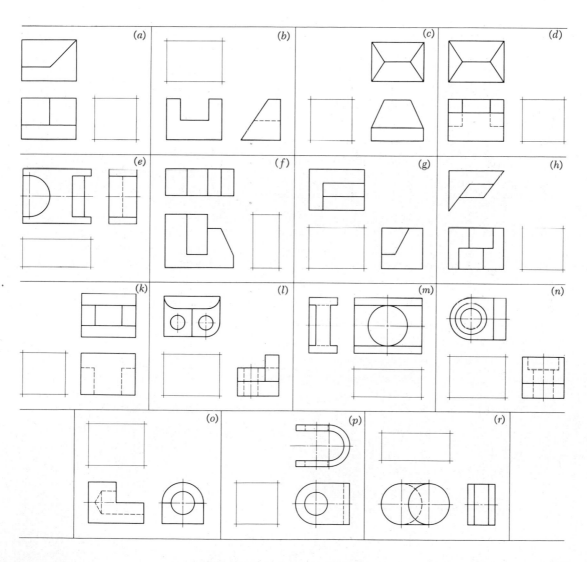

PROB. 55 Views to be supplied freehand.

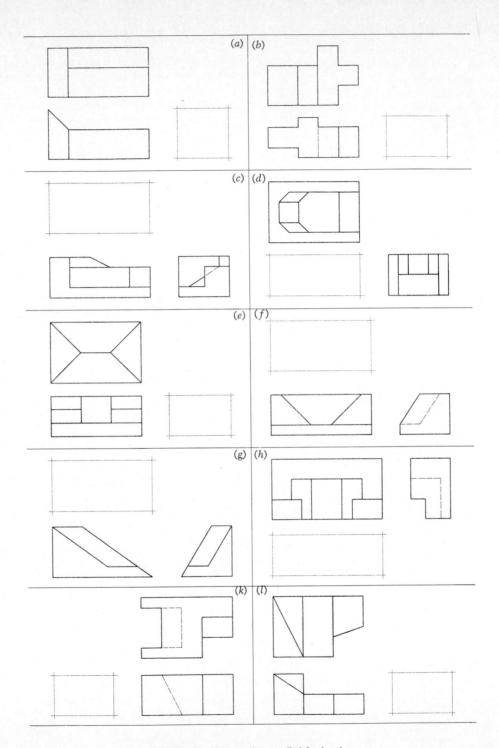

PROB. 56   Views to be supplied freehand.

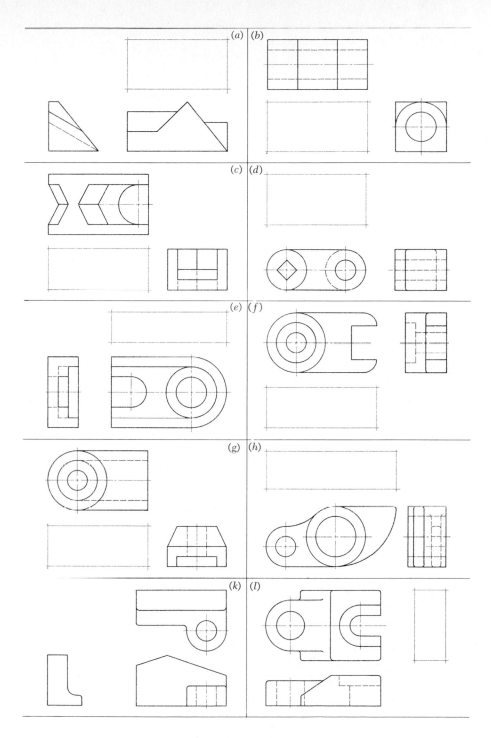

PROB. 57 Views to be supplied freehand.

58. Given top and front views, add side view. Find at least three solutions. Use tracing paper for the second and third solutions.

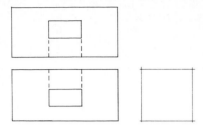

PROB. 58

59. Given front and side views, add top view. Find at least three solutions. Use tracing paper for the second and third solutions.

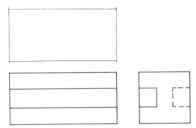

PROB. 59

60. Given top and front views, add side view. Find two solutions. Use tracing paper for the second solution.

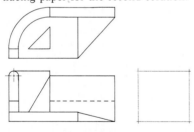

PROB. 60

GROUP 5. VIEWS TO BE SUPPLIED

61. Given front and right-side views, add top view.

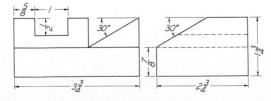

PROB. 61 Projection study.

62. Given front and right-side views, add top view.

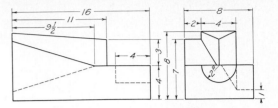

PROB. 62 Bit-point forming die.

63. Given front and top views, add right-side view.

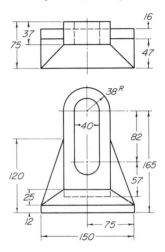

PROB. 63 **METRIC.** Rabbeting-plane guide.

64. Given top and front views, add right-side view.

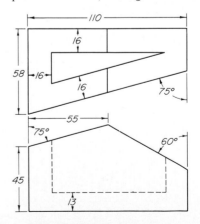

PROB. 64 **METRIC.** Wedge block.

65. Given front and left-side views, add top view.

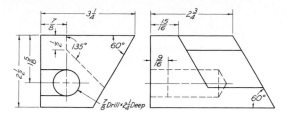

PROB. 65 Burner-support key.

66. Given front and right-side views, add top view.

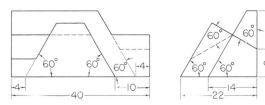

PROB. 66 METRIC. Abutment block.

67. Given front and top views, add right-side view.

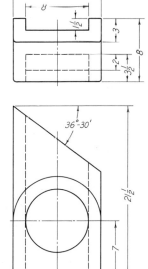

PROB. 67 Sliding port.

68. Assume this to be the right-hand part. Draw three views of the left-hand part.

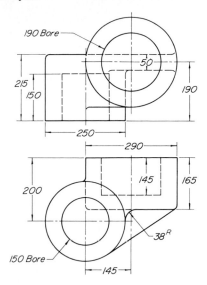

PROB. 68 METRIC. Bumper support and post cap.

69. Given front and top views, add side view.

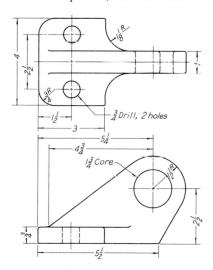

PROB. 69 Anchor bracket.

70. Given front and top views, add side view.

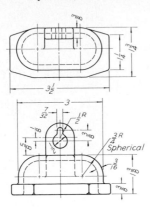

PROB. 70 Entrance head.

72. Given front and top views, add side view.

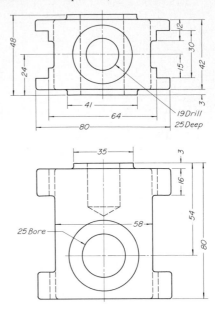

PROB. 72 **METRIC.** Crosshead.

71. Given top and front views, add left-side view.

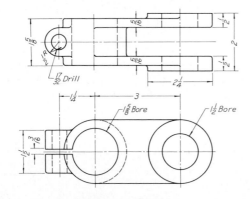

PROB. 71 Yoked link.

73. Given front and top views, add side view.

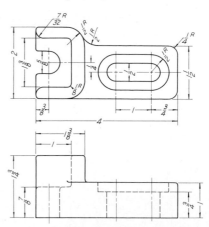

PROB. 73 Tool holder.

74. Given top and front views, add left-side view.

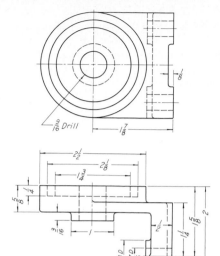

PROB. 74 Bevel-gear mounting.

75. Given top and front views, add side view.

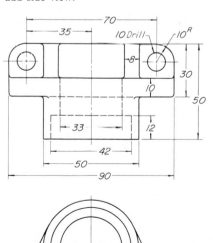

PROB. 75 **METRIC.** Cylinder support.

76. Given front and top views, add side view.

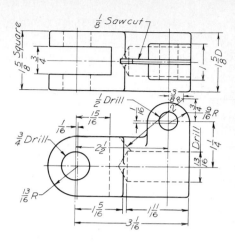

PROB. 76 Rod yoke.

77. Given front and right-side views of electric-motor support, add top view.

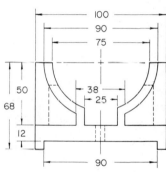

PROB. 77 **METRIC.** Electric-motor support.

78. Given front and right-side views of end frame for engine starter, add top view.

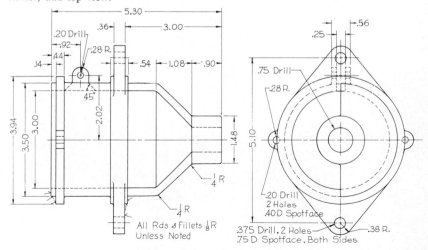

PROB. 78 End frame for engine starter.

GROUP 6. VIEWS TO BE CHANGED

These problems (79 to 82) are given to develop the ability to visualize the actual piece in space and from this mental picture to draw the required views as they would appear if the object were looked at in the directions specified.

In addition to providing training in reading orthographic views and in orthographic projection, these problems are valu-able exercises in developing graphic technique. Note that all the problems given are castings containing the usual features found on such parts, that is, fillets, rounds, runouts, etc., on unfinished surfaces. Note also that sharp corners are formed by the intersection of an unfinished and finished surface or by two finished surfaces. After finishing one of these problems, check the drawing carefully to make sure that all details of construction have been represented correctly.

79. Given front, left-side, and bottom views, draw front, top, and right-side views.

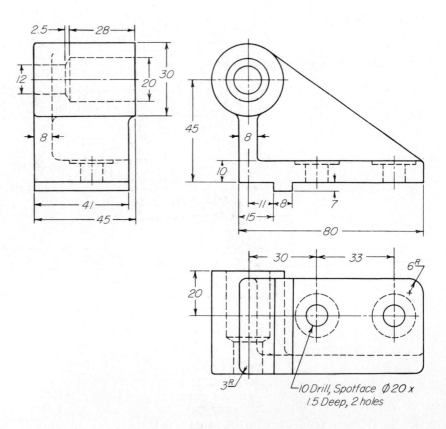

PROB. 79 METRIC. Plunger bracket.

80. Given front, right-side, and bottom views, draw front, top, and left-side views.

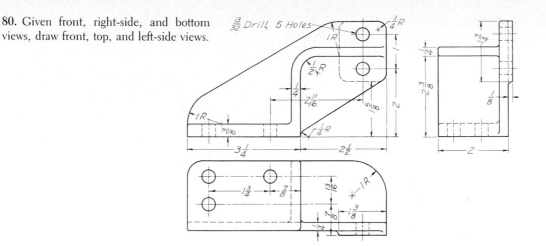

PROB. 80 Offset bracket.

81. Given front, left-side, and bottom views, draw front, top, and right-side views.

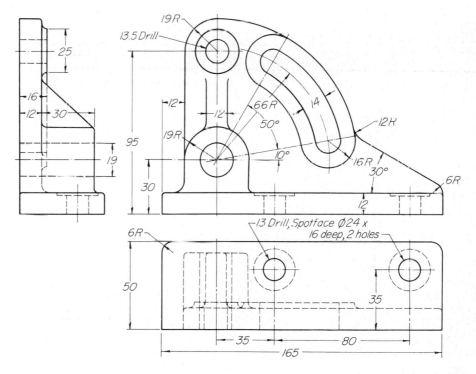

PROB. 81 METRIC. Sector bracket.

82. Given top and front views of jet-engine hinge plate. Add right- and left-side views.

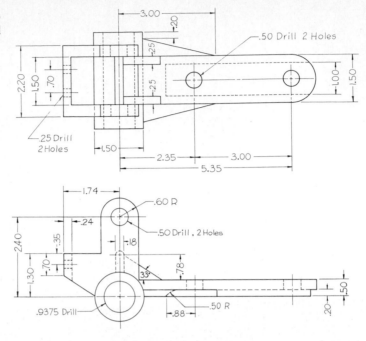

PROB. 82 Jet-engine hinge plate.

GROUP 7. DRAWING FROM MEMORY

One of the valuable assets of an engineer is a trained memory for form and proportion. A graphic memory can be developed to a surprising degree in accuracy and power by systematic exercises in drawing from memory. It is well to begin this training as soon as you have a knowledge of orthographic projection.

Select an object not previously used; look at it with concentration for a certain time (from 5 sec to ½ min or more), close the book, and make an accurate orthographic sketch. Check with the original and correct any mistakes or omissions. Follow with several different figures. The next day, allow a 2-sec view of one of the objects, and repeat the orthographic views of the previous day.

1. PURPOSE OF SECTIONAL VIEWS

A primary purpose of an engineering design drawing is to communicate clearly to the user of the drawing information about size and shape. Showing information clearly can be difficult if an object has many hidden lines. Figure 1 shows a part that has a confusing front view (Fig. 1A) because of the numerous hidden lines.

A portion of the part can be cut away to reveal the interior, as in Fig. 1B. This imaginary cut is called a *section*. A view in which the section is seen is called a *sectional view*. Sectioning also can be used to show a small cross section of the part, such as section C-C in Fig. 1C. Sectioning obviously helps considerably in clarifying the interior shape of a part.

2. HOW SECTIONS ARE SHOWN

The place from which the section is taken must be identifiable on the drawing, and the solid portions and voids must be distinguished on the sectional view. The place from which the section is taken is in many cases obvious, as it is for the sectional view B in Fig. 1; the section is quite evidently taken through the center. In such cases no further description is needed.

If the place from which a section is taken is not obvious, as at C, a *cutting plane*, directional arrows, and identification letters are used to identify it. Whenever there is any doubt, the cutting plane should be shown. A cutting plane is the imaginary medium used to show the path of cutting an object to make a section. Cutting planes for each kind of section will be discussed in the following paragraphs.

A sectional view must show which portions of the object are solid material and which are spaces. This is done by section lining, sometimes called "crosshatching," the solid parts with lines, as shown at B and C of Fig. 1.

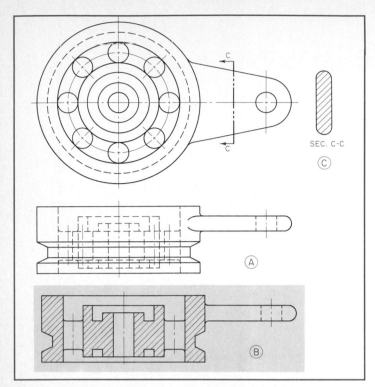

SEC. C-C

FIG. 1 Advantage of sectional views. *(A)* Orthographic view with hidden edges indicated by dashed lines; *(B)* the same view but made as a section to clarify the shape; *(C)* cross-sectional shape of lug shown by removed section.

3. TYPES OF SECTIONS

Several types of sections are available for use in clarifying the interior of parts. The engineer or technician may select from among these types the sectioning format that best aids clarity of shape. Let us look over the most common types of sections.

A. Full Section   A full section is one in which the cutting plane passes entirely across the object, as in Fig. 2, so that the resulting view is completely "in section." The cutting plane may pass straight through, as at A, or be offset, changing direction forward or backward, to pass through features it would otherwise have missed, as at B and C. Sometimes *two* views are drawn in section on a pair of cutting planes, as at A in Fig. 3. In such cases each view is considered separately, without reference to what has been removed for another view. Thus B shows the portion remaining and the cut surface for one sectional view, and C shows them for the other sectional view.

Figure 4 shows a full section made on an offset cutting plane. Note that the *change in plane direction* is *not* shown on the sectional view, for the cut is purely imaginary and *no edge* is present on the object at this position.

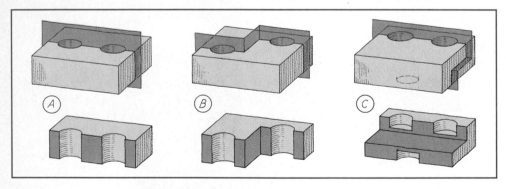

FIG. 2 Cutting planes for a full section. The plane may cut straight across *(A)* or change direction *(B and C)* to pass through features to be shown.

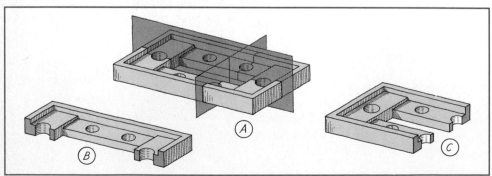

FIG. 3 Two cutting planes. The two planes at *(A)* will produce sections *(B)* and *(C)*. Each section is considered separately.

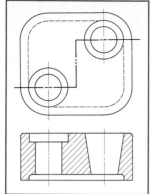

FIG. 4 A full section.

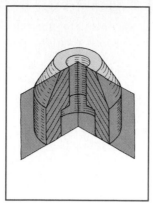

FIG. 5 Cutting plane for a half section.

B. Half Section This is a view sometimes used for symmetrical objects in which one half is drawn in section and the other half as a regular exterior view. The cutting plane is imagined to extend halfway across, then forward, as in Fig. 5. A half section has the advantage of showing both the interior and exterior of the object on one view without using dashed lines, as at A in Fig. 6. However, a half section thus made is difficult to dimension without ambiguity, and so, if needed for clarity, dashed lines may be added, as at B.

Note particularly that a *center line* separates the exterior and interior portions on the sectional view. This is for the same reason that the change in plane direction for the offset of the cutting plane of Fig. 4 is not shown—no edge exists *on the object* at the center.

C. Broken-Out Section Often an interior portion must be shown but a full or half section cannot be used because the cutting plane would remove some feature that must be included. For this condition the cutting plane is extended only so far as needed. Figure 7 is an example. Note the irregular break line, which limits the extent of the section.

D. Rotated Section This type of section is useful for showing cross sections of parts while saving space. Figure 8 shows a typical cutting plane for a rotated section. When rotated 90° about a vertical axis, the resulting section is termed a "rotated section" (or "revolved section").

The example of Fig. 9 gives the rotated section resulting from swinging the cross section 90°, as indicated in the top view. The depth of the cross section seen in the top view becomes the width in the front view. Height is

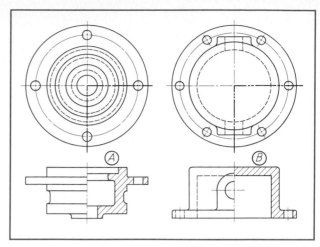

FIG. 6 Half sections. Dashed lines are rarely necessary *(A)*, but may be used for clarity or to aid in dimensioning *(B)*.

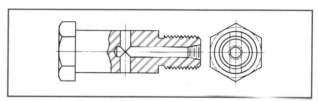

FIG. 7 A broken-out section.

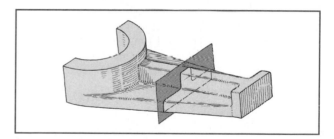

FIG. 8 The cutting plane for a rotated or a removed section.

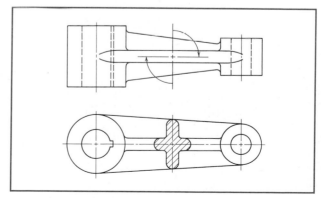

FIG. 9 A rotated section.

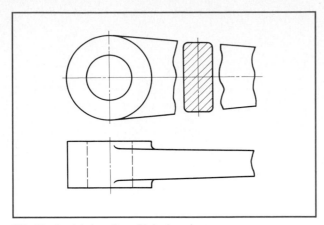

FIG. 10   A rotated section with broken view.

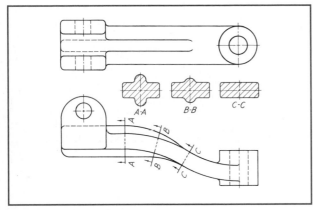

FIG. 11   Removed sections.

determined from the actual height that exists in the front view at the position selected for the rotated section. If the view outline interferes with the section, the view is broken, as in Fig. 10.

E. Removed Section   Removed sections are used for the same purpose as rotated sections, but instead of being drawn *on* the view, they are removed to some adjacent place on the paper (Fig. 11).

The cutting plane with reference letters should always be indicated unless the place from which the section has been taken is obvious. Removed sections are used whenever restricted space for the section or the dimensioning of it prevents the use of an ordinary rotated section. When the shape of a piece changes gradually or is not uniform, several sections may be required (Fig. 12). It is often an advantage to draw the sections to larger scale than that of the main drawing in order to show dimensions more clearly.

It is recommended that, if possible, a removed section be drawn in its natural projected position. This practice is followed in Fig. 12.

F. Auxiliary Section   An auxiliary section conforms to the principles of orthographic projection discussed in Chap. 5. As seen in Fig. 13, an auxiliary section shows the true surface of a cutting plane that is in a position on an inclined feature. (For a full discussion of auxiliary views, see Chap. 9.)

All types of sections—full, half, broken-out, rotated, and removed—are used on auxiliaries. Figure 14 shows

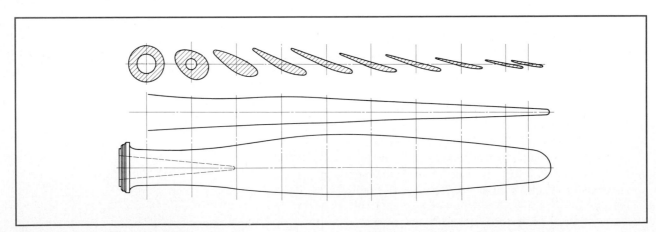

FIG. 12   Removed sections in projection.

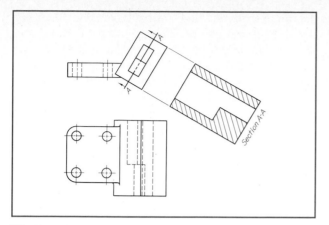

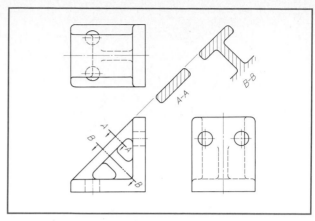

FIG. 13 Auxiliary section. The section is a normal view of the cutting plane.

FIG. 14 Auxiliary sections (partial). Identifications of the cutting planes and mating sections are necessary.

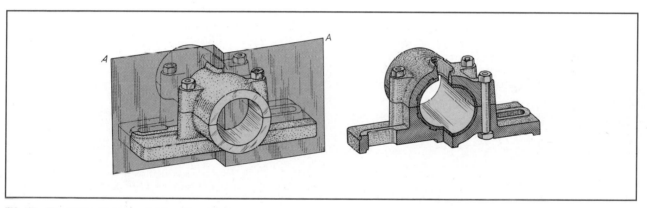

FIG. 15 The cutting plane for an assembly section.

auxiliary partial sections, also known as removed sections in auxiliary position.

G. Assembly Section As the name implies, an assembly section is made up of a combination of parts. All the previously mentioned types of sections may be used to increase the clarity and readability of assembly drawings. The cutting plane for an assembly section is often offset, as in Fig. 15, to reveal the separate parts of a machine or structure.

The purpose of an assembly section is to reveal the interior of a machine or structure so that the separate parts can be clearly shown and identified, but the separate parts do not need to be completely described. Thus only such hidden details as are needed for part identification or dimensioning are shown. Also, the same amount of clearance between mating or moving parts is not shown because, if shown, the clearance would have to be greatly exaggerated, thus confusing the drawing. Even the clearance between a bolt and its hole, which may be as much as $1/16$ in, is rarely shown.

4. DRAWING PRACTICES FOR SECTIONAL VIEWS

In general, the rules of projection are followed in making sectional views. Figure 16 shows the picture of a casting intersected by a cutting plane, giving the appearance that the casting has been cut through by the plane A-A and

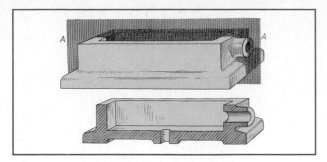

FIG. 16 Cutting plane for a sectional view.

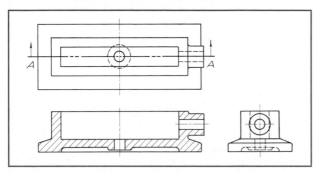

FIG. 17 A drawing with a sectional view. This is the same object as in Fig. 16.

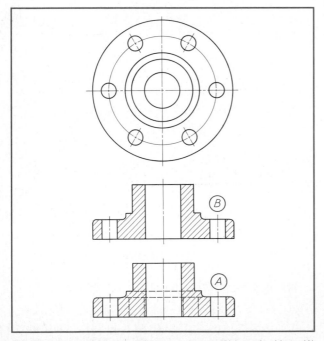

FIG. 18 Hidden edges and surfaces not shown. *(B)* is preferable to *(A)* because the dashed lines are not necessary for description of the part.

the front part removed, exposing the interior. Figure 17 shows the drawing of the casting with the front view in section. The edge of the cutting plane is shown in the top view by the cutting-plane symbol, with reference letters and arrows to show the direction in which the view is taken. It must be understood that the portion assumed to be removed is not omitted in making other views. Therefore, the top and right-side views of the object in Fig. 17 are full and complete, and only in the front view has part of the object been removed.

The American National Standards Institute (ANSI) recommends certain practices to assist in making effective sections. These practices are outlined in paragraphs A to G.

A. The Cutting-Plane Symbol The cutting-plane symbol may be shown on the orthographic view where the cutting plane appears as an edge. It may be more completely identified with reference letters along with arrows to show the direction in which the view is taken. The cutting-plane line symbol is shown in the alphabet of lines, Fig. 39, Chap. 2.

Often when the position of the section is evident, the cutting-plane symbol is omitted (Fig. 18). It is not always desirable to show the symbol through its entire length; so in such cases the beginning and ending of the plane are shown. Removed sections usually need the cutting-plane symbol with arrows for the direction of sight and letters for the resulting sectional view (Fig. 11).

B. Unnecessary Hidden Detail Hidden edges and surfaces are not shown unless they are needed to describe the object. Much confusion may result if all detail behind the cutting plane is drawn. Figure 18A shows a sectional view with all the hidden edges and surfaces shown by dashed lines. These lines complicate the view and do not add any information. The view at *B* is preferred because it is simpler and less time-consuming to draw, and more easily read than the view at *A*. The holes lie on a circular center line, and where similar details repeat, all may be assumed to be alike.

C. Necessary Hidden Detail Hidden edges and surfaces are shown if necessary for the description of the object. In Fig. 19, view *A* is inadequate since it does not show the thickness of the lugs. The corrected treatment is in view *B*, where the lugs are shown by dashed lines.

D. Visible Detail Shown in Sectional Views Figure 20 shows an object pictorially with the front half removed, thus exposing edges and surfaces behind the cutting plane. At A a sectional view of the cut surface only is shown, with the visible elements omitted. This treatment should *never* be used. The view should be drawn as at B, with the visible edges and surfaces behind the cutting plane included in the sectional view.

E. Visible Detail Not Shown in Sectional Views Sometimes confusion results if all visible detail behind the cutting plane is drawn, and it may be omitted if it does not aid in readability. Omission of detail should be carefully considered and may be justified as time saved in drawing. This applies mainly to assembly drawings, in which it is desired to show how the pieces fit together rather than to give complete information for making the parts.

F. Section Lining Wherever material has been cut by the section plane, the cut surface is indicated by section lining done with fine lines generally at 45° to the principal lines in the view and spaced uniformly to give an even tint. These lines are spaced entirely by eye except when some form of mechanical section liner is used. The pitch, or distance between lines, is governed by the size of the surface. For ordinary working drawings, it will not be much less than 1/16 in and rarely more than 1/8 in. Take care in setting the pitch by the first two or three lines, and glance back at the first lines often to see that the pitch does not gradually increase or decrease. Nothing mars the appearance of a drawing more than poor section lining. The alphabet of lines gives the weight of crosshatch lines.

Two adjacent pieces in an assembly drawing are crosshatched in opposite directions. If three pieces adjoin, one of them may be sectioned at other than 45° (usually 30 or 60°, Fig. 21), or all pieces may be crosshatched at 45° by using a different pitch for each piece. If a part is so shaped that 45° sectioning runs parallel, or nearly so, to its principal outlines, another direction should be chosen (Fig. 22).

Very thin sections, as of gaskets, sheet metal, or structural-steel shapes to small scale, may be shown in solid black, with white spaces between the parts where thin pieces are adjacent (Fig. 23).

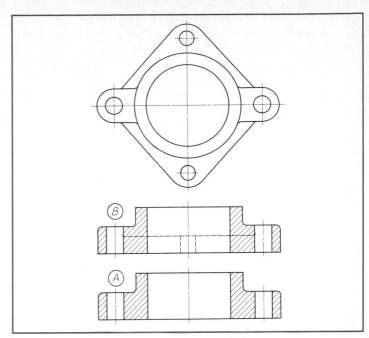

FIG. 19 Hidden edges and surface shown. *(B) must* be used instead of *(A)* because the dashed lines are necessary for description of the part.

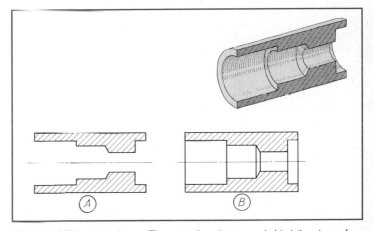

FIG. 20 Visible edges shown. These are the edges seen behind the plane of the section and must be shown as at *(B)*.

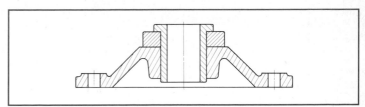

FIG. 21 Crosshatching of adjacent parts.

G. Symbols for Materials in Section It is sometimes helpful to show a distinction between materials sectioned by using symbols depicting the various materials. One application of this technique is in assembly sections. Commonly used symbols appear in Fig. 24.

Using symbols for various materials is only an aid in reading a drawing. The limited number of symbols available could never include all materials and their numerous alloys. Exact specification of the material for each piece appears within notes on the drawing.

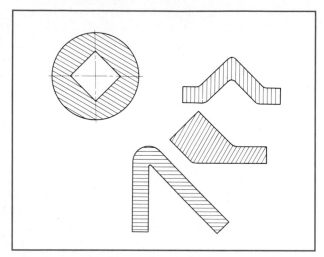

FIG. 22 Section-line directions for unusual shapes.

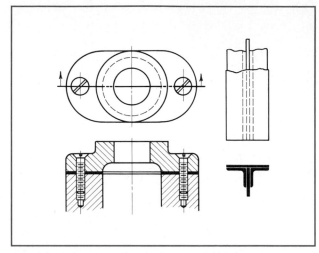

FIG. 23 Thin material in section (drawn solid for lack of room to cross-hatch).

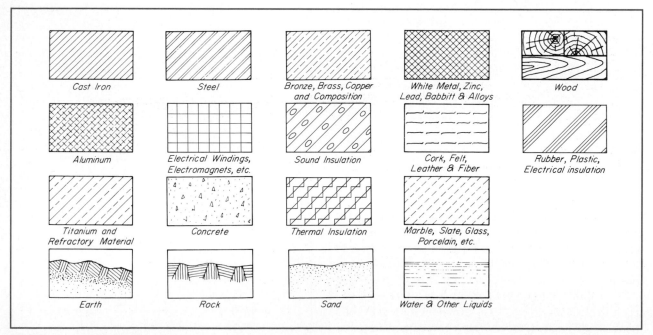

FIG. 24 Symbols for materials (section).

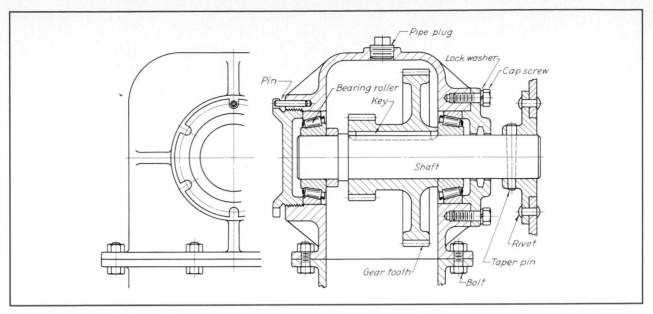

FIG. 25 Part of a sectional assembly.

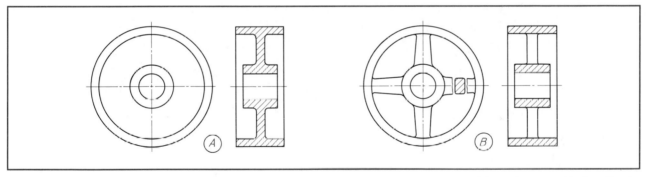

FIG. 26 Spokes in section. The wheel with spokes *(B)* is treated as though the cutting plane were in front of the spokes, to avoid misreading the section as a solid web line *(A)*.

5. PARTS NOT SECTIONED

Many parts are more easily recognized by their exterior views than by sections. These parts include solid bolts, pins, and shafts that may lie in the path of a sectioning plane. Such parts should be left in full view and should not be sectioned. Figure 25 shows several nonsectional parts. If the shaft, bolts, nuts, pins, and rivets were sectioned, the drawing would be confusing and difficult to read.

6. PORTIONS OF PARTS NOT SECTIONED

A basic principle for sectioning circular parts is that any element not continuous (not solid) around the axis of the part should be drawn without crosshatching in order to avoid a misleading effect.

A. Spokes in Section For example, consider the two pulleys in Fig. 26. Pulley A has a solid web connecting the hub and rim. Pulley *B* has four spokes. Even though

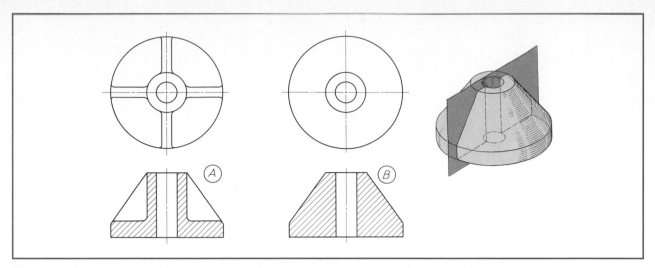

FIG. 27 Ribs in section. Ribs at *(A)* are treated as though the cutting plane were in front of them, to avoid misreading the section as a solid *(B)*.

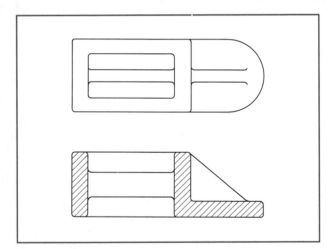

FIG. 28 Ribs in section.

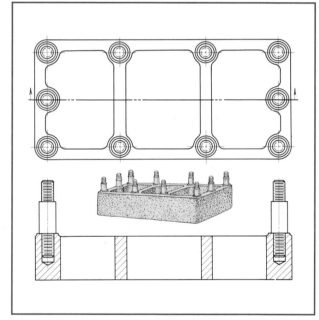

FIG. 29 Ribs at right angles to the cutting plane. Note the omission of detail behind the sectioning plane.

the cutting plane passes through two of the spokes, the sectional view of *B* must be made without crosshatching the spokes in order to avoid the appearance of a solid web, as in pulley A.

Other machine elements treated in this manner are teeth of gears and sprockets, vanes and supporting ribs of cylindrical parts, equally spaced lugs, and similar parts.

B. Ribs in Section When the cutting plane passes longitudinally through the center of a rib or web, as in Fig. 27A, the crosshatching is eliminated from the ribs as if the cutting plane were just in front of them.

A true sectional view with the ribs crosshatched gives a heavy, misleading effect suggesting a cone shape, as shown in Fig. 27B. The same principle applies to ribs cut longitudinally on rectangular parts (Fig. 28). When the cutting plane cuts a rib transversely, that is, at right angles to its length or axis direction (the direction that shows its thickness), it is always crosshatched (Fig. 29).

C. Lugs in Section A lug or projecting ear (Fig. 30A), usually of *rectangular* cross section, is not crosshatched; note that crosshatching either of the lugs would suggest a circular flange. However, the somewhat similar condition at B should have the projecting ears crosshatched as shown because these ears *are* the base of the part.

7. ALTERNATE CROSSHATCHING

In some cases omitting the crosshatching of ribs or similar parts gives an inadequate and sometimes ambiguous treatment. To illustrate, Fig. 31A shows a full section of an idler pulley. At B four ribs have been added. Note

that the top surfaces of the ribs are flush with the top of the pulley. Without crosshatching, the section at B is identical with A and the ribs of B are not identified at all on the sectional view. A better treatment in this case is to use alternate crosshatching for the ribs, as at C, where half (alternating) the crosshatch lines are carried through the ribs. Note that the line of demarcation between rib and solid portions is a *dashed* line.

8. USE OF ALIGNMENT

Alignment is a technique in which features of parts can be shown in their true radial positions. Clarity of shapes

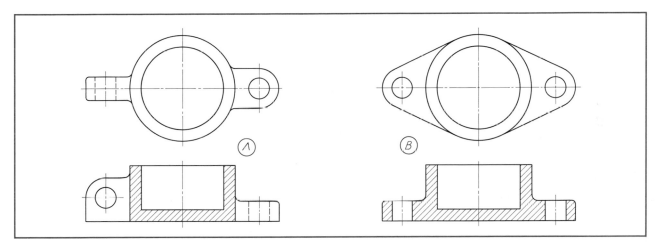

FIG. 30 Lugs in section. Small lugs *(A)* are treated like spokes and ribs. Large lugs *(B)* are considered as the solid base of the part.

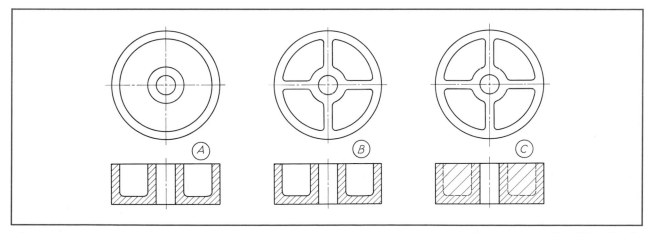

FIG. 31 Alternative crosshatching. Section *(B)*, with ribs flush at the top, looks like section *(A)*, with no ribs. Alternate crosshatching *(C)* identifies the ribs.

is greatly increased. The following cases illustrate the principle of alignment.

A. Aligned Spokes Any part with an odd number of spokes or ribs (three, five, seven, etc.) will give an unsymmetrical and misleading section if the principles of true projection are strictly adhered to, as illustrated by the drawing of a handwheel in Fig. 32. The preferred projection is shown in the second sectional view, where one spoke is drawn as if aligned or, in other words, the spoke is rotated to the path of the vertical cutting plane and then projected to the side view. Note that neither spoke should be sectioned, for the reasons given in paragraph 6A.

This practice of alignment is well justified because a part with an odd number of equally spaced elements is just as symmetrical as a part with an even number and therefore should be shown by a symmetrical view. Moreover, the symmetrical view shows the true *relationship* of the elements, whereas the true projection does not.

B. Aligned Ribs, Lugs, and Holes Ribs, lugs, and holes often occur in odd numbers and should be aligned to show the true relationship of the elements. In Fig. 33,

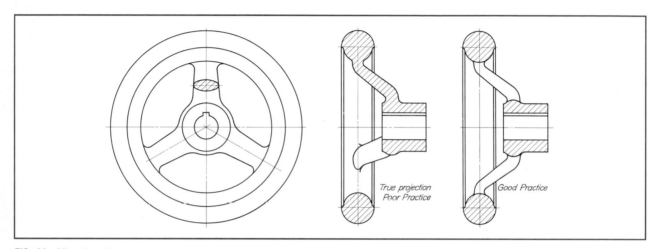

FIG. 32 Aligned spokes.

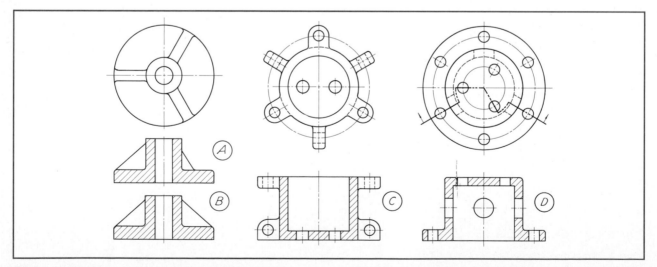

FIG. 33 Aligned ribs, lugs, and holes. True projection of ribs *(A)* is misleading. Alignment *(B)* gives a symmetrical section for a symmetrical part. The same is true for lugs *(C)*. Offset sectioning aids clarity in case *D*.

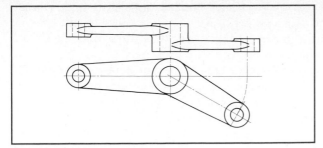

FIG. 34 Aligned view.

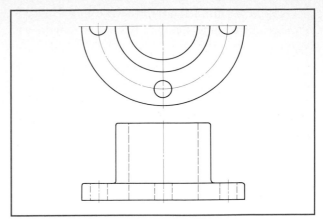

FIG. 35 Half view. The front half is drawn when the mating view is external.

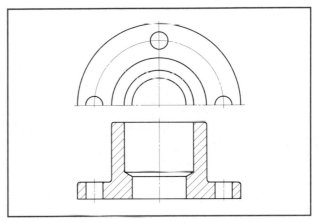

FIG. 36 Half view. The rear half is drawn when the mating view is a full section.

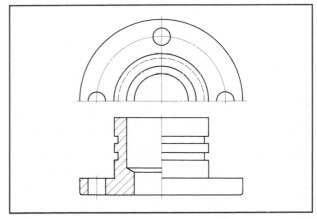

FIG. 37 Half view. The rear half is drawn when mating view is a half section.

true projection of the ribs would show the pair on the right foreshortened, as at *A*, suggesting in the sectional view that they would not extend to the outer edge of the base. Here, again, the alignment shown at *B* gives a symmetrical section of a symmetrical part and shows the ribs in their true relationship to the basic part. To illustrate further, at *C* the lugs and holes are aligned, thus showing the holes at their true radial distance from the axis, and, incidentally, eliminating some difficult projections.

In all cases of alignment, the element can be thought of as being swung around to a common cutting plane and then projected to the sectional view. Note at *C* that because an offset cutting plane is used, each hole is brought separately into position on a common cutting plane before projection to the sectional view.

C. Aligned Views Pieces that have elements at an angle to one another, as the lever of Fig. 34, may be shown straightened out or aligned in one view.

9. HALF VIEWS

When space is limited, it is allowable to make the top or side view of a symmetrical piece as a half view. If the front is an exterior view, the *front* half of the top or side view would be used, as in Fig. 35; but if the front view is a sectional view, the *rear* half would be used, as in Fig. 36. Figure 37 shows another space-saving combination of a half view with a half section.

10. CONVENTIONAL PRACTICES

We have seen several cases in which the strict projection of features can cause confusion. Clarity is enhanced by using alignment and by not sectioning such features as spokes and ribs. Further aids to clarity can be applied to certain cases for intersections, fillets and rounds, and breaks, as seen in the following examples.

A. Conventional Practice for Intersections There are occasions when the true lines of intersection are of no value as aids in reading and should be ignored. Some typical examples are shown in Fig. 38. It must be noted, however, that in certain cases where there is a major difference in line position when the true projection is given, as compared with conventional treatment, the true line of intersection should be shown. Compare the treatment of the similar objects in Figs. 38 and 39. It would not be good practice to conventionalize the intersections on the objects in Fig. 39 because the difference between true projection and the convention is too great.

B. Conventional Practice for Fillets and Rounds Designers rarely leave sharp internal corners in a casting because of possible stress cracks arising from these sharp corners.

A fillet is used to provide a blending radius instead of a sharp corner. Also, external corners may be rounded for safety and appearance. Small fillets and rounds are often sketched in freehand.

Runouts are also seen. They are filleted intersections where there is no line because there is no abrupt change of direction. Figure 40 shows conventional practice for fillets and rounds with runouts.

C. Conventional Breaks In making the detail of a long bar or piece with a uniform cross section, it is rarely necessary to draw its whole length. It may be shown to a

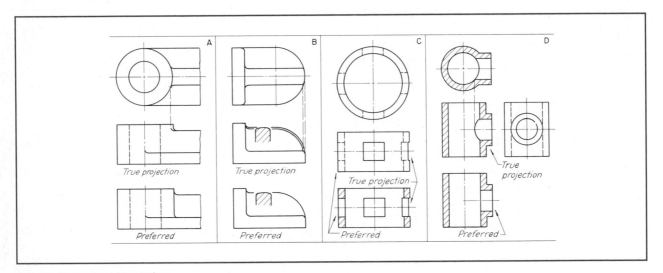

FIG. 38 Conventional intersections.

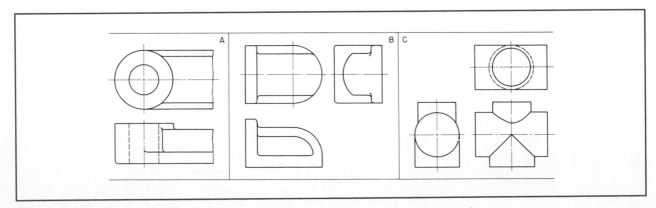

FIG. 39 True intersections.

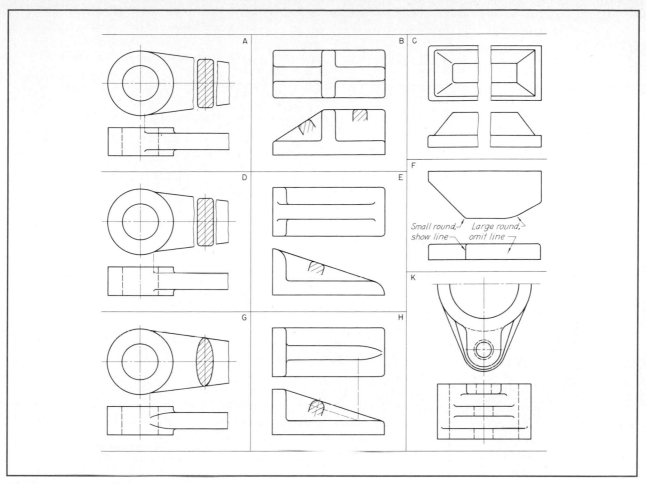

FIG. 40 Conventional fillets, rounds, and runouts.

larger and therefore better scale by breaking out a piece, moving the ends together, and giving the true length by a dimension, as in Fig. 41. The shape of the cross section is indicated by a rotated section or by a semipictorial break line, as in Fig. 42.

11. OTHER CONVENTIONAL SYMBOLS

Throughout this book you will see the applications of various symbols not discussed in this chapter. Examples are symbols for screw threads, springs, pipe fittings, and electrical devices.

Designations for screw threads are one of the important uses of conventional symbols that will be discussed

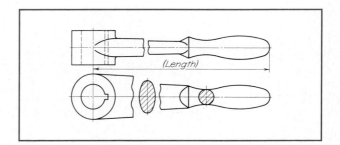

FIG. 41 Broken view with rotated sections.

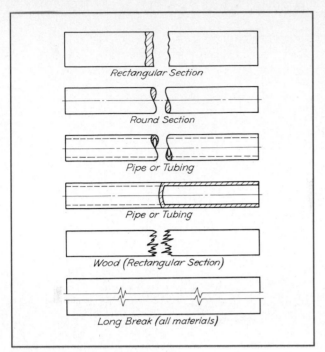

FIG. 42 Conventional breaks.

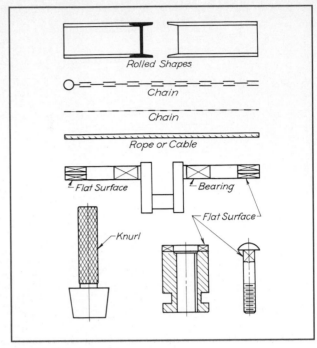

FIG. 43 Various symbols.

in Chap. 16. Such symbols are used primarily on machine drawings. Architectural drawing also uses many conventional symbols. Topographic drawing is made up almost entirely of symbols. Electrical diagrams are completely symbolic.

The symbol of two crossed diagonals is used for two distinct purposes:

1. To indicate on a shaft the position of finish for a bearing.
2. To indicate that a certain plane is flat. See Fig. 43.

PROBLEMS

GROUP 1. SINGLE PIECES

The following problems can be used for practice in shape description only or, by adding dimensions, in making working drawings.

1. Draw the top view, and change the front and side views to sectional views as indicated.

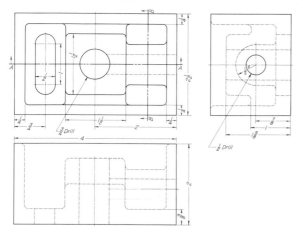

PROB. 1 Section study.

2. Draw the top view, and make the front and two side views in section on cutting planes as indicated. Scale to suit.

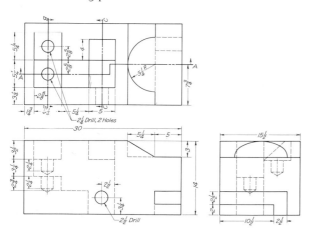

PROB. 2 Section study.

3 and 4. Given the side view, draw full front and side views in section. Scale to suit.

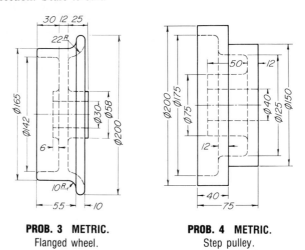

PROB. 3 METRIC.
Flanged wheel.

PROB. 4 METRIC.
Step pulley.

5 and 6. Change the right-side view to a full section.

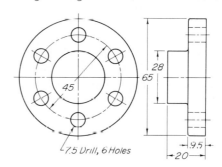

PROB. 5 METRIC. Cap.

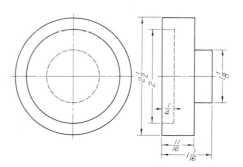

PROB. 6 Flanged cap.

7 and 8. Change the right-side view to a full section.

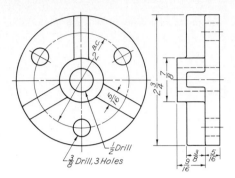

PROB. 7 Pump-rod guide.

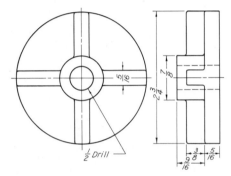

PROB. 8 Face plate.

9. Change the right-side view to a full section.

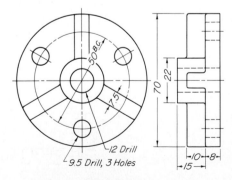

PROB. 9 METRIC. Ribbed support.

10. Change the right-side view to a sectional view as indicated.

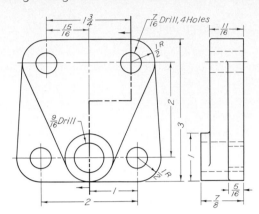

PROB. 10 Housing cover.

11 and 12. Change the front view to a full section.

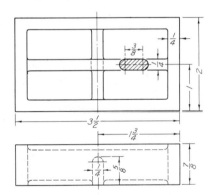

PROB. 11 Filler block.

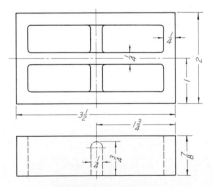

PROB. 12 Filler block.

13 and 14. Change the front view to a full section.

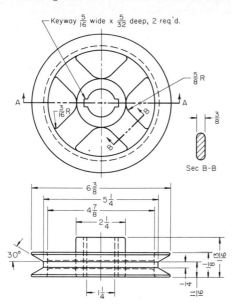

PROB. 13   V-belt pulley.

14 and 15. Select views that will best describe the piece.

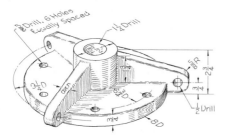

PROB. 14   End plate.

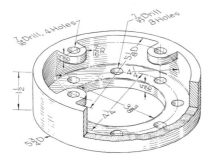

PROB. 15   Piston cap.

16. Draw the top view as shown in the illustration and the front view as a full section.

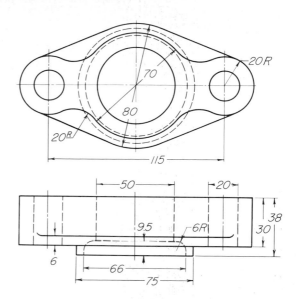

PROB. 16   METRIC. Pump flange.

17. Draw the top view as illustrated and the front view in half section on A-A.

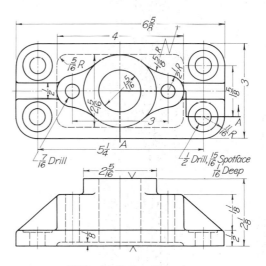

PROB. 17   Brake-rod bracket.

18. Draw the top view and sectional view (or views) to best describe the object.

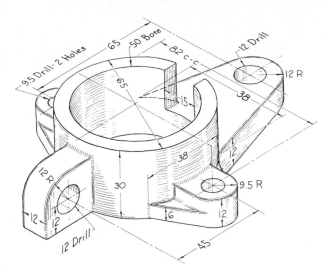

PROB. 18 METRIC. Column collar.

19. Draw the top view and front view in section.

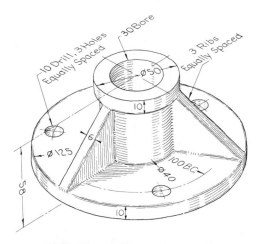

PROB. 19 METRIC. Stem support.

20. Turn the object through 90°, and draw the given front view as the new top view; then make the new front view as section *B-B* and auxiliary section *A-A*. Refer to Sec. 3 for instructions on making auxiliary sections. See also Chap. 9 for the method of projection for the auxiliary section. Note in this case that the new top view, front-view section *B-B*, and auxiliary section *A-A* will completely describe the object. However, if desired, the side view may also be drawn, as shown or as an aligned view (described in Sec. 8).

20A. As an alternative for Prob. 20, draw views as follows: with the object in the position shown, draw the front view as shown, draw the left-side view as section *B-B*, and draw the new top view as an aligned view.

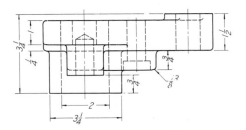

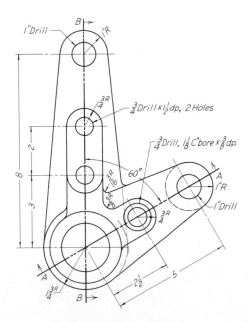

PROB. 20 Compound bell-crank.

21 and 22. Draw the views and add the sectional views indicated.

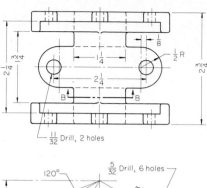

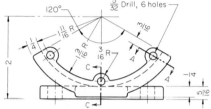

PROB. 21 Spool base.

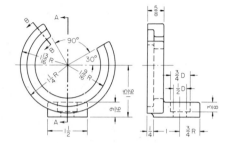

PROB. 22 Saddle collar.

23 and 24. Draw a view and sectional view to best describe the piece.

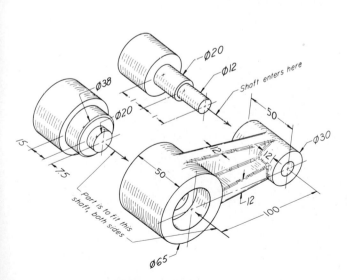

PROB. 23 METRIC. Actuator link.

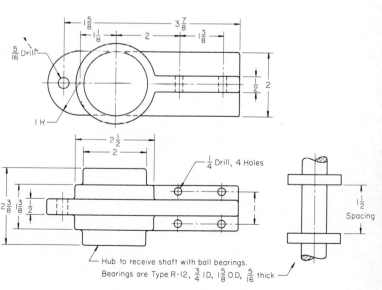

PROB. 24 Vibrator-drive bearing support.

25. Draw the top view and necessary sectional view (or views) to best describe the object.

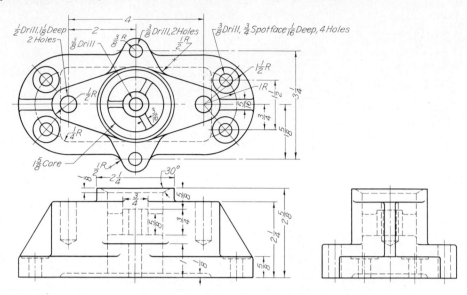

PROB. 25   Cover and valve body.

26. Draw three views, making the side view as a section on *B-B*

26A. Draw three views, making the top view as a half section on A-A.

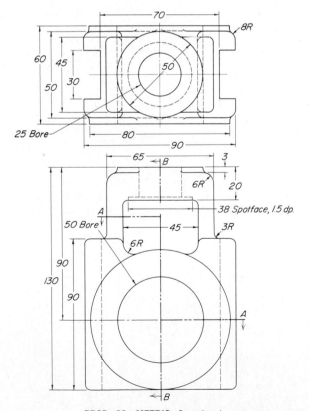

PROB. 26   METRIC. Crosshead.

GROUP 2. ASSEMBLIES

27. Draw a half end view and a longitudinal view as full section.

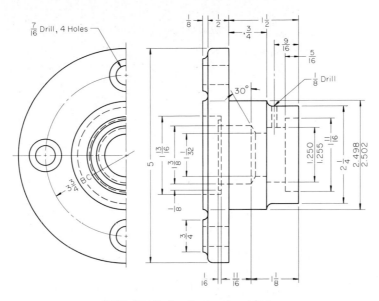

PROB. 27 Push rod cap and seal body.

28. Draw the top view as shown in the figure and new front view in section. Show the shape (right section) of the link with rotated or removed section. The assembly comprises a cast-steel link, two bronze bushings, steel toggle pin, steel collar, steel taper pin, and part of the cast-steel supporting lug.

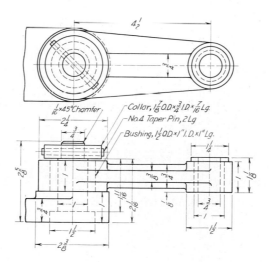

PROB. 28 Link assembly.

29. Draw the front view and longitudinal section. The assembly comprises a cast-iron base, a bronze bushing, a bronze disk, and two steel dowel pins.

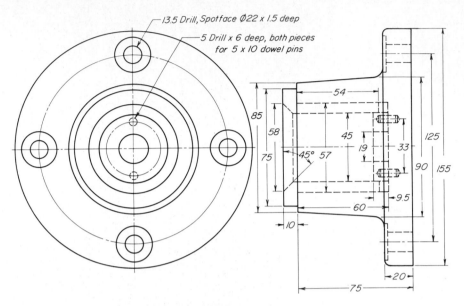

PROB. 29 METRIC. Step bearing.

30. Draw two half end views and a longitudinal section. The assembly consists of cast-iron body, two bronze bushings, steel shaft, cast-iron pulley, and steel taper pin.

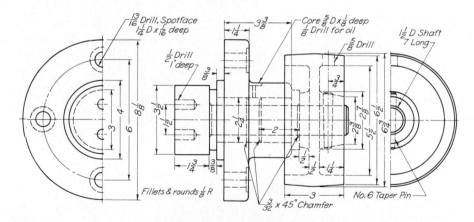

PROB. 30 Pulley-bracket assembly.

31. Make an assembly drawing in section. The bracket is cast iron, the wheel is cast steel, the bushing is bronze, and the pin and taper pin are steel. Scale: full size.

31A. Make a drawing of the bracket with one view in section. Material is cast iron. Scale: full size.

31B. Make a drawing of the wheel with one view in section. Material is cast steel. Scale: full size.

32. Make an assembly drawing in section. The assembly comprises two cast-iron brackets, two bronze bushings, steel shaft, cast-steel roller, and cast-iron base. The bushings are pressed into the roller, and the shaft is drilled for lubrication. Scale: full size.

32A. Make a drawing of the roller and bushing assembly that gives one view in section. See Prob. 32 for materials. Scale: full size.

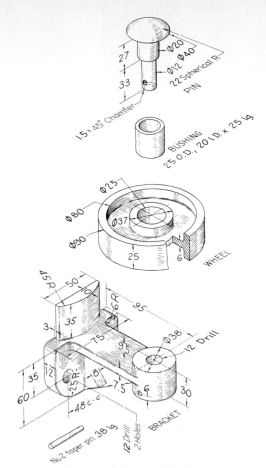

PROB. 31 METRIC. Sliding-door guide.

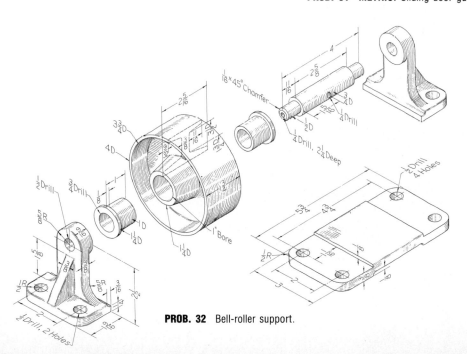

PROB. 32 Bell-roller support.

7 PICTORIAL DRAWING AND SKETCHING

1. THE REASON FOR PICTORIAL DRAWINGS

The orthographic views we discussed in Chap. 5 are very valuable in providing information as to true heights, widths, and depths of objects. However, the formats of object representation in Chap. 5 are two-dimensional. It can be helpful to see objects in three-dimensional forms.

A three-dimensional drawing or sketch shows the entire object in one view. The object is not scattered among top, front, and side views. A three-dimensional pictorial greatly enhances one's ability to visualize the object, especially if one does not understand orthographic projection. There is often a need to illustrate for a nontechnical person some object under discussion; the three-dimensional format is probably the best way to present the object.

Also, pictorial views can be used well as technical illustrations. These include Patent Office drawings, layouts, pipe plans, and the like. Pictorial methods are useful also in making freehand sketches, and this is one of the most important reasons for learning them.

2. PICTORIAL METHODS

There are three main divisions of pictorial drawing: (1) axonometric, with its divisions into trimetric, dimetric, and isometric; (2) oblique, with several variations; and (3) perspective. These methods are illustrated in Fig. 1.

The trimetric form gives an effect more pleasing to the eye than the other axonometric and oblique methods and allows almost unlimited freedom in orienting the object, but is difficult to draw. With the dimetric method the result is less pleasing and there is less freedom in orienting the object, but execution is easier than with trimetric. The isometric method gives a result less pleasing than dimetric or trimetric, but it is easier to draw and has the distinct advantage of being easier to dimension.

The oblique method is used principally for objects with circular or curved features only on one face or on parallel faces. For objects of this type the oblique is easy to draw and dimension. Perspective drawing gives a result most pleasing to

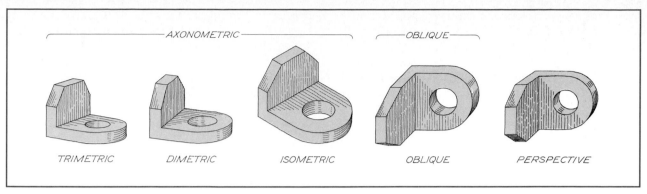

FIG. 1 Pictorial methods.

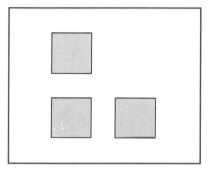

FIG. 2 The object face is parallel to the picture plane. One face only is seen in the front view.

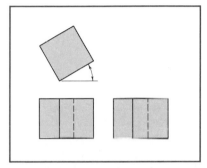

FIG. 3 The object is rotated about a vertical axis. Two faces are seen.

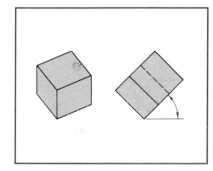

FIG. 4 The object is rotated about both a vertical and a profile axis. Three faces are seen.

the eye, but it is of limited usefulness because many lines are unequally foreshortened. Isometric and oblique are the forms most commonly used.

3. AXONOMETRIC PROJECTION

Axonometric projection is theoretically orthographic projection in which only one plane is used, the object being turned so that three faces show. Imagine a transparent vertical plane with a cube behind it, one face of the cube being parallel to the plane. The projection on the plane, that is, the front view of the cube, will be a square (Fig. 2). Rotate the cube about a vertical axis through any angle less than 90°, and the front view will now show two faces, both foreshortened (Fig. 3). From this position, tilt the cube forward (rotation axis perpendicular to profile) any amount less than 90°. Three faces

will now be visible on the front view (Fig. 4). There can be an infinite number of axonometric positions, depending upon the angles through which the cube is rotated. Only a few of these positions are ever used for drawing. The simplest is the isometric (equal-measure) position, in which the three faces are foreshortened equally.

4. ISOMETRIC PROJECTION

If the cube in Fig. 5A is rotated about a vertical axis through 45°, as shown at (*b*), and then tilted forward, as at (*c*), until the edge *RU* is foreshortened equally with *RS* and *RT*, the front view of the cube in this position is said to be an "isometric projection." (The cube has been tilted forward until the body diagonal through *R* is perpendicular to the front plane. This makes the top face slope approximately 35°16′.) The projections of the three mutually perpendicular edges *RS, RT, RU* meeting at the front corner *R* make equal angles, 120°, with each other and are called "isometric axes." Since the projections of parallel lines are parallel, the projections of the other edges of the cube will be, respectively, parallel to these axes. Any line parallel to the edge of the cube, whose projection is thus parallel to an isometric axis, is called an "isometric line." The planes of the faces of the cube and all planes parallel to them are called "isometric planes."

The isometric axes *RS, RT,* and *RU* are all foreshortened equally because they are at the same angle to the picture plane.

5. ISOMETRIC DRAWING

In nearly all practical use of the isometric system, the foreshortening of the lines is disregarded, and *their full lengths are laid off on the axes.* This gives a figure of exactly the same shape but larger in linear proportion of 1.23 to 1. The effect of increased size is usually of no consequence, and since the advantage of measuring the lines directly is of great convenience, isometric drawing is used almost exclusively rather than isometric projection.

In isometric projection the isometric lines have been foreshortened to approximately $81/100$ of their length, as seen in Fig. 6. An isometric scale to this proportion can be made graphically, as shown in Fig. 7, if it is necessary to make an isometric projection instead of an isometric drawing.

An isometric drawing is not difficult to begin. If the object is rectangular (Fig. 8), start with a point representing a front corner, shown at A with heavy lines, and draw from it the three isometric axes 120° apart: one vertical, as at B; the other two with the 30° triangle. On

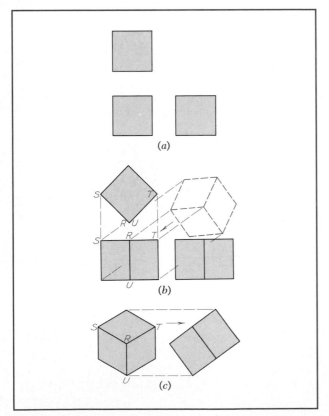

FIG. 5 The isometric cube. Rotated from position *(a)* to *(b)* then to *(c)*, the three perpendicular edges are now equally foreshortened.

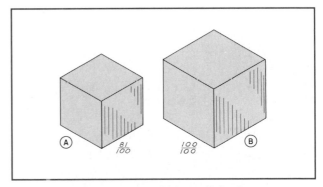

FIG. 6 *(A)* Isometric projection; *(B)* isometric drawing.

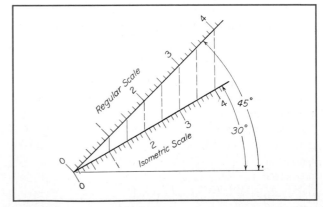

FIG. 7 To make an isometric scale.

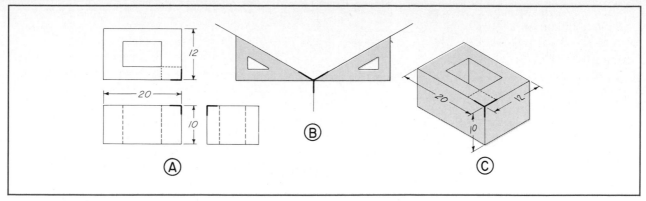

FIG. 8 METRIC. Isometric axes.

these three lines measure the height, width, and depth of the object, as indicated at C. Through the points so determined draw lines parallel to the axes, completing the figure. Hidden lines are omitted except when they are needed to describe the piece.

6. NONISOMETRIC LINES

Edges whose projections or drawings are not parallel to one of the isometric axes are called "nonisometric lines." The one important rule is that measurements can be made only on the drawings of *isometric* lines; conversely, measurements *cannot* be made on the drawings of *nonisometric* lines. For example, the diagonals of the face of a cube are nonisometric lines; although equal in length, their isometric drawings will not be at all of equal length on the isometric drawing of the cube.

There are two basic methods for drawing nonisometric lines: the boxing method and the offset method.

A. Boxing Method When an object contains many nonisometric lines, it is drawn by the "boxing method" or the "offset method." When the boxing method is used, the object is enclosed in a rectangular box, which is drawn around it in orthographic projection. The box is then drawn in isometric and the object located in it by its points of contact, as in Fig. 9. It should be noted that the isometric views of lines that are parallel on the object are parallel. This knowledge can often be used to save a large amount of construction, as well as to test for accuracy. Figure 9 might be drawn by putting the top face into

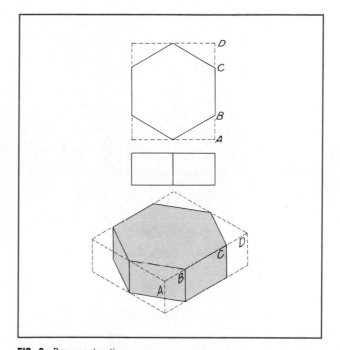

FIG. 9 Box construction.

isometric and drawing vertical lines equal in length to the edges downward from each corner. It is not always necessary to enclose the whole object in a rectangular "crate." The pyramid (Fig. 10) would have its base enclosed in a rectangle and the apex located by erecting a vertical axis from the center.

B. Offset Method When an object is made up of planes at different angles, it is better to locate the ends of

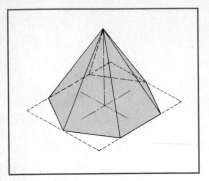

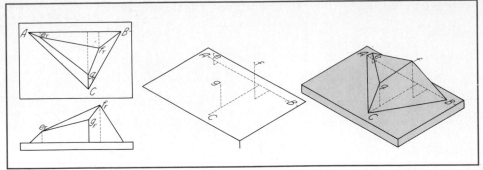

FIG. 10 Semibox construction. Points on the base are transferred by boxing.

FIG. 11 Offset construction.

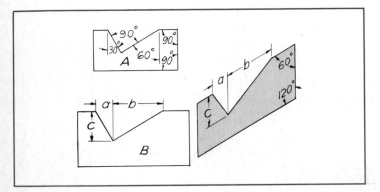

FIG. 12 Angles in isometric.

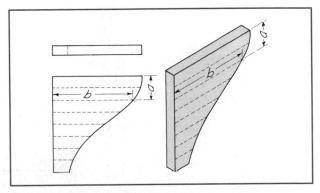

FIG. 13 Curves in isometric.

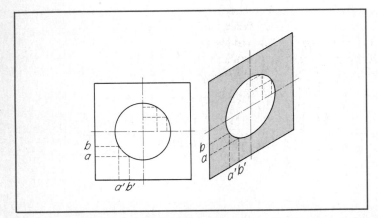

FIG. 14 Isometric circle, points plotted.

the edges by the offset method rather than by boxing. When the offset method is used, perpendiculars are extended from each point to an isometric reference plane. These perpendiculars, which are isometric lines, are lo-

cated on the drawing by isometric coordinates, the dimensions being taken from the orthographic views. In Fig. 11, line *AB* is used as a base line and measurements are made from it as shown, first to locate points on the base; then verticals from these points locate *e*, *f*, and *g*.

7. ANGLES IN ISOMETRIC

The three isometric axes, referred back to the isometric cube, are mutually perpendicular but in an isometric drawing appear at 120° to each other. For this reason, angles specified in degrees do not appear in their true size on an isometric drawing and must be laid off by coordinates that will be parallel to the isometric axes. Thus if an orthographic drawing has edges specified by angular dimensions, as in Fig. 12A, *a view to the same scale as the isometric drawing* is made as at *B*; from this view the coordinate dimensions *a*, *b*, and *c* are transferred with dividers or scale to the isometric drawing.

8. CURVES IN ISOMETRIC

A circle or any other curve will not show in its true shape when drawn in isometric. A circle on any isometric plane will be an ellipse, and a curve will be shown as the isometric projection of the true curve.

Any curve can be drawn by plotting points on it from isometric reference lines (coordinates) that are parallel to the isometric axes, as shown in Fig. 13. A circle plotted in this way is shown in Fig. 14. Note that in both these figures coordinates *a* and *b* are parallel to the isometric axes and the coordinate distances must be obtained from an orthographic view drawn to the same scale as the isometric.

A. Isometric Circles Circles occur so frequently that they are usually drawn by a four-centered approximation, which is sufficiently accurate for ordinary work. Geometrically, the center for any arc tangent to a straight line lies on a *perpendicular from the point of tangency* (Fig. 15A).

In isometric, if perpendiculars are drawn from the middle point of each side of the circumscribing square, the intersections of these perpendiculars will be centers for arcs tangent to two sides, Fig. 15B. Two of these intersections will evidently fall at the corners A and C of the isometric square, because the perpendiculars are altitudes of equilateral triangles. Thus the construction at B to D can be made by simply drawing 60° lines (horizontals also at C and D) from the corners A and C and then arcs with radii R and R_1, as shown.

If a true ellipse is plotted as in Fig. 14 in the same square, it will be a little longer and narrower than this four-center approximation, but in most drawings the difference is not sufficient to warrant the extra expenditure of time required in execution.

The isometric drawing of a *sphere* is a circle with its diameter equal to the long axis of the ellipse that is inscribed in the isometric square of a great circle of the sphere. It would thus be 1.23/1.00 of the actual diameter (the isometric *projection* of a sphere would be a circle of the actual diameter of the sphere).

B. Isometric Circle Arcs To draw any circle arc, draw the isometric square of its diameter in the plane of its face, with as much of the four-center construction as is necessary to find centers for the part of the circle needed, as illustrated in Fig. 16. The arc occurring most frequently is the quarter circle. Note that in illustrations D and E only two construction lines are needed to find the center of a quarter circle in an isometric plane. Measure the true radius R of the circle from the corner on the two isometric lines as shown, and draw *actual* perpendiculars from these points. Their intersection will be the required center for radius R_1 or R_2 of the isometric quadrant. Figure 16F illustrates the construction for the two vertical isometric planes.

9. REVERSED ISOMETRIC

It is often desirable to show the lower face of an object by tilting it *back* instead of *forward*, thus reversing the usual position so as to show the underside. The construction is the same as when the top is shown. Figure 17 shows the reference cube and the position of the axes, as well as the application of reversed-isometric construction to circle arcs.

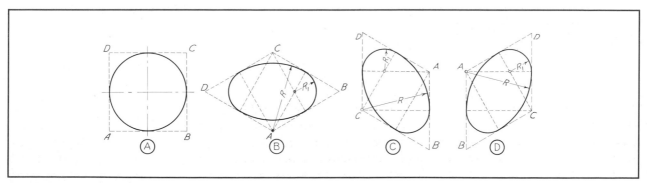

FIG. 15 Isometric circles, four-centered method.

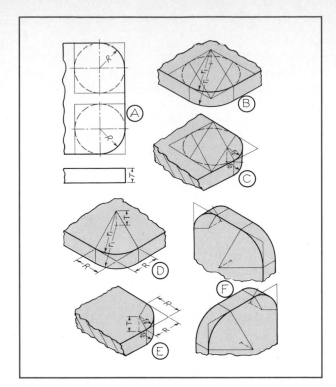

FIG. 16 Isometric quarter circles.

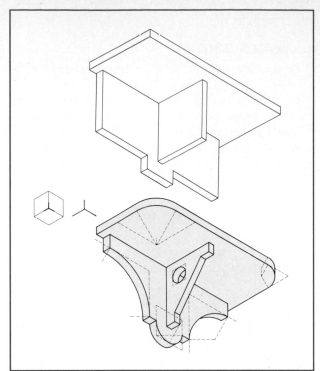

FIG. 17 Isometric with reversed axes. The bottom and two sides are shown. Construction methods are the same as for regular position.

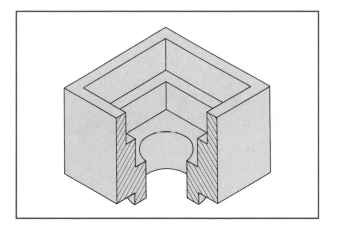

FIG. 18 Isometric half section. One-fourth of the object is removed to reveal interior construction.

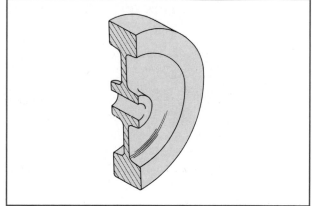

FIG. 19 Isometric full section. Half of object is removed to reveal object shape.

10. ISOMETRIC SECTIONS

Isometric drawings are, from their pictorial nature, usually outside views, but sometimes a sectional view is used to good advantage to show a detail of shape or interior construction. The cutting planes are taken as isometric planes, and the section lining is done in the direction that gives the best effect; this is, in almost all cases, the direction of the long diagonal of a square drawn on the surface. As a general rule, a half section is made by outlining the figure in full and then cutting out the front quarter, as in Fig. 18; for a full section, the cut face is drawn first and then the part of the object behind it is added (Fig. 19).

11. DIMETRIC PROJECTION

The reference cube can be rotated into any number of positions in which two edges are equally foreshortened, and the direction of axes and ratio of foreshortening for any one of these positions might be taken as the basis for a system of dimetric drawing. A simple dimetric position is one with the ratios 1 to 1 to ½. In this position the tangents of the angles are ⅛ and ⅞, making the angles approximately 7° and 41°. Figure 20 shows a drawing in this system. Dimetric is seldom used because of the difficulty of drawing circles in this projection.

12. TRIMETRIC PROJECTION

Any position in which all three axes are unequally foreshortened is called "trimetric." Compared with isometric and dimetric, distortion is reduced in trimetric projection, and even this effect can be lessened with some positions. However, because it is slower to execute than is isometric or dimetric, it is seldom used except when done by projection.

13. OBLIQUE PROJECTION

Oblique projection can yield three-dimensional drawings just as can isometric projection. However, the process for creating an oblique projection is virtually the op-

posite of that for making an isometric projection. Remember that in making an isometric projection, one looks perpendicularly through the projection plane at an object which has been rotated and tilted.

To create an oblique projection the object is not rotated and tilted. Rather, the object remains with a major face parallel to the projection plane, as in Fig. 21. Now, instead of looking perpendicularly through the projection plane as in isometric projection, one looks through the projection plane with any *oblique* angle.

As Fig. 21 shows, the result is a three-dimensional

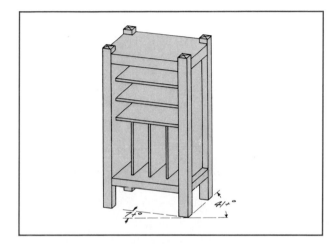

FIG. 20 Dimetric drawing. It is used principally for rectangular objects.

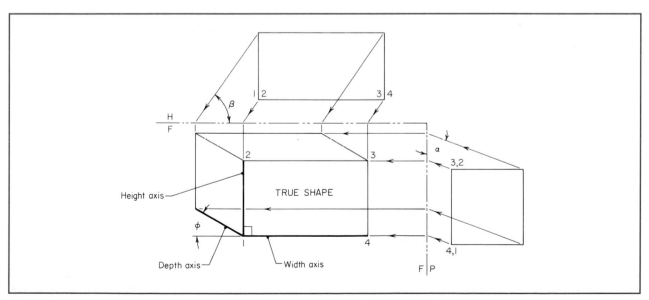

FIG. 21 Oblique projection.

pictorial. The axes of height and width remain at 90° to each other. Any plane that is parallel to the frontal projection plane will keep its true shape, such as plane 1-2-3-4. This true-shape feature is the chief virtue of oblique projection. Therefore, any curve, circle, or irregular shape in a plane parallel to the frontal plane will keep its shape, which can greatly aid in the drawing of oblique pictorials.

The axis of depth seen in Fig. 21 can vary in angle ϕ and in length, depending on the particular oblique angle originally selected (which in Fig. 21 gave projected angles α and β). Angle ϕ and the length of the depth axis can be virtually anything, but in practice angle ϕ is usually set at 30 or 45° because each is easy to use.

The length of the depth axis is usually set to be one of two possibilities: one-half the scale used on the height and width axes (called "cabinet" drawing) or the same scale as used on the height and width axes (called "cava-lier" drawing). Cabinet drawing is considered to have a more pleasing appearance in many cases than does cavalier drawing.

Figure 22 compares the separate effects of cabinet vs. oblique drawing. The object is the same for both Fig. 22A and B. Only the ratio varies for depth to height to width for the axis scales. For cabinet oblique, the ratio is ½ to 1 to 1; for cavalier oblique the ratio is 1 to 1 to 1. Do you think the cabinet oblique is more pleasing?

Cabinet drawing is popular because of the easy ratio, but the effect is often too thin. Other oblique drawing ratios, such as 2 to 3 or 3 to 4, may be used with pleasing effect.

14. TO MAKE AN OBLIQUE DRAWING

Oblique drawing is similar to isometric drawing in that it has three axes which represent three mutually perpendicular edges and upon which measurements can be made. To draw a rectangular object (Fig. 23) start with a point representing a front corner (A) and draw from it the three oblique axes, one vertical, one horizontal, and one at an angle. On these three axes measure the height, width, and depth of the object. In this case the width is made up of the 65-mm distance and the 35-mm radius. Locate the center of the arc, and draw it as shown. The center for the arc of the hole in the figure will be at the same point as the center for the outside arc on the front face. The center for the rear arc of the hole will be 30 mm rearward on a depth-axis line through the front center.

15. OBJECT ORIENTATION FOR OBLIQUE

Any face parallel to the picture plane will evidently be projected without distortion. In this, oblique projection has an advantage over isometric that is of particular value in representing objects with circular or irregular outline.

The *first rule* for oblique projection is to *place the object with the irregular outline or contour parallel to the picture plane*. Note in Fig. 24 the greater distortion at (*b*) and (*c*) than at (*a*).

One of the greatest disadvantages in the use of isometric or oblique drawing is the effect of distortion produced by the lack of convergence in the receding lines—a violation of perspective. In some cases, particularly with

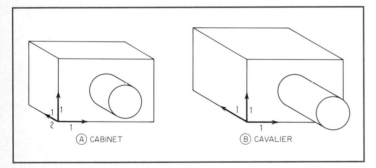

FIG. 22 Obliques: cabinet vs. cavalier.

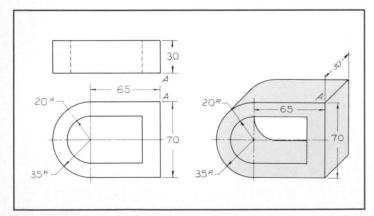

FIG. 23 **METRIC.** Oblique drawing.

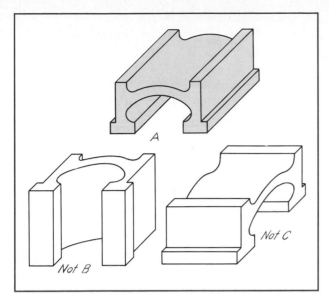

FIG. 24 Illustration of the first rule.

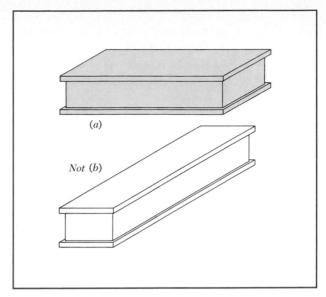

FIG. 25 Illustration of the second rule. Note the exaggerated depth at (b).

large objects, this becomes so painful as practically to preclude the use of these methods. This is perhaps even more noticeable in oblique than in isometric and of course increases with the length of the depth dimension.

Hence the *second rule: preferably, the longest dimension should be parallel to the picture plane.* In Fig. 25, (a) is preferable to (b).

16. STARTING PLANE

Note that as long as the front of the object is in one plane parallel to the plane of projection, the front face of the oblique projection is *exactly the same as in the orthographic front view.* When the front is made up of more than one plane, take care to preserve the relationship between the planes by selecting one as the starting plane and working from it. In a piece such as the link in Fig. 26, the front bosses can be imagined as cut off on the plane A-A, and the front view, that is, the section on A-A, drawn as the front of the oblique projection. Then lay off depth axes through the centers C and D, the distances, for example, CE behind and CF in front of the plane A-A.

When an object has no face perpendicular to its base, it can be drawn in a similar way by cutting a right section

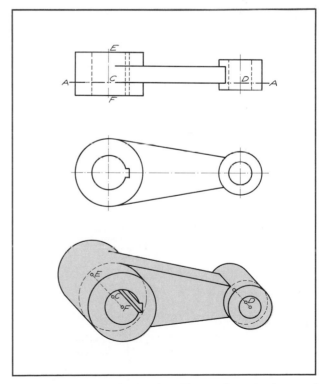

FIG. 26 Offsets from reference plane. Distances forward and rearward are measured from the fontal plane.

and measuring offsets from it, as in Fig. 27. This offset method, previously illustrated in the isometric drawings in Figs. 11, 12 and 13, is a rapid and convenient way of drawing almost any figure, and it should be studied carefully.

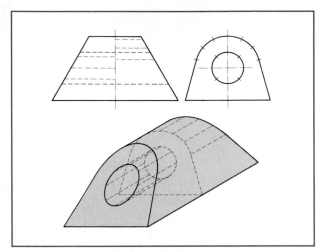

FIG. 27 Offsets from right section. Measurements forward and rearward are made from the frontal plane.

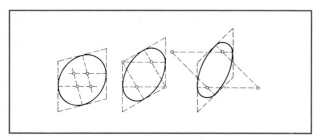

FIG. 28 Oblique circle construction. Note that the tangent points of arcs must be at the midpoints of the enclosing oblique square.

17. CIRCLES AND ARCS IN OBLIQUE

When it is necessary to draw circles that lie on oblique faces, they can be drawn as circle arcs, with the compasses, on the same principle as the four-center isometric approximation shown in Fig. 15. In isometric it happens that two of the four intersections of the perpendiculars from the middle points of the containing square fall at the corner of the square, and advantage is taken of the fact. In oblique, the position of the corresponding points depends on the angle of the depth axis. Figure 28 shows three squares in oblique positions at different angles and the construction of their inscribed circles. The important point to remember is that the circle arcs *must* be tangent at the midpoints of the sides of the oblique square.

Circle arcs representing rounded corners, etc., are drawn in oblique by the same method given for isometric arcs in Sec. 8B. The only difference is that the angle of the sides tangent to the arc will vary according to the angle of the depth axis chosen.

18. CONCEPTS OF PERSPECTIVE DRAWING

Perspective drawing represents an object as it appears to an observer stationed at a particular position relative to it. The object is seen as the figure resulting when visual rays from the eye to the object are cut by a picture plane. In a technical way, perspective is used more in architecture and in illustration than in other fields, but every engineer will find it useful to know the principles of the subject.

Imagine an observer standing on the sidewalk of a city street, as in Fig. 29, with the picture plane erected between him and the street scene ahead. Visual rays from

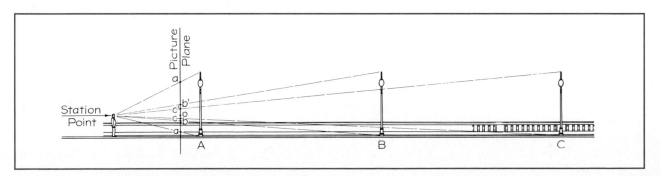

FIG. 29 Theory of perspective illustrated.

FIG. 30 The perspective drawing. This is the image formed on the picture plane of Fig. 29.

his eye to the ends of lamppost *A* intercept a distance *aa'* on the picture plane. Similarly, rays from post *B* intercept *bb'*, a smaller distance than *aa'*. This apparent decrease in the size of like objects as the distance from the objects to the eye increases agrees with our everyday experience and is the keynote of perspective drawing. It is evident from the figure that succeeding lampposts will intercept shorter distances on the picture plane than the preceding ones, and that a post at infinity would show only as a point *O* at the level of the observer's eye.

In Fig. 30 the plane of the paper is the picture plane, and the intercepts *aa'*, *bb'*, etc., show as the heights of the respective lampposts as they diminish in their projected size and finally disappear at a point on the horizon. In a similar way the curbings appear to converge at the same point *O*. Thus a system of parallel horizontal lines will vanish at a single point on the horizon, and all horizontal planes will vanish on the horizon. Verticals such as the lampposts and the edges of the buildings, being parallel to the picture plane, pierce the picture plane at an infinite distance and therefore show as vertical lines in the picture.

19. DEFINITIONS

Figure 31 illustrates perspective theory and names the points, lines, and planes used. An observer in viewing an object selects his or her *station point* and thereby determines the *horizon plane*, as the horizontal plane is at eye level. This horizon plane is normally above the horizon-

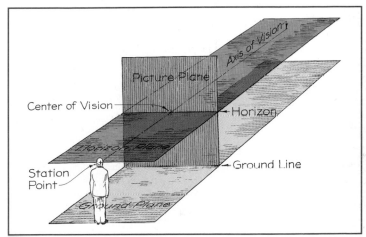

FIG. 31 Perspective nomenclature.

tal *ground plane* upon which the object is assumed to rest. The *picture plane* is usually located between the station point and the object being viewed and is ordinarily a vertical plane perpendicular to the horizontal projection of the line of sight to the object's center of interest. The *horizon line* is the intersection of the horizon plane and picture plane, and the *ground line* is the intersection of the ground plane and picture plane. The *axis of vision* is the line through the station point which is perpendicular to the picture plane. The piercing point of the axis of vision with the picture plane is the *center of vision*.

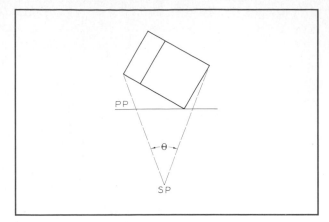

FIG. 32   Lateral angle of view.

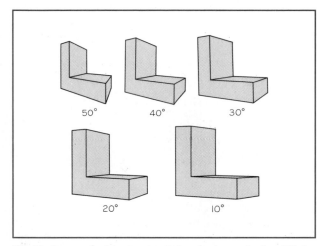

FIG. 33   Comparative lateral angles of view. Angles greater than 30° give an unpleasing perspective.

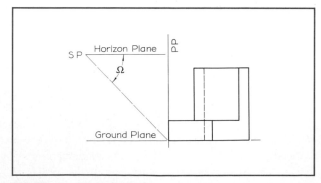

FIG. 34   Elevation angle of view.

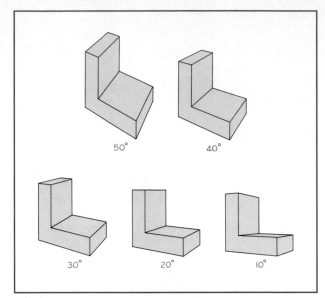

FIG. 35   Comparative elevation angles of view. Angles greater than 30° give an unpleasing perspective.

20. SELECTION OF THE STATION POINT

In beginning a perspective drawing, take care in selecting the station point, as an indiscriminate choice may result in a distorted drawing. If the station point is placed to one side of the drawing, the same effect is obtained as when a theater screen is viewed from a position close to the front and well off to one side: heights are seen properly but not horizontal distances. Therefore, *the center of vision should be somewhere near the picture's center of interest.*

Wide angles of view result in a violent convergence of horizontal lines and so should be avoided. The angle of view is the included angle θ between the widest visual rays (Fig. 32). Figure 33 shows the difference in perspective foreshortening for different lateral angles of view. In general, an angle of about 20° gives the most natural picture.

The station point should be located at the point from which the object is seen to best advantage. For this reason, for large objects such as buildings, the station point is usually taken at a normal standing height of about 5 ft above the ground plane; for small objects, the best representation demands that the top, as well as the lateral surfaces, be seen, and the station point must be elevated accordingly. Figure 34 shows the angle of elevation Ω between the horizon plane and the extreme visual ray. By illustrating several different angles of elevation (Ω), Fig. 35 shows the effect of elevation of the station point. In general, the best picturization is obtained at an angle of about 20° to 30°.

21. PERSPECTIVE BY PROJECTION

Perspective projection is based on the theory that visual rays from the object to the eye pierce the picture plane and form an image of the object on the plane. Thus in Fig. 36, the image of line YZ is formed by the piercing points y and z of the rays. Several projective methods may be used. The simplest method, basically, but the most laborious to draw, is illustrated by the purely orthographic method of Fig. 37, in which the top and side views are drawn in orthographic. The picture plane (edge view) and the station point are located in each

view. Assuming that the line YZ in Fig. 36 is one edge of the L-shaped block in Fig. 37, visual rays from Y and Z will intersect the picture plane in the top view, thus locating the perspective of the points laterally.

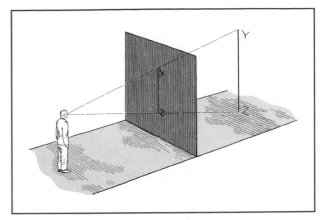

FIG. 36 Perspective of a line.

FIG. 37 Perspective drawing (orthographic method). Points are plotted from the intersection of rays with the picture plane.

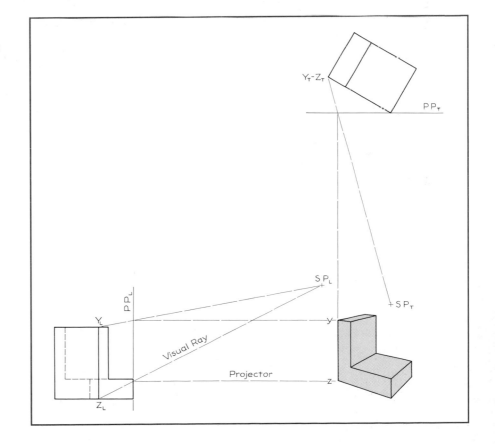

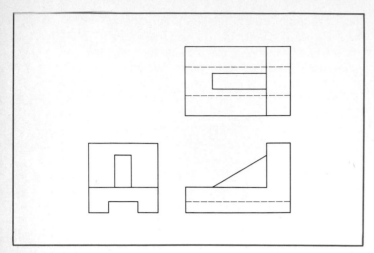

FIG. 38   Sliding block. This object is drawn in perspective in Fig. 39.

Similarly, the intersections of the rays in the side view give the perspective heights of Y and Z. Projection from the top and side views of the picture plane gives the perspective of YZ, and a repetition of the process for the other lines will complete the drawing. Note that *any* point such as Y or Z can be located on the perspective, and thus the perspective is actually plotted, by projection, point after point.

22. PERSPECTIVE BY USE OF VANISHING POINTS

The steps required to make a perspective of the sliding block in Fig. 38 are as follows. The edge view of the picture plane (plan view) is drawn (Fig. 39), and behind it the top view of the object is located and drawn. In this case, one side of the object is oriented at 30° to the pic-

FIG. 39   Use of vanishing points and measuring lines.

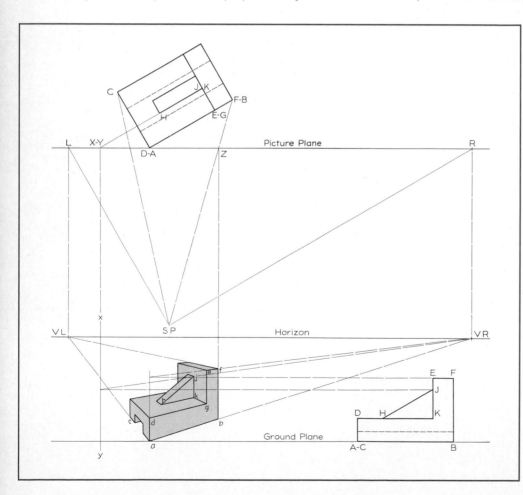

ture plane in order to emphasize the L shape more than the end of the block. The station point is located a little to the left of center and far enough in front of the picture plane to give a good angle of view. The ground line is then drawn, and on it is placed the front view of the block from Fig. 38. The height of the station point is then decided—in this case, well above the block so that the top surfaces will be seen—and the horizon line is drawn at the station-point height.

To avoid the labor of redrawing the top and front views in the positions just described, the views can be cut from the orthographic drawing, oriented in position, and fastened with tacks or tape.

To visualize the location of a vanishing point, imagine that a horizontal line, such as AB in Fig. 39, is extended back an infinite distance toward the horizon. We know that infinitely long horizontal lines disappear across the horizon. To see the end of an infinitely long line, we would, in fact, look parallel to it.

Therefore, to see the vanishing point of horizontal line AB, we look parallel to it. Where our sight pierces the picture plane at point R is where the vanishing point occurs. Point R is then projected to the horizon line, locating VR (vanishing point right). Point VR is the vanishing point not only for line AB but for all lines parallel to AB, since all parallel lines vanish at the same point. The vanishing point VL (vanishing point left) for line AC and lines parallel to AC is found similarly, as shown.

Point A lies in both the picture plane and the ground plane and will therefore be shown in the perspective at a, on the ground line, and in direct projection with the top view. The perspective of AB is determined by drawing a line from a to VR and then projecting the intercept Z (of the visual rays SP to B) to the line, thus locating b.

All lines behind the picture plane are foreshortened in the picture, and only those lying in the picture plane will appear in their true length. For this reason, all measurements must be made in the picture plane. Since AD is in the picture plane, it will show in its actual height as ad.

A *measuring line* will be needed for any verticals such as BF that do not lie in the picture plane. If a vertical is brought forward to the picture plane along some established line, the true height can be measured in the picture plane. If, in Fig. 39, BF is imagined as moved forward along ab until b is in coincidence with a, the true

height can be measured vertically from a. This vertical line at a is then the measuring line for all heights in the vertical plane containing a and b. The height of f is measured from a, and from this height point, a vanishing line is drawn to VR. Then from Z (the piercing point in the picture plane of the visual ray to F), f can be projected to the perspective.

The measuring line can also be thought of as the intersection of the picture plane with a vertical plane that contains the distance to be found. Thus ad, extended, is the measuring line for all heights in surface ABFEGD. The triangular rib in Fig. 39 is located by continuing surface HJK until it intersects the picture plane at XY, thereby establishing xy as the measuring line for all heights in HJK. In the figure, the height of J is measured on the measuring line xy, and j is found as described for f.

Note that heights can be measured with a scale on the measuring line or they can be projected from the front view, as indicated in Fig. 39.

23. PERSPECTIVE FOR PLANES PARALLEL TO THE PICTURE PLANE

Objects with circles or other curves in a vertical plane can be oriented with their curved faces parallel to the picture plane. The curves will then appear in true shape. This method, often called "parallel perspective," is also suitable for interiors and for street vistas and similar scenes where considerable depth is to be represented.

The object in Fig. 40 has been placed so that the planes containing the circular contours are parallel to the picture plane. The horizontal edges parallel to the picture plane will appear horizontal in the picture and will have no vanishing point. Horizontals perpendicular to the picture plane are parallel to the axis of vision and will vanish at the center of vision CV. Except for architectural interiors, the station point is usually located above the object and either to the right or left, yet not so far in any direction as to cause unpleasant distortion. For convenience, one face of the object is usually placed in the picture plane and is therefore not reduced in size in the perspective.

In Fig. 40, the end of the hub is in the picture plane; thus the center o is projected from O in the top view, and the circular edges are drawn in their true size. The cen-

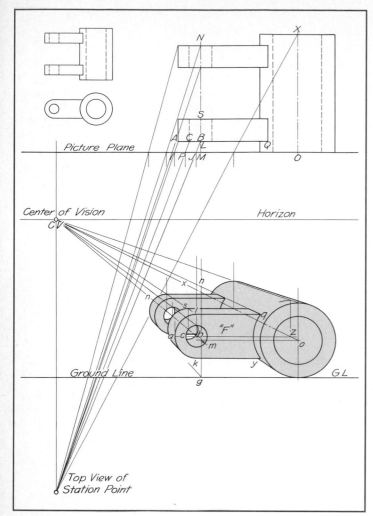

FIG. 40 Planes parallel to the picture plane. Compare the position of this object with that of Fig. 39.

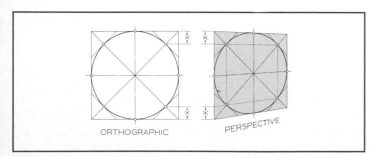

FIG. 41 Perspective of a circle.

ter line *ox* is vanished from *o* to CV. To find the perspective of center line *MN*, a vertical plane is passed through *MN* intersecting the picture plane in measuring line *gh*. A horizontal line from *o* intersecting *gh* locates *m*, and *m* vanished to CV is the required line.

By using the two center lines from *o* and *m* as a framework, the remaining construction is simplified. A ray from the station point to *B* pierces the picture plane at *J*, which, projected to *mn*, locates *b*. The horizontal line *bz* is the center line of the front face of the nearer arm, and the intercept *IJ* gives the perspective radius *ab*. The circular hole having a radius *CB* has an intercept *PJ*, giving *cb* as the perspective radius. The arc *qy* has its center on *ox* at *z*. On drawing the tangents *lq* and *ky*, the face "F" is completed.

The remaining construction for the arms is exactly the same as that for "F." The centers are moved back on the center lines, and the radii are found from their corresponding intercepts on the picture plane.

24. CIRCLES AND CURVES IN PERSPECTIVE

The perspective of a circle is a circle only when its plane is parallel to the picture plane. In all other positions the circle projects as an ellipse whose major and minor diameters are not easily found. It should be noted that in all cases the center of the circle is not coincident with the center of the ellipse representing the circle and that concentric circles are not represented by concentric ellipses. The major and minor diameters of the ellipses for concentric circles are not even parallel except in special cases.

The perspective of a circle can be plotted point by point, but the most rapid solution is obtained by enclosing the circle in a square, as shown in Fig. 41, and plotting points at the tangent points and at the intersections of the diagonals. The eight points thus determined are usually sufficient to give an accurate curve. The square, with its diagonals, is first drawn in the perspective. From the intersection of the diagonals, the vertical and horizontal center lines of the circle are established; where these center lines cross the sides of the square are four points on the curve. In the orthographic view, the measurement X is made, then laid out *in the picture plane* and vanished, crossing the diagonals at four additional points.

The perspectives of irregular curves can be drawn by projecting a sufficient number of points to establish the curve, but if the curve is complicated, the method of graticulation may be used to advantage. A square grid is overlaid on the orthographic view, as shown in Fig. 42; then the grid is drawn in perspective and the outlines of the curve are transferred by inspection from the orthographic view.

25. MEASURING POINTS

It has been shown that all lines lying in the picture plane can be scaled directly on the perspective drawing. The adaptation of this principle has an advantage in laying off a series of measurements, such as a row of pilasters, because it avoids a confusion of intercepts on the picture plane and the inaccuracies due to long projection lines.

In the measuring points method, a surface, such as the wall between A and B in Fig. 43, is rotated into the picture plane for the purpose of making measurements, as shown at AB'. While in the picture plane, the entire surface can be laid out directly to the same scale as the top view; therefore, ab' and other horizontal dimensions of the surface are established along the ground line as shown. The counterrotation of the wall to its actual position on the building and the necessary projections in the perspective are based on the principle that the rotation has been made about a vertical axis and that any point has traveled in a horizontal plane. By drawing, as usual, a line parallel to BB', from the station point to the picture plane, and then projecting to the horizon, the vanishing point MR is found. This vanishing point is termed a *measuring point* and may be defined as the vanishing point for lines joining corresponding points of the actual and rotated positions of the face considered. The divisions on ab' are therefore vanished to MR; where this construction intersects ab (the perspective of AB), the lateral position of the pilasters in the perspective is determined. Heights are scaled on the vertical edge through a, since this edge lies in the picture plane. The perspective of the wall between A and B is completed by the regular methods previously described. For work on the end of the building, the end wall is rotated as indicated, measuring point ML is found, and the projections are continued as described for the front wall.

Measuring points can be more readily located if the drafter recognizes that the triangles ABB' and R-O-SP

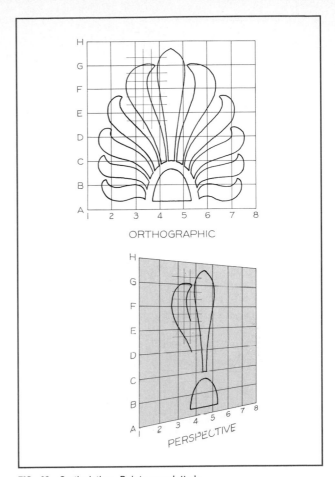

FIG. 42 Graticulation. Points are plotted.

are similar. Therefore, a measuring point is as far from its corresponding vanishing point as the station point is from the picture plane, measuring the latter parallel to the face concerned. MR can then be found by measuring the distance from the station point to R and laying off RO equal to the measurement, or by swinging an arc, with R as center, from the station point to O, as shown. The measuring point MR is then projected from O.

26. INCLINED LINES IN PERSPECTIVE

Any line that is not parallel or perpendicular to either the picture plane or the horizon plane is termed an inclined line. Any line may have a vertical plane passed through it, and if the vanishing line of the plane is found, a line

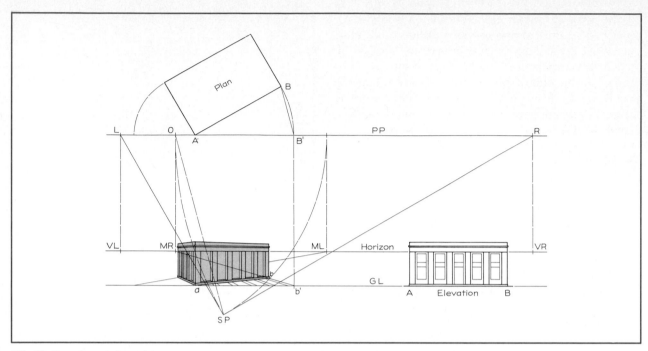

FIG. 43 Use of measuring points.

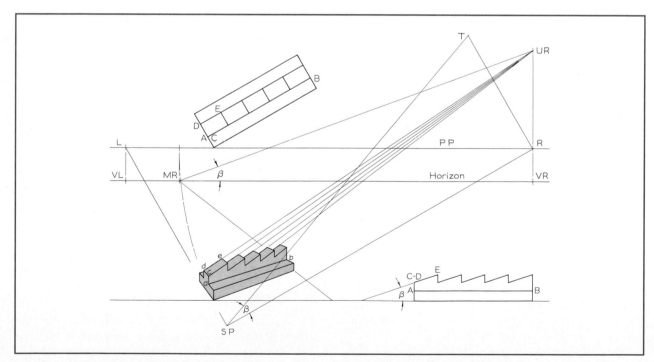

FIG. 44 Vanishing point of inclined lines. This simplifies construction when there are many parallel inclined lines.

in the plane will vanish at some point on the vanishing line of the plane. Vertical planes will vanish on vertical lines, just as horizontal planes vanish on a horizontal line, the horizon.

In Fig. 44, the points *a* to *e* have all been found by regular methods previously described. The vanishing point of the horizontal *ab* is VR. The vertical line through VR is the vanishing line of the plane of *abc* and all planes *parallel to abc*. This vanishing line is intersected by the extension of *de* at UR, thereby determining the vanishing point for *de* and all edges *parallel to de*.

The vanishing point for inclined lines can also be located on the theory that the vanishing point for any line can be determined by moving the line until it appears as a point, while still retaining its original angle with the picture plane. The vanishing point of *de* can therefore be located by drawing a line through the station point parallel to *DE* and finding its piercing point with the picture plane. This is done by laying out SP T at the angle β to SP R and erecting RT perpendicular to SP R. Then RT is the height of the vanishing point UR above VR.

27. INCLINED PLANES IN PERSPECTIVE

An inclined plane is any plane not parallel or perpendicular to either the picture plane or the horizon plane. The vanishing line for an inclined plane can be found by locating the vanishing points for any two systems of parallel lines in the inclined plane. To determine the vanishing line of plane ABCD in Fig. 45, the vanishing point VL of the horizontal edges AD and BC is one point, and the vanishing point UR for the inclined edges AB and DC gives a second point on the vanishing line VL UR for plane ABCD.

It is often necessary to draw the line of intersection of two inclined planes. The intersection will vanish at the point of intersection of the vanishing lines of both planes. The intersection J of the two vanishing lines of the roof planes in Fig. 45 is the vanishing point of the line of intersection of the two planes.

28. PICTORIAL SKETCHING

In designing and inventing, the first ideas come into the mind in pictorial form, and sketches made in this form

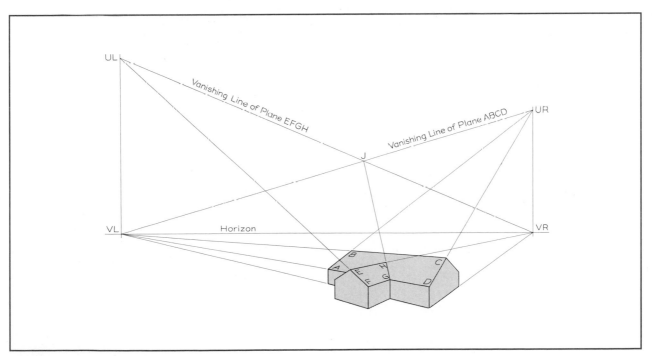

FIG. 45 Vanishing lines for inclined planes and vanishing point for the line of intersection of two inclined planes.

preserve the ideas as visualized. From this record the preliminary orthographic design sketches are made. A pictorial sketch of an object or of some detail of construction can often be used to explain it to a client or worker who cannot read the orthographic projection intelligently. One of the best ways of reading a working drawing that is difficult to understand is to start a pictorial sketch of it.

It should be clearly understood at the outset that pictorial sketching means the making of a pictorial drawing *freehand*. The same construction that is used for locating points and lines and for drawing circles and arcs with instruments will be used in pictorial sketching. From this standpoint, a knowledge of the constructions already given is necessary before attempting pictorial sketching.

Although this is not a complete classification, there may be said to be three pictorial methods of sketching: axonometric, oblique, and perspective. The mechanical construction has been explained in detail.

29. CHOICE OF VIEWS IN PICTORIAL SKETCHING

After a clear visualization of the object, the first step is to select the type of pictorial sketch—axonometric, oblique, or perspective—to be used.

Isometric is the simplest axonometric position, and it will serve admirably for representing most objects. Although dimetric or trimetric may be definitely advantageous for an object with some feature that is obscured or misleading in isometric, it is best to try isometric first.

Oblique forms (cavalier or cabinet) may be used to advantage for cylindrical objects or for objects with a number of circular features in parallel planes. Nevertheless, a true circle, representing a circular feature parallel to the picture plane in oblique, is much harder to sketch than an ellipse, representing the same feature in isometric, because the slightest deviation from a circle is evident, whereas the same deviation in an ellipse is unnoticed.

Perspective is the best form for pictorial sketching because it is free from any distortion. A perspective is not much more difficult to sketch than an axonometric or an oblique, but attention must be paid to the convergence of the lines and to keeping good proportion.

Choose carefully the *direction* in which the object is to be viewed. There are many possibilities. The object may be turned so that any lateral face will be represented on the right or the left side of the pictorial. Orient the object so that the two *principal* faces will show to advantage.

30. SKETCHING THE AXES

After the type of pictorial and the position of the object have been decided upon, the first step in making the sketch is to draw the axes.

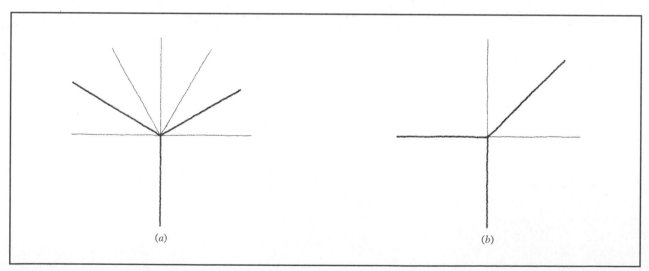

(a)   *(b)*

FIG. 46 Locating the axes. *(a)* Isometric; *(b)* oblique at 45°.

In isometric, the three axes should be located as nearly as possible 120° from one another (one vertical and two at 30° with the horizontal). Because no triangles are used in sketching, the angles must be located by judgment. Figure 46A shows a satisfactory method of judging the position of the lines. First draw a *light* horizontal and vertical and then divide both upper quadrants into thirds. The lines at the top of the lower thirds are then the two axes at 30° to the horizontal, and the third axis is the vertical. This method is simple and accurate because it is easy to estimate equal thirds of a quadrant.

In dimetric, the standard angles are 7° and 41°. For the 7° axis, again referring to Fig. 46A, draw the bisector of the 30° axis for isometric to get 15°, and then bisect again to get 7½°, which will be quite satisfactory and can be done fairly accurately. For the 41° axis, take the midline of the second 30° section, which is 45°, and shade it a little to approximate 41°.

In trimetric, almost any combination representing three mutually perpendicular lines is possible. However, a pronounced distortion will occur if the two transverse axes are more than 30° from the horizontal.

In oblique, the common angles are 30° and 45° although theoretically any angle is possible. To prevent violent distortion, never make the depth axis greater than 45°. To locate a 45° axis, sketch a vertical and horizontal as in Fig. 46b, and then sketch the bisector of the quadrant (see oblique positions) where the axis is wanted. For an axis at 30°, proceed as in Fig. 46a for isometric.

For perspective, take care to have a reasonable included angle of view; otherwise a violent convergence will occur. In order to prevent difficulty, first locate two vanishing points, as in Fig. 47, as widely separated as the paper will allow (attach extra paper with transparent tape if necessary). Then sketch the *bottom* edges of the object (the heavier lines in Fig. 47) and on these sketch a rectangular shape, as shown. It will be immediately evident either that the arrangement is satisfactory or that the vanishing points must be moved *or* that the base lines must be altered. Remember that the two vanishing points *must* be at the same level, the horizon.

31. SKETCHING THE PRINCIPAL LINES

Almost without exception, the first lines sketched should be those that box in the whole object, or at least its major

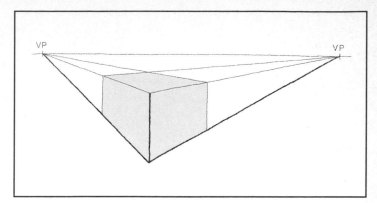

FIG. 47 Perspective layout with vanishing points.

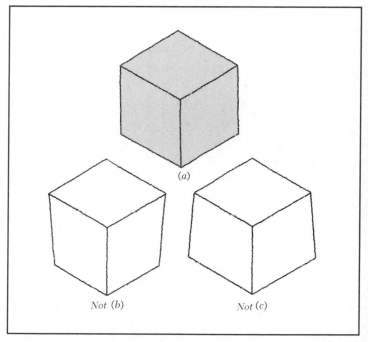

FIG. 48 Sketching vertical lines. These must be accurately vertical to define object shape.

portion. These first lines are all-important to the success of the sketch, for mistakes at the outset are difficult to correct later. Observe carefully the following three points:

1. *Verticals must be parallel to the vertical axis.* In Fig. 48 a cube is sketched correctly at (*a*). At (*b*) and (*c*) the same cube is sketched but at (*b*) the

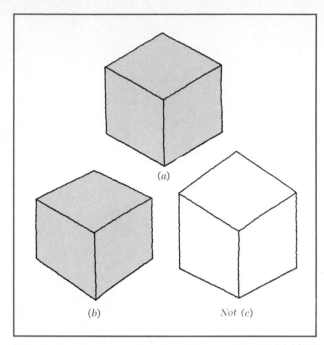

FIG. 49 Sketching the receding edges. These must be parallel *(a)*; or converging *(b)*; *never* separating *(c)*.

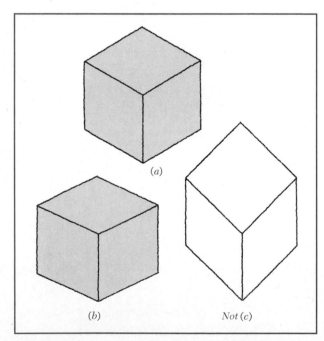

FIG. 50 Angle of axes. Isometric position *(a)* or flattened *(b)* gives a natural appearance. Distortion is inherent in steep axes as at *(c)*.

verticals converge downward and at (*c*), upward. It is evident that (*b*) and (*c*) do not look cubical at all, but like frustums of pyramids. This is proof that verticals *must be kept accurately vertical*. Be critical of the verticals throughout the construction. Accurate verticals add a stability and crispness not attained in any other way.

2. *Transverse lines must be parallel or converging.* In Fig. 49 a cube is sketched at (*a*) with the transverse lines (receding right and left edges) made accurately parallel. At (*b*) these lines are made to converge as they recede. Note that (*b*) looks more natural than (*a*) because of the effect of perspective foreshortening. The monstrosity at (*c*) is produced by the separating of the receding lines as they recede. In attempting to get the lines parallel, the beginner often makes a mistake like that at (*c*). Converge the lines deliberately, as in (*b*), to avoid the results of (*c*)!

3. *Axes must be kept flat to avoid distortion.* In Fig. 50 an accurate isometric sketch is shown at (*a*). At (*b*) the axes have been flattened to less than 30° with the horizontal. Possibly, (*b*) looks more natural than (*a*). At (*c*), however, the axes are somewhat more than 30° to the horizontal. Note the definite distortion and awkward appearance of (*c*). Therefore, especially in isometric, but also in other forms, keep the axes at their correct angle or flatter than the normal angle.

The foregoing three points must be kept constantly in mind. Hold your sketch at arm's length often during the work, so that you will see errors that are not so evident in the normal working position. Become critical of your

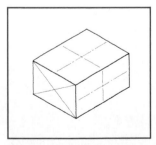

FIG. 51 Location of centers. Use diagonals or judge the position of center lines.

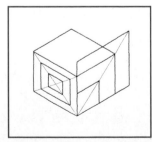

FIG. 52 Use of diagonals for increasing or decreasing rectangular shapes.

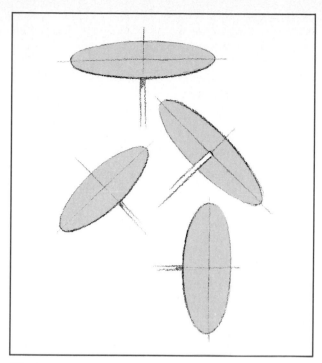

FIG. 53 Circles in pictorial. The major diameter of the ellipse is perpendicular to the axis of rotation.

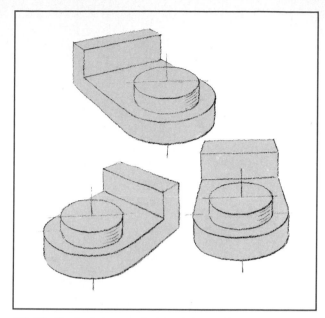

FIG. 54 Circular features on horizontal planes. The major diameter of the ellipse is horizontal.

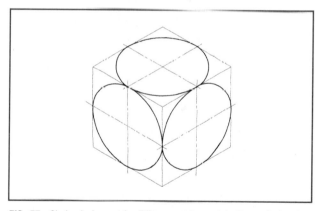

FIG. 55 Circles in isometric. Ellipses are tangent to the enclosing isometric squares at the mid-points of the sides.

own work and you will soon develop confidence and a good sense of line direction.

32. USES OF THE DIAGONAL

The two diagonals of a square or rectangle will locate its geometric center, as indicated on the left face of Fig. 51. The center is also readily located by drawing two center lines, estimating the middle of the space as shown on the top and right faces.

The diagonals of a rectangular face can also be used to increase or decrease the rectangle symmetrically about the same center and in proportion, as shown on the left face of Fig. 52, or with two sides coincident, as shown on the right face. To increase or decrease by equal units, the distance (or space) between lines must be judged.

33. SKETCHING CIRCULAR FEATURES

A circle in pictorial is an ellipse whose major diameter is always perpendicular to the *rotation axis*. Thus its minor diameter coincides on the drawing with the rotation axis (Fig. 53). These facts can be used to advantage when drawing an object principally made up of cylinders on the same axis. Note particularly from the above that *all* circles on horizontal planes are drawn as ellipses *with the major diameter horizontal*, as shown in Fig. 54.

Most objects, however, are made up of combinations of rectangular and circular features, and for this reason it is best to draw the enclosing pictorial square for all circular features. Figure 55 shows circles on all three axono-

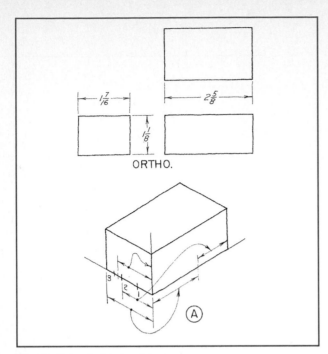

FIG. 56 Proportioning the distances.

metric planes. Note particularly that the ellipses must be tangent to the sides of the pictorial square at the midpoints of the sides; accordingly, it is best always to draw center lines, also shown. Always sketch the enclosing pictorial square for *all* circular features because by this method the size of the ellipse and the thickness of the cylindrical portion are easily judged.

34. PROPORTIONING THE DISTANCES

The ability to make divisions into equal units is needed in proportioning distances on a sketch. The average object does not have distances that are easily divisible into even inches, millimeters, etc., but since sketches are *not* made to scale, only to good proportions, great accuracy is not necessary.

Figure 56 shows a simple rectangular object, but the distances are not multiples of any simple unit. The best way to proportion this object is to lay off on *one* axis a distance that is to represent one side of the object. This

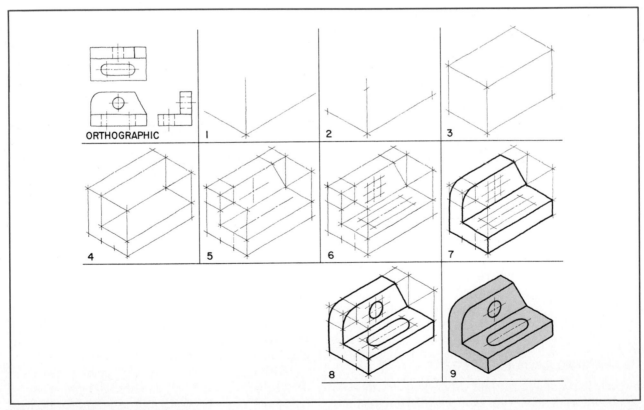

FIG. 57 Steps in making a pictorial sketch.

sets the size of the sketch. Then divide this first side into some unit that can be used easily to proportion other distances. At A the *left* side has been laid off and divided into three parts. Each of these parts now represents *approximately* ½ in. To get the distance for the vertical axis, the last third (at the rear) has been divided in half, and again in half, so that the dimension shown is approximately 1⅛ in. Transfer this distance to the vertical axis by (1) judging by eye, (2) measuring with the finger on the pencil, or (3) marking the distance on a piece of scratch paper.

The method to be used will suggest itself according to the relative accuracy needed. The right-axis distance is obtained similarly, by first transferring the whole left distance (representing approximately 1½ in) and then adding two-thirds of the left distance (approximately 1 in), which gives a total of 2½ in, close enough to the actual 2⅝-in dimension of the object.

35. STEPS IN MAKING A PICTORIAL SKETCH

Because a variety of objects is sketched, the order of procedure will not always be the same, but the following will serve as a guide:

1. Visualize the shape and proportions of the object from the orthographic views, a model, or other source.
2. Mentally picture the object in space and decide the pictorial position that will best describe its shape.
3. Decide on the type of pictorial to use—axonometric, oblique, or perspective.
4. Pick a suitable paper size.
5. Proceed as shown in Fig. 57.

Numbers 1 to 6 show light construction; 7 to 9, completion to final weight.

36. OBLIQUE SKETCHING

The methods presented thus far have been directed toward the making of axonometric sketches because they are the type most used. However, the practices given apply to oblique and perspective sketching, for which additional helps are given in the paragraphs that follow.

The advantage of oblique projection in preserving one face without distortion is of particular value in

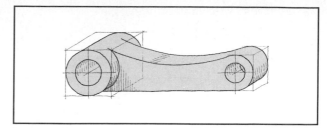

FIG. 58 An oblique sketch.

sketching, as illustrated by Fig. 58. The painful effect of distortion in oblique drawing that is done instrumentally can be greatly lessened in sketching by foreshortening the depth axis to a pleasing proportion.

37. PERSPECTIVE SKETCHING

A sketch made in perspective gives a better effect than a sketch made in axonometric or oblique. For constructing a perspective drawing of a proposed structure from its plans and elevations, a knowledge of the principles of perspective drawing is required, but for making a perspective sketch from the object, you can get along by observing the ordinary phenomena of perspective which affect everything we see: the fact that objects appear proportionately smaller as their distance from the eye increases, that parallel lines appear to converge as they recede, and that horizontal lines and planes appear to "vanish" on the horizon.

Make the sketch lightly, with free sketchy lines, and do not erase any lines until the whole sketch has been blocked in. Do not make the mistake of making the sketch too small.

In starting a sketch from the object, set it in a position to give the most advantageous view, and sketch the directions of the principal lines, running them past the limits of the figure toward their vanishing points. Block in the enclosing squares for all circles and circle arcs and proceed with the figure, drawing the main outlines first and adding details later; then brighten the sketch with heavier lines.

The drawing in Fig. 59 shows the general appearance of a "one-point" perspective sketch before the construction lines have been erased. Figure 60 is an example showing the object turned at an angle to the picture plane.

38. PICTORIAL ILLUSTRATION: RENDERING

Pictorial illustration combines any one of the regular pictorial methods with some method of shading or "rendering." In considering a specific problem, decide upon the pictorial form—axonometric, oblique, or perspective—and then choose a method of shading that is suited to the method of reproduction and the general effect desired.

A. Light and Shade The conventional position of the light in light-and-shade drawing is the same as that used for orthographic line shading, that is, a position to the left, in front of, and above the object. Any surface or portion of a surface perpendicular to the light direction and directly illuminated by the light would receive the greatest amount of light and be lightest in tone on the drawing; any face not illuminated by the light would be "in shade" and darkest on the drawing. Other surfaces, receiving less light than the "high" light but more than a shade portion, would be intermediate in tone.

An understanding of the simple one-light method of illumination is needed at the outset, as well as some artistic appreciation for the illumination on various surfaces of the object. Figure 61 shows a sphere, cylinder, cone, and cube illuminated as described and shaded accordingly. Study the tone values in this illustration.

B. Shade Lines Shade lines, by their contrast with other lines, add some effect of light and shade to the drawing. These lines used alone, without other shading, give the simplest possible shading method. Usually the best effect is obtained by using heavy lines only for the left vertical and upper horizontal edges of the dark faces (Fig. 62). Holes and other circular features are drawn with heavy lines on the shade side. Shade lines should be used sparingly, as the inclusion of too many heavy lines simply adds weight to the drawing the does not give the best effect.

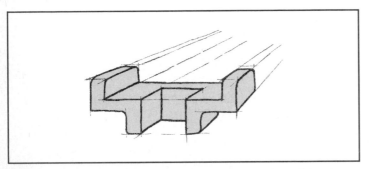

FIG. 59 A perspective sketch. The front face of the object is parallel to the picture plane.

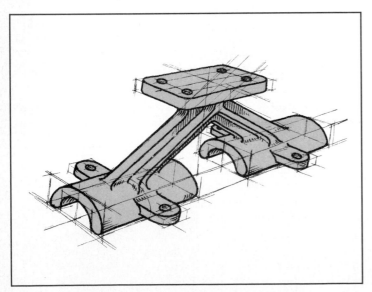

FIG. 60 A perspective sketch. The object is in an angular position.

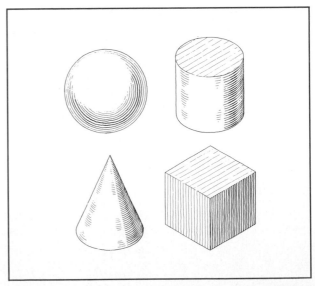

FIG. 61 Light and shade. The light source is from the upper left front.

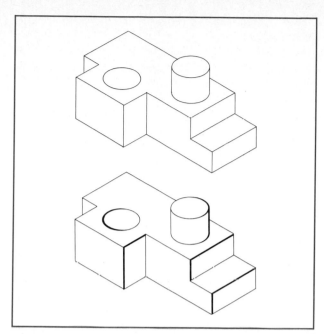

FIG. 62 Outline and shade lines. The side away from the light source is drawn with heavier lines.

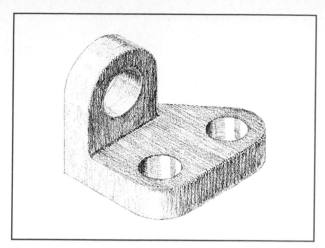

FIG. 63 Continuous-tone shading.

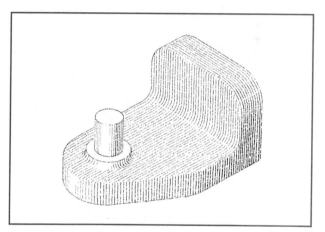

FIG. 64 Line-tone shading.

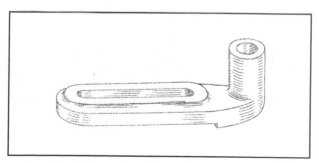

FIG. 65 Line-shading technique in pencil.

C. Pencil Rendering There are two general methods of pencil shading—continuous-tone and line-tone. Continuous-tone shading is done with a fairly soft pencil with its point flattened. A medium-rough paper is best for the purpose. Start with a light, overall tone and then build the middle tones and shade portions gradually. Figure 63 is an example. Clean highlights with an eraser.

Line-tone shading requires a little more skill, as the tones are produced by line spacing and weight. Light lines at wide spacing produce the lightest tone, and heavy lines at close spacing make the darkest shade. Leave highlights perfectly white. Pure black may be used sparingly for deep shade or shadow. Figure 64 is an example, drawn with only a light outline.

Complete overall shading is somewhat heavy, and a lighter, more open treatment is usually desired. To achieve this, leave light portions of the object with little or no shading, and line middle tones and shade sparingly. The few lines used strongly suggest light, shade, and surface finish (Fig. 65). There are many variations that can be made in this type of rendering.

D. Pen-and-Ink Rendering Pen-and-ink methods follow the same general pattern as work in pencil, with the exception that no continuous tone is possible. However, there are some variations not ordinarily used in pencil work. Figure 66 shows line techniques. As in pencil work, the common and usually the most pleasing method is the partially shaded, suggestive system.

E. Special shading methods There are several methods of representing textures, in addition to those just discussed.

Smudge shading is a rapid method often used for smooth surfaces (Fig. 67). Graphite from a soft pencil, powdered graphite, charcoal, or crayon sauce is first rubbed on a piece of paper and then picked up with a piece of cotton or an artist's stump and applied to the drawing. Highlights are easily brought out with a sharpened eraser.

Stippling with pen, pencil, brush, or sponge is an effective method of indicating rough-texture surfaces. In pen and pencil stippling, a good effect of light and shade is achieved with a multitude of dots, widely spaced for light surfaces, more closely for dark surfaces. In brush and sponge stippling, printer's ink or artist's oil color (drier added) is first worked out smoothly on a palette; then the medium is picked up from the palette with a bristle brush or sponge and applied to the drawing with a dabbing motion. Sharp edges of shaded areas can be maintained by the use of masks.

Small areas are easily lightened with a razor blade or "scratcher." Highlights are brought out with an eraser after the ink is dry. Figure 68 is an example of brush

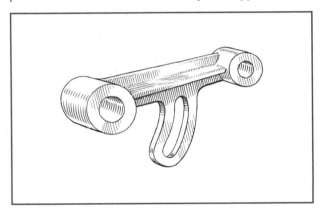

FIG. 66 Line-shading technique in ink.

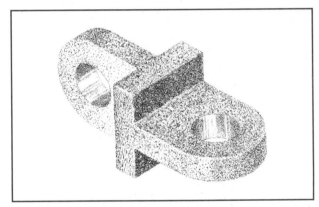

FIG. 68 Brush-stipple shading. This type of shading can also be done with a pen, pencil, or sponge.

FIG. 67 Smudge shading. This method gives an appearance similar to airbrush work and photography.

stippling with the smooth surfaces rendered by smudging.

Shading screens of clear cellulose with printed line or dot patterns provide a simple and effective shading method. Such screens are of clear cellulose printed with a shading pattern and backed with a special adhesive. The screen is applied to the drawing and rubbed down lightly wherever shading is wanted. The shaded sections are then outlined with a cutting needle and the unwanted pieces stripped off. The portions left on the drawing are then rubbed down firmly with a burnisher.

Scratchboard. Drawing paper with a chalky surface is often used for commercial illustrations because of the ease with which white lines can be produced on a black background, as well as black lines on a white background.

For ink work, the drawing is penciled on the board in the usual way and then inked by line-shading the lighter areas, working gradually to the darker areas. The darker areas are painted in with a brush, and when dry, white lines, dots, etc., are produced by scratching off the ink with a sharp-pointed knife, sharp stylus, or a needle point. Corrections are easily made by scratching off unwanted ink. Scratched areas can be reinked if necessary. Figure 69 is an example of scratchboard technique.

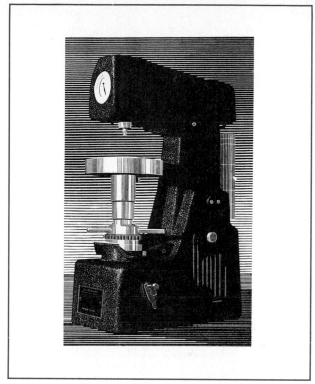

FIG. 69 A scratchboard drawing. The effect is similar to a woodcut.

PROBLEMS

The following problems are intended to furnish practice (1) in the various methods of pictorial representation and (2) in reading and translating orthographic projections.

In reading a drawing, remember that a line on any view always means an edge or a change in direction of the surface of the object; always look at another view to interpret the meaning of the line.

GROUP 1. ISOMETRIC DRAWINGS

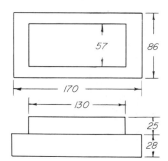

PROB. 1 METRIC. Jig block.

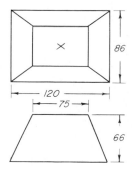

PROB. 2 METRIC. Frustum of pyramid.

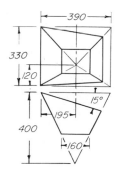

PROB. 3 METRIC. Hopper.

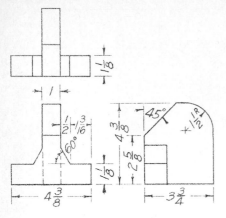

PROB. 4 Stop block.

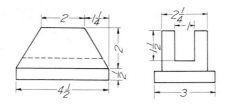

PROB. 5 Guide block.

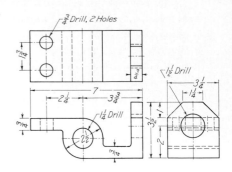

PROB. 6 Bracket.

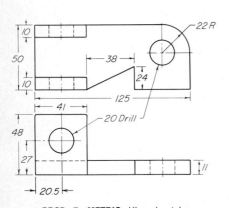

PROB. 7 METRIC. Hinged catch.

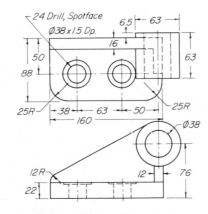

PROB. 8 METRIC. Offset bracket.

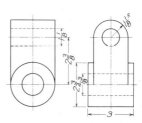

PROB. 9 Cross link.

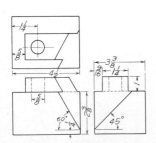

PROB. 10 Wedge block.

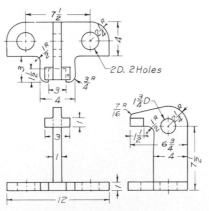

PROB. 11 Slide stop.

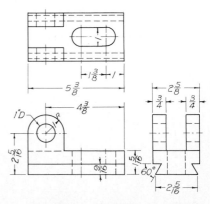

PROB. 12 Dovetail bracket.

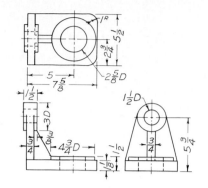

PROB. 13 Cradle bracket.

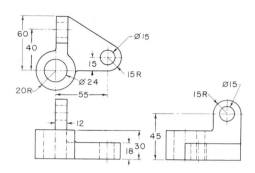

PROB. 14 METRIC. Cable clip.

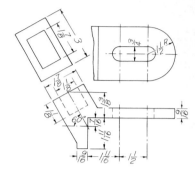

PROB. 15 Strut anchor.

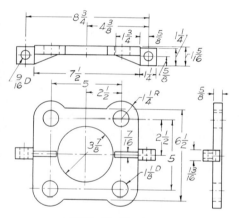

PROB. 16 Tie plate.

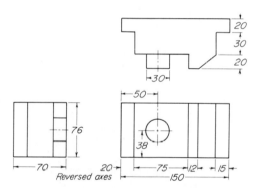

Reversed axes

PROB. 17 METRIC. Forming punch.

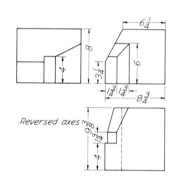

Reversed axes

PROB. 18 Springing stone.

GROUP 2. ISOMETRIC SECTIONS

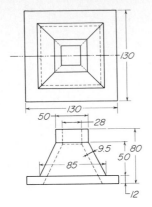

PROB. 19 METRIC. Column base.

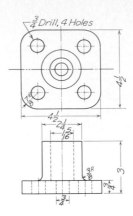

PROB. 20 Base plate.

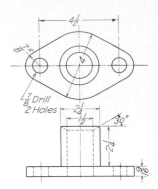

PROB. 21 Gland.

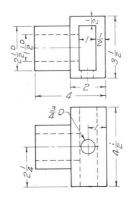

PROB. 22 Squared collar.

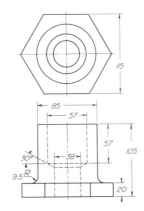

PROB. 23 METRIC. Blank for gland.

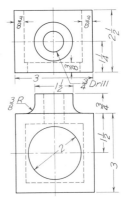

PROB. 24 Sliding cover.

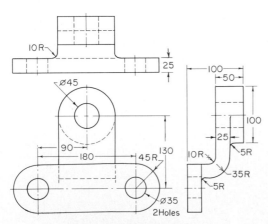

PROB. 25 METRIC. J bracket.

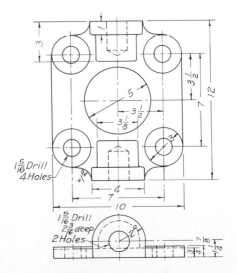

PROB. 26 Trunnion plate.

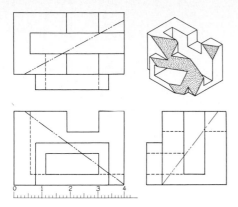

PROB. 27 Section study.

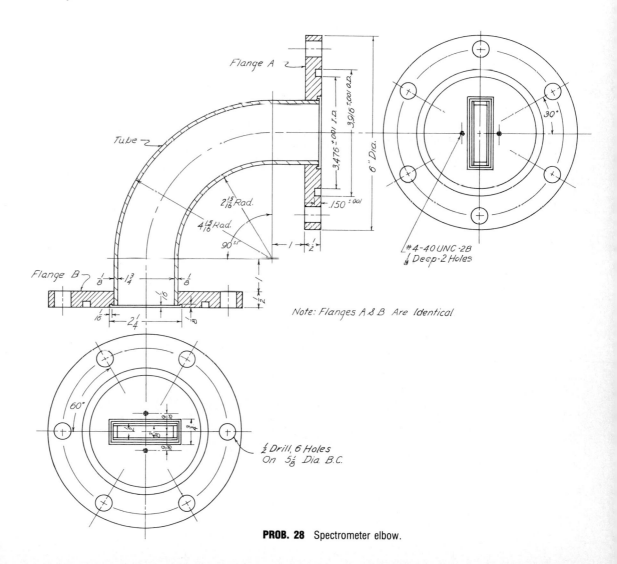

Flange A

Tube

$2\frac{15}{16}$ Rad.

$4\frac{15}{16}$ Rad.

$90^{\pm 1}$

$3.476^{+.001}$ I.D.

$3.916^{+.001}$ O.D.

$6''$ Dia.

$.150^{\pm .001}$

1 — $\frac{1}{2}$

$30°$

#4-40 UNC-2B
$\frac{1}{8}$ Deep-2 Holes

Flange B

$\frac{1}{8}$ $1\frac{3}{4}$ $\frac{1}{8}$

$\frac{1}{16}$

$\frac{1}{16}$ $2\frac{1}{4}$ $\frac{3}{8}$

$\frac{1}{2}$ — $\frac{1}{4}$

Note: Flanges A & B Are Identical

$60°$

$\frac{1}{2}$ Drill, 6 Holes
On $5\frac{1}{8}$ Dia. B.C.

PROB. 28 Spectrometer elbow.

GROUP 3. OBLIQUE DRAWINGS

Work problems in this group as either cavalier or cabinet drawings as the instructor indicates.

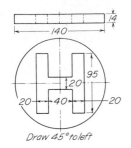

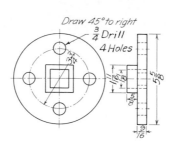

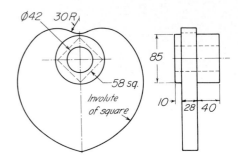

PROB. 29 METRIC. Letter die.

PROB. 30 Guide plate.

PROB. 31 METRIC. Heart cam.

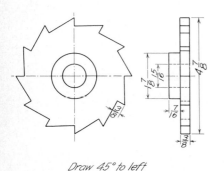

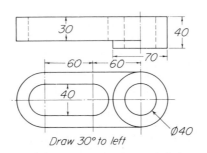

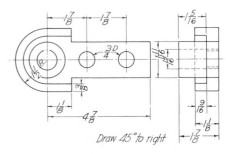

PROB. 32 Ratchet wheel.

PROB. 33 METRIC. Slotted link.

PROB. 34 Swivel plate.

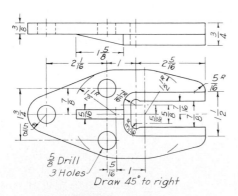

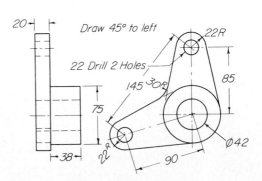

PROB. 35 Jaw backet.

PROB. 36 METRIC. Bell crank.

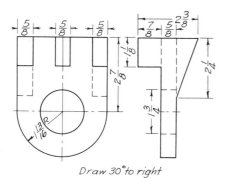

Draw 30° to right

PROB. 37 Stop plate.

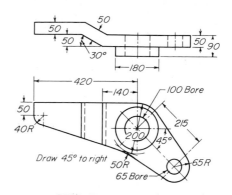

PROB. 38 METRIC. Pawl.

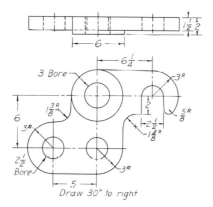

Draw 30° to right

PROB. 39 Hook brace.

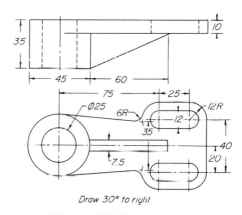

Draw 30° to right

PROB. 40 METRIC. Adjusting-rod support.

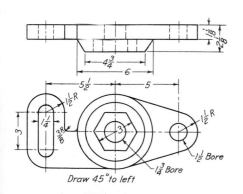

Draw 45° to left

PROB. 41 Link.

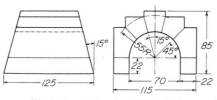

Offsets from right section, 30° to right

PROB. 42 METRIC. Culvert model.

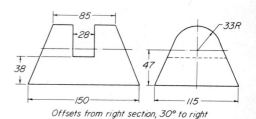

Offsets from right section, 30° to right

PROB. 43 METRIC. Slotted guide.

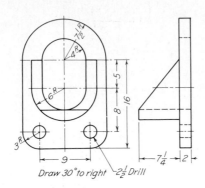

PROB. 44 Support bracket.

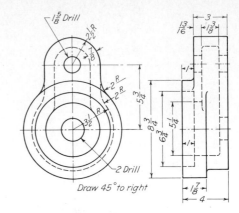

PROB. 45 Port cover.

GROUP 4. OBLIQUE SECTIONS

46. Oblique full section of sliding cone.

46A. Oblique half section of sliding cone.

48. Oblique half section of the base plate of Prob. 20.

49. Oblique half section of the gland of Prob. 21.

50. Oblique half section of regulator level.

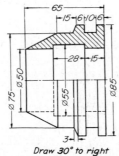

Draw 30° to right

PROB. 46 **METRIC.** Sliding cone.

47. Oblique full section of conveyor-trough end.

47A. Oblique half section of conveyor-trough end.

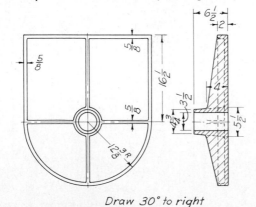

Draw 30° to right

PROB. 47 Conveyor-trough end.

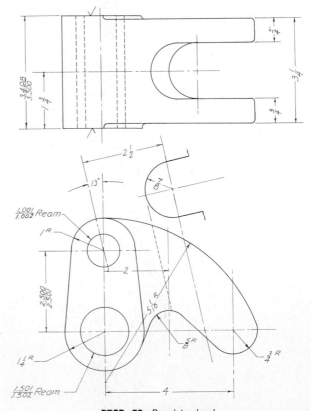

PROB. 50 Regulator level.

51. Oblique full section of anchor plate.

52. Oblique drawing of shaft guide. Use sections, as needed, to describe the part.

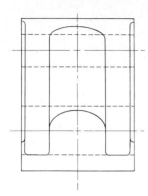

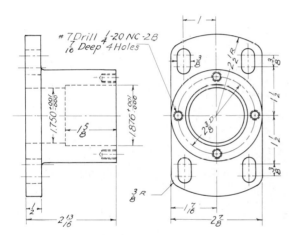

PROB. 51 Anchor plate.

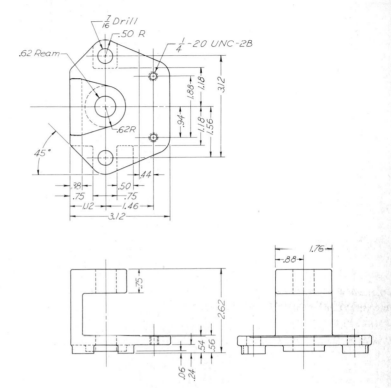

PROB. 52 Shaft guide.

53. Oblique half section of rod support. Partial sections and phantom lines can also be used to describe the part.

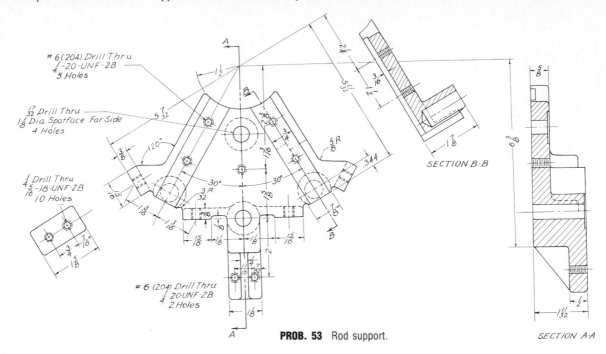

PROB. 53 Rod support.

GROUP 5. PERSPECTIVE DRAWINGS

The following are a variety of different objects to be drawn in perspective. A further selection can be made from the orthographic drawings in other chapters.

54. Double wedge block.

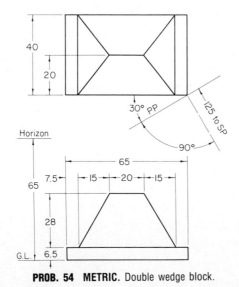

PROB. 54 METRIC. Double wedge block.

55. Notched holder.

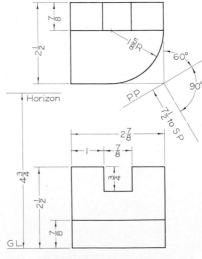

PROB. 55 Notched holder.

56. Crank.

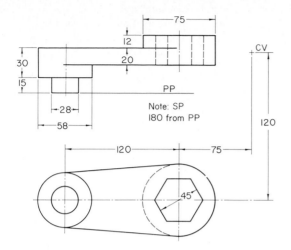

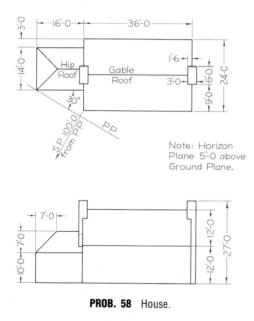

PROB. 56 METRIC. Crank.

57. Corner lug.

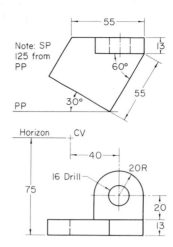

PROB. 57 METRIC. Corner lug.

58. House.

Note: Horizon Plane 5'-0 above Ground Plane.

PROB. 58 House.

59. Church.

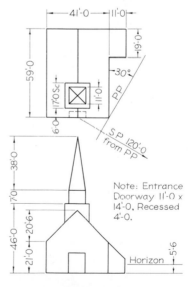

Note: Entrance Doorway 11'-0 x 14'-0, Recessed 4'-0.

PROB. 59 Church.

GROUP 6. PICTORIAL SKETCHING

The following problems are planned to develop skill not only in pictorial sketching but also in reading orthographic drawings. Make the sketches to suitable size on 8½- by 11-in. paper, choosing the most appropriate form of representation—axonometric, oblique, or perspective—with partial, full, or half sections as needed. Small fillets and rounds may be ignored in these problems and shown as sharp corners.

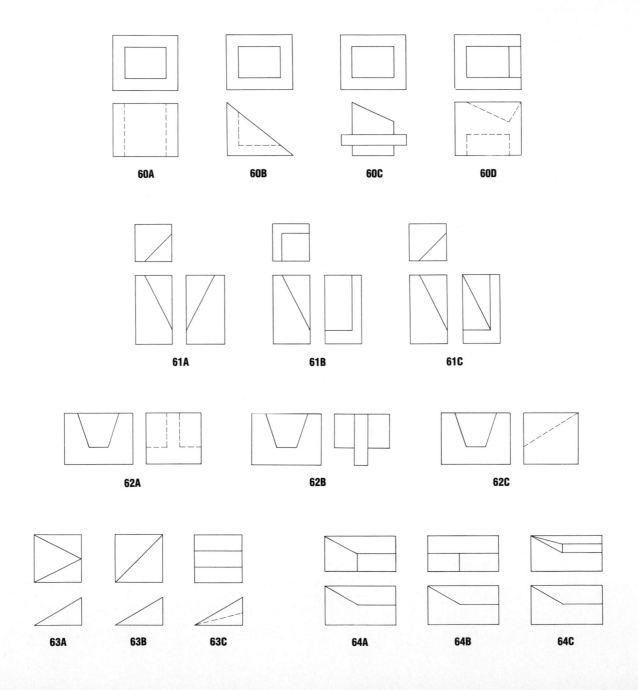

60A   60B   60C   60D

61A   61B   61C

62A   62B   62C

63A   63B   63C   64A   64B   64C

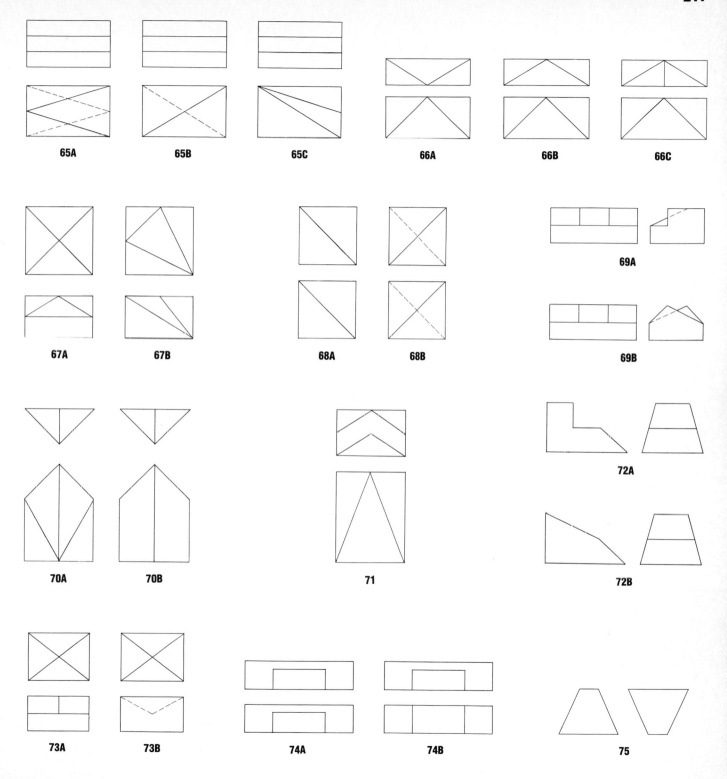

65A 65B 65C 66A 66B 66C

67A 67B 68A 68B 69A

69B

70A 70B 71 72A

72B

73A 73B 74A 74B 75

76

77

78A

78B

78C

79A

79B

80A

80B

80C

81

82

83

84

85

86

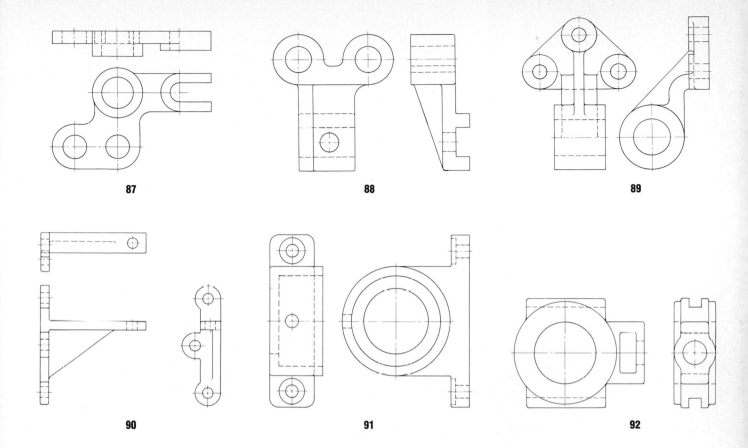

87

88

89

90

91

92

GROUP 7. OBLIQUE SKETCHING

Select an object from Group 3 not previously drawn with instruments, and make an oblique sketch.

GROUP 8. PERSPECTIVE SKETCHING

Select an object not previously drawn with instruments, and make a perspective sketch.

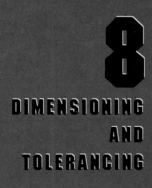

8

DIMENSIONING AND TOLERANCING

1. PURPOSE OF DIMENSIONING

To this point we have concentrated on describing the shapes of parts, both in two-dimensional and in three-dimensional formats. It is now necessary to learn how to dimension a part. Only then can a person responsible for producing the part have exact information on the size and position of each feature of the part.

One needs to learn the whys of dimensioning, or its theory. Also, one needs information on the hows of dimensioning, or its techniques. Included must be knowledge of how to apply permitted variation (tolerance) to sizes and positions of part features.

This chapter is divided into four sections to aid understanding of the material: theory of dimensioning, techniques and conventions, tolerancing systems, and standardized tolerances.

THEORY OF DIMENSIONING

2. SIZE VS. POSITION DIMENSIONS

Dimensions can be classified as those of size or position. Figure 1 shows a part where each geometric shape is accounted for in terms of size and position. The letter S indicates size and P indicates position.

Any part can be broken down into a combination of basic geometric shapes, primarily prisms, cylinders, cones, and pyramids. Any of these basic shapes can be positive or negative; for example, a hole is a negative cylinder. In our study we will learn how to systematically describe all common shapes.

In all our work two vital rules will be followed:

1. Each feature is dimensioned and positioned only once.
2. Each feature is dimensioned and positioned where its shape shows.

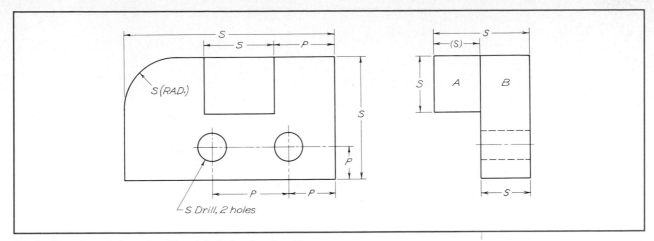

FIG. 1 Dimensions of size and position. *S* indicates size, *P* position.

These rules help to clarify the reading of any drawing and reduce the chance of error and confusion for the person making the part. Even when the part is produced on a computer-controlled machine, the critical setup of the computer program is simplified when dimensional instructions are clear and concise. The occasional exceptions to these rules will be distinctly pointed out.

3. BASIC EXAMPLES OF DIMENSIONS OF SIZE AND POSITION

Since every solid has three dimensions, each of the geometric shapes making up the object must have its height, width, and depth indicated in the dimensioning.

The *prism* is the most common shape and requires three dimensions for square, rectangular, or triangular (Fig. 2A). For regular hexagonal or octagonal types, usually only two dimensions are given, the length and either the distance "across corners" or the distance "across flats."

The *cylinder* is the second most common shape. A cylinder obviously requires only two dimensions, diameter and length (Fig. 2B). Partial cylinders, such as fillets and rounds, are dimensioned by radius instead of diameter. A good general rule is to dimension complete circles with the diameter and circle arcs (partial circles) with the radius.

Right cones can be dimensioned by giving the altitude and the diameter of the base. They usually occur as frus-

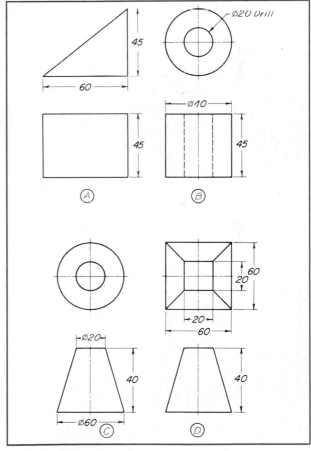

FIG. 2 METRIC. Size dimensions. *(A)* prism; *(B)* cylinder; *(C)* cone; *(D)* pyramid.

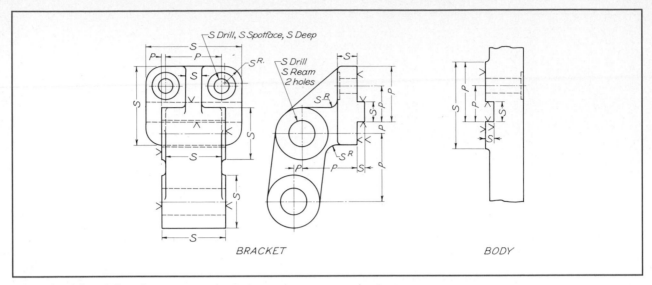

FIG. 3 Correlation of dimensions.

tums, however, and require the diameters of the ends and the length (Fig. 2C).

Right pyramids are dimensioned by giving the dimensions of the base and the altitude. Right pyramids are often frustums, requiring dimensions of both bases (Fig. 2D).

Oblique cones and *pyramids* are dimensioned in the same way as right cones and pyramids but with an additional dimension parallel to the base to give the offset of the vertex.

Spheres are dimensioned by giving the diameter and other surfaces of revolution by dimensioning the generating curve.

After the basic geometric shapes have been dimensioned for size, the position of each relative to the others must be given. Position must be established in height, width, and depth directions. Rectangular shapes are positioned with reference to their faces, cylindrical and conic shapes with reference to their center lines and their ends.

One basic shape will often coincide or align with another on one or more of its faces. In such cases the alignment serves partially to locate the parts. Thus in Fig. 1, prism *A* requires only one dimension for complete positioning with respect to prism *B*, because two surfaces are in alignment and two in contact.

Coincident center lines often eliminate the need for

position dimensions. In the cylinder in Fig. 2B the center lines of the hole and of the cylinder coincide, and no dimensions of position are needed. The two holes of Fig. 1 are on the same center line, and the dimension perpendicular to the common center line positions both holes in that direction.

4. SELECTION OF DIMENSIONS

The *selection of dimensions of size* arrived at by shape breakdown will usually meet the requirements of a particular shop (e.g., machine shop, shop involved in pattern making, forging, sheet-metal or steel work, die casting, welding, etc.) since the basic shapes result from the fundamental shop operations. However, a shop often prefers to receive dimensions of size in note form rather than as regular dimensions when a shop process in involved, such as drilling, reaming, counterboring, and punching.

Selecting dimensions of position ordinarily requires more consideration than selecting dimensions of size because there are usually several ways in which a position might be given. In general, positional dimensions are given between finished surfaces, center lines, or a combination thereof (Fig. 3). Remember that rough castings or forgings vary in size; so do not position machined surfaces from unfinished surfaces.

The position of a point or center by offset dimensions from two center lines or surfaces (Fig. 4) is preferable to angular dimensions (Fig. 5) unless the angular dimension is more practical from the standpoint of construction.

5. THE CONTOUR PRINCIPLE

In all cases of selecting dimensions the important consideration is clarity. One view of a part will usually describe the shape of some feature better than other views. Using the view that best shows shape involves use of the contour principle: A feature is best dimensioned in the view showing the shape of the feature.

It is natural to look for the dimensions of a given feature wherever that feature appears most characteristic, and an advantage in clarity and in ease of reading will certainly result if the dimension is placed there. In Fig. 6 the rounded corner, the drilled hole, and the lower notched corner are all characteristic in, and dimensioned on, the front view. The projecting shape on the front of the object is more characteristic in the top view and is dimensioned there.

Use of the contour principle implies that duplicate and/or unnecessary dimensions are to be avoided. When a drawing is changed or revised, a duplicate dimension may go unnoticed. As a result, a particular dimension could have two different values, one incorrect!

An unnecessary dimension, other than a duplicate, is a dimension that is not essential to make a part.

Unnecessary dimensions always occur when all the individual dimensions are given in addition to the overall dimension (Fig. 7A). One dimension of the series must be omitted if the overall dimension is used, thus allowing only one possible positioning from each dimension (Fig. 7B).

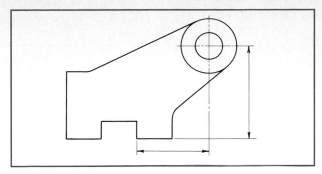

FIG. 4 Position by offsets.

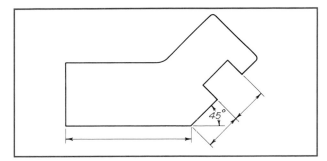

FIG. 5 Position by angle.

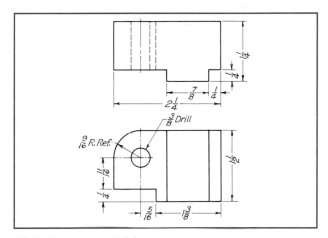

FIG. 6 The contour principle.

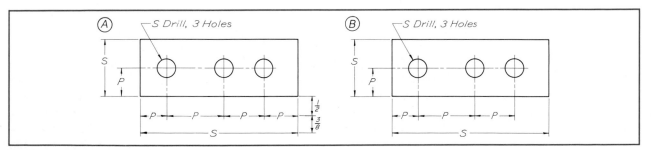

FIG. 7 Unnecessary dimensions.

223

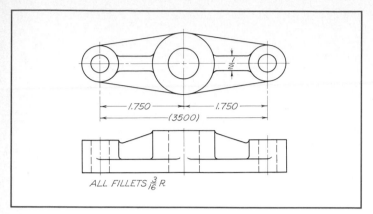

FIG. 8 One reference dimension.

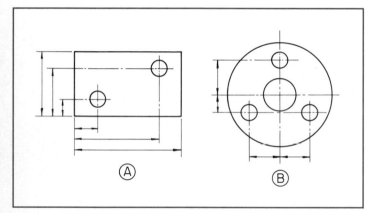

FIG. 9 Dimensions from datum. *(A)* Datum surface; *(B)* datum center.

In architectural and structural work, where the interchangeability of parts is usually irrelevant, unnecessary dimensions cause no difficulty and all dimensions are given.

There are several aids for the placement of dimensions that make the contour principle easier to apply. These aids are as follows:

1. Dimensions outside the view are preferred, unless added clearness, simplicity, and ease of reading will result from placing some of them inside. For good appearance, dimensions should be kept off the cut surfaces of sections.

2. Dimensions between the views are preferred unless there is some reason for placing them elsewhere.

3. Dimension lines should be spaced, in general, ½ in away from the outlines of the view. This applies to a single dimension or to the first dimension of several in a series. See Fig. 7A.
    Parallel dimension lines should be spaced uniformly with at least ⅜ in between lines. See Fig. 7A.

4. Always place a longer dimension line outside a shorter one to avoid crossing dimension lines with the extension lines of other dimensions. Thus an overall dimension (maximum size of piece in a given direction) will be outside all other dimensions.

Occasionally, for reference and checking, all dimensions in a series are given as well as the overall dimension. In such cases one dimension to be made the reference dimension is enclosed within parentheses, as in Fig. 8. The former practice was to use the abbreviation REF, as in 3.500 REF.

6. DIMENSIONAL CLARITY USING A DATUM

Datum points, lines, and edges of surfaces of a part are features that are assumed to be exact for purposes of computation or reference, and *from* which the positions of other features are established. In Fig. 9A the left side and bottom surfaces of the part are the datum surfaces, and at *B* the center lines of the central hole in the part are datum lines. Where positions are specified by dimensions from a datum, different features are always positioned from this datum and *not* with respect to one another.

A feature selected to serve as a datum must be clearly identified and readily recognizable. Note in Fig. 9 that the datum lines and edges are obvious. On an actual part, the datum features must be accessible during manufacture so that there will be no difficulty in making measurements.

A datum surface (on a physical part) must be more accurate than the allowable variation on any dimension of position that is referred to the datum. Thus, it may be necessary to specify the perfection of datum surfaces for flatness, straightness, roundness, etc. (See sections on geometric tolerancing.)

TECHNIQUES AND CONVENTIONS

7. PRACTICAL CONSIDERATIONS: USE OF SCALE

Much of the basic theory behind dimensioning has been presented. We now need practical information about techniques and conventions used in dimensioning—lines, notes, leaders, and lettering—so that the user is provided dimensional information that is as clear as possible.

Special mention must be made concerning the *scale* of the particular part that is to be dimensioned. (See Chap. 2, Sec. 10.) A part may be drawn to any scale convenient to the drafter. However, the scale must be indicated in an easily seen area of the working drawing, often in the title block (see Chap. 13).

It is important to realize that dimensions placed on a part are always *actual size* dimensions. The number for a size or a position dimension never changes, regardless of the scale to which the part is drawn. Therefore, if a certain length is 50 mm in actual size, the value 50 will appear on the drawing no matter what scale is used in drafting the part.

8. DIMENSIONS AND NOTES AS RELATED TO ANSI STANDARDS

The two basic methods used to give distance on a drawing are a dimension (Fig. 10) and a note (Fig. 11). Please notice the use of R15 and ϕ12 DRILL in these two figures.

The use of R15 and ϕ12 DRILL conforms to metric notation as used in Europe and Asia. Such notation was adopted as an acceptable standard in the United States by the American National Standards Institute in ANSI Y14.5M—1982, Dimensioning and Tolerancing. This excellent standard is used extensively throughout this chapter. However, one may find the particular formats R15 and ϕ12 DRILL given as 15R and 12 DRILL on many drawings in industry as well as within this book.

Referring to Fig. 10 one sees that a dimension is used to give the distance between two points, lines, planes, or some combination. Extension lines lead the eye to the particular feature described by the dimension.

Notes are word statements giving information that cannot be given by the views and dimensions. They al-most always specify some standard shape, operation, or material, and are classified as *general* or *specific*. A general note applies to the entire part, and a specific note applies to an individual feature. Occasionally a note will save making an additional view, for example, by indicating right- and left-hand parts.

General notes do not require the use of a leader and should be grouped together above the title block. Examples are "Finish all over," "Fillets ¼R, rounds ⅛R, unless otherwise specified," "All draft angles 7°," "Remove burrs," etc.

Specific notes almost always require a leader and should therefore be placed fairly close to the feature to which they apply. Most common are notes giving an operation with a size, as "ϕ15 thru, 4 holes."

Notes often use abbreviations to save space. Many persons are familiar with common abbreviations such as MAX, MIN, DIA, and THD. Uncommon abbrevia-

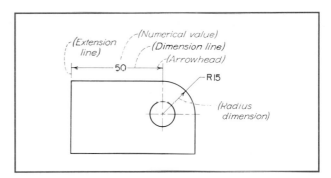

FIG. 10 METRIC. Dimensions.

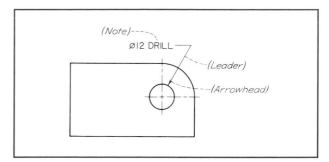

FIG. 11 METRIC. A note.

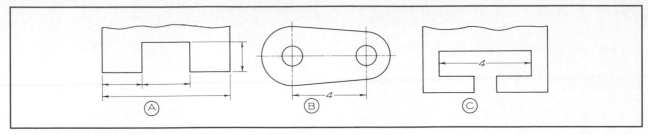

FIG. 12 Dimension placement.

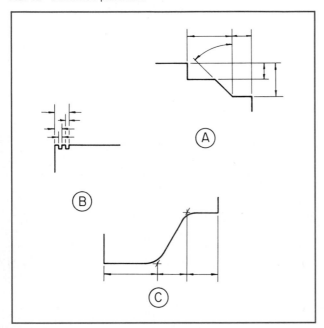

FIG. 13 Technique of drawing extension lines.

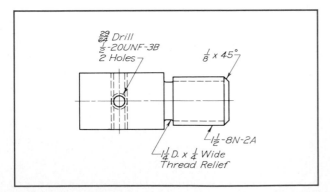

FIG. 14 Leaders for notes.

tions such as PH BRZ (phosphor bronze) should be used sparingly because of possible misinterpretation. Abbreviations should conform to accepted standards, such as those with ANSI Y1.1—1972.

9. EXTENSION LINES AND LEADERS

A. Extension Lines Extension lines should not touch the outline of the view but should start about $\frac{1}{16}$ in from it and extend about $\frac{1}{8}$ in beyond the last dimension line (Fig. 12A). This example is printed approximately one-half size.

Dimensions may also terminate at *center lines* or *visible outlines of the view*. Where a measurement between centers is to be shown, as at *B*, the center lines are continued to serve as extension lines, extending about $\frac{1}{8}$ in beyond the last dimension line. Usually the outline of the view becomes the terminal for arrowheads, as at *C*, when a dimension must be placed inside the view.

Extension lines for an angular dimension are shown in Fig. 13A, with one of the extension lines used for a linear dimension.

Extension lines should not be broken where they cross each other or an outline of a view, as shown in Figs. 12A and 13A. However, when space is restricted and extension lines come close to arrowheads, the extension lines may be broken for clarity, as in Fig. 13*B*.

Where a point is located by extension lines alone, the extension lines should pass through the point, as at *C*.

B. Leaders Leaders are *straight* (not curved) lines leading from a dimension value or an explanatory note to the feature on the drawing to which the note applies (Fig. 14). An arrowhead is used at the pointing end of the leader. The note end of the leader should terminate with a short horizontal bar at the midheight of the lettering and should run to the beginning or the end of the note, never to the middle.

Leaders should be drawn at an angle to contrast with the principal lines of the drawing, which are mainly horizontal and vertical. Thus leaders are usually drawn at 30, 45, or 60° to the horizontal.

A "dot" may be used for termination of a leader, as in Fig. 15A, when the dot is considered to be a clearer representation than an arrowhead. The dot should fall within the outline of the object.

When dimensioning a circular feature with a note, make the leader *radial*, as shown at B. A leader directed to a flat surface should meet the surface at an angle of *at least* 30°; an angle of 45 or 60° as at C is best.

If possible *avoid* crossing leaders; *avoid* long leaders; *avoid* leaders in a horizontal or vertical direction; *avoid* leaders parallel to adjacent dimension lines, extension lines, or crosshatching; and *avoid* small angles between leaders and the lines they terminate on.

10. LINE WEIGHTS AND ARROWHEADS

Line weight for dimension and extension lines should be the same as those for center lines, that is, fine, so as to contrast with the heavier outlines of the views (refer to the alphabet of lines in Chap. 2).

Arrowheads are drawn freehand, as shown in Figs. 16 and 17. All arrowheads on the same drawing should be the same type, open or solid, and the same size, except in restricted spaces. Arrowhead lengths vary somewhat depending upon the size of the drawing. A good general length for small drawings is ⅛ in and for larger drawings ³⁄₁₆ in.

11. LETTERING PRACTICE

A. Figures Figures must be carefully lettered in vertical or inclined style. In an effort to achieve neatness, the beginner often gets them too small; ⅛ in for small drawings and ⁵⁄₃₂ in for larger drawings are good general heights.

The general practice is to leave a space in the dimension line for the dimension value (Fig. 18). It is universal in structural practice and common in architectural practice to place the value above a continuous dimension line (Fig. 18).

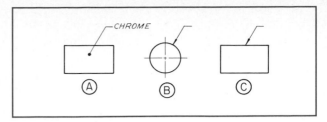

FIG. 15 Leaders. *(A)* special terminal; *(B)* and *(C)* proper angle.

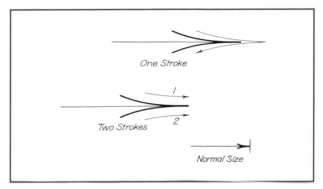

FIG. 16 Open-style arrowhead strokes.

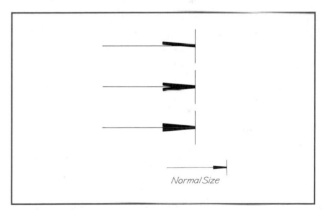

FIG. 17 Solid-style arrowhead strokes.

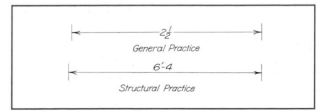

FIG. 18 Placement of dimension values.

B. Common Fractions Common fractions should be made with the fraction bar parallel to the guide lines for making the figure and with the numerator and denominator each somewhat smaller than the height of the whole number so that the *total* fraction height is twice that of the integer (Fig. 19). Avoid the incorrect forms shown.

C. Feet and Inches Indicate feet and inches as 9'-6". When there are no inches, indicate as 9'-0", 9'-0½". When dimensions are all in inches, the inch mark is preferably omitted from all the dimensions and notes unless there is some possibility of misunderstanding; if that is the case, "1 bore," for example, should be given for clarity as "1" bore."

In structural drawing, length dimensions should be given in feet and inches. Plate widths, beam sizes, etc., are given in inches. Inch marks are omitted, even though the dimension is in feet and inches (Fig. 18).

D. Decimal Systems As with figures, ⅛ in [3 mm] is an effective height for lettering in decimal systems. If two lines are involved, as seen in Fig. 20, 1/16 in [1.5 mm] is a proper distance between lines. The method of expressing metric numbers varies, however, from that of expressing numbers in the English decimal system. Table 1 compares the two systems.

By way of general observation, note that in the metric system, millimeters (mm) are used in drawings of nearly

TABLE 1   COMPARISON OF ENGLISH DECIMAL AND METRIC SYSTEMS

English decimal	Metric	Comment
3,214,759	3 214 759 0.754 76 2476 *or* 2 476	Decimal system separates groups of three by commas; the metric system, by space.
.75 *or* 0.75	0.75	Metric system always uses a zero before decimal point, a feature that is optional in the English decimal system.
0.4010 *or* 0.40100	0.401	Common practice in the English decimal system is to keep a consistent number of figures to the right of the decimal point, including zeros; the metric system drops useless zeros.

all machine parts. Centimeters (cm) are used most often for larger items, such as measurements of the human body. Meters (m) are commonly used for measurements of medium distance on land, such as lengths of swimming pools and distances calculated in surveying. Kilometers (km) are used for considerable distances, as on maps. Note also that a space is always placed between a figure and the metric unit of measure, as 25 mm, not 25mm.

12. NUMERICAL SYSTEMS FOR DIMENSIONING

The preceding section indicates that fractional and decimal systems may be used in dimensioning. This is indeed so. There are three systems in use: (1) the common-fraction system, (2) the common-fraction, decimal-

2½ 2½ 2I/2 2 ½

Avoid

FIG. 19 Technique for lettering values with common fractions.

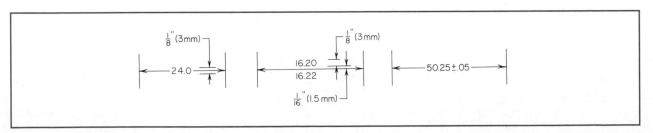

FIG. 20 Size of decimals.

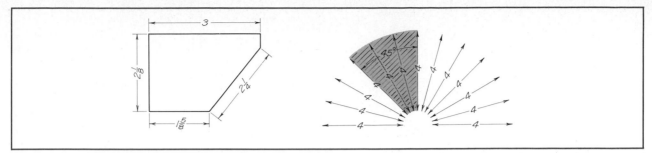

FIG. 21 Reading direction of values (aligned system).

fraction system, and (3) the decimal system. The system chosen depends on the part being designed.

A. The Common-Fraction System This system, used in general drawing practice, including architectural and structural work, has all dimension values written as units and common fractions, as 3½, 1¼, ⅜, 1/16, 3/32, 1/64. Values thus written can be laid out with a steel tape or scale graduated in sixty-fourths of an inch.

B. The Common-Fraction, Decimal-Fraction System This system is used principally in machine drawing whenever the degree of precision required calls for fractions of an inch smaller than those on the ordinary steel scale. To continue the use of common fractions below 1/64, such as 1/128 or 1/256, is considered impractical. The method followed is to give values (1) in units and common fractions for distances not requiring an accuracy closer than 1/64 in; and (2) in units and decimal fractions, as 2.375, 1.250, 0.1875, etc., for distances requiring greater precision. The decimal fractions are given to as many decimal places as needed for the degree of precision required.

C. The Decimal System A decimal system can be either an English decimal or metric system. Both systems are widely used, more so than types A and B above, especially for machine parts. The metric system is gaining steadily on the decimal inch system, however, because metric units have universal application throughout the world.

D. Rounding It is often necessary to round off a decimal to a smaller number. The procedure is identical for both inch units and millimeter units. The standard described in ASTM (E380) 41.1981 is as follows:

When the figure beyond the last figure to be retained is less than 5, the last figure retained should not be changed. Example: 3.46325, if cut off to three places, should be 3.463.

When the figure beyond the last figure to be retained is more than 5, the last figure retained should be increased by 1. Example: 8.37652, if cut off to three places, should be 8.377.

When the figure beyond the last place to be retained is exactly 5 with only zeros following, the preceding number, if even, should be unchanged; if odd, should be increased by 1. Example: 4.365 becomes 4.36 when cut off to two places. Also, 4.355 becomes 4.36 when cut off to two places.

13. READING DIRECTION

Engineering drawings use either of two systems for reading direction, the aligned or the unidirectional.

In the aligned system the figures are oriented to be read from a position *perpendicular* to the dimension line; thus the guide lines for the figures will be parallel to the dimension line, and the fraction bar in line with the dimension line (Fig. 21). The figures should be arranged so as to be read from the *bottom* or *right side* of the drawing. Avoid running dimensions in the directions included in the shaded area of Fig. 21; if this is unavoidable, they should read downward with the line.

The *unidirectional system* originated in the automotive and aircraft industries, and is sometimes called the "horizontal system." All figures are oriented to read from the bottom of the drawing. Thus the guide lines and fraction bars are horizontal regardless of the direction of

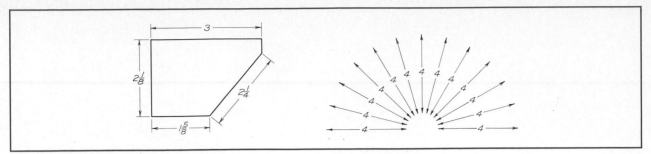

FIG. 22 Reading direction of values (unidirectional system).

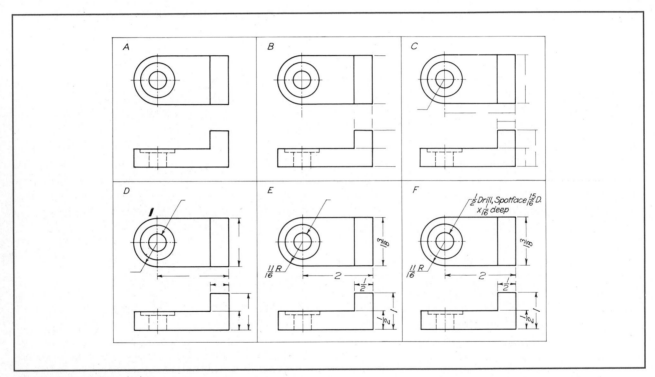

FIG. 23 Order of dimensioning.

the dimension (Fig. 22). The "avoid" zone of Fig. 21 has no significance with this system.

Notes must be lettered horizontally and read from the bottom of the drawing in either system.

14. ORDER OF DIMENSIONING

A systematic order of working is a great help in placing dimensions. Figure 23 illustrates the procedure. First complete the shape description (A). Then place the extension lines and extend the center lines where necessary (B), thus planning for the location of both size and position dimensions; study the placement of each dimension and make alterations if desirable or necessary. Add the dimension lines (C). Draw arrowheads and leaders for notes (D). Then add values and letter notes (E and F).

It is desirable to add the notes *after* the dimensions have been placed. If the notes are placed first, they may occupy a space needed for a dimension. Because of the freedom allowed in the use of leaders, notes may be given in almost any available space.

Dimensions should never be crowded into space too small to contain them. One of the methods in Fig. 24 can be used where space is limited. Sometimes a note is appropriate. If the space is small and crowded, an enlarged removed section or part view can be used (Fig. 25).

15. DIMENSIONING AN AUXILIARY VIEW

In placing dimensions on an auxiliary view, the same principles of dimensioning apply as for any other drawing, but special attention is paid to the contour principle given in Sec. 5. An auxiliary view is made for the purpose of showing the normal view (size) of some inclined or oblique face, and for this reason the dimensioning of the face should be placed where it is easiest to read, *which will he on the normal view*. Note in Fig. 26 that the spacing and size of holes as well as the size of the inclined face are dimensioned on the auxiliary view. Note also that the angle and dimension of position tying the inclined face to the rest of the object could not be placed on the auxiliary view.

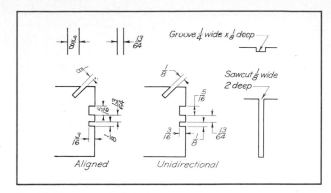

FIG. 24 Dimensions in limited space.

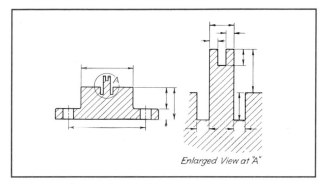

FIG. 25 Use of enlarged view to clarify dimensions.

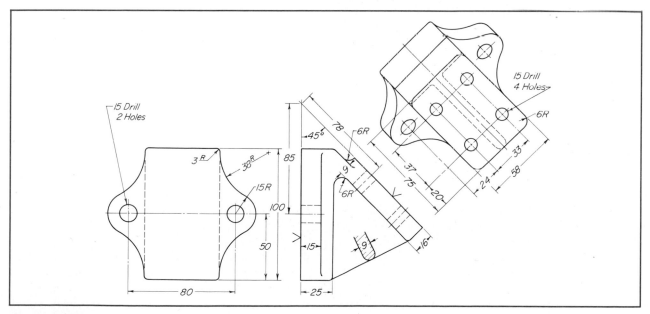

FIG. 26 METRIC. Dimensioning on an auxiliary (normal) view.

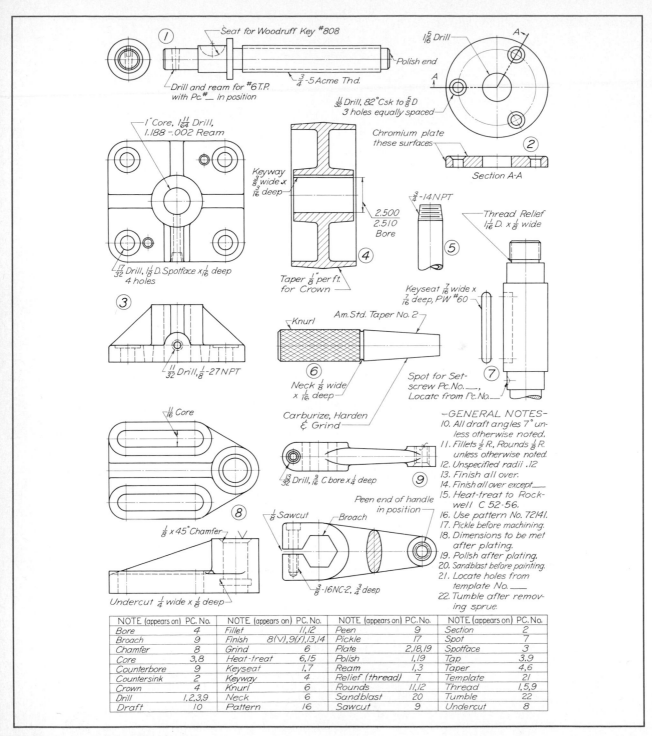

FIG. 27 Recommended working of notes.

16. DIMENSIONING TECHNIQUES FOR COMMON FEATURES

We have arrived at a stage in this chapter where it would be helpful to give some specific information on the dimensioning of common shapes and features. You have studied indirectly many of these shapes already. Now we can concentrate on such shapes so that proper dimensioning practices can be used with ease in future work.

A. Notes for Standard Features   Notes were discussed in Sec. 8 and their values indicated. A comprehensive example of the use of notes is seen in Fig. 27, in which both specific and general notes are seen. Use of notes is a powerful way to compress much information into a small area of the drawing.

B. Arcs and Curves   Arcs should be dimensioned by giving the radius on the view that shows the true shape of the curve. The dimension line for a radius should always be drawn as a radial line at an angle (Fig. 28), never horizontal or vertical; and only one arrowhead is used.

A position on a curved part may be misconstrued unless there is a specification showing the *surface* of the part to which the dimension applies, as in Fig. 29.

The numerical value is followed by the letter R when using the inch system. This practice is still common when the metric system is used in the United States, though other countries usually give the letter R first, as in R10, meaning a radius of 10 mm, as endorsed by ANSI Y14.5M-1982. Mention of the standard was made in Sec. 8.

When the center of an arc lies outside the limits of the drawing, the center is moved closer along a center line of the arc and the dimension line is jogged to meet the new center (Fig. 30). The portion of the dimension line adjacent to the arc is a radial line of the true center.

Curves for which great accuracy is not required are dimensioned by offsets, as in Fig. 31. For greater accuracy, dimensions from datum features, as in Fig. 32, are recommended.

Round-end shapes should be dimensioned according to their method of manufacture. Figure 33 shows several similar contours and the typical dimensioning for each. The link A, to be cut from thin material, has the radius of the ends and the center distance given as it would be laid out. At *B* is shown a cast pad dimensioned as at A, with the dimensions most usable for the patternmaker.

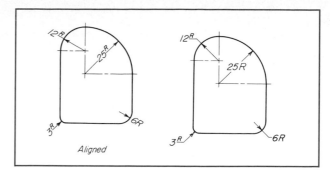

FIG. 28   METRIC. Dimensioning of arcs.

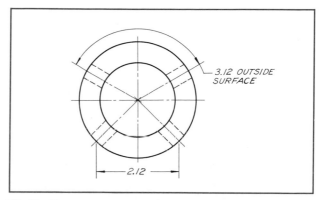

FIG. 29   Dimensioning a position on a curved surface.

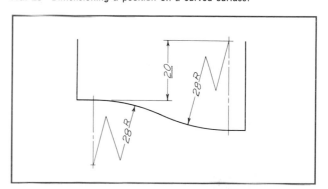

FIG. 30   METRIC. Dimensioning of objects having inaccessible centers.

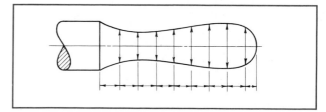

FIG. 31   A curve dimensioned by offsets.

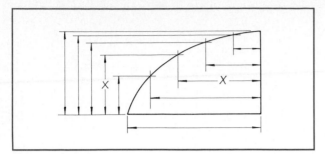

FIG. 32   A curve dimensioned from datum edges.

The drawing at C shows a slot machined from solid stock with an end-milling cutter. The dimensions give the diameter of the cutter and the travel of the milling-machine table. The slot at D is similar to that at C but it is dimensioned for quantity production, where overall length, not table travel, is wanted for gaging purposes. Pratt and Whitney keys and key seats are dimensioned by the method shown at D.

C. Holes   Holes that are made by being drilled, reamed, bored, or punched are normally specified by a

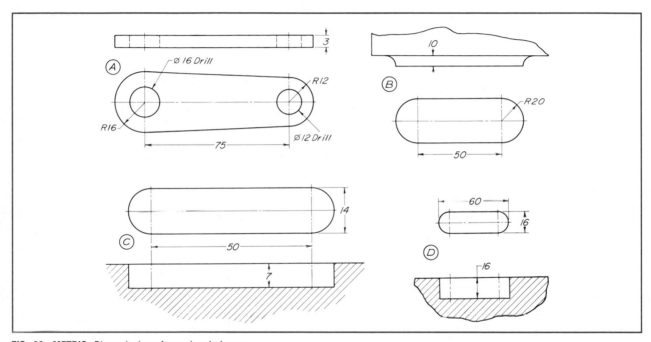

FIG. 33   **METRIC.** Dimensioning of round-end shapes.

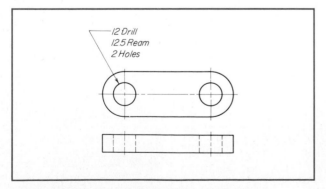

FIG. 34   **METRIC.** Dimensioning of drilled and reamed holes.

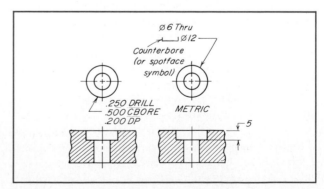

FIG. 35   Dimensioning of counterbored holes.

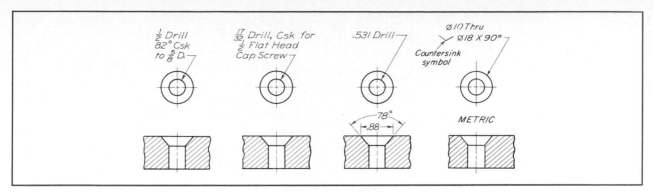

FIG. 36 Dimensioning of countersunk holes.

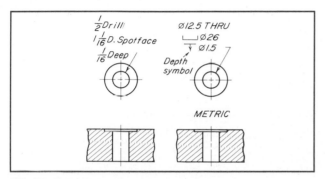

FIG. 37 Dimensioning of spot-faced holes.

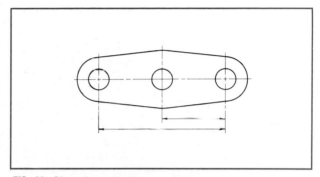

FIG. 38 Dimensions of position for holes.

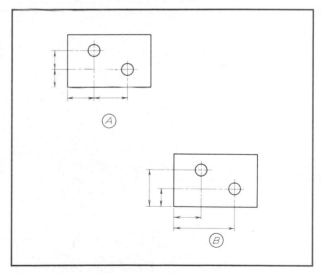

FIG. 39 Dimensions of position for holes.

note giving the diameter, operation, and depth. If there is more than one hole of the same kind, the leader needs to point to but one hole, and the number of holes is stated in the note (Fig. 34). Several operations involving one hole may be grouped in a common note. Figures 35 to 37 show typical dimensioning practice for drilled and reamed, counterbored, countersunk, and spot-faced holes.

The leader to a hole should point to the *circular* view if possible. The pointing direction is toward the center. With concentric circles, the arrowhead should touch the inner circle (usually the first operation) unless an outer circle would pass through the arrowhead. Then, the arrow should be drawn to touch the outer circle.

Positioning of holes can be influenced by how parts are held together. Mating parts secured by screws, rivets, etc., must have holes positioned from common datum surfaces or lines to ensure matching of the holes. When two or more holes are on an established center line, the holes will require a dimension of position in one direction only (Fig. 38). If the holes are not on a common center line, they will require positioning in two directions, as in Fig. 39. The method at *B* is preferred when it is important to have the positions of both holes estab-

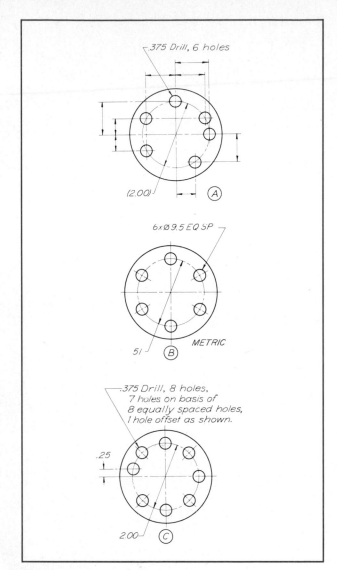

FIG. 40 Positioning of holes on "hole circles." *(A)* from datum lines; *(B)* equally spaced; *(C)* one hole offset.

lished from datum features (left and lower sides of the part).

The coordinate method for the positioning of holes (Fig. 40A) is preferred in precision work. The hole circle is often drawn and its diameter given for reference purposes, as in the figure. The diameter of a hole circle is invariably given on the circular view. The datum lines in this case are the center lines of the part.

Hole circles are circular center lines, often called "bolt circles," on which the centers of a number of holes are located.

One practice is to give the diameter of the hole circle and a note specifying the size of the holes, the number required, and the spacing, as in Fig. 40B. If one or more holes are not in the regular equally spaced position, their location may be given by an offset dimension, as shown at C.

Figure 41 shows holes located by polar coordinates.

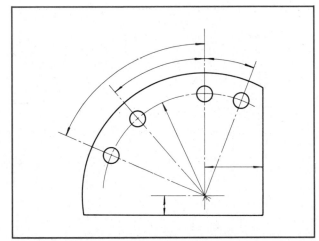

FIG. 41 Holes positioned by angle from datum line.

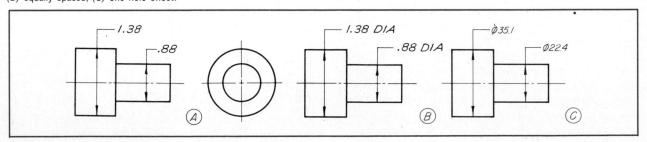

FIG. 42 Dimensioning of cylinders.

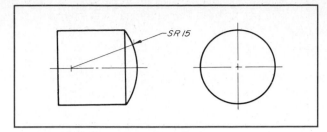

FIG. 43 Dimensioning a spherical surface.

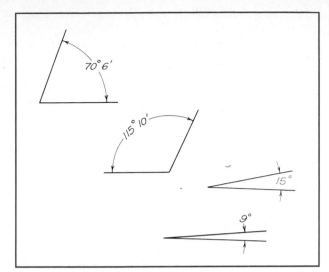

FIG. 44 Dimensioning of angles.

This method should be used when accurate shop equipment is available for locating holes by angle.

D. Cylinders and Spherical Surfaces Cylinders having an end view may be dimensioned as in Fig. 42A. If there is no end view, DIA should be used in the inch system, as in Fig. 42B. In the metric system diameter is given by the symbol ϕ as seen in Fig. 42C.

Spherical surfaces should be dimensioned as in Fig. 43 where SR stands for spherical radius.

E. Angles The dimension line for an angle is a circle arc with its center at the intersection of the sides of the angle (Fig. 44). The value is placed to read horizontally, with the exception that in the aligned system large arcs have the value aligned with the dimension arc. Angular values should be written in the form 35°7′ with no dash between the degrees and the minutes.

F. Chamfers Chamfers may be dimensioned by note, as in Fig. 45A, if the angle is 45°. The linear size is understood to be a short side of the chamfer triangle; the dimensioning without a note shown at B is in conformity with this. If the chamfer angle is other than 45°, it is dimensioned as at C. Internal chamfers may be dimensioned as in Fig. 46.

G. Tapers The term "taper" as used in machine work usually means the surface of a cone frustum. If a standardized taper is used, the specification should be accompanied by one diameter and the length as shown in Fig. 47A. The general method of giving the diameters of both ends and the taper per foot is illustrated at B. An alternative method is to give one diameter, the length, and the taper per foot. Taper per foot is defined as the difference in diameter in inches for 1 ft of length. A metric method is seen at C.

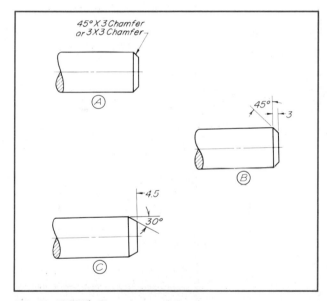

FIG. 45 METRIC. Dimensioning of chamfers.

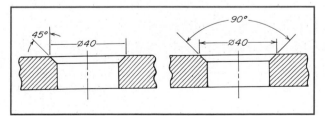

FIG. 46 METRIC. Dimensioning internal chamfers.

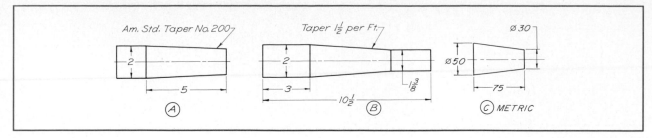

FIG. 47   Dimensioning of tapers.

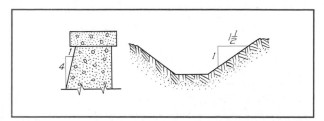

FIG. 48   Dimensioning of batters and slopes.

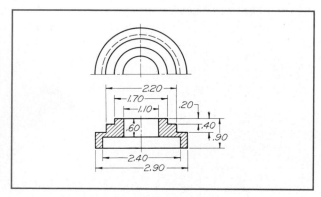

FIG. 49   Dimensions for a sectional view.

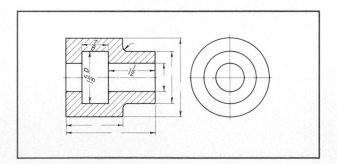

FIG. 50   Dimensions inside the view.

H. Batters, Slopes, Grades *Batter* is a deviation from the vertical, such as is found on the sides of retaining walls, piers, etc., and *slope* is a deviation from the horizontal. Each is expressed as a ratio with one factor equal to unity, as illustrated in Fig. 48. *Grade* is identical with slope but is expressed in percentage, the inclination in feet per hundred feet. In structural work angular measurements are shown by giving the ratio of run to rise, with the larger side 12 in.

17. DIMENSIONING SIZES, PARTS, AND TOOLS

For dimensioning common fasteners, wire sizes, and sheet-metal gages, standards have been established to assist both the drafter and the designer. Sometimes there are prevailing company standards that may be used. Also, manufacturers' standards are available. The ANSI standards are quite useful as an excellent alternative. There are numerous tables in the appendix that will be of assistance to you in the dimensioning of various parts.

18. DIMENSIONING A SECTIONAL VIEW

Dimensions that must be placed on sectional views are usually placed outside the view so as not to be crowded within crosshatched areas. See Fig. 49.

However, sometimes a dimension *must* be placed across a crosshatched area. When this is the case, the crosshatching is left out around the dimension figures, as illustrated in Fig. 50.

19. DIMENSIONING A HALF SECTION

In general, the half section is difficult to dimension clearly without some possibility of crowding and giving misleading or ambiguous information. Generous use of notes and careful placement of dimension lines, leaders, and figures will in most cases make the dimensioning clear; but if a half section cannot be clearly dimensioned, an extra view or part view should be added on which to describe the size.

Inside diameters should be followed by the letter *D* and the dimension line carried over the center line, as in Fig. 51, to prevent the possibility of reading the dimension as a radius. Sometimes the view and the dimensioning can both be clarified by showing the dashed lines on the unsectioned side. Dimensions of internal parts, if placed inside the view, will prevent confusion between extension lines and the outline of external portions.

20. DIMENSIONING A PICTORIAL VIEW

Pictorial drawings are often more difficult to dimension than orthographic drawings because there is one view instead of several, and the dimensioning may become crowded unless the placement is carefully planned. In general, the principles of dimensioning for orthographic drawings should be followed whenever possible. The following rules should be observed:

1. Dimension and extension lines should be placed so as to lie *in* or *perpendicular to* the face on which

the dimension applies. See Prob. 27, Chap. 5.

2. Dimension numerals should be placed so as to lie in the plane in which the dimension and extension lines lie. See Prob. 16, Chap. 5.

3. Leaders for notes and the lettered note should be placed so as to lie in a plane parallel or perpendicular to the face on which the note applies. See Prob. 28, Chap. 5.

4. Lettering of dimension values and of notes should be made so that the lettering appears to lie in or parallel to one of the principal faces of the pictorial drawing. See Prob. 31, Chap. 5.

The American Standard permits lettering of pictorial drawings according to the unidirectional system, with either vertical or slant lettering. This is principally to make possible the use of mechanical lettering devices. If the unidirectional system is employed,

1. Dimension values are lettered to read from the bottom of the sheet.

2. Notes are lettered so that they lie *in the picture plane*, to read from the bottom of the sheet. Notes should be kept off the view, if possible.

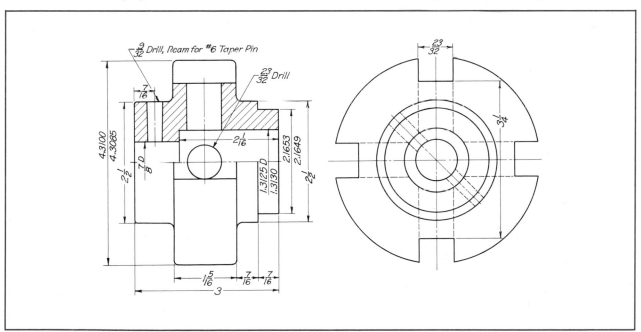

FIG. 51 Dimensioning a half section.

TOLERANCING SYSTEMS

21. NEED FOR DIMENSIONAL CONTROL

Most products in use today consist of an assembly of two or more parts. The individual parts are made separately and then are brought together in the assembly process. One hopes that all of the parts go together in the manner the designer intended, that the degree of "closeness" or "tightness" desired between mating parts is indeed achieved. Mass production techniques demand that careful control measures exist as to part size, location, and assembly precision.

Precision is the degree of accuracy necessary to ensure the functioning of a part as intended. As an example, a cast part usually has two types of surfaces: mating surfaces and nonmating surfaces. The mating surfaces are machined to achieve the proper smoothness and to be at the correct distance from each other. The nonmating surfaces, with no important relationship to other parts or surfaces, are left in their original rough-cast form. Thus mating surfaces ordinarily require much greater manufacturing precision than nonmating surfaces. The dimensions on a drawing must indicate which surfaces are to be finished and what degree of precision is required in finishing. However, because it is impossible to produce any distance to an absolute size, some variation must be allowed in manufacture.

22. FORMS OF SIZE CONTROL

As a preview of concepts yet to be presented, a brief historical overview should help clarify the scope of size control. Early in the industrial revolution of the 1800s engineers began to realize the importance of controlling size and location so that parts could be assembled properly. Controls placed on lengths and diameters of parts were among the first controls. As machine tools developed to give more precision, size control became easier to manage.

By the 1930s, interest had grown in devising formal ways to control the shape and form of parts, for example, in the areas of parallelism between surfaces, perpendicularity, and concentricity of diameters. From these concerns has evolved the concept of geometric tolerancing, an area that will be covered in this chapter.

Standards of practice for size control have become even more vital with the emergence of computer-aided manufacturing. The programs used by the computers require specific instructions, but to be cost-effective the programs must have wide application by adhering to standards.

Essentially all size control is applied to dimensions in decimal form. Decimals may be in inches or millimeters. However, since the inch system evolved mainly in Great Britain and the United States, and the metric system in continental Europe, some minor variations exist between standards of the inch and the millimeter. In recent years, considerable effort has been given to reaching compatible compromises between inch and millimeter systems. The International Organization for Standardization (ISO) and the American National Standards Institute (ANSI) are among major groups involved in the modernization of standards.

23. NOMENCLATURE

The terms used in size control must be understood before a detailed study is done. The following terms are adapted from ANSI definitions:

1. *Nominal size.* The nominal size is the designation that is used for the purpose of general identification.

2. *Basic size.* The basic size is that size from which the limits of size are derived by the application of allowances and tolerances.

3. *Design size.* The design size is that size from which the limits of size are derived by the application of tolerances. When there is no allowance, the design size is the same as the basic size.

4. *Dimension.* A dimension is a geometric characteristic such as diameter, length, angle, or center distance.

5. *Tolerance.* A tolerance is the total permissible variation of a size. The tolerance is the difference between the limits of size.

6. *Unilateral tolerance.* A unilateral tolerance is a tolerance in which variation is permitted only in one direction from the design size.

7. *Bilateral tolerance.* A bilateral tolerance is a tolerance in which variation is permitted in both directions from the design size.

8. *Fit.* Fit is the general term used to signify the range of tightness that may result from the application of a specific combination of allowances and tolerances in the design of mating parts.

9. *Actual fit.* The actual fit between two mating parts is the relation existing between them with respect to the amount of clearance or interference that is present when they are assembled.

10. *Clearance fit.* A clearance fit is one having limits of size so prescribed that a clearance always results when mating parts are assembled.

11. *Interference fit.* An interference fit is one having limits of size so prescribed that an interference always results when mating parts are assembled.

12. *Transition fit.* A transition fit is one having limits of size so prescribed that either a clearance or an interference may result when mating parts are assembled.

13. *Allowance.* An allowance is an intentional difference between the maximum material limits of mating parts. (See definition of "fit.") It is a minimum clearance (positive allowance) or maximum interference (negative allowance) between mating parts.

14. *Maximum material limit.* A maximum material limit is the maximum limit of size of an external dimension or the minimum limit of size of an internal dimension.

15. *Minimum material limit.* A minimum material limit is the minimum limit of size of an external dimension or the maximum limit of size of an internal dimension.

16. *Basic-hole system.* A basic-hole system is a system of fits in which the design size of the hole is the basic size and the allowance is applied to the shaft.

17. *Basic-shaft system.* A basic-shaft system is a system of fits in which the design size of the shaft is the basic size and the allowance is applied to the hole.

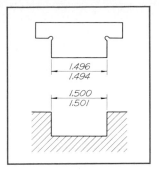
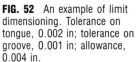
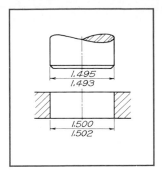

FIG. 52 An example of limit dimensioning. Tolerance on tongue, 0.002 in; tolerance on groove, 0.001 in; allowance, 0.004 in.

FIG. 53 A clearance fit. The tightest fit is 0.005 in clearance; the loosest, 0.009 in clearance.

In an illustration of some of these terms, a pair of mating parts is dimensioned in Fig. 52. In this example the *nominal size* is 1½ in. The *basic size* is 1.500. The *allowance* is 0.004. The *tolerance* on the tongue is 0.002, and on the slot it is 0.001. The *limits* are, for the tongue, 1.496 (maximum) and 1.494 (minimum) and, for the slot, 1.501 (maximum) and 1.500 (minimum).

24. EXAMPLES OF GENERAL FITS

The clearance, interference, and transition fits referred to in the previous section can be further explained by an example of each. The fits established on machine parts are classified as follows:

A *clearance fit* is the condition in which the internal part is smaller than the external part, as illustrated by the dimensioning in Fig. 53. In this case the largest shaft is 1.495 in and the smallest hole 1.500 in, leaving a clearance of 0.005 for the tightest possible fit.

An *interference fit* is the opposite of a clearance fit, having a definite interference of metal for all possible conditions. The parts must be assembled by pressure or by heat expansion of the external member. Figure 54 is an illustration. The shaft is 0.001 in larger than the hole for the loosest possible fit. The allowance in this case is 0.003 in interference.

A *transition fit* is the condition in which either a clearance fit or an interference fit may be had. Figure 55 illustrates a transition fit; the smallest shaft in the largest hole results in 0.0003 in clearance, and the largest shaft in the smallest hole results in 0.0007 in interference.

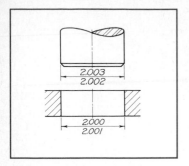

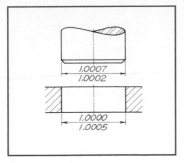

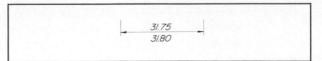

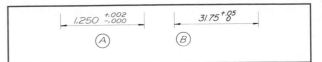

FIG. 56 A tolerance expressed as limits in millimeters.

FIG. 54 An interference fit. The loosest fit is 0.001 in. interference; the tightest, 0.003 in. interference.

FIG. 55 A transition fit. The loosest fit is 0.0003 in. clearance; the tightest, 0.0007 in. interference.

FIG. 57 Tolerances plus and minus.

25. GENERAL VS. SPECIFIC TOLERANCES

Tolerance may be *general*, as given as a note in the title block, or *specific*, as given with a dimension value.

A. General Tolerance Notes   The general tolerances apply to all dimensions not carrying a specific tolerance. The general tolerance should be allowed to apply whenever possible, using specific tolerances only when necessary. If no tolerances are specified, the value usually assumed for fractional dimensions is $\pm\frac{1}{64}$ in; for angular dimensions $\pm\frac{1}{2}°$; and for decimal inch dimensions plus or minus the nearest significant figure, as, for example, ±0.01 in for a two-place decimal and ±0.001 in for a three-place decimal.

As in the inch system, a general note in the metric system may be given in the following form, for example: *"Unless otherwise specified all untoleranced dimensions are ±0.6 mm."* Such a note is used only when the magnitudes of the dimensions do not vary widely.

The standard given in ANSI B4.3—1978 suggests the use of a chart (Table 2) to provide appropriate general notes. In Fig. 51 the fine and medium series apply to machined parts and the coarse series to other objects. As an example, a length of 50 mm given a medium-series general tolerance would have a size of 50 ± 0.3 mm.

B. Specific Tolerance Notes   There are several methods of expressing tolerances. The method preferred in quantity-production work is to write the two limits representing the maximum and minimum acceptable sizes, as in Fig. 56. An internal dimension has the *minimum* size above the line, and an external dimension has the *maximum* size above the line. This arrangement is for convenience in machining.

Another method is to give the basic size followed by the tolerance, plus and minus, as in Fig. 57. Inch units are shown in Fig. 57A, while the millimeter equivalent is shown in Fig. 57B.

TABLE 2   RECOMMENDED GENERAL TOLERANCES—METRIC LINEAR

Basic dimensions, mm	VARIATIONS IN MILLIMETERS						
	0.5 to 3	Over 3 to 6	Over 6 to 30	Over 30 to 120	Over 120 to 315	Over 315 to 1000	Over 1000 to 2000
Permissible variations:							
Fine series	±0.05	±0.05	±0.1	±0.15	±0.2	±0.3	±0.5
Medium series	±0.1	±0.1	±0.2	±0.3	±0.5	±0.8	±1.2
Coarse series	...	±0.2	±0.5	±0.8	±1.2	±2	±3

(ANSI B4.3—1978.)

Figure 57 is an example of a *unilateral* tolerance, in which variation occurs in only one direction from basic size. Unilateral tolerance can also be applied to fractions and angles. For fractional dimensions,

$$\tfrac{1}{2} - \tfrac{1}{32} \quad \text{or} \quad \tfrac{1}{2}\, {}^{+\ 0}_{-\ \tfrac{1}{32}}$$

For angular dimensions,

$$64°15'30'' + 0°45'0''$$

$$\text{or} \quad 64°15'30''\, {}^{+\ 0°45'0''}_{-\ 0°0'0''}$$

Bilateral tolerances are expressed by giving the basic value followed by the divided tolerances, both plus and minus (commonly equal in amount), as

$$1.500\, {}^{+\ 0.002}_{-\ 0.002} \quad \text{or} \quad 1.500 \pm 0.002$$

The millimeter equivalent would be $38.1 + 0.051$. Bilateral tolerances can be applied to fractions and angles. For fractional dimensions,

$$1\tfrac{1}{2}\, {}^{+\ \tfrac{1}{64}}_{-\ \tfrac{1}{64}} \quad \text{or} \quad 1\tfrac{1}{2} \pm \tfrac{1}{64}$$

For angular dimensions,

$$30°0'\, {}^{+\ 0°10'}_{-\ 0°10'} \quad \text{or} \quad 30°0' \pm 0°10'$$

26. CUMULATIVE TOLERANCES

Tolerances are said to be cumulative when a position in a given direction is controlled by more than one tolerance. In Fig. 58A the holes are positioned one from another. Thus, the distance between two holes separated by two, three, or four dimensions will vary in position by the sum of the tolerances on all the dimensions. This difficulty can be eliminated by dimensioning from *one* position, which is used as a datum for all dimensions, as shown in Fig. 58B. This system is commonly called base-line dimensioning.

Figure 59 is a further example of the effect of cumulative tolerances. The position of surface Y with respect to surface W is controlled by the additive tolerances on dimensions A and B. If it is important, functionally, to hold surface Y with respect to surface X, the dimensioning used is good. If, however, it is more important to hold surface Y with respect to surface W, the harmful effect of cumulative tolerances can be avoided by dimensioning as in Fig. 60. Cumulative tolerance, however, is always present; in Fig. 60 the position of surface Y with respect to surface X is now subject to the cumulative tolerances of dimensions A and C.

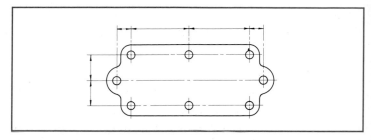

FIG. 58A Successive dimensioning. Tolerances accumulative.

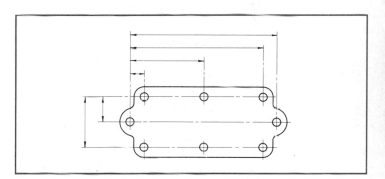

FIG. 58B Dimensions from datum.

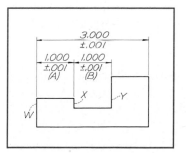

FIG. 59 Control of surface position through different toleranced dimensions. Here one dimension is successive.

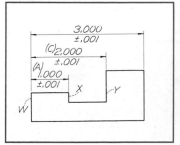

FIG. 60 Control of surface position through different toleranced dimensions. All dimensions are from datum.

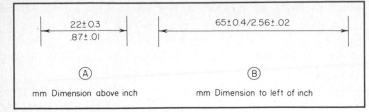

FIG. 61　Dual dimensioning: position method.

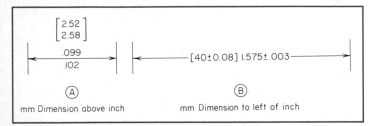

FIG. 62　Dual dimensioning: bracket method.

27. DUAL DIMENSIONING

Until a particular company is fully converted to the metric system, a format known as dual dimensioning is sometimes used. Dual dimensioning uses both inch units and millimeter units on the same drawing. The two decimals (inches and millimeters) for a given dimension must properly round off so that the same type of fit is obtained with either decimal. One good source of information on rounding and conversion is ASTM (E380) 41.1981.

Three methods are used for dual dimensioning: (1) the position method, (2) the bracket method, and (3) the chart method.

A. The Position Method　In this method the millimeter units are placed above the inch units, *or* the millimeter units are placed to the left of the inch units, with a slash separating them. Figure 61 illustrates each case.

B. The Bracket Method　The most common form of this method encloses the millimeter units in brackets, as in Fig. 62. Position of the millimeter units is not critical, but they are often placed above or to the left of the inch units.

C. Chart Method　This method uses only one decimal format on the drawing, typically millimeter units. The conversion chart is then provided at a convenient loca-

Operation	Tolerance
Casting	
Sand	
Small	±0.030
Medium	±0.060
Large	±0.015+
Die	
Small	±0.002
Medium	±0.008
Large	±0.010
Forging, size	
½–2 lb	±0.030
2–10 lb	±0.060
10–60 lb	±0.120
Drilling, diameter	
No. 60–No. 30	0.002
No. 29–No. 1	0.004
¼–½	0.006
½–¾	0.008
¾–1	0.010
1–2	0.015
Reaming, diameter	
0.2–0.5	0.0005
0.5–1.0	0.0010
1.0+	0.0015
Lath work	
Rough, diameter	
¼–½	0.005
½–1	0.007
1–2	0.010
2+	0.015
Finish, diameter	
¼–½	0.002
½–1	0.003
1–2	0.005
2+	0.007
Broaching, diameter	
½–1	0.001
1–2	0.002
2–4	0.003
Grinding	0.0005
Milling	0.002–0.005
Planing	0.005–0.010

FIG. 63　Typical tolerances available with specific machining operations (Values in inches. Multiply by 25.4 for millimeters.).

tion on the drawing. Within the conversion chart each dimension on the drawing has listed its equivalent value, whether inches or millimeters. In this way the actual drawing is less crowded with dimensions than when using the position or bracket methods.

It is common practice when converting from millimeters to inches to round off to one more digit to the right of the decimal point than given in the millimeter dimension. For example, using the conversion factor 25.4 mm/in, 40.5 mm = 1.59 in.

Conversely, when converting from inches to millimeters, round off the millimeters to one less digit to the right of the decimal point than given in the inch dimension. For example, 1.25 in = 31.8 mm.

28. AVAILABLE PRECISION IN MANUFACTURING

An engineer must have knowledge of manufacturing processes before he or she can decide on the precision needed for a designed part. Only then can proper fits and tolerances be specified. Engineering experience is helpful, and critical tolerances are sometimes determined by exhaustive testing of experimental models. The different manufacturing processes all have inherent minimum possible accuracies, depending upon the size of the work, the condition of the equipment, and, to some extent, the skill of the workers. Figure 63 gives typical tolerances to be used as a guide; they are based on the assumption that the work is to be done on a quantity-production basis with equipment in good condition. Greater precision can be attained by highly skilled workers on a unit-production basis.

29. SURFACE QUALITY

Simple dimensioning of the height, width, and depth of a part indicates little of the surface condition of that part. A surface could be in an as-cast condition or it could be machined by various methods giving varying degrees of roughness. A designer must be able to specify surface quality so that the part can function as intended. Normally a designer would leave a surface (internal or external) as rough as feasible to minimize production costs.

A. Finish Marks   The most basic designation that may be placed on a surface to indicate that it has been finished is the finish mark. Figure 64 shows a part having finish marks. Finish marks indicate only that a surface has been machined. Degrees of roughness or waviness of the surface are not revealed.

Finish marks always appear on the *edge view* of the surface to be machined, and in all views where edge view of the particular surface shows. The V symbol serves as a finish mark. It is made about ⅛ in [3 mm] high.

When dimensioning notes are given for drilling, boring, reaming, and other operations requiring machining, finish marks are omitted because they are redundant. Finish marks are also omitted if a part is made from machined stock, such as a cold-rolled rod. If an entire part is machined, finish marks may be omitted and a note substituted: "Finish all over."

B. Control of Surface Texture   Information is needed in addition to a finish mark if the exact nature of surface texture is to be known. Figure 65 illustrates the terms associated with surface texture. ANSI Standard B46.1—1978 defines the factors of surface texture. Any surface

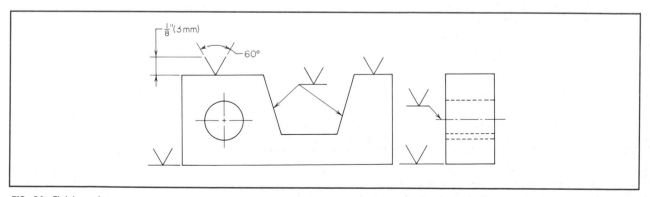

FIG. 64 Finish marks.

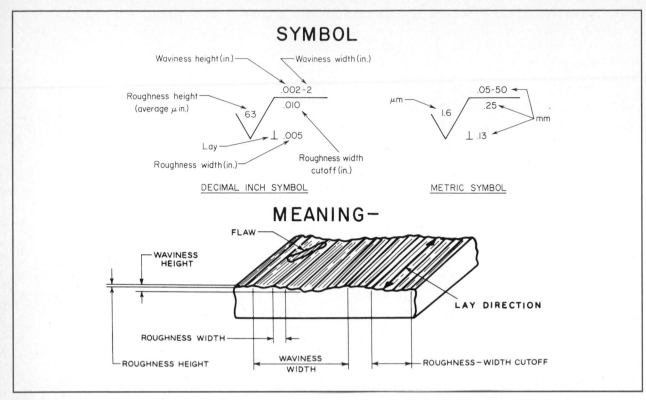

FIG. 65 Interpretation of surface texture symbols.

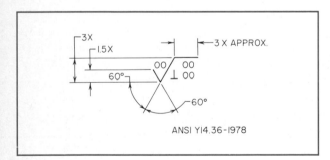

FIG. 66 Size of surface texture symbol.

TABLE 3 PREFERRED SERIES OF ROUGHNESS AVERAGE VALUES

MICROMETERS		MICROINCHES	
μm	μin	μm	μIn
0.012	0.5	1.25	50
0.025*	1*	1.60*	63*
0.050*	2*	2.0	80
0.075	3	2.5	100
0.10*	4*	3.2*	125*
0.125	5	4.0	160
0.15	6	5.0	200
0.20*	8*	6.3*	250*
0.25	10	8.0	320
0.32	13	10.0	400
0.40*	16*	12.5*	500*
0.50	20	15	600
0.63	25	20	800
0.80*	32*	25*	1000*
1.00	40		

*Recommended   (ANSI Y14.36—1978.)

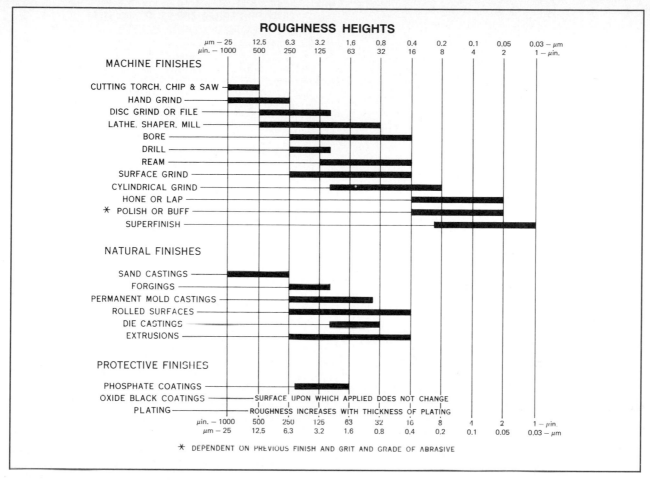

FIG. 67 Roughness heights as related to machine finishes.

has minute peaks and valleys, the height of which is termed "surface roughness." Surface roughness may be superimposed on a more general "waviness." The direction of tool marks is called "lay." The relative size of the surface-texture symbol is given in ANSI Standard Y14.36—1978 and is shown in Fig. 66.

30. DESIGNATION OF SURFACE TEXTURE

The terms to which the surface-texture symbol relates are shown in Fig. 65 and are defined below.

1. *Roughness*, produced by tool cutting edges and feed, is expressed as the arithmetical average [in micrometers (μm) or microinches (μin)].

2. *Roughness height* is the average variation from the average height line of the cut (in micrometers or microinches). ANSI Y14.36—1978 offers a chart of preferred roughness average values. See Table 3. Roughness heights actually produced by common tooling are shown in Fig. 67.

3. *Roughness width* is the maximum permissible spacing between repetitive units of the surface pattern (in inches or millimeters).

4. *Roughness-width cutoff* is the maximum width of surface irregularities to be included in the measurement of roughness height (in inches or millimeters).

5. *Waviness* designates irregularities of greater spac-

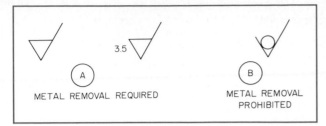

FIG. 68 Modifications to surface texture symbol.

Lay Symbol	Meaning	Example Showing Direction of Tool Marks
—	Lay approximately parallel to the line representing the surface to which the symbol is applied.	
⊥	Lay approximately perpendicular to the line representing the surface to which the symbol is applied.	
X	Lay angular in both directions to line representing the surface to which the symbol is applied.	
M	Lay multidirectional.	
C	Lay approximately circular relative to the center of the surface to which the symbol is applied.	
R	Lay approximately radial relative to the center of the surface to which the symbol is applied.	
P[3]	Lay particulate, non-directional, or protuberant.	

FIG. 69 Lay symbols. *(ANSI Y14.36—1978.)*

ing than roughness, resulting from tool deflection and vibration. Height is the peak-to-valley distance; width is the distance between peaks of two adjoining waves (in inches or millimeters). ANSI

TABLE 4 PREFERRED SERIES OF MAXIMUM WAVINESS HEIGHT VALUES

Millimeters	Inches	Millimeters	Inches	Millimeters	Inches
0.0005	0.00002	0.008	0.0003	0.12	0.005
0.0008	0.00003	0.012	0.0005	0.20	0.008
0.0012	0.00005	0.020	0.0008	0.25	0.010
0.0020	0.00008	0.025	0.001	0.38	0.015
0.0025	0.0001	0.05	0.002	0.50	0.020
0.005	0.0002	0.08	0.003	0.80	0.030

(ANSI Y14.36—1978.)

Y14.36—1978 provides a chart of preferred maximum waviness height values. See Table 4.

The symbol shown in Fig. 66 is the basic symbol for surface texture. It is modified as needed to convey specific information. One way in which the symbol may be modified is to indicate whether or not material is to be removed from the part to achieve the desired surface texture. Figure 68A gives the symbol used when material removal by machining is required. A numeral, 3.5 mm in this example, placed to the left of the symbol is the material removal allowance, and indicates the amount to be removed (in inches or millimeters). Figure 68B is used when metal removal is prohibited, meaning that the part should be produced by a process leaving the surface with an appropriate texture. Such a process could be forging, hot or cold finishing, casting, or powder metallurgy. Realize that a surface texture may or may not indicate the need to remove metal. However, a *finish* mark (Sec. 29) always implies material removal.

The application of surface-texture notation to a complete symbol was seen in Fig. 65. Note the portion of the symbol that indicates that the lay is to be perpendicular. The various lay symbols (ANSI Y14.36—1978) are illustrated in Fig. 69.

One can appreciate that the use of surface-texture symbols is a well-standardized area. Use of such symbols is best left to well-qualified persons who will use them correctly and sparingly. Production costs rise as more and more surface requirements are placed on a part. In general, only those specifications of surface texture that are essential to the proper functioning of the part are designated.

STANDARDIZED TOLERANCES

31. OVERVIEW

The general concepts of tolerancing were presented in Secs. 21 to 26. Although instructive, our work in tolerancing so far has not dealt with specific standards for fits of parts. A designer could presumably make up a system of fits, but in terms of interchangeability of parts with other manufacturers, a confusing situation might result. Conversely, use of a standard fit for a particular part would enable a person in London to communicate effectively with a person in San Francisco regarding the exact tolerances permitted.

The application of standards is a broad field that would fill several textbooks. However, a brief study of the more common applications is appropriate. We will deal with the following areas:

1. Cylindrical fits, decimal inch system, based on ANSI Standard B4.1—1967 (1979).

2. Cylindrical fits, millimeter (SI metric) system, based on ANSI Standard B4.2—1978.

3. Geometric tolerancing, including tolerances of form and location, based on ANSI Standard Y14.5M—1982.

32. CYLINDRICAL FITS: DECIMAL INCH SYSTEM

A. Designation of ANSI Fits   The fits in ANSI Standard B4.1—1967 (1979) are designated by symbols that facilitate reference for educational purposes. These symbols are not to be shown on manufacturing drawings; instead, sizes should be specified. The letter symbols used are as follows:

RC, running or sliding fit

LC, locational clearance fit

LT, transition fit

LN, locational interference fit

FN, force or shrink fit

These letter symbols are used in conjunction with numbers for the class of fit; thus FN 4 represents a class 4 force fit. Each symbol (two letters and a number) represents a complete fit, for which the minimum and maximum clearance or interference and the limits of size for the mating parts are given in the appendix.

B. Description of ANSI Fits

▪ **(1) Running and Sliding Fits** (Appendix 8A)   These are classified as follows:

RC 1, *close sliding fits*, accurately locate parts that must assemble without perceptible play.

RC 2, *sliding fits*, are for accurate location, but with greater maximum clearance than RC 1. Parts move and turn easily but do not run freely, and in the larger sizes may seize with small temperature changes.

RC 3, *precision running fits*, are about the closest fits expected to run freely, and are for precision work at slow speeds and light journal pressures. They are not suitable under appreciable temperature differences.

RC 4, *close running fits*, are chiefly for running fits on accurate machinery with moderate surface speeds and journal pressures, where accurate location and minimum play are desired.

RC 5 and RC 6, *medium running fits*, are higher running speeds, heavy journal pressures, or both.

RC 7, *free running fits*, are for use where accuracy is not essential, where large temperature variations are likely, or under both these conditions.

RC 8 and RC 9, *loose running fits*, are for materials such as cold-rolled shafting and tubing, made to commercial tolerances.

▪ **(2) Locational Fits** (Appendix 8B to D)   These fits determine only the location of mating parts and may provide rigid or accurate location—as in interference fits—or some freedom of location—as in clearance fits. They fall into three groups:

LC, *locational clearance fits*, are for normally stationary parts that can be freely assembled or disassembled. They run from snug fits for parts requiring accuracy of location through the medium clearance fits for parts such as spigots, to the looser fastener fits where freedom of assembly is important.

LT, *locational transition fits*, fall between clearance and interference fits for application where accuracy of location is important, but a small amount of clearance or interference is permissible.

LN, *locational interference fits*, are used where accuracy of location is of prime importance, and for parts needing rigidity and alignment with no special requirements for bore pressure. Such fits are not for parts that transmit frictional loads from one part to another by virtue of the tightness of fit; these conditions are met by force fits.

■ **(3) Force Fits** (Appendix 8E)   A force fit is a special type of interference fit, normally characterized by maintenance of constant bore pressures throughout the range of sizes. Thus the interference varies almost directly with diameter, and to maintain the resulting pressures within reasonable limits, the difference between its minimum and maximum value is small.

FN 1, *light-drive fits*, require light assembly pressures and produce more or less permanent assemblies. They are suitable for thin sections, long fits, or cast-iron external members.

FN 2, *medium-drive fits*, are for ordinary steel parts or for shrink fits on light sections. They are about the tightest fits that can be used with high-grade cast-iron external members.

FN 3, *heavy-drive fits*, are suitable for heavier steel parts or for shrink fits in medium sections.

FN 4 and FN 5, *force fits*, are for parts that can be highly stressed, or for shrink fits where heavy pressing forces are impractical.

C. Example of Dimensioning ANSI Fits   Assume that a 1-in shaft is to run with moderate speed but with a fairly heavy journal pressure. The fit class chosen is RC 6.

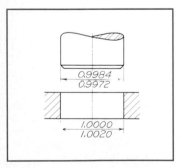

FIG. 70 An ANSI clearance fit. This one is class RC 6. Tolerance on shaft, 0.0012 in; tolerance on hole, 0.0020 in; min. clearance, 0.0016 in, max. clearance, .0048 in.

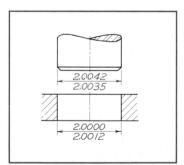

FIG. 71 An ANSI interference fit. This one is class FN 4. Tolerance on shaft, 0.0007 in; tolerance on hole, 0.0012 in; allowance, 0.0042 in interference.

From Appendix 8A the limits given are:

$$Hole: \begin{matrix} +2.0 \\ -0 \end{matrix}$$
$$Shaft: \begin{matrix} -1.6 \\ -2.8 \end{matrix}$$

The dimensioned parts are shown in Fig. 70. These values are thousandths of an inch. The basic size is 1.0000. Therefore, for the hole, 1.0000 + 0.0020 gives 1.0020 as the maximum limit, and 1.0000 + 0.0000 gives 1.0000 for the minimum limit. For the shaft, 1.0000 − 0.0016 gives 0.9984 for the maximum limit, and 1.0000 − 0.0028 gives 0.9972 for the minimum limit.

To analyze the fit, the tightest condition (largest shaft in smallest hole) is 1.0000 − 0.9984, or 0.0016, clearance. This is the *allowance*. The loosest condition (smallest shaft in largest hole) is 1.0020 − 0.9972, or 0.0048, clearance. The values of these two limits are given in the table under "limits of clearance."

To illustrate further, suppose a 2-in shaft and hub are to be fastened permanently with a drive fit. The hub is high-grade steel. Fit FN 4 has been chosen. Appendix 8E gives:

$$Hole: \begin{matrix} +1.2 \\ -0 \end{matrix}$$
$$Shaft: \begin{matrix} +4.2 \\ +3.5 \end{matrix}$$

The dimensioned parts are shown in Fig. 71.

These values are in thousandths of an inch. The basic size is 2.0000. Thus, for the hole 2.000 + 0.0012 gives 2.0012 for the maximum limit, and 2.000 − 0.0000 gives 2.0000 as the minimum limit. For the shaft, 2.0000 + 0.0042 gives 2.0042 for the maximum limit, and 2.0000 + 0.0035 gives 2.0035 for the minimum limit. The limits of fit are 0.0023 minimum interference and 0.0042 maximum interference. The allowance is 0.0042.

Note in both the above cases that, in dimensioning, the maximum limit of the hole and the minimum limit of the shaft are placed below the dimension line. This is for convenience in reading the drawing and prevention of mistakes by the machinist.

The ANSI fits are based on standard hole practice. Note in the tables (Appendix 8A to E) that the minimum hole size is always the basic size.

D. Basic-Shaft Fits The basic-shaft system could be used if a common shaft of constant diameter were to be used in conjunction with a series of holes. Recall that in the basic-shaft system (Sec. 23) the design size of the shaft is the basic size and the allowance is applied to the hole. The ANSI tables used for the basic-hole system still apply in terms of the given limits of clearance or interference.

Symbols for basic-shaft fits are the same as for basic-hole fits except that an S is added. For example, an RC 3 would become an RC 3S in the basic-shaft system.

To convert from a basic-hole system to a basic-shaft system, do the following:

1. For clearance fits: *Increase* the limits for hole and shaft by the amount needed to change the maximum shaft to the basic size.

2. For transition or interference fits: *Decrease* the limits for hole and shaft by the amount needed to change the maximum shaft to basic size. As an illustration, consider the examples shown in Table 5.

33. CYLINDRICAL FITS: SI METRIC SYSTEM

A. Background Nations of the world have been involved for many years in developing standards for metric cylindrical fits. In the United States, General Motors was one company that saw a need in 1973 for metric standards and developed an interim standard. In 1975 the B4 Standards Committee of ASME (American Society of Mechanical Engineers) proposed a standard for

TABLE 5 COMPARING BASIC-HOLE AND BASIC-SHAFT SYSTEMS

	BASIC-HOLE SYSTEM			BASIC-SHAFT SYSTEM	
	Hole	Shaft		Hole	Shaft
RC 6	1.000 1.002	0.9984 0.9972	RC 6S*	1.0016 1.0036	1.000 0.9988
FN 4	2.000 2.0012	2.0042 2.0035	FN 4S	1.9958 1.9970	2.0000 1.9993

*RC 6S limits were obtained by bringing the maximum limit on the shaft, 0.9984, up to 1.000, and adding this factor of 0.0016 to all other limits.
FN 4S limits were obtained by bringing the maximum limit on the shaft, 2.0042, down to 2.000, and subtracting this factor of 0.0042 from all other limits.

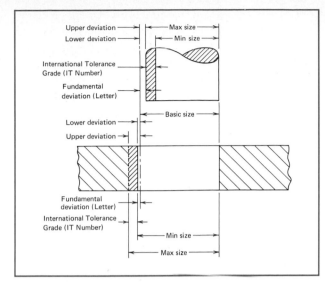

FIG. 72 Illustration of definitions. *(ANSI B14.2—1978.)*

preferred metric limits and fits, based on eight standards from countries worldwide. The proposed standard was accepted in 1978. This standard is highly compatible with the ISO standards used by numerous countries, and products produced to ANSI Standard B4.2—1978 are more cost-effective in world markets because of interchangeability of parts.

B. Definitions While general concepts of tolerancing are the same for both the decimal inch and metric systems, some specific definitions vary. We have already discussed those definitions relating to the decimal inch system (Sec. 23). Definitions within the metric system that *vary* from the decimal inch system are given below. Figure 72 is keyed to these definitions.

1. *Deviation* is the difference between any given size and the basic size for that diameter (or for any space between two parallel faces, such as a keyway, slot, etc.).

2. *Upper deviation* is the difference between the *maximum* permitted size and the basic size.

3. *Lower deviation* is the difference between the *minimum* permitted size and the basic size.

4. *Fundamental deviation* equals either the upper or lower deviation, whichever is closer to the basic size.

5. *Tolerance zone* is a zone that represents tolerance and its position relative to the basic size.

6. *International tolerance (IT) grade* is a group of tolerances that gives the same general level of accuracy within a given grade. The tolerances themselves vary in accordance with basic size.

Other definitions are identical to those within the decimal inch system, i.e., basic size, tolerance, hole basis, shaft basis, clearance fit, interference fit, and transition fit.

C. Symbols An appropriate symbol has been developed to express the maximum and minimum metric limits of a part. Let us give an example of a symbol and then explain it. For a hole, one could have the symbol

25H7

where 25 = the basic diameter, mm
   H = the tolerance position letter, which is an expression of the fundamental deviation desired
   7 = the grade number, which relates to the amount of tolerance desired (smaller grade numbers are for smaller tolerance zones; larger grade numbers for larger tolerance zones) and which correlates with international tolerance grades, such as IT6, IT9, etc.

For a shaft, one could have the symbol

25s6

where 25 = the basic diameter, mm
   s = the tolerance position letter
   6 = the grade number

Notice that the tolerance position letter for a hole is uppercase, for a shaft, lowercase.

A fit is shown by combining symbols for the hole and shaft. For our example above, the fit becomes

25H7/s6

The basic size is followed by the hole symbol. The shaft symbol comes last.

D. Preferred Sizes To simplify tooling setups, a system of preferred basic sizes is given in ANSI B4.2—1978. First and second choices are provided, as seen in

Table 6. Sizes lower than 1 mm or greater than 1000 mm may be selected by multiplying or dividing the sizes of Fig. 7.71 by 1000 or its multiples (such as 2000, 3000, etc.).

E. Preferred Fits Many combinations are possible for toleranced holes and toleranced shafts. To bring some order into the system, ANSI B4.2—1978 recommends a particular set of preferred fits. These fits are based on first-choice tolerance zones (although second and third choices do exist). The fits are also based on the first choice of preferred sizes (Table 6). The set of preferred fits is listed in the chart of Fig. 73. If one must use a nonpreferred size, its calculated fit is available in ANSI B4.2—1978.

As an example, let us say we wish to find the limit dimensions for a close running fit, hole basis, for a 60-mm diameter. Figure 73 indicates that an H8/f7 fit is

TABLE 6 PREFERRED SIZES

BASIC SIZE, mm		BASIC SIZE, mm		BASIC SIZE, mm	
First choice	Second choice	First choice	Second choice	First choice	Second choice
1		10		100	
	1.1		11		110
1.2		12		120	
	1.4		14		140
1.6		16		160	
	1.8		18		180
2		20		200	
	2.2		22		220
2.5		25		250	
	2.8		28		280
3		30		300	
	3.5		35		350
4		40		400	
	4.5		45		450
5		50		500	
	5.5		55		550
6		60		600	
	7		70		700
8		80		800	
	9		90		900
				1000	

(ANSI B4.2—1978.)

ISO symbol		Description
Hole basis	**Shaft* basis**	
H11/c11	C11/h11	*Loose running* fit for wide commercial tolerances or allowances on external members
H9/d9	D9/h9	*Free running* fit not for use where accuracy is essential, but good for large temperature variations, high running speeds, or heavy journal pressures
H8/f7	F8/h7	*Close running* fit for running on accurate machines and for accurate location at moderate speeds and journal pressures
H7/g6	G7/h6	*Sliding fit* not intended to run freely, but to move and turn freely and locate accurately
H7/h6	H7/h6	*Locational clearance* fit provides snug fit for locating stationary parts; but can be freely assembled and disassembled
H7/k6	K7/h6	*Locational transition* fit for accurate location, a compromise between clearance and interference
H7/n6	N7/h6	*Locational transition* fit for more accurate location where greater interference is permissible
II7/p6	P7/h6	*Locational interference* fit for parts requiring rigidity and alignment with prime accuracy of location but without special bore pressure requirements
H7/s6	S7/h6	*Medium drive* fit for ordinary steel parts or shrink fits on light sections, the tightest fit usable with cast iron
H7/u6	U7/h6	*Force* fit suitable for parts which can be highly stressed or for shrink fits where the heavy pressing forces required are impractical

On the left, a bracket spans the first four to five rows labeled "Clearance fits", then "Transition fits", then "Interference fits". On the right, arrows labeled "More clearance" and "More interference".

*The transition and interference shaft basis fits shown do not convert to exactly the same hole basis fit conditions for basic sizes in range from 0 through 3 mm. Interference fit P.7/h6 converts to a transition fit H7/p6 in the above size range.

FIG. 73 Description of preferred fits. *(ANSI B4.2—1978.)*

the proper one. Appendix 9A gives the information shown in Fig. 74.

Therefore the minimum clearance is 0.030 mm and the maximum clearance is 0.106 mm. We have, then, an easy-to-use set of tables available, thanks to considerable effort on the part of the persons on the B4 Standards Committee.

34. GEOMETRIC TOLERANCING: BACKGROUND

Traditional tolerancing methods which we have presented have covered the control of linear dimensions and diameters. A knowledge of such methods is vital to the practicing design engineer. Nevertheless, a person skilled in using only control of linear dimensions and diameters will be unable to deal with other types of con-

CLOSE RUNNING FIT		
Hole H8	**Shaft f7**	**Fit**
60.046	59.970	0.106
60.000	59.940	0.030

FIG. 74 Example of H8/f7 fit.

trol needed for parts. Control is needed for the *position* of features within parts, such as holes. Control is also needed for the *form* or shape of features, such as straightness, angularity, or parallelism of specific portions of the part.

Let us take an example of the implication of control of *position*. The use of traditional coordinate dimensioning of the center of the hole in Fig. 75 would yield a *square* tolerance zone. The designer might think that he or she is controlling the location of the center of the

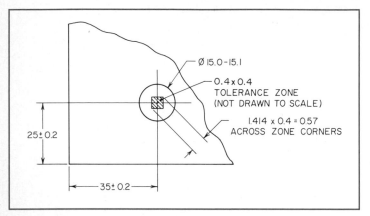

FIG. 75 METRIC. Effect of coordinate dimensioning.

15-mm hole within a 0.4-mm boundary. Actually, the center of the hole could vary across the diagonal of the square tolerance zone, yielding a maximum tolerance of not 0.4 mm but 0.57 mm.

This discrepancy suggests that there might be a better system to locate centers of holes. The natural tendency would be to locate the center of a hole within a circular tolerance zone rather than a square one. If the part could be dimensioned to provide a circular tolerance zone, the dimensioning would provide a more functional and better-controlled location of the center.

Fortunately for the engineer, a system called "geometric tolerancing" has been developed to aid in the control of tolerancing of position and form for parts. Much work has been done since the 1950s in standardizing practices in geometric tolerancing. One result of the effort is ANSI Standard Y14.5M—1982. Work in geometric tolerancing will reflect use of this standard, which provides for a closer understanding among nations in the

FIG. 76 Symbols for geometric tolerancing. *(ANSI Y14.5M—1982.)*

	TYPE OF TOLERANCE	CHARACTERISTIC	SYMBOL
FOR INDIVIDUAL FEATURES	FORM	STRAIGHTNESS	—
		FLATNESS	▱
		CIRCULARITY (ROUNDNESS)	○
		CYLINDRICITY	⌭
FOR INDIVIDUAL OR RELATED FEATURES	PROFILE	PROFILE OF A LINE	⌒
		PROFILE OF A SURFACE	⌓
FOR RELATED FEATURES	ORIENTATION	ANGULARITY	∠
		PERPENDICULARITY	⊥
		PARALLELISM	//
	LOCATION	POSITION	⊕
		CONCENTRICITY	◎
	RUNOUT	CIRCULAR RUNOUT	↗ *
		TOTAL RUNOUT	↗↗ *

*Arrowhead(s) may be filled in.

TERM	SYMBOL
AT MAXIMUM MATERIAL CONDITION	Ⓜ
REGARDLESS OF FEATURE SIZE	Ⓢ
AT LEAST MATERIAL CONDITION	Ⓛ
PROJECTED TOLERANCE ZONE	Ⓟ
DIAMETER	∅
SPHERICAL DIAMETER	S∅
RADIUS	R
SPHERICAL RADIUS	SR
REFERENCE	( )
ARC LENGTH	⌒

FIG. 77 Abbreviations for modifying symbols. *(ANSI Y14.5M—1982.)*

area of tolerancing, facilitating international use of tolerancing standards.

35. SYMBOLS USED IN GEOMETRIC TOLERANCING

Figure 76 gives a complete list of all necessary symbols to be used in our study of geometric tolerancing, while Fig. 77 gives common abbreviations to be used (examples will be given shortly). The proper size of feature control frames to be discussed is given in Fig. 78.

36. USE OF DATUMS

A datum is a feature of a part that acts as a master reference used to locate other features of the part. A datum can be a point, a line, or a plane. In Fig. 75, two planes are used as datum planes, the left-hand edge and the bottom surface. Datums are chosen for their functional nature. Any datum should be a readily available feature, such as a finished surface on a bench vise.

It may seem that establishing datums is an unnecessary complication for the designer. Actually, proper datums can make designs more clear, with increased ease of manufacture and subsequent inspection. Datums that increase the clarity of the design can reduce errors in production and thereby be very cost-effective. We will

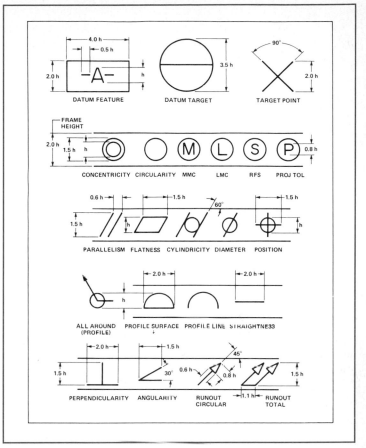

FIG. 78 Relative sizes for geometric tolerancing symbols. *(ANSI Y14.5M—1982.)*

soon see how datums can be represented by symbols on a drawing.

37. DATUM PLANES IN SPACE

The production of a part often relies on three mutually perpendicular directions to define space. We know these directions as width, height, and depth, which describe the x, y, and z axes, respectively, as seen in Fig. 79. A datum plane may be created along each of the three axes. The designer decides the priority of importance for the three datum planes.

The most important plane is termed datum A, chosen in Fig. 79 to be the horizontal plane. Datum A is known as the primary datum. Datums B and C of Fig. 79 have

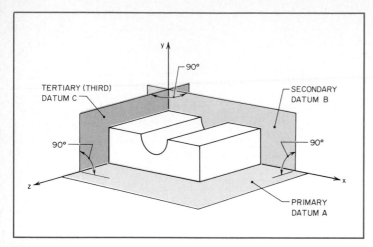

FIG. 79 Datum planes in space.

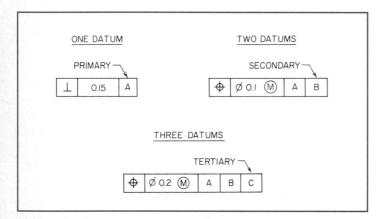

FIG. 80 Datums within the feature control frame (metric).

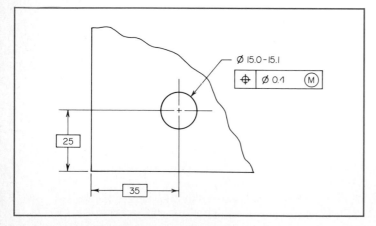

FIG. 81 **METRIC.** Positional tolerancing for Fig. 75.

been chosen to be the secondary and tertiary (third) datums, meaning that datum *B* is the second most important, and datum *C* the least important among our arbitrary choices.

Three points of contact with the part are needed to set the position of datum *A*. However, only two contact points are needed to set datum *B* since datum *B* is also 90° to datum *A*. Just a single point of contact of datum *C* with the part is needed, since datum *C* is at 90° to both datums *A* and *B*.

The number of datums needed to control positions on a part is dependent on the function of the part and how the designer wishes it to be made. Use of a single datum is usually essential. Sometimes two or even three datums are required. The feature control frame can be expanded to include one, two, or three datums as shown in Fig. 80.

38. TOLERANCES OF LOCATION

We have spent some time discussing the location (position) of features, beginning with Fig. 75. Let us now discuss control symbols that can help describe specific types of location. There are two types of locational tolerances: position and concentricity. (Refer to the list of Fig. 76.) Of these two, positional tolerancing is the more widely used. It is also called "true-position tolerancing." Positional tolerancing will help resolve the difficulty created in Fig. 75 with the square tolerance zone.

A. Positional Tolerancing Figure 81 converts Fig. 75 from that of traditional coordinate tolerancing to that of positional tolerancing. Note that the 25- and 35-mm locations have been boxed in. This technique indicates that these two dimensions are basic, that is, the perfect, exact locators of the center of the hole.

Of course, no center can be perfectly located. Therefore, within the feature control frame is given the permitted variation of the position of the center, a 0.4-mm diameter at MMC (maximum material condition). Figure 82 gives the positional interpretation as a cylinder with a 0.4-mm diameter, as opposed to a 0.4-mm square as was the case with coordinate tolerancing of Fig. 75. The location of the hole is within tolerance so long as its center stays within the 0.4-mm-diameter cylinder. Note that the tolerance on the *location* of the hole is com-

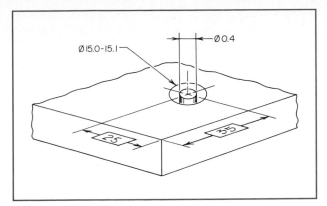

FIG. 82 METRIC. Interpretation of positional tolerancing.

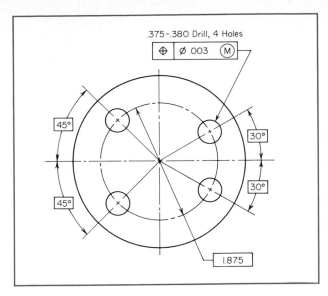

FIG. 83 Angular location by positional tolerancing.

plctely independent of the tolerance on the *size* of the hole, 0.1 mm.

Positional tolerancing is often used to provide angular location of holes on a fixed diameter. Note in Fig. 83 that the four holes are located both by the diameter on which the holes' centers lie, 1.875 in, and by angular deviation from the horizontal center line. The 1.875-in diameter and the 30 and 45° angular locations have been made basic by boxing in these dimensions. The four holes may have their centers move within 0.003-in diameters, as seen by the information within the feature control frame.

B. Concentricity Concentricity controls the extent to which the axis of one cylinder is collinear with the axis on an adjoining cylinder. Figure 84 shows an example of concentricity control using a proper feature control frame. The diameter of the larger cylinder has been made the datum diameter A. The center line of the smaller diameter is to be concentric to the center line of datum A within a diameter of 0.1 mm.

The interpretation of the feature control frame is also given in Fig. 84. The offset of the center line of surface B, relative to the center line of surface A, may not exceed 0.05 mm. The tolerance zone is therefore a 0.1-mm diameter.

Concentricity controls are used less often than positional controls because of the difficulty of locating the actual centers. It is recommended that a runout control be used instead of a concentricity control, where feasible. For runouts, see Sec. 39J.

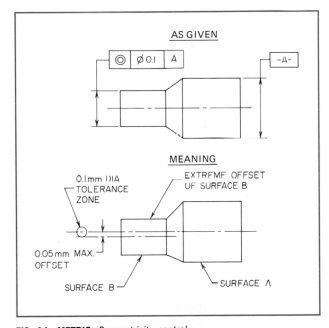

FIG. 84 METRIC. Concentricity control.

39. TOLERANCES OF FORM

Our discussion so far has been limited to tolerances of location: position and concentricity. These two types of tolerance control help the designer, but in themselves are insufficient. Tolerance control of the form, profile, and orientation of parts is vital. We will give examples for each of the nine tolerances given in Fig. 76. Runout tolerances will also be included since they are a type of form tolerance.

A. Straightness A straightness tolerance controls the extent to which an element on the surface of a part is

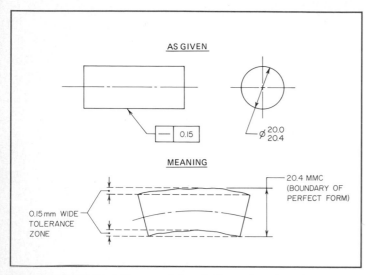

FIG. 85 METRIC. Straightness symbol controlling surface elements.

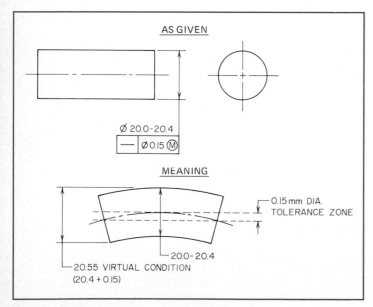

FIG. 86 METRIC. Straightness symbol controlling axis.

indeed a straight line. A tolerance zone of constant width is created within which must lie all points of the surface element being controlled, as shown in Fig. 85. The tolerance zones are parallel lines sharing a common plane with the axis. The straightness tolerance would apply to each longitudinal element of the cylinder. The straightness tolerance may not be greater than the size tolerance, however, The symbol is placed separately from the size control, 20.0 to 20.4 mm.

One variation of the example of Fig. 85 may be noted. The diameter symbol may be included *with* the size control, as in Fig. 86. In this way, the axis is controlled rather than the surface elements. This straightness control is not dependent on the size of the outside diameter, and is therefore applied in the condition of RFS (regardless of feature size).

The dimension of 20.55 mm possible in Fig. 86 is the *virtual condition*. Virtual condition is the most extreme case that can occur while the part is at MMC (maximum material condition). Adding the MMC size and the form tolerance gives virtual condition. Virtual condition can occur for many types of form control, not just straightness.

Note that the feature control frame for straightness contains only two parts. A datum is not referenced. Straightness refers only to itself, as do three other types of form control: flatness, roundness, and cylindricity. The remaining types of form tolerances require reference to datums.

B. Flatness Flatness is easily defined. While straightness relates to a *line*, flatness relates to a plane *surface*. A flatness tolerance is defined by two parallel planes within which all portions of the surface must lie (see Fig. 87). Flatness exists on an RFS basis. Note that the high points of the surface are controlled by the upper plane of the tolerance zone. The low points are controlled by the lower plane of the tolerance zone. The tolerance zone for flatness must be contained within the size tolerance.

C. Roundness A roundness tolerance controls only those points on a surface intersected by any plane perpendicular to the part's axis. Refer to Fig. 88. The tolerance zone lies between two concentric circles on any plane perpendicular to the axis. Note that roundness does not control taper. As was true with flatness, a

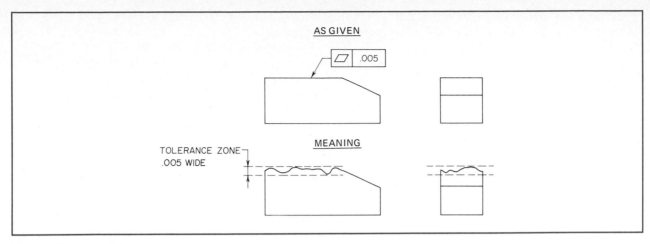

FIG. 87 Flatness symbol.

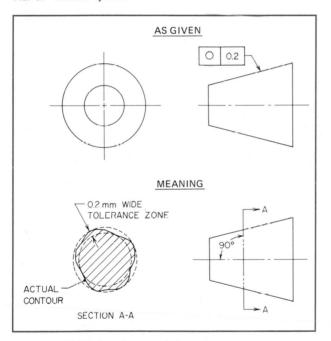

FIG. 88 METRIC. Roundness symbol.

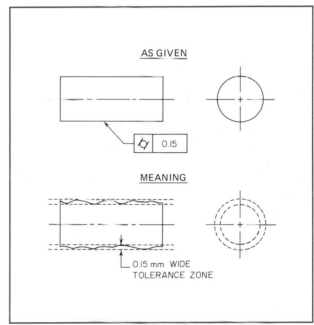

FIG. 89 Cylindricity symbol.

roundness tolerance must stay within the size tolerance of the part.

D. Cylindricity It is interesting to compare the similarities between straightness and roundness and between flatness and cylindricity. Both straightness and roundness relate to a line condition. Straightness relates to a straight line, while roundness wraps the line around a surface. Flatness and cylindricity both involve areas, but the area for cylindricity wraps around a surface.

As seen in Fig. 89, a cylindricity tolerance defines a tolerance zone consisting of two concentric cylinders. Cylindricity tolerance is used for cylinders only, and really controls taper, roundness, and straightness. Cylindricity tolerance may not exceed size tolerance.

E. Profile of a Line The profile of any part is its outline as seen in any two-dimensional view. Profiles contain combinations of lines and arcs. A true profile is described by the use of basic (boxed) dimensions without tolerances. A feature control frame contains the tolerance zone to be applied to the basic dimensions.

Figure 90 gives an example of line profile tolerancing. The feature control frame applies to each and every vertical cross section that may be taken in the left-side view. The shape of each cross section must lie within the 0.15-mm-wide tolerance zone specified in the feature control symbol. The tolerance zone is in relation to datum A.

The control exists between points X and Y shown in the front view. Each line element of the surface between points X and Y, along any typical cross section, must stay within the two profile boundaries which are 0.15 mm apart in relation to the datum planes A and B. Also the surface must stay within the size limits of 39.6 mm and 40.4 mm.

F. Profile of a Surface Profile tolerancing applied to a surface is actually more common than the case just discussed, line profile tolerancing. Profile tolerancing of a surface is also easier to describe than line profile tolerancing. Refer to Fig. 91.

Notice that the dashed control line is shown on the *inside* of the profile of the upper surface. Placing the control on the inside of the profile implies that unilateral tolerance exists downward from the profile. If the dashed line were outside the profile, an outside unilateral tolerance would apply. If *no* dashed line were used, a bilateral tolerance would apply on both sides of the given profile.

If it is desired to have the tolerance zone go entirely around the profile of the part, one should include the symbol needed to indicate "all around." The symbol is placed on the leader as shown in Fig. 91.

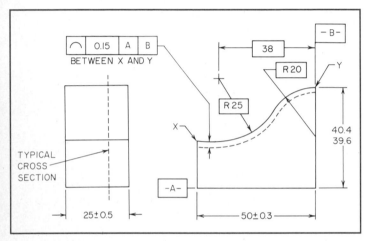

FIG. 90 METRIC. Line profile symbol.

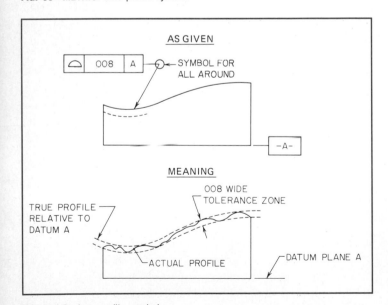

FIG. 91 Surface profile symbol.

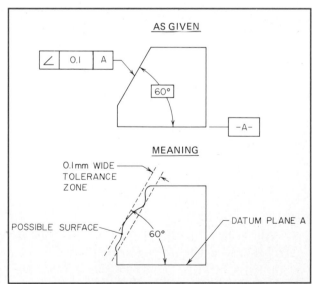

FIG. 92 METRIC. Angularity symbol.

G. Angularity Angularity relates to an axis or surface at some specified angle to a datum. The angle may not be 90°, however, as such a case is covered by a perpendicularity control. Figure 92 illustrates angularity control.

The tolerance zone lies between two parallel planes 0.1 mm apart. The parallel planes have a specified angle of 60° to datum plane A. The actual surface between the two parallel planes may have any shape whatsoever. In a real sense, an angularity control also controls flatness. This is a case of overlap between types of form tolerancing, a situation that can happen in other instances.

One can control the angularity of an axis as well as that of a surface. While not illustrated, the angle of an inclined hole may be controlled, using exactly the same format as in Fig. 92.

H. Perpendicularity A feature control frame for perpendicularity can control several forms of perpendicularity: a plane with a plane, an axis to a plane, an axis to an axis. Two examples will be given, a plane with a plane and an axis to a plane. Other examples may be found in ANSI Y14.5M—1982.

As a first example, consider Fig. 93. The perpendicularity of the left-hand vertical plane is controlled relative to the horizontal datum plane A. Notice that the control of perpendicularity is also controlling flatness of

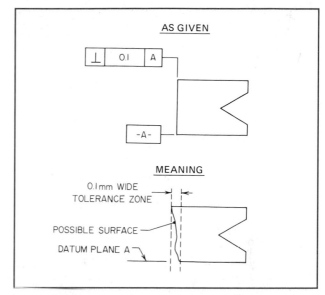

FIG. 93 METRIC. Perpendicularity symbol applied to two planes.

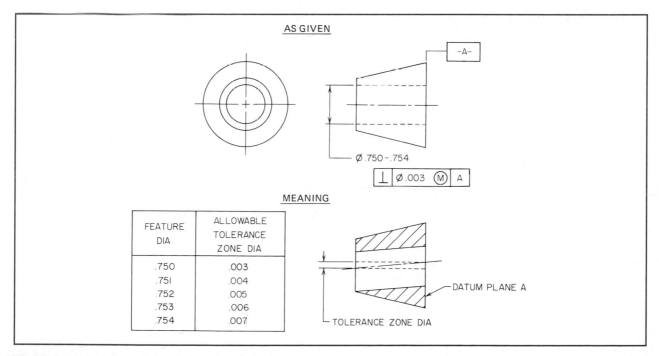

FIG. 94 Perpendicularity symbol applied to axis and plane.

the vertical plane. This plane must be flat at least within a 0.1-mm-wide tolerance zone.

A second example is seen in Fig. 94. Here the axis of the hole is being controlled relative to the vertical datum plane A. Note that tolerance zone diameter of 0.003 in applies at MMC, or when the hole is at its smallest size (maximum material condition). The effect of adding the MMC modifier is seen in the accompanying chart. The tolerance zone is permitted to increase in diameter in direct relation to the increase in the feature (hole) diameter. Therefore when a 0.754-in-diameter hole exists, the allowable tolerance zone diameter is 0.007 in. This situation can permit less precision and lower costs in manufacturing, assuming the function intentions are still met.

I. Parallelism Parallelism exists when a feature is at a constant distance from a datum. The feature could be an axis or a plane, as could be the datum. Two examples will be shown, one for a plane parallel to a plane, and another for an axis parallel to an axis.

Figure 95 shows the control of a plane parallel to plane. The tolerance zone 0.1 mm wide must be perfectly parallel to datum plane A. The actual surface may assume any shape within the tolerance zone. Like perpendicularity, parallelism also controls flatness.

Let us use Fig. 96 as an example of control of parallelism for two axes. One can see that the axis of the larger hole may vary at will so long as it stays within a tolerance zone of 0.25 mm diameter. However, the cylinder defining the tolerance zone must be parallel to datum axis A. Since no MMC modifier is given, the parallelism of the two axes exists in an RFS condition. If an MMC condition were specified, the situation given in Fig. 94 would apply. That is, the allowable tolerance zone diameter could increase by an amount equal to the tolerance placed on the size of the hole. If the hole were, say, 10.0 to 10.3 mm in diameter, the maximum allowable zone tolerance would be 0.25 + 0.3, or 0.55 mm diameter.

J. Runout Runout is a measure of the deviation from perfect form, determined as a part is revolved 360° around its axis. Runout tolerance is a truly composite tolerance that incorporates variations in straightness, roundness, and parallelism. Surface features are related to a datum axis. Features controlled may be those wrapped around the datum axis, or may be those perpendicular to the datum axis.

Figure 97 shows that runout may be found by reading the net change on a dial indicator while a part is revolved one full turn. The reading found is called FIM (full indicator reading) or TIR (total indicator reading).

The two types of runout control are circular and total. An example of each will be given. It is a designer's decision as to which runout control, if any, is to be chosen. Either of the runout types operates on an RFS basis.

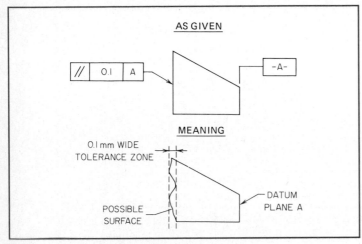

FIG. 95 METRIC. Parallelism symbol applied to two planes.

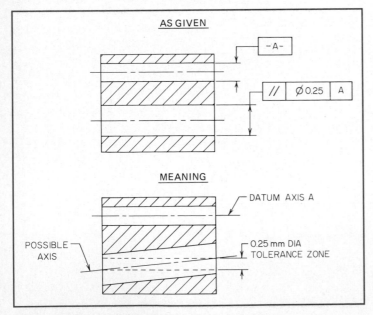

FIG. 96 METRIC. Parallelism symbol applied to two axes.

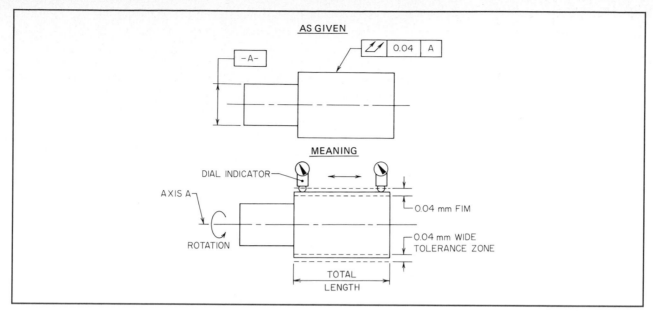

FIG. 97 METRIC. Using dial indicators to read total runout.

Circular runout provides control at any single fixed position along the axis of a part. Refer to Fig. 98. We see that the dial indicator may not exceed 0.03 FIM while kept normal to the surface and while *not* moved parallel to datum axis A. Therefore the indicator checks runout within the circular element at that position.

Total runout varies from circular runout in that control is given for an entire surface, not a circular element alone. Note in Fig. 97 that total runout control is desired because of the double arrow within the feature control frame. The dial indicator *is* moved parallel to datum axis A as the FIM of 0.04 mm maximum is checked. Both profile and circular elements are therefore controlled.

Runout is a powerful type of form control because it incorporates control of roundness, straightness, and cylindricity, to mention three of the possible single types overseen. When surface (as opposed to coaxial) relationships are to be controlled, runout control is an ideal choice.

40. CONCLUDING COMMENTS

Through study of geometric tolerancing one can appreciate the power that the various feature controls have

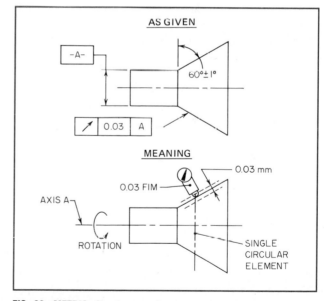

FIG. 98 METRIC. Circular runout.

over the location and form of parts. A designer would never use feature controls in an excessive way merely to obtain unnecessary precision. Every use of a feature control implies the cost of producing the desired control and then inspecting for that control feature.

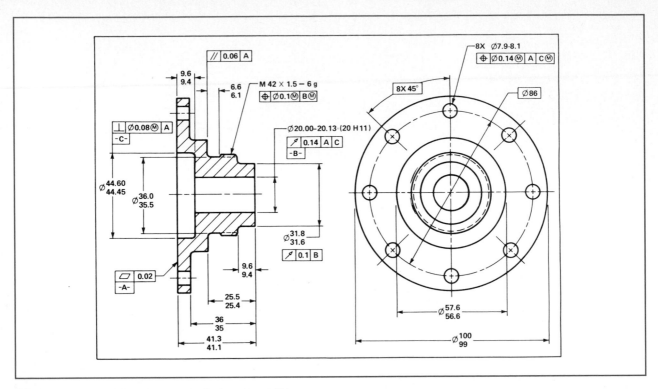

FIG. 99 Use of geometric tolerancing. *(ANSI Y14.5M—1982.)*

However, when a control of location or form is legitimately needed, the availability of feature controls is very useful to the designer. An example of the use of many control features on a part is seen in Fig. 99. Although it would be costly to produce this part, the figure illustrates the many ways in which feature controls can be applied.

PROBLEMS

GROUP 1. DIMENSIONED DRAWINGS FROM PICTORIAL VIEWS

The following problems will give you practice in applying principles of dimensioning. The various shapes and locations of features should be given single dimensions, unless tolerancing is requested. For each group of problems follow the specific instructions.

A selection of problems is offered, using examples seen in Chap. 5.

1 to 6. Use the pictorials given in Probs. 1–6, Chap. 5 using case C or D for each problem. Draw and dimension three orthographic views for each problem selected. English decimals or fractions may be used, allowing each scale graduation on the object to be 0.25 (¼) in. Alternatively, each scale graduation may represent 1 cm.

7 to 10. Use Prob. 12, Chap. 5. For all four pictorials, draw and dimension three orthographic views. Assume the objects are one-third size. Therefore, triple the sizes shown to give full size, using dividers to "walk off" the distances. Give dimensions in millimeters.

11 to 15. Use Prob. 13, Chap. 5, portions A, B, D, F, and G. For each example, draw and dimension three orthographic views. Triple the sizes shown. Give full-size dimensions in decimal inches.

GROUP 2. DIMENSIONED DRAWINGS FROM ORTHOGRAPHIC VIEWS

16 to 22. Refer to Prob. 55, Chap. 5, portions E to K. Draw the missing view, if not done previously, and dimension in millimeters. Draw triple the given sizes to give full size.

23 to 28. Refer to Prob. 56, Chap. 5, all portions. Draw the missing views; dimension in fractions. Triple the sizes given to give full size.

GROUP 3. DIMENSIONED DRAWINGS FROM DIMENSIONED PICTORIAL DRAWINGS

29. Refer to Prob. 28, Chap. 5. Dimension three orthographic views, converting to millimeters of three significant figures (e.g., ⅞ converts to 22.2 mm). Draw full size.

30. Refer to Prob. 31, Chap. 5. Dimension three orthographic views, using the decimal inch. Draw full size.

31. Refer to Prob. 34, Chap. 5. Using the decimal inch, dimension three orthographic views, full size. Dimension the larger hole to an RC 7 fit and the smaller hole to an FN 2 fit. Indicate that the base is to be flat within 0.007 in, using a feature control symbol.

32. Refer to Prob. 36, Chap. 5. Dimension three orthographic views, using millimeters as given. Draw full size. Indicate that the three holes on the 82-mm bolt circle are to have a position tolerance for angular location within a 0.15-mm diameter at MMC, using feature control symbols. The size of these three holes is to be to an H8/f7 fit. Indicate that the sides of the slot are to be parallel within 0.08 mm, using a feature control frame.

GROUP 4. MACHINE PARTS TO BE DIMENSIONED

33. Given: A stud shaft shown half size, machined from steel-bar stock. Give the proper dimensions for diameter A to conform to an H8/f7 fit. The thread is coarse metric.

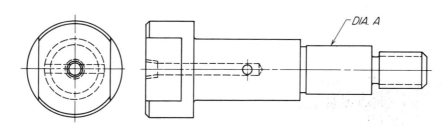

PROB. 33 METRIC. Stud shaft.

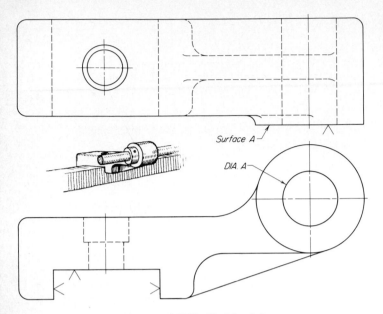

PROB. 34 METRIC. Shaft bracket.

34. Given: A shaft bracket shown half size, made from malleable iron. The slot is to be given a surface-texture symbol, with roughness appropriate to a milling operation. Diameter A is to have an H7/h6 locational clearance fit. The diameter of the H7/h6 hole is to be perpendicular to surface A within a 0.1-mm diameter.

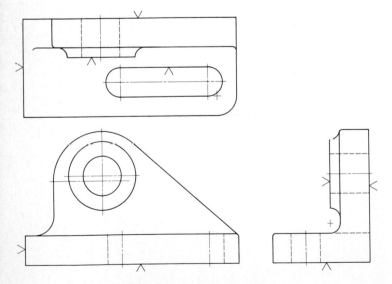

PROB. 35 Idler bracket, left-hand.

35. Given: An idler bracket shown one-third size, made of cast iron. The hole is bored and reamed; the slot is milled. Dimension the hole to an RC 6 fit. Give a feature control frame so that the base is flat within 0.008 in. Give a surface-texture symbol for all indicated finished surfaces consistent with roughness common to milling operations. Dimension with decimals.

36. Given: A filter flange shown half size, made of cast aluminum. The small holes are drilled. Dimension the large diameter to an LC 4 fit. The small holes are to be given a feature control frame so that the holes are positionally located for angularity within a 0.006-in diameter at MMC. Dimension with decimals.

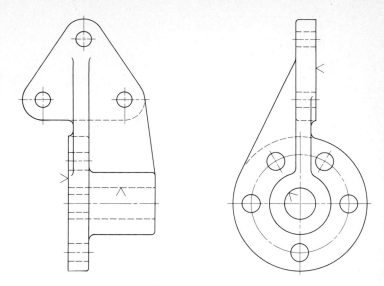

PROB. 36 Filter flange.

37. Given: A boom-pin rest made from a steel drop forging, shown half size. A top view may be added if desired. Draft angles are 7°. Holes are drilled; corner notches are milled. Assume the part is being dimensioned for the machinist.

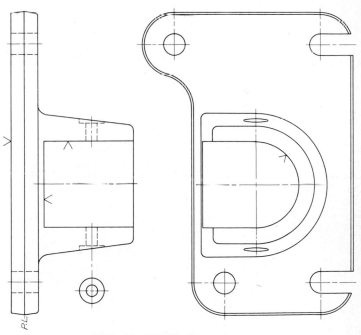

PROB. 37 METRIC. Broom-pin rest.

38. Given: A clutch-lever drop-forged from aluminum. The part is shown half size. Add a top view if helpful. Holes are drilled and reamed to an H8/f7 fit. The ends of the hub are ground to a 1.6-μm finish. Draft angles are 7°. Show machining allowance with alternative position lines.

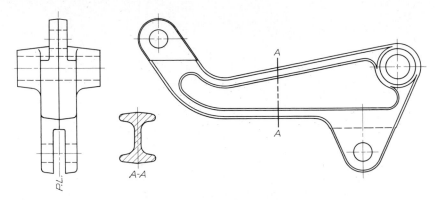

PROB. 38 METRIC. Clutch lever.

39. Given: A radiator mounting clip made of 1.5-mm thick steel, shown half size. The holes and slot are punched. See Fig. 39, Chap. 13, for reference.

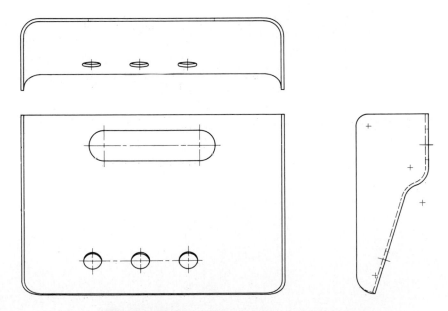

PROB. 39 METRIC. Radiator mounting clip, left-hand.

40. Given: A pulley bracket shown half size, made of 24-ST aluminum sheet, 0.032 in thick. Reference: Fig. 39, Chap. 13.

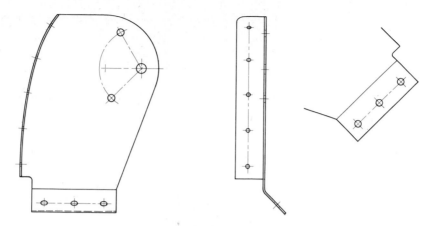

PROB. 40 Pulley bracket.

41. Using the scale shown, draw and dimension the check-valve body.

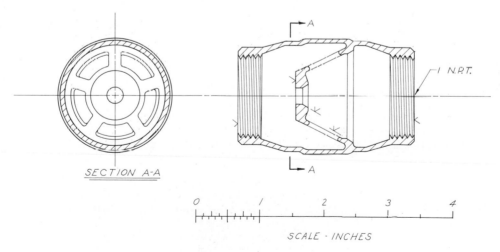

PROB. 41 Check-valve body.

42. Given: A hydrostatic pressure housing at the size shown. Use the decimal inch system. The eight tapped holes are to be given a position tolerance for angular location within a 0.004-in diameter at MMC, using feature control frames. The two ends are to be parallel within a 0.005-in-wide tolerance zone. The right-hand end face is to be perpendicular to the longitudinal center line within a 0.010-in diameter. Feature control frames are needed.

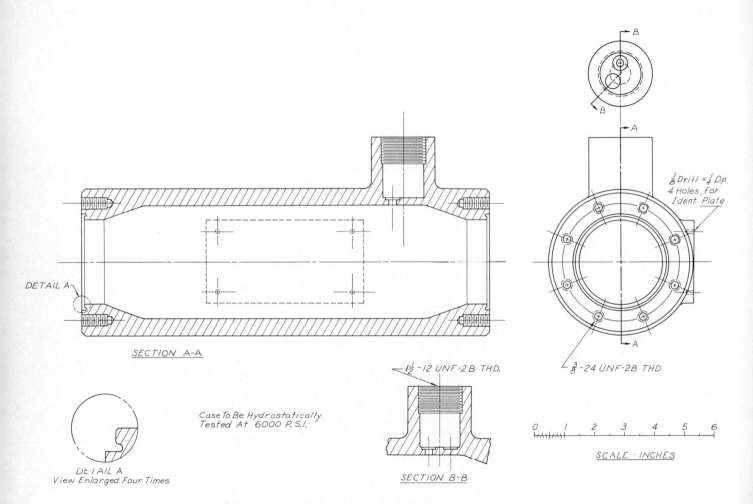

SECTION A-A

DETAIL A

DETAIL A
View Enlarged Four Times

Case To Be Hydrostatically
Tested At 6000 P.S.I.

$1\frac{1}{2}$-12 UNF-2B THD.

SECTION B-B

$\frac{1}{8}$ Drill $\times\frac{1}{4}$ Dp
4 Holes, for
Ident. Plate

$\frac{3}{8}$-24 UNF-2B THD

0 1 2 3 4 5 6
SCALE - INCHES

PROB. 42 Hydrostatic pressure housing. *(Courtesy of Westinghouse Air Brake Co.)*

43. Given: A universal-joint housing made of steel, shown one-third size. Although the design is fairly complex, a careful analysis will allow you to move along in a steady manner. The center line of the spline should be perpendicular to the finished base within a 0.15-mm diameter, using a feature control frame. Tolerance the four drilled holes to an H7/h6 fit.

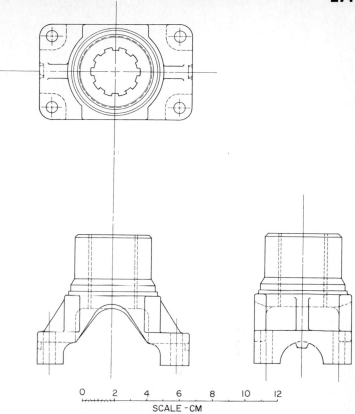

0 2 4 6 8 10 12
SCALE - CM

PROB. 43 METRIC. Universal-joint housing.

44. Given: A connecting link made of cast iron. Using decimals, dimension the piece. The center lines of the two holes are to be concentric within a 0.010-in-diameter tolerance zone. Use a feature control frame. The holes are to be an FN 1 fit.

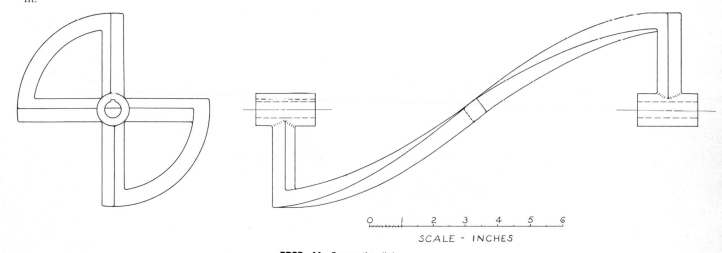

0 1 2 3 4 5 6
SCALE - INCHES

PROB. 44 Connecting link.

PART B

ELEMENTS OF SPACE GEOMETRY

9

AUXILIARY VIEWS: LINES AND PLANES EXPLORED

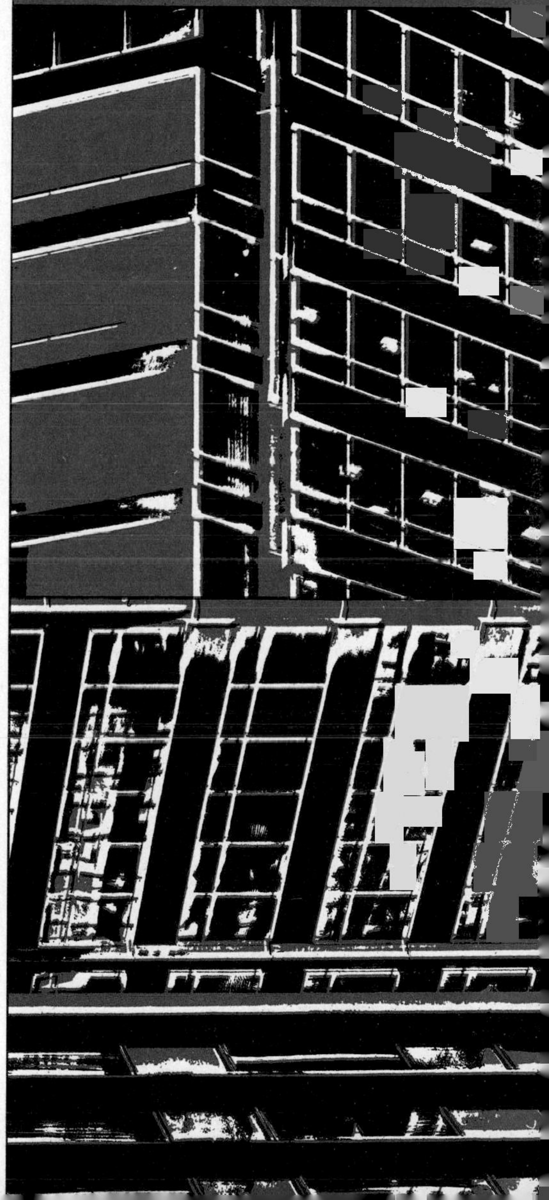

1. THE NEED FOR AUXILIARY VIEWS

So far we have used only horizontal, frontal, and profile views to describe shapes. The information on orthographic projection (Chap. 5) was very helpful in allowing us to show heights, widths, and depths of objects while keeping within the six principal planes of projection (see Fig. 11 of Chap. 5).

As long as shapes can be seen clearly there is no need to go beyond the six principal views. An object such as in Fig. 1 is well defined using three of the six principal views, for example. However, for an object such as seen in Fig. 2 the top and front views need the supplementary *auxiliary* views to show the normal (true surface) views of the inclined tabs.

An auxiliary view is merely any orthographic view that is *not* a horizontal, frontal, or profile view. We will learn the whys and hows of making useful auxiliary views. Using auxiliary views takes us into an area study known as space geometry or descriptive geometry.

Basic, useful problems in space geometry can be solved by using appropriate auxiliary views. The views we will study are:

1. The normal view of a line
2. The point view of a line
3. The edge view of a plane
4. The normal view of a plane

The term "normal" means perpendicular to or at right angles. The term is here applied to a view in which the direction of sight is perpendicular to a line or plane.

The point view of a line is that view of a line which shows the line as a single point, obtained by looking parallel to the line. The edge view of a plane, on the other hand, is that view in which all lines of the plane appear in a continuous single line, obtained by looking parallel to the plane.

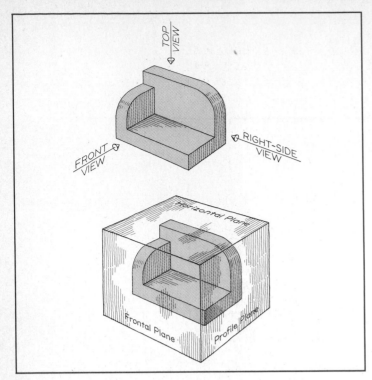

FIG. 1 An object needing only three principal views to define shape.

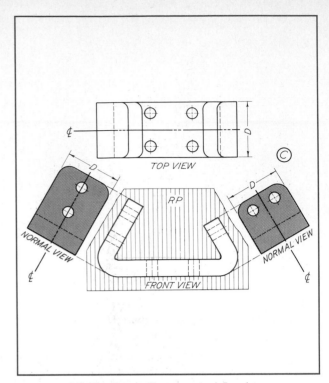

FIG. 2 An object needing auxiliary views to define shape.

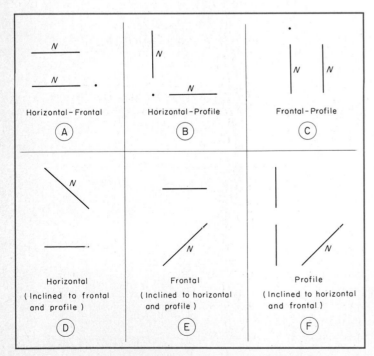

FIG. 3 Normal views of lines lying in *two* principal planes of projection and in *one* principal plane of projection.

2. DEFINING A LINE AND A PLANE

It is helping to begin by defining a line and a plane so we will be on common ground.

A line may be located in space by specifying *two points* on it. If the two points are located in two adjacent orthographic views, the points are fixed in space and the line passing through these two points therefore has its direction in space determined. A line is considered to be infinite in length, and the portion between any two points on it simply specifies a segment. The *space direction* (bearing and slope) and *one point* will also locate a line.

A plane, on the other hand, can be defined by three points, one point and one line, two parallel lines, or two intersecting lines. Planes are thought often to be infinite in size. The definition of a plane simply sets its orientation in three-dimensional space.

3. CLASSIFICATION OF LINES AND PLANES

It is necessary to classify lines and planes before we can intelligently discuss them.

A. Line Positions A line may lie in two, one, or none of the principal planes of projection.

1. A line that lies in *two* principal planes of projection will appear *normal* (showing its true length) in *two* views and is named for the planes it lies in (see Fig. 3). A line lying in *two* principal planes will appear as a point in the third principal plane.

2. A line in *one* principal plane of projection will appear normal in *one* view and is named for the plane it lies in (see Fig. 3). A line lying in *one* principal plane of projection may be inclined to the other two principal planes. To show the end view of the line an auxiliary view is required.

3. A line in *none* of the principal planes of projection, called an "oblique line," will not appear normal in any of the principal views. Auxiliary views are needed to show the normal and end views of an oblique line.

B. Plane Positions A plane may be parallel to one of the principal planes of projection, inclined to two principal planes of projection, or inclined to all three principal planes.

1. A plane *parallel* to *one* of the principal planes receives its name from that principal plane and will appear normal in *that* one principal view (see Fig. 4).

2. A plane *inclined* to *one* principal plane of projection will also be inclined to another principal plane and will not appear normal in any view, but will appear as an edge in one view (see Fig. 4).

3. A plane *inclined* to *all three* principal planes of projection, called an *oblique plane*, will not appear

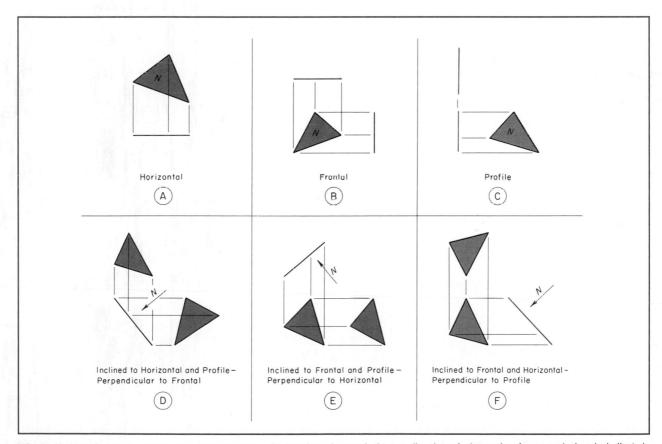

FIG. 4 Horizontal, frontal, profile, and inclined planes. Designation of normal view or direction of observation for normal view is indicated.

normal or as an edge in any of the principal views. Auxiliary views are required to show the edge and normal views.

C. Terms of Reference Terms of reference are necessary for identification of the views and constructions needed in the solution of space problems. This is to ensure accuracy of description both in the textbook and in class discussion. The following terms of reference are used in this text:

LINES

1. Aligned with two principal planes
   a. Horizontal-frontal
   b. Horizontal-profile
   c. Frontal-profile
2. Aligned with one principal plane
   a. Horizontal (inclined to frontal and profile)
   b. Frontal (inclined to horizontal and profile)
   c. Profile (inclined to horizontal and frontal)
3. Oblique (inclined to all principal planes)

PLANES

1. Principal
   a. Horizontal
   b. Frontal
   c. Profile
2. Inclined
   a. Inclined to frontal and profile (perpendicular to horizontal)
   b. Inclined to horizontal and profile (perpendicular to frontal)
   c. Inclined to horizontal and frontal (perpendicular to profile)
3. Oblique (inclined to all principal planes)

PROJECTORS

1. Always aligned with (parallel to) the viewing direction; description the same as for viewing direction

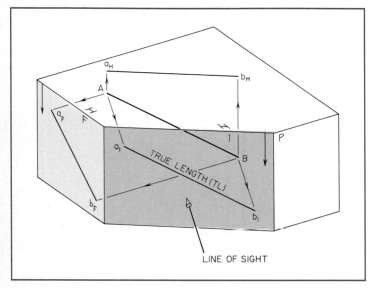

FIG. 5 Projection box to show a normal view of an oblique line.

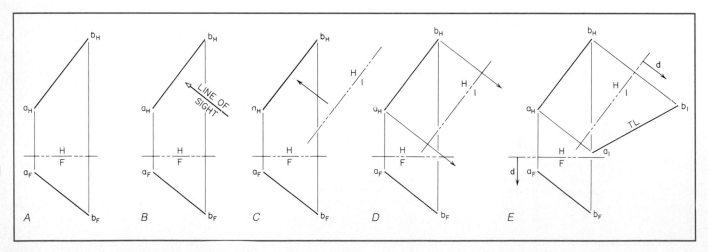

FIG. 6 *A,* line as given; *B,* step 1; *C,* step 2, *D,* step 3, *E,* step 4.

REFERENCE PLANES

1. Horizontal
2. Frontal
3. Profile
4. Auxiliary

4. NORMAL VIEWS OF A LINE

One of the first cases studied in space geometry is that of locating the normal view of a line. Since a normal view shows a line's true length, we have valuable information. Every design requires knowledge of true lengths of its parts. The general case involves finding the true length of an oblique line (not parallel to any of the three principal planes of projection).

Note in Fig. 5 that line *AB* is an oblique line contained within a projection box. The horizontal plane is labeled *H*, the frontal plane *F*, the right-side profile plane *P*, and a new plane is called 1. Plane 1 is a first auxiliary plane about which we will say much. Note that plane 1 has been placed parallel to line *AB*, which forces the line of sight to be perpendicular to *AB*. (Remember that in orthographic projection the line of sight is always perpendicular to the plane of projection.) With the line of sight perpendicular to *AB*, *AB* must show its true length in plane 1. Plane 1 shows a normal (true-length) view of *AB*.

It is important to see how the true length of line *AB* may be constructed in two-dimensional orthographic multiview. If one is confronted with the horizontal and frontal views of *AB*, as seen in Fig. 6A, how can the normal view, and therefore true length, be found? The process by which this new view is constructed is the keystone for making any and all views in space geometry. The process involves the following four basic steps used in constructing any new view in orthographic projection.

Step 1. Determine the Desired Line of Sight In Fig. 6B this is seen to be perpendicular to line *AB*. One has essentially walked around in space until the desired line of sight can be obtained.

Step 2. Establish a New Projection Plane Perpendicular to the Line of Sight (Fig. 6C) Since the new plane (auxiliary plane 1) is perpendicular to the line of sight, it is also parallel to line *AB*. The new plane 1 may be

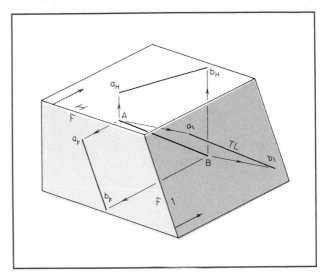

FIG. 7 Normal view of a line using auxiliary view attached to the frontal plane.

placed at any convenient spot on the paper so long as it is perpendicular to the line of sight.

Step 3. Project all Points into the New View (Fig. 6D) For line *AB* there are, of course, only two points, A and B. The points must project perpendicular to auxiliary plane 1 because we are in an orthographic system.

Step 4. Locate Points *A* and *B* along the Projections in Plane 1 This action will give line *AB* in its normal (true-length) position. To locate A and B in plane 1, transfer along direction *d* by going perpendicularly away from the same plane in space. Going perpendicularly away from the horizontal plane into plane 1 is exactly the same direction in space as going perpendicularly away from the horizontal plane into the frontal plane. Note in Fig. 5 that the solid arrows are perpendicular to the horizontal plane. These same arrows are shown in Fig. 6E.

Step 4, as just explained, highlights the most profound concept in space geometry: A *direction perpendicular to the same plane is always the same direction in space.* This concept can be reinforced by reworking the same problem for line *AB*, but by using different views. As first solved, the auxiliary plane 1 was attached perpendicular to the horizontal view. One could have also obtained the same true length by attaching an auxiliary plane perpendicular to the frontal view. See Fig. 7. The

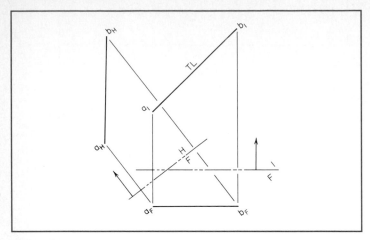

FIG. 8 Orthographic solution for Fig. 7.

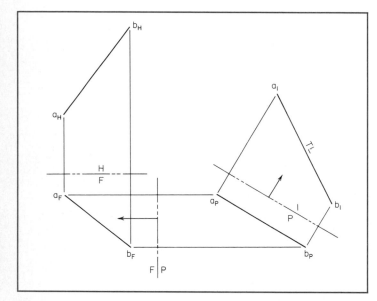

FIG. 9 Normal view of a line using auxiliary view attached to a profile plane.

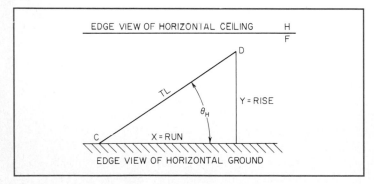

FIG. 10 Relation of a line to horizontal planes.

process of solution would involve the same four steps as before. The full solution is seen in Fig. 8. Alternatively, one could solve for the true length of line *AB* by attaching an auxiliary plane perpendicular to the side (profile) plane. See Fig. 9. The three-dimensional projection box has been omitted, but the two-dimensional multiview construction is shown in full.

5. SLOPE AND GRADE OF A LINE

One may obtain several important items of information from the relationship of a line to the horizontal plane. One such item is the slope of a line. In Fig. 10 the slope of line *CD* is defined as follows:

$$\frac{\text{Rise}}{\text{run}} = \frac{Y}{X} = \tan \theta_H$$

The angle θ_H is the slope angle of line *CD*. One can also speak of the grade of a line, which is $(\tan \theta_H)(100\%)$. The slope, slope angle, and grade of a line are considered positive in sign if the line rises upward as one travels along it from beginning to end. The beginning of a line is said to be at the first letter or number used to signify a line, i.e., from *C* to *D*, from *X* to *Y*, from 1 to 2. For line *CD* the sign is positive. Slope, slope angle, and grade are negative in sign if one travels downward from the beginning of the line to the end. Line *CD* would be negative in sign if the line were *DC*, that is, if one were to start at *D* and travel downward to *C*.

Be aware of one vital factor regarding slope, slope angle, and grade. They must be measured only where the line is seen as true length and only in a view showing height. Such a view occurs only in a view adjacent to (90° from) the horizontal view. If a line is not true length in the front view, one must first find the line's true length in a view adjacent to the horizontal view. This is the case for an oblique line such as *AB* in Fig. 11, previously discussed. Notice that in Fig. 6E we are prepared to measure slope, slope angle, and grade. Figure 11 is Fig. 6E with slope angle indicated. Slope angle is negative because one travels downward away from the horizontal as the line is traveled from *A* to *B*. The slope is *Y/X*. The calculation of slope is made easy if the denominator *X* is made to equal 10, using any convenient scale on an engineers' or metric scale. The numerator *Y* can then be read in a direction perpendicular to *X*, and the ratio Y/X can be determined in one's head. For line AB,

$$\text{slope} = \frac{Y}{X} = \frac{-4.9}{10} = -0.49$$

The grade of AB is therefore $-0.49 \times 100\% = -49\%$.

6. BEARING AND AZIMUTH OF A LINE

Engineers, pilots, and others who work with maps often use the term "bearing" of a line. Bearing is the deviation from the north or south point of a compass. Bearing can be seen in the horizontal view only, since maps are in themselves horizontal views. In Fig. 12 are seen several bearings. Note that the acute angle (less than 90°) is always used. The special cases of N0°E, N90°E, S0°W, and S90°W are called due north, due east, due south, and due west, respectively. The bearings of the oblique line of Fig. 11 used earlier can be found in the horizontal view. If you assume that the north arrow is straight up the paper and that this north arrow passes directly through point A (the beginning of the line), you will find the bearing to equal N40°E.

Azimuth of a line is an alternative way to express bearing. Azimuth is always read clockwise from the north arrow and uses only the letter N together with the clockwise angle. The bearings given in Fig. 12 would convert to the azimuths of Fig. 13.

7. POINT VIEW OF A LINE

The point view of a line occurs when the line of sight is parallel to the line. The resulting view shows the line as a single point, since both ends of the line are coincident. Point views of lines will be very useful in the solving of many problems in space geometry.

To make a point view of an oblique line, let us go back and use Fig. 11, redrawn in Fig. 14. To get the point view the line of sight must be parallel to the true length of AB. Line AB is in true length in plane 1. Therefore, place the line of sight parallel to AB in plane 1, as indicated.

The point view of AB can now be constructed, using the concepts expressed in Sec. 4. The four steps for constructing any view work for us here as well:

1. Determine the line of sight (this has already been done).

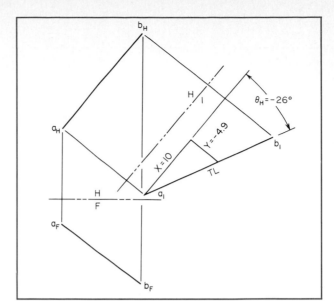

FIG. 11 Slope of an oblique line.

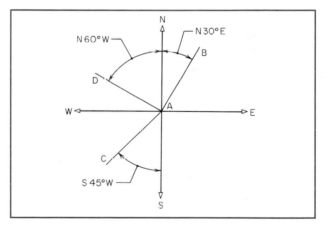

FIG. 12 Bearing of a line.

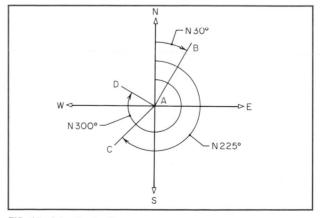

FIG. 13 Azimuth of a line.

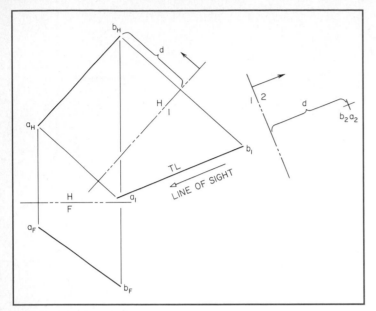

FIG. 14   Point view of a line.

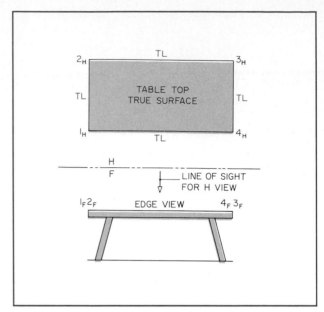

FIG. 15   Edge view of a table.

2. Establish a projection plane perpendicular to the line of sight. Figure 14 shows this new projection plane as plane 2, attached 90° to plane 1. Again, the position of plane 2 on the paper does not matter so long as plane 2 is perpendicular to the line of sight.

3. Project all points into the new view. Points *A* and *B* project perpendicular to plane 2, sharing a single projector.

4. Locate points along the projector. The distance along the projector into plane 2 can be found as before: Use a transfer distance measured perpendicularly away from the same plane. The solid arrows show that the direction perpendicularly away from plane 1 into plane 2 is exactly the same in space as the direction perpendicularly away from plane 1 back into plane *H*. Line *AB* appears as a point in plane 2 because the transfer measurements for *A* and *B* in plane *H* are identical (shown as distance *d*).

8. EDGE VIEW OF A PLANE

The edge view of a plane is seen when all lines of the plane appear to coincide in a single line. An edge view is

guaranteed if one turns 90° off a normal view (true surface) of a plane. Take as an example a tabletop, representing a horizontal plane. A turn 90° from the top view of the table will show the table in edge view. See Fig. 15, in which the particular 90° turn off the top is a front view which shows the edge view of the tabletop.

The edge view of a plane can be seen by a second means. If one sees any line of the plane in point view, the plane itself is in edge view. This effect shows itself in Fig. 15, where lines 1-2 and 3-4 are point views in the front view because the line of sight for the front view is parallel to the true lengths of 1-2 and 3-4 of the horizontal view.

A more general case is seen in Fig. 16. An oblique plane 1-2-3 is given in the *H* and *F* views. None of the boundary lines 1-2 or 2-3 or 3-1 is seen in true length. Therefore one cannot look parallel to any of these three lines to get a point view. You must add some line that will be in true length within the *H* or *F* view. Line 2-4 serves this purpose. By adding it to the *F* view so that it is parallel to the *H* plane, line 2-4 must be in true length in the *H* view. Then you need only look parallel to 2-4 in the *H* view to construct an auxiliary view giving the point view of line 2-4, with the total plane being in edge view. Note that the slope angle θ_H is also available in the auxiliary view. The slope angle of a plane is the angle between the edge view of the plane and the edge of the horizontal plane, much like the case for a line.

Edge views of a plane may be constructed adjacent to any plane desired. If you wish an edge view adjacent to the frontal plane, the line added within the plane must show its true length in the frontal plane. Similarly, for a plane to appear in edge view adjacent to the profile plane, the line added within the plane must show its true length in the profile view. The example of Fig. 16 has the edge view in a view adjacent to the horizontal view because the true length within the plane occurs in the horizontal plane. For any case, no more than one auxiliary view ever is required to obtain the edge view of a plane.

9. NORMAL VIEW OF A PLANE

The normal (true-surface) view of a plane is seen when the line of sight is perpendicular to the plane. The table-top of Fig. 15 shows this clearly. The horizontal view is obtained by looking perpendicularly down in the front view, and in so doing the line of sight is perpendicular to the edge view of the tabletop.

If no edge view is available within the existing views, one must be constructed, because no true surface can be seen unless there is an edge view which can be viewed perpendicularly. Figure 16, in which the edge view of plane 1-2-3 was constructed, can be used again to build a true surface. One has only to look perpendicularly at the edge view and construct the auxiliary view to see the true surface. The procedures for constructing an auxiliary view remain unchanged. The four-step method used earlier for the point view of a line (Fig. 14) must be used. Therefore, the true surface of plane 1-2-3 is completed by locating points 1, 2, and 3 by transferring distances perpendicularly away from plane A into plane H and using these distances in plane B (note the solid arrows of Fig. 16). Transferring point 4 into the true surface is purely optional. as it is not needed to describe the plane.

The angles shown within the true surface of plane 1-2-3 of Fig. 16 are all true angles because all lines within the true surface are in true length. Therefore, if

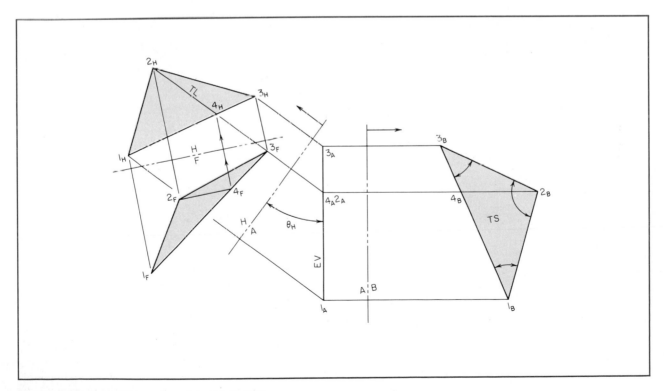

FIG. 16 Edge view and true surface of an oblique plane.

you ever need the angle between any two lines, simply construct a true surface of the plane described by the two lines.

10. AUXILIARY VIEWS APPLIED TO PLANES OF SOLIDS

It is interesting and worthwhile to apply to solids the concepts of space geometry for lines and planes since the actual world contains many objects that are solids. Sur-

faces may occur at any of the positions shown on the solids of Fig. 17. At A all the surfaces are parallel to the principal planes of projection, and therefore each of the principal views is a normal (true-surface) view. At B the shaded surface is at an angle to *two* of the principal planes but perpendicular to one plane and is called an *inclined surface*. At C the shaded surface is at an angle to *all three* principal planes of projection and is known as an *oblique surface*.

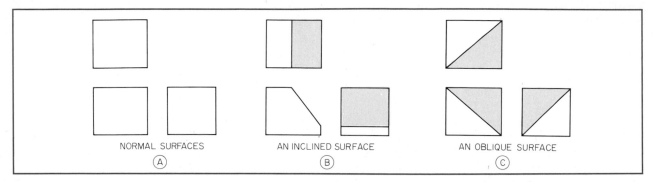

FIG. 17   Classification of surfaces.

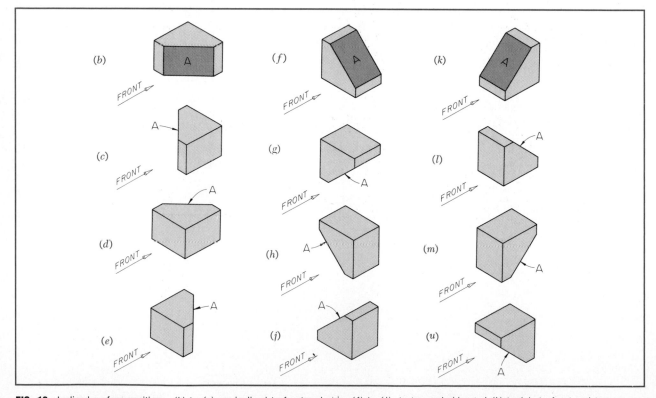

FIG. 18   Inclined-surface positions. *(b)* to *(e)* are inclined to front and side; *(f)* to *(j)*, to top and side; and *(k)* to *(n)*, to front and top.

11. THE NORMAL VIEW OF AN INCLINED SURFACE ON A SOLID

Inclinced surfaces may occur anywhere on an object, and because other features of the object must also be represented, the inclined surface may be in any one of the twelve positions shown in Fig. 18. The first column, (*b*) to (*e*), shows surfaces inclined to the front and side so that the surface may be on the (*b*) right front, (*c*) left front, (*d*) left rear, or (*e*) right rear.

The second column, (*f*) to (*i*), shows surfaces inclined to the top and side, so that the surface may be on the (*f*) upper right, (*g*) lower right, (*h*) lower left, or (*i*) upper left.

The third column, (*k*) to (*n*), shows surfaces inclined to the top and front, so that the surface may be on the (*k*) upper front, (*l*) upper rear, (*m*) lower rear, or (*n*) lower front.

Inclined surfaces occur at *angles* of inclination differing from those shown in Fig. 18, but for general position no other locations are possible.

Figure 19 shows an object with one inclined face, *ABDC*. This face is inclined to the horizontal and profile planes and perpendicular to the frontal plane. The face *ABDC* therefore appears as an edge in the front view, but none of the principal views shows the true size and shape of the surface. To show the true size and shape of *ABDC*, a view is needed that has a direction of observation perpendicular to *ABDC* and is projected on a plane parallel to *ABDC*, as shown in Fig. 20. This view is known as a *normal* view.

Figure 21 is the orthographic counterpart of Fig. 20. To get the normal view of surface *ABDC*, a projection is made perpendicular to *ABDC*. This projection is made *from* the view where the surface shows as an edge, in this case, the front view, and thus perpendicularity from the surface is seen in true relationship.

Figure 22 is another pictorial drawing of a solid that has two inclined surfaces, one of which is plane 1-2-3-4. While plane 1-2-3-4 is a vertical surface, it is also inclined to the frontal and profile planes. Plane 1-2-3-4 appears as an edge view (EV) in the horizontal view. We have seen that it is necessary to look perpendicularly at an edge view to cause a plane to become a true surface (normal) in the next view. The next view should be auxiliary view A. The pictorial shows that plane 1-2-3-4 will project as a true surface in plane A.

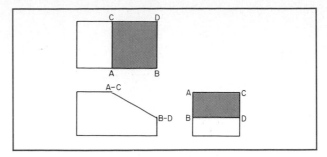

FIG. 19 One object face inclined to two principal planes.

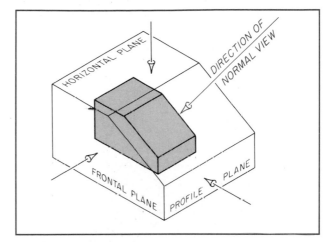

FIG. 20 Pictorial of object.

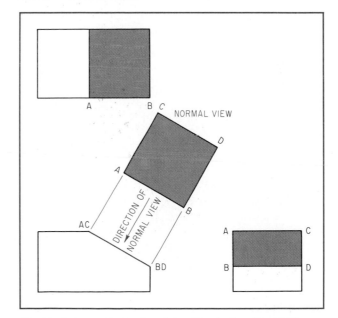

FIG. 21 Orthographic drawing of the object in Figs. 19 and 20.

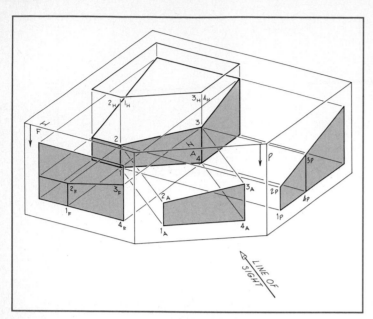

FIG. 22 Pictorial of a solid having an inclined surface.

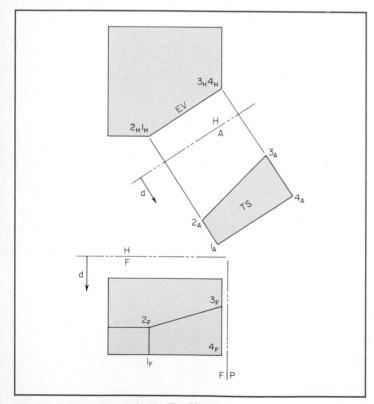

FIG. 23 Orthographic solution for Fig. 22.

The pictorial drawing has been converted into an orthographic drawing in Fig. 23. Complete horizontal and frontal views are shown in the figure. The true surface of plane 1-2-3-4 has been constructed in auxiliary view A, using the now familiar four steps. Note that the transfer of points 1, 2, 3, and 4 into plane A is possible because the common transfer direction d is available in the front view to be placed into plane A. Remember, the direction perpendicularly away from the same plane is always the same direction in space. In this case, the direction perpendicularly away from the horizontal plane is seen in the front view, and this direction may be transferred with confidence into plane A. Both the frontal view and plane A are perpendicular to the horizontal view. The arrows pointing perpendicularly away from the horizontal plane are shown. The arrows show height for this particular problem.

The process of constructing a true-surface view from an inclined plane is the correct process regardless of the position of an inclined plane in space. A plane may be inclined to a frontal plane, as is the case of Fig. 22, or it may be inclined to a horizontal, as in Fig. 20. The process of constructing the true surface from the given edge view, however, will always remain the same.

12. NORMAL VIEWS OF INCLINED SURFACES ON PRACTICAL OBJECTS

So far we have presented procedures for drawing a normal view, using a simple object as an illustration. Practical objects are, of course, made up of numerous basic shapes: rectangular prisms, conic shapes, cylinders, etc. Some might consider this a complication, but it need not be.

Figure 24 shows a part with a surface inclined to front and side, about in the same position as the inclined surface in Fig. 22. At A the object is shown pictorially, surrounded by the planes of projection, and at B the projection planes are opened up into the plane of the paper. Note that the reference plane is placed, in this case, at the *base* of the object because this is a natural reference surface. Thus at C the reference plane is drawn at the base in the front and normal views, and measurements are made *upward* (dimensions H and H') to needed points. Note that the top view is the normal view

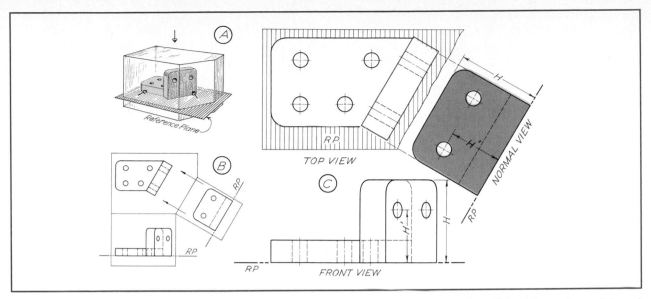

FIG. 24 Machine part with surface inclined to front and side. *(A)* planes of projection; *(B)* the planes opened up; *(C)* front, top, and normal views of the inclined surface.

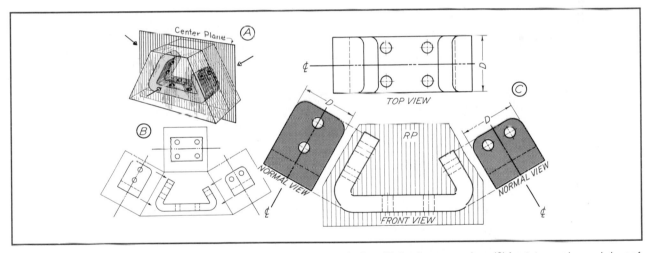

FIG. 25 A machine part with surfaces inclined to top and side. *(A)* planes of projection; *(B)* the planes opened up; *(C)* front, top, and normal views of the inclined surfaces.

of the reference plane, which appears as an edge in the front and normal views.

Figure 25 shows an object with two inclined surfaces. These surfaces are inclined with respect to the horizontal and profile planes (not shown). The object has symmetry about the center plane. This symmetry makes the center plane a convenient reference plane because distance in direction D (a depth) is equal on either side of the center

plane. Direction D is perpendicular to the front view, incidentally. Note that the normal views are only partial views, because the base of the object is fully described in the top and front views. It is common practice in auxiliary views to show only portions of complete views sufficient to provide new information, not to repeat known information.

Figure 26 shows another case in drawing a normal

view of a face inclined to the top and side. However, one cannot use the advantage of symmetry that was present in Fig. 25. We shall use the techniques of a reference plane as was done for Fig. 24.

1. Draw the partial top and front views, as at (b) and locate the view direction by drawing projectors perpendicular to the inclined surface, as shown.

2. Locate the reference plane in the top view. The reference plane is here located at the rear flat surface of the object because of convenience in measuring. The reference plane in the normal view will be perpendicular to the projects already drawn, and is located at a convenient distance from the front view, as shown at (c).

3. As shown at (d), measure the distance (depths) from the reference plane (of various points needed), and transfer these measurements with dividers to the normal view, measuring from the reference plane in the normal view. Note that the points are in front of the reference plane in the top view and are therefore measured toward the front in the normal view.

4. From specifications, complete the normal view as shown at (e).

5. Complete the drawing, as shown at (f). In this case the top view could have been completed before the

normal view was drawn. However, it is considered better practice to lay out the normal view before completing the view that will show the surface foreshortened.

A part might also be such that a surface is inclined to the horizontal plane. Figure 27 shows such a part. Again, a center plane has been used as a reference plane so as to take advantage of the part's symmetry. Direction W (a width) is equal on either side of the center plane. Direction W is perpendicular to the profile plane. A partial auxiliary view is again used because projecting the entire part into the normal (true-surface) view would add nothing to our knowledge and would require considerable extra work. In fact, the two small holes seen in the front view would become ellipses in the auxiliary view, something a drafter should avoid if possible.

13. PRACTICAL USES OF NORMAL VIEW

In practical work the chief reason for using a normal view is to show the true shape of an inclined surface.

In a normal view, the inclined surface will be shown in its true shape, but the other faces of the object appearing in the view will be foreshortened. In practical work these foreshortened parts are usually omitted, as in Fig. 28. Views thus drawn are called *partial views*.

Another important use of a normal view is in the case

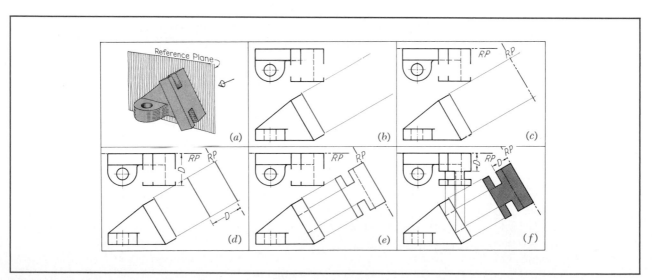

FIG. 26 Stages in drawing a machine part with a surface inclined to top and side.

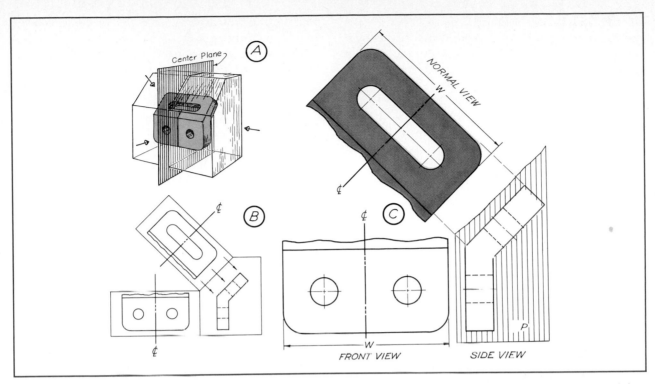

FIG. 27 A machine part with a surface inclined to top and front. *(A)* planes of projection; *(B)* the planes opened up; *(C)* front, side, and normal views of the inclined surface.

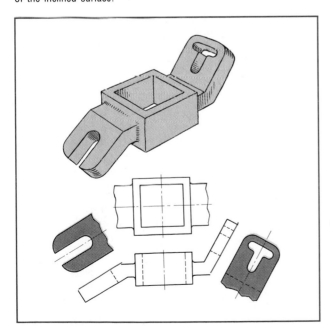

FIG. 28 Use of partial views.

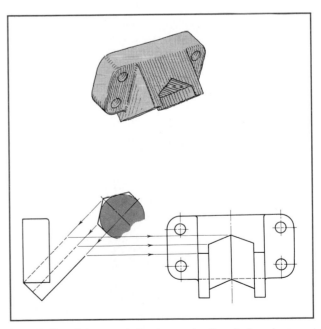

FIG. 29 Use of the normal view for construction of other views.

289

where a principal view has a part in a foreshortened position that cannot be drawn without first constructing a normal view in its true shape, from which the part can be projected back to the principal view. Figure 29 illustrates this procedure. Note in this figure that the view direction is set up looking along the semihexagonal slot and perpendicular to the face that is at a right angle to the slot. From the normal view showing the true semihexagonal shape, the side view and then the front view can be completed.

In most cases the normal view cannot be projected

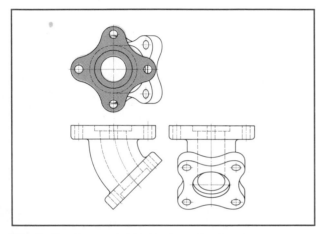

FIG. 30 Top, front, and right-side views of an irregular part.

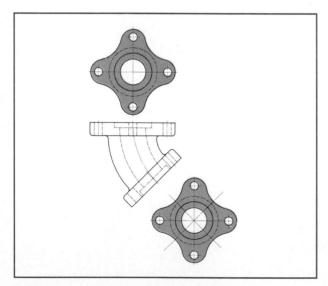

FIG. 31 Front and two normal views of an irregular part.

from the principal views but *must be drawn from dimensional specifications of the surface shape.*

Another practical example is the flanged 45° elbow of Fig. 30, a casting with an irregular inclined face, which not only cannot be shown in true shape in any of the principal views but also is difficult to draw in its foreshortened position. An easier and more practical selection of views for this piece is shown in Fig. 31, where normal views looking in directions perpendicular to the inclined faces show the true shape of the surfaces and allow for simplification. This is because each view can be laid out independently from specifications, and there is not even need for a reference plane.

14. NORMAL VIEWS OF OBLIQUE SURFACES

An oblique surface exists on an object if the surface never shows an edge view in any frontal, horizontal, or profile projection plane. An oblique surface *ABC* exists on the object in Fig. 32. Since surface *ABC* does not appear as an edge view in the principal view (*H*, *F*, *P*), an auxiliary view is necessary to establish an edge view.

Frontal line *AC* is in true length in the frontal view, and is so labeled. Therefore an edge view may be constructed by looking parallel to the true length (as described in Sec. 8). The open arrows indicate that the direction perpendicularly away from the frontal plane is the same direction in space. Therefore transfer measurements are taken from the horizontal plane and are placed in plane 1.

In this example the auxiliary view was constructed adjacent to the front view since a true length of a line was available in the front view. However, an auxiliary view could have been taken adjacent to the horizontal view just as readily, since horizontal line *AB* is in true length in the horizontal view. The view selected for making an edge view of plane *ABC* was done simply for the sake of variety.

Constructing the normal (true-surface) view of plane *ABC* was accomplished exactly as in Sec. 9. The line of sight was perpendicular to the edge view. The second auxiliary plane 2 was placed parallel to the edge view. All projected points crossed from plane 1 into plane 2 by going perpendicularly across the fold (reference) line 1-2. Transfers of points *A*, *B*, and *C* from plane 1 into plane 2 were located by going perpendicularly away from the same plane, that is, perpendicularly from plane 1

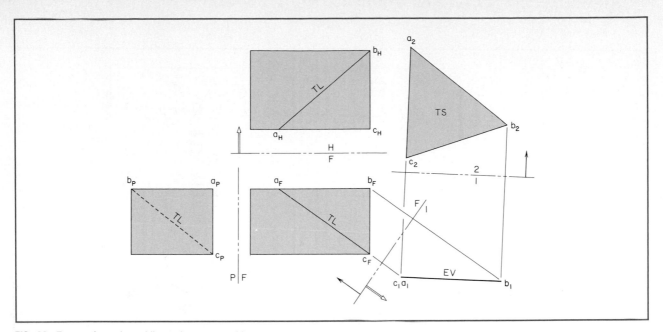

FIG. 32 True surface of an oblique plane on an object.

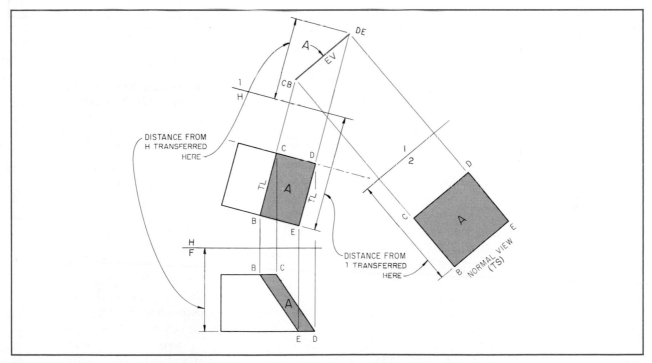

FIG. 33 Normal view of an oblique surface. *BCDE* is an oblique surface because of the position of the object.

into plane 2 in the same direction as perpendicularly away from plane 1 into plane *F*. The solid arrows indicate this common direction.

Another example involving an oblique surface is offered in Fig. 33. On the shaded oblique surface A one sees that none of the given edges are true length in the front view. However, both lines *BC* and *DE* are true length in the horizontal view because they are horizontal

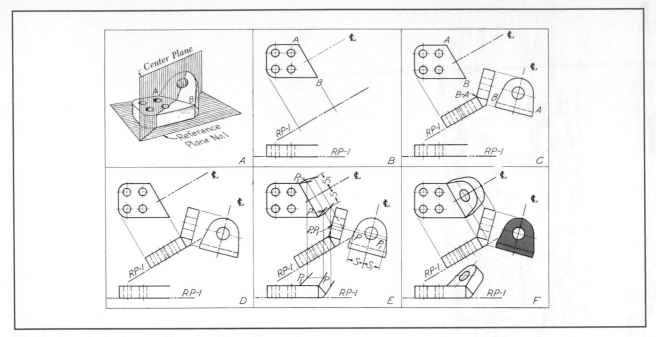

FIG. 34 Stages in drawing a machine part with an oblique surface.

lines. The procedure to find the normal view of surface A follows exactly that of previous examples. The proper transfers are given here with leaders and notes to aid understanding.

15. NORMAL VIEWS OF OBLIQUE SURFACES ON PRACTICAL OBJECTS

Figure 34 shows the successive steps in constructing a normal view. The pictorial illustration A shows a typical part with an oblique surface. The line of intersection between the oblique surface and the horizontal base in line AB. An edge view of the oblique surface may be obtained by looking parallel to line AB. A reference plane (RP-1) may be used. It is actually a horizontal plane placed under the base rather than between front and horizontal views, as is usually done. Plane RP-1 is used simply as an alternative position for the usually H-F fold line.

The line of sight for the normal (true-surface) view will be perpendicular to the oblique surface. A center plane for the normal view will be placed through the

single large hole and perpendicular to the base, so as to use to good advantage the symmetry within the oblique surface. The center plane is shown in illustration A.

At B, partial top and front views are shown. The projectors and reference plane for the required edge view are also shown. Note the centerline symbol.

At C, the edge view has been drawn. Note that line AB appears as a point in the edge view. The angle that the oblique surface makes with the base is laid out in this view from specifications.

At D, the normal view is added. The projectors for the view are perpendicular to the edge view. The centerline is drawn perpendicular to the projectors for the normal view and at a convenient distance from the edge view. This is the same centerline seen in the top view. The normal view is drawn from specifications of the shape. The projection back to the edge view can then be made.

The views thus completed at D describe the object, but the top and front views may be completed for illustrative purposes or as an exercise in projection. The method is illustrated at *E* and *F*. Any point, say *P*, may be selected and projected back to the edge view. From this view a projector is drawn back to the top view. Then the distance *S* from the normal view is transferred to the reference place in the top view. A number of points so located will complete the top view of the circular portion, and the straight-line portion can be projected in similar manner. The front view is found by drawing projectors to the front view for the points needed, measuring the heights from the reference plane in the edge view, and transferring these distances to the front view. Note that this procedure for completing the top and front views is the same as that for drawing the views originally but in reverse order.

It is hoped that a careful study of this chapter will help you understand both the theory and technique of making auxiliary views. This chapter can serve as an introduction to the study of space geometry, whose principles can solve almost any problem involving three dimensional relationships among lines, planes, and solids. Other texts are available to take you further into the area of space geometry, should you desire.

PROBLEMS

GROUP 1. NORMAL VIEWS OF INCLINED SURFACES

1 to 7. Draw given views and add normal views to the inclined surface using the reference plane indicated.

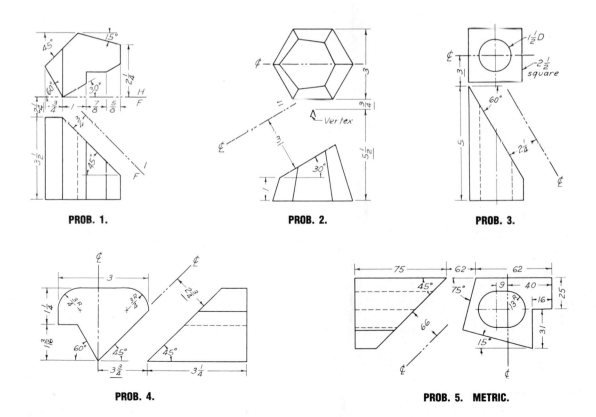

PROB. 1.   PROB. 2.   PROB. 3.

PROB. 4.   PROB. 5.   METRIC.

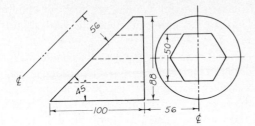

PROB. 6. METRIC.

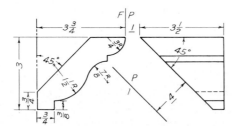

PROB. 7.

8. Draw the front view, partial top view, and normal view of the inclined surface.

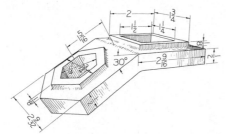

PROB. 8 Holder.

9. Draw the top view, partial front view, and normal view of the inclined surfaces.

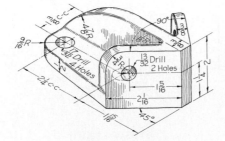

PROB. 9 Push plate.

10. Draw the front view, partial top view, and normal view of the inclined surface.

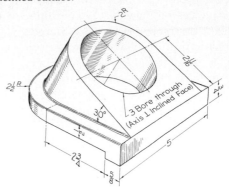

PROB. 10 Bevel washer.

11. Draw the front view, partial top view, and normal view of the inclined surfaces.

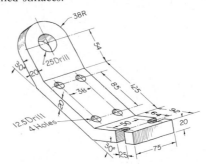

PROB. 11 METRIC. Connector strip.

12. Draw the front view, partial right-side view, and normal view of the inclined surface. Draw the normal view before completing the front view.

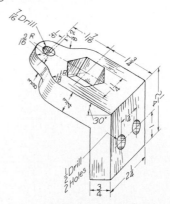

PROB. 12 Jig angle.

13. Draw the front view, partial top view, and normal view of the inclined surface.

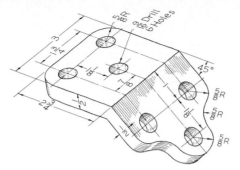

PROB. 13   Angle clip.

14. Draw the front view, partial top and left-side views, and normal view of the inclined surface.

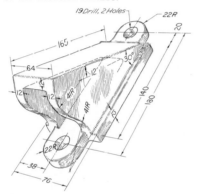

PROB. 14   METRIC. Channel support.

15. Draw the front view, partial top view, and normal view of the inclined surface.

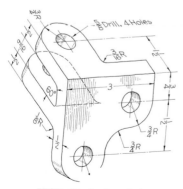

PROB. 15   Angle swivel.

16. Draw the front view, partial top and right-side views, and normal view of the inclined surface.

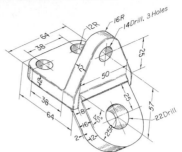

PROB. 16   METRIC. Corner tie.

17. Determine what views and partial views will best describe the part. Sketch the proposed views before making the drawing with instruments.

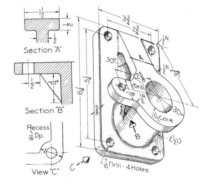

PROB. 17   Angle-shaft base.

18. Draw the front view, partial bottom view, and normal view of the inclined surface.

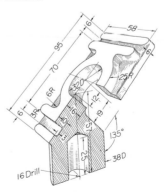

PROB. 18   METRIC. Catenary clip.

19 and 20. Determine what views and partial views will best describe the part.

21. Draw the front view, partial left-side and bottom views, and normal view of the inclined surface.

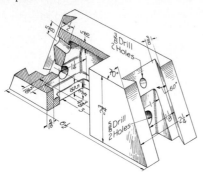

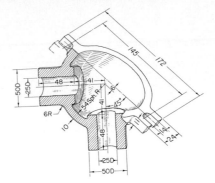

PROB. 19 Slide base.

PROB. 21 METRIC. Bevel-gear housing.

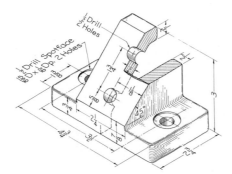

PROB. 20 Idler bracket.

22. Draw the given front view and add the views necessary to describe the part.

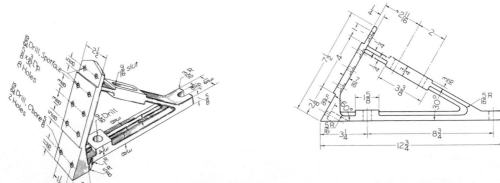

PROB. 22 Corner brace.

23 and 24. This pair of similar objects has the upper lug in two different positions. Layouts are for 11- × 17-in paper. Draw the views and partial views as indicated on the layouts.

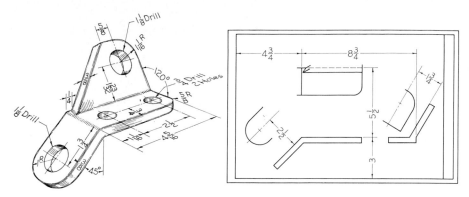

PROB. 23 Spar clip, 90°.

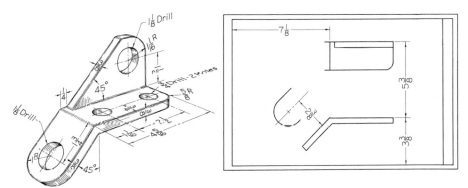

PROB. 24 Spar clip, 120°.

25. Draw the front view; partial top, right-side, and left-side views; and normal view of the inclined surface. Use decimal scale for layout.

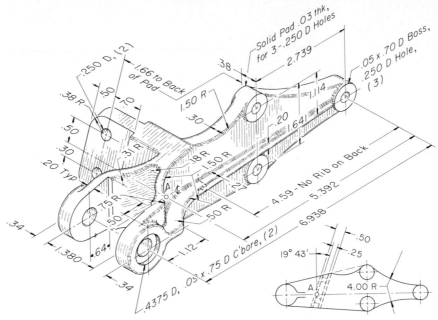

PROB. 25 Seat release.

26. Draw the views and partial views that will best describe the part.

27. Draw top and front views and a normal view of the inclined face. Will the normal view of the inclined face show the true cross section of the square hole?

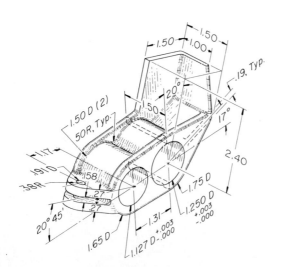

PROB. 26 Idler link.

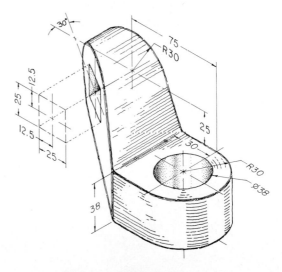

PROB. 27 **METRIC.** Actuator bracket.

28. Front and right-side views of the shaft-locator wedge are shown on a layout for 11- × 17-in paper. Add normal view of the inclined faces at centerline position shown.

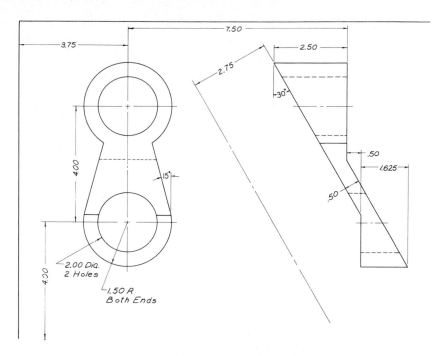

PROB. 28 Shaft-locator wedge.

GROUP 2. NORMAL VIEWS OF OBLIQUE SURFACES

29. Draw the partial front and top views, edge view showing the contour of the slot, and normal view of the oblique surface.

30. Draw the partial front and top views, and edge and normal views of the oblique surface. Draw the normal view before completing the edge view.

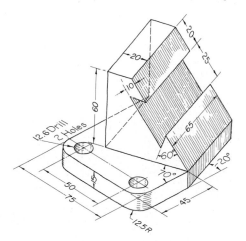

PROB. 29 METRIC. Dovetail clip.

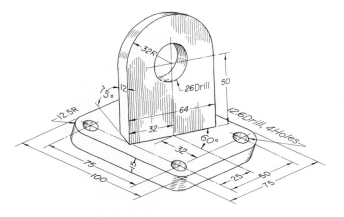

PROB. 30 METRIC. Anchor base.

31. Draw the partial front and top views, and edge and normal views of the oblique surface.

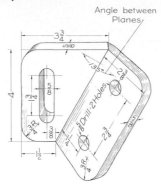

PROB. 31 Adjusting clip.

32. Draw the views given, omitting the lugs in the top view. Add normal views to describe the lugs.

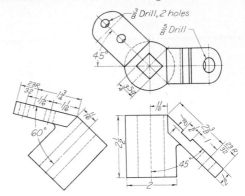

PROB. 32 Bar-strut anchor.

33. Draw the views given, using edge and normal views to obtain the shape of the lugs.

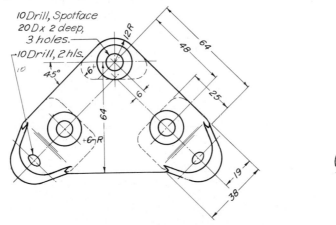

PROB. 33 **METRIC.** Cable anchor.

34. Draw the partial top, front, and side views. Add edge and normal views to describe the lugs.

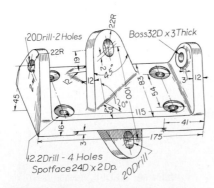

PROB. 34 **METRIC.** Transverse connection.

35. Draw the top and front views and use edge and normal views to describe the slots and oblique surfaces. The part is symmetrical about the main axis.

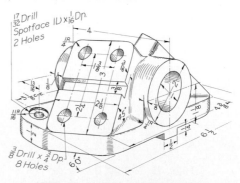

PROB. 35 **METRIC.** Chamber tool base.

36. Draw the spar clip, using the layout shown for 11- × 17-in paper. Note that an edge and two normal views are required.

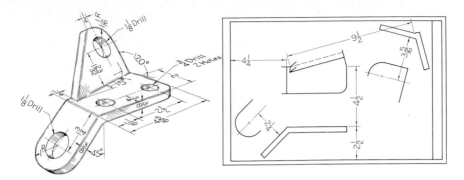

PROB. 36 Spar clip.

37. Draw the views given, using edge and normal views to describe the lugs.

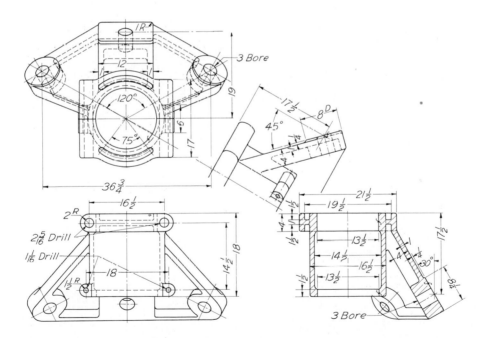

PROB. 37 Crane-masthead collar and cap.

38. Draw top and front views and auxiliary views that will describe the oblique surface.

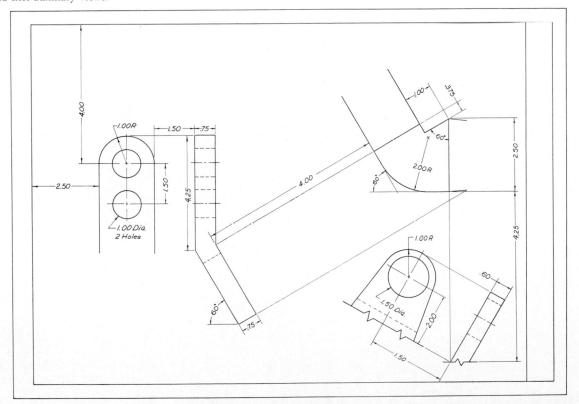

PROB. 38 Bus-bar connector.

39. Shown on layout for 11- × 17-in paper are partial front and right-side views and partial auxiliary views that describe the position and shape of the oblique surface. Complete front, right-side, and first auxiliary views.

PROB. 39 Valve control-shaft bracket.

40. Shown on layout for 11- × 17-in paper are partial top and front views and auxiliaries that describe the position and shape of the clamp portion of the part. Complete the top and front views.

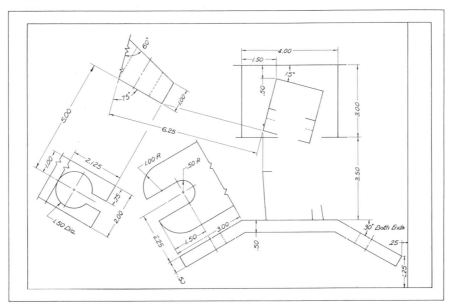

PROB. 40 Bipod shaft clamp.

GROUP 3. NORMAL VIEW OF A LINE

41. Find, by the use of an auxiliary projected from the top view, the true length of AB. Scale: $1'' = 1'-0''$.

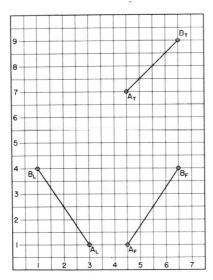

PROB. 41.

42. The line CL slopes upward from C with a 60 percent grade. Draw the front view of CL. Find the distance from C to L. Scale: $1'' = 50'$.

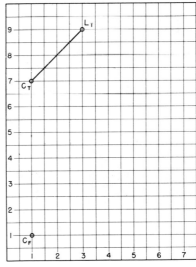

PROB. 42.

43. Line *AB* is to be established with a bearing of N60°E with a 60 percent positive grade from *A*. The distance from *A* to *B* is 336 ft. Draw the top and front views of *AB*. Scale: $1'' = 100'$.

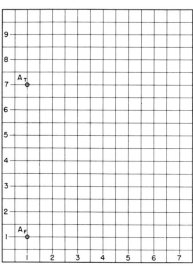

PROB. 43.

44. A power-line pole is supported by two guy wires. *E* and *D* are the points on the ground where the guy wires are to be attached. Find the length of the guy wires from point *C* in the pole. Disregard lengths for attaching. Scale: $1'' = 10'$.

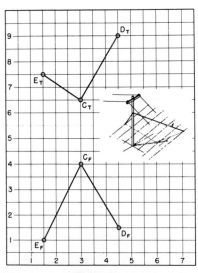

PROB. 44.

GROUP 4. NORMAL AND END VIEWS OF A LINE

45. Using the layout of Prob. 41, show the end view of *AB* by projecting from a first auxiliary view projected from the top view. Identify all reference planes, and label all points.

46. Using the layout of Prob. 41, show the end view of *AB* by projecting from a first auxiliary view projected from the front view. Identify all reference planes, and label all points.

47. Using the layout of Prob. 41, consider *AB* as the centerline of a 3½-in (nominal size) American Standard pipe. Show the end view of the pipe by projecting from a normal view of the centerline *AB* projected from the side view. Scale: half size.

GROUP 5. EDGE VIEW OF A PLANE

48. Show the edge view of *RSP* by projecting parallel to a horizontal line of *RSP*.

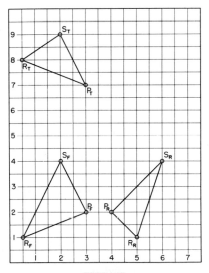

PROB. 48.

49. Using the layout given for Prob. 48, determine the edge view of *RSP* by projecting parallel to a frontal line of *RSP*.

50. Using the layout of Prob. 48, draw the edge view of *RSP* when seen from the rear along a profile line of sight.

GROUP 6. EDGE AND NORMAL VIEWS OF A PLANE

51. Using the layout of Prob. 48, draw the normal view of *RSP* by projecting from an edge view of *RSP* projected from the top view. Label all points and identify the reference planes used.

52. Using the layout of Prob. 48, determine the true size and shape of *RSP* when projecting from an edge view of *RSP* projected from the front view. Label all points, and identify the reference planes used.

53. Using the layout of Prob. 48, first show the edge view of *RSP* in an edge view projected from the side view; then draw the normal view of *RSP*. Label all points, and identify all reference planes used.

10
SURFACE INTERSECTIONS

1. INTERSECTIONS IN THE PRACTICAL WORLD

Intersections occur very frequently in the design world. Examples are the joining of ducts in heating and ventilation work, the hookup of pipes in various installations, and the placement of braces within structures. The applications are endless.

This chapter will cover the basic types of intersections using simple geometric forms. Since all designs involve combinations of simple shapes, we will study the concepts of intersections applied to basic shapes. By learning a few fundamental concepts, you will be able to handle all types of intersections simply as applications of the concepts.

2. TYPES OF INTERSECTIONS TO BE STUDIED

1. Line with line	9. Line with cylinder
2. Line with plane	10. Line with cone
3. Line with solid	11. Plane with cylinder
4. Plane with plane	12. Plane with cone
5. Plane with prism	13. Cylinder with cylinder
6. Prism with prism	14. Cone with cylinder
7. Prism with pyramid	15. Cone with cone
8. Pyramid with pyramid	16. Line with surface of revolution

At first, this may appear to be a lengthy list of types for consideration. However, a pattern of strategy for solutions will appear early, and this pattern will assist you in all intersection problems.

3. INTERSECTION OF TWO LINES

Lines that pass through the same point are intersecting lines. The common point is the point of intersection. A single orthographic view does not give sufficient

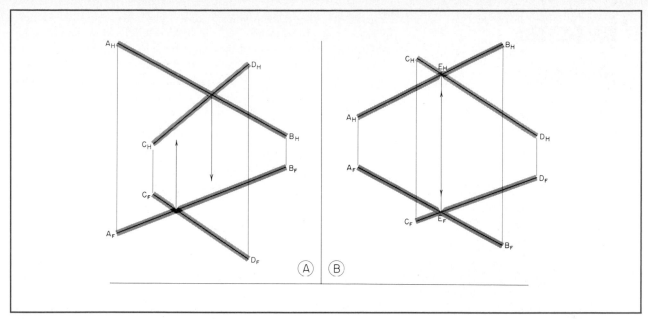

FIG. 1 Nonintersecting lines *(A)* and intersecting lines *(B)*.

information to determine whether existing lines do or do not intersect. The front view only of Fig. 1A does not justify the conclusion that *AB* and *CD* intersect, because the crossing of the lines is only the projection in the front view. The top view shows a crossing not in projection with the crossing in the front view; therefore there is no common point, and the lines do not intersect. On the other hand, if both the front and the top views of Fig. 1B are read by following along the projector from E_F, it can be seen that point *E* lies on both *AB* and *CD*, and the lines therefore intersect. Thus if two views (or more) show the lines to have a common point, the lines intersect. This intersecting-line principle is an important tool for solving many space problems.

Mention should be made of nonintersecting lines. Nonintersecting lines are lines that do not share a common point. As already indicated, lines *AB* and *CD* of Fig. 1A do not have an intersection point in common and therefore are nonintersecting lines.

Lines at an angle to each other are called "skew lines." The lines of Fig. 1 are examples of skew lines. Note that skew lines may or may not intersect.

4. INTERSECTION OF LINE WITH PLANE

We first shall look at the general case of the intersection of a line and a plane. The general case can be solved by two methods: edge-view method and cutting-plane method. The general case is represented by any oblique line and any oblique plane as seen in Fig. 2.

A. Intersection of Line with Plane—Edge-View Method The edge-view method requires no new concepts. One has only to remember how to construct an edge view of a plane. Any line then passing through that edge view will reveal the intersection point. Figure 2 illustrates the edge-view method. Any edge view could be constructed. In this case it was decided arbitrarily to build the edge view of plane 1-2-3 adjacent to the top view. Line 4-5 was carried into auxiliary view A along with the edge view of the plane. The intersection is found and is projected back into the top and front views. Realize that line 4-5 need never be in true length to find an intersection point. An edge view of the plane is fully sufficient.

Before our solution is finished we must consider one more item. That item is the visibility of line 4-5 relative

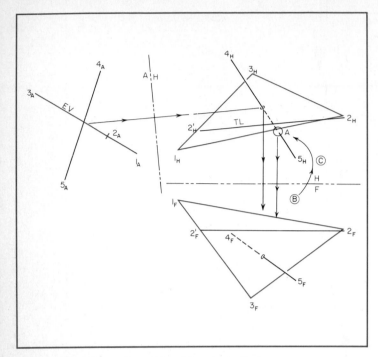

FIG. 2 Intersection of line and plane-edge view method.

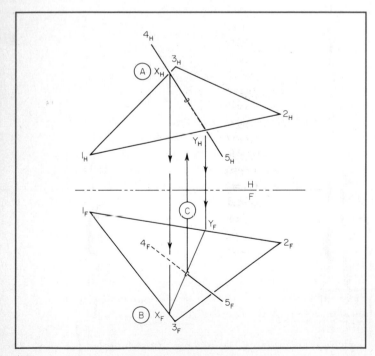

FIG. 3 Intersection of line and plane-cutting plane method.

to the plane in each view. "Visibility" refers to whether a line appears visible or hidden relative to a plane. Visibility in this solution must be determined for the H and F views, but visibility is not necessary for view A because the plane is in edge view. The following steps will determine visibility not only for this solution but for any problem you might encounter.

▪ **Step A** Select an *apparent* intersection of two lines in the view for which visibility is desired. In Fig. 2 in the top view the apparent intersection of lines 1-2 and 4-5 were selected. (One could have just as well used lines 1-3 and 4-5.)

▪ **Step B** Look into any immediately adjacent view to see which of the two selected lines is seen first. That line will be visible in the original view; the second line seen will be hidden in the original view.

For the case at hand, for lines 1-2 and 4-5, line 1-2 is seen first in adjacent view F, and line 4-5 is seen second, along the projector from the apparent intersection. Alternatively, one could look into adjacent view A and obtain identical results.

▪ **Step C** By knowing which of the two apparently intersecting lines is first and which is second, you can conclude visibility in the original view. With line 4-5 being second in this case, it is drawn hidden in the original (top) view, up to the intersection point. Visibility always reverses at the intersection point between a line and a plane. Therefore line 4-5 becomes visible in the rearward portion of the top view.

The visibility for the front view would be determined in the same manner—that is, select a pair of apparently intersecting lines, see which of the two is seen first and second in the adjacent view, and use this knowledge to complete the visibility back in the front view.

B. Intersection of Line with Plane—Cutting-Plane Method The cutting-plane method is a powerful, efficient method for finding intersections that do not require an auxiliary view. When using the edge-view method, create an auxiliary view to provide an edge view of a plane when no edge view exists in the given views. With the cutting-plane method, only the existing views are required.

Three major steps are needed for the cutting-plane method. The steps are demonstrated in Fig. 3. The problem solved by the edge-view method is repeated so that the difference between the edge-view method and the cutting-plane method will be apparent.

▪ **Step A** The cutting-plane method is based on the premise that any intersection between a line and a plane must occur along the path that the line travels as it crosses the plane. For example, in the top view of Fig. 3, as line 4-5 travels across the plane, it does so along line XY on the plane. This means that if line 4-5 is to "hit" the plane in the top view, the hit (or intersection) must occur along line XY. If we know that the intersection must occur along line XY, point X on line 1-2 and point Y on line 1-3 are so labeled. One could say that line XY has been "cut" across the plane—hence the term "cutting-plane method."

▪ **Step B** Placing line XY on the plane in step A is useful but not yet conclusive. Lines 4-5 and XY lie on top of each other. The actual point of intersection cannot be seen. However, when line XY is projected to the front view, lines XY and 4-5 are separate from one another in the front view and the intersection between them can be seen. Since line XY is on the plane, line 4-5 goes through the plane when it goes across line XY.

▪ **Step C** The intersection point found in the front view is easily projected onto line 4-5 on the top view. Determining visibility concludes the solution.

The problem just solved could have been just as well started in the front view rather than in the top view. Selecting the view to start the problem is a matter of personal choice. The overall process of using the cutting-plane method will be helpful in all types of intersection problems, as we will see.

5. INTERSECTION OF A LINE WITH A SOLID: GENERAL CASE

This type of intersection has applications ranging in size from making a model to boring a train tunnel through a mountain. One example is seen in Fig. 4, where line 4-5 intersects a pyramid. It is possible to have just one point of intersection if a line enters but does not exit a solid. In Fig. 4, however, there are two points of intersection.

Using the cutting-plane method for solids yields a more direct solution than does the edge-view method, since no time-consuming auxiliary views are required.

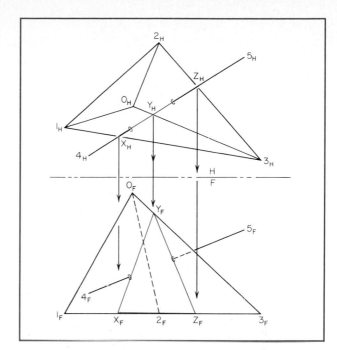

FIG. 4 Intersection of line and solid-cutting plane method.

The concept used in Sec. 4B for finding the intersection of a line and a plane by the cutting-plane method will be demonstrated again. The intersection of the line and the solid must occur along the path covered by line 4-5 as it crosses the solid. Again the solution is begun in the top view (an arbitrary decision). If line 4-5 is to intersect the pyramid, the intersections can take place only along lines XY, YZ, or XZ, which are labeled in the top view.

Then points X, Y, and Z are projected to the front view, giving lines XY, YZ, and XZ in the front view. If line 4-5 is to hit the pyramid, the intersection will show in the front view. Two intersection points do occur, one on line XY and one on line YZ. These two intersection points are returned to the top view and placed on line 4-5. Completing visibility of line 4-5 relative to the pyramid finishes the solution. Note that it is not necessary to show line 4-5 inside the solid.

6. INTERSECTION OF A LINE WITH A SOLID: SPECIFIC CASES

You may think that the examples so far in this chapter are rather abstract. The general cases discussed so far are valuable in that all specific cases relate directly to the general case and can be solved by the techniques used to solve the general case.

We will illustrate several specific cases to show that they are really just applications of the general case. The cases will get more specific as we move from example to example.

A. Oblique Line Intersecting a Solid Having an Oblique Surface Figure 5 shows an example on this case, a case that is quite similar to the general case seen in Fig. 2. The difference between the two figures is simply that a true length of line *RS* is available in the horizontal view of Fig. 5, but a horizontal line had to be added in the general case of Fig. 2. Otherwise, the two solutions are the same.

B. Oblique Line Intersecting a Solid Having an Inclined Surface We see in Fig. 6 three examples, all of which involve an oblique line intersecting an inclined surface. In case 6a the plane intersected is inclined to the horizontal and profile planes and shows itself edge-view in the frontal view. The intersection is point *P*. Case 6b shows the plane inclined to the frontal and profiles planes (edge-view in the horizontal view). Finally, case 6c gives the edge view in the profile view, meaning that the plane is inclined to the horizontal and frontal planes.

C. Oblique Line Intersecting Principal Planes of a Solid Figure 7 provides the three possibilities of intersection with principal planes. Case 7a shows an intersection with the horizontal plane. Case 7b reveals intersections with both the frontal and profiles planes in that the line *AB* is forced entirely through the solid.

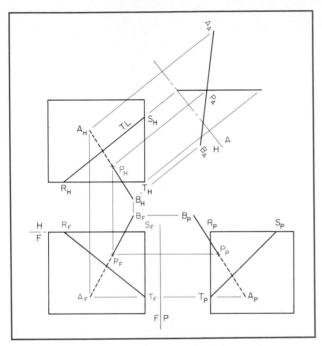

FIG. 5 Intersection of an oblique line with a solid having an oblique surface.

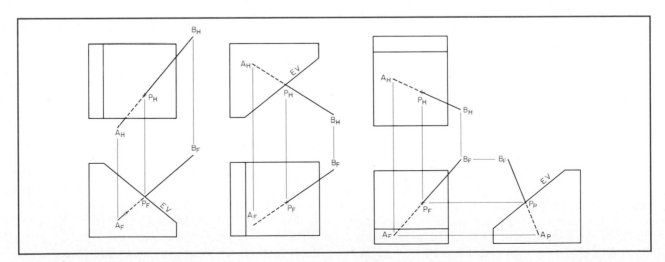

FIG. 6 Intersection of an oblique line with a solid having an inclined surface.

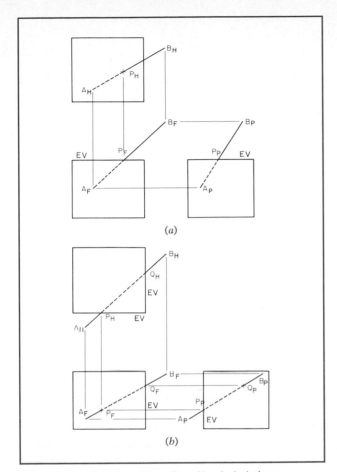

(a)

(b)

FIG. 7 Intersection of an oblique line with principal planes.

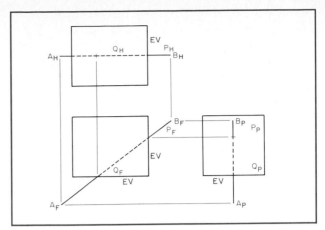

FIG. 8 Intersection of an inclined line with a principal plane.

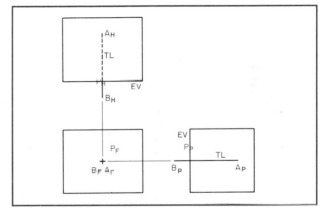

FIG. 9 Intersection of a principal line with a principal plane.

D. Inclined Line Intersecting Principal Planes of a Solid This specific case is given in Fig. 8. Note that the horizontal and profile planes are intersected by line AB.

E. Principal Line Intersecting a Principal Plane of a Solid This is the most specific, restricted case of a line intersecting a solid. Figure 9 shows line AB to be a horizontal-profile line, intersecting a frontal plane of the solid at point P. Note that line AB appears point-view in the frontal view.

7. INTERSECTION OF TWO PLANES: GENERAL CASE

The world of design is filled with examples of the intersection of two planes. Whether it be the intersection of two roof planes on a building or of two sides of an air duct, this type of intersection occurs frequently. The edge-view method of solution and the cutting-plane method are both used. Each will be illustrated.

A. Intersection of Two Planes—Edge-View Method The edge-view method implies that intersection will be revealed in a view showing the edge view. This idea worked well for a line and a plane (Sec. 4A). It will work just as well here. The example of Fig. 10 desires the intersection between plane 1-2-3-4 and plane 5-6-7. Planes always intersect along a line, and a line of intersection is also expected here.

Plane 1-2-3-4 was selected to be made in edge view. Plane 5-6-7 could just as well have been selected. The

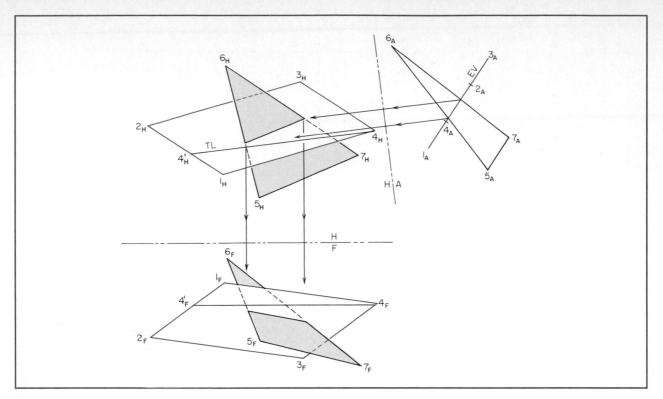

FIG. 10 Intersection of two planes-edge-view method.

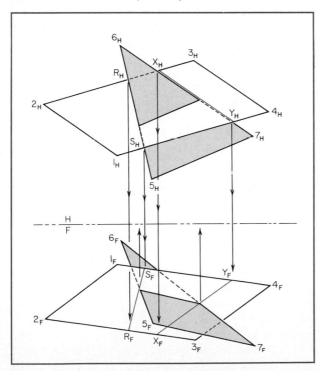

FIG. 11 Intersection of two planes-cutting-plane method.

horizontal line 4-4' added to plane 1-2-3-4 shows in true length in the horizontal view. Auxiliary view A is constructed to give line 4-4' in point view and therefore plane 1-2-3-4 in edge view. Plane 5-6-7 is transferred into view A. The line of intersection is then available and is projected back into the horizontal view onto lines 5-6 and 6-7. The front view receives the line of intersection by projection from the horizontal view.

The line of intersection between any two planes is always visible. On the other hand, the visibility between the two planes is enhanced by shading of one of them (plane 5-6-7 in this case).

B. Intersection of Two Planes—Cutting-Plane Method

The problem of Fig. 10 is repeated in Fig. 11, using the cutting-plane method. Having used this method for a line and a plane, and a line and a solid, you should be reasonably familiar with the process. The problem is somewhat complicated, however, in that with this method one does not know which lines are actually intersecting. One could argue that possibly lines 5-6 and 6-7 are intersecting the parallelogram 1-2-3-4, or that the solution is a combination of these two possibilities (e.g., lines 2-3 and 5-6 might intersect).

Do not be concerned; if you try a line for possible intersection and it doesn't intersect, simply try another line. For Fig. 11, lines 5-6 and 6-7 are found to work. Cutting along line 5-6 in the top view gives cut RS on plane 1-2-3-4. Projecting RS into the front view yields an intersection point on line 5-6. Similarly, cutting along line 6-7 in the top view gives cut XY on plane 1-2-3-4. Projecting XY into the front view gives the second intersection point. The two intersection points determine the line of intersection, which can then be projected back into the top view. Correct visibility completes the solution.

8. INTERSECTION OF TWO PLANES: SPECIFIC CASES

We shall discuss three specific cases in which an oblique plane intersects a plane of a solid. You will recognize that all three cases are merely applications of the general case.

A. Oblique Plane Intersecting an Oblique Plane of a Solid

Refer to Fig. 12. The oblique plane $RSTU$ is intersected by the other oblique plane along the line of intersection PQ. The solution is obtained by the cutting plane method discussed in the general case (Sec. 7). Line $A_F A_F'$ is extended to give cut WX. Line $B_F B_F'$ is extended to give cut YZ. The cut lines are projected into

the horizontal view where $A_H A_H'$ and $B_H B_H'$ are extended to intersect the cut lines. The result is line PQ, the line of intersection.

B. Oblique Plane Intersecting an Inclined Plane of a Solid

Figure 13 shows clearly that lines $A_F A_F'$ and $B_H B_H'$ are extended in the frontal view to intersect the

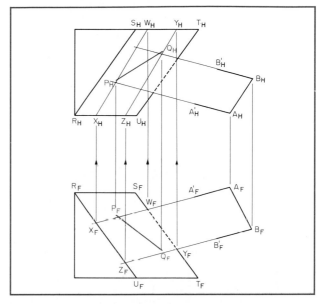

FIG. 12 Intersection of an oblique plane with an oblique plane of a solid.

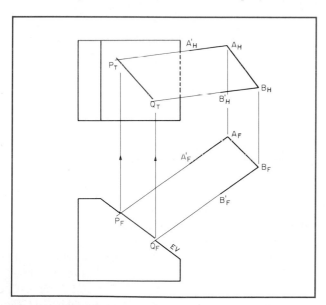

FIG. 13 Intersection of an oblique plane with an inclined plane of a solid.

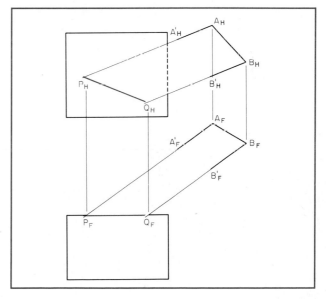

FIG. 14 Intersection of an oblique plane with a principal plane of a solid.

edge view of the inclined plane. The intersection provides the line of intersection *PQ* which is projected into the horizontal view.

C. Oblique Plane Intersecting a Principal Plane of a Solid

The most specific case is seen in Fig. 14. The edge view of a horizontal plane of the solid is intersected in the frontal view by lines $A_F A_F'$ and $B_F B_F'$ extended. The line of intersection is again line *PQ*.

9. INTERSECTION OF A PLANE AND A PRISM

The process for solving the problem of an intersection of a plane and a prism may be applied to the many areas of design in which a plane intersects a solid. Figure 15 combines both edge-view and cutting-plane methods. A step-by-step analysis is provided.

Step A. Determine the Expected Number of Intersection Points

Looking at the top view one sees that line 1-4 hits edge views of the prism at two points. Line 2-3 also hits the prism at two points, giving a subtotal of four expected intersection points so far. Vertical lines 9-14 and 7-12 will hit plane 1-2-3-4 somewhere, adding two more in-tersection points. This gives a total of six expected points of intersection.

Step B. Locate Intersection Points

When an edge view exists and a line hits it, the intersection point is immediately available without the need for any construction. Such is the case for lines 1-4 and 2-3. Each of these two lines hits the prism twice where edge views exist, giving four intersection points which can be projected to the same respective lines in the front view.

Lines 9-14 and 7-12 do not hit edge views. Therefore two options exist: Either make plane 1-2-3-4 into an edge view and carry lines 9-14 and 7-12 into that view, or use the cutting-plane method, which is definitely easier.

Lines *RS* and *XY* are cut in the top view through the point views of lines 9-14 and 7-12, respectively. Lines *RS* and *XY* are projected to the front view. This reveals the intersection of cut lines *RS* and *XY* with lines 9-14 and 7-12, respectively, yielding two more (and final) intersection points. Note that lines *RS* and *XY* in the top view are arbitrary cut lines. Any lines cutting through corners 9-14 and 7-12 would be usable.

Step C. Connect the Intersection Points in a Logical Manner

For example, move clockwise from intersection point to intersection point. Connect a line from where line 1-4 hits plane 9-8-13-14 to where line 9-14 passes through plane 1-2-3-4 (seen in the front view). Then go from this last point up to where line 2-3 hits plane 5-9-14-10. These operations complete the path of intersection on the left-hand side of the prism.

Repeat the process on the right-hand side of the prism. Move from where line 2-3 hits plane 6-7-12-11 down to where line 7-12 hits plane 1-2-3-4. Finish by moving from the last point down to where line 1-4 hits plane 8-7-12-13.

Step D. Determine Visibility

There are two visibilities that must be considered. One visibility needed is that of the plane in regard to the prism. The other visibility required is that for the paths of intersection. For visibility between the plane and prisms, realize that no visibility concerns exist in the top view since the prism shows itself in a series of edge views. Only the front view has to be resolved in terms of visibility. Then realize that lines 5-10 and 6-11 of the prism are seen to be at the very rear of the picture (in the top view) and are therefore hidden

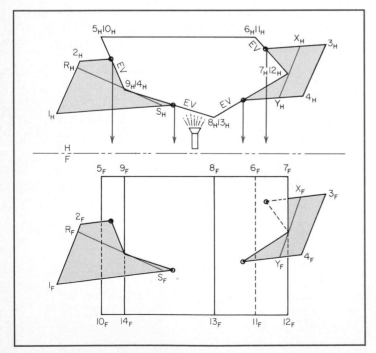

FIG. 15   Intersection of plane and prism.

behind plane 1-2-3-4 in the front view. Conversely, line 1-4 of the plane is in front of the prism (in the top view), making line 1-4 visible in the front view. These analyses give us the proper visibility.

Finally, the visibility of path of intersection is needed. A quick way to determine this is to use a "flashlight" technique. In the top view a flashlight is shown, lighting up three surfaces of the prism (5-9-14-10, 9-8-13-14, 8-7-12-13) but leaving two surfaces in darkness (5-6-11-10 and 6-7-12-11). Any lines on visible surfaces are visible, and any lines on hidden (dark) surfaces are hidden. This analysis leads to three lines of intersection being visible and one hidden, as shown.

10. INTERSECTION OF TWO PRISMS

The same processes used to find the intersection of a plane and a prism can be used effectively to find the intersection of two prisms. Recall from Sec. 9 that four steps are needed:

A. Determine the expected number of intersection points.

B. Locate the intersection points.

C. Connect the intersection points to form lines of intersection.

D. Determine visibility.

Step A Looking at Fig. 16 one can realize that the two prisms intersect in some kind of closed path made up of straight-line segments, since neither solid contains curved segments. To estimate the number of intersection points, observe that the triangular prism will have lines E, F, and G hit the vertical prism, giving three intersection points. It also appears that lines NS, RS, and ST of the vertical prism will each hit the triangular prism once, giving three additional intersection points for a total of six points. (The actual intersection points will be labeled in this problem, whereas such points were not labeled in earlier, simpler problems.)

Step B To locate the intersection points we again will make use of both the edge-view and cutting-plane methods. Use the edge-view method first because it is the most direct method if lines hit edge views. Three lines hit edge views. Line RS hits the vertical edge of the triangular prism in the top view to yield point 4. Line F of the

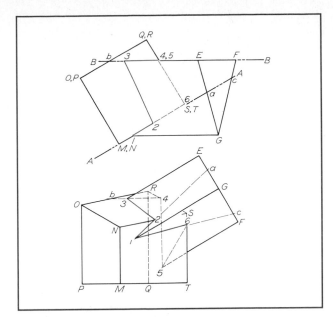

FIG. 16 Intersection of two prisms.

triangular prism hits edge view $QRST$ in the top view to give point 5. Line G of the triangular prism hits edge view $MNST$ in the top view to give point 1. Intersection points 4, 5, and 1 are then projected onto the proper lines in the front view.

With the information gained from edge views now exhausted, we turn to the cutting-plane method. To locate where lines NS and ST of the vertical four-sided prism hit the triangular prism, use is made of cutting plane A-A. In the top view cutting plane A-A is placed coincident with lines NS and ST. We will know where lines NS and ST hit the triangular prism if we know where the cutting plane cuts across this triangular prism. The top views show that cutting plane A-A cuts the triangular prism from point 1 to a (on line EG) to c (on line FG).

The lines cut by cutting plane A-A are projected to the front view, showing themselves as lines $1a$, $1c$, and ac. If lines NS and ST hit any of these three lines, NS and ST hit the triangular prism, because lines $1a$, $1c$, and ac are on the triangular prism. Intersection does occur at point 2 for line NS and at point 6 for line ST.

Five of the six intersection points have now been found. Only the intersection of line E of the triangular prism is needed. This intersection is located by using

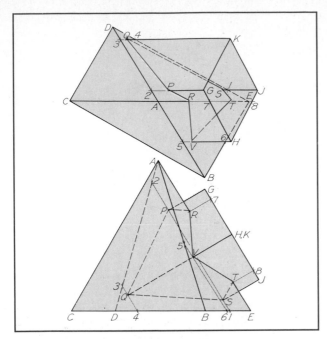

FIG. 17 Intersection of a pyramid and a prism.

cutting plane *B-B* in the top view to cut across plane *NORS* of the vertical prism. The line cut is *b*4 between lines *OR* and *RS*. Line *b*4 is projected to the front view, where line *E* does hit at point 3. All six intersection points have been located.

Step C To form lines of intersection, take an imaginary walk around the bounded area common to both prisms. Specifically, start in the top view where the vertical prism has all vertical surfaces showing in edge view. Walk from point 1 to 2 and then mark this path in the front view. Walk on from point 2 to 3; mark it the front view. Walk from 3 to 4, from 4 to 5, 5 to 6, and finally arrive back where you started by walking from 6 to 1. Connecting all points in this logical clockwise fashion in the top view gives the complete bounded area of intersection between the two prisms.

Step D Visibility is needed to complete the solution. One needs the visibility of the solids with respect to one another and also the visibility of the paths of intersection. Visibility is not a major concern in the top view because of the many edge views. The base plane *MPQT* of the vertical prism is the only hidden feature. In the front view line *QR* is hidden because it is at the rear, as seen in the top view.

Visibility of the intersection path can be determined readily when one realizes that planes *MNST* and *NORS* of the vertical prism are visible in the front view. Therefore intersection lines 6-1, 1-2, and 2-3 on these two planes are visible. Planes *OPQR* and *QRST* are hidden in the front since these planes are on the "dark" back side of the vertical prism. Therefore intersection lines 3-4, 4-5, and 5-6 are hidden in the front view. The problem is done.

11. INTERSECTION OF A PRISM AND A PYRAMID

The problem is quite similar in technique of solution to that used for intersecting prisms. In Fig. 17 it can be seen that no edge views are helpful. Therefore the solution must rely on the cutting-plane method.

A vertical plane through edge *G* cuts line 1-2 from surface *AED*, and a vertical plane through edge *K* cuts line 3-4 from surface *AED*, giving the two piercing points *P* and *Q* on surface *AED*. A vertical plane through edge *AE* cuts elements 7 and 8 from the prism and gives piercing points *R* and *T*. Continue in this manner until the complete line of intersection *PQS-TVR* is found.

12. INTERSECTION OF TWO PYRAMIDS

This case is a variation on the theme already expressed in Figs. 16 and 17. Refer to Fig. 18. Find where edge *AD* pierces plane *EHG*, by assuming a vertical cutting plane through edge *AD*. This plane cuts line 1-2 from plane *EHG*, and the piercing point is point *P*, located first on the front view and then projected to the top view. Next, find where *AD* pierces plane *EFG*, by using a vertical cutting plane through *AD*. This plane cuts line 3-4 from plane *EFG*, and the piercing point is point *Q*.

After finding point *P* on plane *EHG* and point *Q* on plane *EFG*, locate the piercing point of edge *EG* with plane *ABD* in order to draw lines of intersection. Thus edges of the "first" pyramid pierce surfaces of the "second," and edges of the second pierce surfaces of the first. Continue in this manner until the complete lines of intersection *PRQSTV* has been found.

The use of a vertical cutting plane to obtain the piercing points is perhaps the simplest method and the easiest to visualize. Nevertheless, it should be noted that a plane receding from the frontal or profile plane could also be used. As an example of the use of a plane receding from the frontal, consider that such a plane has been passed through line CA in the front view. This plane cuts line 7-8 from plane *EFG*, and the point of intersection is *S* on line *CA*.

13. CLASSIFICATION OF SURFACES

We shall soon study intersections involving surfaces. It is helpful to learn first how surfaces are classified.

A surface may be considered to be generated by the motion of a line: the generatrix. Surfaces are thus divided into two general classes: (1) those that can be generated by a moving *straight* line, and (2) those that can be generated by a moving *curved* line. The first are called *ruled surfaces*; the second, *double-curved surfaces*. Any position of the generatrix is called an *element* of the surface.

Ruled surfaces are divided into *(a) the plane, (b) single-curved surfaces*, and *(c) warped surfaces*.

The *plane* is generated by a straight line moving so as to touch two other intersecting or parallel straight lines or a plane curve.

Single-curved surfaces have their elements parallel or intersecting. In this class are the cylinder and the cone and also a third surface, which we shall not consider, known as the "convolute," in which only consecutive elements intersect.

Warped surfaces have no two consecutive elements that are parallel or intersecting. There are a great variety of warped surfaces. The surface of a screw thread and that of an airplane wing are two examples.

Double-curved surfaces are generated by a curved line moving according to some law. The commonest forms are *surfaces of revolution*, made by revolving a curve about an axis in the same plane, such as the sphere, torus or ring, ellipsoid, paraboloid, hyperboloid. Illustrations of various surfaces may be found in Fig. 87, Chap. 3.

14. INTERSECTION OF CURVED SURFACES

Before you work a specific example, you will find it helpful to learn some general facts about intersections of curved surfaces.

As was the case for plane surfaces, the line of intersection of two curved surfaces is a line all points of which are common to both surfaces. These may be more than one line of intersection between two curved surfaces. For example, a small pipe that passes completely through a larger pipe will intersect the larger pipe in two lines, one line as it *enters* and another as it *leaves*.

The line of intersection of any two surfaces may be found by either one of the following methods:

- Selected-Line Method   Select a sufficient number of lines of one surface. Find the point where each one of these lines pierces the other surface. A line joining these piercing points will be the line of the intersection of the two surfaces.

- Cutting-Plane Method   Pass a sufficient number of cutting planes through each of the given surfaces simultaneously. Each plane will "cut" a line (straight or curved) from each of the given surfaces. These lines will intersect in a point or points common to the two given surfaces. A line connecting these points will be the line of intersection of the given surfaces.

In attacking any problem of surface intersection, always examine the problem carefully to discover the *simplest* lines possible to cut from each surface.

In applying either of the general methods outlined above, exercise care, or the solution of the problem may become more complicated than necessary. In applying

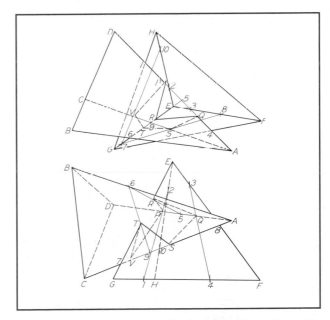

FIG. 18 Intersection of two pyramids. The edges of one, intersecting the faces of the other, determine the intersection.

the selected-line method, for example, select lines from one surface by giving primary consideration to the method that will be used to find where these lines pierce the other surface. In applying the cutting-plane method, select planes that will cut *simple* lines (either straight lines of circles) from each of the given surfaces.

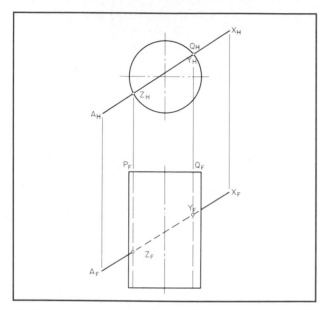

FIG. 19 Intersection of a line and cylinder. The edge view of the cylinder determines the two piercing points.

15. INTERSECTION OF A LINE AND A CYLINDER

The common case occurs when the cylinder is in a simple position, as in Fig. 19, with one view the end view—the *edge* view of the surface—and the other view showing the normal view of all the elements. Any line such as AX is therefore seen to intersect the cylinder where the surface appears as an edge, in this case the top view at Y_H and Z_H. These two points lie on elements of the cylinder P and Q, which are projected to the front view thus locating Y_F and Z_F.

16. INTERSECTION OF A LINE AND A CONE

Figure 20A illustrates pictorially the problem of finding the points where a line pierces a cone. A cutting plane VXY, containing the given line *AB* and passing through the apex V of the cone, is selected. A plane passing through the apex of a cone will intersect the cone in straight lines. The lines of intersection of the cutting plane and the cone are lines V-1 and V-2. These lines are determined by finding the line of intersection *RS* between the cutting plane and the *base plane* of the cone and then finding the points 1 and 2 where this line crosses the *base curve*. Line *AB* intersects both lines V-1 and V-2, establishing the points *P* and *P′* where line *AB* pierces the cone.

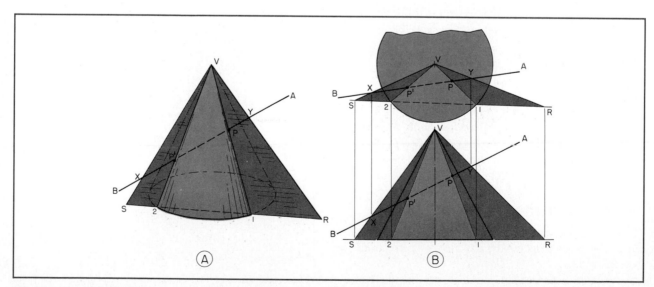

FIG. 20 Intersection of a line and a cone. (A) Pictorial aid, (B) orthographic construction.

The orthographic solution is illustrated in Fig. 20B. Convenient points X and Y on line AB are selected so that plane VXY (a plane containing the apex of the cone and line AB), extended, will intersect the base plane of the cone in the line RS. In the top view, RS cuts the base curve at 1 and 2, thus locating elements V-1 and V-2, which then intersect BA at P and P′, the piercing points of line AB and the cone.

17. INTERSECTION OF A PLANE AND A CYLINDER

In Fig. 21, a vertical cutting plane is shown. This plane, parallel to the cylinder axis, will cut elements of the cylinder at T and U and will also cut a line RS from the plane ABC. Points Z and X are then points common to the cutting plane, given plane, and cylinder and are two points on the line of intersection sought. A series of planes similar to the one shown will give other points to complete the line of intersection.

18. INTERSECTION OF A PLANE AND A CONE

Figure 22A illustrates pictorially the determination of the line of intersection between a cone and a plane by the selected-line method. Lines (elements) of the cone, such as V-1, V-2, etc., will intersect the plane at a, b,

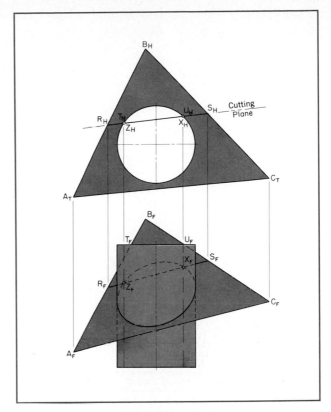

FIG. 21 Intersection of a plane and a cylinder.

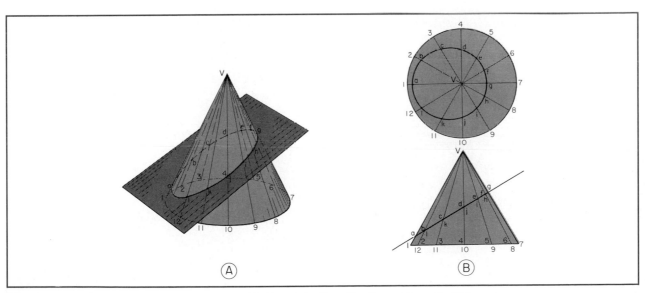

FIG. 22 Intersection of a plane and a cone. (A) Pictorial aid to explanation, (B) orthographic construction.

etc., respectively. If the edge view of the plane appears in one of the views, as in 22*B*, the solution is quite simple. Selected elements of the cone, V-1, V-2, etc., are drawn in both views. These elements are seen to intersect the plane in the front view at *a*, *b*, etc. Projection of these

points to the top view then gives points through which a smooth curve is drawn to complete the solution.

If the plane intersecting the cone is in an oblique position, as in Fig. 23, the cutting-plane method would probably be preferred because of the simplicity of the solution. Any plane passed through the apex of the cone will cut straight lines from the cone. Therefore, in Fig. 23, if vertical cutting planes are employed, all passing through the apex V, these planes will cut lines such as V-2 and V-8 from the cone. At the same time, such a cutting plane will cut XZ from the plane. XZ then intersects V-2 at *b* and V-8 at *h*, giving two points on the line of intersection. A series of planes thus passed and points found will complete the solution.

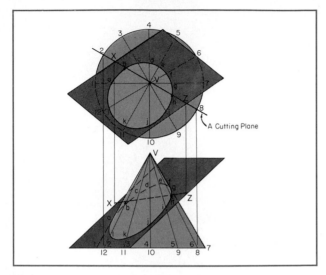

FIG. 23 Intersection of a plane and a cone.

19. INTERSECTION OF TWO CYLINDERS

A.   Case for Axes in Vertical Planes In Fig. 24, cutting planes parallel to the axis of a cylinder will cut straight-line elements from the cylinder. The frontal cutting planes A, B, C, and D, *parallel to the axis of each cylinder*, cut elements from each cylinder, the intersections of which are points on the curve. The pictorial sketch shows a slice cut by a plane from the object,

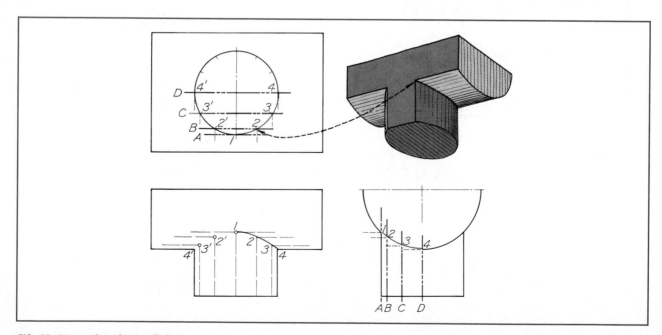

FIG. 24 Intersection of two cylinders.

which has been treated as a solid in order to illustrate the method more easily.

When the axes of the cylinders do not intersect, as in Fig. 25, the same method is used. Certain "critical planes" give the limits and turning points of the curve. For the position shown, planes A and D give the depth of the curve; the plane B, the extreme height; and the plane C, the tangent or turning points on the contour element of the vertical cylinder. After the critical points have been determined, a sufficient number of other cutting planes are used to give an accurate curve.

B. Case for Axes Not in Vertical Planes

Figure 26 illustrates pictorially at A and orthographically at B the determination of the line of intersection of two cylinders. As explained before, a plane passed *parallel to the axis* of a cylinder will cut straight-line elements from the cylinder. Logic then dictates that a plane parallel to the axes

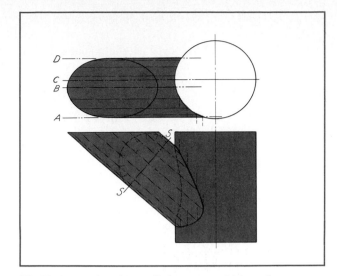

FIG. 25 Intersection of two cylinders, axes not intersecting.

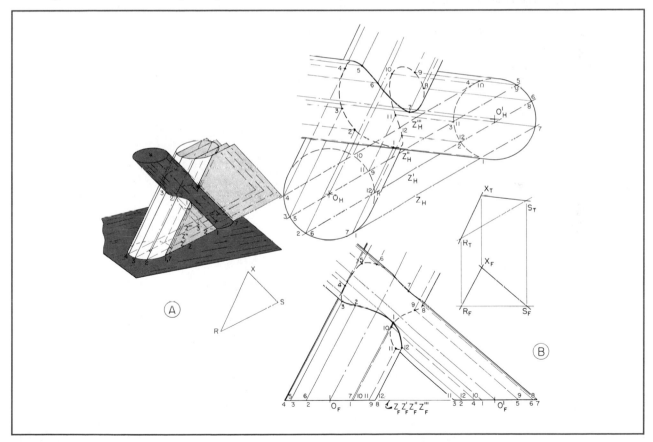

FIG. 26 Intersection of two cylinders; edge view of bases given.

of *two* cylinders will cut straight-line elements from both.

In Fig. 26A, line XS is made parallel to one cylinder axis and line XR parallel to the other cylinder axis. *RS* is located so that it will be *in the plane of the bases of both cylinders*. Thus a line parallel to *RS* will cut the base curves of both cylinders and locate points, such as 1 through 7 (shown), that will fix the location of elements of each cylinder that lie in the same plane and will therefore determine points on line of intersection. The details of projection are shown in Fig. 26B.

First, for the method to be applied, *both cylinder bases must lie in the same plane*, preferably horizontal, frontal, or profile. If the original views do not include

bases in one plane, find the cylinder intersections with a common plane and they will have them. In this case (Fig. 26B), the bases are in a horizontal plane.

Second, a plane must be established parallel to both cylinder axes. This construction is shown at XRS, where point X has been assumed at any convenient place, and then XR is drawn parallel to the left cylinder axis and XS parallel to the right cylinder axis. *RS* is made horizontal *because the bases are horizontal*, thus producing a line of intersection RS of plane XRS with a horizontal plane. The direction of $R_H S_H$ will fix the direction of the intersection with horizontal planes of *any* plane parallel to both cylinders. Lines Z, Z', Z'', and Z''', drawn parallel to *RS*, are therefore the intersections of planes parallel to both cylinders with the horizontal base plane of the cylinders.

Third, the horizontal intersection of any plane passed as described will cut the base curves of the cylinders. For example, plane Z' cuts the base of the left cylinder at points 2 and 8 and the base of the right cylinder at 6 and 12. The elements of the cylinders then emanate from these points and intersect in four points, 2, 6, 8, and 12, which are points on the line of intersection. The top view is drawn first, and then the elements and points are projected down to the front view.

Notice in this case that 2 and 12, cut by plane Z', are on the bottom side of the cylinders and that points 6 and 8 are on the upper side. Note also that plane Z is the foremost plane that will cut both cylinders and that Z''' is the rearmost. Except for a plane tangent to one of the cylinders, four points on the line of intersection are obtained by each cutting plane. The visibility is determined by inspection after the line of intersection is plotted.

20. INTERSECTION OF A CONE AND A CYLINDER

In Fig. 27, cutting planes may be taken, as at A, so as to pass through the vertex of the cone and parallel to the axis of the cylinder, thus cutting the straight-line elements from both cylinder and cone; or, as at B, when the cylinder's axis is parallel or perpendicular to the cone's axis, cutting planes may be taken parallel to the base so as to cut circles from the cone. Both systems of planes are illustrated in the figure. The pictorial sketches show slices taken by each plane through the objects, which have been treated as solids in order to illustrate the

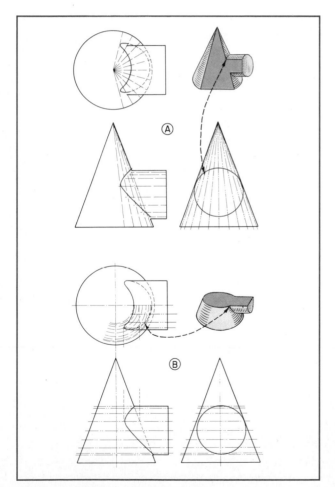

FIG. 27 Intersection of a cylinder and cone.

method more easily. Some judgment is necessary in the selection of both the direction and the number of cutting planes. More points need to be found at the places of sudden curvature or change of direction of the projections of the line of intersections.

In Fig. 27A the cutting planes appear as edges in the right-side view, where the cylinder surface also appears as an edge. The observed intersections in the side view are then projected to the cone elements formed by the cutting planes, in the top and front views, to complete the solution.

At B, horizontal cutting planes cut circles from the cone and straight-line elements from the cylinder. The top view then reveals intersections of cone circles and cylinder elements. Projection to the front view will complete the solution.

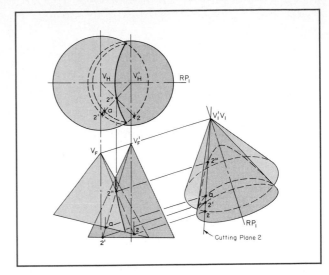

FIG. 28 Intersection of two cones.

21. INTERSECTION OF TWO CONES

A. Case for Axes in Vertical Planes Figure 28 illustrates another variation of the intersection of single-curved surfaces, this time for two cones. Employing only top and front views (given), no series of planes either horizontal, frontal, profile, or receding in any view will cut simple lines from the surfaces. Yet if a plane is passed through the apex of a cone, it will cut straight-line elements from the cone. It follows then that if a plane is passed through the apexes of *both* cones, straight-line elements will be cut from both cones. Therefore, a plane passed through a line connecting the apex points of both cones should give a relatively simple position.

A point view of the line connecting the apex points of the cones of Fig. 28 is given in the auxiliary view shown at V'V; and this view, having the point view of a line through which a plane is passed, gives the edge view of the plane. One such plane (2) is shown, which cuts an element V-2 from one cone and an element V'-2' from the other. These two elements intersect at 2", giving one point on the line of intersection of the cones. A series of planes passed as described will give other points to plot the complete line of intersection.

B. Case for Axes Not in Vertical Planes In Fig. 28, the edge views of the bases of both cones appear in the front view, but when the point view of the line connecting the apexes is made in this new view, the base curves must be

plotted, Depending, to some extent, on the position of the cones, a better solution might be to pass planes through the apexes of both cones, determining where the planes intersect the base planes of the cones. Figure 29A illustrates the principle pictorially. Line V-V' connecting the apex points is extended to P and P', where the line intersects the base planes. Then a point such as X, chosen on the line of intersection of the base planes, will make a plane PP'X which contains a line P'X cutting the base curve of one cone and a line PX cutting the base curve of the other. Elements of each cone thus located (by cutting the base curves) will intersect in four points on the line of intersection.

The orthographic drawing at Fig. 29B will illustrate the details of construction. Apex line V-V' is extended to intersect the base planes at P and P'. The line of intersection of the base plane appears as a point in the front view and as a line in the top view. A point X selected on the base-plane intersection in the top view produces P'X, points 2 and 8 on the base of the left cone, and PX and points 6 and 12 on the right cone. These elements 2-V', 8-V', 6-V, and 12-V, intersect at 2, 6, 8, and 12 in the cutting plane and on both cones. Elements 2-V', 8-V', 6-V, and 12-V then located in the front view allow projection of 2, 6, 8, and 12 to the front view. Other points are located by shifting the plane to another position, such as PP'X' and PP'X". In the illustration, ele-

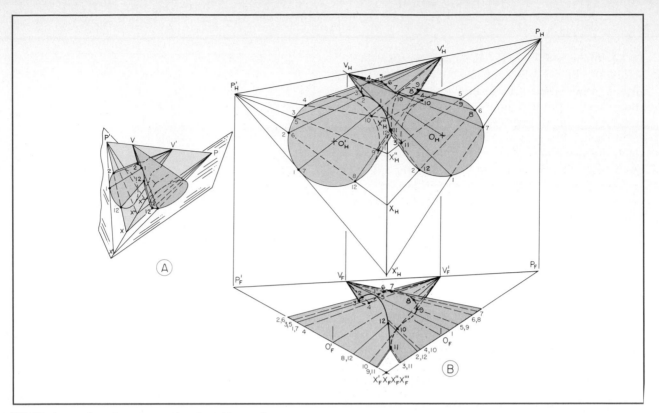

FIG. 29 Intersection of two cones; edge view of bases given.

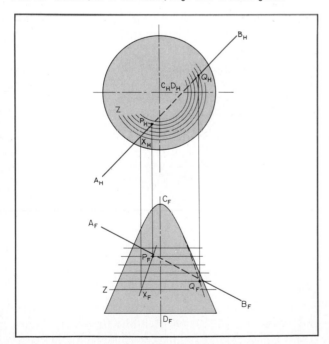

FIG. 30 Intersection of a double-curved surface (hyperboloid) and a line.

ments cut from each cone by each cutting plane are numbered alike so that identically numbered elements intersect at a point designated by the number. This method of determining the line of intersection of two cones is sometimes called "the swinging-plane method."

22. INTERSECTION OF A LINE AND A SURFACE OF REVOLUTION

Double-curved surfaces of revolution present problems somewhat more complicated than do single-curved surfaces because straight lines cannot be cut from a double-curved surface. However, the geometric double-curved surfaces—the ellipsoid, paraboloid, etc.—are surfaces of revolution, and circles may therefore be cut by planes perpendicular to the axis of revolution.

Because no straight line can be drawn on a surface of revolution, any plane passed through the surface will be a curve. In finding the points of intersection of a line and a surface of revolution, it is necessary to pass a plane through the line. The curve resulting from the intersection of the cutting plane and the surface of revolution will then intersect the line to locate the points of intersection of line and surface.

Figure 30 illustrates the intersection of an oblique line *AB* and a surface of resolution (hyperboloid). A vertical cutting plane is passed through the line *AB*. To find the intersection of the cutting plane and the surface of revolution a number of horizontal planes are passed through both. Plane *Z*, for example, cuts a circle from the surface of revolution. This circle in the top view is seen to intersect the edge view of the cutting plane at point X_T. Point X_F, then projected from the top view, locates one point on one curve of intersection between

the vertical plane through *AB* and the hyperboloid. Other points so located will plot the curves, shown in the front view, that intersect the line *AB* at P_F and Q_F, the points of intersection of *AB* and the hyperboloid.

You have now completed a rather technical and involved chapter detailing the specifics of the most usual forms of intersections. The overall concepts of solution are not unduly difficult, although the practice of these concepts requires patience. In many ways this chapter represents the peak of perseverance needed for space analysis; the next chapter is less abstract.

PROBLEMS

In the study of intersections, a purely theoretical approach may be used without any reference whatever to practical utility where the lines, planes, and surfaces form pipes, ducts, or other real objects. On the other hand, some may prefer to teach the subject by applying the theory on useful objects. For these reasons, the problems on intersections are given first from the standpoint of pure theory and then from a more practical standpoint. Note that even though the surfaces are geometrical, they are the ducts, hoppers, transitions, etc., that will often be found in practical work.

GROUP 1. INTERSECTION OF LINES AND PLANES

1. Locate the top and front views of *P*, the point where the line *JK* pierces the plane *EFG*.

2. Lines *AB* and *CD* are to be connected by a single straight line passing through point *P*. Draw top and front views of the line.

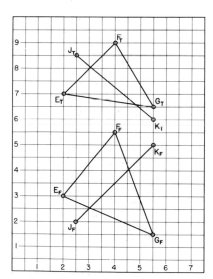

PROB. 1.

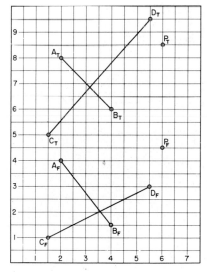

PROB. 2.

3. *AB* is the axis of a duct of irregular cross section. The duct is to be cut off in the direction determined by plane *KLM*. Show the top and front views of the cut. Determine the true shape of the cap that will fit this cut. (Neglect bend-overs, etc.)

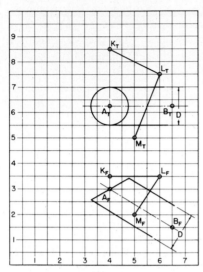

PROB. 3.

GROUP 2. INTERSECTION OF PLANES

4. Find the line of intersection between the planes *ABC* and *DEF*. Show visibility.

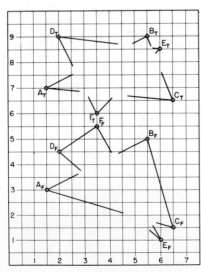

PROB. 4.

5. Find the line of intersection between the planes *MNO* and *RST*.

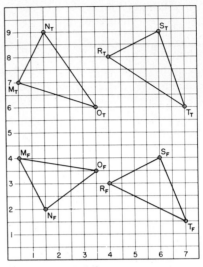

PROB. 5.

6. Find the line of intersection and complete the views of the two hollow sheet-metal shapes shown. The top of the right-hand shape is enclosed.

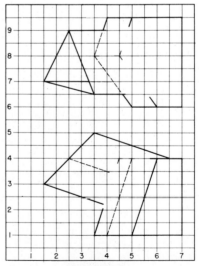

PROB. 6.

GROUP 3. INTERSECTIONS OF PRISMATIC DUCTS

7 and 8. Find the line of intersection, considering the prisms as pipes opening into each other. Use care in indicating visible and invisible parts of the line of intersection.

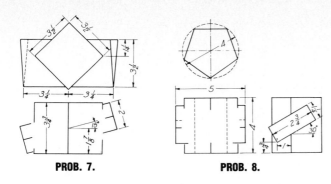

PROB. 7.   **PROB. 8.**

9. Find the line of intersection, indicating visible and invisible parts and considering prisms as pipes opening into each other.

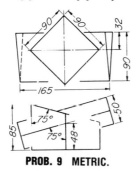

PROB. 9 METRIC.

10. Find the intersection between the two prismatic ducts.

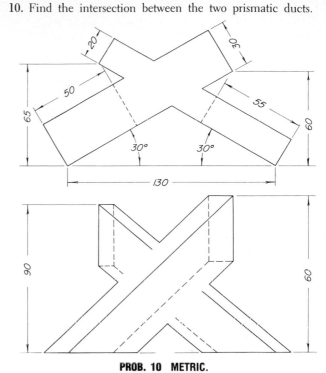

PROB. 10 METRIC.

11. The layout as illustrated shows a tower and proposed conveyor-belt galleys. Find the intersection between the tower and the main supply galley.

PROB. 11 Tower and conveyor-belt galleys.

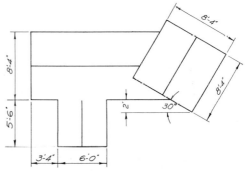

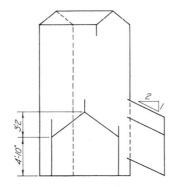

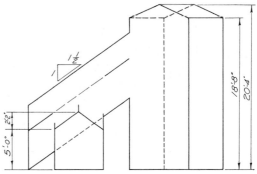

11A. Same layout as Prob. 11. After solving Prob. 11, find the intersection between the two galleys.

GROUP 4. INTERSECTIONS OF PYRAMIDAL OBJECTS

12 to 15. Find the lines of intersection.

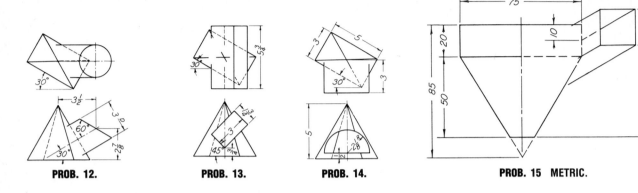

PROB. 12.   **PROB. 13.**   **PROB. 14.**   **PROB. 15 METRIC.**

GROUP 5. INTERSECTIONS OF SINGLE-CURVED SURFACES

16. Find the points X and Z where line AB pierces the 2-in-diameter right-circular cylinder OP. Find the points R and S where line CD pierces this cylinder.

17. Draw the top view of the line of intersection between plane A and the 2-in-diameter right-circular cylinder OP. Draw the front view of the line of intersection between plane B and this cylinder.

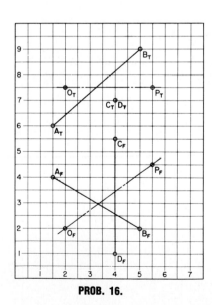

PROB. 16.

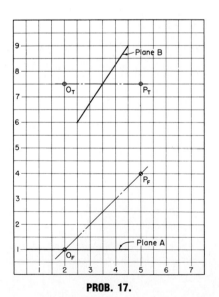

PROB. 17.

18. Find the points X and Z where line *AB* pierces the cone *VO*. Show visibility of line *AB*.

20. Draw the top and front views of the line of intersection between plane *ABC* and cone *VO*. Complete the cone and show visibility. Draw the top view of the line of intersection between plane *B* and cone *VO*.

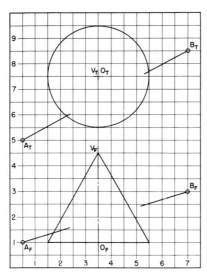

PROB. 18.

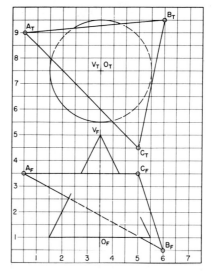

PROB. 20.

19. Draw the front view of the line of intersection between plane *A* and cone *VO*.

21 and 22. Complete the views of cylinders *AB* and *CD*, showing their line of intersection.

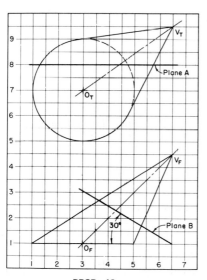

PROB. 19.

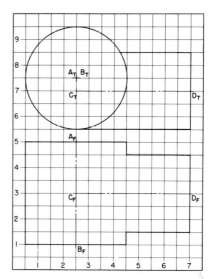

PROB. 21.

23. Complete the views of cones *VO* and *XP*, showing their line of intersection.

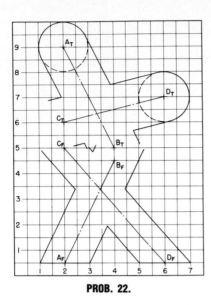

PROB. 22.

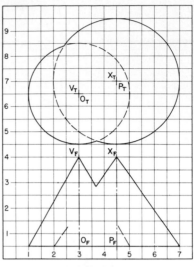

PROB. 23.

GROUP 6. INTERSECTIONS OF CYLINDRICAL DUCTS

24 and 25. Find the line of intersection, indicating visible and invisible portions and considering cylinders as pipes opening into each other.

26. The layout shows a pump casing composed of a cylinder and elbow. Find the intersection between the cylinder and elbow.

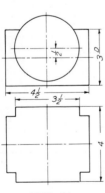

PROB. 24.

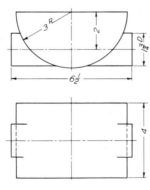

PROB. 25.

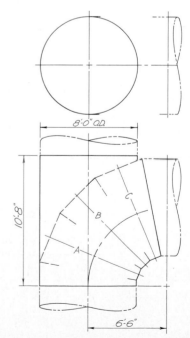

PROB. 26 Pump casing.

27. The layout shows a portion of a liquid oxygen pumping system consisting of pipes A, B, C, and D. Find the intersection of the pipes.

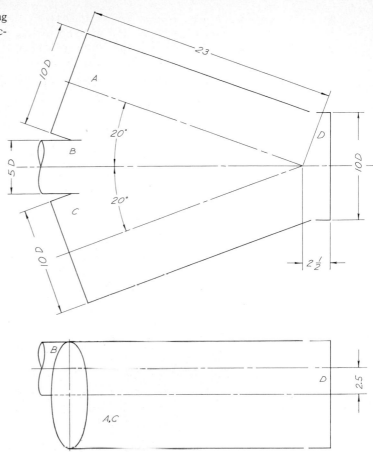

PROB. 27 METRIC. Lox pump outlet.

GROUP 7. INTERSECTIONS OF CONIC OBJECTS AND DUCTS

28 to 32. Find the line of intersection.

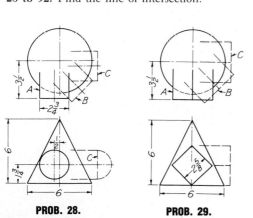

PROB. 28.

PROB. 29.

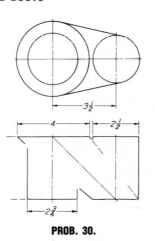

PROB. 30.

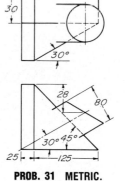

PROB. 31 METRIC.

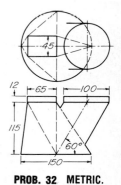

PROB. 32 METRIC.

33. The layout shows the plan view of an intake manifold consisting of a conical connector intersected by two cylindrical pipes. Find the intersection of the pipe that enters the side of the conical connector. The axes of the two shapes intersect. Solve this problem, using only the plan view.

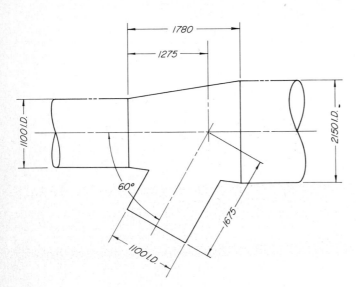

PROB. 33 METRIC. Discharge manifold.

34. The layout shows a right circular cylinder and an elliptical-base cone. Find the intersection of the two surfaces.

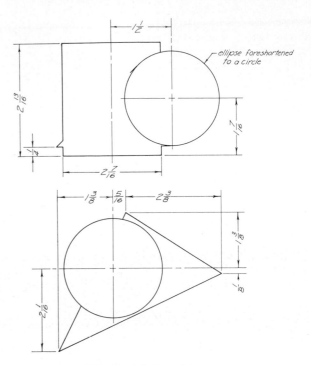

PROB. 34 Cylinder and cone.

1. DEFINING DEVELOPMENTS

In many different kinds of construction full-size patterns of some or all of the faces of an object are required; for example, in sheet-metal work, a pattern to which a sheet may be cut so that when rolled, folded, or formed it will make the object. The complete surface laid out in a plane is called the "development" of the surface.

Surfaces about which a thin sheet of flexible material (such as paper or tin) can be wrapped smoothly are said to be "developable"; these include objects made up of planes and single-curved surfaces only.

Warped and double-curved surfaces are nondevelopable, and when patterns are required for their construction, they can be made only by methods that are approximate; but, assisted by the ductility or pliability of the material, they give the required form. Thus while a ball cannot be wrapped smoothly, a two-piece pattern developed approximately and cut from leather can be stretched and sewed on in a smooth cover, or a flat disk of metal can be die-stamped, formed, or spun to hemispherical or other shape.

2. BASIC CONSIDERATIONS

We have learned the method of finding the true size of a plane surface by projecting its normal view. If the true sizes of all the plane faces of an object are found and the faces joined in order at their common edges so that all faces lie in a common plane, the result will be the developed surface. Usually this may be done to the best advantage by finding the true length of the edges.

The development of a right cylinder is evidently a rectangle whose width is the altitude and length the rectified circumference (Fig. 1); and the development of a right-circular cone is a circular sector with a radius equal to the slant height of the cone and an arc equal in length to the circumference of its base (Fig. 1).

As illustrated in Fig. 1, developments are drawn with the inside face up. This is primarily the result of working to inside rather than outside dimensions of

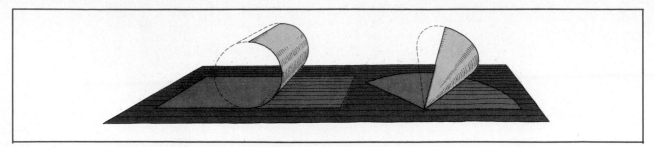

FIG. 1 Theory of development. The surface is composed into a plane by unfoldment.

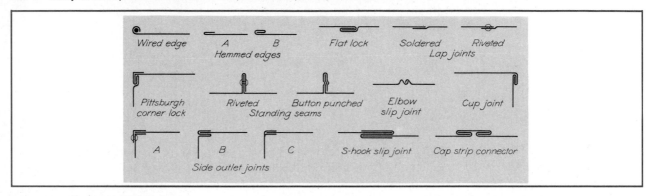

FIG. 2 Joints and finished edges.

ducts. This procedure also facilitates the use of fold lines, identified by punch marks at each end, along which the metal is folded in forming the object. In laying out real sheet-metal designs, an allowance must be made for seams and lap and, in heavy sheets, for the thickness and crowding of the metal; there is also the consideration of the commercial size of material as well as the question of economy in cutting. In all this some practical shop knowledge is necessary.

There are numerous joints used in seaming sheet-metal ducts and in connecting one duct to another. Figure 2 illustrates some of the more common types, which may be formed by hand on a break or by special seaming machines. No attempt to dimension the various seams and connections has been made here because of the variation in sizes for different gages of metal and in the forming machines of manufacturers.

Hemming is used in finishing the raw edges of the end of the duct. In wire hemming, an extra allowance of about 2½ times the diameter of the wire is made for wrapping around the wire. In flat hemming, the end of the duct is bent over once or twice to relieve the sharp edge of the metal.

3. AN AID TO DEVELOPMENTS: REVOLUTION

A development exists on a flat surface. Therefore all lines within the development appear in true length when we look at the flat surface as a true (normal) surface. It follows that developments can be created if true lengths of lines are known for the surfaces to be developed.

Sometimes it is easy to find true lengths of lines of surfaces, as is the case for vertical prisms, where all vertical lines are in true length in the front view. Often, however, we will need to determine true lengths of lines for use in developments.

A powerful method for finding true lengths of lines is the method of revolution, an additional method to that of using auxiliary views studied in Chap. 9. Revolution is useful also for finding edge views and true surfaces of planes. In fact, revolution has many applications, not all of which will be considered here.

Revolution is used to save constructing as many additional views as are needed with the auxiliary-view method. Using the method of revolution saves constructing one auxiliary view. That is, if three auxiliary views were needed for a solution, revolution would require only two auxiliary views. If two auxiliary views were needed, revolution allows us to get by with one auxiliary view. Those cases requiring only one auxiliary view, such as the true length of a line, would need no auxiliary view.

To become familiar with revolution as a tool and to facilitate the discussion of developments, let us look at four common examples of revolutions: true length of a line, edge view of a plane, true surface of a plane, and a point revolved about an oblique axis. The first of these applies directly to developments. The remaining examples are useful elsewhere.

A. True Length of a Line by Revolution   The basic concept of revolution is shown in Fig. 3A. Revolution is so named because something is revolved or rotated. That something can be a point, a line, a plane, or a solid. In Fig. 3, a line is being revolved to become in true length in a front view.

In Fig. 3A, note several conditions concerning revolution:

1. Revolution requires an axis and a path of revolution.

2. The path of revolution is perpendicular to its axis and generates a circular path around the axis. In Fig. 3A, point X is on the axis and does not move, but point Y revolves in a circular path around the axis. If point Y revolved 360°, a right cone would be generated.

3. The angle that a line makes with a plane will be maintained if the axis of revolution is perpendicular to the plane. In Fig. 3A, the axis is perpendicular to the horizontal plane H. Therefore the angle that line XY makes with the H plane (the slope angle, actually) will not change as line XY revolves.

Finding the true length of line XY is seen in Fig. 3B. Line XY is revolved in the horizontal view about the vertical axis until line XY becomes parallel to the frontal plane. This revolved position of point Y_H is noted as Y_{HR}. With revolved line XY_{HR} being parallel to the frontal plane, line XY_{HR} must be in true length in front view. Note that point Y_F in the front view has moved in a perpendicular path to the axis as point Y_F moved out to become point Y_{HR}. The slope angle θ_H is available if desired.

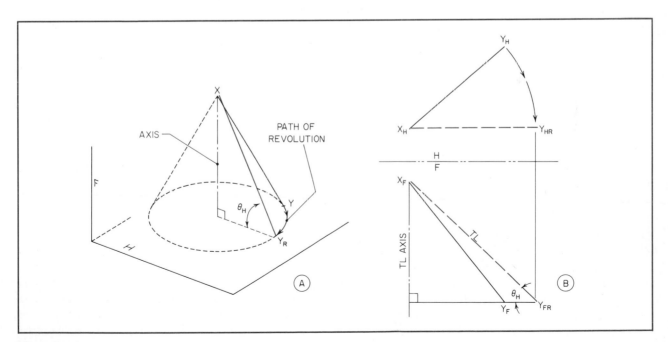

FIG. 3   True length of a line by revolution.

The axis of revolution may be made perpendicular to any plane. If you wish the axis to be perpendicular to the frontal plane, simply turn Fig. 3B upside down and reverse the H-F notation. Angle θ_H will become θ_F, the angle line YX makes with the frontal plane. Similarly, by rotating Fig. 3B 90° counterclockwise, and relabeling plane H as F and plane F as P, you can make the axis perpendicular to the frontal plane. Line XY will then be in true length in the profile plane. Angle θ_H will be θ_F.

B. Edge View of a Plane by Revolution

In Chap. 9, finding the edge view of a plane using the auxiliary method required one auxiliary view. Using revolution should require no auxiliary view.

In Fig. 4, instead of looking parallel to the true length

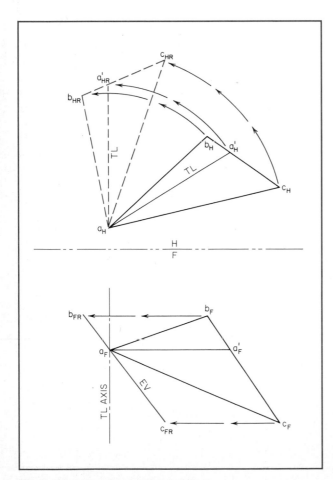

FIG. 4 Edge view of a plane by revolution.

of added line AA′ to make line AA′ in point view in the first auxiliary view, we revolve line AA′ in the top view until it is perpendicular to the frontal plane. Line AA′ will now be forced to be in point view in the front view. With line AA′ in point view, the plane is in edge view. Note in the top view that all points must revolve through the same angle.

C. True Surface of a Plane by Revolution

Finding the true (normal) surface of a plane by the auxiliary-view method usually requires two auxiliary views, one for the edge view and one for the true surface itself. The method of revolution should require only one auxiliary view. Figure 4 gave plane ABC in edge view in the front view. Let us say we wish the same plane to appear as a true surface in the front view.

Figure 5 shows how revolution can assist in making plane ABC a true surface. An edge view is first constructed adjacent to the front view, using added line CC′, which is in true length in the front view. (For review, see Chap. 9.) The edge view is then revolved until it is parallel to the frontal plane. The true surface of plane ABC must then appear in the front view. Point A did not revolve in the front view because it is on the axis, whereas points B and C did move perpendicular to the axis.

D. Point Revolved about an Oblique Axis

In design work one must on occasion revolve a point around an axis. An example would be that of checking for clearance in a machine as a wrench is swung around a bolt. Even though revolving a point about an oblique axis does not relate directly to our needs in development work, this application is presented briefly to consolidate our understanding of revolution.

Figure 6 shows the general procedure. To see a circular path move perpendicularly around any axis, the axis must be seen in point view, meaning that the axis must first be made in true length. The true length of line RS is constructed in auxiliary view 1, and the point view is made in auxiliary view 2. Point O is to revolve 90° clockwise around axis RS. Therefore in view 2 the 90° revolution can be done and point O may be relocated in view 1, realizing that point O moves in a path perpendicular to the true-length axis. Point O could be transferred back into the H and F views if desired. Point O would generate an elliptical path in the H and F views.

4. TO DEVELOP A TRUNCATED HEXAGONAL PRISM

Developments should proceed well with our newly learned knowledge of revolution, which will be of benefit in most of the following examples. The first example, however, has all necessary lines given in true length and will therefore not require the use of revolution to find true lengths. Refer to Fig. 7.

First draw two projections of the prism: (1) a normal view of a right section (a section or cut obtained by a plane perpendicular to the axis) and (2) a normal view of the lateral edges. The base *ABCDEF* is a right section shown in true size in the bottom view. Lay off the perimeter of the base on line *AA* of the development. This line is called by sheet-metal workers the "stretchout" or "girth" line. At points *A*, *B*, *C*, etc., erect perpendiculars called "measuring lines" or "bend lines," representing the lateral edges along which the pattern is folded to form the prism. Lay off on each of these its length *A-1*, *B-2*, *C-3*, etc., as given on the front view. Connect the points 1, 2, 3, etc., in succession, to complete the development of the lateral surfaces.

Note on the pattern that the inside of the lateral faces is toward the observer. For the development of the entire surface in one piece, attach the true sizes of the upper end and the base as shown, finding the true size of the upper end by an auxiliary (normal) view as described in Chap. 9. For economy of solder or rivets and time, it is customary to make the seam on the shortest edge or surface. In seaming along the intersection of surfaces whose dihedral angle is other than 90°, as is the case here, the lap seam lends itself to convenient assembling. The flat lock could be used if the seam were made on one of the lateral faces.

5. TO DEVELOP A TRUNCATED RIGHT PYRAMID

In this example, which is reasonably straightforward, you will use revolution in finding true lengths of lines. Refer to Fig. 8. Draw the projections of the pyramid that show (1) a normal view of the base or right section and (2) a normal view of the axis. Lay out the pattern for the pyramid, and then superimpose the pattern of the truncation.

Since this is a portion of a right regular pyramid, the lateral edges are all of equal length. The lateral edges *OA* and *OD* are parallel to the frontal plane and conse-

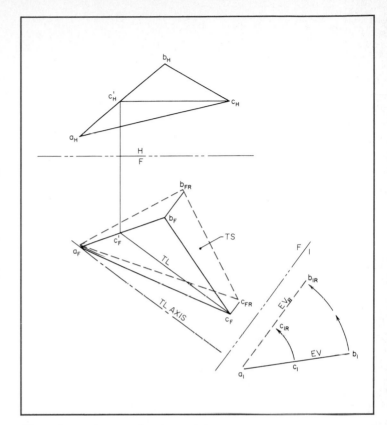

FIG. 5 True surface of a plane by revolution.

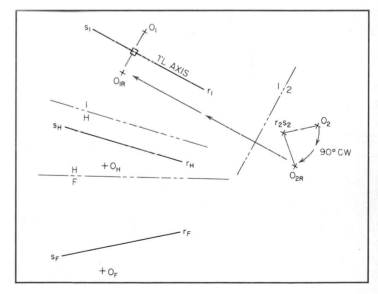

FIG. 6 Revolution of a point about an oblique axis.

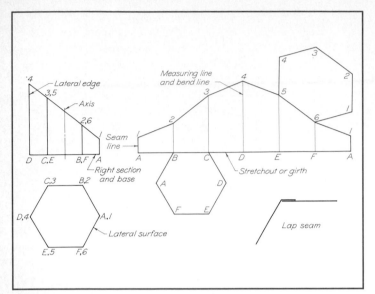

FIG. 7 Development of a prism. The true size of each side is laid out in successive order.

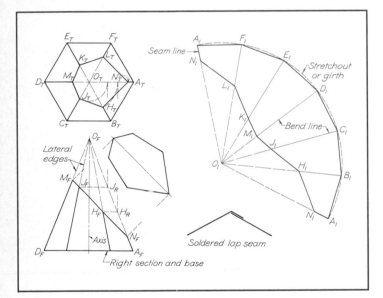

FIG. 8 Development of a pyramid.

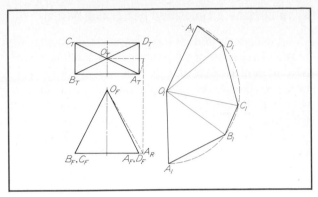

FIG. 9 Development of a pyramid. The true length from vertex to base is found by rotating one edge until frontal.

The intersection of the cutting plane and lateral surfaces is developed by laying off the true length of the intercept of each lateral edge on the corresponding line of the development. The true length of each of these intercepts, such as OH and OJ, is found by rotating it about the axis of the pyramid until they coincide with O_FA_F. The path of any point, as H, will be projected on the front view as a horizontal line. To obtain the development of the entire surface of the truncated pyramid, attach the base; also find the true size of the cut face and attach it on a common line.

The right-rectangular pyramid, Fig. 9 is developed in a similar way, but as the edge OA is not parallel to the plane of projection, it must be rotated to O_FA_R to obtain its true length.

6. TO DEVELOP AN OBLIQUE PYRAMID

This variation of a pyramid is seen in Fig. 10. Since the lateral edges are unequal in length, the true length of each must be found separately by rotating it parallel to the frontal plane. With O_1 taken at any convenient place, lay off the seam line O_1A_1 equal to O_FA_R. With A_1 as center and radius A_1B_1 equal to the true length of AB, describe an arc. With O_1 as center and radius O_1B_1 equal to O_FB_R, describe a second arc intersecting the first in vertex B_1. Connect the vertices O_1, A_1, and B_1, thus forming the pattern for the lateral surface OAB. Similarly, lay out the patterns for the remaining three lateral surfaces, joining them on their common edges. The stretchout is equal to the summation of the base edges. If the complete development is required, attach

quently show in their true length on the front view. With center O_1, taken at any convenient place, and a radius O_FA_F, draw an arc that is the stretchout of the pattern. On it step off the six equal sides of the hexagonal base, obtained from the top view, and connect these points successively with each other and with the vertex O_1, thus forming the pattern for the pyramid.

the base on a common line. The lap seam is suggested as the most suitable for the given conditions.

7. TO DEVELOP A TRUNCATED RIGHT CYLINDER

The development of a cylinder as seen in Fig. 11 is similar to the development of a prism.

Draw two projections of the cylinder: (1) a normal view of a right section and (2) a normal view of the elements. In rolling the cylinder out on a tangent plane, the base or right section, being perpendicular to the axis, will develop into a straight line. For convenience in drawing, divide the normal view of the base, here shown in the bottom view, into a number of equal parts by points that represent elements. These divisions should be spaced so that the chordal distances closely enough approximate the arc to make the stretchout practically equal to the periphery of the base or right section.

Project these elements to the front view. Draw the stretchout and measuring lines as in Fig. 7, the cylinder

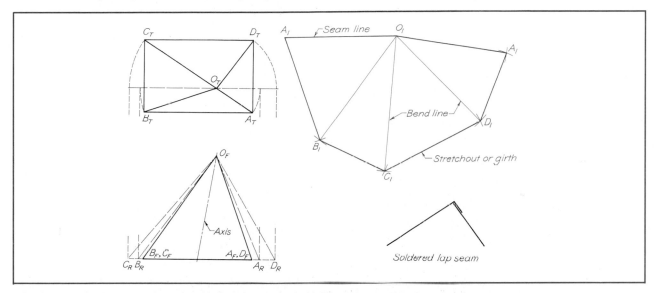

FIG. 10 Development of an oblique pyramid. Each lateral edge is of different length. The true length for each must be determined.

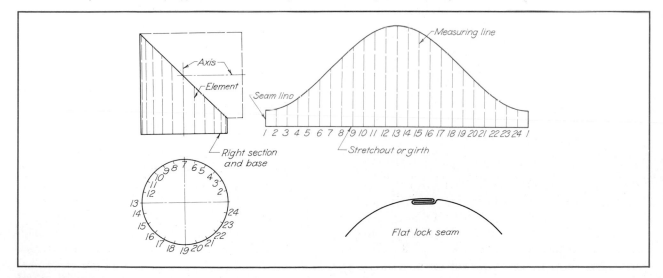

FIG. 11 Development of a cylinder. The cylinder is treated as a many-sided prism.

now being treated as a many-sided prism. Transfer the lengths of the elements in order, by projection or with dividers, and join the points thus found by a smooth curve, sketching it in freehand very lightly before fitting the french curve to it.

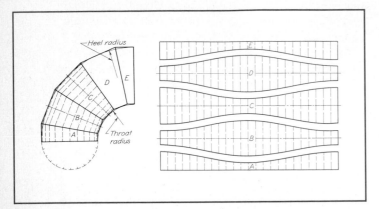

FIG. 12 Development of an elbow (cylinders). Note the identity of *A*, *E*, *B*, and *D*.

This development might be the pattern of one-half of a two-piece elbow. Three-piece, four-piece, or five-piece elbows can be drawn similarly, as illustrated in Fig. 12. As the base is symmetrical, only one-half of it need be drawn. In these cases, the intermediate pieces, as *B*, *C*, and *D*, are developed on a stretchout line formed by laying off the perimeter of a right section. If the right section is taken through the middle of the piece, the stretchout line becomes the centerline of the development.

Evidently any elbow could be cut from a single sheet without waste if the seams were made alternately on the long and short sides. The flat lock seam is recommended for Figs. 11 and 12, although other types could be used.

The octagonal dome (Fig. 13) illustrates an application of the development of cylinders. Each piece is a portion of a cylinder. The elements are parallel to the base of the dome and show in their true lengths in the top view. The true length of the stretchout line for sections A and A′ shows in the front view at $O_F H_F$. By

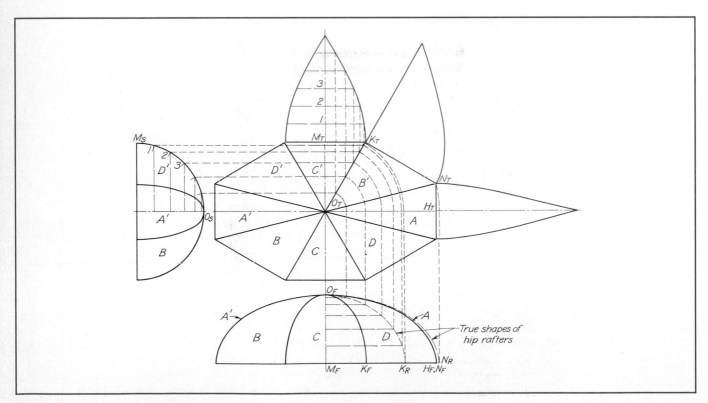

FIG. 13 Development of a dome (cylinders). Each cylindrical portion is unrolled separately.

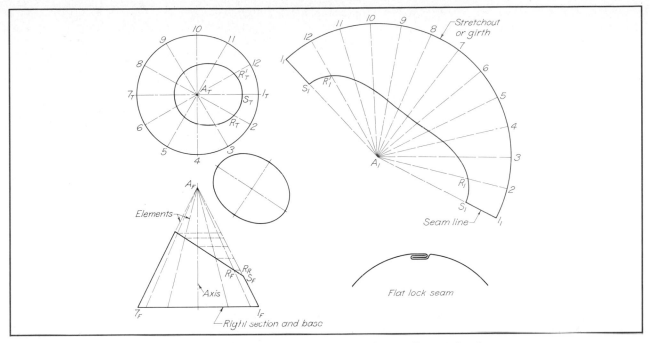

FIG. 14 Development of a cone. This is a right cone. All elements from vertex to base are the same length.

considering $O_T H_T$ as the edge of a plane cutting a right section, the problem is identical with the preceding problem.

Similarly, the stretchout line for sections B, B', D, and D' shows in true length at $O_F K_R$ in the front view, and for section C and C' at $O_S M_S$ in the side view.

The true shape of hip rafter ON is found by rotating it until it is parallel to the frontal plane, as at $O_F N_R$, in the same manner as in finding the true length of any line. A sufficient number of points should be taken to give a smooth curve.

8. TO DEVELOP A TRUNCATED RIGHT-CIRCULAR CONE

Draw the projections of the cone in Fig. 14 that will show (1) a normal view of the base or right section and (2) a normal view of the axis. First develop the surface of the complete cone and then superimpose the pattern for the truncation.

Draw the projections of the cone that will show (1) a normal view of the base or right section and (2) a normal view of the axis. First develop the surface of the complete cone and then superimpose the pattern for the truncation.

Divide the top view of the base into a sufficient number of equal parts so that the sum of the resulting chordal distances will closely approximate the periphery of the base. Project these points to the front view, and draw front views of the elements through them. With center A_1 and a radius equal to the slant height $A_F 1_F$, which is the true length of all the elements, draw an arc, which is the stretchout, and lay off on it the chordal divisions of the base, obtained from the top view.

Connect these points, 1_1, 2, 3, etc., with A-1, thus forming the pattern for the cone. Find the true length of each element from vertex to cutting plane by rotating it to coincide with the contour element A-1, and lay off this distance on the corresponding line of the development. Draw a smooth curve through these points. The flat lock seam along element S-1 is recommended, although other types could be employed. The pattern for the inclined surface is obtained from the auxiliary (normal) view.

9. UTILIZING TRIANGULATION

Nondevelopable surfaces are developed approximately by assuming them to be made of narrow sections of

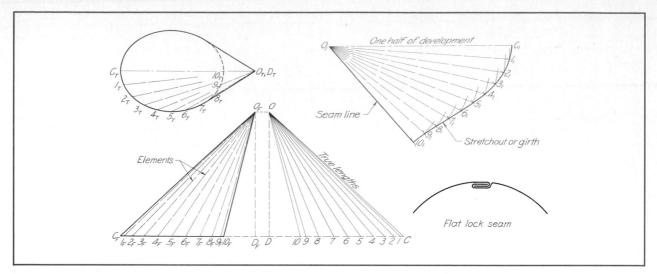

FIG. 15 Development of an oblique cone. A true-length diagram is used to obtain the lengths of the elements.

developable surfaces. The commonest and best method for approximate development is that of triangulation, that is, the surface is assumed to be made up of a large number of triangular strips, or plane triangles with very short bases.

This method is used for all warped surfaces and also for oblique cones. Oblique cones are single-curved surfaces and thus are theoretically capable of true development, but they can be developed much more easily and accurately by triangulation, a simple method which consists merely of dividing the surface into triangles, finding the true lengths of the sides of each, and constructing them one at a time, joining these triangles on their common sides.

The use of triangulation will be shown in the next section, development of an oblique cone.

10. TO DEVELOP AN OBLIQUE CONE

An oblique cone differs from a cone of revolution in that the elements have different lengths. The development of the right-circular cone is, practically, made up of a number of equal triangles which meet at the vertex and whose sides are elements and whose bases are the chords of short arcs of the base of the cone. In the oblique cone each triangle must be found separately.

In Fig. 15 draw two views of the cone showing (1) a

normal view of the base and (2) a normal view of the altitude. Divide the true size of the base, here shown in the top view, into a sufficient number of equal parts so that the sum of the chordal distances will closely approximate the length of the base curve. Project these points to the front view of the base.

Through these points and the vertex, draw the elements in each view. Since the cone is symmetrical about a frontal plane through the vertex, the elements are shown only on the front half of it. Also, only one-half of the development is drawn. With the seam on the shortest element, the element OC will be the centerline of the development and can be drawn directly at O_1C_1, as its true length is given at O_FC_F. Find the true length of the elements by rotating them until parallel to the frontal plane or by constructing a "true-length diagram."

The true length of any element would be the hypotenuse of a triangle in which one leg is the length of the projected element as seen in the top view and the other leg is equal to the altitude of the cone. Thus, to make the diagram, draw the leg OD coinciding with or parallel to O_FD_F. At D and perpendicular to OD, draw the other leg, on which lay off the lengths D-1, D-2, etc., equal to D_T1_T, D_T2_T, etc., respectively. Distances from O to points on the base of the diagram are the true lengths of the elements.

Construct the pattern for the front half of the cone as

follows: With O_1 as center and radius O-1, draw an arc. With C_1 as center and radius C_T-1_T, draw a second arc intersecting the first at 1_1; then O_1-1_1 will be the developed position of the element O-1. With 1_1 as center and radius 1_T-2_T, draw an arc intersecting a second arc with O_1 as center and radius O-2, thus locating 2_1. Continue this procedure until all the elements have been transferred to the development. Connect the points C_1, 1_1, 2_1, etc., with a smooth curve, the "stretchout line," to complete the development. The flat lock seam is recommended for joining the ends to form the cone.

11. A CONIC CONNECTION BETWEEN TWO PARALLEL PIPES

This is shown in Fig. 16. The method used in drawing the pattern is an application of the development of an oblique cone. One-half of the elliptical base is shown in true size in an auxiliary view, here attached to the front view. Find the true size of the base from its major and minor diameters, divide it into a number of equal parts so that the sum of these chordal distances closely approximates the periphery of the curve, and project these points to the front and top views. Draw the elements in each view through these points, and find the vertex O by extending the contour elements until they intersect.

The true length of each element is found by using the vertical distance between its ends as the vertical leg of the diagram and its horizontal projection as the other leg. As each true length from vertex to base is found, project the upper end of the intercept horizontally across from the front view to the true length of the corresponding element to find the true length of the intercept. The development is drawn by laying out each triangle in turn, from vertex to base, as in Sec. 10, starting on the centerline O_1C_1 and then measuring on each element its intercept length. Draw smooth curves through these points to complete the pattern. Join the ends with a flat lock seam.

12. TRANSITION PIECES

Transition pieces are used to connect pipes or openings of different shapes of cross section. Figure 17, showing a transition piece for connecting a round pipe and a rectangular pipe, is typical. Transition pieces are always developed by triangulation. The piece shown in Fig. 17 is made up of four triangular planes, whose bases are the sides of the rectangle, and four parts of oblique cones, whose common bases are arcs of the circle and whose vertices are at the corners of the rectangle. To develop it, make a true-length diagram as in Fig. 15. When the true length of O-1 is found, all the sides of triangle A will be known. Attach the development of cones B and B', and then those of triangles C and C', and so on.

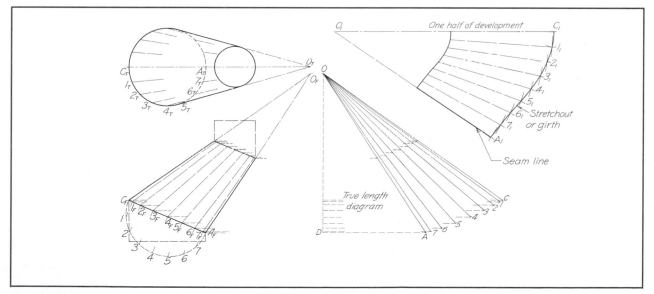

FIG. 16 Development of a conic section.

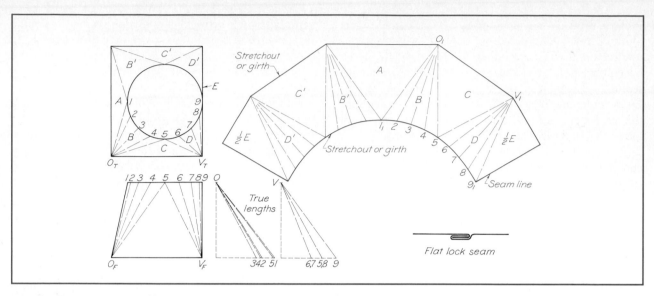

FIG. 17 Development of transition.

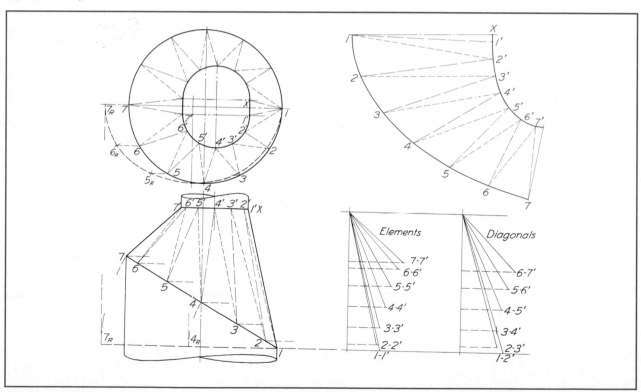

FIG. 18 Development of a transition (warped surface). This transition connects an oval and a cylindrical pipe. Triangulation of surface portions is necessary.

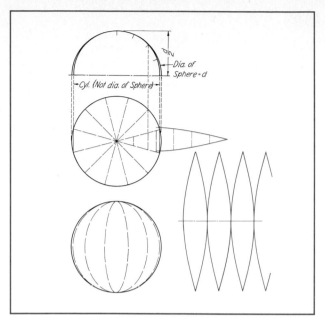

FIG. 19 Development of a sphere (gore method). This is a double-curved surface and development is only approximate.

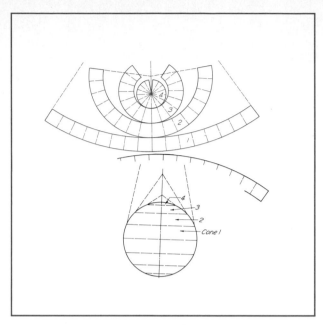

FIG. 20 Development of a sphere (zone method). This is a double-curved surface and development is only approximate.

13. TRIANGULATION OF WARPED SURFACES

The approximate development of a warped surface is made by dividing it into a number of narrow quadrilaterals and then splitting each of these quadrilaterals into two triangles by a diagonal, which is assumed to be a straight line, although it is really a curve. Figure 18 shows a warped transition piece to connect an ovular (upper) pipe with a right-circular cylindrical pipe (lower).

Find the true size of one-half the elliptical base by rotating it until horizontal about an axis through 1, when its true shape appears on the top view. The major diameter is $1\text{-}7_R$, and the minor diameter through 4_R will equal the diameter of the lower pipe.

Divide the semiellipse into a sufficient number of equal parts, and project these to the top and front views. Divide the top semicircle into the same number of equal parts, and connect similar points on each end, thus dividing the surface into approximate quadrilaterals. Cut each into two triangles by a diagonal. On true-length diagrams find the lengths of the elements and the diagonals, and draw the development by constructing the true sizes of the triangles in regular order.

14. TO DEVELOP A SPHERE

The sphere may be taken as typical of double-curved surfaces, which can be developed only approximately. It may be cut into a number of equal meridian sections, or "lunes," as in Fig. 19, and these may be considered to be sections of cylinders. One of these sections, developed as the cylinder in Fig. 19 will give a pattern for the others.

Another method is to cut the sphere into horizontal sections, or "zones," each of which may be taken as the frustum of a cone whose vertex is at the intersection of the extended chords (Fig. 20).

We hope the material of this chapter has proved to be understandable and interesting. The world of design contains many uses of developments, and their many applications should now become apparent. Some problems follow that should strengthen comprehension.

PROBLEMS

GROUP 1. DEVELOPMENTS OF PRISMS

1 to 6. Develop lateral surfaces of the prisms.

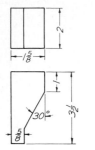

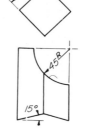
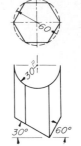
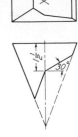

PROB. 1.   **PROB. 2 METRIC.**   **PROB. 3.**   **PROB. 4 METRIC.**   **PROB. 5 METRIC.**   **PROB. 6.**

GROUP 2. DEVELOPMENTS OF PYRAMIDS

7 to 10. Develop lateral surfaces of the hoppers.

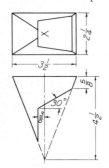
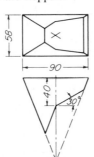

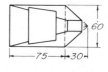
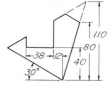

PROB. 7.   **PROB. 8 METRIC.**   **PROB. 9.**   **PROB. 10 METRIC.**

GROUP 3. DEVELOPMENTS OF CYLINDERS

11 to 15. Develop lateral surfaces of the cylinders.

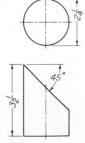
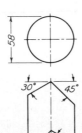

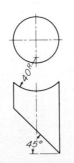
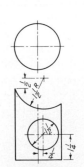

PROB. 11.   **PROB. 12 METRIC.**   **PROB. 13.**   **PROB. 14 METRIC.**   **PROB. 15.**

GROUP 4. DEVELOPMENTS OF COMBINATIONS OF PRISMS AND CYLINDERS

16 to 18. Develop lateral surfaces.

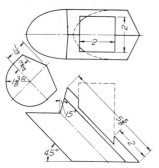

PROB. 16.

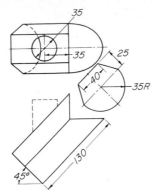

PROB. 17 METRIC.

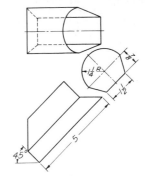

PROB. 18.

GROUP 5. DEVELOPMENTS OF CONES

19 to 22. Develop lateral surfaces.

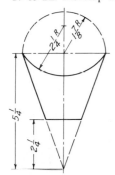

PROB. 19.

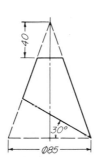

PROB. 20 METRIC.

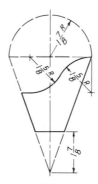

PROB. 21.

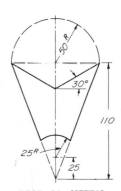

PROB. 22 METRIC.

GROUP 6. DEVELOPMENTS OF COMBINATIONS OF SURFACES

23 to 26. Develop lateral surfaces of the objects.

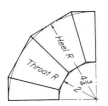

PROB. 23.

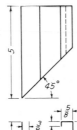

PROB. 24.

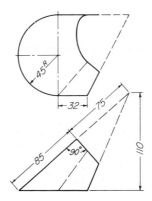

PROB. 25 METRIC.

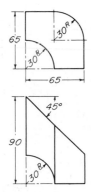

PROB. 26 METRIC.

GROUP 7. DEVELOPMENTS OF CONES AND TRANSITION PIECES

27 to 33. Develop lateral faces of the objects.

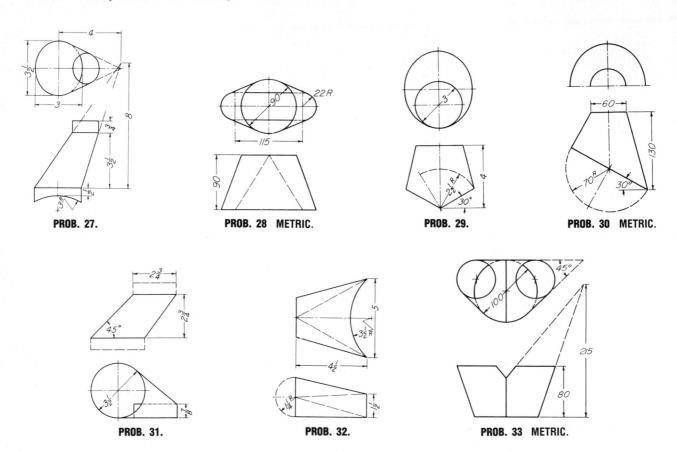

PROB. 27.

PROB. 28 METRIC.

PROB. 29.

PROB. 30 METRIC.

PROB. 31.

PROB. 32.

PROB. 33 METRIC.

GROUP 8. DEVELOPMENTS OF FURNACE-PIPE FITTINGS

34 to 39. Develop surfaces and make paper models.

PROB. 34.

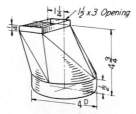

PROB. 35 METRIC.

PROB. 36.

PROB. 37 METRIC.

PROB. 38.

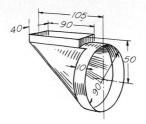

PROB. 39 METRIC.

GROUP 9. SPECIALTIES

There are innumerable combinations of geometric shapes and, sometimes, nongeometric forms used in modern engineering work that, fundamentally, are problems of layout and intersection and then either development or representation. The following two problems are typical examples.

40. The diverter-duct system for a VTOL jet consists essentially of a formed elbow and a cylinder. Find the intersection between the cylinder and elbow. Develop the cylinder and make an accurate representational drawing of the elbow.

41. The microwave horn reflector antenna consists essentially of three surfaces: a cone, a cylinder, and a paraboloid of revolution, which, if chosen correctly, should intersect one another on the same curve. To verify this construction, lay out the complete horn according to the dimensions shown in the illustration. Then find the intersection between the cone and cylinder and between the cylinder and paraboloid. Develop all parts that can be developed.

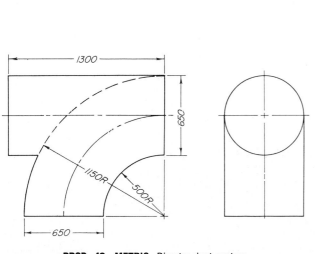

PROB. 40 METRIC. Diverter-duct system.

PROB. 41 Reflector horn.

PART C

APPLIED GRAPHICS AND DESIGN

1. OVERVIEW OF DESIGN

The word "design" covers a broad range of meanings and objects. Design is involved certainly in such things as houses, bridges, and automobiles. Design is involved also in items such as furniture, toasters, and swimming pools. In every case, however, design implies the creation of ideas and the transmitting of these ideas to others for use.

In engineering, design refers to a creative solution to a problem that leads to the production of a useful device. A design may be represented by drawings, models, or specifications. Representations may be done manually or by computer, but in every case each detail needed for production must be given. Details include size and location dimensions, materials, finishes, and assembly procedures.

A design often begins with sketches. This highly creative mode has been used for centuries, as seen in the sketches of Leonardo da Vinci shown in Fig. 1. One can also "sketch" by interacting with the computer, as seen in Fig. 2. Whatever the medium of expression, a person who designs must know a reasonable amount about what is being attempted in the design.

As an example, suppose that a person who has never fished and knows nothing about the sport attempts to design a fishing reel. Because of ignorance of such aspects as weight, balance, line capacity, drag characteristics, and overall performance, such a person is not capable of producing a good design. Nevertheless, most good engineering designers are capable of designing a wide variety of devices because of their knowledge of materials, processes, production methods, and other related aspects.

2. TYPES OF DESIGN

A. Design As Related to Practicality   Design may be classified according to its practical value. We will consider three such classifications of design: aesthetic, abstract, and pure functional.

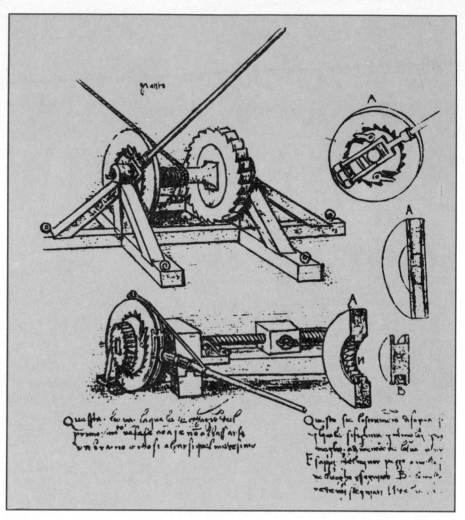

FIG. 1 A historical design sketch. Leonardo da Vinci (about A.D. 1500).

Aesthetic design is not strictly practical since it is used primarily for decorative purposes. Figure 3 shows an example. Aesthetic design has application far beyond engineering, of course. Such design should be pleasing to the eye and should enrich one's living environment. Most objects such as furniture, rooms, and appliances, incorporate aesthetic design.

Any completely new and original creation can usually be designed with freedom because there is no other similar entity to compare or compete with it. The choice of configuration, colors, contrasts, materials, and other similar factors offers extensive opportunity for experiment. The designer may try a number of combinations, build mock-ups, and then choose the most appealing.

The design of any product presents a difficult aesthetic problem. The project requires artistic ability, a keen color sense, good judgment of configurations and combinations, and an instinct for supplying the factors that appeal strongly to most buyers and users. This may sound difficult to persons who think that they have no artistic ability. A measure of success can be attained, however, by diligent study, observation, and work.

An *abstract design*, on the other hand, has no necessary relationship to reality, but may have visual interest or perhaps some future use. Abstract design may have use in research areas of engineering, such as Fig. 4, which shows the computer output for a set of mathematical equations giving a three-dimensional format.

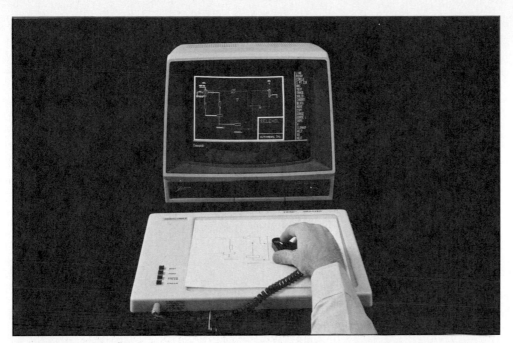

FIG. 2 Interactive sketching using a graphics tablet. The tablet interfaces with a computer, which displays output on the monitor shown. *(Courtesy Heath Company.)*

FIG. 3 Aesthetic design. The design has no function other than that of decoration.

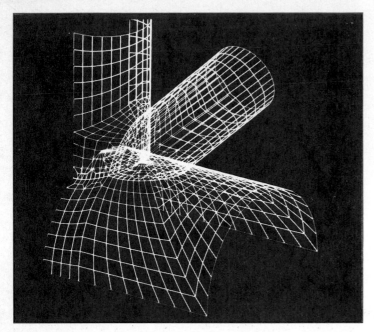

FIG. 4 Abstract design from a computer.

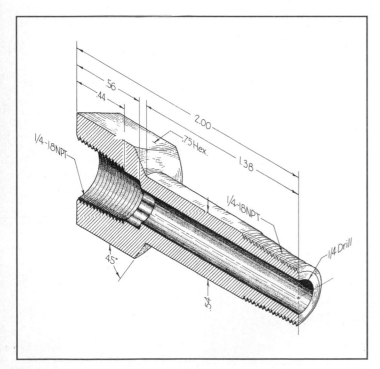

FIG. 5 Purely functional design for a gage connector.

Purely functional design is any design where function is completely dominant, with aesthetics not considered. An example is shown in Fig. 5. This part is completely functional; aesthetics plays no part in the design. Other examples are machines such as lathes, boring machines, motors, power tools, conveyors, material-handling equipment, and the great bulk of manufacturing and production equipment.

Some of the best examples to include all design aspects are to be found in automotive and commercial aircraft design. Purely aesthetic features are evidenced in color used, elegance of fabric—plastic or leather in upholstery, finish and appearance of appointments, body lines, and artistic configuration. Functional aesthetic features are obvious in such aspects as artful and functional controls, dials, instruments, glass areas, seat comfort, and safety features. Purely functional design is clearly manifested by examination of the power supply and all of its auxiliaries and components. Figure 6 represents a product incorporating all elements of design.

B. Design Related to Physical Form So far we have mentioned three types of design: aesthetic, abstract, and functional. These classifications relate a design to its degree of practicality. It is also possible to clarify design by its *physical form*, of which there are at least three types.

If the physical form of the design involves many variables, all interrelated, such design may be termed *system design*. One example is a subway system, involving a host of variables, including economic, social, and engineering considerations. System design involves much teamwork and is seldom covered in a single book or course. A designer becomes part of a team and contributes a small but vital individual portion to the design under the supervision of a project engineer.

Hardware design is the second type of design named for its physical form. Such design is so labeled because the product is in a "hard" form. Many products fall into this category: for example, stereo equipment, tractors, toasters, and air conditioners. Hardware design is often called "product design."

Software design is a third type of design labeled for its physical form. The product is in a "soft" form, often paper containing a solution to a problem. Computer output on paper is one of the most common forms of software design. The design may be the plot of the de-

FIG. 6 Functional design involving all aspects of design. *(U.S. Air Force Photo.)*

sired orbit of a communications satellite or the mathematical equations for automotive spring analysis. There is no limit to the possibilities of software design except the limits of the human mind.

3. BACKGROUND NEEDED FOR DESIGN WORK

There are certain elements common to all successful designers. These elements are (1) knowledge of the particular design field, (2) experience, (3) creativity, (4) a knowledge of materials and processes of manufacture, and (5) the ability to represent the design to others.

It is in the ability to represent the design to others that all elements come together. The design may be represented manually by hand-drawn plans, or may be generated by computer, with instructions provided by the designer. Regardless of the mode of representation, designing is personally very satisfying because of its creative aspects.

Designing requires a thorough knowledge of all the elements involved in a particular problem. This means that for every particular field of endeavor, the background required will be dictated by that field. Nevertheless, no field today is quite pure. For example, in mechanical engineering there are many cases where an electrical application is an important part of the mechanical device, and vice versa.

Therefore, a diversified background is a distinct advantage. Furthermore, the necessary background will vary considerably. For example, a designer working in

one of the aerospace fields must have extensive study in physics, chemistry, mathematics, etc., while a designer working for a company that manufactures small home appliances probably need not have anything beyond basic courses.

All designers *must* have a thorough training in graphics. Without it, a designer would fail, because as the design conception proceeds, the designer's own thinking must be recorded in the form of sketches and drawings. Furthermore, as the design is developed, it must be discussed with and approved by such people as the chief designer, chief engineer, and management executives, which means that clear and concise communication accomplished through sketches and drawings made by the designer is necessary. Design drawings will often be augmented and supported by mathematical data and diagrams, including computer data.

4. DESIGN PROCEDURE

This section discusses how a designer carries forward the work of designing. A chart has been provided in Fig. 7 that summarizes a series of 13 steps or phases in generating a design. These 13 phases are not the only system of design procedures that could be devised, but they are representative of how a design progresses.

Phase 1: The Project In this first phase the need is recognized for a solution to an existing problem. Information is gathered that relates to the problem. Management discusses with the client what criteria are to be met

DESIGN PROCEDURE

PHASE ONE—THE PROJECT

A. Discussion with management, engineering, client.
B. Statements and specification of the design problem.
C. Collection of all pertinent information.

PHASE TWO—FORMULATION

A. Recognition of requirements.
B. Definition of requirements.
C. Consideration of previous designs.
D. Assembly of all original data needed—mathematical, graphical, mechanical, electrical, etc.

PHASE THREE—CONCEPTS

A. Preliminary design sketches.
B. Preliminary design data giving materials, methods, construction details, and projected characteristics.

PHASE FOUR—ANALYSIS

A. Critical analysis of all design concepts.
B. Selection of most promising design or designs.

PHASE FIVE—DESIGN CONFERENCE

A. Discussion of preliminary designs with engineering, management, client.
B. Approval of design or designs.

PHASE SIX—REFINEMENT

A. More complete drawings and specifications of selected design or designs.
B. More complete data supporting projected design.

PHASE SEVEN—DESIGN CONFERENCE

A. Discussion of refined design or designs.
B. Approval of most promising design.

PHASE EIGHT—SYNTHESIS

A. Projected design supported by mathematical, graphical, and computer-aided and combined systems data.
B. Investigation of all physical aspects and proof of soundness of the design.

PHASE NINE—MODELS

A. Components.
B. Mock-ups.
C. Models of critical features.

PHASE TEN—TESTING

A. Proof of operating characteristics of components.
B. Proof of soundness of complete entity.

PHASE ELEVEN—CONFERENCE

A. Final discussion with originating authority.
B. Approval of final design.

PHASE TWELVE—FINAL PREPARATION

A. Final design drawings.
B. Final specifications.

PHASE THIRTEEN—TRANSMITTAL

A. Transmittal of final design drawings and specifications to originating authority.

NOTE: In the outline above three conferences (Phases 5, 7, and 11) are scheduled with the originating authority, engineering, and management. This is usual, but it does not mean that these are the only conferences during the development of the project. The designer confers frequently with colleagues, and with engineers, components suppliers, materials experts, and others, to obtain information and confirm design features.

FIG. 7 Chart for design procedure.

in order to have a satisfactory design. The criteria are usually expressed in a set of specifications that outline the specific items needed in the design, such as, for example, particular speeds, pressures, temperatures, etc., in which the final product is to operate.

Phase 2: Formulation The search for solutions is started, and here the ability to create shows itself. In the beginning, conventional solutions will come to mind, and there may be several of them. Then as the investigation continues, newer, more modern, and previously unheard-of answers may appear. At this stage the designer should let the imagination really run rampant. Every unusual physical, chemical, and electrical application and use of material, combination of elements, and their relationships should be carefully studied. A number of possible solutions will probably emerge.

Phase 3: Concepts While the solutions are being composed, the designer must make sketches for personal use. To be sure, these may be very rough and may lack detailed information, but they will probably include many written notes giving information for later use. This is where real ability and facility in the use of the graphic language play a very important role. An active mind and fluency at recording the mind's products of originality on paper combine to produce successful solutions.

Phase 4: Analysis All the solutions should be evaluated. This analysis must include feasibility of the design from every standpoint—from every engineering detail to economics and aesthetics. This must be done carefully and honestly because it is principally self-evaluation. Personal idiosyncrasies and preferences should be subservient to open-mindedness.

Phase 5: Design Conference This phase represents an ongoing process of review by project engineers and management. Rather than a single conference, there will probably be a series of frequent checks by management about the progress of the design. At some point in time, a "go" signal by management is needed to permit continuing development of the design. Approval to continue means that the company is willing to commit additional money and resources to the evolution of a design.

Phase 6: Refinement Further development takes place. The design begins to reflect its final format with respect to size, shape, cost. Prototype assembly drawings,

known as "design layouts," are produced. These drawings help to show the feasibility of the final production version of the product, as yet not completed. Design layouts are sometimes called "preliminary design drawings."

Phase 7: Design Conference Another conference occurs here in the design sequence to indicate the continual dialogue that is taking place between engineers and management. Choices are being narrowed by now. Commitments are firmer.

Phase 8: Synthesis Supporting information is brought into the design, in the form of mathematical, graphical, and computer aids and resource materials of all kinds. At this point the probable soundness of the proposed design should be known. The engineers and management have on hand all basic information to finalize the design.

Phase 9: Modeling This phase need not occur precisely in this sequence, but if modeling is to be part of the design, it is done at this time. New automobiles have clay mock-ups done full size. Small parts may be tried out in wood or plastic instead of the metal of the final form. Modeling can do much to resolve packaging problems that show up in three-dimensional space as opposed to the two-dimensional space of a design on paper.

Phase 10: Testing Testing is a necessary phase in any design process and should begin early so that vital data can be given the designer, allowing refinement of the design. While indicated here as phase 10, testing is actually a continuous activity. However, it tends to peak in the later stages of the design process. A product made up of well-tested components should prove to be reliable.

Phase 11: Conference Another conference is included at this point to indicate ongoing involvement between engineering and management. Final approvals are secured at this stage.

Phase 12: Final Preparation The design is fully set by this phase. The design layouts of phase 6 have now evolved into final versions ready for manufacture. All component parts needing design are fully dimensioned in what are called "detail drawings." Components of the design from previous products have been purchased for inclusion in the new design.

Phase 13: Transmittal The design drawings are checked, given final approval, and transmitted to the production engineers. The actual product is then manufactured, shipped to the customer, and installed. Feedback on the product's performance is given to the original designers so that any needed revisions can be made.

5. RELATIONSHIPS OF DESIGN, MATERIALS, AND CONSTRUCTION

A prime consideration in design is the materials to be used, because the assembly of parts, their detailed shapes, and their aesthetic appearance are affected by the materials used. More important reasons, however, are the conditions under which the finished product is expected to operate.

For example, a machine to be used in both tropical and extremely cold climates will have to be made of materials not adversely affected by either climate. In the hot and humid climate, parts deteriorate because of heat, corrosion due to humidity, and damage by fungus growths. On the other hand, in an extremely cold climate contraction causes reduction of clearances, and parts may "freeze" in position. Also, in extreme cold some materials become brittle and crack or break; there are operational difficulties caused by cold, "gummy" lubricants; and hazards are created by frost and ice. The same conditions may affect many aircraft parts. For example, in the summertime an airplane may move in a very short time from the ground, where the temperature is 100°F, to an altitude of 30,000 ft, where the temperature may be of the order of −40°F. Even if a device will still operate at two temperature extremes, the operational characteristics, especially in delicate instruments, may change radically, making either insulation or maintenance at a constant temperature (or both) a necessary part of the design.

The actual details of construction also have much to do with the successful operation of a machine. Inadequate bearing surfaces or insufficient lubrication will cause the machine to have a short operating life before repair or replacement. Protective seals against contaminants, abrasive materials, or corrosive liquids and gases are frequently necessary.

Another consideration in design is the need to keep the user in mind, particularly if the user is a person not technically familiar with the product. Such is the case with most products designed for use in the home or car.

Control knobs that come off and get lost; levers that break; screws and nuts that readily loosen and drop off; multiple ("stacked") controls so made that the careless person can inadvertently move two or more while adjusting one; finishes that rust, corrode, pit, and wear off; parts that are difficult to keep clean; and many other similar items all cause the user to take a dim view of the device. A principal rule in design is to provide as many guards against harm to the user and the device as are reasonable within the constraints of the design.

6. GOOD PRACTICE IN HARDWARE DESIGN

Offered in this section are some practical principles of design related to the shape of parts. These principles have been determined by geometry, experience in production, and testing. Examples are given in the area of hardware design, as opposed to software design, because hardware by nature involves shapes and proportions of firm material such as metal. Principles of software design

FIG. 8 Sharp corner. Poor design.

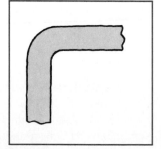

FIG. 9 Rounded corner. Good design.

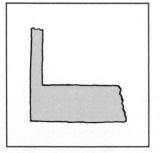

FIG. 10 Wide difference in metal thickness. Poor design.

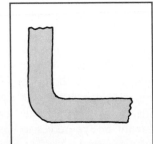

FIG. 11 Uniform section. Good design.

may be found in texts devoted to computer-aided design (CAD) and in certain math and physics books with design orientations.

The examples to be given are helpful in design work, but you should realize that for critical applications stress analyses are needed. For example, high-speed turbine blades for jet engines cannot be designed by "sight" or by what "looks good." Exact analyses involving strength of materials must be done, often by using a computer for complicated solutions. The illustrations that follow apply mainly to static (nonmoving) applications.

Corners Sharp corners on single parts are not recommended. In cast parts, cracks may occur at a sharp corner because of cooling strains. In other parts (forged, molded, etc.) breaks caused by overstressing will always start at a sharp corner. A corner such as that shown in Fig. 8 is poor design. Design with a rounded corner as shown in Fig. 9 if at all possible. The minimum inside radius is usually about one-half the metal thickness, making the minimum outer radius about 1½ times the metal thickness.

Differences in Metal Thickness When joined sections vary greatly in metal thickness, especially in castings and forgings, cooling strains may produce cracks and warping. Figure 10 is poor design because of the wide difference in metal thickness and also because of the sharp corners. Keep joined sections as uniform as possible and avoid sharp corners, as in Fig. 11.

Bracing Ribs or Supports Internal bracing should be designed according to geometrical principles. The ribs of Fig. 12 lend very little stiffness to the part because the geometry of the part is based on the square, which is an unstable shape. Stiffening ribs based on the triangle, a stable shape, should be employed if possible, as shown in Fig. 13. These designs are for torsioned stiffening only and are not applied to beam stiffening.

Size of Corner Radii We have seen before that corners should be rounded for several reasons. However, if the corner radius is too large for intersecting members, as in Fig. 14, too much material is left at the intersection and the design becomes poor because of wide differences in metal thickness. Keep the radius to a reasonable (minimum) value of about half the metal thickness, as in Fig. 15.

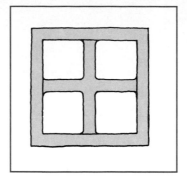

FIG. 12 Stiffening ribs. Poor design because of an unstable geometry.

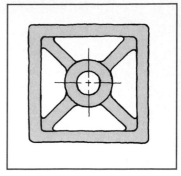

FIG. 13 Stiffening ribs. Good design because of stable geometric shapes.

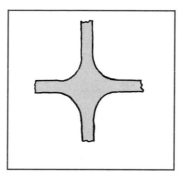

FIG. 14 Corner radius too large. Poor design.

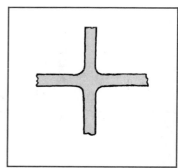

FIG. 15 Proper corner radius. Good design.

When several (five or more) members intersect, as in Fig. 16, even a small corner radius will not solve the problem and the difference in metal thickness becomes critical. The solution of the problem is to relieve the center of the part with a hole, as in Fig. 17, regardless of whether or not the hole has any function for the part.

Good Proportions Many combinations of hubs, ribs, holes, corner sections, flanges, and other elements occur in machine parts. It is impossible to describe all combinations, but the following examples will cover the bulk of the elements usually combined. Please note that the values given for corner radii are *maximum* values. Also note that the proportioning is based on the theory that differences in metal thickness should be "blended" into each other. This produces the best insurance against dis-

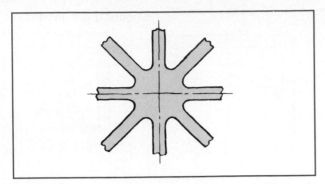

FIG. 16 Several intersecting members produce a great difference in metal thickness. Poor design.

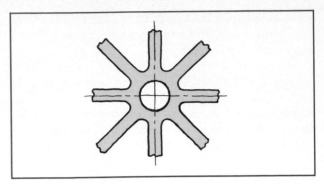

FIG. 17 Great difference in metal thickness relieved with a hole. Good design.

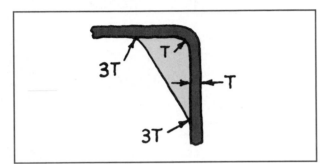

FIG. 18 Good proportioning for a corner rib. Radius values are maximum.

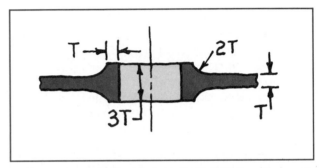

FIG. 19 Good proportioning for a centralized hub. Radius value is maximum.

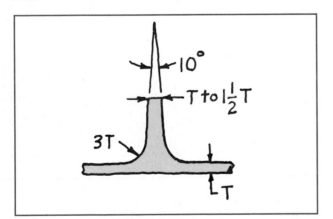

FIG. 20 Good proportioning for a support rib. Radius value is maximum.

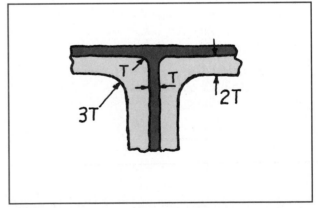

FIG. 21 Good proportioning for a tee section.

tortion and cracking during cooling and develops the maximum strength capability of the part.

Figure 18 shows good proportioning for a corner rib.

Figure 19 shows good proportioning for a centralized hub.

Figure 20 shows good proportions for a support rib. Note that the tapered section of the rib provides a good blend of metal thickness from the body section.

Figure 21 shows good proportions for a ribbed tee section.

Figure 22 shows good proportions for a flanged rib. The flanged section provides extra strength.

Figure 23 shows good proportions for extra material needed for a tapped hole in a thin wall section.

Figure 24 shows good proportions for a hub. The sizes given are based on the diameter of the hole needed.

Figure 25 shows good proportions for a symmetrical enlarged section necessary to provide a hole equal to, or larger than, a wall section.

Figure 26 shows a simple cylindrical part machined from solid bar stock. The sharp edges weaken the part, and even though it may not be highly stressed, they may cause difficulties in assembly and malfunctioning later. A better design is shown in Fig. 27, where the edges are chamfered.

Parts can be severely weakened by poorly chosen placement of keyseats, grooves, and other similar elements. Note in Fig. 28 that the keyway severely weakens

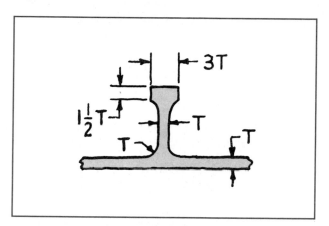

FIG. 22 Good proportioning for a flanged rib.

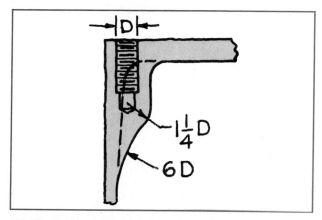

FIG. 23 Good proportioning for extra material needed for a threaded hole in a thin wall.

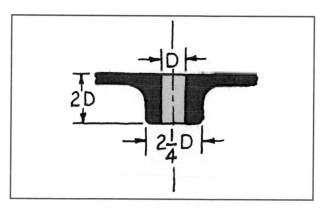

FIG. 24 Good proportioning based on the diameter of hole needed.

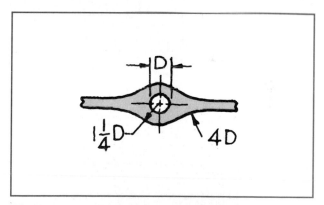

FIG. 25 Good proportioning for an enlarged portion.

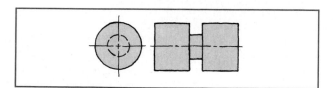

FIG. 26 A machined part with sharp edges. Poor design.

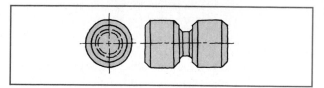

FIG. 27 A machined part with rounded edges. Good design.

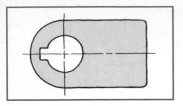

FIG. 28 Keyway in a position that severely weakens the part. Poor design.

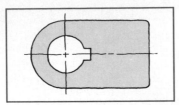

FIG. 29 Keyway position does not weaken the part. Good design.

the rounded end portion of the part. The difficulty is easily eliminated by placing the keyway as shown in Fig. 29.

In good machinery, finished-surface seats are always provided for fasteners. A simple unfinished surface, like that shown in Fig. 30, is poor design. Surfaces are finished either by adding a boss with machined surface as in Fig. 31 or by spot-facing the surface as in Fig. 32.

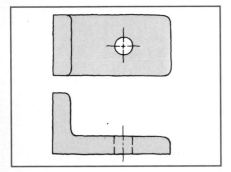

FIG. 30 No finished surface for bolt or screw seat. Poor design.

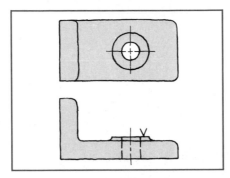

FIG. 31 Seat for bolt or screw provided by a finished boss. Good design.

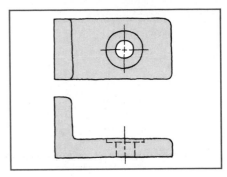

FIG. 32 Seat for bolt or screw provided by spot-face. Good design.

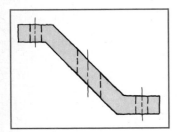

FIG. 33 Drilled hole at an angle to the part surface. Poor design.

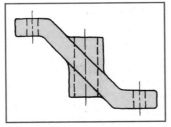

FIG. 34 Normal surface provided for drilled hole. Good design.

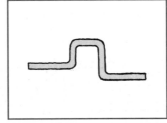

FIG. 35 Formed sheet-metal part with 90° bends and small bend radii. Poor design.

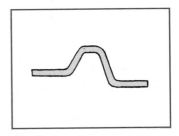

FIG. 36 Formed sheet-metal part with bends less than 90° and bend radii (internal) greater than the metal thickness. Good design.

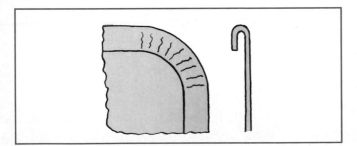

FIG. 37 Bent flange on round corner causes crowding of the metal. Poor design.

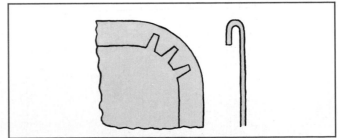

FIG. 38 Notches are used to relieve the crowding of metal on a bent corner flange. Good design.

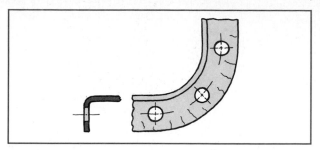

FIG. 39 Formed outside flange on rounded corner produces breaks in the material. Poor design.

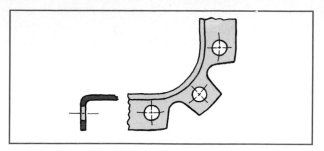

FIG. 40 Notches on outside flange prevent material from stretching. Good design.

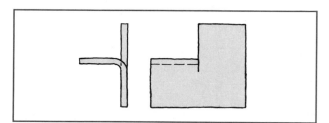

FIG. 41 Bent portion of part formed by shearing and bending. Poor design.

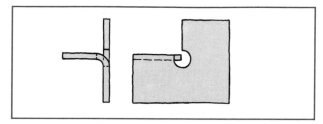

FIG. 42 Bent portion of part formed by shearing to a terminal hole and bending. Good design.

Parts should never be designed so that a hole must be drilled at an angle to a surface, as in Fig. 33. Even though a drill jig is used, the design will cause drill breakage, inaccuracy, and severe wear on the drill bushing. Provide a normal surface for the hole by building up the piece as in Fig. 34.

In parts formed from sheet metal two practices should be avoided if possible. First, full 90° bends will cause undue stretching of the material by the die and the part may be difficult to strip from the die. Second, small radii less than the metal thickness (internal) may cause cracking of the material at the bend. Such a poorly designed part is shown in Fig. 35. Unless full 90° bends are needed for some functional reason, a far better design is to use bends of less than 90° and radii equal to the metal thickness (internal) or larger, as shown in Fig. 36.

Bent stiffening flanges on round corners of sheet-metal parts must be specially designed or crowding and crinkling of the metal will occur, as shown in Fig. 37. A design with notches to relieve the crowding of the material will solve the problem, as indicated in Fig. 38.

Formed flanges on the outside rounded corners of sheet-metal parts are almost impossible to make without undue stretching, weakening, and cracking of the mate-

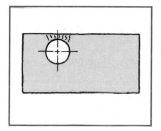

FIG. 43 Hole too close to the edge causes distortion and possible breakage. Poor design.

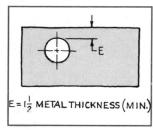

$E = 1\frac{1}{2}$ METAL THICKNESS (MIN.)

FIG. 44 Proper setback for a punched hole in sheet metal. Good design.

rial, as indicated in Fig. 39. The solution of the problem is to provide notches, as in Fig. 40.

When a portion of a sheet-metal part must be made by shearing or cutting and bending over as in Fig. 41, breakage because of overstressing or fatigue will start from the end of the sheared or cut surface. The sheared or cut line should be terminated by a hole (drilled or punched), as in Fig. 42. The hole contour will distribute stresses and prevent breakage.

Punched holes too close to the edge of a sheet-metal part may cause distortion and possible breakage of the material, as indicated in Fig. 43. The hole should be set back with the minimum edge distance shown in Fig. 44.

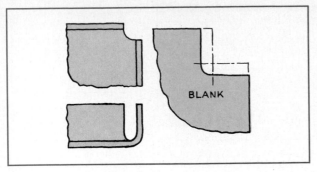

FIG. 45 Corner relief for a formed sheet-metal pan. Economical design.

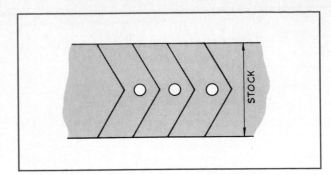

FIG. 46 Maximum use of stock. Good design.

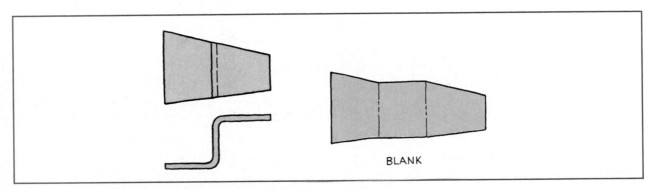

FIG. 47 Irregular blank. Poor design.

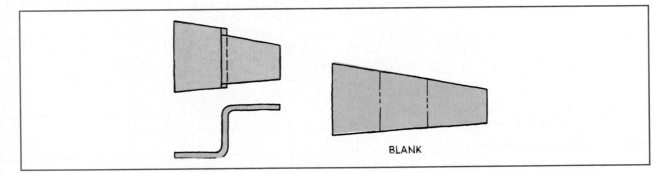

FIG. 48 Simplified shape of blank. Good design.

If the corner of a sheet-metal pan does not have to be closed for functional purposes, an economical part can be made by relieving the corner as shown in Fig. 45.

If possible, sheet-metal parts should be designed so that maximum use can be made of the stock without excessive scrap, as indicated in Fig. 46.

Sheet-metal parts should be designed so as to simplify the shape of the blank unless there is some functional reason for not doing so. The design of the part shown in Fig. 47 produces an irregular blank and excessive scrap from the blanking operation.

A more economical design is shown in Fig. 48, where

the blank has straight sides. By reversal in the blanking operation excessive scrap is eliminated and maximum use of the stock is attained.

7. GOOD PRACTICE IN JOINING PARTS

There are many methods of joining parts. The designer should be familiar with all possible methods and should select the method that will satisfactorily hold the parts together and also be economical from the standpoints of cost, repair, and replacement. Several different methods may be used depending somewhat on whether the fastening is to be permanent or removable. The following list describes the most common methods.

Mechanical Fasteners These include bolts and screws, rivets, special springs, snap rings, O-rings, rollpins, and keys.

Adhesives Adhesives are classified by *bonding type* (heat, pressure, time, catalyst, vulcanizing); *form* (liquid, paste, powder, mastic); *flow* (flowable or nonflowable); *vehicle* (solvent dispersion, water emulsion, or complete solids); or chemical composition, which is the most significant.

A joint like that shown in Fig. 49 is weak in tension because as the parts joined are stressed, the metal will distort and will be peeled away from the adhesive bond.

A simple butt joint such as that shown in Fig. 50 is not strong in tension because of the weakness of the bond itself, and also because in most joints of this type the actual bond area is limited.

A better joint may be made by lapping the parts as in Fig. 51. This puts the bond in shear, but as the parts are stressed, some distortion due to the offset will occur and the adhesive may peel.

Probably the best joint in tension is that shown in Fig. 52, where tension on the joint will not distort the members and the adhesive remains in shear.

When an adhesive bond is used for a right-angle member as in Fig. 53, the member should be flared as shown to get a greater bonding area.

When a joint is designed as in Fig. 54, for a right-angle corner, the adhesive bond will be in peel and will be weak. The joint should be designed as in Fig. 55, where the adhesive bond is in shear.

Solders Many combinations of solder metals are

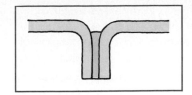

FIG. 49 Adhesive-bonded joint. Weak in tension because of peel.

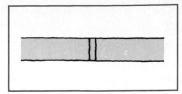

FIG. 50 Adhesive-bonded butt joint. Weak in tension because of bond strength and lack of bone area.

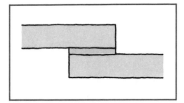

FIG. 51 Adhesive lap joint. Better in tension than the joints of Figs. 49 and 50.

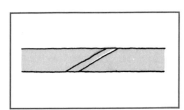

FIG. 52 Adhesive-bonded joint. Better in tension than the joints of Figs. 49, 50, or 51.

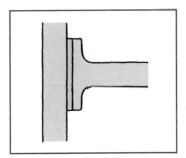

FIG. 53 Right-angle member flared for greater area of bonded joint.

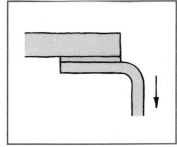

FIG. 54 Adhesive-bonded joint for right angle. Poor design.

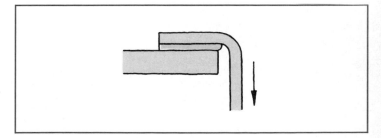

FIG. 55 Adhesive-bonded joint for right angle. Good design.

commercially available. *Tin-lead* solders are used for general-purpose applications on ferrous and nonferrous metals for joining and coating. *Tin-lead-antimony* solders are used for machine and torch soldering and coat-

ing of metals, except galvanized iron. *Tin-antimony* solders are used for joints on copper in electrical, plumbing, and heating applications. *Tin-silver* solders are used principally for work on fine instruments. *Tin-*

Joinability of Materials*

	Arc welding	Oxyacet-ylene welding	Resistance welding	Brazing	Soldering	Adhesive bonding	Threaded fasteners	Riveting and stitching
Cast iron	×	×				×	×	×
Carbon and low-alloy steels	×	×	×	×		×	×	×
Stainless steels	×	×	×	×		×	×	×
Aluminum magnesium	×	×	×	×		×	×	×
Copper and its alloys	×	×	×	×	×	×	×	×
Nickel and its alloys	×	×	×	×	×	×	×	×
Titanium	×		×			×	×	
Lead, zinc	×	×			×	×	×	×
Thermoplastics	×	×	×			×	×	×
Thermosets						×	×	×
Elastomers						×		×
Ceramics						×	×	
Glass		×				×	×	
Wood						×	×	×
Leather						×		×
Fabric						×		×
Dissimilar metals				×	×	×	×	×
Metals to nonmetals						×	×	×
Dissimilar nonmetals						×		×

* The crossmark (×) in the table indicates common practice. Difficult and unreliable methods are not given.

FIG. 56 Joinability of materials.

zinc solders are used to join aluminum. *Lead-silver* solders are used on copper, brass, and similar metals (with torch heating). *Cadmium-zinc* solders are used for soldering aluminum. *Cadmium-silver* solders are used for joining aluminum to aluminum and to dissimilar metals. *Zinc-aluminum* solders are used for high-strength aluminum joints.

Brazing Brazing is the operation of joining two similar or dissimilar metals by partial fusion with a metal or alloy applied under very high temperature. Brazing alloys include aluminum-silicon, copper-phosphorus, silver, gold, copper, copper-zinc, magnesium, and nickel.

Welding Welding is the operation of joining two similar metals with a fused joint of the same or very similar metal. Steel, cast iron, copper and copper alloys, nickel alloys, and aluminum can be welded.

The accompanying table of Fig. 56 lists joinability of common materials by various procedures.

8. EVOLUTION OF A DESIGN: AN EXAMPLE

In this section an example of various drawings accompanying the evolution of a design is offered for your study. Design drawings in the early phases of design supply the information from which the final assembly and detail drawings are made (see Chap. 13). Design drawings may be divided into two classes, preliminary and final.

Preliminary design drawings, for the most part, are sketches. Figures 57 to 59 illustrate preliminary design drawings. These are made freehand. While working up the sketches, the designer is keenly attentive to the needs of the problem.

As an example of how the design thinking and the preliminary drawings are brought forward together, Figs. 57 to 60 illustrate a series of design drawings made for a nutcracker. The first drawing, Fig. 57, shows that the designer conceived a lever and base with fixed and pivoted pads between which the nut would be placed. It is immediately obvious that when the shell breaks, the release of counterforce will allow the lever to descend at once and crush the nut meat. The second design, Fig. 58, is a refinement over the first design, with the linkage arranged so that when the actuating lever descends to its maximum, the holding pads will come to a minimum separation, but no closer together.

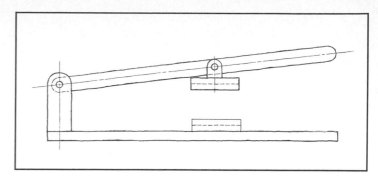

FIG. 57 Design sketch of nutcracker. This is the first of a series.

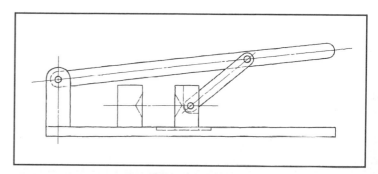

FIG. 58 Design sketch of nutcracker. This is a refinement of the first sketch (Fig. 57).

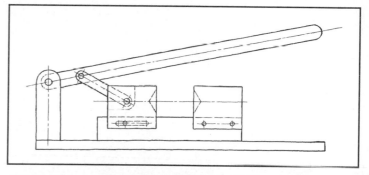

FIG. 59 Design sketch of nutcracker. This is the third in a series, a refinement of the second (Fig. 58).

A further refinement is seen in Fig. 59, where the linkage is reversed to place it away from the hand position on the operating lever. Finally, in Fig. 60, the basic arrangement of Fig. 59 is refined to include an adjustable distance between the holding pads. Also at this time a more accurate and complete drawing is made, and some basic dimensions and details, including material types, are added.

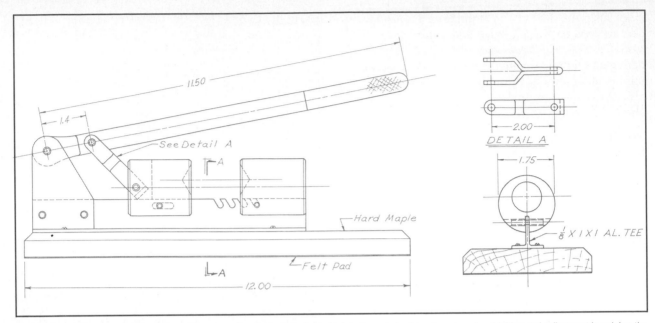

FIG. 60 Design drawing of nutcracker. This represents the final design, which is the culmination of progressive thinking and refinement in solving the problem.

FIG. 61 An elevation view of a house is being generated on a high-speed plotter by this automated drafting system. *(Courtesy Bausch & Lomb.)*

Note that Fig. 60 is not a sketch, but is drawn with instruments. This last drawing could also have been done on a computer-driven plotter. Computer assistance would be particularly helpful if many drawings were needed, each having slight variations from the others.

One example of a drawing done on a computer-driven plotter is seen in Fig. 61. The design shown on the monitors is being plotted on the large sheet using a pen positioned by computer command. While costly to purchase, such an automated drafting system can be cost-effective if used during a high proportion of working hours. Speed and accuracy over manual methods can more than offset the initial cost.

PROBLEMS

A series of design projects is presented on the following pages. The projects are designed to tax the ingenuity of the student without requiring any special knowledge of engineering mechanics or higher mathematics.

Each project has an accompanying suggested solution. These solutions are deliberately left undimensioned and incomplete so that the minimum requirement for the student is to complete the design and make a set of working drawings. They do not necessarily represent the best possible design for the project. The student is urged to consider as many other ideas as possible, weighing such factors as cost and reliability before making a final selection. Also see Chap. 13.

Use of standard purchased parts such as rollpins, retainer rings, screws, etc., is encouraged. Refer to the appendixes for specific information.

1. Design a device to be used in a drill press to cut circular gaskets from sheet cork, rubber, etc. It must cut gaskets to varying sizes, as large as 4 in outside diameter, with the interior hole adjusting down as small as practicable.

The design indicated makes use of a spring-loaded centering pin to keep the material from slipping during the cutting.

Torque must be transmitted from the spring tube to the cutter bar by some fastening device not shown. If this design is used, the means incorporated to make this connection must permit the parts to be disassembled for spring replacement and sharpening of the center pin.

2. An industrial dolly made from steel angle is to be rebuilt to carry delicate electronics instruments.

Design a new caster mount to absorb shock incurred when the dolly is used on rough floors. A design load of about 100 lb per caster (including dolly) is to be carried.

The suggested design makes use of a compression spring having a spring constant of 100 lb/in with a total deflection of 2 in.

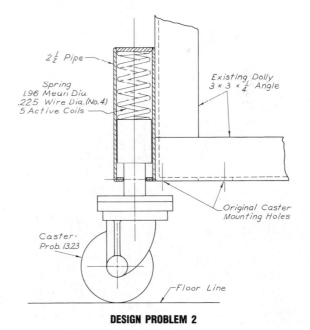

DESIGN PROBLEM 1

DESIGN PROBLEM 2

3. Design a rust-resistant portable stand suitable for holding serving trays, campers' cookstoves, laundry baskets, etc. This stand is to be folded into as small and neat a package as possible.

The suggested solution makes use of aluminum angles and flat aluminum bar stock. The fasteners are threadless pins held in place by pushnuts.

4. Design a portable solar stove suitable for survey parties and campers, for use in relatively sunny climates. It must be lightweight and disassemble to pack into a small box. A parabolic reflector at least 1 m in diameter must be used, and its focal point must not be too precise or it will tend to burn a hole in a small container. The reflector must be on an adjustable mounting so that it can be focused on the sun at different times of the day and year, and at different latitudes.

The suggested solution consists of a reflector made of eight properly developed segments of polished sheetmetal. These segments lie flat when not in use, and are bent to the proper shape by fasteners holding the segment edges together.

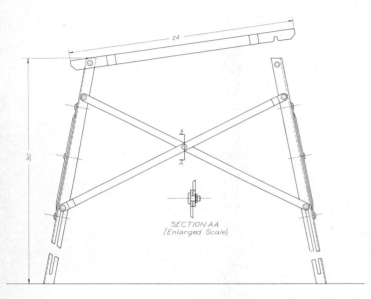

SECTION AA
(Enlarged Scale)

DESIGN PROBLEM 3

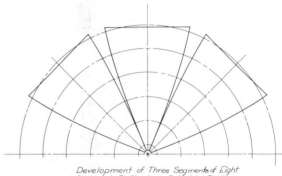

Development of Three Segments of Eight
Segment Reflector - Suitable Fasteners
Will Pull Flat Pieces To Parabolic Curve

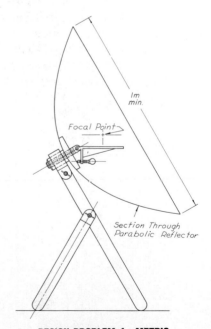

1m
min.

Focal Point

Section Through
Parabolic Reflector

DESIGN PROBLEM 4 METRIC.

5. A small brass shell is fed from a hopper into the existing supply tube shown, which in turn supplies a press that further processes the part. All parts must have the closed end down.

Design a device to turn those pieces that have the open end down. The hopper feeds the pieces at such a rate that they do not touch each other.

One possible solution is shown; it utilizes an adjustable pin which enters the open end of those pieces that need turning. Gravity will then flip the pieces as indicated. The closed ends of the other pieces will hit the pin, and the pieces will bounce up and fall without turning.

The design must be mounted on the existing 6-in l-section column shown.

6. An industrial drying process is supplied with warm air through a 200-mm-diameter round duct. A variable-diameter orifice is to be installed in the duct to control the airflow. To aid in maintaining laminar flow the opening is to be kept approximately round and concentric in the duct. The diameter must reduce down to approximately one-half of the original diameter.

Design a suitable device for this service.

The solution indicated is based on the idea of six curved blades pivoted outside the periphery of the duct and moved in and out of the airstream by a control ring which is concentric with the duct.

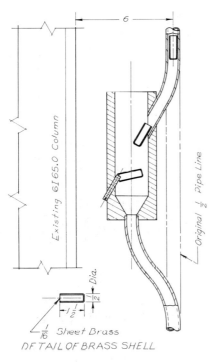

DESIGN PROBLEM 5

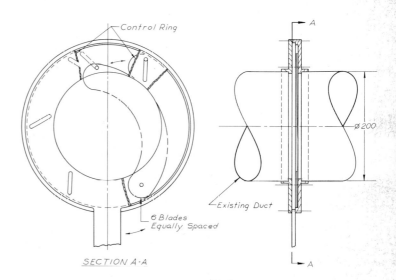

SECTION A·A

DESIGN PROBLEM 6   METRIC.

7. Many mechanical devices are lubricated for life at the time of assembly. This requires that a metered amount of oil be placed in the machine.

Design a device that will meter out 2 fluid ounces of oil with a single "push-pull" motion of the operator (1 U.S. fluid ounce = 1.805 cubic inches).

Some adjustment is desirable so that precise control is possi-ble. Oil must not flow when the device is not in use.

The partial solution indicated uses two pistons, mounted on a single rod, sliding back and forth in a carefully finished cylin-der.

Air trapped in the cylinder may impede movement of either the oil or the piston. The plastic float valve indicated will per-mit air to enter and leave the cylinder, but will trap the oil.

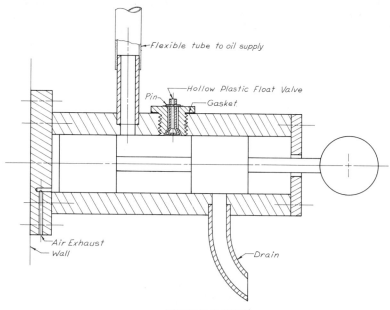

DESIGN PROBLEM 7

8. A "tin"-can food-container manufacturing plant is supply-ing cans directly to a food processor. The flow of cans must be equally divided and supplied to the two conveyor belts which will be installed as shown. Design a device and make the nec-essary assembly, working, and installation drawings for this.

The solution indicated consists of a "tee" member pivoted as shown. Each can will rotate the member in one direction or the other, evenly dividing the flow.

The cans are 4 in in diameter and 4¾ in high. They are supplied spaced out so that they do not touch each other.

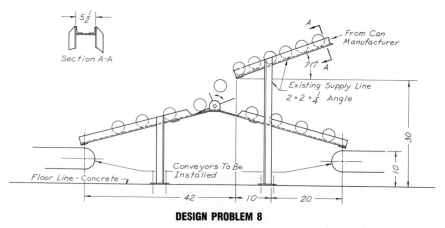

DESIGN PROBLEM 8

9. Design a device to check the compression-type coil spring specified below for spring rate (pounds to deflect spring 1 inch); maximum free length; maximum length when compressed solid.

Specification of spring: 1 in mean diameter, .0625-in.-diameter wire, five active coils, ends squared and ground, 2-in maximum free length, 0.44-in maximum solid length, and 4.4 lb/in spring rate. A load of 4.4 lb must deflect spring from 0.9 to 1.1 in.

The device indicated is one possible solution. A cam turned by hand deflects the master spring 1 in. (It is carefully set to give a force of 4.4 lb at 1-in deflection.) If the two shoes are equidistant from the center pivot, the spring under test will carry a 4.4-lb load. Lines scribed on the base then will show, if the spring is within tolerance. The other lines scribed on the base will show, if spring meets length-tolerance specifications.

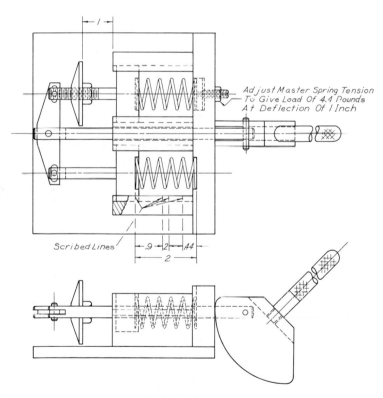

DESIGN PROBLEM 9

13
PRODUCTION
DRAWINGS

1. THE RELATIONSHIP OF DRAWINGS TO PRODUCTION

In the previous chapter we discussed how designs are created within an engineering office. Now we will study how drawings expressing a design relate to production methods. There are different forms of drawings for different production methods. Figure 1 shows two such drawings. The computer-generated drawing in the foreground is for an electrical circuit. The drawing in the background is an assembly drawing for a mechanical component.

First, basic methods of production will be surveyed to assist in understanding of fundamental machining operations. Operations such as turning, milling, and boring are essential in the production of parts. Machines doing the operations may be controlled manually or, as is more and more common, may be automatically controlled by computer. Computer-aided manufacturing (CAM) is a vital ingredient in obtaining high productivity in today's highly competitive markets.

This chapter will concentrate on basic forms of production and the drawings that support them. Other methods, such as CAM, will be covered only lightly, as will be computer-aided drafting (CAD). Both CAD and CAM build upon the *basic* principles of drafting and production. Excellent texts in the CAD/CAM areas should be consulted for in-depth coverage of these advanced techniques.

2. THE ROLE OF QUANTITY IN PRODUCTION

Production methods can be classified as (1) unit production, when one or only a few devices or structures are to be built, and (2) quantity or mass production, when a large number of practically identical machines or devices are to be made with the parts interchangeable from one machine to another.

Unit production often applies to custom-made products. The products may be large, such as a made-to-order steam turbine for an electrical power generation plant. Or they may be relatively small, such as a specially fitted artificial arm. By its nature unit production yields limited numbers of pieces. Cost may be rather high per piece because the design and production costs must be written off across a low total number of units.

Quantity production, commonly known as *mass production*, is used whenever a great many identical products are made. Examples are easy to cite: toasters, TVs, refrigerators, most cars. The fact that a product is mass-produced need in no way diminish its value or quality compared to unit-production items. Much top engineering talent can be expended to ensure a consistent, well-designed mass-produced product. The consumer can benefit in terms of good value per dollar spent.

Quality control is needed in quantity production to ensure a consistent standard. Each and every part produced may be inspected where the design dictates such and where the added cost is permitted. On the other hand, random sampling techniques may be used for products such as light bulbs and spark plugs, both very high production items.

3. PERSONS INVOLVED IN PRODUCTION

The setting up of production processes and sequences can seldom be left to the judgment of a single person. In preparing drawings intended for production, the engineering department must assume full responsibility for the success of the machine by making the drawings so exact and complete that the resulting parts should not fail to be satisfactory.

The drafter should be well acquainted with methods of parts production, because the manufacturing processes selected will determine the way in which the supporting drawings are dimensioned. Persons involved with particular production processes expect the drawings for those processes to be dimensioned in particular ways. These persons work with specific machines. For example, the person to keep in mind when dimensioning a part cut from a solid material is the machinist.

For parts produced by casting, the persons to be considered are the patternmaker (for sand castings) or the diemaker (for die castings) and, for finishing, the machinist. Forged parts subject to quantity production are dimensioned for the diemaker and machinist. For parts produced from sheet stock, the template maker, the diemaker, and the machinist must be considered; information for making the template and for forming the blank is obtained from a detail drawing showing the part as it should be when completed. In every case one drawing, appropriately dimensioned, must show the finished part.

Each of the different methods produces a characteristic detailed shape and appearance of the parts, and these features must be shown on the drawing. Figure 2 shows and lists typical features of some methods and indicates the differences in drawing practice.

4. BASIC MANUFACTURING PROCESSES

In this section we will give an overview of the various basic machining tools and fabrication processes available to the production engineer. We will look at the tools in their simplest formats. Realize that basic tools in their pure forms, such as simple drills, can be incorporated into sophisticated special tooling of semiautomatic and automatic types. Commanded by programmed instruc-

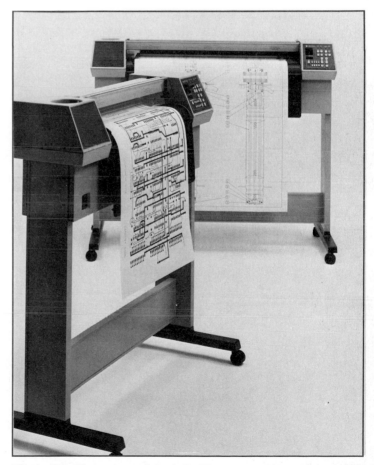

FIG. 1 Two different types of drawings. The computer-driven plotters provide highly accurate, quickly drawn results. *(Courtesy Hewlett-Packard Company.)*

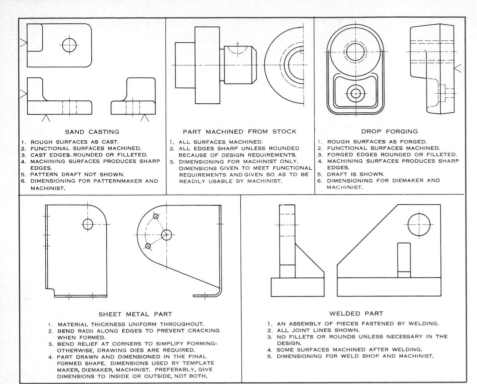

SAND CASTING

1. ROUGH SURFACES AS CAST.
2. FUNCTIONAL SURFACES MACHINED.
3. CAST EDGES ROUNDED OR FILLETED.
4. MACHINING SURFACES PRODUCES SHARP EDGES.
5. PATTERN DRAFT NOT SHOWN.
6. DIMENSIONING FOR PATTERNMAKER AND MACHINIST.

PART MACHINED FROM STOCK

1. ALL SURFACES MACHINED.
2. ALL EDGES SHARP UNLESS ROUNDED BECAUSE OF DESIGN REQUIREMENTS.
3. DIMENSIONING FOR MACHINIST ONLY. DIMENSIONS GIVEN TO MEET FUNCTIONAL REQUIREMENTS AND GIVEN SO AS TO BE READILY USABLE BY MACHINIST.

DROP FORGING

1. ROUGH SURFACES AS FORGED.
2. FUNCTIONAL SURFACES MACHINED.
3. FORGED EDGES ROUNDED OR FILLETED.
4. MACHINING SURFACES PRODUCES SHARP EDGES.
5. DRAFT IS SHOWN.
6. DIMENSIONING FOR DIEMAKER AND MACHINIST.

SHEET METAL PART

1. MATERIAL THICKNESS UNIFORM THROUGHOUT.
2. BEND RADII ALONG EDGES TO PREVENT CRACKING WHEN FORMED.
3. BEND RELIEF AT CORNERS TO SIMPLIFY FORMING- OTHERWISE, DRAWING DIES ARE REQUIRED.
4. PART DRAWN AND DIMENSIONED IN THE FINAL FORMED SHAPE, DIMENSIONS USED BY TEMPLATE MAKER, DIEMAKER, MACHINIST. PREFERABLY, GIVE DIMENSIONS TO INSIDE OR OUTSIDE, NOT BOTH.

WELDED PART

1. AN ASSEMBLY OF PIECES FASTENED BY WELDING.
2. ALL JOINT LINES SHOWN.
3. NO FILLETS OR ROUNDS UNLESS NECESSARY IN THE DESIGN.
4. SOME SURFACES MACHINED AFTER WELDING.
5. DIMENSIONING FOR WELD SHOP AND MACHINIST.

FIG. 2   Drawing requirements for different manufacturing methods.

FIG. 3   This numerically controlled milling machine is designed as an instructional unit in a student laboratory. *(Courtesy Feedback, Inc.)*

tions, these machines are basically the same as ordinary lathes, grinders, etc., but contain mechanisms to control the movements of cutting tools and produce identical parts with little attention from the operator once the machine has been "tooled up." Figure 3 shows a machine tool controlled by a computer.

5. THE MACHINE SHOP

We will soon look at specific machines within the area, or shop, used to produce parts. First we need to make some general comments regarding the operations within a machine shop.

The machine shop produces parts machined from stock material and finishes castings, forgings, etc., requiring machined surfaces. Cylindrical and conic surfaces are machined on a lathe. Flat or plane surfaces are machined on a planer, shaper, milling machine, broaching machine, or in some cases (facing) a lathe. Holes are drilled, reamed, counterbored, and countersunk on a drill press or lathe; holes are bored on a boring mill or lathe. For exact work, grinding machines with wheels of abrasive material are used. Grinders are also

coming into greatly increased use for operations formerly made with cutting tools. In quantity production many special machine tools and automatic machines are in use. The special tools, jigs, and fixtures made for the machine parts are held in the toolroom ready for the machine shop.

All machining operations remove metal, either to make a smoother and more accurate surface, as by planing, facing, milling, etc., or to produce a surface not previously existing, as by drilling, punching, etc. The metal is removed by a hardened steel, carbide, or diamond-cutting tool (matching) or an abrasive wheel (grinding), the product, or "work piece," as well as the tool or wheel, being held and guided by the machine.

All machining methods are classified according to the operating principle of the machine performing the work:

1. The surface may be *generated* by moving the work with respect to a cutting tool or the tool with respect to the work, following the geometric laws for producing the surface.

2. The surface may be *formed* with a specially shaped cutting tool, moving either work or tool while the other is stationary.

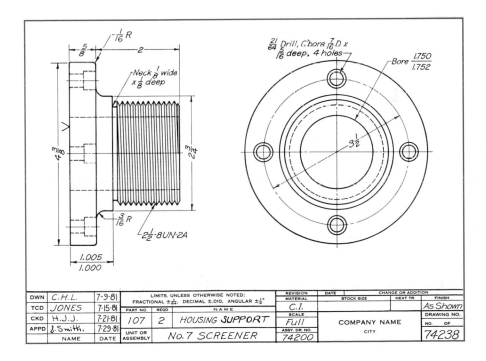

FIG. 4 A working drawing of a cast part.

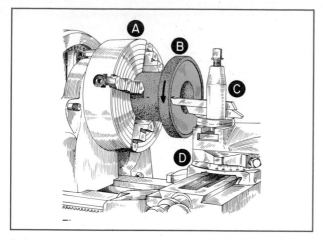

FIG. 5 Facing. The chuck (A) holds the part (B) and revolves it in the direction shown. The tool in the holder (C) is brought against the work surface by the lathe carriage, and the motion of the tool across the face is controlled by the cross slide (D).

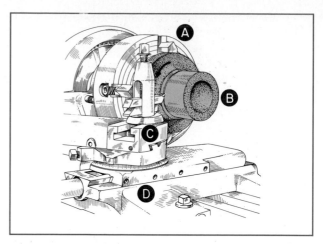

FIG. 6 Turning. The chuck (A) holds the part (B) and revolves it in the direction shown. The tool in the holder (C) is brought against the work surface by the cross slide, and the motion of the tool is controlled by the lathe carriage (D).

The forming method is, in general, less accurate than the generating method, inasmuch as any irregularities in the cutter are reproduced on the work. In some cases a combination of the two methods is used.

6. SPECIFIC MACHINE TOOLS

It is helpful to discuss tools used in manufacture so that a drafter can dimension more effectively the drawing of a part to be made by a particular tool. We will start with the lathe, a common and vital machine tool.

A. The Lathe Called the "king of machine tools," the lathe is said to be capable of producing all other machine tools. Its primary function is machining cylindrical, conic, and other surfaces of revolution, but with special attachments it can perform a great variety of operations.

Figure 4 shows the working drawing of a cast part. Complete information is given for all operations to make the part.

Figure 5 shows the casting made from the drawing of Fig. 4 held in the lathe chuck. As the work revolves, the cutting tool is moved across perpendicular to the axis of revolution, removing metal from the base and producing a plane surface by generation an operation called *facing*.

After being faced, the casting is turned around, and the finished base is aligned against the face of the chuck, bringing the cylindrical surface into position for *turning*

to the diameter indicated in the thread note on the drawing. The neck shown at the intersection of the base with the body is turned first, running the tool into the casting to a depth slightly greater than the depth of the thread. The cylindrical surface is then turned (generated) by moving the tool parallel to the axis of revolution (Fig. 6). Figure 7 shows the thread being cut on the finished cylinder. The tool is ground to the profile of the thread space, carefully lined up to the work, and moved parallel to the axis of revolution by the lead screw of the lathe. This operation is a combination of the fundamental processes, the thread profile being formed while the helix is generated.

The hole through the center of the casting, originally cored, is now finished by *boring*, as the cutting of an interior surface is called (Fig. 8). The tool is held in a boring bar and moved parallel to the axis of revolution, thus generating an internal cylinder.

Note that in these operations the dimensions used by the machinist have been (1) the finish mark on the base and thickness of the base, (2) the thread note and outside diameter of the thread, (3) the dimensions of the neck, (4) the distance from the base to the shoulder, and (5) the diameter of the bored hole.

B. The Drill Press The partially finished piece of Fig. 4 is now taken to the drill press for drilling and counterboring the holes in the base according to the dimensions

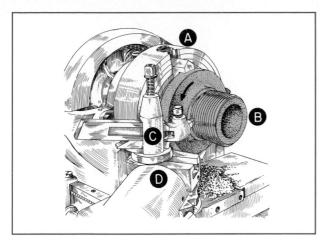

FIG. 7 Threading. The chuck (A) holds the part (B) and revolves it in the direction shown. The tool, ground to the profile of the thread space, in the holder (C) is brought into the work surface by the angled cross slide (D), and the rate of travel is controlled by gears and a lead screw that moves the lathe carriage.

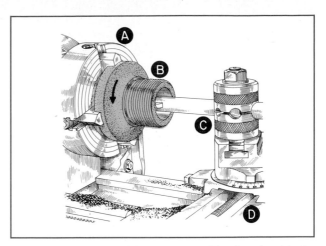

FIG. 8 Boring. The chuck (A) holds the part (B) and revolves it in the direction shown. The boring bar with the tool in the holder (C) is brought against the work surface by the cross slide, and advance is controlled by the lathe carriage (D).

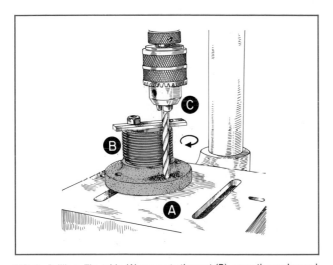

FIG. 9 Drilling. The table (A) supports the part (B), sometimes clamped as shown. The drill in the chuck (C) revolves in the direction shown, and is forced downward by a gear and rack in the drill-press head, to make the hole.

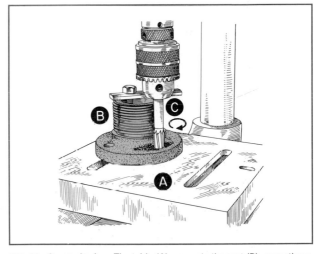

FIG. 10 Counterboring. The table (A) supports the part (B), sometimes clamped as shown. The counterboring tool with piloted end to fit the previously drilled hole is held in the chuck (C) and revolves in the direction shown. The gear and rack in the drill-press head bring the tool into the work.

on the drawing. These dimensions give the diameter of the drill, the diameter and depth of the counterbore, and the location of the holes. The casting is clamped to the drill-press table (Fig. 9) and the rotating drill is brought into the work by a lever operating a rack and pinion in the head of the machine.

The cutting is done by two ground lips on the end of the drill (see Fig. 20). Drilling can also be done in a lathe, the work revolving while the drill is held in and moved by the tailstock. In Fig. 10 the drill has been replaced by a counterboring tool (Fig. 20) of which the diameter is the size specified on the drawing and which

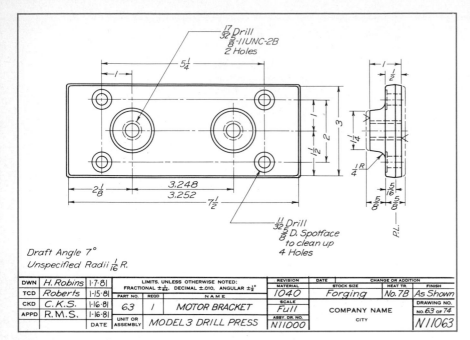

Draft Angle 7°
Unspecified Radii $\frac{1}{16}$ R.

$\frac{17}{32}$ Drill
$\frac{5}{8}$-11UNC-2B
2 Holes

$\frac{11}{32}$ Drill
$\frac{5}{8}$ D. Spotface
to clean up
4 Holes

DWN	H. Robins	1·7·81	LIMITS, UNLESS OTHERWISE NOTED: FRACTIONAL ±$\frac{1}{64}$, DECIMAL ±.010, ANGULAR ±$\frac{1}{2}$°				REVISION	DATE	CHANGE OR ADDITION		
TCD	Roberts	1-15-81					MATERIAL	STOCK SIZE		HEAT TR.	FINISH
			PART NO.	REQD	NAME		1040	Forging		No. 7B	As Shown
CKD	C.K.S.	1-16-81	63	1	MOTOR BRACKET		SCALE		COMPANY NAME		DRAWING NO.
APPD	R.M.S.	1-16-81					Full				NO. 63 OF 74
		DATE	UNIT OR ASSEMBLY		MODEL 3 DRILL PRESS		ASSY. DR. NO. N11000		CITY		N11063

FIG. 11 A working drawing of a forged part.

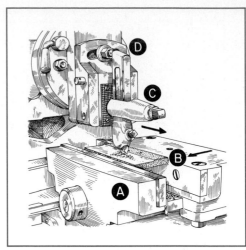

FIG. 12 Shaping. The vise (A) holds the part (B). The tool in the holder (C) is forced across the work surface by the ram (D) in the forward-stroke direction shown. The tool lifts on the return stroke by the pivot head on the ram. To make successive cuts, the table holding (A) and (B) moves in the direction shown.

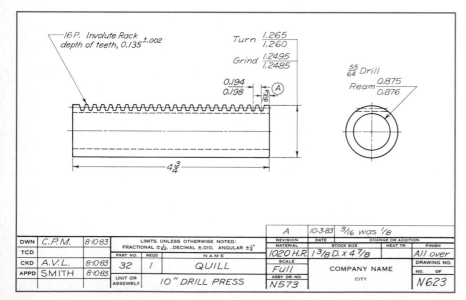

16 P. Involute Rack
depth of teeth, 0.135 ±.002

Turn $\frac{1.265}{1.260}$

Grind $\frac{1.2495}{1.2485}$

$\frac{0.194}{0.198}$ Ⓐ

$\frac{55}{64}$ Drill
Ream $\frac{0.875}{0.876}$

$4\frac{3}{4}$

DWN	C.P.M.	8-10-83	LIMITS, UNLESS OTHERWISE NOTED: FRACTIONAL ±$\frac{1}{64}$, DECIMAL ±.010, ANGULAR ±$\frac{1}{2}$°			A	10-3-83	$\frac{3}{16}$ was $\frac{1}{8}$		
TCD						REVISION	DATE	CHANGE OR ADDITION		
			PART NO.	REQD	NAME	MATERIAL	STOCK SIZE		HEAT TR.	FINISH
CKD	A.V.L.	8-10-83	32	1	QUILL	1020 H.R.	1$\frac{3}{8}$ D. x 4$\frac{7}{8}$			All over
APPD	SMITH	8-10-83				SCALE Full		COMPANY NAME		DRAWING NO.
		UNIT OR ASSEMBLY			10" DRILL PRESS	ASSY. DR. NO. N573		CITY		NO. OF N623

FIG. 13 Working drawing of a part machined from stock.

has a cylindrical pilot on the end to fit into the drilled hole, thus ensuring concentricity. This tool is fed in to the depth shown on the drawing.

C. The Shaper and the Planer The drop forging of Fig. 11 requires machining on the base and boss surfaces.

Flat surfaces of this type are machined on a shaper or a planer. In this case the shaper (Fig. 12) is used because of the relatively small size of the part. The tool is held in a ram that moves back and forth across the work, taking a cut at each pass forward. Between the cuts, the table moves laterally so that closely spaced parallel cuts are made until the surface is completely machined.

The planer differs from the shaper in that its bed, carrying the work, moves back and forth under a stationary tool. It is generally used for a larger and heavier type of work than that done on a shaper.

7. PARTS MACHINED FROM STANDARD STOCK

The shape of a part will often lend itself to machining directly from standard stock, such as bars, rods, tubing, plates, and blocks, or from extrusions and rolled shapes, such as angles and channels. Hot-rolled (HR) and cold-rolled (CR) steel are common materials.

Parts produced from stock are usually finished on all surfaces, and the general note "Finish all over" on the drawing eliminates the use of finish marks. Figure 13 is the drawing of a part to be made from bar stock. Note the specification of material, stock size, etc., in the title.

Figure 14 illustrates the shop interpretation in Fig. 13 and indicates the order of machining operations to be performed.

8. ADDITIONAL MACHINE TOOLS

Let us look at additional tools used in the production of the part just seen in Fig. 13. Figure 14 gives the sequence of operations needed plus the machines needed. More detail is now offered on the turret lathe, milling machine, grinder, and broach.

A. The Turret Lathe The *quill* of Fig. 13, produced in quantity, may be made on a turret lathe, Fig. 15, except

for the rack teeth and the outside-diameter grinding. The stock is held in the collet chuck of the lathe. First the end surface is faced, and then the cylindrical surface (OD) is turned. The work piece is then ready for drilling and reaming. The turret holds the various tools and swings them around into position as needed. A center drill starts a small hole to align the larger drill, and then the drill and reamer are brought successively into position. The drill provides a hole slightly undersize, and

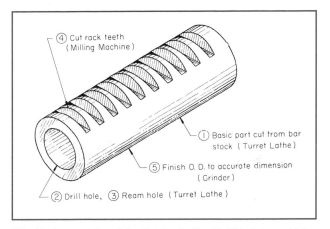

FIG. 14 Interpretation of the drawing in Fig. 13. This lists machining operations.

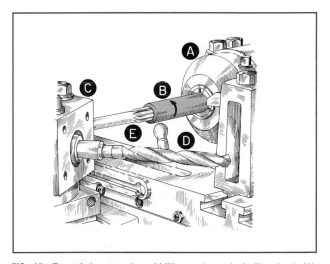

FIG. 15 Turret-lathe operations (drilling and reaming). The chuck (A) holds, indexes (for length), and rotates the stock (B) in the direction shown. The turret (C) swings the drill (D) and reamer (E) successively into position.

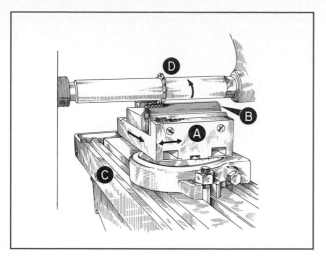

FIG. 16 Milling. The vise (A) holds the part (B) to the table (C), which moves laterally to index cuts and longitudinally, in the direction shown, for individual cuts. The milling cutter (D) revolves in the direction shown.

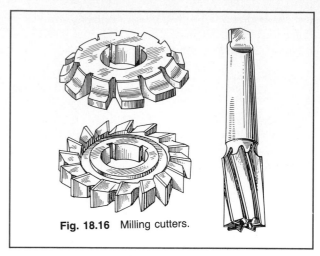

Fig. 18.16   Milling cutters.

FIG. 17 Milling cutters.

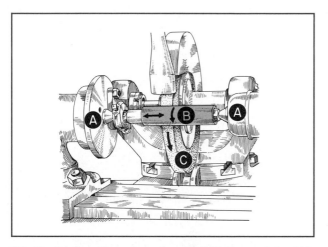

FIG. 18 Grinding. The cone centers (A) hold the mandrel, which mounts and rotates the part (B) in the direction shown. The wheel (C) moves laterally to traverse the entire surface of (B).

then the reamer, cutting with its fluted sides, cleans out the hole and gives a smooth surface finished to a size within the dimensional limits on the drawing. Figure 15 shows the turret indexed so that the drill is out of the way and the reamer is in position. At the right is seen the cutoff tool ready to cut the piece to the length shown on the drawing.

B. The Milling Machine   The dimensions of the rack teeth (Fig. 13) give the depth and spacing of the cuts and

the specifications for the cutter to be used. This type of work may be done on a milling machine. The work piece is held in a vise and moved horizontally into the rotating milling cutter, which, in profile, is the shape of the space between the teeth (Fig. 16). The cuts are spaced by moving the table of the machine to correspond with the distance shown on the drawing. Note that this operation is a forming process, in that the shape depends upon the contour of the cutter. With several cutters mounted together ("gang milling"), a number of teeth can be cut at the same time.

There are many types of milling cutters made to cut on their peripheries, their sides, or their ends, for forming flat, curved, or special surfaces. Three milling cutters are shown on Fig. 17.

C. The Grinder   The general purpose of grinding is to make a smoother and more accurate surface than can be obtained by turning, planing, milling, etc. The limit dimensions for the outside diameter of the quill (Fig. 13) indicate a grinding operation on a cylindrical grinder (Fig. 18). The abrasive wheel rotates at high speed, while the work piece, mounted on a mandrel between conic centers, rotates slowly in the opposite direction. The wheel usually moves laterally to cover the surface of the work piece. The machine for flat surfaces, called a *surface grinder*, holds the work piece on a flat table moving back and forth under the abrasive wheel. The table "indexes" laterally after each pass under the work.

D. Surface Finishing Lapping, honing, and super-finishing are three methods of producing smooth surfaces after grinding. All three methods use fine abrasives (1) powdered and and carried in oil on a piece of formed soft metal (lapping) or (2) in the form of fine-grained compact stones (honing and superfinishing) to rub against the surface to be finished and reduce scratches and waviness.

E. The Broach A broach is a long, tapered bar with a series of cutting edges (teeth), each successively removing a small amount of material until the last edge forms the shape desired. For flat or irregular external surfaces, the broach and work piece are held by the broaching machine, and the broach is passed across the surface of the work piece. For internal surfaces, the broach is pulled or pushed through a hole to give the finished size and shape.

Some machined shapes can be more economically produced by broaching than by any other method. Figure 19 shows several forms of broaches and the shapes they produce.

9. SMALL TOOLS FOR MODIFYING HOLES

The shop uses a variety of small tools, both powered machines and hand tools. Figure 20 shows a *twist drill*, available in a variety of sizes (numbered, lettered, fractional and metric) for producing holes in almost any material; a *reamer*, used to enlarge and smooth a previously existing hole and to give greater accuracy than is

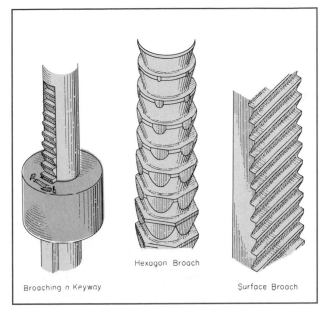

Hexagon Broach

Broaching a Keyway

Surface Broach

FIG. 19 Broaches. Each tooth takes its small cut as the broach is forced across or through the work piece.

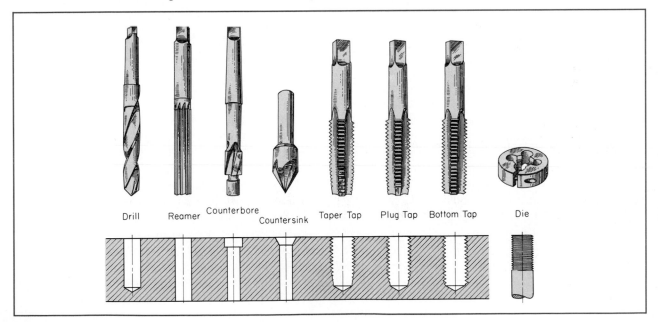

Drill · Reamer · Counterbore · Countersink · Taper Tap · Plug Tap · Bottom Tap · Die

FIG. 20 Small tools.

possible by drilling alone; and a *counterbore* and a *countersink*, both used to enlarge and alter the end of a hole (usually for screwheads). A *spot-facing tool* is similar to a counterbore. *Taper*, *plug*, and *bottoming taps* for cutting the thread of a tapped hole and a die for threading a rod or shaft are also shown.

10. HEAT TREATMENT OF PARTS

"Heat treatment" is a general term applied to the processing of metals by heat and chemicals to change the physical properties of the material.

The glossary of shop terms (Appendix 1) gives definitions of such heat-treatment processes as annealing, carburizing, case hardening, hardening, normalizing, and tempering.

The specification of heat treatment may be given on the drawing in several ways: (1) by giving a general note listing the steps, temperatures, and baths to be used, (2) by listing a standard heat-treatment number (SAE or company standard) in the space provided in the title block, (3) by giving the Brinell or Rockwell hardness number to be attained, or (4) by noting the tensile strength, in pounds per square inch, to be attained through heat treatment. Figure 11 illustrates the need for heat treatment within the title block where a heat treatment no. 7B is requested.

11. REQUIREMENTS FOR DRAWINGS

The machines used to produce parts have been illustrated in their most basic formats. Realize that these machines and their operations can be combined into multiple-machine setups where the additional costs are overcome by the increased productivity of multiple machining. Regardless of the machining sequencing desired, specific drawing requirements are necessary. For any machining operation, the dimensions to be a result of the operation must appear on a working drawing.

The drawing must also give the complete shape and size description of the part and any machining operations needed. Dimensions must be clear, without requiring addition or subtraction of dimensions on the drawing.

Two general practices are followed: (1) the "single-drawing" system, in which only one drawing, showing the finished part, is made to be used by all the shops involved in producing the part; and (2) the "multiple-drawing" system, in which different drawings are prepared, one for each shop, giving only the information required by the shop for which the drawing is made.

The second practice is recommended, because the drawings are easier to dimension without ambiguity, somewhat simpler and more direct, and therefore easier for each shop to use.

12. WORKING DRAWINGS NEEDED FOR PRODUCTION

A working drawing is any drawing used to give information for the manufacture or construction of a machine or the erection of a structure. Complete knowledge for the production of a machine or structure is given by a *set* of working drawings conveying all the facts fully and explicitly so that further instructions are not required.

The description given by the set of drawings will include:

1. The graphic representation of the shape of each part (shape description)
2. The dimensions of each part (size description)
3. Explanatory notes, general and specific, on the individual drawings, giving the specifications of material, heat treatment, and finish
4. A descriptive title on each drawing
5. A description of the relationship of each part to the others (assembly)
6. A parts list or bill of material

A set of drawings will include, in general, two classes of drawings: *detail drawings*, giving the information included in items 1 to 4; and an *assembly drawing*, giving the location and relationship of the parts, item 5.

Recall that working drawings used for purposes of production are preceded by design layouts. (Refer to Chap. 12.3, phase 6 of the design procedure.) Design layouts are prototype assembly drawings from which working drawings evolve. Figure 21 is an example of a fairly complicated design layout.

If a machine is to be produced in quantity, "operation" or "job" sheets will be prepared describing the sep-

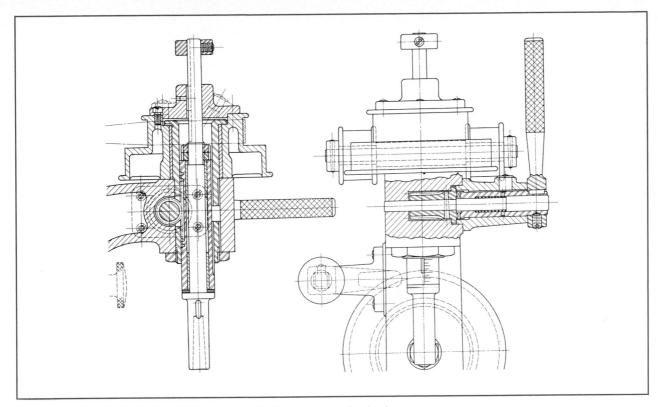

FIG. 21 A portion of a design layout. Notes and specifications accompany the drawing.

arate manufacturing steps required and indicating the use and kinds of special tools. The tool-design group, working from the detail drawings and operation sheets, designs and makes the drawings for the special tools needed.

13. SPECIFIC TYPES OF WORKING DRAWINGS

We will now turn our attention to particular types of drawings that are used in the *production* of parts. Design-layout drawings are part of the initial *design* procedure and were discussed in Chap. 12.

    A. Detail Drawings A *detail* drawing is the drawing of a single piece, giving a complete and exact description of its form, dimensions, and construction. A successful detail drawing will tell the worker *simply* and *directly* the shape, size, material, and finish of a part; what shop

operations are necessary; what limits of accuracy must be observed; etc. It should be so exact in description that if followed, a satisfactory part will result. Figures 22 to 24 illustrate commercial detail drawings.

    Detail drawings are made by studying carefully the initial design-layout drawing. Use is made of the scale of the design layout, dimensions that may be given, and all notes provided. Detail drawings include approved standards for the specific company involved with respect to lettering style, dimensioning techniques, position of notes, etc.

    When a working drawing is finished, it must be *checked* by an experienced person, who becomes responsible for any errors. This is the final "proofreading" and cannot be done by the one who has made the drawing nearly so well as by another person. All notes, computations, and checking layouts should be preserved for future reference.

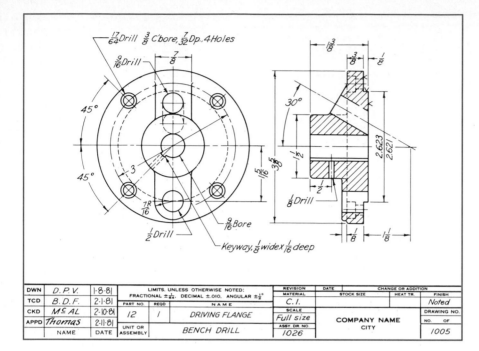

FIG. 22 A detail drawing. This gives complete information for the production of a single part.

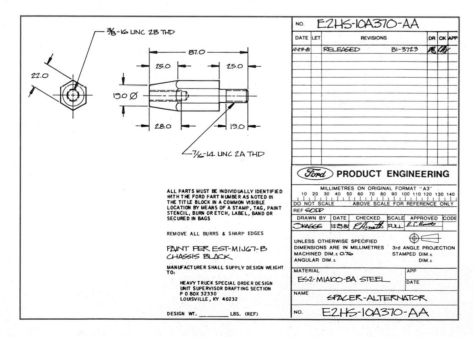

FIG. 23 A detail drawing. Metric dimensions are used for all features except the threads in this particular application. (*Courtesy Kentucky Truck Plant, Ford Motor Company.*)

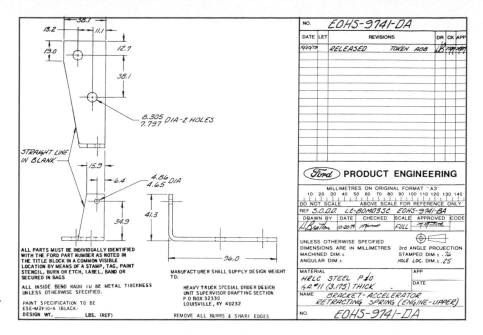

FIG. 24 A detail drawing. Note the thorough use of engineering notes and specifications. *(Courtesy Kentucky Truck Plant, Ford Motor Company.)*

B. Assembly Drawings A *complete assembly drawing* is a drawing of the product put together, showing all parts in their operational positions.

The finished separate pieces come to the assembly department to be put together according to the assembly drawings. Sometimes it is desirable or necessary to perform some small machining operation during assembly, often drilling, reaming, or hand finishing. In such cases the assembly drawing should carry a note explaining the required operation and give dimensions for the alignment or location of the pieces.

The assembly drawing can be produced by several methods, one being to simply trace from the design layout. This method lacks the accuracy in checking obtained if a person draws the assembly from dimensions of the detail drawings. This second method by nature is very time-consuming. Where feasible, computer-aided drafting (CAD) can be a tremendous timesaver when an assembly drawing is needed. This third and quite modern method is illustrated in Fig. 25. The method using CAD is beneficial when the high initial costs can be recovered with heavy usage of the CAD equipment.

The assembly drawing sometimes gives the overall dimensions and the distances between centers or from

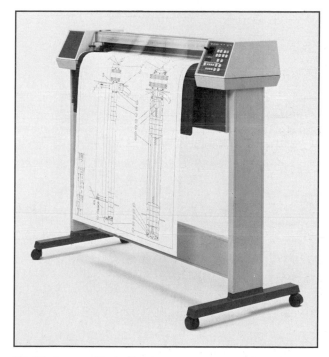

FIG. 25 An assembly drawing done on an automated drafting plotter. Such equipment permits rapid and exact drafting unobtainable by manual methods. *(Courtesy Hewlett-Packard Co.)*

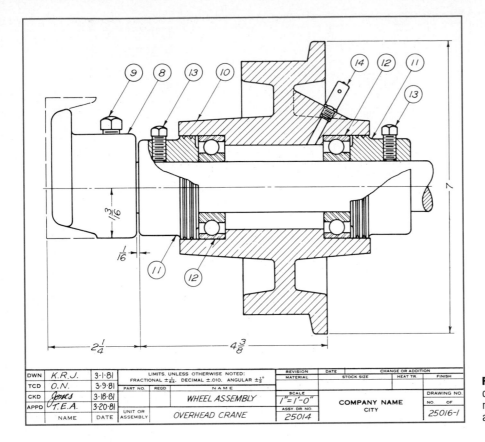

DWN	K.R.J.	3·1·81	LIMITS, UNLESS OTHERWISE NOTED: FRACTIONAL ±1/64. DECIMAL ±.010. ANGULAR ±1/2°		REVISION	DATE	CHANGE OR ADDITION		
TCD	O.N.	3·9·81			MATERIAL	STOCK SIZE	HEAT TR.	FINISH	
CKD	Jones	3·18·81	PART NO.	REQD	NAME				
APPD	T.E.A.	3·20·81			WHEEL ASSEMBLY	SCALE 1"=1'-0"	COMPANY NAME	DRAWING NO.	
	NAME	DATE	UNIT OR ASSEMBLY		OVERHEAD CRANE	ASSY. DR. NO. 25014	CITY	NO. OF 25016-1	

FIG. 26 A unit assembly drawing. For complex machines, a number of units may be used in place of one complete assembly drawing.

part to part of the different pieces, thus fixing the relation of the parts to each other and aiding in erecting the machine. However, many assembly drawings need no dimensions. An assembly drawing should not be overloaded with detail, particularly hidden detail.

Assembly drawings usually have reference letters or numbers designating the different parts. These "piece numbers," sometimes enclosed in circles with a leader pointing to the piece (Fig. 26), are used in connection with the details and bill of material (parts list).

A *unit assembly drawing* or subassembly, Fig. 26, is a drawing of a related group of parts used to show the assembly of complicated machinery where it would be practically impossible to show all the features on one drawing. Thus, in the drawing of a lathe, there would be included unit assemblies of groups of parts such as the headstock, tailstock, gearbox, etc.

An *outline assembly drawing* is used to give a general idea of the exterior shape of a machine or structure, and contains only the principal dimensions (Fig. 27). When it is made for catalogs or other illustrative purposes, dimensions are often omitted. Outline assembly drawings are frequently used to give the information required for the installation or erection of equipment and are then called *installation drawings*.

An *assembly working drawing* gives complete information for producing a machine or structure on one drawing. This is done by providing adequate orthographic views together with dimensions, notes, and a descriptive title. The figure for Prob. 30 is an example.

A *diagram drawing* is an assembly showing, symbolically, the erection or installation of equipment. Erection and piping and wiring diagrams are examples. Diagram drawings are often made in pictorial form.

The *bill of material* is a tabulated list placed either on the assembly drawing or on a separate sheet. Figure 30 in the problems shows the list on the assembly drawing. The list gives the part number, name, quantity, material, sometimes the stock size of raw material, detail drawing numbers, weight of each piece, etc. A final column is usually left for remarks. The term "bill of material" is ordinarily used in structural and architectural drawing. The term "parts list" applies more accurately in machine-drawing practice. In general, the parts are listed in the order of their importance, with the larger parts first and the standard parts such as screws, pins, etc., at the end.

C. Tabular Drawings A *tabular drawing*, either assembly or detail, is one on which the dimension values are replaced by reference letters, and an accompanying table lists the corresponding dimensions for a series of sizes of the machine or part, thus making one drawing serve for the range covered. Some companies manufacturing parts in a variety of sizes use this tabular system of size description, but a serious danger with it is the possibility of misreading the table. Figure 27 shows a tabular assembly drawing.

D. Standardized Drawings To avoid the difficulties experienced with tabular drawings, some companies make a "standard drawing," complete except for the actual figured dimensions. This drawing is reproduced by offset printing or black-and-white reproduction on vellum paper, and the reproductions are dimensioned separately for the various sizes of parts. This method gives a separate complete drawing for each size of part, and when a new size is needed, the drawing is easily and quickly made. Figure 28 shows a standard drawing, and Fig. 29 shows the drawing filled in, making a completed working drawing.

E. Set of Drawings A *complete set* of working drawings consists of detail sheets and assembly sheets, the former giving all necessary information for the manufacture of each of the individual parts that cannot be purchased and the latter showing the parts assembled as a finished unit or machine. The set includes the bill of material or parts list and may also contain special drawings for the purchaser, such as foundation plans or oiling diagrams.

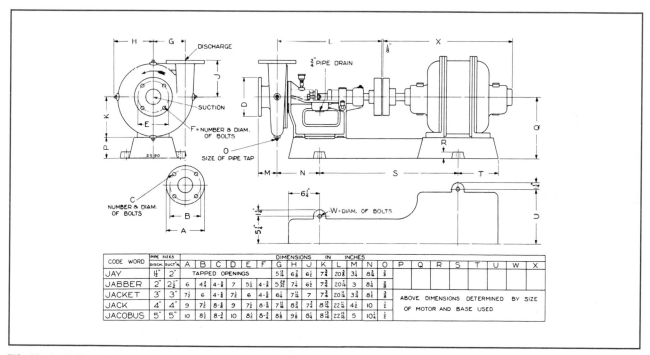

FIG. 27 An outline assembly drawing (tabular). Tabulation gives the dimensions for different-sized units.

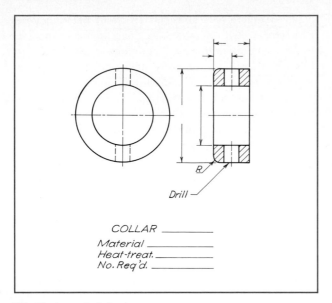

FIG. 28 A standard drawing.

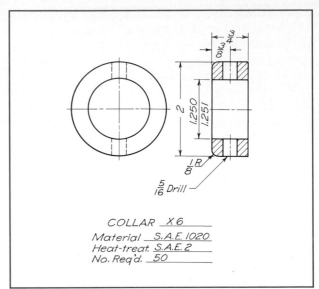

FIG. 29 A filled-in standard drawing.

14. USE OF STANDARD PARTS

Purchased or company standard parts are specified by name and size or by number and, consequently, need not be detailed. All standard parts, such as bolts, screws, antifriction bearings, etc., are shown on the assembly drawing and are given a part number. The complete specifications for their purchase are given in the parts list.

Sometimes, however, a part is made by *altering* a standard or previously produced part. In this case a detail drawing is made, showing and specifying the original part with changes and dimensions for the alteration.

15. TITLE BLOCKS

All drawings used in production need title blocks. The title of a working drawing is usually placed in the lower right-hand corner of the sheet, the size of the space varying with the amount of information to be given. The spacing and arrangements are designed to provide the information most helpful in a particular line of work.

In general, the title of a machine drawing should contain the following information:

1. Name of company and its location

2. Name of machine or unit

3. Name of part (if a detail drawing)

4. Drawing number

5. Part number (if a detail drawing)

6. Number of parts required (for each assembly)

7. Scale

8. Assembly-drawing number (given on a detail drawing to identify the part in assembly)

9. Drafting-room record: names or initials of drafter, tracer, checker, approving authority, each with date

10. Material

To these, depending upon the need for the information, may be added:

11. Stock size

12. Heat treatment

13. Finish

14. Name of purchaser, if special machine

15. Drawing "supersedes" and "superseded by"

Form of Title Every drafting room has its own standard form for titles. In large offices the blank form is often printed in type on the paper or vellum to be used. Figure 30 is a characteristic example.

A form of title that is used to some extent is the *record strip*, a strip marked off across the lower part or right end of the sheet, containing the information required in the title and space for the recording of orders, revisions, changes, etc., that should be noted, with the date, as they occur. Figure 31 illustrates one form.

16. REVISIONS TO DRAWINGS

Once a drawing has been reproduced and released for production, any changes should be recorded on the drawing and new prints issued. If the changes are extensive, the drawing may be made *obsolete* and a new drawing made that *supersedes* the old drawing. Many drawing rooms have "change-record" blocks printed in conjunction with the title, where minor changes are recorded (Fig. 23). The change is identified in the record and on the face of the drawing by a letter.

New designs may be changed so often that the alterations cannot be made fast enough to reach the shop when needed. In this case sketches showing the changes are rapidly made, reproduced, and sent to the shop, where they are fastened to each print of the drawing. These sketches, commonly known as "engineering orders," are later incorporated on the drawing. If computer-aided drafting is being used, output from the computer printer can often be used due to the speed of the process, and so sketches can be eliminated.

17. WORKING SKETCHES

Working sketches can be very useful for a part or assembly that does not need, for some reason, a fully developed set of working drawings. Sometimes a part must be made very quickly, and time doesn't permit a drawing to be made. On the other hand, working sketches can precede design layouts at the stage where the designer is thrashing out ideas on paper.

Working sketches can be divided into two general classes: (1) those made before the structure is built and (2) those made after the structure is built.

In the first class are included the sketches made in connection with the designing of the structure. These can be classified as (*a*) *idea sketches*, used in studying and developing the arrangement and proportion of parts; (*b*) *computation sketches*, made in connection with the

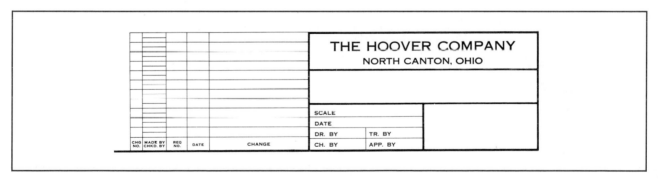

FIG. 30 A printed title form.

FIG. 31 A strip title. This type extends across one side of the drawing.

figured calculations for motion and strength; (c) *executive sketches*, made by the chief engineer, inventor, or consulting engineer to give instructions for special arrangements or ideas that must be embodied in the design; (d) *design sketches*, used in working up the ideas into suitable form so that the design drawing can be started; and (e) *detail sketches*, made as substitutes for detail drawings.

The second class includes (a) *detail sketches*, drawn from existing models or parts, with complete notes and dimensions, from which duplicate parts can be constructed directly or from which working drawings can be made (Fig. 32); (b) *assembly sketches*, made from an assembled machine to show the relative positions of the various parts, with center and location dimensions, or sometimes for a simple machine, with complete dimensions and specifications; and (c) *outline or diagrammatic sketches*, generally made for the purpose of location: sometimes, for example, to give the size and location of pulleys and shafting, piping, or wiring, that is, information for use in connection with the setting up of machinery.

Note that a sketch contains dimensions just like any drawing. Add all remarks and notes that are of value.

Include a title. Remember to *date* a sketch so that valid documentation can be complete.

18. DRAWINGS FOR SPECIFIC MANUFACTURING PROCESSES

We now turn our attention to drawings used in certain processes of manufacture. The drawings to be illustrated are representative of the more widely used processes in parts production. Realize that in every case the drafter strives to give clear and complete information that will be immediately helpful to the person doing the particular process of manufacture.

A. Drawings for a Sand Casting The dimensions required for sand castings can be classified as those used by the patternmaker and those used by the machinist. Since a cast part has two distinct phases in its manufacture, we will discuss a particular case first, one for the patternmaker (Fig. 33) and one for the machinist (Fig. 34).

The *casting drawing* gives the shape of the unmachined casting and carries dimensions for the patternmaker only. Shape breakdown will show that each geometric shape has been dimensioned for size and then

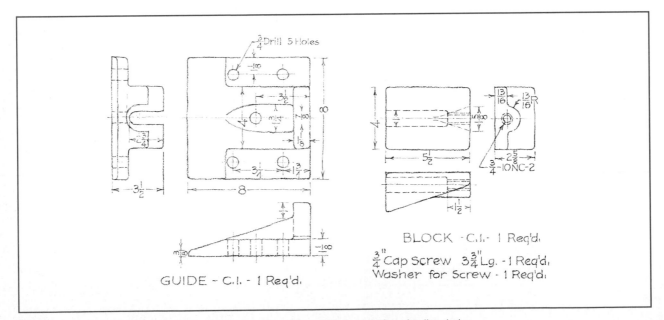

FIG. 32 A detail sketch. This is a sketch of two parts, the guide and block, parts for a leveling device.

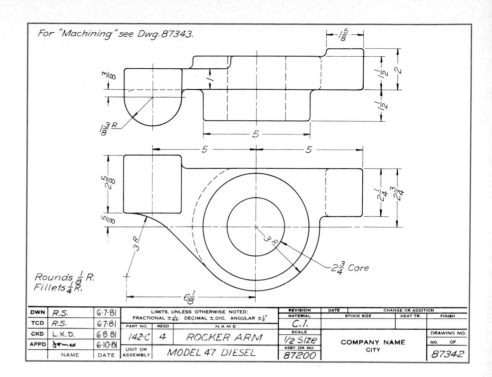

FIG. 33 Dimensioning an unmachined casting.

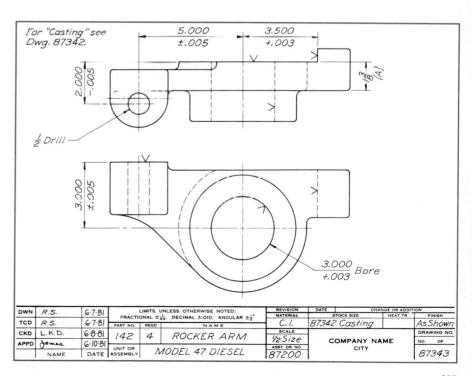

FIG. 34 Dimensions for machining a casting.

positioned, thus providing dimensions easily usable by the worker.

Note also that several of the dimensions have been selected to agree with required functional dimensions of the machined part, even though the dimensions employed have been selected so as to be directly usable by the patternmaker; they also achieve the *main objective*, which is to state the sizes that the unmachined casting must fulfill when produced.

The *machining drawing* shows only the dimensions required by the machinist. These are almost all functional dimensions and have been selected to correlate with mating parts. It is important to note that a starting point must be established in each of the three principal directions for machining the casting. In this case a starting point is provided by (1) the coincidence of the centerlines of the large hole and cylinder (positioning in two directions) and (2) dimension A to position the machined surface on the back, from which is positioned the drilled hole. Dimension A is a common fraction carrying the broad tolerance of $\pm\frac{1}{64}$ in, because there is no functional reason for working to greater precision.

Figure 35 is a drawing of the same part used in Figs. 33 and 34, but with the casting drawing dispensed with and the patternmaker's dimensions incorporated in the drawing of the finished part. In combining the two drawings, we have eliminated some dimensions, because the inclusion of all the dimensions of both drawings would result in overdimensioning.

Therefore the patternmaker must make use of certain machining dimensions in his or her work. In working from the drawing in Fig. 35, the patternmaker provides for machining allowance, being guided by the finish marks. In the drawing in Fig. 33, the engineering department provides for the machining allowance by showing and dimensioning the rough casting oversize where necessary for machining, and no finish marks are used.

B. Shops Using Drawings of a Sand Casting

■ **1. The Pattern Shop** The drawing is first used by the patternmaker, who makes a pattern, or "model," of the part in wood. From this, if a large quantity of castings is required, a metal pattern, often of aluminum, is made. The patternmaker provides for the shrinkage of the cast-

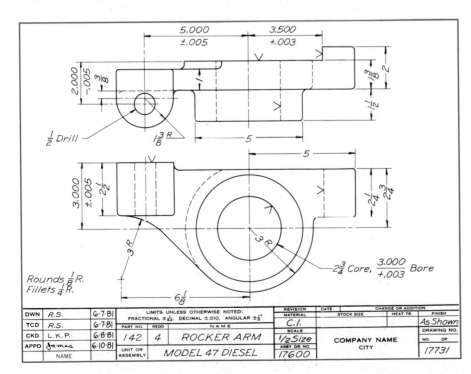

FIG. 35 All dimensions for a casting.

ing by making the pattern oversize, and also provides additional metal (machining allowance) for the machined surfaces, indicated on the drawing by (1) finish marks, (2) dimensions indicating a degree of precision attainable only by machining, or (3) notes giving machining operations. The patternmaker also provides the "draft," or slight taper, not shown on the drawing, so that the pattern can be withdrawn easily from the sand. A "core box," for making sand cores for the hollow parts of the casting, is also made in the pattern shop. A knowledge of patternmaking is a great aid in dimensioning, since almost all dimensions are used by the patternmaker, while only those for finished features are used by the machine shop.

■ **2. The Foundry** The pattern and core box or boxes are sent to the foundry, and sand molds are made so that molten metal can be poured into the molds and allowed to cool, forming the completed rough casting. Figure 36 is a cross section of a two-part mold, showing the space left by the pattern and the core in place. Only in occasional instances do the foundry workers turn to the drawing for assistance, inasmuch as their job is simply to reproduce the pattern in metal.

Permanent molds, made of cast iron coated on the molding surfaces with a refractory material, are sometimes an advantage in that they can be used over and over again, thus saving the time to make an individual sand mold for each casting. This method is usually limited to small castings.

Die castings are made by forcing molten metal under pressure into a steel die mounted in a special die-casting machine. Alloys with a low melting point are used in order to avoid damaging the die. Because of the accuracy possible in making a die, a fine finish and accurate dimensions of the part can be obtained; thus machining may be unnecessary.

C. Drawings for Forgings Forgings are made by heating metal to make it plastic and then forming it to shape on a power hammer with or without the aid of special steel dies. Large parts are often hammered with dies of generalized all-purpose shape. Smaller parts in quantity may warrant the expense of making special dies. Some small forgings are made with the metal cold.

Drop forgings are the most common and are made in

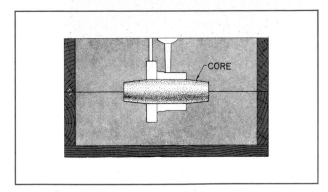

FIG. 36 Cross section of a two-part mold.

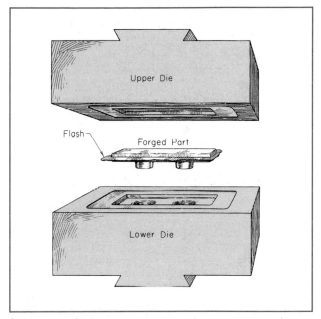

FIG. 37 Drop-forging dies and the forged part. These are the dies for the part shown in Fig. 11.

dies of the kind shown in Fig. 37. The lower die is held on the bed of the drop hammer, and the upper die is raised by the hammer mechanism. The hot metal is placed between the dies, and the upper die is dropped several times, causing the metal to flow into the cavity of the dies. The slight excess of material will form a thin fin, or "flash," surrounding the forging at the parting plane of the dies (Fig. 37). This flash is then removed in

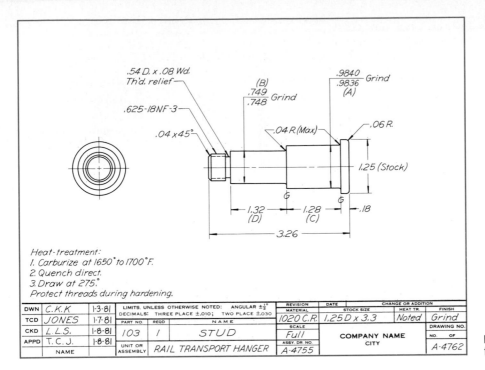

Heat-treatment:
1. Carburize at 1650° to 1700°F.
2. Quench direct.
3. Draw at 275.°
Protect threads during hardening.

DWN	C.K.K	1-3-81	LIMITS, UNLESS OTHERWISE NOTED: ANGULAR ±½° DECIMALS: THREE PLACE ±.010; TWO PLACE ±.030				REVISION	DATE		CHANGE OR ADDITION		
							MATERIAL		STOCK SIZE		HEAT TR.	FINISH
TCD	JONES	1-7-81	PART NO.	REQD	NAME		1020 C.R.	1.25 D x 3.3		Noted	Grind
CKD	L.L.S.	1-8-81	103	1	STUD		SCALE		COMPANY NAME		DRAWING NO.	
APPD	T.C.J.	1-8-81	UNIT OR ASSEMBLY				Full		CITY		NO. OF	
	NAME			RAIL TRANSPORT HANGER			ASSY. DR. NO. A-4755				A-4762	

FIG. 38 Dimensioning a part machined from stock.

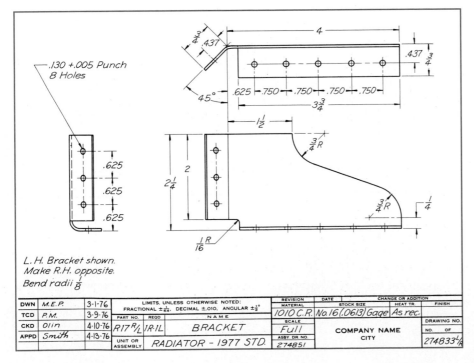

.130 +.005 Punch
8 Holes

L.H. Bracket shown.
Make R.H. opposite.
Bend radii ⅛

DWN	M.E.P.	3-1-76	LIMITS, UNLESS OTHERWISE NOTED: FRACTIONAL ±1/64. DECIMAL ±.010. ANGULAR ±½°				REVISION	DATE		CHANGE OR ADDITION		
							MATERIAL		STOCK SIZE		HEAT TR.	FINISH
TCD	P.M.	3-9-76	PART NO.	REQD	NAME		1010 C.R.	No.16 (.0613) Gage		As rec.	
CKD	Olin	4-10-76	R17 R/L	1R-1L	BRACKET		SCALE		COMPANY NAME		DRAWING NO.	
APPD	Smith	4-13-76	UNIT OR ASSEMBLY				Full		CITY		NO. OF	
				RADIATOR - 1977 STD.			ASSY. DR. NO. 274851				274833½	

FIG. 39 Dimensioning a sheet-metal part.

a "trimming" die made for the purpose. Considerable draft must be provided for release of the forging from the dies.

Forging drawings are prepared according to the multiple-drawing system, one drawing for the diemaker and one for the machinist; or the single-drawing system, one drawing for both (Fig. 11). In either case the parting line and draft should be shown and the amount of draft specified. On the single drawing (Fig. 11), the shape of the finished forging is shown in full outline, and the machining allowance is indicated by "alternate-position" lines, thus completing the shape of the rough forging. This single drawing combines two drawings in one, with complete dimensions for both diemaker and machinist.

D. Drawings for Parts Machined from Stock Figure 38 is a detail drawing of a stud. The stud is produced by machining on a lathe. Cold-rolled steel stock, 1¼ in in diameter, is used. The stock diameter is the same as the large end of the stud, thus eliminating one machining operation.

Shape breakdown of the part results in a series of cylinders each requiring two dimensions—diameter and length. Many of the dimensions selected are those that best suit shop requirements. Note that the thread length and overall dimension cannot both be given, or the part would be overdimensioned.

E. Drawings for Sheet-Metal Parts Parts to be made of thin materials are usually drawn showing the part in its finished form, as in Fig. 39. The template maker first uses the drawing to lay out a flat pattern of the part. If only a few parts are to be made, this template will serve as a pattern for cutting the blanks. Then the part will be formed and completed by hand. If a large number of parts is to be made, the diemaker will use the template and drawing in making up the necessary dies for blanking, punching, and forming. The work of both template maker and diemaker is simplified by giving the dimensions to the same side of the material inside or outside, whichever is more important from the functional standpoint, as shown in Fig. 39. Dimensions to rounded edges (bends) are given to the theoretical sharp edges, which are called *mold lines*. The thickness of the material is given in the "stock" block of the title strip. Note in the figure that the holes are positioned in groups (because of functional requirements) and that important functional dimensions are three-place decimals.

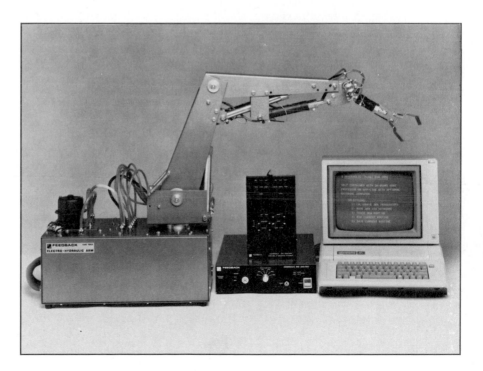

FIG. 40 This robot can handle five-pound loads with its five-axis revolute arm. It is set up for instructional use. *(Courtesy Feedback, Inc.)*

19. NUMERICALLY CONTROLLED MACHINING

In the foregoing discussion of drawings used in certain specific manufacturing processes, it was assumed that the worker set up the tooling manually. For example, a part to be machined on the lathe was inserted into the lathe, and the cutting operations were set up by hand.

Now we will briefly consider machining operations that are not manually performed. We will see how an automated machining process can affect the way one dimensions a drawing for the part to be produced.

A. Background Computer-aided manufacturing (CAM) relieves a person of much of the need to interact manually with tooling. Once tooling has been set up, machining can proceed automatically, often untouched by human hands, thus reducing the risk of human error.

Even assembly processes can be computer-controlled. The use of robots is now widespread in industry for per-

forming tasks such as welding automobile bodies. Smaller robots are useful in areas such as the assembling of electronic circuits. Figure 40 shows a small robot designed for demonstrations.

One part of CAM is the process called *numerical control* (NC), in which coded instructions to a machine automatically guide the machining sequence. Most coded instructions for NC are generated through computers that are appropriately programmed for a particular machine tool.

A simplified schematic diagram for NC is seen in Fig. 41. The diagram shown is that of a closed-loop system, in which a feedback loop is provided. Basically, the overall process is that coded data generated by the computer are fed to the machine-control unit, which decodes the data into a format that instructs the actual machine tool to make specific cuts. The feedback loop senses any errors that may creep into the operation and tells the machine-control unit to adjust the operation so that the machine tool will cut properly.

Computer language has been developed to control machines. Beginning in 1962, a language known as APT (automatically programmed tools) has been updated and refined to become one of the most common of the computer languages for machine tools. Professional groups such as the Society of Manufacturing Engineers (SME) and the Numerical Control Society (NCS) are continually at work to incorporate the latest technology into standards that aid both computer programmers and manufacturing engineers.

B. Spatial Control Modes Machine tools operate in three-dimensional space. We have become accustomed to expressing information in three-dimensional terms throughout our studies. Figure 42 shows a conventional representation of the three axes in space, *x*, *y*, and *z*. Appropriate plus and minus signs have been applied to each axis. Note also that three rotational modes are available (*a*, *b*, and *c*) with the plus rotational modes given. (Minus would be opposite.) An engineer may use any or all of the three directional axes and three rotational modes for programming the desired motion of a machine tool.

Not all machining needs a three-dimensional format. Only two dimensions, such as the *xy* axis format of Fig. 42, would be needed in the case of milling a flat horizontal plate. Adding the *z*-axis control, however, makes

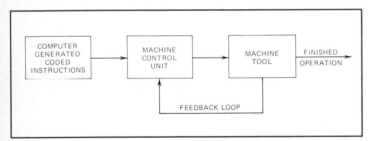

FIG. 41 Schematic diagram for an NC system.

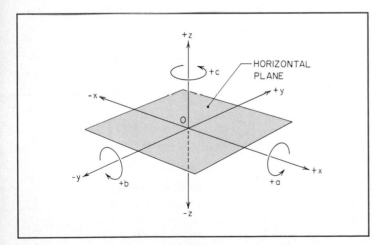

FIG. 42 Axial and rotational modes available for NC programming.

a system more versatile. Rotational modes are very useful when machining complex shapes such as propellers.

C. Control-Path Systems Machine tools can be programmed to move in three paths, depending on the type of NC system used. The three control-path systems are point to point, positioning/straight cut, and contouring.

The *point-to-point* system is often the least costly and is concerned only with properly positioning the tool at desired points. The tool is not cutting as it travels from point to point. Figure 43 shows that any path involving 90° turns is feasible; the exact path is not critical. Examples of machining that might use point-to-point NC control are drilling, spot welding, and riveting.

The *positioning/straight-cut* path seen in Fig. 43 requires a more complex path than does the point-to-point system. However, precise straight lines can be generated that are not possible with point-to-point control. Milling and lathe work usually need positioning/straight-cut control.

Contouring systems, seen in Fig. 43, can cut at any angle. This is because the feed-rate driving motors of the machine are independently controlled along the *x, y,* and *z* axes. Driving motors running at equal speeds give a 45° diagonal path (usually the only case for positioning/straight cut), but differing drive speeds give different angles. Contouring systems, also known as "continuous-path" systems, are the most costly but also the most versatile of the three systems.

20. EFFECT OF NUMERICAL CONTROL ON DRAWINGS

At this point you may be wondering how the use of numerical control affects drawings. Structurally a part should be the same whether it's made by hand setting all tooling or whether the tooling is automatically controlled via NC. The difference in drawings for manual control versus NC control lies chiefly in the way a part is dimensioned. In turn, the way a part is dimensioned is dependent on which of the two-dimensional modes is available for use, the *absolute mode* or the *incremental mode*.

A. Absolute Mode The absolute mode for dimensioning is used in cases where all the dimensions in a given direction are taken from the same baseline. Refer to Fig. 44, where case A shows the absolute mode. All

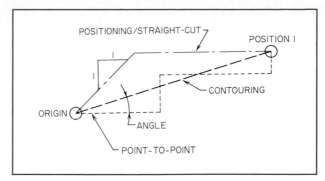

FIG. 43 Three control-path systems for NC.

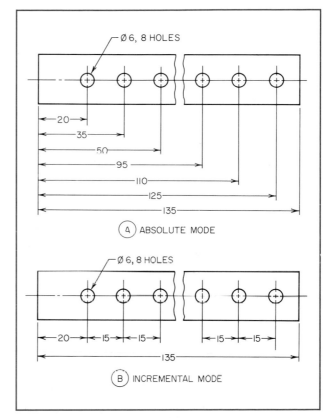

FIG. 44 Dimensioning modes for NC.

dimensions are taken from the left-hand edge. This left-hand position is also called the *zero point*. Absolute mode is used widely in NC. A person always knows where the tool is located because all dimensions relate to the baseline.

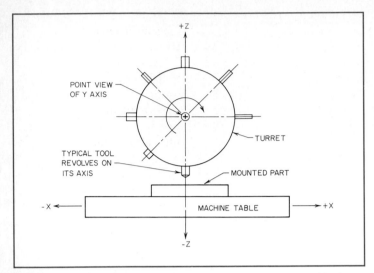

FIG. 45 Basic layout of a turret drill. Seven tools have been mounted in the indexing turret.

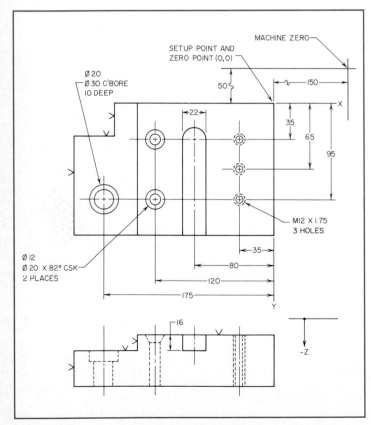

FIG. 46 METRIC. Part dimensioned for NC-controlled turret drill operation.

B. Incremental Mode It is possible to move incrementally from one point to the next, as seen in case B of Fig. 44. The movement of a tool is described in terms of where it was last positioned. A system operating in this way operates in the incremental mode, a mode that can be used in NC as readily as the absolute mode.

C. Example of Dimensioning for NC Operations Let us use a turret drill as an example to highlight several operations that can employ NC. Figure 45 shows the basic operations of a turret drill. The figure shows a frontal view consisting of the x and z axes. (Refer to Fig. 42.) The turret drill features an indexing turret that rotates around the horizontal y axis. In the turret various tools can be mounted, such as drills, counterbores, taps, and face mills.

The turret can be programmed by NC to index a tool upon demand and to perform a machining operation by moving the turret in the $-z$ direction into the mounted part. Most turret drills that are numerically controlled can move along all three axes (x, y, z). Position may be controlled in either the absolute or the incremental mode.

Figure 46 shows a part that has been dimensioned for operations to be done on a numerically controlled turret drill. Note that only operations done by the turret drill are dimensioned. Operations done on another machine, such as the notch in the back left-hand corner, are not dimensioned. Dimensioning is in the absolute mode in this case, and is taken from the zero point indicated. The zero point is at an arbitrary position relative to the permanent machine zero point.

With the part mounted on the machine's table, the table would move in the x and y directions to bring a particular position under the centerline of a specifically indexed tool. Motion in the $-z$ direction would then permit machining to take place.

In using basic tooling, seven turret positions would be required, as follows:

1. Drill to precede tapping operation.
2. Tap for the M12 × 17.5 threaded holes.
3. Drill for the 12-mm-diameter holes.
4. Countersink (CSK).
5. Drill for the 20-mm hole preceding the counterbore (CBORE).

6. Counterbore.

7. End mill of 22-mm diameter for the 16-mm-deep slot.

It would be possible to combine some of the above tooling if high-volume production were desired. A special drill-countersink tool could occupy a single tooling-station position, for example, as could a drill-counterbore tool. In this way seven turret positions could be reduced to five. The extra cost of special tools would have to be weighed against savings in time.

21. DIMENSIONING GUIDELINES FOR NUMERICAL CONTROL

The following guidelines are general but should be helpful when planning dimensioning for parts to be made via numerical control.

1. Think of the part as aligned with principal surfaces parallel to the x, y, z axes of the tool machine selected.

2. Place the zero point (origin) at an easily located and used position.

3. If the NC input system will permit, use mathematical equations to define desired geometric contours, such as a circle, ellipse, parabola.

4. Use decimal dimensioning with bilateral tolerancing (e.g., $\varnothing 20 \pm 0.03$).

5. Give angles by degree and decimal parts of a degree.

6. Use geometric tolerancing where feasible.

7. Use tolerances in keeping with economical design. Do not call out excessively close tolerances even though the machine tool can theoretically produce them.

22. USING NUMERICAL CONTROL TO PRODUCE DRAWINGS

Our discussion of numerical control has been limited to how it affects machining and dimensioning of drawings used to set up such machining. We now touch briefly on another aspect of NC: the generation of drawings themselves using NC.

Information that can be put into digital form can be converted directly into a drawing. The drawing is created by an NC drafting machine which operates a pen over an XY plotter or drum. The actual drawing may be done on paper, vellum, or film. Speed and accuracy of NC-automated drafting machines far exceed that possible by any human drafter.

One interesting example of automated drafting is that done by General Motors Corporation in the creation of the 1982 Camaro automobile. A full-scale clay mock-up of the body shell was scanned by a digitizer in paths around the body. A separate drawing could be generated for the body contour for each path selected. The saving in time alone was considerable compared with previous methods involving human interpretation of data. Figure 47 should be of interest to illustrate automated drafting.

Automated drafting is part of the larger field of computer-aided design (also known as CAD, as is computer-aided drafting). A person may enter data points for a geometric configuration or part, display the part, change the part as desired, and have the part drawn on a plotter. CAD systems vary widely in size, features, and cost. One modestly sized system is shown in Fig. 48. This system incorporates a computer with supporting display, light pen, and dual-disk drive. Also included is a digitizer pad and plotter. Data may be entered as X, Y, and Z coordinate points, and resulting objects can be rotated to any desired viewing angle. Permanent copy on the XY plotter is available in varying colors.

23. REPRODUCTION OF DRAWINGS

A finished working drawing represents a considerable effort on the part of the drafter. Time has been spent, time that translates into dollars spent for salary and overhead. Protection of the original drawing is naturally given high priority. Therefore, copies are made of the original for use by the engineering office, factory, and other authorized users. The original drawing is kept safely stored, except when copies must be made or when revisions must be made to the part or parts shown on the drawing.

Several processes are used to reproduce drawings. We will touch on several of the better known processes: blueprinting, diazo printing, Vandyke printing, xerography, photostating, and microfilming.

A. Blueprinting This once common method is now seldom used because of faster and more convenient

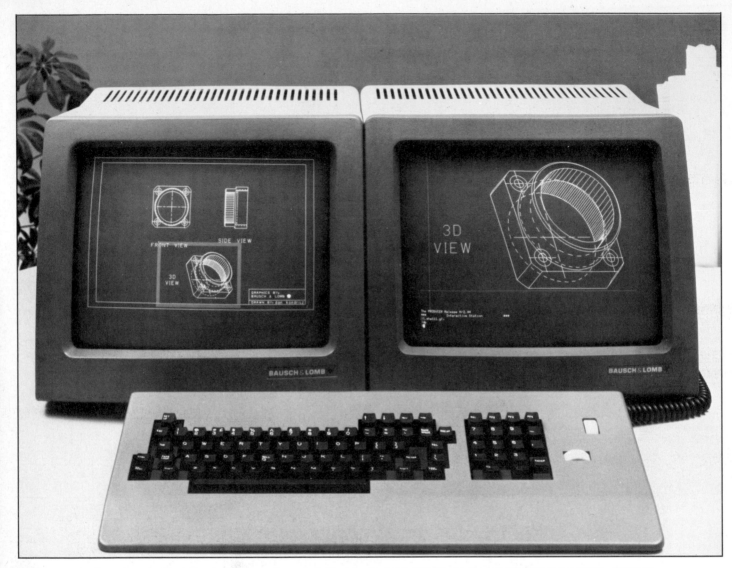

FIG. 47 The drawings shown on this dual raster display can be converted to paper hard copy with the use of a computer-driven pen-plotter. *(Courtesy Interactive Graphics Division, Bausch & Lomb.)*

FIG. 48 Computer-aided design is possible on this integrated system, from data entry to plotted output. *(Houston Instrument.)*

methods now available. However, the term "blueprint" is still found in the language of drawing reproduction, even though the process is seldom used today.

Blueprinting is so named because the process results in white lines on a deep blue background. The print is made by exposing to light a chemically sensitized sheet and the original drawing, which has been done on a semitransparent medium such as vellum, tracing paper, or film. The sensitized sheet and drawing are in close surface contact with each other. On exposure to light, a chemical reaction takes place that gives a blue color when fixed by washing in water. The parts protected from the light by the black lines of the drawing wash out, leaving white lines.

Blueprinting has been used for decades, and archives of design contain innumerable blueprints. The inconvenience of washing the sheets in water and then drying them has led to a dramatic decrease in the use of blueprinting, inasmuch as newer methods have been developed to reproduce drawings.

B. Diazo Printing This very popular process uses light-sensitive paper, as does blueprinting. However, the diazo process is a completely dry process.

The original drawing must be on a semitransparent medium. It is placed directly against a chemically treated sheet, which appears yellow before exposure to light. The original drawing and diazo sheet are run together through a diazo-process machine, which has a built-in light source. The light bleaches out all exposed yellow areas, leaving a white background. These portions of the sheet beneath the protective black lines of the original drawing are left as yellow lines.

FIG. 49 A modern diazo printing machine. *(Courtesy Ozalid Corp.)*

The diazo sheet is then passed through a developer containing ammonia fumes. The remaining yellow lines are developed into permanent lines that are usually blue, but can be black, brown, or red, depending on the particular type of diazo sheet selected. An example of a diazo machine is given in Fig. 49. Such machines are often called "whiteprinters" or, in the case of machines by the GAF Corporation, "ozalid whiteprinters." The term "ozalid" is sometimes used for prints made by the diazo process.

C. Vandyke Printing Vandyke printing utilizes a thin sensitized paper that turns dark brown when exposed to light and is "fixed." Lines of the original drawing remain as white lines. In this form the negative Vandyke can be used for special effects in presentation of drawings, such as architectural drawing displays, by mounting them on poster boards.

More often, though, the negative Vandyke is used to make another print, a positive. The Vandyke negative is placed next to the chemically sensitized side of another sheet. That sheet is exposed to light, and the result is blue lines on white, sometimes called a "blue-line blueprint." In this manner the Vandyke negative acts as an original, leaving the actual original untouched when further copies are needed.

D. Xerography Printing Developed by the Xerox Corporation, this process has seen widespread adoption in recent years. Prints are positives made directly on plain unsensitized paper, a distinct advantage over other processes requiring special papers. Xerox lines are black on whatever background color the print paper has.

Xerography involves an electrostatic process in which an electrically charged powder is transferred to the paper supplied, using the drawing as the original image. By using heat the powder is baked onto the paper supplied to give the final copy.

The original drawing need not be on a semitransparent medium. Opaque paper works well. Also, enlargements and reductions can be made, and high-run multiple copies are feasible. Provisions are available to transmit a copy of an original through telephone lines to distance points.

E. Photostating This process has evolved through many years and uses a special camera that may enlarge or reduce the original drawing. No darkroom is needed; all portions of the process take place within the machine.

A print with white lines on a dark background is made directly from the drawing. This print can again be photostated, giving a brown-black line on a white background. Xerography has given photostating considerable competition, so that photostating has a smaller share of the market than was formerly the case.

F. Microfilming When storage space for multiple drawings becomes a problem, microfilming provides a welcome and useful solution. The information explosion of the last several decades has given microfilming a firm acceptance among all frequent users of drawings.

By photographic processes, microfilming converts drawings to film, usually 16 or 35 mm. The film may be on rolls or aperture cards. Film rolls and aperture cards are viewed in a reader. Storage and retrieval of information is conveniently assigned to a computer, since the aperture card is actually a standard data card.

Retrieval may take the form of reading the film on a built-in screen or reading the information off a hard-copy printout. The printout can be an enlargement or reduction of the original size. Duplicate aperture cards are readily available for distribution to persons needing original-drawing information. Considerable design effort has gone into developing sophisticated retrieval systems based on microfilm, yielding a broad-based technology in its own right.

24. USE OF PHOTOGRAPHY FOR DRAWINGS

For certain applications special techniques may be used to create usable drawings. Photography is one such technique that can reduce drafting time or replace conventional drawings.

Individual applications of photo-drawing technique vary widely in industry, depending on the type of business in which the company is engaged. There are two principal areas in which photo drawings are of value: (1) revising and adding to existing equipment, (2) recording and supplementing model designs. Many process companies constantly change their facilities, to make a new product or to improve and expand; photography can play a major role in such engineering changes. Some companies have manufacturing operations that call for adding completely new units. Photo drawings of models are used to record the design.

Photo drawings have two main advantages:

1. They lower engineering costs. The camera records information that would have to be drawn if photodrawings were not utilized.

2. They provide better visualization of the design intent than is obtained from a drawing. The camera records an image approximately as the eye sees it—in three dimensions. In contrast, when conventional plan and elevation drawings are made, the reader has to mentally combine the various views to visualize the design intent.

A. Using Photo Drawings for Revisions The cost of engineering revisions of existing facilities can be reduced through the use of photo drawings. The original and main use of photo drawings is in the field of piping rework, but new applications are being found in the electrical, mechanical, structural, and the site-clearance fields.

No matter what phase of rework is being engineered, preparation of a photo drawing is similar and involves four basic steps: (1) taking the photographs, (2) making the "reproducible" (the transparent base for all drawing effort), (3) drawing on the reproducible to show the design intent of the revision, and (4) making prints from the photo drawing for distribution to the construction forces.

The highest-quality reproducibles, and therefore the best photo drawings, are obtained when the picture is taken by a skilled photographer working in close association with the engineer or designer responsible for the project.

A reproducible is made in several ways, depending on the company's facilities. It is a screened image on which highlights and shadows have been obtained by varying the concentration of white and black dots (see Fig. 50). Screened images, with their tonal characteristics, give much superior and more realistic prints than are available with printing processes that record only in solid black and pure white.

The drawing technique is the same no matter what material is used to prepare the reproducible. There are no unusual mechanical problems in drawing on photo drawings. By using a soft pencil, which produces heavy black lines, and by controlling the printing speed, it is possible to make additions stand out in contrast to the existing photographic background, as shown in Fig. 51.

Frequently it is necessary to clear a site before new equipment can be installed. Instructions for site clearance are easily transmitted by crossing out on a photo drawing the items that are to be removed, as suggested in Fig. 52. A second photo drawing of the same area is then made with the existing equipment removed from the picture so that it does not interfere with the drawing of new equipment. Any part of the photographic image can be removed from a photo-drawing reproducible with a wet eraser or with chemicals. Note the removed and drawn-in portions in Fig. 53.

Usually photo drawings convey design intent with only the important control dimensions indicated. It would be difficult, but not impossible, to include all of the detailed dimensions normally shown on conventional drawings. Field workers are accustomed to revision work and to "field checking" the drawings they receive, so the lack of complete dimensioning is not critical if the design intent is clear. Note the dimensioning and extensive use of notes in Fig. 54.

When more than one photograph is required to portray a change, each photo drawing is referenced to the other. This is done when there are multiple images on one photo drawing as well as when there are several sin-

FIG. 50 Illustration of screening on a reproducible. The drawing at the bottom shows a normal, screened positive image. The portion at the top is the small section enlarged ten times to show the screen pattern of dots. *(Courtesy of Procter & Gamble.)*

gle-image photo drawings. When there are several photo drawings for a project, a "key plan" is made. The key plan is a simple sketch of the area involved on which are shown the numbers of the various photo drawings with an arrow indicating the direction in which each picture was taken.

B. Using Photography for Models The camera can also record a design shown by a model, thus giving the engineer or designer a drawing base to which he or she can add information.

The basic steps followed in making photo drawings of

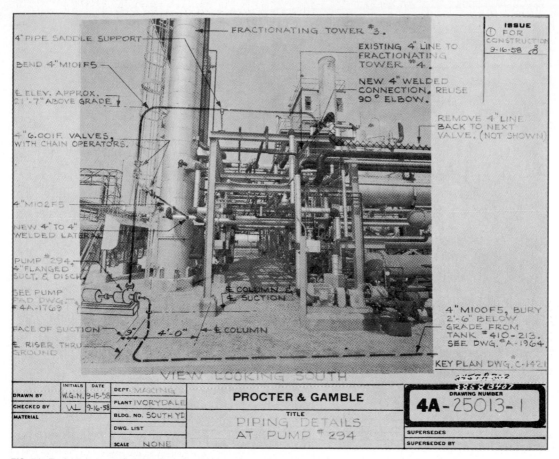

FRACTIONATING TOWER #3.

4" PIPE SADDLE SUPPORT

BEND 4" M101F5

&. ELEV. APPROX.
21'-7" ABOVE GRADE

4" 6.00IF. VALVES,
WITH CHAIN OPERATORS.

4" M102F5

NEW 4" TO 4"
WELDED LATERAL

PUMP #294,
4" FLANGED
SUCT. & DISCH.

SEE PUMP
PAD DWG.
#4A-1769

FACE OF SUCTION

&. RISER THRU
GROUND

EXISTING 4" LINE TO
FRACTIONATING
TOWER #4.

NEW 4" WELDED
CONNECTION, REUSE
90° ELBOW.

REMOVE 4" LINE
BACK TO NEXT
VALVE. (NOT SHOWN)

&. COLUMN &.
&. SUCTION

9"   4'-0"   &. COLUMN

4" M100F5, BURY
2'-6" BELOW
GRADE FROM
TANK #410-213.
SEE DWG. #A-1964.

KEY PLAN DWG. C-1421

VIEW LOOKING SOUTH

DRAWN BY	INITIALS W.G.N.	DATE 9-15-58	DEPT. MACHINE	**PROCTER & GAMBLE**	DRAWING NUMBER **4A-25013-1**
CHECKED BY	WL	9-16-58	PLANT IVORYDALE		
MATERIAL			BLDG. NO. SOUTH YD	TITLE	
			DWG. LIST	PIPING DETAILS	SUPERSEDES
			SCALE NONE	AT PUMP #294	SUPERSEDED BY

FIG. 51 Typical photo drawing of piping for the revision of an existing facility. Observe the extensive notes and directions added to the reproducible in pencil. *(Courtesy of Procter & Gamble.)*

FIG. 52 Typical drafting-room scene when work is being done on a photo drawing.

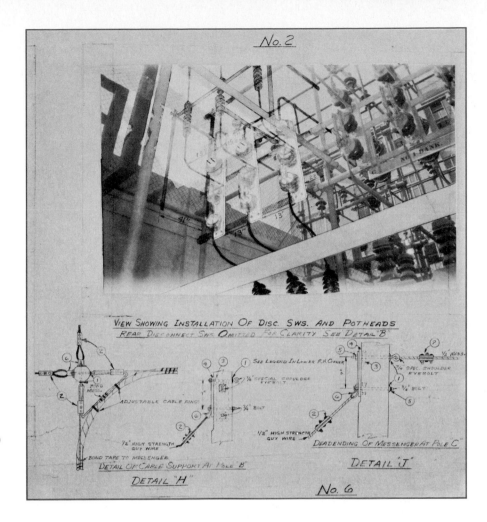

FIG. 53 Typical completed photo drawing for piping revision of an existing facility. (Compare with Fig. 51.) In this case the photo portion shows design intent, and the necessary details are given below in conventional orthographic views. *(Courtesy of Procter & Gamble.)*

existing areas (taking the picture, preparing the reproducible, drawing in the engineering information, making the prints) apply also in making model photo drawings. The principal difference between the two uses is in the taking of the pictures.

Model photo drawings are used for (1) equipment layouts, (2) location drawings, (3) information for people remote from the model, and (4) permanent records of the job once the model is destroyed. All these applications require that the photo drawings be shadowless, so that the equipment and piping on the model stand out clearly. An example of a shadowless model photo drawing is seen in Fig. 55.

The problems of perspective are more difficult to solve. The ultimate in model photo drawings is a complete absence of perspective, that is, purely orthographic photo drawings, representing all dimensions to true scale. Most companies use the more easily attainable minimum perspective, which is obtained through use of a long-focal-length camera lens. Photo drawings in this perspective are not unreasonably difficult to dimension and understand, but are still not scalable. Development work has been done on equipment that will take orthographic photographs—devices that utilize segments of parabolic mirrors combined to form a telecentric optical system. Such devices can take scalable photographs of a

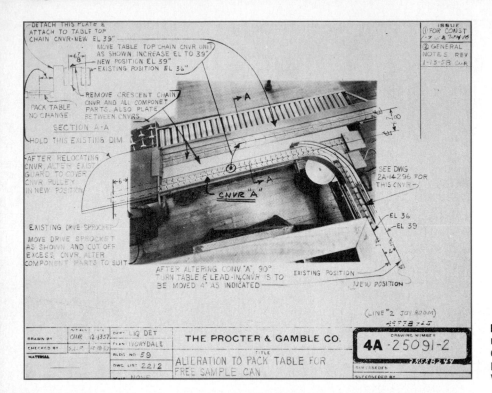

FIG. 54 Typical photo drawing for revision of mechanical equipment. Note that this drawing employs dimensioning and notes applied to the photo portion, and also uses small explanatory views. *(Courtesy of Procter & Gamble.)*

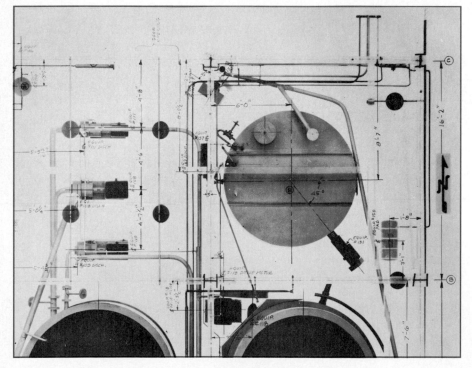

FIG. 55 A portion of a shadowless model photo drawing. *(Courtesy of Procter & Gamble.)*

FIG. 56 Comparison of methods of taking photographs of a model. The lower picture was taken with a conventional camera; the upper picture, with a device employing a telecentric system to produce an orthographic (nonperspective) view. *(Courtesy of Procter & Gamble.)*

model or other three-dimensional subject. An example is shown in Fig. 56.

Once the picture has been taken, under conditions that will produce either orthographic or minimum-perspective results, the reproducible, which is the base for the model photo drawing, is made in the same manner as previously outlined for existing-area photo drawings.

25. SPECIALIZED DRAWINGS USED FOR ILLUSTRATION

We will conclude our study of production drawings by looking at several types of specialized drawings that can enhance an understanding of how a product is assembled. Assembly drawings, as such, were discussed back in Sec. 13. However, it is possible to use drawings that make more use of the three-dimensional format than do most conventional assembly drawings. Such 3-D drawings are often called *illustration drawings*. While not all illustration drawings are three-dimensional, all seek to make assembly easier to understand.

A. Use of Illustration Drawings

Illustration drawings can be designated by the use to which they are put. These drawings often combine artistry with mechanical drawing and may be done by commercial illustrators as opposed to engineering drafters. Drawings may also be created in the three-dimensional illustrative mode by sophisticated computer programs, as seen in Fig. 57.

Examples of illustration drawings by use are listed below.

Advertising illustrations are usually shaded pictorial drawings, often with color added to make the presentation as forceful as possible.

Catalogue illustrations, orthographic or pictorial, are often colored or shaded in pencil or ink, by stippling, airbrush, etc.

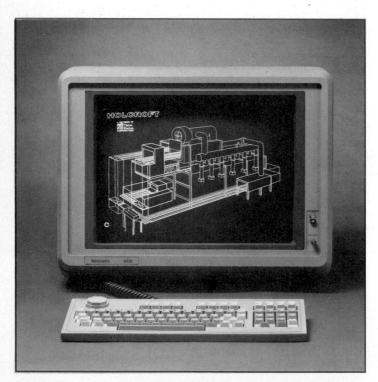

FIG. 57 A three-dimensional output of an industrial process unit that can be converted to paper format by use of a pen plotter. *(Courtesy Tektronix, Inc.)*

Operation, service, and repair charts are drawings showing the working parts of a machine with appropriate directions for the purpose intended. These drawings are effective in shaded or colored pictorial form.

Piping, wiring, and installation diagrams are easy to read when made in shaded or colored pictorial form.

Architectural and engineering presentation drawings, sometimes shaded in pencil or ink, often in watercolor, show the building or structure as it will appear when completed.

Textbook illustrations in pictorial form, usually in black and white but sometimes in two colors or full color, add visual clarity and give emphasis not achieved by words alone.

Patent drawings are usually shaded to bring out and clarify every feature of the invention.

Production drawings, from original design sketches to the final details, subassemblies, and assemblies, are frequently made in pictorial form, often shaded. Such drawings are particularly useful when persons not trained in reading orthographic drawings are employed for manufacturing.

B. Illustrated Working Drawings

Illustration drawings can make drawings more understandable for persons not trained to read two-dimensional orthographic drawings. The illustration drawings can be used in every phase of production, from the original design to the final operating instructions. In general, the drawings can be classified as design drawings, assembly illustrations, and operation and maintenance illustrations.

Design drawings include a variety of pictorial illustrations that show the machine or structure broken down into small, workable units, and then give details of construction, location of equipment, structural features, function of parts and equipment, tooling methods, etc. These drawings are used to study the complete production job and to plan and correlate the work. As the design progresses they are altered, corrected, or redrawn. The preliminary production breakdown shown in Fig. 58 is an example of the type of illustration used in the design stage of the work.

Assembly illustrations give detailed information regarding the breakdown of the machine, from which a

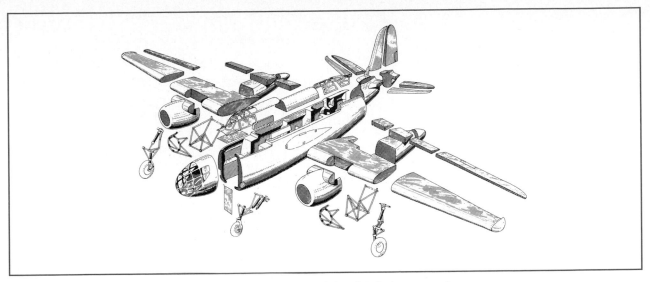

FIG. 58 A design illustration. This shows the breakdown of a vintage airplane into basic components.

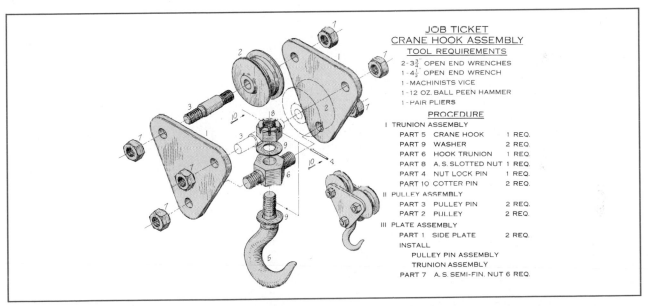

JOB TICKET
CRANE HOOK ASSEMBLY
TOOL REQUIREMENTS
2 - 3¾" OPEN END WRENCHES
1 - 4½" OPEN END WRENCH
1 - MACHINISTS VICE
1 - 12 OZ. BALL PEEN HAMMER
1 - PAIR PLIERS

PROCEDURE
I TRUNION ASSEMBLY
  PART 5 CRANE HOOK   1 REQ.
  PART 9 WASHER   2 REQ.
  PART 6 HOOK TRUNION   1 REQ.
  PART 8 A. S. SLOTTED NUT 1 REQ.
  PART 4 NUT LOCK PIN   1 REQ.
  PART 10 COTTER PIN   2 REQ.
II PULLEY ASSEMBLY
  PART 3 PULLEY PIN   2 REQ.
  PART 2 PULLEY   2 REQ.
III PLATE ASSEMBLY
  PART 1 SIDE PLATE   2 REQ.
  INSTALL
     PULLEY PIN ASSEMBLY
     TRUNION ASSEMBLY
  PART 7 A. S. SEMI-FIN. NUT 6 REQ.

FIG. 59 A manufacturing illustration. It specifies the parts and tools needed for an assembly. Drawings of this type are used for original manufacture and in service manuals.

multitude of separate illustrations are made to show the location of subassemblies, parts, and equipment; give directions for performing operations; provide detailed information for the manufacture of parts; and furnish directions for assembly. Figure 59 is an example of an assembly illustration.

Operation and maintenance illustrations give directions for disassembly, repair and replacement of parts, lubrication, inspection and care of equipment, etc. In many cases the drawings that were originally made for manufacturing can be used in a service manual. Figure 60 is an example.

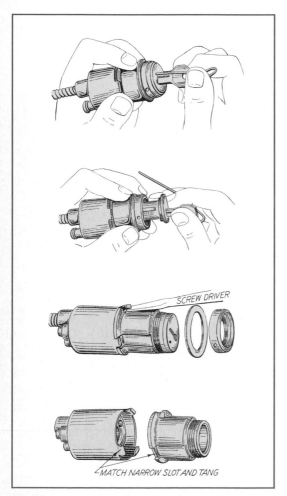

FIG. 60 A maintenance illustration. This type of drawing is used in a service manual to illustrate repair by replacement of components.

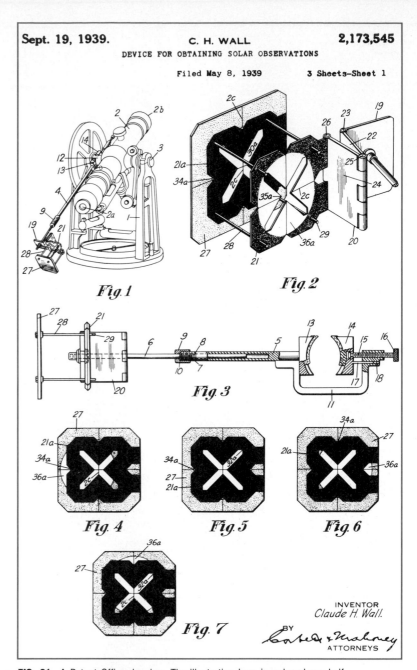

FIG. 61 A Patent Office drawing. The illustration here is reduced one-half.

C. Patent Drawings In an application for letters patent on an invention a written description, called the "specification," is required, and for a machine or device a drawing showing every feature of the invention must also be supplied. A high standard of execution and conformity to the rules of the Patent Office must be observed. A pamphlet called "Rules of Practice," giving full information and rules governing Patent Office procedure in applying for a patent, can be obtained from the Commissioner of Patents, Washington, D.C.

Patent Office drawings are not working drawings. They are descriptive and pictorial rather than structural; hence they have no centerlines, dimensions, notes, or names of views. The views are lettered with figure numbers, and the parts are designated by reference numbers through which the invention is described in the specification.

The drawings may be made in orthographic, axonometric, oblique, or perspective projection. The pictorial system is used extensively, for all or part of the views. Surface shading is used whenever it will aid readability.

Figure 61 is an example of a Patent Office drawing.

D. Cautions for Drawing for Reproduction Specialized drawings for illustrations in books, periodicals, catalogues, or other printed materials must be made with an understanding of reproduction processes.

The drawings are usually made larger than the final size of the illustration and reduced photographically in the reproduction process. Consequently the work must be done with visualization of the line weights, contrast, size of lettering, and general effect in reduced size. To preserve the hand-drawn character of the original, the reduction should be slight. For a very smooth effect, the drawing may be as much as three or four times larger than the reproduction. The best general size is one and a half (for one-third reduction) to two times (for one-half reduction) the linear size of the cut.

The reduction is usually an even proportion, such as one-fourth, one-third, one-half, etc., although odd reductions may be used. For a drawing marked "Reduce ⅓," the reproduction will be two-thirds the linear size of the original. Figure 62 illustrates the appearance of an original drawing, and Fig. 63 shows the same drawing reduced one-half. The coarse appearance, open shading,

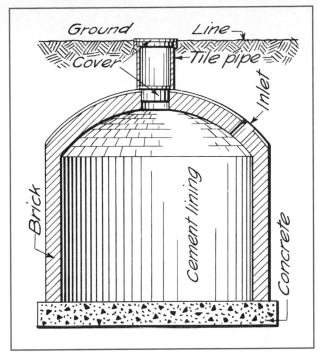

FIG. 62 A drawing for reproduction. This drawing is made for reduction to half size. The reduced reproduction is shown in Fig. 63.

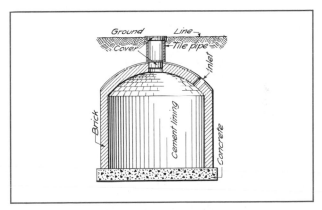

FIG. 63 The drawing shown in Fig. 62 reduced to half size and reproduced.

and lettering size of the original should be noted. The line work must be kept fairly "open," for if lines are drawn close together the space between them may choke in the reproduction and mar the effect.

PROBLEMS

GROUP 1. DETAIL DRAWINGS

This group of problems gives practice in making drawings with sectional views, auxiliaries, and conventional representation. Several methods of part production are included. See Appendixes for information on thread symbols and notes.

1. Make complete working drawing with necessary sectional views. Cast iron.

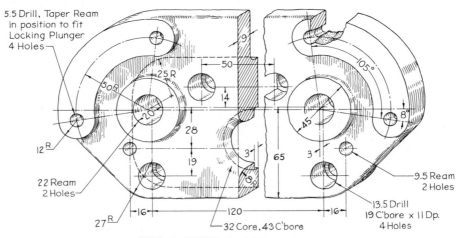

PROB. 1 METRIC. Gear-shifter bracket.

2. Working drawing of mixing-valve body. Cast brass.

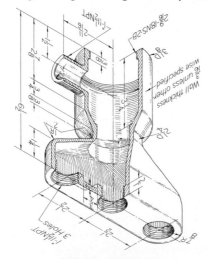

PROB. 2 Mixing-valve body.

3. Working drawing of conveyor hanger. Determine what views and part views will best describe the piece. Cast steel.

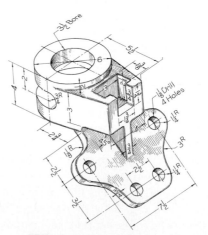

PROB. 3 Conveyor hanger.

4. Working drawing of valve cage. Cast bronze.

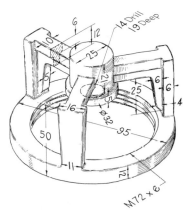

PROB. 4　METRIC. Valve cage.

5. Determine what views and part views will most adequately describe the piece. Malleable iron.

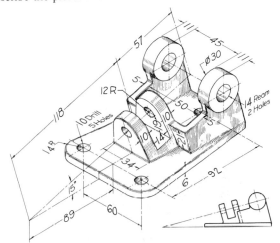

PROB. 5　METRIC. Brace plate.

6. Make working drawing of shifter fork. Cast steel.

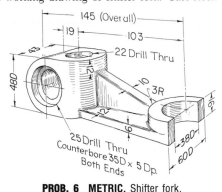

PROB. 6　METRIC. Shifter fork.

7. Make working drawing of rubber support anchor.

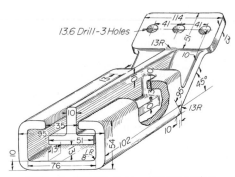

PROB. 7　METRIC. Rubber support anchor.

8. Make working drawing of breaker. Steel.

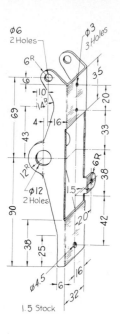

PROB. 8 METRIC. Breaker.

9. Make working drawing of relief-valve body. Cast brass.

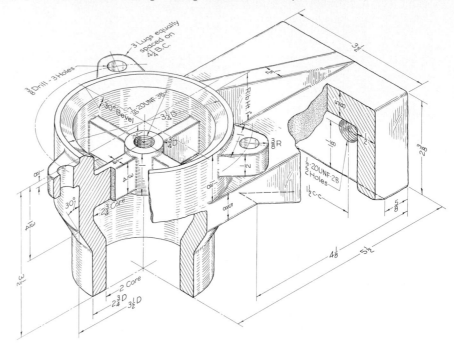

PROB. 9 Relief-valve body.

10. Make working drawing of selector ring. Brass.

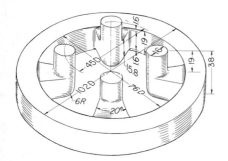

PROB. 10 METRIC. Selector ring.

11. Make working drawing of impeller. Aluminum.

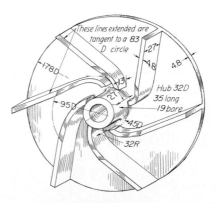

PROB. 11 METRIC. Impeller.

12. (*a*) Make detail working drawings on same sheet, one for *rough* forging and one for *machining*; or (*b*) make one detail drawing for forging and machining. Alloy steel.

PROB. 12 Steering knuckle.

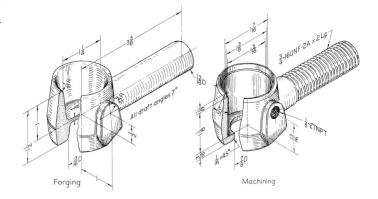

Forging   Machining

13. Working drawing of automotive connecting rod. Drop forging, alloy steel. Drawings same as Prob. 12.

PROB. 13   METRIC.
Automotive connecting rod.

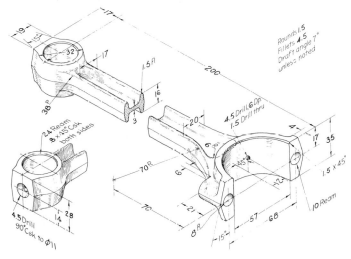

14. Make working drawing of buffer stand. Steel drop forging. Drawings same as Prob. 12.

15. Make working drawing of flap link. High-strength aluminum alloy.

PROB. 14 Buffer stand.

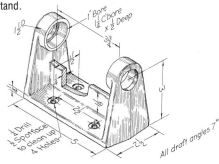

PROB. 15 Flap link.

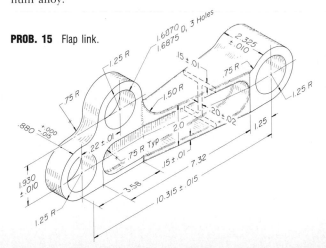

16. Make working drawing of torque-tube support. Drop forging, aluminum alloy. Drawings same as Prob. 12.

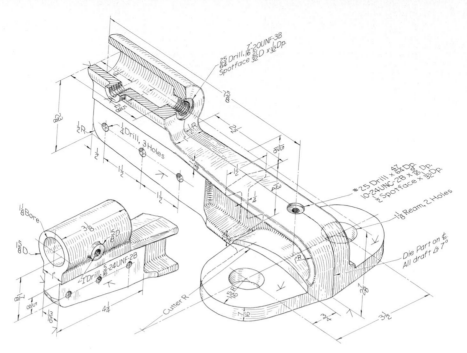

PROB. 16 Torque-tube support.

17. Make working drawing of bearing block, cast iron.

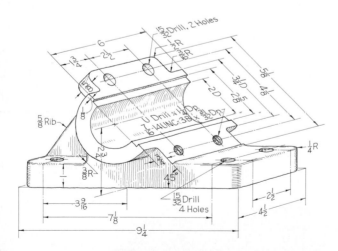

PROB. 17 Bearing block.

18. Make working drawing of elevator control bracket. Aluminum alloy, welded.

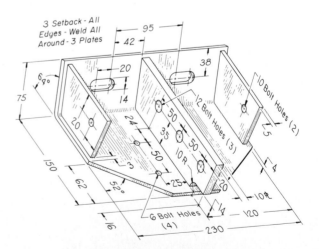

PROB. 18   METRIC. Elevator control bracket.

19. Make working drawing of wing fitting. High-strength aluminum alloy.

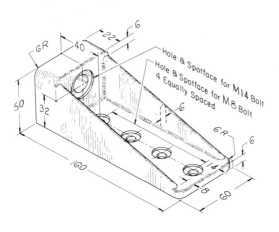

PROB. 19   METRIC. Wing fitting.

20. Make working drawing of microswitch bracket. Aluminum sheet, 0.15 thick.

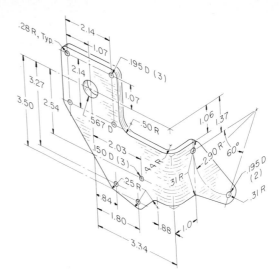

PROB. 20   Microswitch bracket.

21. Make working drawing of tab link. High-strength aluminum alloy.

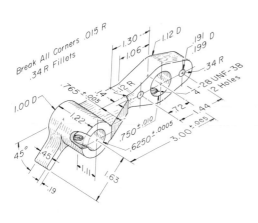

PROB. 21   Tab link.

22. Make a complete working drawing of the desurger loading-valve gage connector. This part is made of ¾-in hexagon steel (AISI C1212).

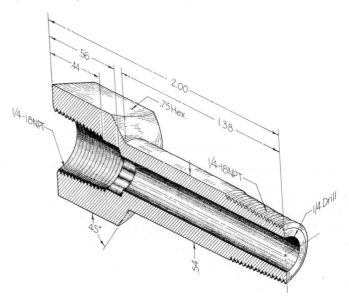

PROB. 22   Desurger loading-valve gage connector. *(Courtesy of Westinghouse Air Brake Co.)*

23. Make a complete working drawing of the needle-valve
needle. Material is stainless steel.

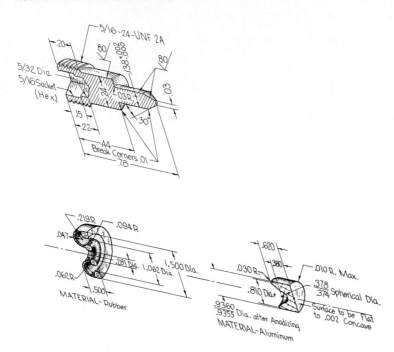

PROB. 23   Needle-valve needle. *(Courtesy of Westinghouse Air Brake
Co.)*

GROUP 2. AN ASSEMBLY DRAWING FROM THE DETAILS

24. Make an assembly drawing of sealed shaft unit. Bracket and gland are cast iron. Shaft, collar, and studs are steel. Bushing is bronze. Drill bushing in assembly with bracket.

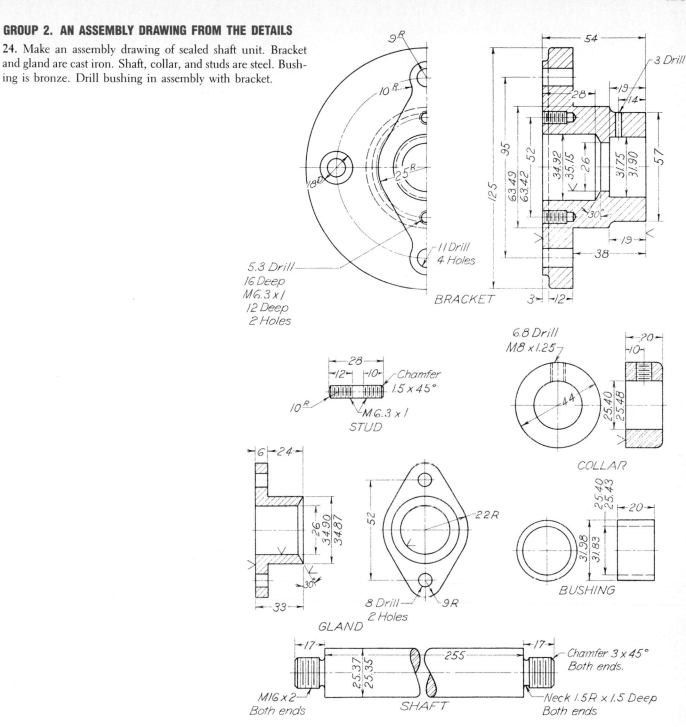

PROB. 24   METRIC. Sealed shaft unit.

25. Make an assembly drawing of the crane hook from details given. Standard parts 7 to 10 are not detailed; see handbook for sizes.

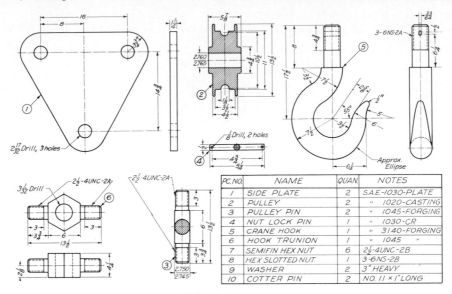

PC. NO.	NAME	QUAN.	NOTES
1	SIDE PLATE	2	S.A.E.-1030-PLATE
2	PULLEY	2	" 1020-CASTING
3	PULLEY PIN	2	" 1045-FORGING
4	NUT LOCK PIN	1	" 1030-CR
5	CRANE HOOK	1	" 3140-FORGING
6	HOOK TRUNION	1	" 1045 "
7	SEMIFIN HEX NUT	6	2½-4UNC-2B
8	HEX SLOTTED NUT	1	3-6NS-2B
9	WASHER	2	3" HEAVY
10	COTTER PIN	2	NO. 11 × 1" LONG

PROB. 25 Crane hook.

26. Make an assembly drawing, front view in section, of caster.

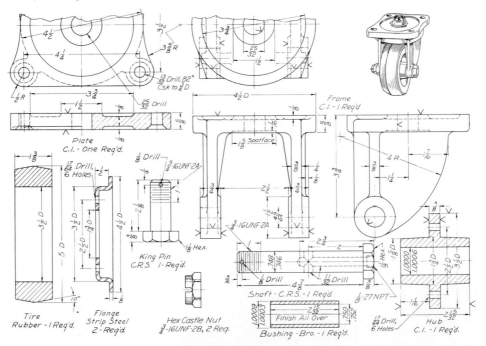

PROB. 26 Caster.

GROUP 3. DETAIL DRAWINGS FROM THE ASSEMBLY

27. Make detail drawings of the jig table. Parts, cast iron.

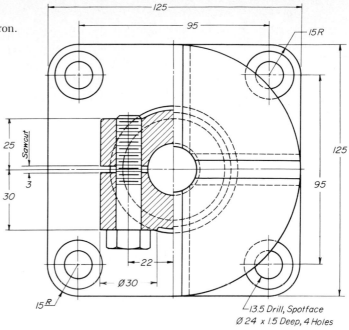

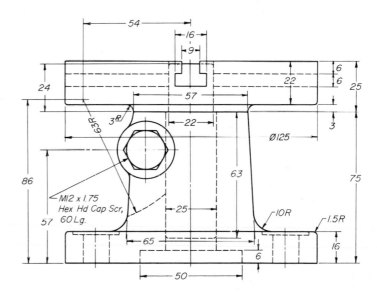

PROB. 27 METRIC. Jig table.

28. Make detail drawings of the door catch.

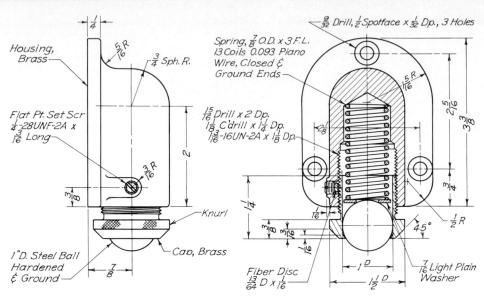

PROB. 28 Door catch.

29. Make detail drawings of belt drive. The pulley and bracket are cast iron; the gear and shaft, steel. The bushing is bronze.

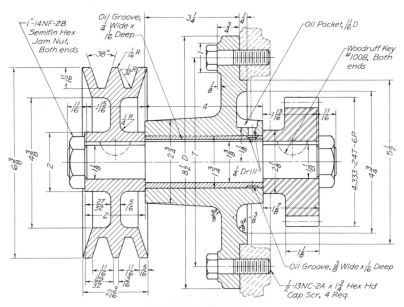

PROB. 29 Belt drive.

30. Make detail drawings of ball-bearing idler pulley.

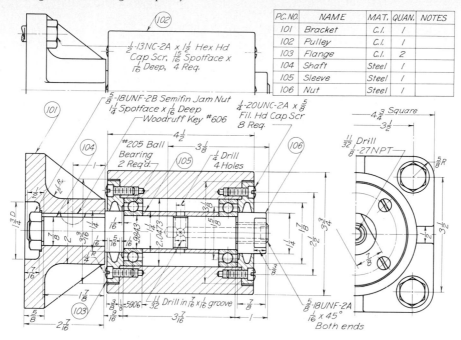

PC.NO.	NAME	MAT.	QUAN.	NOTES
101	Bracket	C.I.	1	
102	Pulley	C.I.	1	
103	Flange	C.I.	2	
104	Shaft	Steel	1	
105	Sleeve	Steel	1	
106	Nut	Steel	1	

PROB. 30 Ball-bearing idler pulley.

31. Make detail drawings for V-belt drive.

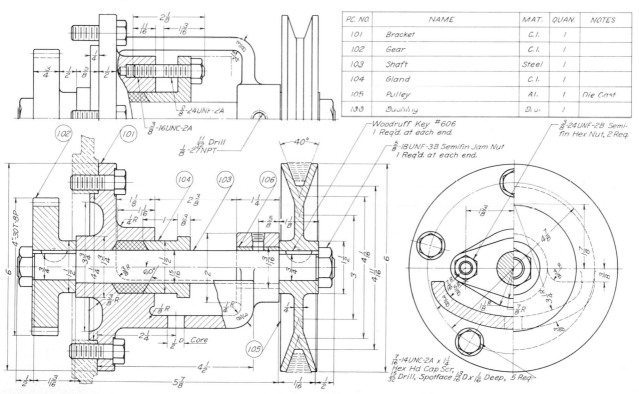

PC. NO.	NAME	MAT.	QUAN.	NOTES
101	Bracket	C.I.	1	
102	Gear	C.I.	1	
103	Shaft	Steel	1	
104	Gland	C.I.	1	
105	Pulley	Al.	1	Die Cast
106	Bushing	Br.	1	

PROB. 31 V-belt drive.

32. Make detail drawings of sealed ball joint.

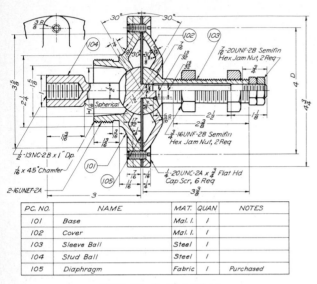

PC. NO.	NAME	MAT.	QUAN.	NOTES
101	Base	Mal. I.	1	
102	Cover	Mal. I.	1	
103	Sleeve Ball	Steel	1	
104	Stud Ball	Steel	1	
105	Diaphragm	Fabric	1	Purchased

PROB. 32 Sealed ball joint.

33. Make detail drawings of belt tightener. The bracket, pulley, and collar are cast iron. The bushing is bronze; the shaft, steel.

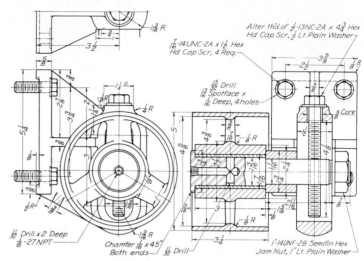

PROB. 33 METRIC. Belt tightener.

34. Make detail drawings of hydraulic punch. In action, the punch assembly proper advances until the cap, piece 109, comes against the work. The assembly (piece 106 and attached parts) is then stationary, and the tension of the punch spring (piece 114) holds the work as the punch advances through the work and returns.

PC. NO.	NAME	MAT.	QUAN.	NOTES
101	Bracket	C.I.	1	
102	Cylinder	C.I.	1	
103	Piston	C.I.	1	
104	Cylinder Head	C.I.	1	
105	Sleeve Stop	C.I.	1	
106	Sleeve	Steel	1	
107	Sleeve Nut	Steel	1	
108	Punch	Steel	1	
109	Punch Cap	Steel	1	
110	Packing Plate	Steel	1	
111	Packing	Leather	1	Purchase
112	Piston Rod	Steel	1	
113	Sleeve Spring	Steel	1	Purchase
114	Punch Spring	Steel	1	Purchase

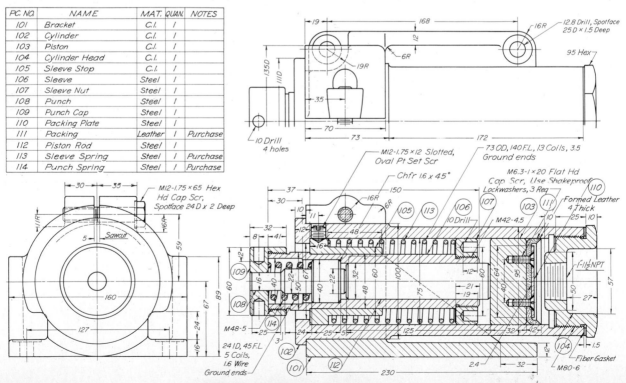

PROB. 34 METRIC. Hydraulic punch.

35. Make detail drawings of double-acting air cylinder. Length of stroke to be assigned. Fix length of cylinder to allow for clearance of 25 mm at ends of stroke. Note that pieces 101 and 102 are identical except for the extra machining of the central hole in piece 101 for the shaft, packing, and gland. Make separate drawings for this piece, one for the pattern shop and two for the machine shop.

PC.NO.	NAME	MAT.	QUAN.	NOTES
101	Cylinder Head, Front	C.I.	1	
102	Cylinder Head, Rear	C.I.	1	Use Cast. 101
103	Packing Gland	C.I.	1	
104	Piston	C.I.	1	
105	Piston Rod	Steel	1	
106	Piston Plate	Steel	2	
107	Cylinder	Steel	1	Shelby Tubing
108	Tie Rod	Steel	4	
109	Stop Collar	Steel	1	
110	Piston Packing	Leath.	2	Purchase
111	Gasket	Fiber	2	Purchase

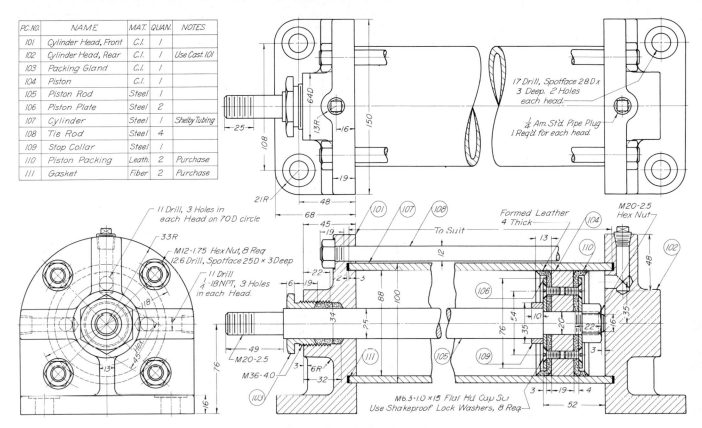

PROB. 35 METRIC. Double-acting air cylinder.

36. Make detail drawings of rail-transport hanger. Rail is 10-lb ASCE.

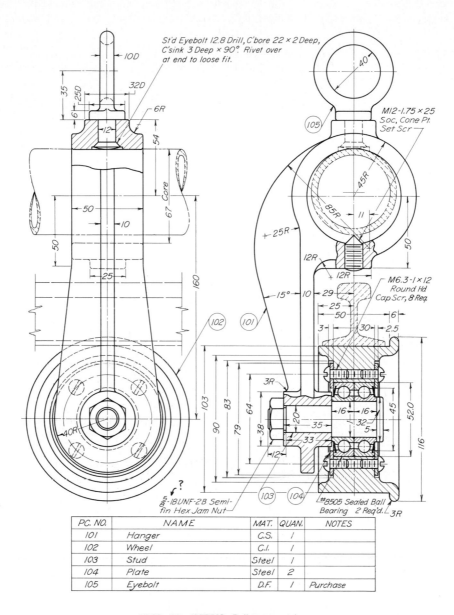

PC. NO.	NAME	MAT.	QUAN.	NOTES
101	Hanger	C.S.	1	
102	Wheel	C.I.	1	
103	Stud	Steel	1	
104	Plate	Steel	2	
105	Eyebolt	D.F.	1	Purchase

PROB. 36 METRIC. Rail-transport hanger.

GROUP 4. A SET OF DRAWINGS FROM AN EXPLODED PICTORIAL

The problems that are given in this group have been arranged as complete exercises in making a set of working drawings. Remember that the dimensions given on the pictorial views are to be used only to obtain distances or information needed. In some cases the data needed for a particular part may have to be obtained from the mating part.

The detail drawings should be made with each part on a separate sheet. Drawings of cast or forged parts may be made in the single-drawing or the multiple-drawing system.

The assembly drawing should include any necessary dimensions, such as number, size, and spacing of mounting holes, that might be required by a purchaser or needed for checking with the machine on which a subassembly is used.

For the style and items to be included in the parts list, see Prob. 30. For information on threads and notes, see Appendixes.

37. Make a complete set of working drawings for the antivibration mount.

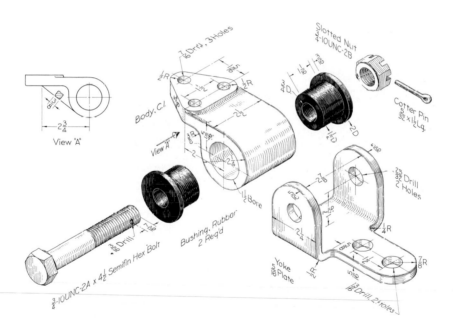

PROB. 37 Antivibration mount.

38. Make a complete set of drawings for pivot hanger, including detail drawings, assembly drawing, and parts list. All parts are steel. This assembly consists of a yoke, base, collar, and standard parts.

39. Make a complete set of drawings for pump valve. The valve seat, stem, and spring are brass; the disk is hard-rubber composition. In operation, pressure of a fluid upward against the disk raises it and allows flow. Pressure downward forces the disk tighter against the seat and prevents flow.

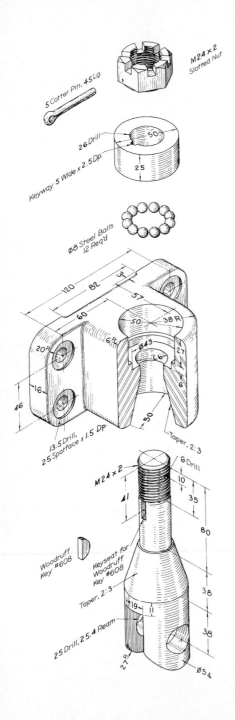

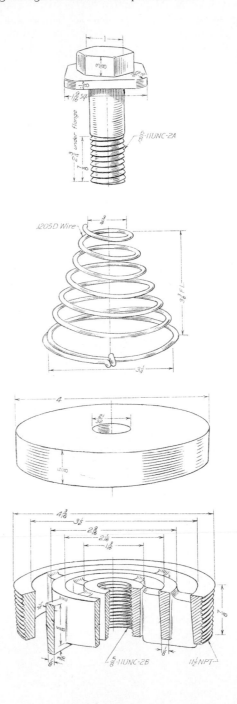

PROB. 38 METRIC. Pivot hanger.

PROB. 39 Pump valve.

40. Make a complete set of working drawings of the boring-bar holder. All parts are steel. Note that the *body* is made in one piece, then split with a ⅛-in-wide cut (exaggerated in the picture).

41. Make a complete set of working drawings of the ratchet wrench.

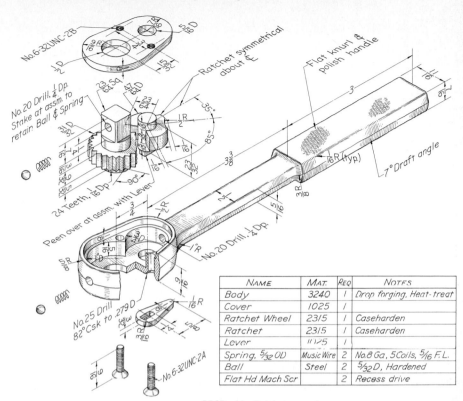

NAME	MAT.	REQ	NOTES
Body	3240	1	Drop forging, Heat-treat
Cover	1025	1	
Ratchet Wheel	2315	1	Caseharden
Ratchet	2315	1	Caseharden
Lever	1025	1	
Spring, 5/32 OD	Music Wire	2	No. 8 Ga., 5 Coils, 5/16 F.L.
Ball	Steel	2	5/32 D, Hardened
Flat Hd Mach Scr		2	Recess drive

PROB. 41 Ratchet wrench.

42. Make a complete set of working drawings for access door support.

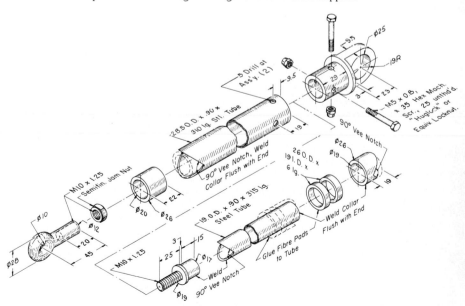

PROB. 42 METRIC. Access-door support.

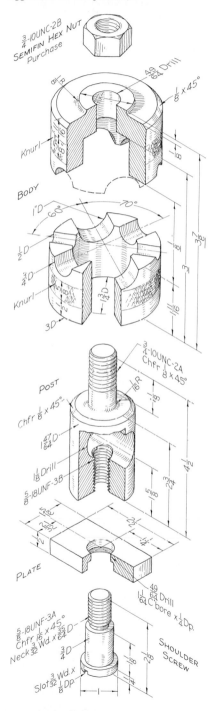

PROB. 40 Boring-bar holder.

43. Make detail and assembly drawings of the pressure-cell piston. Body parts are high-strength aluminum alloy. O rings are Teflon.

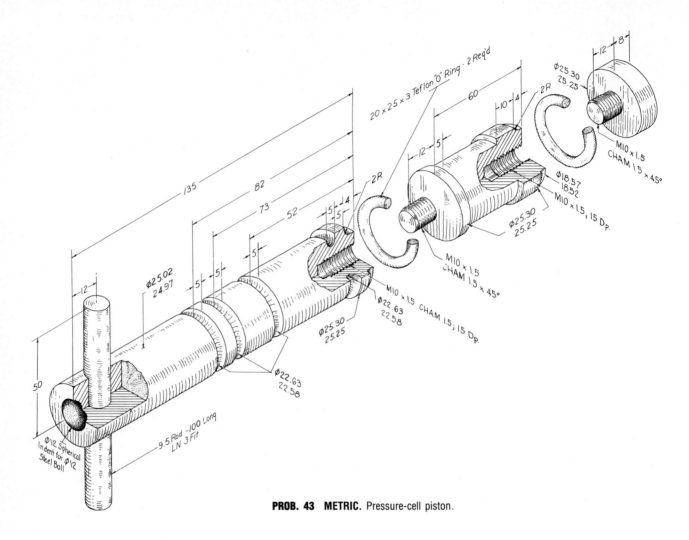

PROB. 43   METRIC. Pressure-cell piston.

GROUP 5. A SET OF DRAWINGS FROM A PICTORIAL ASSEMBLY

44. Make a complete set of working drawings of the high-tension coil mount.

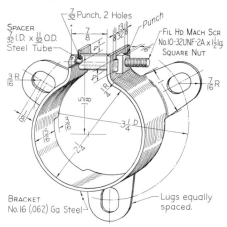

PROB. 44 High-tension coil mount.

46. Make a complete set of working drawings of the pipe clamp. The flange is cast steel.

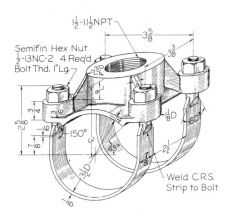

PROB. 46 Pipe clamp.

45. Make a complete set of working drawings of the stay-rod pivot. Parts are malleable iron.

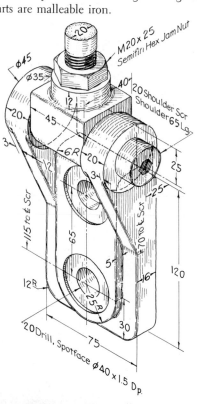

PROB. 45 METRIC. Stay-rod pivot.

47. Make a complete set of working drawings of the diaphragm assembly. Draw four times actual size.

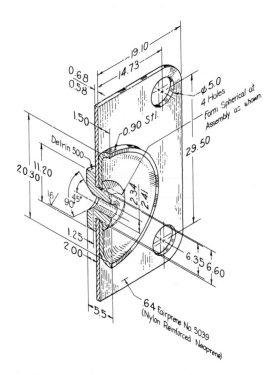

PROB. 47 METRIC. Diaphragm assembly. *(Courtesy of Bellows-Valvair, Division of International Basic Economy Corporation.)*

48. Make a complete set of working drawings of the tool post. All parts are steel.

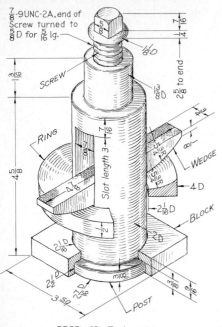

PROB. 48 Tool post.

49. Make a unit-assembly working drawing of the wing-nose rib.

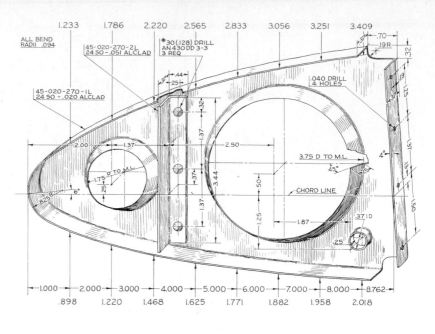

PROB. 49 Wing-nose rib.

GROUP 6. WORKING DRAWINGS FROM PHOTO DRAWINGS

Photo drawings are coming into use commercially, especially in the processing industries. These problems give practice in making detail and assembly drawings, and illustrate the possibilities of drawings produced from photographs of models or test parts.

50. Detail drawing of conveyor link. SAE 1040 steel.

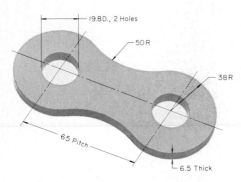

PROB. 50 METRIC. Conveyor link.

51. Detail drawing of hydrocylinder support. Aluminum sheet.

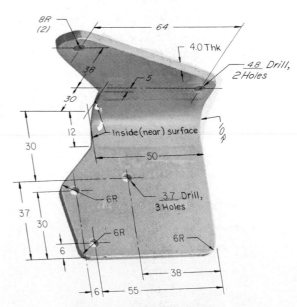

PROB. 51 METRIC. Hydrocylinder support.

52. Make detail and unit assembly of pivot shaft. Shaft is bronze; nuts are steel.

53. Make detail drawing of third terminal (temperature control). Material is red brass (85% CU).

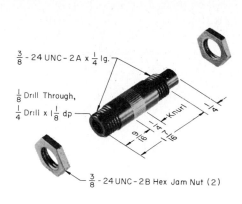

$\frac{3}{8}$ - 24 UNC - 2A x $\frac{1}{4}$ lg.

$\frac{1}{8}$ Drill Through,
$\frac{1}{4}$ Drill x 1$\frac{1}{8}$ dp

Knurl

$\frac{9}{16}$ $\frac{1}{16}$ $\frac{1}{4}$

$\frac{3}{8}$ - 24 UNC - 2B Hex Jam Nut (2)

PROB. 52 Pivot shaft.

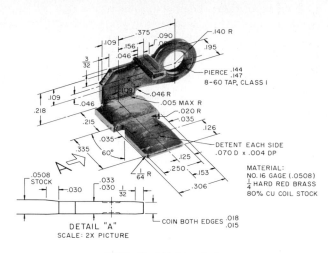

.375   .090   .140 R
.109   .156   .195
.046
$\frac{3}{32}$ PIERCE .144 / .147
8-60 TAP, CLASS 1
.109   .046 R
.005 MAX R
.020 R
.218   .046   .035
.215   .126
.035
.335   60°   .125   DETENT EACH SIDE
.250   .153   .070 D x .004 DP
$\frac{1}{64}$ R   .306

MATERIAL:
NO. 16 GAGE (.0508)
$\frac{1}{4}$ HARD RED BRASS
80% CU COIL STOCK

.0508 STOCK
-.030
.033
.030 $\frac{1}{32}$

COIN BOTH EDGES .018 / .015

DETAIL "A"
SCALE: 2X PICTURE

PROB. 53 Third terminal.

54. Create a set of working drawings of conveyor link unit. All parts are steel.

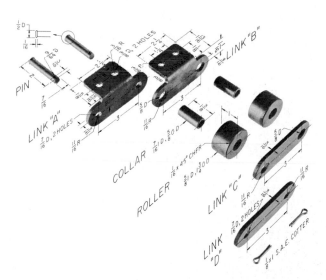

PROB. 54 Conveyor link unit.

55. Marking machine. From design drawing, shown half size, make complete set of drawings. Base, piece 1, is malleable-iron casting. Frame, piece 2, is cast iron. Ram, piece 3, is 1020 HR; bushing, piece 4, is 1020 cold-drawn tubing; heat treatment for both is carburize at 1650 to 1700°F, quench direct, temper at 250 to 325°F. Spring, piece 5, is piano wire, no. 20 (0.045) gage, six coils, free length 50 mm, heat treatment is "as received." Marking dies and holders are made up to suit objects to be stamped.

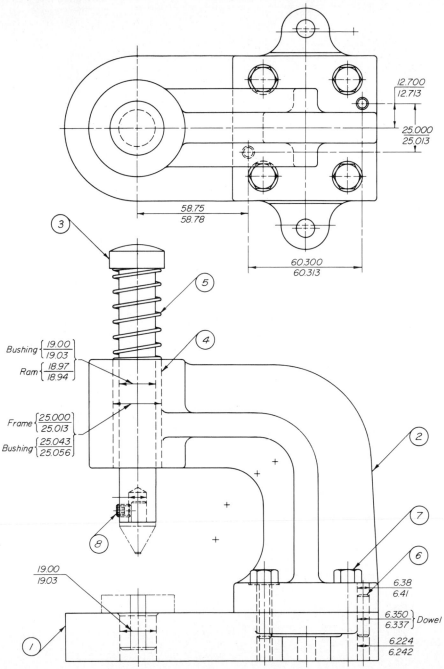

PROB. 55 METRIC. Marking machine (one-half size).

56. Arbor press. From design drawing, shown one-quarter size, make complete set of drawings. All necessary information is given on the design drawing.

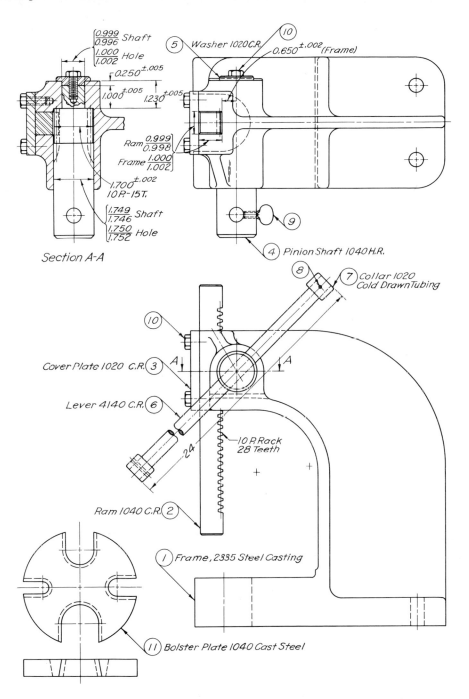

$\dfrac{0.999}{0.996}$ Shaft

$\dfrac{1.000}{1.002}$ Hole

$0.250^{\pm.005}$

$1.000^{\pm.005}$ $1.230^{\pm.005}$

⑤ Washer 1020 C.R.   ⑩

$0.650^{\pm.002}$ (Frame)

Ram $\dfrac{0.999}{0.998}$

Frame $\dfrac{1.000}{1.002}$

$1.700^{\pm.002}$
10 P.-15 T.

$\dfrac{1.749}{1.746}$ Shaft

$\dfrac{1.750}{1.752}$ Hole

Section A-A

⑨

④ Pinion Shaft 1040 H.R.

⑧   ⑦ Collar 1020
Cold Drawn Tubing

⑩

Cover Plate 1020 C.R. ③

A   A

Lever 4140 C.R. ⑥

2A

10 P. Rack
28 Teeth

Ram 1040 C.R. ②

① Frame, 2335 Steel Casting

⑪ Bolster Plate 1040 Cast Steel

PROB. 56   One-ton arbor press (one-quarter size).

57. Number 2 flanged vise. From design drawing, make complete set of drawings, including details, parts list, and assembly. Design drawing is shown one-third size. All necessary information will be found on the design drawing.

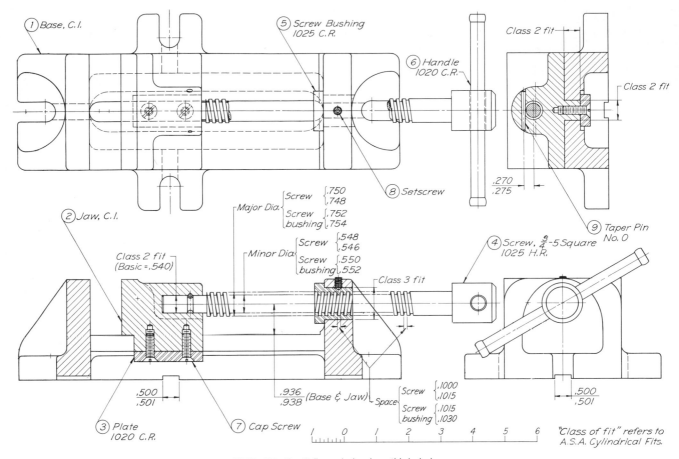

PROB. 57   No. 2 flanged vise (one-third size).

58. Bench shears. Design drawing is half size. Make a complete set of drawings. For dimensions where close fits are involved, either the decimal limit or the tolerance to be applied to the scaled basic size is given on the design drawing. Heat treatments should be specified as follows: for base and jaw, "normalize at 1550°F"; for eccentric, "as received"; for blade, "to Rockwell C57-60"; for handle, none. Finish for blades, "grind." Washers are special but may be specified on the parts list by giving inside diameter, outside diameter, and thickness. The key may be specified on the parts list by giving width, thickness, and length.

Following are some specifications for limits and tolerances, in millimeters.

Diameter of shoulder screw and hole in both base and jaw:

$$\text{Screws (as manufactured): } \frac{15.82}{15.77}$$

$$\text{Hole: } \frac{15.85}{15.90}$$

Diameter of eccentric and width of slot in jaw:

Eccentric: $\dfrac{22.20}{22.15}$

Jaw: $\dfrac{22.23}{22.28}$

Width of keyway in both handle and eccentric:

Key (as purchased): $\dfrac{6.35}{6.32}$

Key seat and keyway: $\dfrac{6.35}{6.38}$

Tolerance on shoulder screw and eccentric hole locations:
For base and jaw (two dimensions on each part)

± 0.05

Limits for eccentric offset:

Center to center: $\dfrac{6.30}{6.40}$

Length of handle:

Center of hub to center of ball: 300 mm

Diameter of eccentric and hole in handle:

Eccentric: $\dfrac{25.37}{25.32}$

Handle: $\dfrac{25.37}{25.42}$

Diameter of eccentric and hole in base:

Eccentric: $\dfrac{34.912}{34.900}$

Base: $\dfrac{34.925}{34.938}$

Depth of keyway and key seat in both handle and eccentric:

2.50–2.54

Control of clearance between blades:

Thickness of blade: $\dfrac{3.175}{3.162}$

Dimension A: $\dfrac{14.30}{14.27}$

Dimension B: $\dfrac{20.50}{20.53}$

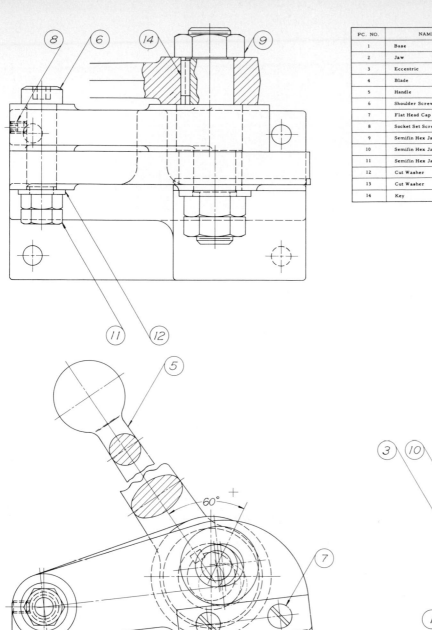

PC. NO.	NAME	MAT.	REQ	NOTES
1	Base	2335	1	Steel Casting
2	Jaw	2335	1	Steel Casting
3	Eccentric	2340 HR	1	
4	Blade	1095 HR	2	
5	Handle	1040	1	Drop Forging
6	Shoulder Screw		1	
7	Flat Head Cap Screw		4	
8	Socket Set Screw		1	Flat Point
9	Semifin Hex Jam Nut		1	
10	Semifin Hex Jam Nut		2	
11	Semifin Hex Jam Nut		2	
12	Cut Washer	1112 CR	1	
13	Cut Washer	1112 CR	1	
14	Key	Key Stock	1	

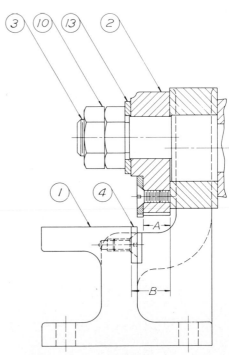

PROB. 58 METRIC. Bench shears (one-half size).

59. The overriding clutch is shown half size. From this design drawing make a complete set of working drawings. The frame is made of cast iron and the shaft is cold-rolled steel. Other parts are as indicated.

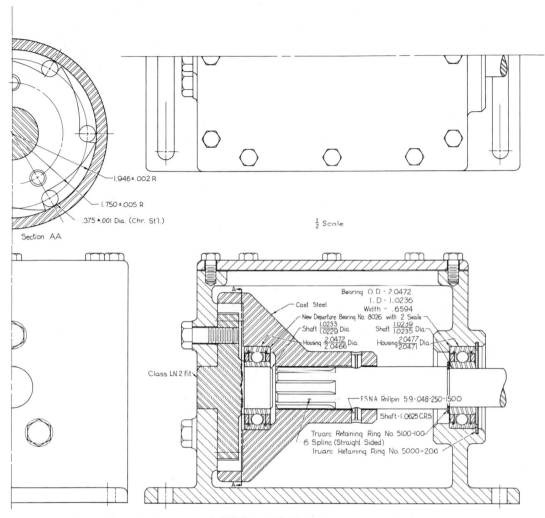

PROB. 59 Overriding clutch.

GROUP 8. WORKING SKETCHES

60. Select one of the problems in Group 1, and make a detail working sketch.

61. Select one of the problems in Group 2, and make an assembly sketch.

62. Select a single part from one of the assemblies of Groups 3 to 5, and make a working sketch.

63. Select a single part from one of Probs. 1 to 43 and make a pictorial working sketch.

64. Select one of the problems that are given in Group 3, and make detail working sketches.

65. Select one of the problems in Group 4 or 5, and make a complete set of working sketches.

14

PRESENTING DATA: CHARTS AND GRAPHS

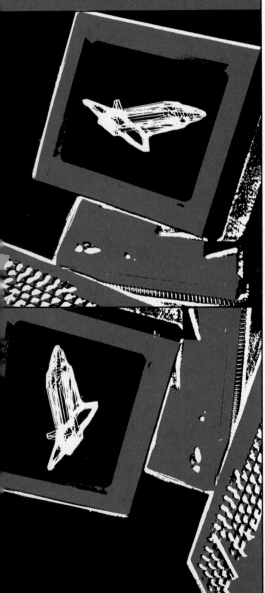

1. THE NEED FOR CHARTS AND GRAPHS

The creation of data is a routine part of the technological world in which we live. Data can be recorded as the interaction of variables. Some variables may be easily noted, such as the number of airplanes landing and taking off at a given airport each hour. Other variables may be less readily measured, such as the relationship between attitude and fuel consumption in a jet engine.

The way in which variables are recorded as data leads to various forms of charts and graphs. Some charts and graphs are designed to be understood by the nontechnical person; others are created for the technician or engineer. The various forms of data expression have their advantages and disadvantages. In this chapter we will review forms of charts and graphs that can be of use to a person wishing to express data.

2. CLASSIFICATION OF CHARTS AND GRAPHS

Charts, graphs, and diagrams fall roughly into two classes: (1) those used for purely technical purposes and (2) those used in advertising or in presenting information in a way that will have popular appeal. The engineer is concerned mainly with those of the first class, but he or she should be acquainted also with the preparation of those of the second and understand their potential influence.

3. TYPES OF CHARTS AND GRAPHS

A. Rectilinear Graphs   Because the preliminary chart work in experimental engineering is done on rectilinear graph paper, the student should become familiar with this form of graph.

Figure 1 shows two forms of rectilinear graphs such as might be made for inclusion in a written report.

A rectilinear graph is made on a sheet ruled with equispaced horizontal lines crossing equispaced vertical lines. The spacing is optional. One commercial graph paper is divided into squares of $\frac{1}{20}$ in, with every fifth line heavier, to aid in plotting and reading. Sheets are available with various other rulings, such as 4, 6, 8, 10, 12, and 16 divisions per inch.

It is universal practice to use the upper right-hand quadrant for plotting experimental-data curves, making the lower left-hand corner the origin. In case both posi-

tive and negative values of a function are to be plotted, as occurs with many mathematical curves, the origin must be placed so as to include all desired values.

B. Logarithmic Graphs Logarithmic graphs are charts in which both abscissas and ordinates are spaced

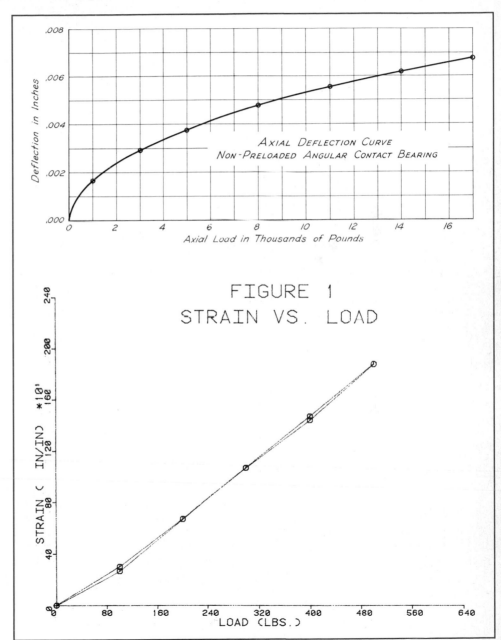

FIG. 1 Rectilinear graphs. The form at top is hand-done. The form at bottom is done on a computer-driven pen plotter.

logarithmically. Any equation of the form $y = ax^b$, in which one quantity varies directly as some power of another, will plot as a straight line on logarithmic coordinates. Thus, multiplication, division, powers, and roots are examples. Figure 2 shows a logarithmic plot of sound intensity (and the power required to produce it), which varies as a power of the distance from the source. Other examples are the distance-time relationship of a falling body and the period of a simple pendulum.

Logarithmic graph paper is sold in various combinations of ruling. It is available in one, two, three, or more cycles; in multiples of 10; and also in part-cycle and split-cycle form. In using logarithmic paper, interpolations should be made logarithmically; arithmetical interpolation with coarse divisions might lead to considerable error.

C. Semilogarithmic Graphs Semilogarithmic graphs have equal spacing on one axis and logarithmic spacing on the other axis. The semilogarithmic graph is often called a "rate-of-change" graph or "ratio chart" because the slope of the curve at any point is the measure of the rate of increase or decrease in the data plotted. It is therefore very useful in statistical work, since it shows at a glance the rate at which a variable changes.

Refer to Fig. 3 for an example. In Fig. 3A data are plotted on an ordinary rectilinear scale. It is difficult to tell the rate of increase of curve 1 compared with curve

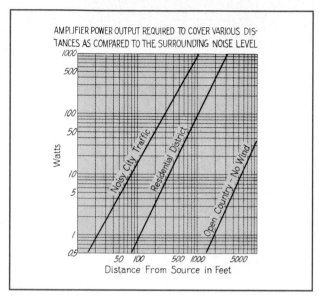

FIG. 2 A logarithmic chart.

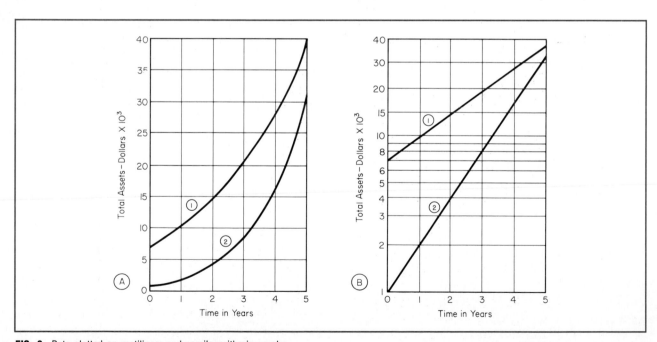

FIG. 3 Data plotted on rectilinear and semilogarithmic graphs.

2. The actual rates of increase show nicely on the semi-log paper seen in Fig. 3*B*. There we see that curve 2 has an increase in assets of 100 percent from one year to the next (e.g., $1000 goes to $2000 in the first year, $2000 goes to $4000 during the second year, etc.). Curve 1, on the other hand, has a much slower rate of increase, about 45 percent per year, or less than half the rate for curve 2. This difference is not readily apparent from looking at the rectilinear plot in Fig. 3*A*.

One must realize, however, that a semilogarithmic graph has certain disadvantages. Some persons may think mistakenly that they are looking at a rectilinear graph. Also, semilogarithmic graphs cannot be used for data having zero or negative values.

In choosing between rectilinear ruling and semilogarithmic ruling, the important point to consider is whether the chart is to represent *numerical* increases and decreases or *percentage* increases and decreases. In many cases it is desirable to emphasize the percentage, or rate, change, not the numerical change; for these a semilogarithmic chart should be used.

Exponential equations of the form $y = ae^{bx}$ plot on semilogarithmic coordinates as a straight line. This represents the type of relationship in which a quantity increases or decreases at a rate proportional to the amount present at any time. For example, the passage of light through a translucent substance varies (in intensity) exponentially with the thickness of material.

D. Nomograms Nomograms allow us to find specific values of variables, using specially constructed charts. Figure 4 shows a fairly simple nomogram. If one variable is known, the second can be found by placing a straight-edge from the known variable through pivot point *P* to read the corresponding variable on the second scale. Construction of nomograms requires a knowledge of scale formulation, nomogram types, and the necessary mathematics.

E. Polar Charts Polar coordinate paper is often used for representing intensity of heat and illumination. An example is seen in Fig. 5, where the area of illumination from a reflector is shown. Polar charts consist of a set of concentric circles with the origin located at the center. Radial lines are drawn from the center. Data can be plotted through any angular sector desired, up to 360°, a full circle.

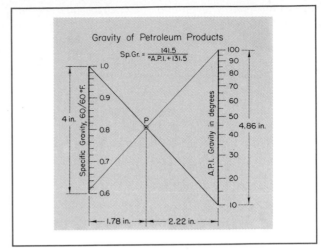

FIG. 4 A nomogram. This one is a conversion chart for specific gravity to degrees API.

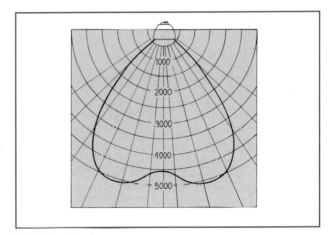

FIG. 5 A polar chart.

Other uses of polar charts are for stresses and loads that occur over a 360° sweep, such as loads on engine bearings. By studying polar charts, engineers can determine the angular positions of peak loads on bearings.

F. Trilinear Charts The trilinear chart, or "triaxial diagram," as it is sometimes called, affords a valuable means of studying the properties of chemical compounds consisting of three elements, alloys of three metals or compounds, and mixtures containing three variables. The chart has the form of an equilateral triangle, the altitude of which represents 100 percent of each of the three constituents. Figure 6, showing the ultimate tensile strength of copper-tin-zinc alloys, is a typical example of its application. The usefulness of such diagrams depends upon the geometric principle that the sum of the perpendiculars to the sides from any point within an equilateral triangle is a constant and is equal to the altitude.

G. Bar Charts The bar chart is a type of chart easily understood by nontechnical readers. One of its simplest forms is the *100 percent bar* for showing the relations of the constituents to a given total. Figure 7 is an exam-

ple. The different segments should be crosshatched, shaded, or distinguished in some other effective manner; the percentage represented should be placed on the related portion of the diagram or directly opposite; and the meaning of each segment should be clearly stated. Bars may be placed vertically or horizontally, the vertical position giving an advantage for lettering and the horizontal an advantage in readability, since the eye judges horizontal distances readily.

Figure 8 is an example of a *multiple-bar chart*, in which the length of each bar is proportional to the magnitude of the quantity represented. Means should be provided for reading numerical values represented by the bars. If it is necessary to give the exact value represented by the individual bars, the values should not be lettered at the ends of the bars, because this would increase their apparent length. This type of chart is made horizontally, with the description at the base, and vertically.

H. Pie Charts The "pie diagram," or 100 percent circle (Fig. 9), is much inferior to the bar chart but is used constantly because of its popular appeal. It is a simple form of chart and, except for the lettering, is easily constructed. It can be regarded as a 100 percent bar bent into circular form. The circumference of the circle is divided into 100 parts, and sectors are used to represent percentages of the total.

To be effective, this diagram must be carefully lettered and the percentages marked on the sectors or at the circumference opposite the sectors. For contrast, it is best to crosshatch or shade the individual sectors. The percentage notation should always be placed where it can be read without removing the eyes from the diagram.

I. Strata Charts Strata charts are commonly used to graphically depict data by superimposing one variable's plot over the plot of one or more other variables. The area between the curves is often shaded to emphasize each variable shown. Figure 10 is a typical strata chart.

The relative position or strength of one variable to another is seen well. However, it must be realized that only the area *under* a curve has meaning, not the area *between* curves.

J. Volume Charts Volume is by nature three-dimensional. Therefore, volume charts show three-dimensional configurations. One common case for volume

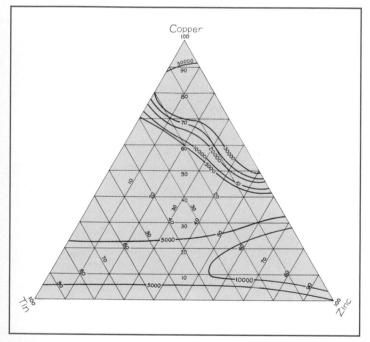

FIG. 6 A trilinear chart. This type is used to represent properties of three combined elements.

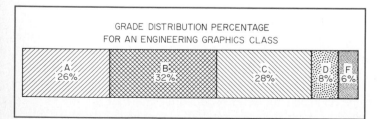

FIG. 7 A 100 percent bar chart.

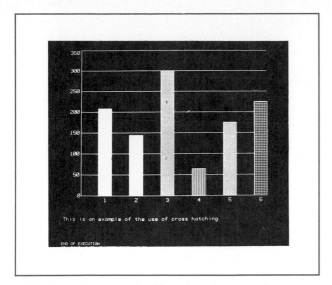

FIG. 8 A bar chart with textured bars drawn by a computer program.

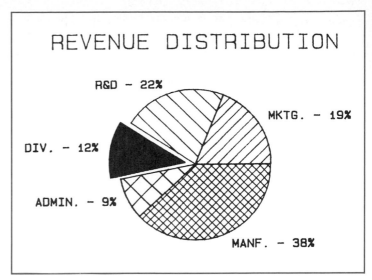

FIG. 9 A pie chart as drawn on a computer-driven plotter. *(Courtesy Hewlett-Packard.)*

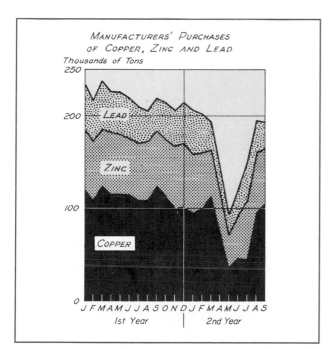

FIG. 10 A strata chart.

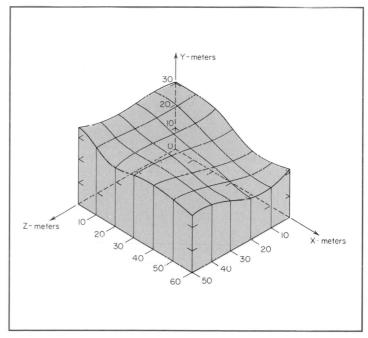

FIG. 11 Volume chart: Model for computation of earth removal.

charts involves the three mutually perpendicular axes x, y, and z.

Figure 11 shows a typical use of the three axes for generating a volume chart. The variables, in this case linear measurements of soil, are plotted on the x, y, z coordinates. It is possible to compute the volume of soil displayed. Problems of this type lend themselves to solution by computer graphics.

K. Classification Charts and Flowcharts The uses to which these two types of charts can be put are widely different, but their underlying principles are similar and so they have been grouped together for convenience.

A *classification chart*, illustrated in Fig. 12, is intended to show the subdivisions of a whole and the interrelation of its parts with one another. Such a chart often takes the place of a written outline, because it gives a better visualization of the facts than can be achieved with words alone. A common application is an organization chart of a corporation or business. It is customary to enclose the names of the divisions in rectangles, although circles or other shapes may be used. The rectangle has the advantage of being convenient for lettering,

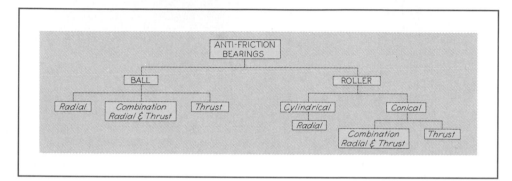

FIG. 12 A classification chart.

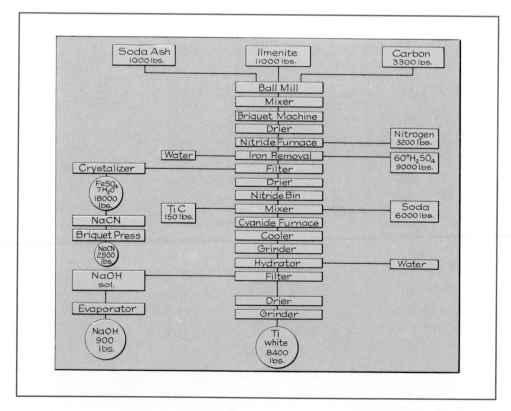

FIG. 13 A flowchart.

while the circle can be drawn more quickly and possesses greater popular appeal. Often a combination of both is used.

The *flowchart* illustrated in Fig. 13 is an example of a route chart applied to a chemical process. Charts of this type show in a dynamic way facts that might require considerable study to be understood from a written description.

4. CHOICE OF GRAPH OR CHART

The function of a chart is to reveal facts. It may be entirely misleading if wrong paper or coordinates are chosen. For example, the growth of an operation, if plotted on a rectilinear chart, might entirely mislead an executive analyzing the trend of a business, while if plotted on a semilogarithmic chart, it would give a true picture of conditions. Intentionally misleading charts have been used many times in advertising; the commonest form is the chart with a greatly exaggerated vertical scale. In engineering work it is essential to present the facts with honesty and scientific accuracy.

5. PLOTTING OF A GRAPH

A. General Procedures So far we have discussed various ways in which data can be presented. In many cases representation alone gives an accurate description of the value of one variable relative to another. It is important to be familiar with sound principles for plotting data in order to provide representation of data that are accurate and easily understandable to the reader.

To plot data effectively follow these general procedures:

1. Assemble and compute all necessary data.
2. Determine the size and type of graph or chart needed to represent the data in the format desired.
3. Determine the scales for the x and y axes to give the best effect to the resulting curve.
4. Lay off the independent variable (abscissa) on the x axis and the dependent variable (ordinate) on the y axis.
5. Plot points from the data.
6. Pencil or ink the curve or curves, as desired.
7. Compose and letter the title, axes, and any other

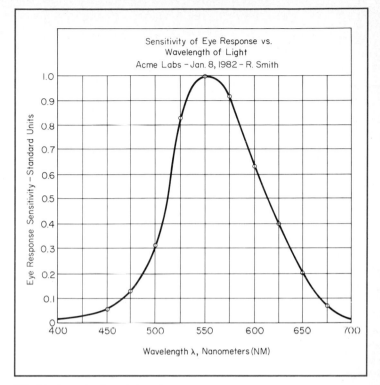

FIG. 14 Example of graph layout.

items needed to make the graph fully understandable.

B. Specific Advice for Plotting a Graph Additional comment is needed to help create a graph that gives a proper total effect. A good graph is like a well-composed picture. It is complete, neat, carefully balanced, and easily readable. Major elements of a graph are the scales, the plotting and drawing of curves, and designations and titles. Each of these areas will be discussed. Figure 14 may be used as an appropriate example.

▪ 1. **Scales** The range of scales should ensure efficient use of the coordinate area. In Fig. 14 the x axis has values from 400 to 700 nm. Values outside this range would not contribute to the graph. The y axis runs from 0 to unity (1) because the values of eye sensitivity have been set in relation to a maximum of 1 unit, which occurs at 550 nm.

For arithmetical scales, scale numbers should correspond to 1, 2, or 5 units of measurement, multiplied or

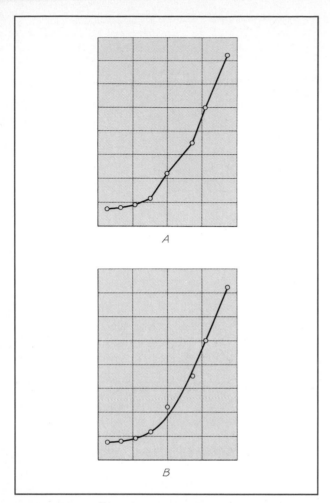

FIG. 15 Methods of drawing curves (A) for data not supported by theory; (B) for data known to produce a smooth curve.

divided by 1, 10, 100, etc. Avoid the use of many digits in the scale numbers. For example, if you are plotting pressures of 0 to 60,000 lb/in^2 on the y axis, you could plot numerals 0 to 6 and within the scale caption indicate "Pressure, lb/in^2 × 10^4." The scale caption should always include the variable measured and the units of measurement.

Scale values and captions should be placed outside the grid area, as seen in Fig. 14. It is also normal to place values and captions at the bottom for the horizontal scale and at the left side for the vertical scale. The overall effect of the scale captions, curves, and open areas should be balanced and aesthetically pleasing.

■ **2. Plotting and Drawing Curves** Observed data points are commonly designated by circles. If several curves are to be placed on the same grid, squares and triangles are often used in addition to circles to tell one curve from the other.

In drawing graphs from experimental data, it is often a question whether the curve should pass through all the points plotted or strike a mean between them. In general, observed data not backed up by definite theory or mathematical law are shown by connecting the points plotted with straight lines, as in Fig. 15A. An empirical relationship between curve and plotted points may be used, as at B, when, in the opinion of the engineer, the curve should exactly follow some points and go to one side of others.

If one curve is shown, it is best to use a solid line. If several curves are shown, you may use alternate forms of lines for additional curves, such as dashed or dotted lines.

■ **3. Designations and Titles** If there is more than one curve on the same graph a brief label should be placed close to each curve. Occasionally curves may be indicated by letters or numbers that require a key or code. This method is convenient if the labels themselves are lengthy or if space is tight near the curves. If a key is used it should be placed within the grid area in an open area, often closed by the border of straight lines. The different curves should be drawn so as to be easily distinguishable. This can be done by varying the character of the lines, using full, dashed, and dot-and-dash lines, with a tabular key for identification.

The title is an important part of a chart, and its wording should be clear and concise. In every case, it should contain sufficient description to tell the nature of the chart, the source or authority, the name of the observer, and the date. Approved practice places the title at the top of the sheet, arranged in phrases symmetrically about a centerline. If it is placed within the ruled space, a border line or box should set it apart from the sheet.

6. PLOTTING OF DATA TO BE A STRAIGHT LINE

Data should be plotted on coordinates that will cause the curve to appear as a straight line if possible. The equation of a straight line is then easily found. Equations found from data points are known as "empirical equations."

The usual procedure is first to plot the data on rectangular coordinates. If a straight line results, the equation is of the first degree (linear). If a curve still results from the plot, the equation is of a second or higher degree. Semilogarithmic or logarithmic coordinates should be tried. If the data then appear as a straight line, the equation can be determined. If not, more sophisticated methods, aided by computer programs, can assist.

7. LINEAR EQUATIONS

Empirical data that plot as a virtually straight line on rectilinear coordinates are known to have an equation of the general form $y = a + bx$, where a and b are constants. This equation is in the slope-intercept form, with a as the y intercept and b as the slope of the line.

Two methods by which the constants of the equation can be found will be illustrated. The first method is known as the slope-intercept method; the second method is that of selected points.

A. Slope-Intercept Method Figure 16 illustrates the graphic procedure for determination of the equation from observed data recorded in an experiment. These data are plotted as shown in the figure, using light, sharp, intersecting lines for the location of points. Then the best straight-line representative of the data is drawn through as many of the points as possible, striking an average between points not on the line.

The y intercept of this representative line will be the value of a in the equation and may be measured directly from the y scale (at $x = 0$). In this case, the value is seen to be 5.0. To complete the equation, the slope b is measured.

In Fig. 16, two selected points on the line (shown by the dashed line extending to the axes) will have the coordinates of x_1, y_1 and x_2, y_2. The tangent is found from the equation

$$\frac{y_2 - y_1}{x_2 - x_1} = b$$

$$\frac{43.5 - 13.8}{35.0 - 8.0} = \frac{29.7}{27} = 1.1$$

The equation of the line is

$$y = 5 + 1.1x$$

Note especially that this method as well as the following one requires the slope and the intercept to be calculated from the representative line rather than from the experimental data. This point is often neglected, and the equation obtained by use of the data can be extremely inaccurate. The purpose of the representative line is to average graphically all the data values and thereby eliminate the effects of various inaccuracies.

B. Method of Selected Points The equation of a straight line plotted on rectilinear coordinates may be evaluated by reading coordinates of two selected points on the line, x_1, y_1 and x_2, y_2, and substituting in the equation

$$\frac{y - y_1}{x - x_1} = \frac{y_2 - y_1}{x_2 - x_1}$$

This equation simply states the fact that the slope of a straight line is everywhere constant and sets up a proportion between corresponding changes in x and y values.

From the graph of Fig. 16, coordinates of two selected points are

$$x_1 - 8 \qquad x_2 = 35.0$$
$$y_1 = 13.8 \qquad y_2 = 43.5$$

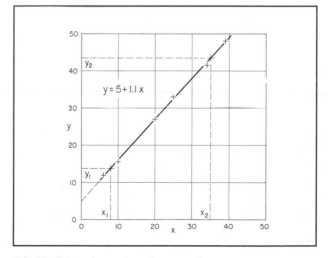

FIG. 16 Data and curve for a linear equation.

Substituting in the equation, we get

$$\frac{y - 13.8}{x - 8.0} = \frac{43.5 - 13.8}{35.0 - 8.0}$$

$$\frac{y - 13.8}{x - 8.0} = 1.1$$

$$y = 5.0 + 1.1x$$

It should be noted that the coefficient 1.1 is the slope of the line

$$b = \frac{y_2 - y_1}{x_2 - x_1}$$

and that the constant term 5.0 is the ordinate-axis intercept a.

Figure 17 shows a practical example of the determination of an equation by the method of selected points. Brinell hardness and tensile strength of a number of wrought-aluminum alloys in the soft or annealed condition are tabulated. These data are plotted on rectangular coordinates in Fig. 17, and a representative straight line is drawn through the points. Coordinates of two selected points on the line are

$$B_1 = 25 \qquad B_2 = 55$$
$$T_1 = 14,700 \qquad T_2 = 31,500$$

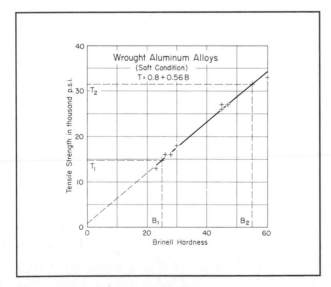

FIG. 17 Data and curve for a linear equation.

Writing the equation of the straight line, we get

$$\frac{T - 14,700}{B - 25} = \frac{31,500 - 14,700}{55 - 25}$$

$$T = 700 + 560B$$

It seems reasonable to make a comparison between the slope-intercept method and the method of selected points. In the slope-intercept method, the y-intercept value may be read from the graph, but the slope must be obtained by a pair of selected points; in the method of selected points, the equation is determined solely from the selected points. Also, in many cases, the y intercept will not appear on the graph, making the method of selected points the better choice.

8. LOGARITHMIC PLOTS: POWER EQUATIONS

Power relationships occur frequently in nature. The distance-time relationship of a falling body, the period of a simple pendulum, and the velocity of free discharge of water from an orifice are power relationships.

The power equation $y = ax^b$, in which one quantity varies directly as some power of another, plots on the uniform rectangular coordinates of Fig. 18A as a family of parabolic and hyperbolic curves passing through the point $(1, a)$. When the exponent b is equal to unity, the curve is a straight line through the origin. For other positive values of b, the curves are parabolic, symmetric with respect to the y axis for values of b greater than 1, and symmetric with respect to the x axis for values of b less than 1.

Negative values of b result in hyperbolic curves having their coordinate axes as asymptotes, one variable decreasing as the other increases. When placed in logarithmic form, the equation becomes $\log y = \log a + b \log x$, in which $\log y$ and $\log x$ are variable and $\log a$ and b are constant terms. Hence, curves of this type may be rectified either by plotting the logarithms of x and y on rectilinear coordinates or by plotting y versus x directly on logarithmic coordinates as in Fig. 18B.

An example of data giving a power equation is shown in Fig. 19, where a test resulted in tabulated values of two variables, as shown.

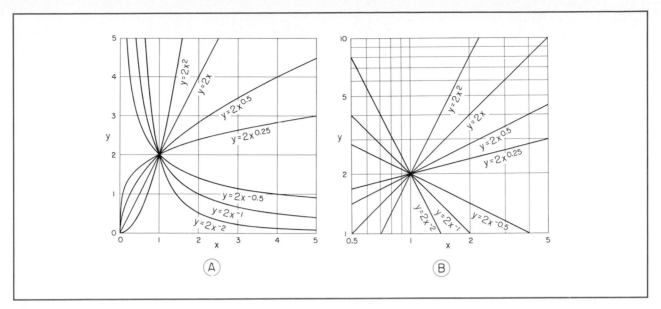

FIG. 18 Plots of power equations.

These data plot on uniform coordinates as a curved line but when plotted on logarithmic coordinates as in Fig. 19, they rectify to a straight line.

The representative straight line is drawn through the points. Coordinates of two selected points on the line are

$$x_1 = 2.5 \qquad x_2 = 55.0$$
$$y_1 = 3.8 \qquad y_2 = 26.5$$

The equation of a straight line by the method of selected points is

$$\frac{\log y - \log 3.8}{\log x - \log 2.5} = \frac{\log 26.5 - \log 3.8}{\log 55.0 - \log 2.5}$$

Evaluating logarithms, using base 10, we have

$$\log y = 0.330 + 0.628 \log x$$
or
$$y = 2.14x^{0.628}$$

Note that the coefficient 2.14 is the ordinate-axis intercept obtained by extending the line to $x = 1$ and that the exponent 0.628 is the slope of the rectified curve. The y intercept occurs at the coordinate line $x = 1$ since $\log 1 = 0$. The value 1 is therefore the origin of the logarithmic scale.

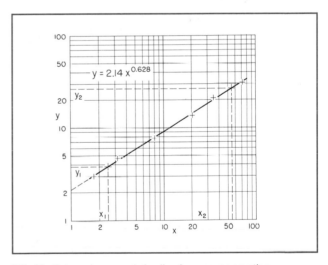

FIG. 19 Data and representative line for a power equation.

9. SEMILOGARITHMIC PLOTS: EXPONENTIAL EQUATIONS

In nature many physical quantities increase or decrease at a rate proportional to the amount present at any time. As time varies arithmetically by a constant difference, the quantity changes geometrically by a constant factor.

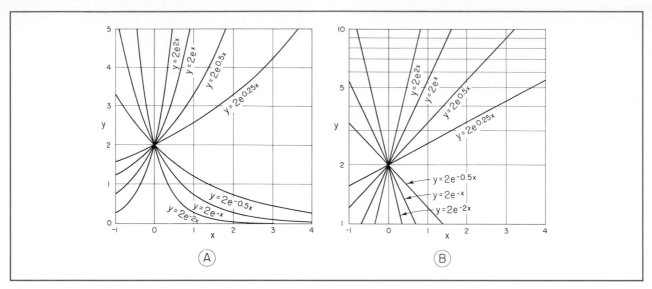

FIG. 20 Plots of exponential equations.

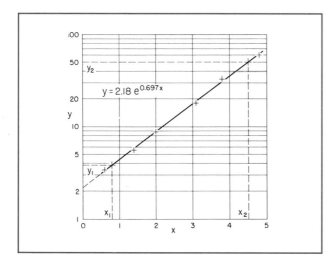

FIG. 21 Data and representative line for an exponential equation.

Exponential relationships involving time change are found in formulas for transient currents in an inductive electric circuit, for the decay of natural radioactivity, and for the increase in a sum of money at compound interest. Variables other than time may also be involved. The intensity of light passing through a substance varies exponentially with the thickness of the material.

The exponential equation $y = ae^{bx}$ or $y = a10^{bx}$ plots on rectilinear coordinates of Fig. 20A as a family of curves passing through the point $(0, a)$ and having the x axis as asymptote. When placed in logarithmic form, the equation $y = ae^{bx}$ becomes

$$\ln y = \ln a = bx \ln e$$

Since $\ln e = 1$,

$$\ln y = \ln a + bx$$

Similarly the equation $y = a10^{bx}$ becomes

$$\log y = \log a + bx$$

Curves of this type may be rectified by plotting either $\ln y$ versus x or $\log y$ versus x on rectilinear coordinates or by plotting y versus x directly on semilogarithmic coordinates in which the y scale is logarithmic and the x scale uniform, as shown in Fig. 20B.

An example of data giving an exponential equation is shown in Fig. 21, where a test resulted in tabulated values of the two variables. These data plot on uniform coordinates as a curved line. However, when the data are plotted on semilogarithmic coordinates, as in Fig. 21, with a logarithmic y scale and a uniform x scale, a representative straight line may be drawn through the points.

Coordinates of two selected points on the line are

$$x_1 = 0.8 \qquad x_2 = 4.5$$
$$y_1 = 3.8 \qquad y_2 = 50$$

Writing the equation of the straight line gives

$$\frac{\ln y - \ln 3.8}{x - 0.8} = \frac{\ln 50 - \ln 3.8}{4.5 - 0.8}$$
$$\ln y = 0.777 + 0.697x$$
$$y = 2.18e^{0.697x}$$

If the use of logarithms to the base 10 is preferred, the equation of the straight line is written

$$\frac{\log y - \log 3.8}{x - 0.8} = \frac{\log 50 - \log 3.8}{4.5 - 0.8}$$
$$\log y = 0.338 + 0.302x$$
$$y = 2.18(10)^{0.302x}$$

10. AESTHETIC QUALITIES IN PRESENTING DATA

Our intent to this point has been to get a graph or chart technically correct and readable, and to find the equation for data that can plot to a straight line. We have seen the need for information on a graph to be accurate, complete, and not misleading. Beyond these critical factors a graph should also be aesthetically pleasing and reasonably dynamic in its format. Aesthetic qualities are particularly important if a graph or chart is to be used in a presentation, such as to an audience of persons interested in your subject.

A. Special Effects in Lettering and Line Work If you need assistance in creating dynamic special effects, consult an industrial illustrator. Large companies usually have personnel skilled in developing high-quality layouts used for advertising, staff seminars, and service literature, and such illustrators are available for consultation if needed.

You also should be aware of several aids that can provide useful enhancement of technical data, for example, lettering materials and transfer films. There are readily available today a wide selection of special inking pens, including felt-tip markers, which can give various line widths and colors. Commercial lettering guides can provide mechanically drawn letters in a wide variety of styles.

Transfer films offer an almost limitless means of creating interesting effects for data presentation. One may use dry transfer letters, which are transferred from the film to the graph by rubbing over the letters, or transfer lines in different colors. Also available are colored acetate strips in different widths that will adhere directly to the graph. One can also obtain sheets of different hues and tints to be used as background on which the graph is created. Another variation, the gummed overlay, is seen in Fig. 22.

B. Use of Color and Tints Color can add to the readability and emphasis of a graph. If only black and white are used, various line codes (dashes, dots, etc.) must be employed and areas can be emphasized only by crosshatching. In color, lines stand out and areas are easily distinguished. The use of color requires artistic judgment, and definitive rules are difficult to provide; however, the following principles may serve as a guide.

▪ **1. Backgrounds** A pure white background for a chart is not as effective as a lightly tinted background, which can add cohesiveness and vigor to graphs and charts. White lines and areas can be produced only if a tinted background is used. Figure 23 shows a tinted background and use of white areas. Tints of a color are made either by diluting the color with white or, in printing, by use of screens that reduce the color intensity. In printing, screens are rated in percentages: a 75 percent screen will print (in tiny dots) 75 percent color and 25 percent white (blank area).

▪ **2. Contrast** Contrast may be defined as the degree of dissimilarity. High contrast occurs when a color of strong intensity is placed alongside a color of weak intensity. Low contrast results when two colors of about the same intensity are placed side by side. High contrast emphasizes adjacent areas, while low contrast diminishes the impact of adjacent areas.

Adjacent areas are well defined and easily read if they are made of different colors. Note in Fig. 24 that adjacent areas are different colors and also that different values of the colors have been used. Maximum separation of areas occurs when a strong color and value are placed next to a weak color and value. In some cases maximum separation is needed, in others a more gradual change is desirable. For example, in Fig. 25 the stronger colors are

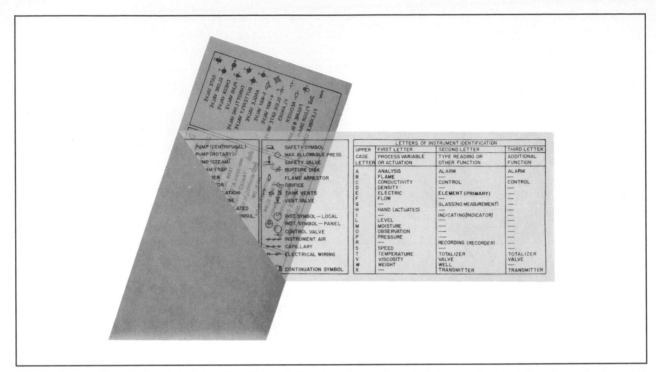

FIG. 22 A gummed overlay for tabular information. The illustration shows the overlay partially in place, to indicate how the overlay is applied. *(Courtesy of Stanpat Co.)*

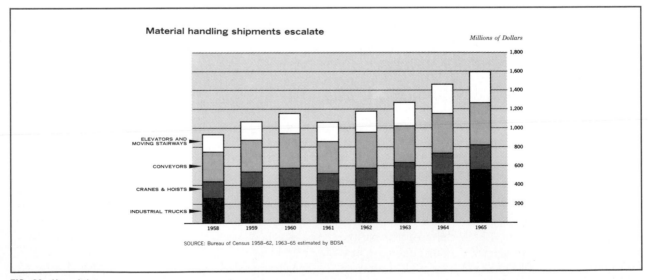

FIG. 23 Use of tints.

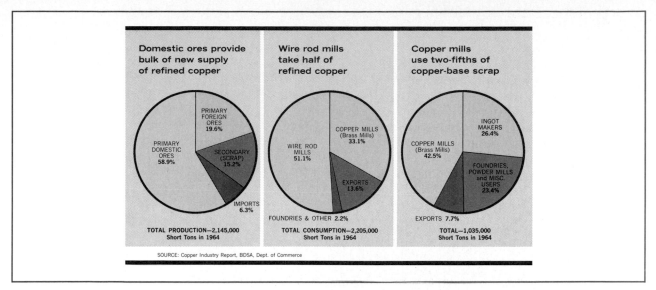

FIG. 24 A series of pie charts in color. Readability and appeal is superior to single-color rendition.

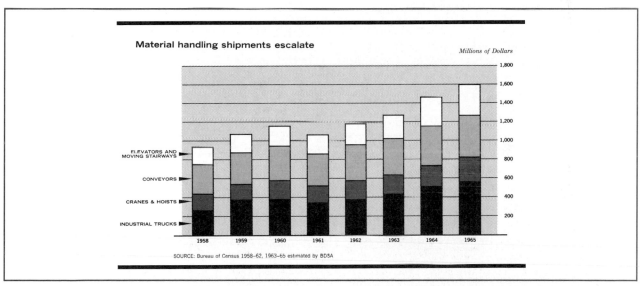

FIG. 25 A multiple-bar chart in color. Two basic colors and tints of each are used. Note the difference in effect between this chart and the one of Fig. 23.

used at the bottom of the bars and the weaker colors at the top. This is done not only for the sake of stability of the bars, but because, if adjacent areas of the bars are made as high-contrast areas, the bars become "spotty" in appearance and the value of the continuous bar is lost.

The best separation of *lines* occurs when adjacent lines are contrasty, in either color or value, or both. Note the use of alternate lines in Fig. 26 and the contrasting white line at the top.

▪ **3. Choice of Colors** The actual choice of colors is almost entirely a matter of artistic judgment, but this is

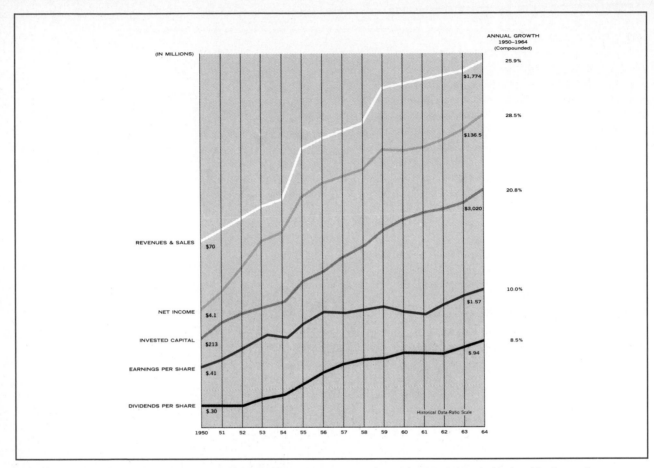

FIG. 26 A graph in color. Note that one more colored line is produced by employing a tinted background with the white line on it.

often tempered by other factors. Colors at the high (wavelength) end of the spectrum are known as "cold" colors. Colors at the low end of the spectrum are known as "warm" colors. The colors of nature are the blue of the sky; the greens of trees, grass, and shrubbery; and the browns of tree trunks, soil, and rock. Note that these colors comprise a cold color, a medium color, and a warm color. However, even though we view the colors of nature daily and are not disturbed by them, such colors are not commonly put together in a chart. Also, studying the colors of the spectrum will be of little help in choosing colors to be used for a chart, graph, or diagram. This is because the *only* important considerations are those of pleasing and visually effective combinations, contrasts, and color values.

Colors that are not pleasingly compatible should not be used together unless a startling or garish effect is wanted. For example, violet, purple, and yellow are almost never used for charts. The colors used most are blues, greens, reds, browns, white, and black. The symbolic colors of a company or organization may also play a part in the choice, as for example, the blue and white of the Pure Oil Company or the red, white, and blue of the Exxon Corporation. In addition, many of the charts and graphs in such magazines as *Scientific American* have been printed in black, shades of black (gray), red, blue, and green, with occasionally some yellow, but black, gray, and red are predominant.

The best practice is to lay out the chart and then try several colors by holding colored sheets or strips over the chart to judge finally the most pleasing and effective combination.

When the chart is made exclusively for reproduction, the basic chart is outlined and drawn in india ink as would be done for a black-and-white chart, with the exception that areas are not blacked in. Then, an overlay sheet of tracing paper or drafting film such as Mylar is placed over the chart and fastened with drafting (or transparent) tape. On this overlay sheet directions are lettered or written, designating the colors to be used for lines and areas. The engraver, following these directions, will then make plates for the printing of each color. In giving the directions for color and color value, either an approximate value must be given, such as *light blue*, or an exact value must be designated, such as 30 percent blue.

This brief section has only touched on some of the major factors important for increasing the effectiveness of a graph or chart. Industrial illustration studios and libraries, as well as helpful persons in the field, can give you valuable additional assistance in generating the special emphasis desired in presenting data. Make good use of these available resources.

PROBLEMS

GROUP 1. PLOTTING OF DATA

1. The data given below were obtained in a tension test of a machine-steel bar. Plot the data on rectangular coordinates, using the elongation as the independent variable, the applied load as the dependent variable.

Applied load, lb/in^2	Elongation per inch of length
0	0
3,000	0.00011
5,000	0.00018
10,000	0.00033
15,000	0.00051
20,000	0.00067
25,000	0.00083
30,000	0.00099
35,000	0.00115
40,000	0.00134
42,000	0.00142

2. In testing a small 1-kW transformer for efficiency at various loads, the following data were obtained: Watts delivered: 948, 728, 458, 252, 000. Losses: 73, 62, 53, 49, 47.

Plot curves on rectangular coordinate paper, showing the relation between percentage of load and efficiency, using watts delivered as the independent variable and remembering that efficiency = output ÷ (output + losses).

3. A test of the corrosive effect of 5 percent sulfuric acid, both air-free and air-saturated, on 70 percent nickel, 30 percent copper alloy (Monel) over a temperature range from 20 to 120°C resulted in the data tabulated below. Plot these data on rectangular coordinates, with corrosion rate as ordinate vs. temperature as abscissa.

Temperature, °C	CORROSION RATE, MDD (mg/dm^2 PER DAY)	
	Acid saturated with air	Acid air-free, saturated with N$_2$
20	195	35
30	240	45
40	315	50
50	425	63
50	565	70
70	670	74
80	725	72
83	715	70
85	700	68
92	580	57
95	470	42
101	60	12

4. The data given below were obtained from a test of an automobile engine.

Plot curves on rectangular coordinate paper, showing the relation between fuel used per brake horsepower–hour and brake horsepower developed. Show also the relation between thermal efficiency and brake horsepower developed, assuming the heat value of the gasoline to be 19,000 Btu/lb.

r/min	Length of run, min	Fuel per run, lb	Bhp
1006	11.08	1.0	5.5
1001	4.25	0.5	8.5
997	7.25	1.0	13.0
1000	5.77	1.0	16.3
1002	2.38	0.5	21.1

5. A test of the resistance of alloy steels to high-temperature steam resulted in the data tabulated below. Represent the results of the text graphically by means of a multiple-bar chart.

CORROSION OF STEEL BARS IN CONTACT WITH STEAM AT 600°C FOR 2000 h

Steel	Average penetration, cm
SAE 1010	0.00432
1.25 Cr-Mo	0.00297
4.6 Cr-Mo	0.00243
9 Cr 1.22 Mo	0.00176
12 Cr	0.00011
18-8-Cb	0.00003

6. From the data below, plot curves showing the "thinking distance" and "braking distance." From these curves, plot the sum curve "total distance." Title "Automobile Minimum Travel Distances When Stopping—Average Driver."

mph	ft/sec	Thinking distance, ft	Braking distance, ft
20	29	22	18
30	44	33	40
40	59	44	71
50	74	55	111
60	88	66	160
70	103	77	218

7. Put the data given in Fig. 9 into 100 percent bar form.

8. Put the data of Fig. 7 into pie-chart form.

9. On trilinear coordinate paper plot the data given below. Complete the chart, identifying the curves and lettering the title below the coordinate lines.

FREEZING POINTS OF SOLUTIONS OF GLYCEROL AND METHANOL IN WATER

Water	WEIGHT, % Methanol	Glycerol	Freezing points, °F
86	14	0	+14
82	8	10	+14
78	6	16	+14
76	24		−4
74	2	24	+14
71	17	12	−4
70	0	30	+14
68	32	0	−22
65	10	25	−4
62	23	15	−22
60	40	0	−40
58	5	37	−4
57	17	26	−22
55	0	45	−4
54	28	18	−40
53	47	0	−58
52	10	38	−22
50	20	30	−40
47	31	22	−58
45	0	55	−22
42	7	51	−40
39	12	49	−58
37	0	63	−40
36	8	56	−58
33	0	67	−58

10. Thermal conductivities of a particular insulating material at various mean temperatures are tabulated as shown.

Plotting the data on uniform rectangular coordinates with thermal conductivity as ordinate and mean temperature as abscissa, evaluate the constants a and b in the linear equation $K = a + bT$ relating the variables.

Mean temperature (hot to cold surface) T, °F	Thermal conductivity K, Btu/(h)(ft²)(°F) temperature difference, in thickness
100	0.365
200	0.415
320	0.470
400	0.520
460	0.563
600	0.628
730	0.688
900	0.780

11. Corrosion tests on specimens of pure magnesium resulted in values for weight increase in pure oxygen at 525°C tabulated as shown.

Plotting the data on uniform rectangular coordinates with weight increase as ordinate and elapsed time as abscissa, evaluate the constant a in the linear equation $W = aT$ relating the variables.

Elapsed time T, h	Weight increase W, mg/cm²
0	0
2.0	0.17
4.0	0.36
8.2	0.65
11.5	1.00
20.0	1.68
24.1	2.09
30.0	2.57

GROUP 2. EMPIRICAL DATA OF KNOWN EQUATION FORM

The following problems are identified with respect to the form of equation that fits the given data. Preliminary plots of the data should be made on rectilinear coordinates in order to illustrate the characteristics of each type of equation.

Completed plots of the rectified data should include careful calibration of the scales, properly labeled scale captions, and the equation of the data with the constants evaluated.

12. Measurement of the electrical resistance of a 1-ft length of no. 0 AWG standard annealed copper wire at various temperatures resulted in the data shown in the accompanying table.

After plotting the data on uniform rectangular coordinates with resistance as ordinate and temperature as abscissa, evaluate the constants a and b in the linear equation $R = a + bT$ relating the variables.

Temperature T, °C	Resistance R, $\mu\Omega$
14.0	96.00
19.2	97.76
25.0	100.50
30.0	102.14
36.5	105.00
40.1	105.75
45.0	108.20
52.0	110.60

13. Creep-strength tests of a high-chromium (23 to 27 percent) ferritic steel used in high-temperature service resulted in the stress values to produce a 1 percent deformation in 10,000 h at various temperatures, tabulated as shown.

After plotting the data on semilogarithmic coordinates with stress as ordinate on a logarithmic scale and temperature as abscissa on a uniform scale, evaluate the constants a and b in the exponential equation $S = a10^{bT}$ relating the variables.

Temperature T, °C	Stress S, kPa
540	35,500
590	14,400
650	8,600
700	3,900
760	2,200

14. Approximate rates of discharge of water under various heads of fall through a 300-m length of 10-cm pipe with an average number of bends and fittings tabulated as shown.

After plotting the data on logarithmic coordinates with discharge as ordinate and head as abscissa, evaluate the constants a and b in the power equation $Q = aH^b$ relating the variables.

Head of fall H, m	Discharge Q, L/min
0.3	134
0.6	188
1.2	264
1.8	322
3.0	425
3.5	452
5.0	545
6.0	596
7.5	672
9.0	732
12.0	849
15.0	948
23.0	1162
30.0	1340

Gap length L, cm	Peak voltage V, kV
5	0.16
10	0.32
15	0.48
20	0.64
25	0.81
30	0.98
35	1.15
40	1.32
45	1.49
50	1.66
60	2.01
70	2.37
80	2.74
90	3.11
100	3.49

16. Capacities of horizontal conveyer belts, carrying materials with a density of 50 lb/ft^3 at a speed of 200 ft/min and running on standard 20° troughing idlers, are tabulated as shown.

Belt width W, in	Capacity C, short tons (2000 lb)/h
12	29
18	68
24	125
30	200
36	290
42	406
48	550
54	716
60	900

GROUP 3. EMPIRICAL DATA OF UNKNOWN EQUATION FORM

Preliminary plots should be made on rectilinear coordinates for data in all the following problems so that the equation form may be determined. Further preliminary plots of the data should be made on the correct coordinate system required for rectification. This will avoid loss of time and waste of large coordinate paper if an improper choice of coordinate has been made. Constants of the equations may be obtained graphically from an accurate rectified plot of the data. The completed plot of the data should include scales with proper graduations, calibrations, legends, and the equation with constants evaluated for the data as part of a descriptive title.

15. Spark-gap breakdown voltages for smooth spherical electrodes 25 cm in diameter and in clean dry air at 25°C and 760 mm are tabulated as shown for various gap lengths.

17. Breaking loads for standard annealed copper wire are given in the accompanying table.

Wire size G, AWG	Breaking load L, N
0	13,260
1	10,800
2	8,590
3	6,800
4	5,385
5	4,280
6	3,390
7	2,695
8	2,140
9	1,690
10	1,410
11	1,105
12	875
13	695
14	550

18. Corrosion tests on specimens of pure magnesium resulted in values for weight increase in pure oxygen at 525°C tabulated as shown.

Elapsed time T, h	Weight increase W, mg/cm²
0	0
2.0	0.17
4.0	0.36
8.2	0.65
11.5	1.00
20.0	1.68
24.1	2.09
30.0	2.57

19. Thermal conductivities of a particular insulating material at various mean temperatures are tabulated as shown.

Mean temperature (hot to cold surface) T, °F	Thermal conductivity K, Btu/(h) (ft²)(°F) temp. diff. in thickness
100	0.365
200	0.415
320	0.470
400	0.420
460	0.563
600	0.628
730	0.688
900	0.780

20. Heat losses from horizontal bare-iron hot-water pipes at 180°F to still ambient air at 75°F are tabulated as shown.

Pipe size D, in	Heat loss H, Btu/ft per day
1	1,896
1¼	2,398
1½	2,746
2	3,430
2½	4,140
3	5,054
3½	5,771
4	6,493
4½	7,211
5	8,020
6	9,530
8	12,442
10	15,528
12	18,418
14	20,140
16	23,088
18	25,998

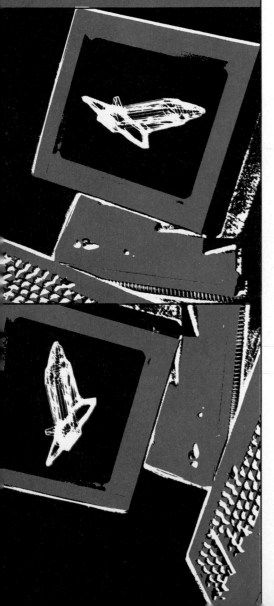

15

COMPUTERS, COMPUTER GRAPHICS, AND COMPUTER-AIDED DESIGN

1. INTRODUCTION

Computers have had two related impacts on graphics. The more obvious effect is the ongoing computerization of the production of the working drawings necessary in the design and production processes of modern industry. The other effect is the enhanced role that graphics itself now has in design, analysis, visualization, control, and representation, not only in industry but in business, education, and cultural pursuits. Even when the quality of the images has been poor, computer-generated graphics have stimulated an increased interest in graphics. One possible reason for this may be that poor line work is no longer an inhibiting factor; everyone's lines look the same in computer graphics.

Computer graphics, like all graphics, is a means of communication. As a means of communication, it can be used to convey information much more rapidly than can the written word. This can only happen, however, where the graphic representation is readily understood. Drawings of an object that are descriptive usually constitute a very rapid means of communication, but they may take a long time to produce. Computer graphics are also time-consuming to produce the first time. One of the great advantages of computer graphics is like the advantage of electronic word processing over writing: the product is very easy to edit and reuse in a variety of ways and places.

Once a drawing has been done by computer it may be filed away just as a traditional vellum drawing may be filed, although a computer file is stored in a tiny fraction of the space that a vellum drawing requires. The computer version may then be copied back into the computer, a modification made, and a completely new file saved in a few minutes. In the case of a vellum drawing, either the old drawing would be modified and you would only have the modified version, or you would have to do a completely new drawing.

The term "computer graphics" refers to any graphics display produced with the aid of a computer, but its usage is sometimes restricted to the more mathematically oriented techniques of geometric modeling as distinct from computer-aided drafting in industry. CAD is used to refer to computer-aided design, com-

This chapter was written by Richard F. Devon, Assistant Professor of Engineering Graphics, The Pennsylvania State University.

puter-aided drafting, or both. CADD is used to refer to computer-aided design and drafting. CAM stands for computer-aided manufacturing. In this chapter we will use CAD to refer to computer-aided drafting.

In CAD a particular shape that has been saved not only can be copied into the computer, but it can be moved around the screen, scaled up or down, distorted, or rotated more or less instantaneously. Sophisticated CAD systems can even do this in three dimensions. Working with files of modular shapes and drawings of frequently used parts, complete with dimensions, industrial working drawings may now be assembled very quickly. Revisions can be made extremely rapidly, and the drawings in their electronic form can be transmitted by computer networks and telephone lines with little or no delay. This means the link between design and manufacture/construction can be greatly speeded up and labor costs reduced. In many large industries these savings in cost have been found sufficient to justify the very large capital expenditures required to acquire and maintain sophisticated CAD systems.

2. COMPONENTS OF A CAD SYSTEM

A computer processes information. For a computer to be of any use there must be a means of giving it information to process (*input*), instructions for the method of processing (*programs*) and a means of retrieving the results (*output*). Because computers handle very large amounts of information, and because their own memory is volatile so the information is lost when the power is lost, it is also highly desirable to have a means of storing information (*storage*).

A typical CAD system will include one or more input, output, and storage devices, as well as the central processing unit. The arrangement at which a drafter or engineer works is known as a "workstation." Workstations are usually networked together so that everyone works with a common and readily accessible catalog of drawings and design data (*database*). An example of a workstation is shown in Fig. 1.

The three functions—input, output, and storage—are discussed more fully below. Included in the discus-

FIG. 1 The Engineering Graphics System EGS/200 of Hewlett-Packard. A typical workstation is shown on the left. The workstation on the right has a large display unit. The processor and storage unit in the center and the plotter in the rear are shared resources. *(Courtesy Hewlett-Packard.)*

sion are many examples of current technologies. However, there are many exciting new technologies which may prove cost-competitive that are not mentioned, such as bubble memory, flat screen devices, ink-jet printers, laser printers, and laser storage on videodisks. Programs will be discussed in the next section of the chapter.

A. Input The computer can only accept electronic signals as information, but there are a surprising number of ways of producing the necessary signals. Some of these devices are listed here.

▪ **Keyboard** The keyboard is usually similar to a typewriter, but it may include other features such as special function keys or a numeric keypad. When a key on the keyboard is depressed, a number which corresponds to that letter is sent to the computer. A number is sent because the computer works only with numbers in its internal processing. The keyboard typically is used in one of two ways. It is used for writing—either programs or word processing. It is also used during the operation of a particular program to allow the user sitting at the keyboard to input further instructions or information by responding to questions that appear on the screen.

▪ **Digitizing Tablet** Also known as a "graphics tablet," the digitizing tablet is a device consisting of a flat operating surface under which is a fine wire grid. When an electronic stylus or cursor is placed over a certain spot and activated, usually by a button, the computer is sent the numerical identity of that spot. It is easiest to think of this identity as the coordinates of the position, and some tablets have a digital display for the coordinates. Tablets are used most in CAD, but popular "pads" exist for inexpensive microcomputers. The signal sent by the tablet may be treated as spatial information and permit the user to draw electronically. It may also be interpreted as an instruction if, in a given program, a certain location on the tablet corresponds to a particular command, e.g., draw a circle. Digitizing tablets used in industry vary from desktop size to a full-sized drawing board. Figure 2 shows an example of a very large digitizer with a cursor pad. A smaller tablet with a stylus can be seen in Figure 3.

▪ **Optical Scanner** The optical scanner runs a head

FIG. 2 CalComp's 9100 Series digitizer can be operated with either a cursor, shown here, or a pen stylus. *(Courtesy CalComp.)*

FIG. 3 The CalComp System 25 interactive design station. *(Courtesy CalComp.)*

with a light beam and a reflected light sensor over a drawing and records the location of the dark object lines. The computer codes and stores this information and reproduces it as required. Although not always error-free, it is incomparably faster than a digitizing tablet for copying a complex drawing into the computer.

▪ **Screen Input** The screen is primarily an output device in a computing system, although it greatly facilitates keyboard input by displaying the keyboard entries and program listings for checking and editing. It may also be used directly for input. An electronic stylus known as a "light pen" may be applied to a screen in much the same manner as it would for the digitizing tablet. Of course, the computer must be suitable for this and the operating program must be designed for its signals. Another type of screen input occurs with a special type of screen called a "touch screen." There are several different technologies in existence that can sense a finger being applied to the screen and transmit a signal to the computer that corresponds to the location touched. This technique is a little too crude to permit electronic drawing, but it works well for making choices between items displayed in boxes on the screen.

▪ **Paddles** These come as a pair of hand-held rotary controls. They are usually used in games, where each paddle corresponds to either vertical or horizontal coor-

dinates, and they can be used to control the movement of an object on the screen. They are rather limited in function, but quite inexpensive.

▪ **Trackball** This is a ball with about the top third exposed, which the user manipulates with the palm of the hand. The trackball is used as the paddles are, but it has a much better facility for achieving diagonal movement.

▪ **Mouse** The mouse is hand-held and is applied to a surface and moved on it. The effect is the same as a trackball, although the mouse is easier to manipulate. It has become quite popular. See Fig. 4.

▪ **Joystick** This consists of a small upright lever in a base weighted or fastened down. The movement of the lever controls an object on the screen. See Fig. 3.

The mouse, trackball, paddles, joystick, and other similar devices usually work with a visible symbol, such as an arrow, on the screen and a button on the device that allows the user to choose a position on the screen. This means that they may be used to create drawings or to make programmed choices. They are all inexpensive devices.

B. Output The output from a computer is usually either on the screen (*soft copy*) or on paper (*hard copy*). Hard copy may be on a printer or plotter. Also, some computers have a built-in camera device known as a

FIG. 4 An Apollo workstation with mouse. *(Courtesy Apollo, Inc.)*

"phototrigger," although a static image on a screen may easily be photographed in the conventional way.

▪ **Screen** The screen or monitor of a computer may be the only one the computer has, as with a microcomputer, or it may be one of a large number of terminals that are linked to a computer powerful enough to serve several different users simultaneously, such as a minicomputer or a mainframe computer. Screens may be monochrome or color. They are usually used in text mode or graphics mode, although some terminals and a few microcomputers do not have a graphics mode. Most microcomputers can be used to emulate a terminal when they are connected to a larger computer.

There is increasing interest in technologies such as LCD (liquid crystal display), which produce low-energy, low-glare, space-efficient flat screens. At present, most screens are still CRTs, (cathode ray tubes). The inside surface of a CRT is coated with phosphor, which glows when excited by a beam of electrons. The two most common ways of exciting the phosphor are *stroke*, or *vector*, *graphics* and *raster-scan graphics*.

In stroke graphics the electron beam is deflected ac-

cording to the shape that is to be traced on the screen, and it sweeps across the screen in a continuous movement. In raster-scan graphics the screen is composed of a discrete number of picture elements known as "pixels." The beam is swept to and fro across the screen, line by line, until the entire screen has been covered, but the beam is only activated for those pixels that the controlling program indicates are part of the display.

Since the phosphor glow will decay rapidly, the screen will be refreshed at least 25 to 30 times a second if flicker is to be avoided. This is true of both raster-scan and stroke graphics, but stroke graphics are sometimes referred to as "vector refresh graphics" or just "refresh graphics." In stroke graphics the refresh rate can be very high for small numbers of vectors (lines) and this is very good for creating animated effects. Also, stroke graphics produce perfect diagonal lines, whereas raster scan produces jagged, stepped diagonal lines that only become insignificantly jagged in more expensive computers with a very high number of pixels. Raster scan also requires a lot of memory, but memory costs have come down and its superior versatility for color and editing have made it increasingly popular.

Another type of screen display is the storage tube. Here the picture is drawn by an electron beam onto a screen grid that stores the image relatively permanently. There is no refresh problem and the image is excellent. However, redrawing is necessary for any editing or additions.

▪ **Plotters** The plotter is the hard-copy equivalent of the stroke graphic monitor. Plotters use pens to draw the lines with great precision, line by line. They may be either flatbed or drum design, and they come in a very wide variety of sizes. The number of pens also vary, and variations in color and width of the pens are typical (Figs. 5 and 1). Plotters are usually expensive, but they produce excellent images.

▪ **Matrix Printers** These are the equivalent of the raster-scan screen, as they compose their images of little dots and sweep across the paper in successive horizontal lines rather than following the object lines of the drawing (Fig. 6). The impact matrix printers have the advantage of being inexpensive and readily able to mix text and drawings. The image quality is poor for the inexpensive matrix printers. However, there are expensive electrostatic matrix printers with 200-dot-per-inch resolution that provide excellent images.

- Line Printers These are matrix printers dedicated to providing text listing at very high speeds.

- Daisy-Wheel Printers These use a typewriter type head to produce high-quality text. They are much slower than line printers.

C. Storage The computer uses primary and secondary storage systems for the information that it processes. The primary system usually consists of solid-state chips which are volatile, that is, the information is stored as electrical charges which are lost when the power is turned off. An exception to this is information which is permanently burnt onto a chip. For example, the Apple II contains Applesoft, a version of BASIC, which is stored in this manner. It is a *read-only memory*, ROM, with fixed information in fixed memory locations, as opposed to the other primary memory, RAM, which can be *randomly accessed* for both reading ("retrieving") and writing ("saving"). Other types of primary memory chips are PROMS and EPROMS. Both are ROM chips, which may be programmed by the user. The EPROM

FIG. 5 An eight-pen turret on CalComp's 1040 plotter. *(Courtesy CalComp.)*

FIG. 6 Bausch & Lomb workstation with a custom configured dot matrix printer/plotter. *(Courtesy Bausch & Lomb.)*

FIG. 7 The DN550 is a 32-bit engineering workstation featuring a dedicated bit-sliced graphics processor, a 1024 × 800 resolution color monitor and a low-profile detachable keyboard. The node can be configured with 1 to 3 megabytes (MB) of main memory and comes standard with Apollo's new graphics software architecture, the Graphics Metafile Resource. *(Courtesy Apollo Computer, Inc.)*

chip may be reprogrammed by the user (*Erasable PRO-grammable Memory*). In Fig. 7 the Apollo computer has the processor and storage unit on the right of the keyboard and screen unit.

The size of the memory is measured in terms of the amount of information that it can store. The smallest unit is a *bit*. A bit may or may not contain an electrical charge (actually, a high or low voltage), that is, it may have either 1 or 0 as a value. A *byte* contains 8 bits (2 "nibbles" each of 4 bits). A *kilobyte* contains approximately 1000 bytes (actually 1024, or 2 to the power of 10). A *megabyte* contains approximately 1,000,000 bytes. A megabyte can store the contents of a 400-page book, and even a 64K microcomputer can run some very substantial programs. In fact, microcomputers are often similar in scope to the mainframe of 10 years ago. (The difference between an 8-bit and a 16-bit computer, incidentally, lies in the amount of information that can be simultaneously transmitted. An 8-bit computer has, in effect, an 8-lane highway for communication. A 16-bit uses a 16-lane highway and is faster than an 8-bit processor when more than 8 lanes are being used.)

An interesting and efficient technique for a raster-scan graphics display is "bit-mapping." In this method the bit values, 1 or 0, stored in a particular area of memory, are used to control the electron gun of the monitor and map in a one-to-one correspondence with the pixels on the screen. Thus, a screen display is actually a bit map of an area of memory with a lit pixel representing a bit value of 1. The Apple II series uses this method. If a variation in color or intensity is required, more than 1 bit per pixel will be required to carry the extra information.

Secondary storage can be done in a variety of different ways. The most common, at present, uses minute magnetic values like a cassette tape. These may be on a tape or, more typically, a disk. In microcomputing the "floppy disk," so called because of its flexibility, is very popular. The significant feature of a floppy disk is that it is easily removed from the disk drive and is therefore highly portable. It is also very cheap. A hard magnetic disk, while much more expensive, will store 10 to 100 times more data than a floppy disk and operate at much higher speeds. Storage technology is a rapidly changing field, however.

The secondary storage system differs from the primary storage by being more or less permanent. Since problems can and do arise, a basic principle of computing is to have a back-up copy of all your important programs and files—even incomplete ones.

Secondary storage is also much larger, and takes longer to access, when compared to primary storage. In microcomputers, the primary storage will usually only handle one program at a time. In mini and mainframe computers, timesharing techniques allow several different users to access the computer using different parts of the primary storage simultaneously. Secondary storage is also shared by different users, of course.

A RAM disk is a hybrid memory form. It is a volatile memory installed on a board in chip form like RAM. Unlike RAM it cannot be directly accessed by the CPU. It is used as a temporary secondary storage because it can be accessed much more quickly than any disk system.

Virtual memory is a technique whereby the computer loads into primary memory only those parts of the program and data currently needed. It is, therefore, a software shuffle which produces the effect of a large RAM capacity that is not actually there.

D. How a Computer Works

The most paradoxical thing about a computer is that although it is a very simpleminded form of intelligence, it is amazingly powerful. It overcomes the restrictions of simplemindedness by the speed with which it operates. The success of the operation also depends on the ability of the people who work with computers. They must handle the complexity of a machine that can solve an enormous array of problems so long as the problems are broken down into the types of simple operations it can process.

The simplicity of a computer is best exemplified by the fact that it works with a binary algebra. This means that it only recognizes two values, 0 or 1. Fortunately, this does produce a number system, although it is a very cumbersome number system, and only the computer's speed of operation makes it manageable. The reason the computer works with a binary system is that it stores all its values electronically in tiny cells which are either on (1) or off (0). A comparison of the binary numbering system with our usual decimal system reveals the clumsy nature of the former.

Binary system	Decimal system
0	0
1	1
10	2
11	3
100	4
101	5
110	6
111	7
1000	8
—	—
10,000	16
—	—
100,000	32
—	—
—	—
100,000,000	256

The correspondence between the two systems can be expressed by the fact that the number of zeros in the binary system corresponds to the power to which 2 is raised in the decimal system. So 2 to the fifth (32) in the decimal system is the same as 10 to the fifth (100,000) in the binary system.

Obviously, the computer will end up handling very large numbers of digits because it works with a binary system. The power of a computer is very much related to its ability to handle these numbers.

The operations performed by a computer are controlled by its CPU (*central processing unit*). This unit, a single chip in a microcomputer, can directly communicate with the primary memory of the computer, the disk storage system, and other input and output peripheral devices. Internally it consists of a control unit and an arithmetic-logic unit (ALU). The control unit issues the instructions to the ALU. The ALU carries out basic functions such as adding, subtracting, and comparing.

Both the control unit and the ALU contain registers that are used to keep track of what the current command is, or, in the case of multiplication, how many multiples of a number still have to be added to yet another register. Like the memory of the computer, the registers are composed of bits. The rate at which the registers work depends on the number of bits that can be simultaneously transferred (*wordsize*). The bigger the communication channel (*data path*), the more data that can be handled at any one instant, and hence the more powerful the computer. Some early computers were 4-bit machines. The Apple II series is based on an 8-bit CPU. The IBM PC uses a 16-bit for the internal CPU operations. The IBM AT and the AT&T personal computers also use a 16-bit data bus. The *bus* is the communication channel between the CPU and the RAM (actually "bus" is used quite generally for information transfer hardware). The Macintosh and an increasing number of desktop minis use 32-bit CPUs, which makes them very fast machines. (A *minicomputer*, despite its name, is a very powerful computer that ranks between microcomputer and mainframe computer in size.)

Another factor in the power of a computer is the *clock*. This sets the frequency of the pulses of electricity that carry the signals through the computer. The higher the frequency, the faster the computer. Higher frequency, however, requires more expensive materials. The clock frequency for the CPU in the Apple II is 1.023 MHz (MHz = megahertz = 1 million cycles/second).

The memory size of microcomputers, like that of all computers, is continually being increased. The amount of memory a CPU can directly address is fixed by the size of the logical address space in the CPU. In the 6502 CPU in the Apple II series and the Commodore 64, the logical address space is 16 bits. If you convert this to decimal, you get 65,536 memory locations (each 1 byte for these computers), which is a nominal 64 kilobytes of directly addressable memory. As the wordsize of the CPU increases it is better able to work with a larger logical address space. A 32-bit machine can address an enormous memory, but the problem is the cost of the memory chip. As the technology for mass-producing 256K, 512K and 1-megabyte memory chips improves, the cost will fall and the memory capacity of desktop machines will become extremely large. The physical size is also quite remarkable. AT&T is currently producing a 1-megabyte chip the size of a fingernail, and Hewlett-Packard produces a 32-bit CPU that is not much bigger.

3. PROGRAMMING THE COMPUTER

A. Introduction There are several different types of programs involved in the operation of the computer. There are internal programs which are built in to operate the CPU and the monitor. There is another program used for addressing the secondary storage system. Often called the *disk operating system* (DOS), this may also be built in, but is not in most microcomputers. DOS is the means of saving and retrieving the files (programs and data files) to and from the storage disks. These files are produced by programming languages.

Programming languages can be low-level or high-level. Low-level languages such as machine language or assembly language are written in, or close to, the numbers and processes the computer uses in its binary operations. These programs are very tedious to write, but they run much more quickly than high-level language programs do because the computer has to convert a high-level language to machine level before it can execute it.

High level-languages such as FORTRAN, PASCAL, or BASIC are much easier to work with since they use a decimal number system and terms and syntax similar to English and algebra. As microprocessors become more powerful, e.g., 32-bit, the need to program in low-level languages is lessened. In fact, speed is not always that important. Even on 8-bit machines, the computer does a lot of waiting for the human brain to catch up. However, speed is often significant in graphics because a display on a screen often represents a very large amount of memory, and because shape manipulation can involve a lot of number crunching. Speed in graphics is particularly important in animation wherein displays must be drawn and erased very rapidly.

In this section we will work through a quick introduction to programming in a high-level language, and then apply what we learn to producing computer graphics. The language used will be the Applesoft version of BASIC. This language works on the Apple II, Apple II+, Apple IIe, and Apple IIc. Other versions of BASIC are usually quite similar so the listings given in the rest of the chapter could be easily modified.

In this section the problems are integrated with the text because it is very important to practice as you learn each new programming technique. In fact, experimenting and doing extra problems that you make up yourself is very helpful if you wish to become a competent programmer.

B. Getting Started The most common message is SYNTAX ERROR. This means that there is an error in the language you are typing in, so the computer cannot understand what you are saying. Just correct it and keep on going.

Turn the computer and the monitor on. The Applesoft language is already in the ROM of the Apple, so you can start using it right away. The DOS is not, however, so if you plan on saving any program that you write, or using programs on a floppy disk, you will need to get DOS loaded into the computer.

DOS is written on the first three tracks of any disk that has been initialized on an Apple II. It can be loaded into the computer in two ways. Place the disk in the disk drive, label face upward and outward, and close the gate. If the computer is off, turn it on and the computer will "boot" (load) the DOS off the disk automatically. If the computer is already on, type PR#6 and press the RETURN key. The RETURN key must always be pressed if the message which you have typed, and which appears on the screen, is to be sent to the computer. The PR#6 sends the computer to SLOT #6 (PR for "peripheral"), which is the sixth of the slots that the Apple has for connecting peripheral devices to. The disk drive(s) is put in slot #6 by convention, not by necessity. It could be connected to any slot, #1 through #7. Slot #1, again by convention, is used for the printer.

When the Apple is turned on it starts searching for a peripheral device, starting with slot #7. Unless you have connected something to slot #7, it will boot (communicate with) the disk drive in slot #6, and if there is no disk in the drive, press RESET to stop the disk drive.

All new disks need to be initialized. To do this boot a working disk and do the following.

Type LOAD HELLO and press RETURN.

Remove the old disk and insert the new disk.

Type INIT HELLO and press RETURN.

Wait for the red operating light on the disk drive to go out.

The disk is now ready for use. Note that if you used the INIT command on an old disk you would delete all the programs on it, so be careful! The LOAD command may be used with any program. It only loads the program named into the primary memory. It does not run it.

To find out what programs are on a working disk, type CATALOG (and press RETURN). This will give you a directory of the programs (files) on the disk that will look something like this:

A 003 HELLO

A 012 PYRAMID

B 064 SPY RING

A 002 CIRCLE

T 015 LETTER

In the first column the letters indicate the type of file it is. An A means the file is a program written in Applesoft BASIC. The B represents a program written in binary (machine language). The T is for text file, which could be a document written by a word processor program or a set of numerical data from, say, an experiment. The numbers in the three middle columns represent the length of the program in sectors. The names of the files come last.

To run an Applesoft program from a disk just type RUN——(put in the program's name) and press RETURN. For binary programs you must use BRUN. Text files may not be run directly; they can only be accessed by a special program. Once a program is in the primary memory you should only type RUN and press RETURN without adding the program name.

If at any time you do not want the computer to be doing what it is doing, you can usually get control back by pressing the RESET button (on some Apples hold down the CONTROL key at the same time).

C. Immediate Execution Immediate execution, like a nonprogrammable calculator, may be performed on a computer. For example, on the Apple II series the PRINT command will execute algebraic and arithmetic expressions. Remember to press RETURN at the end of each statement.

Enter	Response
PRINT 3+3	6
PRINT 3/3	1
PRINT 3*3	9
PRINT 3^3	27
PRINT 3−3	0
PRINT COS(.5)	.877582562
PRINT (12*3.1 − 2*SIN(.2))/3	12.2675538
A=62.5	—
PRINT A	62.5

So the computer works like a calculator, and it will even remember variable names such as A in the example above. These types of functions are useful, but programming is vastly more so.

D. Deferred Execution   Programming is deferred execution because the programs don't run until directed to. In programming we give each statement a line number and the computer stores it away, deferring execution until given the command RUN. Most programs consist of more than one line, and the computer will arrange them in numerical order no matter which order they are typed in. When RUNning a program the computer will execute the statements in numerical order except when the program directs otherwise.

So let's write a program. Type the following lines and remember to press RETURN at the end of each line. (Please note that the ^ is the exponentiation sign in Applesoft and means to the power of.)

    10 A=6
    20 B=2
    30 PRINT A^B,B^A
    RUN

When RUN is the command, this program will produce the values of 6 squared and 2 to the power of 6. If we used a semicolon instead of a comma in the PRINT statement, the output values would be side by side. Note that variable names may have a second character, either a letter or a number, e.g., CD, F3.

Any time that you wish to view your program, type LIST (and press RETURN). If you wish to examine only a particular line, say line 30, type LIST 30.

For editing, either retype a line with an error in it or if you are still typing the line use the backspace key (with the arrow pointing to the left). To delete a line, type the line number and press RETURN. Other editing techniques using the ESC key are described in the Apple II manuals and the reference by Devon.

So far, what has been produced does not look much better than what we can get from a calculator, so let's add some features. Type:

    10 INPUT "A=?";A
    30 PRINT"A^B="A^B,"B^A"B^A

Line 20 has been left as it was, but you have now replaced lines 10 and 30 with new lines. When you run this program, the first thing that will happen will be the execution of line 10. This will put the contents of the quotes A=? on the screen. You must respond by typing in a number, say, 2.5. Line 30 will give the output as before, but this time the output will be identified by the contents of the quotes. Try it by typing RUN (and pressing RETURN).

The program is getting more interesting because now it can be used for any value of A, and it has identification for the input and the output. Obviously we can input a value for B, also:

    10 INPUT "INPUT VALUES FOR A,B";A,B

Now we must get rid of line 20 or any value of B that is input, for line 10 will be superseded by the value of 2 given in line 20, since the computer always works with the most recent value. To do this, type 20 and press RETURN. This replaces the old line 20 with nothing and the computer will not even both to store the line. To check the program type LIST. You should now get a list of the entire program—all two lines! RUN the program and put in two values for A and B separated by a comma.

The INPUT statement is a bit odd in its details. Unlike the PRINT statement, which may have many identifiers, the INPUT statement may have only one. This identifier must precede the input variables and it must be followed by a semicolon. The variables must be separated by commas, as must the values that are input.

The program will work for any values that you input for A and B. Further, by changing line 30, you could make the program do many other things. It may also be written for more than two variables. Let's type in a new program. The NEW command will be used to get rid of the old program.

```
NEW
10 INPUT "INPUT VALUES FOR A,B,C";A,B,C
20 Z = A^2 + B^2 + C^2
30 H = SQR (Z)
40 PRINT "THE ANSWER IS = "H
```

The SQR command finds the square root of the argument (the number in parentheses). You could also do this by raising a number to the power of 0.5. Lines 20 and 30 could have been combined or linked by a colon. If you use either of the following two arrangements, remember to remove line 30.

```
20 H = SQR(A^2 + B^2 + C^2)
20 Z = A^2 + B^2 + C^2 : H = SQR (Z)
```

To save this program is very easy if you have a disk in your disk drive. Just think of a name, e.g., PYTHAGORAS, and type SAVE PYTHAGORAS and press RETURN. To check that it is saved used the CATALOG command. You will be able to run this program at any time the disk it is on is in the disk drive and DOS is loaded. Note that to run a program from a disk you must always add the program name. Using RUN by itself will only run the program in primary memory.

So far we have only used real number variables. There is also an important variable type known as a "string variable." In Applesoft these always have a $ appended, as in A$, or TR$. A string variable is simply a list of letters and numbers. It is always expressed in quotes. Try this modification to the Pythagoras program you are working with now:

```
5 H$ = "THE ANSWER IS = "
40 PRINT H$;H
```

You should get the same result as before. The computer will have no difficulty distinguishing between H and H$. We will do very little more with strings here. For further work with them you may wish to refer to the Apple II manuals or the references by Coan or Devon.

Programs that have INPUT statements in them require responses from the user and are known as "interactive programs." Write interactive programs for some of the following problems to become familiar with what you have learned so far:

1. Find the average of any four numbers.
2. Calculate a 25 percent increase for any value.
3. Find the square root for any number.
4. Add any three numbers together and divide by 33.

E. Looping Since the computer works one small simple step at a time, it is not surprising that the most typical routine in a computer program is a loop in which the computer repeats over and over the same calculations with minor increments in the value of a variable each time.

There are two ways to loop, a good way and a bad way. First, the bad way.

Type NEW and press RETURN to clear the primary memory of whatever program is there—save it first if you want to.

```
10 X = 1
20 Y = X^2
30 PRINT "X = "X,"X^2 = "Y
40 X = X + 1
50 GOTO 20
```

Before you run this program, note that it is an infinite loop and will run endlessly—at least, until the numbers become too large for the computer. So be prepared to stop the program by pressing RESET or holding down CONTROL and C simultaneously. Now run the program for a few moments.

There are better ways to control this program than RESET. For example:

```
35 IF X > 10 THEN GOTO 60
60 END
```

Now when you run the program it will provide output for values of X from 1 to 11 inclusive and then end. The > is called a "relational operator" and all the usual ones may be used.

```
= EQUALS
< LESS THAN
<= LESS THAN OR EQUAL TO
> GREATER THAN
```

$>$ = GREATER THAN OR EQUAL TO

$<>$ or $><$ NOT EQUAL TO

More complex conditional statements than the one in our program may be used. For example, these two statements are acceptable—but don't type them, as we have no need for them:

180 IF X $<>$Y AND P $>$ Q THEN V $=$ 12

200 IF D $>$ 5 OR D $<-$5 THEN X $=$ 0

Now for the proper way to do the loop. Type:

NEW

10 FOR X $=$ 1 TO 11

20 PRINT "X $=$ "X,"X^2 $=$ "X^2

30 NEXT X

This is much more efficient and we don't have GOTO statements to keep track of. Too many GOTO statements and you will end up with what is referred to as a "spaghetti program," and you know how hard it is to figure out what goes where in a pile of spaghetti.

For a variation in output, you might try the following:

20 PRINT TAB(X) "X $=$ "X;" X^2"X^2 : PRINT

The TAB command will make the output start at the column number of its argument (the content of a function's parentheses). In our program this changes with each loop, but it could be a fixed number, such as 5, say. Note that the comma in the PRINT statement has now been changed to a semicolon. This places the two outputs side by side. To compensate for this some spaces have been placed at the beginning of the second identifier. The PRINT at the end of the line will create a blank line between each line of output. Run this to confirm its effects.

The FOR/NEXT loop always increments in steps of one unless otherwise specified, but we can specify otherwise:

10 FOR X $=$ 1 to 11 STEP 3

You will now get loops, and hence output, for X values of 1,4,7, and 10. You will not get output for X values of 2,3,5,6,8,9, and 11. Both negative and decimal loops are possible, also:

10 FOR X $=$ O TO $-$2 STEP $-$.1

Note that since the incrementation is toward a lower value, the step had to be negative. Sometimes, when working with decimals, a rounding process in the computer will prevent the last loop's being completed. In this event you should shift the limit slightly to avoid the problem.

Loops may be used in sequence one after the other. They may also be used "nested," that is, one inside the other. Let's begin again.

NEW

10 FOR A $=$ 1 to 3

20 FOR I $=$ 0 to 18 STEP A

30 PRINT "2*I"2*I

40 NEXT I

50 PRINT

60 NEXT A

The FOR/NEXT loop from line 20 to line 40 is repeated three times, once for each value of A. The PRINT in line 50 separates the three sets of output with a blank line between each set.

Many loops may be nested one inside the other, but each loop must be entirely within the next loop. If line 40 and line 60 were switched our program would not work.

A useful command for illustrating program flow and for debugging a program is TRACE. Just type it in as a direct command and then run your program again. As every statement is executed, its line number is printed on the screen. When you no longer need it type NOTRACE.

▪ **Problem** If an object is dropped from a building of height H, the time, T, taken for it to reach the ground is expressed by:

$$T = SQR(2*H/G)$$

where G is the acceleration due to gravity, and $*$ is the multiplication sign in Applesoft.

We can write a program that, ignoring air resistance, will give us the fall time for a range of different heights:

5 G $=$ 32.2

10 FOR H $=$ 10 TO 200 STEP 20

20 T $=$ SQR(2*H/G)

30 PRINT "FOR HEIGHT "H;" THE FALL TIME
IS "T

40 NEXT H

The units of G given in line 5 are feet per second squared; therefore H will be in feet and T in seconds.

Practice working with loops by doing some of the following problems.

1. Find the area of a circle for radius values of 1,3,5, and 7 inches using a value of 3.14 for pi.

2. If you are paid $500.00 a month and you save half of it, write a program that shows the accumulation, without interest, over a 2-year period.

3. If you gain weight at a rate of 5 percent a year for 5 years, write a program that, starting with your present weight, will calculate your weight for each year.

Now let's do a loop that is variable in length, and introduce a few other new things along the way.

10 REM THIS PROGRAM FINDS THE AVERAGE OF A SET OF NUMBERS

20 INPUT "HOW MANY NUMBERS WILL YOU INPUT? ";N

30 FOR I = 1 to N

40 INPUT"INPUT NEXT NUMBER ";A

50 T = T + A

60 NEXT I

70 PRINT:PRINT "THE AVERAGE IS = "T/N

All variables have the value of 0 until assigned something else, so in line 50, T, having started from 0, is keeping a running total. The extra PRINT at the beginning of line 70 is to separate the output from the input. The REM statement in line 10 is included to identify the program listing and will be ignored by the computer. You may have as many REM statements as you like. Some programmers like to have an explanatory REM for almost every program line. Others prefer to work with a minimum of REMS, one for the whole program and perhaps one for about every 8 to 10 lines.

By going back to the GOTO system of looping we can find a rare good use for it by doing this program for an indefinite number of values.

10 REM AN AVERAGE PROGRAM

20 PRINT "THIS PROGRAM WILL FIND THE AVERAGE OF AN UNLIMITED SET OF NUMBERS ENTERED ONE AT A TIME BY THE USER":PRINT

25 PRINT "ENTER A VALUE OF −2001 to END"

30 INPUT "INPUT YOUR VALUE AND PRESS RETURN";V

40 IF V = −2001 THEN GOTO 80

50 N = N + 1

60 T = T + V

70 GOTO 30

80 PRINT:PRINT "THE NUMBER OF ENTERED VALUES = "N

90 PRINT" THE AVERAGE VALUE ENTERED = "T/N

In this program the program is identified by a PRINT statement as well as the REM statement. This will inform the user by a message on the screen. This type of message helps the program to be "user friendly." Although important, it is possible to be too user friendly and swamp a program in chitchat.

This program will continue until the value −2001 is entered to signal the end of the data. This value is known as a "trailer," or "rogue value," and any number that is extremely unlikely to occur in your data may be used. Line 50 uses the variable N as a countervariable to keep track of the number of data entered.

F. Computer Graphics The Apple II series has two graphics modes, high-resolution and low-resolution.

There is not a great deal of use for low-resolution so we will only use high-resolution graphics.

In high-resolution graphics a picture is created on the screen, but it is also stored in primary memory. There are two places in the memory reserved for creating high-resolution pictures, and these are called "pages." Page one is accessed by the command HGR, page two by HGR2.

HGR and HGR2 differ slightly. Both read X values from left to right, from 0 to 279. Both read Y values from top to bottom. This means the origin is at the top left-hand corner of the screen and Y is measured downward, which is annoying. In HGR2 the Y values range from 0 to 191. In HGR the Y values range from 0 to 159 and the difference at the bottom of the screen is taken up by four lines of text. The text option in HGR is useful in interactive programs.

When entering either high-resolution mode the color must be set by the HCOLOR command. A value of 3 or 7 should be set for monochrome monitors and this is what we will use here. A value of 0 is for black and it may be used for erasure.

HPLOT. This is the basic command in high-resolution graphics.

HPLOT X,Y. This will plot a point (light up a pixel) at point X,Y. Other variable names may be used. The first variable will always be treated as the X coordinate and the second as the Y coordinate.

HPLOT X1,Y1 TO X2,Y2 to X3,Y3 to—. In this manner the HPLOT command can be used to draw a complete shape with lines from point 1 to point 2, and so on.

We will only draw shapes in this section. Putting text on the graphics screen is best done by using shape tables. A "shape table" is a directory of stored shapes which may used at any time and at various scales and orientations. Shape tables are very easy to use and very handy. A table for the keyboard shapes (letters, numbers, etc.) is particularly valuable. Such a table is available through the text by Myers. See the Apple II manuals for the use of shape tables. We will not be working with one.

Here is a simple routine which draws a triangle on the screen:

```
NEW
10 HGR2:HCOLOR=7
20 HPLOT 10,10 TO 200,30 TO 100,130 TO 10,10
```

Once you have gazed at this long enough, either press RESET, or type TEXT and press RETURN, to get back to the text screen.

As you see, the diagonal lines are jagged. For more expensive systems with screen resolutions of, say, 600 by 400 or 1024 by 1024, the diagonal lines look much better.

Do some of the following problems. You will need to use several separate HPLOT statements in some of the problems.

1. Draw a rectangle.
2. Draw the letter A.
3. Draw the letter E.
4. Draw a large *.

By using variables and loops we can easily generate more complex images:

```
NEW
10 HGR2 : HCOLOR=7
20 A=10 : B=40 : C=30
30 FOR P=0 to 48 STEP 4
40 HPLOT   A+P,A+P   TO   B+P,A+P   TO
C+P,C+P TO A+P,A+P
50 NEXT P
```

Run this program and then modify it from a triangle to a four-sided figure. Be careful not to generate a quantity that is off the screen, or your program will end with an ILLEGAL QUANTITY ERROR. If it does this, do a RESET to get back to the text screen. Then PRINT A,B,C,D,P, or whatever your variable names are to see which one(s) caused the problem.

Now let's try interactive graphics on the HGR screen. The VTAB 21 command in line 20 sets the text output at line 21 of the 24 text lines of the screen. This puts the INPUT cue where the user can see it, just below the graphics part of the screen.

```
10 HGR : HCOLOR=7
20 VTAB 21
30 HPLOT 130+A, 80+B TO 150+A, 80+B
40 HPLOT 140+A, 70+B TO 140+A, 90+B
50 INPUT "INPUT SHIFT VALUES X,Y ";A,B
60 GOTO 30
```

This program will draw a cross in the middle of the screen. It will then request an input of X and Y values to translate the image around the screen. The X values input must be in the range $-130 < X < 129$. The Y values input must be in the range $-70 < Y < 69$. Take the time to work out why this is so.

This program draws the next image without erasing the last image. The program may be extended to provide erasure.

```
10 HGR : HCOLOR=7
20 VTAB 21
30 HPLOT 130+A, 80+B TO 150+A, 80+B
40 HPLOT 140+A, 70+B TO 140+A, 90+B
45 PRINT " PRESS ANY KEY TO CONTINUE"
50 GET A$
60 HCOLOR=0
70 HPLOT 130+A, 80+B TO 150+A, 80+B
80 HPLOT 140+A, 70+B TO 140+A, 90+B
90 INPUT "INPUT SHIFT VALUES FOR X,Y
";A,B
100 HCOLOR=7 : GOTO 30
```

Lines 45 and 50 provide a viewing pause that is totally controlled by the viewer. The GET command waits for a single signal from the keyboard. No RETURN is necessary. So just pressing the spacebar will continue the program. The A$ is a string variable which the Apple will not confuse with the real variable, A. A$ plays no other role in the program—although it could if you wanted.

Since the two plotting lines have to be repeated in this program, it is a good candidate for using a subroutine. A "subroutine" is a program routine which is structured away from the main program, and which may be accessed as frequently as desired. It makes sense to use subroutines for routines that are used more than once. We may rewrite the program to include a subroutine, and for a little variety let's draw a diamond this time. When you run this program, the values of A and B must be at least 20.

```
10 HGR : HCOLOR=7
20 VTAB 21
25 A=140 : B=80
30 GOSUB 200
40 PRINT "PRESS ANY KEY TO CONTINUE"
50 GET A$
60 HCOLOR=0
70 GOSUB 200
80 INPUT "INPUT PLACING VALUES FOR
X,Y ";A,B
```

```
90 HCOLOR=7
100 GOTO 30
110 END
200 HPLOT A-20,B TO A,B-20 TO A+20,B TO
A,B+20 TO A-20,B
210 RETURN
```

Either the subroutine must be separated from the rest of the program by an END command, as above, or it can be at the beginning and be protected by a GOTO statement, as in the partial listing shown below for illustrative purposes (don't type it).

```
10 GOTO 40
20 HPLOT X1,Y1 TO X2,Y2
30 RETURN
40 HGR2 : HCOLOR=7
—
```

etc.

There must be a RETURN at the end of the subroutine. After execution of the subroutine the program will return to the next statement after the GOSUB. Organizing a program into a set of clearly defined routines and subroutines is known as "structured programming" and is much to be desired.

G. Animation The movement of an image to create the appearance of motion can be achieved on the Apple if the image is fairly simple. Type in the following program:

```
10 HGR2
20 A=10 : B=40 : C=30
30 P=3
40 FOR I=1 TO 50
45 HCOLOR=0
50 HPLOT A,A TO B,A TO C,C TO A,A
60 A=A+P : C=C+P : B=B+P
70 HCOLOR=7
80 HPLOT A,A TO B,A TO C,C TO A,A
90 NEXT I
100 P=-P
110 GOTO 40
```

After running this program, list it and study it. A good place to begin is with the HPLOT statements. Look to see what is being plotted. In this case we can see that both line 50 and line 80 plot a triangle, because they both join four points with three lines, and the first and last points are the same. This means three distinct points and three distinct lines, and that means a triangle.

Why are lines 50 and 80 the same? Answer: Because we draw the triangle in line 80 and we erase it in line 50.

Why erase it? Answer: Because we want to make the triangle appear to move, so we keep changing its location a little bit and drawing it again. Of course, we must erase the old before we draw the new. How do you keep changing the location? Answer: In line 50, using P as the increment. P is 3 in the program, but we could make it bigger or smaller. If we make it smaller, the image will move more slowly across the screen. If we make P bigger, the image will move more quickly across the screen, but there will come a point when it no longer looks like continuous motion.

The first time you draw the image the color is black. Why do that? It could have been the other way around, and that would seem more logical. The issue here is viewing time. On a simple image like this the viewing time may be too short, resulting in a very blurred image. It may actually be less than the erased time, in which case the continuous-motion effect of a continuously present object will be lost. The Apple takes a finite amount of time to execute each statement, the actual amount depending upon the content of the statement. Sometimes a FOR/NEXT loop is introduced just to create a viewing pause, the length of the pause being controlled by the number of loops set. In this program the looping statements of lines 90 and 40 create the viewing pause, and they could not have if we had plotted the visible lines first.

We can experiment with the viewing pause like this.

85 FOR J = 1 to 5: NEXT J

This little loop is nested inside the one we already have, and it comes right after the triangle has been visibly drawn. The number of loops set is 5. Run the program and then try setting higher and lower numbers until you like the way it looks. You may prefer to do without it entirely.

The program can be enhanced by doing the following changes.

30 P = 3 : Q = 1
60 A = A + P : B = B + Q : C = C + P
100 P = −P : Q = −Q

This will have the effect of changing the size of the triangle as it moves. This, in turn, gives the impression of three-dimensional motion. Unfortunately, the triangle speeds up as it gets smaller (because there is less to draw) and this is a little counterintuitive.

Another dimension you can add is sound. The Apple needs special programming to get much sound out of it, but this line is easy to add:

85 S = PEEK(− 16336)

This simply activates the Apple's speaker by signaling the memory location that controls it.

■ **Problems** Obviously, you can experiment with different values for P and Q so long as you can keep the object on the screen. In fact, you may change the shape drawn and the looping step also. Try at least four different modifications.

H. Curves and Their Families Nothing illustrates the power of computer graphics more than the secondary stage of image production. This occurs after you have produced the first image and you begin to replicate and otherwise manipulate it. Typically, but not always, this will involve nesting the FOR/NEXT loop which produced the first image in another loop that replicates the first image in various ways.

Let's begin with a circle.

10 HGR2 : HCOLOR = 7
20 PI = 3.14159 : C = 2∗PI : DA = C/40
30 R = 20 : REM R = THE RADIUS OF THE CIRCLE
40 Y = R∗SIN(I) : X = R∗COS(I) : HPLOT X + 140, Y + 90
50 FOR I = DA TO C STEP DA
60 Y = R∗SIN(I) : X = R∗COS(I)
70 HPLOT TO X + 140, Y + 90
80 NEXT I

The computer uses radians to measure angles, and there are 2*pi radians in 360 degrees, or one complete revolution. In line 20 we calculate the number of radians in one revolution (C), and then we calculate a small fraction of it (DA) to use as an increment for plotting our way around the circle from 0 to 2*pi radians. In line 40 the initial point is calculated and plotted. Once this is done all we need to do for further plotting is to use HPLOT TO—to draw a line to the next point. This is a very helpful technique in this sort of a program because it saves us from having to keep track of the last point when we calculate the next point.

Line 60 is just a trigonometric method of defining points on a circle from its center. If you don't know this, prove it to yourself with a pencil-and-paper sketch.

Line 40 and line 70 both have X + 140 and Y + 90 as their arguments. The 140 and 90 place the circle on the center of the screen.

Note that this program actually draws a 40-sided polygon. You may change the number 40 in line 20 to draw any polygon that you want.

Now for the fun part. Add these changes:

```
30 FOR R = 10 to 80 STEP 10
35 I = Ø
90 NEXT R
```

Now you have a whole family of circles. Note that I is reset to 0 for each R loop to plot the first point of each circle correctly. Another dramatic effect is to vary the center of the circle:

```
25 CX = 140 : CY = 90
30 FOR R = 10 TO 50 STEP 5
40 Y = R*SIN(I) : X = R*COS(I) : HPLOT X+CX,Y+CY
70 HPLOT TO X+CX,Y+CY
85 CX = CX+5 : CY = CY+5
```

These changes make the position of the center of each circle drawn, CX, CY, a variable. The incremental changes are created in line 85. As always, other increments may be chosen. Line 25 provides the initial values for CX and CY. Note that line 30 had to be changed, because as the center of the circle is moved from the center of the screen the maximum size circle possible

diminishes. Of course you could always start near a corner and move near the center of the screen.

■ **Problems** Try at least three changes in the increments of the loops and/or CX and CY. A really clever change is to move CX and CY along a curve.

I. Windows So far we have avoided illegal quantity errors by planning ahead to keep the coordinates in the proper range. A more standard approach is to use a procedure known as "clipping." This entails program lines that check to see if the values are legal. If they are not the HPLOT is skipped. The circle drawing program may be clipped as follows.

```
10 HGR2 : HCOLOR = 7
20 PI = 3.14159 : C = 2*PI : DA = C/40
30 FOR R = 10 TO 150 STEP 20
40 FOR I = 0 TO C STEP DA
50 Y = R*SIN(I) + 90 : X = R*COS(I) + 140
60 IF X < 0 OR X > 279 THEN FL = 0 : GOTO 100
70 IF Y < 0 OR Y > 191 THEN FL = 0 : GOTO 100
80 IF FL = 0 THEN HPLOT X,Y
85 FL = 1
90 HPLOT TO X,Y
100 NEXT I
105 FL = 0
110 NEXT R
```

This is like our last circle-drawing program, except that the center is fixed and placed in the calculations for X and Y in line 50. The clipping statements are in lines 60 and 70 and are applied to X and Y, the arguments of the HPLOT commands.

The only tricky thing in this program is the function of the flag variable FL. Flag variables control local program flow. In our case the flag variable is used to decide when to use the HPLOT X,Y and when to use the HPLOT TO X,Y. The HPLOT X,Y, which must be used at the start of a line-drawing sequence, is signaled by FL = 0. This must occur at the beginning of a circle and when restarting after clipping, hence the FL=0 in line 105 and in lines 60 and 70, respectively.

A window is created by restricting the legal area to some portion of the screen. Usually the sides of the window are parallel to the sides of the screen and lined in. We can create a window with the clipping program very easily with these changes:

```
25 HPLOT 40,50 TO 160,50 TO 160,150 TO
40,150 TO 40,50

60 IF X < 40 OR X > 160 THEN FL=0 : GOTO
100

70 IF Y < 50 OR Y > 150 THEN FL=0 : GOTO
100
```

▪ **Problems** Change the program to create a different window twice. For an advanced problem, create two windows on the screen at the same time.

J. Finale Another way to manipulate the original image is to use a fixed point to connect it to. Yet another is to turn it upside down, using a technique known as "reflection."

The following program begins with the generation of a sine curve. If you use the line numbers, you will be able to insert the changes easily.

```
20    HGR2 : HCOLOR=7
50    FOR X=0 TO 279
60    Y=50*SIN(X/20)+90
70    IF X=0 THEN HPLOT X,Y
80    HPLOT TO X,Y
100   NEXT X
RUN
```

At present, the program will draw a nice sine curve. The important line is line 60. The value of 90 places the curve at midscreen in terms of height. The argument of the sine function is X/20. Since X goes from 0 to 279, the argument values will go from 0 to almost 14 radians. A complete sine wave is drawn in 2*pi radians, that is, about 6.3 radians. So our program will produce just over two complete sine waves. A value bigger than 20 will give fewer sine waves; a smaller value will give more sine waves. Since the value range of the sine function is between +/− 1, we need to amplify it. In our program the amplitude is 50.

We could draw repeated sine curves as we did for circles, and the results would be very gratifying (try it sometime). Instead we are going to do something new and look at the fixed-point effect. Add this line:

```
85 HPLOT TO 200,150 TO X,Y
RUN
```

The fixed point is 200,150, but it could be anywhere on the screen. For each point plotted on the sine curve, the computer draws a line to it from the last point and then to the fixed point and back again. It looks better if it only occurs every fourth point. This can be arranged:

```
82 IF FL=4 THEN FL=0
84 IF FL > 0 THEN GOTO 90
90 FL=FL+1
RUN
```

Now the HPLOT command in line 85 is only accessed when FL=0, and this only happens every fourth loop. If you want to increase the line spacing, just increase the value of 4 in line 82.

In order to reflect this drawing vertically, we must in some way plot − Y where we are now plotting Y. The Y and the reflected Y will be placed at about the quarter and three-quarter heights, respectively. The amplitude and the frequency will be made interactive.

The whole listing is given here, with the additions and changes marked with an asterisk.

```
*        10 INPUT "A,B? ";A,B
         20 HGR2 : HCOLOR=7
*        40 K=1 : L=1
         50 FOR X=0 TO 279
*        60 Y=K*A*SIN(X/B)+38*L
         70 IF X=0 THEN HPLOT X,Y
         80 HPLOT TO X,Y
         82 IF FL=4 THEN FL=0
         84 IF FL > 0 THEN GOTO 90
*        85 HPLOT TO 200,100 TO X,Y
         90 FL=FL+1
         100 NEXT X
*        105 IF P=1 THEN GOTO 130
```

```
*    110 K = −1:L = 4
*    120 P = 1 : GOTO 50
*    130 GET A$ : TEXT : RUN
```

Line 10 and line 130 make the program usefully interactive. Line 130 allows you to run the program again simply by pressing the spacebar. Line 10 allows you to put in different values for the amplitude and frequency of the sine wave. Larger values of B, for example, will decrease the frequency of the wave. Whereas any value may be used for B, the amplitude value A is controlled by the height of the screen and cannot be greater than 37.

The value of K changes sign in line 110, so Y will be inverted for the second plotting. To separate the two images, the value of L is increased from 1 to 4 in line 110. This places the second image three-quarters of the way down the screen.

In lines 105 and 120, P is used as a flag variable to make sure that the program does not repeat after completing the two images.

In line 85 the fixed point had been moved to be vertically central and horizontally off-center. The fixed point may, of course, be changed to anywhere on the screen.

This program is a nice demonstration of the power of computer graphics. Beginning with a simple sine wave drawing, you now have a program that can swiftly generate an unlimited number of variations of a design motif, many versions of which are strikingly attractive.

■ **Problems** Your task is to play. Vary the input values of A and B. Vary the spacing. Vary the location of the fixed point. For an advanced problem try varying the location of the "fixed" point during a drawing by making it vary in the X loop. The technique can be extended to link two curved lines together.

4. COMPUTER-AIDED DRAFTING

A. Introduction Using a CAD system is quite different from writing programs for computer graphics. On the one hand, you lose the control over the process that lets you produce anything that you want to. On the other hand, you are provided with a ready-made program (set of programs, really) that will do far more things for you than you would ever have time to write the programs for, even if you knew how.

Although CAD systems are much more powerful than traditional instrument drawing, particularly where the replication and manipulation of already recorded (by CAD) drawings are concerned, they have several distinct disadvantages when compared to traditional graphics.

1. They are vastly more expensive. An industrial CAD workstation may cost from $15,000 to $100,000. This is for one worker. A traditional drafter needs $200 to $800 of equipment (the higher figure includes the seat).

2. The training period prior to productive work is much longer on a CAD system. For example, a Computervision system at Penn State came with manuals almost 3 feet thick.

3. CAD expertise has a low portability rate. What you learn on one CAD system will have only general applicability on another. This means that the drafter will have to do a lot of new training with every CAD system, although it will be much quicker after the first time.

4. Because a CAD system is a set of computer programs, there is a limit to the number of things it can do. However, as CAD becomes more and more sophisticated, the limitations become less important and the things that it does better, such as shape manipulation, become critical in the comparison.

5. CAD has two contrary influences on creativity in design. On the one hand, it involves the creation of extensive files of drawings of objects that tend to become standard and modular because of their ease of use. This tends to reduce the creativity of the design. On the other hand, the ease of editing and the power of shape manipulation encourages the designer to do things that were too tedious or too difficult to do before.

These disadvantages aside, CAD continues to take over the production of industrial drawings from traditional graphics because the increased productivity of CAD outweighs the expenditure involved. Anyone currently planning a career as a drafter must acquire training in CAD.

In the ensuing discussion of CAD, please be aware that only a limited amount of material is being presented in order to give the reader some idea of what to expect when using a CAD system.

B. CAD Systems In industry, the drawings produced by CAD are valuable and critically important to the design process. They should not be generally accessible. Most systems will require a password in order to log-on. This is not true of low-cost (under $10,000) systems, but a password can easily be added. Networks almost invariably work with a password system, and since most CAD systems are networked in industry, two passwords may be necessary: one to log-on and one to access particular CAD files.

▪ **The Menu** The menu is a software device that makes the user of the program aware of the structure of the program and what its component parts can do for the user. Whenever the user is presented with a menu, a choice must be made between the various options presented or the program will not proceed.

The menus of a CAD system may appear on the screen, as an overlay on a digitizing tablet, or parts of it may be controlled by special function keys on the keyboard, on the tablet cursor pad, or on a separate function keypad.

A CAD program will contain a great many menus, including some that will present choices between menus. The structure of the menu organization is important to the ease with which it may be used, but it varies from CAD system to CAD system.

In CADAM, which runs on the sophisticated IBM 5080 graphics system, the main function groups are arranged on a function keypad. The keys allow you to choose a major function group, and the various functions in that group will be displayed at the bottom of the screen. The functions will be selected with a screen cursor and any numerical data input or parametric control available for that function will be listed at the top of the screen. The IBM 5080 is shown in Fig. 8, and two typical screen displays can be seen in Figs. 9 and 10.

▪ **Early Choices** Once you have logged-on, your first choice will be whether or not you are going to load existing drawing(s) into the computer. If you intend to edit a drawing or create a new drawing that will use parts of various other drawings, then you will be loading a file

FIG. 8 The IBM 5080 Graphics System showing screen, keyboard, graphics tablet, and function keypad. The screen measures 19 in. diagonally. It has 1024 × 1024 resolution and a refresh rate of 50 Hz. Up to 4096 colors may be used. *(Courtesy IBM.)*

that contains at least some of the material you intend to work with. If you intend to create an entirely new drawing, or to digitize a drawing that, as yet, exists only on paper, you will not need to load any of the existing files.

Other early choices will entail deciding whether or not you will be printing out a drawing, digitizing a paper or vellum drawing, or doing a new drawing. Once these basics have been decided you will be making choices between large function groups, some of which are listed here.

1. Drawing
2. Editing
3. Labeling and other text
4. Transformations
5. Dimensioning

Each one of these may have its own set of submenus, or all the menus may be included in one central menu. We will now discuss these function groups one at a time along with some typical submenus associated with them.

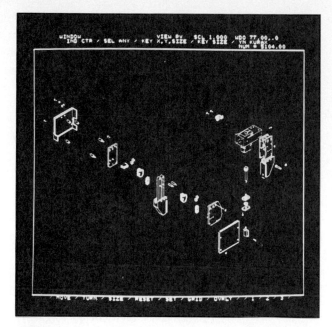

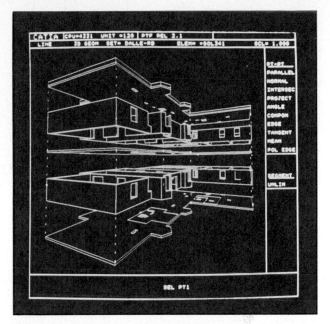

FIG. 9 An exploded isometric of an assembly on the IBM 5080. *(Courtesy IBM.)*

FIG. 10 Color discrimination in a multilevel isometric on the IBM 5080. *(Courtesy IBM.)*

C. Drawing There are choices of mode and function to be made. Some of the choices of mode are positioning, grid choice, zoom, and line quality.

▪ **Positioning** A digitizing tablet may be sensitive to differences of .01 inch or less. Most people's hand-eye coordination is less accurate than this, and very few raster-scan monitors have such a fine resolution. A CAD program will provide a grid constraint system called "snap-lock" or "nearest pen." In this system the drafter will work with a grid set at a chosen spacing and the computer will "snap" the drawn point to the nearest grid point if the drawn point is within a set distance or "gravity field" of that point.

▪ **Grid Choice** Athough a drafter will typically use a square grid for a snap-lock device, an isometric grid may also be used. This is very useful for producing isometric sketches.

▪ **Zoom** As with a camera, the zoom function produces a magnification of a given part—in this case, of a drawing. This part is usually identified by using a window. Zoom, of course, also helps solve the problem of the limited accuracy of hand-eye coordination by enlarging the details of a drawing.

▪ **Line Quality** There are several variations of lines that are usually available: line thickness, line type (solid, dashed, centerline, etc.) and color.

Some of the function choices for drawing lines, rectangles, circles, arcs, and filling, will be described a little further here.

Note that when working with subfunctions when the drawing is on the screen, there will typically be a small flashing arrow or cross on the screen, the location of which is controlled by the drafter. This control will be with a cursor pad or stylus on a graphics tablet, by a joystick or similar hand-held device, or by direction keys on the keyboard. The drafter will "fix," or identify, a point by using a button, making contact with the stylus, or pressing a key on the keyboard.

▪ **Lines** Typically this mode will involve the user

identifying the endpoints of a line and the computer drawing the line that joins them. However, a good CAD system may provide automatic line-drawing functions for horizontal, vertical, perpendicular, and parallel lines. Lines may also be specified as having a certain angle or being tangent to a given curve.

■ **Rectangles** A user may define a rectangle for the computer by providing the locations of two opposite corners. The computer will then draw the rectangle. Actually, this is a very easy thing to program; you might try doing it for coordinate values entered at the keyboard.

■ **Circles** A circle may be drawn in a CAD program in several ways. One way is to enter a center point and any point on the circumference. Another is to enter the center point and provide the value of the radius or the diameter. Yet another is to enter any three points on the circumference. After the information is entered, the computer draws the circle—which may come complete with centerlines.

■ **Arcs** Like circles, these may be defined in more than one way. An obvious way is to specify the center and the endpoints of the arcs on the circumference. An alternative to the endpoints is to specify the radial lines that delimit the arc. A valuable option for surveyors in route layouts is to specify the limiting tangents and the radius of an arc.

■ **Filling** Filling in given areas of a drawing is used for section lining and for shading. In shading, the lines used for filling are often drawn side by side so that the area appears uniform. The user will have the option of varying the quality, angle, and spacing of the lines used for filling. There may also be the option of using symbols for filling.

D. Editing Editing consists of two types of activities. One type is that of using the erase, or delete, function to remove unwanted features from the drawing. The other is that of using other functions (drawing, labeling, transformation) to modify the drawing. Therefore, the edit function usually means erasing, because that is what it does that is unique. Note that erasing will involve the same types of feature specification functions in its sub-menu that are found in drawing and labeling. To be able to erase a line or a circle, you must be able to specify it.

E. Labeling When providing text entries on a drawing, the drafter will be able to choose the size, orientation, and font type of the labels. "Font" refers to the script type, e.g., vertical, sloped, italicized, Gothic.

F. Transformations Once a shape has been drawn it is very convenient to be able to manipulate it in various ways. Some of the more commonly used transformations are translation, scaling (up or down), rotation, reflection, and shear.

■ **Translation** When a shape is moved to a different location on the drawing without being distorted or rotated, it is said to have been translated. In multiview drawing the translation may be considered in terms of an X and a Y coordinate. More generally, when dealing with three-dimensional space, a translation will also have a Z component. The translation consists of adding the X,Y,Z components of the translation to the coordinates of all the corner points and center points that define the shape. That is what the computer does. What the CAD drafter does is to specify translation and then identify the shape to be moved by fixing one point on it. The translation is made after a new location for the point is provided. The whole shape is taken along with the chosen point.

■ **Scaling** Scaling is used for changing the size of the shape. This can be a fairly straightforward process of specifying scaling and identifying the object. However, there are two possibilities to be aware of. One is the opportunity of applying different scale factors for X and Y. This gives you the ability to distort your shape—to turn a square into a rectangle, for example. The other possibility occurs if you are able to specify a point about which the scaling takes place. Consider a square fixed in space. First, increase its size by a factor of 2 about its center point. Now, starting with the original square, increase its size by a factor of 2 about its bottom left-hand corner. The point about which an object is scaled remains fixed at the same point in space. Everything else moves. So in our second scaling there is a translation effect, because the center of the square has moved upward and to the right. The center is now twice as far in X and Y from the fixed bottom left-hand corner. So scaling may involve a translation effect and this must be watched for.

■ **Rotation** Rotation may be a straightforward, clockwise, positive turning of an object about its center. But, like scaling, it becomes translation also when the point of rotation is not the object's center. And, as with scaling, the translation effects become very large when the transformation point is very far from the shape. Care must be taken to consider these effects whenever the rotation function is used.

■ **Reflection** In reflection a shape is drawn again identical to the first, but as a mirror reflection of it. If, for example, a shape is reflected about the X axis, the reflected shape is an upside-down version of the original—just as your reflection in a pond would be. It is common for the reflection axis to be the X or Y axis, but reflection across other lines may be possible in some CAD systems.

■ **Shear** The shear transformation is a type of distortion. Consider two opposite sides of a rectangle, where one side is moved in a direction parallel to those two sides while the other side remains in the same place. The result is a parallelogram in which the two original sides have kept the same orientation, whereas the other two sides have not. This is a shear distortion (see Fig. 11).

■ **Replication** Not strictly a transformation, this involves the reproduction of an existing shape in another location. It is like translation, but the original stays where it was.

There are other forms of manipulation typically available in industrial CAD systems, such as the conversion of circles into ellipses or slots, simple line extension, and the distortion effects of moving around a single corner of a shape.

G. Dimensioning A nice feature of CAD systems is the automatic dimension function. When this function is active, the user identifies two points, between which the dimension is required, and the computer provides the extension lines, the dimension lines, the arrowheads, and the dimension itself. If you intend to use the automatic dimensioning function, it is advisable to use the snap mode of drawing with the gravity field set at a tolerance compatible with the dimensioning accuracy that you intend to use.

If you use the automatic function you will have to choose a number of parametric controls. These include

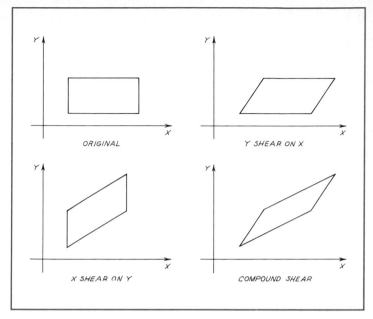

FIG. 11 Shear transformation (sketch).

line quality, dimension orientation (horizontal, vertical, parallel), chain or stack dimensions, dimension type (linear, angular, radial, hole), and text size and location.

CAD systems also permit manual dimensioning, but there will still be parameters to select for extension lines, arrowheads, etc.

The above introduction to industrial CAD systems should be considered as just that. All CAD systems involve many capabilities not even touched on here.

5. INTRODUCTION TO A CAD SYSTEM

A. Introduction The CAD system that will be discussed here works on any 64K Apple II+, IIe or IIc. It was written by Dr. Roy E. Myers, David L. Huggins, and Larry Hampshire of the New Kensington Campus of The Pennsylvania State University. It will be referred to as NKCAD in this chapter. The program is public-domain software, which means that it may be obtained from the authors for a handling fee only. Further, it may then be legally reproduced with no further payments. It comes on a single floppy disk with documentation. For further information write to:

Professor David L. Huggins

The Pennsylvania State University

3550 7th Street Road

New Kensington

Pennsylvania 15068

NKCAD is a 2-D software package that will accept input from the keyboard, paddles, trackball, or graphics tablet. Although originally written for an Apple graphics tablet, a version is available that works with the Houston Instrument HIPAD tablet (parallel interface). It has a graphic dump to compatible dot matrix printers (such as the Epsom MX-80), and it will drive a Bausch & Lomb DMP29 multipen plotter (a version exists for the Hewlett-Packard 7047A two-pen plotter, also). In many cases, modifying it to interface with other printers and plotters would not be hard to do.

The software features of NKCAD include a snap function, square and isometric point grids, automatic box, circle and arc drawing, extension and dimension lines, calculated dimension, labeling, dashed lines, line deletion, zoom, and translation. These are quite enough features to introduce the student to CAD, and, although the screen and printer images are affected by the resolution of the hardware, the results with the plotter can be very nice. In short, NKCAD is a versatile and very inexpensive CAD program. And, since the keyboard version works quite well, using NKCAD means getting started for the price of an Apple. Since most educational institutions cannot acquire sufficient quantities of industrial-standard CAD equipment and software, and most professional CAD drafters will have to be trained on their companies' CAD anyway, it makes sense to consider a free lunch like NKCAD. It may also serve as the base for an introductory course where more advanced equipment is used for more advanced students—and treated with more care by them because of their introductory training.

B. Using NKCAD The first few steps will be as follows:

1. If you are going to be using peripheral devices, these need to be in place. Then insert your NKCAD disk and turn the computer on—or use the PR#6 command.

2. The screen will display an engineering drawing that alternates between inverse and normal displays. Press any key to start the program.

3. The next display will be a CAD menu, from which you will choose either to print or plot a file or to enter a drawing mode. If you choose a drawing mode, you will have to specify which input device you will use.

4. The next message on the screen will ask whether or not you wish to load a file from the disk. This means do you have a previous drawing that you wish to load into the computer so that you may continue to work on it. If yes, you will be prompted for the name of the file.

5. The next display will tell you how many lines, labels, circles, etc., you have space allocated for in the primary memory. You may at this time change the relative proportions of these items, although for most drawings there are plenty of each category. This particular option reflects the limitation of the size of the primary memory. And, since 64K was considered a large memory until the mid 1970s, you can see why CAD has become widespread only since the late 1970s. If you wish to make no adjustment, just press RETURN.

6. Now you will be in the drawing mode, with a mostly blank screen containing two messages at the bottom. At the bottom left the screen will say CHOOSE and at the bottom right it will indicate whether your cursor mode is nearest point (snap function on) or exact pen (snap function off). Changing your cursor mode is just one of the many options available to you.

If you are working with any input device other than a keyboard you will need to initialize it, that is, let the computer know what setting corresponds to 0,0 coordinates (the origin). On the graphics tablet this means pressing the RESET button on the tablet, then pressing the stylus down on the bottom left-hand corner of the pad (or putting the cross-hairs of the cursor pad there and pressing the button), and finally pressing the STREAM button on the tablet.

Now you must choose from the menu that is provided for the keyboard or is an overlay on your graphics tablet. The initial decisions are usually the following:

1. If you have loaded a file, you will probably wish to exercise the REDRAW FILE function. This will draw on the screen the picture stored (in data form) in the file. (If the file has dimensions you will need to use REDRAW DIMENSIONS also.) We will use this option to indicate how most choices are implemented.

   If you have a graphics tablet, you will simply move the cursor or stylus over the box marked REDRAW on the lower right of the overlay and press the stylus down or press the button on the cursor pad. The computer then goes to work. For the other input methods, press the M key on the keyboard. When the command has been executed the CHOOSE prompt will return.

2. You need to decide whether to use the nearest dot function or the exact pen. With nearest dot the computer will snap to the nearest dot if it is close to it, and regardless of whether you have the dots displayed. Nearest dot saves time while you are drawing, but you must decide if the nearest dot will produce the exact shape that you want.

3. The dot grid may be displayed or not. Further, it may be a square grid or an isometric grid. The latter is very useful for isometric sketching. To get the isometric grid you must choose DRAW DOTS and then choose ISOMETRIC DOTS.

4. The scale will be set at 1 (approximately true size in inches on the screen, and it should be exact on a plotter), but this may be changed at the outset. If it is, the dot grid will reflect the new scale.

C. Drawing with NKCAD You are now ready to start drawing, although most of the decisions that you have made so far can be easily changed. We will briefly consider working with the following functions: lines, circles and arcs, labels and dimensions, and scale (zoom) and slide (translation).

■ **Lines** The main choice in line drawing is between DRAW LINE, which allows you to draw one line, and DRAW SEQUENCE OF LINES, which allows you to keep drawing from point to point until you choose CANCEL LAST COMMAND.

To practice, choose DRAW SEQUENCE OF LINES by placing your cursor pad over this function box on the graphics tablet and pressing the button or, for keyboard entry, press the key for the number 4. Now, at the bottom left of the screen, you will be prompted to identify the first point. By moving the cursor pad around the graphics tablet or using the appropriate keys on the keyboard, you can control the cross-hairs on the screen. When the cross-hairs are where you want them press the button on the cursor pad (or trackball), or press the return button on the keyboard if you are using the keyboard mode. A small square will be drawn around the point you identified. Now choose the next point and repeat the process. The computer will join the first two points with a line and ask you for a third point. It will then join the second point to the third point and so on until you select CANCEL LAST COMMAND. When you do this you will then be prompted to choose a function again.

While doing line drawing you have several choices you may wish to use. Try out the following functions:

■ 1. DASHED LINE Choose this before you select the line drawing command.

■ 2. ENLARGE CURSOR Once you have chosen a command, you may enlarge the cross-hairs of the screen cursor. This is helpful if you find it hard to see the regular cross-hairs.

■ 3. DRAW BOX This will draw a rectangle automatically after you have identified two diagonally opposite corners.

■ 4. DRAW DOTS This will give you a square grid of points. Try it with both the NEAREST DOT and the EXACT PEN to get a feel of the size of the gravity field.

■ 5. LINE SEGMENT This will divide any line into two individual line segments, which is useful if you want to delete part of a line.

By now you should have quite a mess on your screen, so it is time to practice some editing.

1. Choose CLEAR SCREEN NOT COMPUTER.

2. Choose REDRAW FILE.

You are now back where you started, but at least you have learned something.

3. Choose DELETE LINE. This works like drawing a line except that you apply it to an existing line.

Delete a few lines, but note that you must identify a line by an endpoint, not any point on it. Where more than one line has the same endpoint, the computer will keep selecting the possible lines until you confirm the correct choice.

4. Choose CLEAR ALL START OVER.

5. Choose REDRAW FILE.

This time you really got rid of your drawing.
Now do the following exercise:

1. DRAW DOTS.

2. ISOMETRIC DOTS.

3. NEAREST DOT.

4. Draw the visible lines of a three-dimensional rectangular block using the dots of the isometric grid.

5. DASHED LINE.

6. Draw in the hidden lines of your block.

▪ **Zoom and Slide** With this rectangular block image, you can practice the following functions.

▪ 1. SLIDE ENTIRE OBJECT Pick any point on your object and provide a new location on the screen for it. The entire object (all the objects if you have more than one) will then move in the same way. Repeat the exercise so that the block is placed more or less centrally on the screen. (With this software you do not actually have to choose a spot on the object. It is enough to place two points anywhere on the screen to represent the desired translation.)

▪ 2. SCALE USING SCALE FACTOR Choose a new scale factor, say, 1.5. This value must be entered at the keyboard even if you are using a graphics tablet. Now choose REDRAW FILE to see the results. Repeat using a scale factor of 0.5.

▪ 3. RETURN TO SCALE 1 This will bring you back to the original scale, but you will need to use REDRAW FILE to see it.

▪ 4. SCALE USING WINDOW This will require you to enter two corners as if you were going to draw a box. The contents of the "box" (window) will be drawn to fill the screen. Do it several times to become familiar with it. Use RETURN TO SCALE 1 to get back to the whole drawing.

▪ NOTE When you scale with SCALE USING SCALE FACTOR or SCALE USING WINDOW you are only changing the appearance of the object on the screen. These are both zoom functions only and they do not change the scale of the object drawing. They are very useful for drawing smallest details.

▪ **Arcs and Circles** The main functions on this topic allow the user to draw and delete arcs and circles. Start with a clear screen, preferably by using CLEAR ALL START OVER.

1. Choose DRAW CIRCLE. You will be asked to identify the center point and any point on the circumference. When you have done this, the computer will draw the circle complete with centerlines. Draw a second circle.

2. Choose DELETE CIRCLE and one of the circles you have drawn. When prompted, identify which circle is to be deleted by its center point. The computer will draw little confirming squares on points on the circumference for the circle it thinks you mean. Enter Y or N at the keyboard for yes or no.

3. Choose SECTOR CIRCLE. This will allow you to eliminate all but a part of the circle, thus leaving an arc. Select the two endpoints of the arc that you wish to end up with. The arc that will be left will be from your first point to your second point in a counterclockwise direction. Once you have entered the two points the computer will erase the entire circle and then draw the arc back in. The centerlines will still be in place. However, if you use CLEAR SCREEN NOT COMPUTER and then REDRAW FILE the centerlines will no longer be there. You may sometimes wish to draw lines to the center of the circle while it is still identified.

4. The command DRAW CENTER-LINE HORIZ/ VERT ONLY may be used for holes as seen from the side.

5. Clear the screen and draw two nonparallel lines to use with the arc commands.

6. Select DRAW ARC OR FILLET. You will be asked to identify which two (tangent) lines will be joined by the arc. The order in which you choose the lines is important. There are two quite differ-

ent arcs that may join any two tangent lines. You probably cannot understand this, but you will soon find out when you start practicing. So, practice away. When prompted for a radius use a value of 1 until you are ready to try other values. When you have tried five or six arcs, clear the screen and draw a box. Now try to round the corners at one end of the box.

7. Practice working with DELETE ARC OR FILLET and DASHED ARC/FILLET.

■ **Labels and Dimensions** For the last introductory session with NKCAD we will look at its options for dimensioning and labeling.

1. Use CLEAR ALL START OVER.

2. Draw a box in the middle of the screen. We will use this to practice on.

3. Select KEYBOARD LABEL. Your label may be up to 12 characters. Type in GEORGE'S BOX or some such nonsense. Then place the label at the bottom center of the screen. If you want to use a longer label, simply repeat the process and put the second label next to the first. Practice by using DELETE LABEL and then KEYBOARD LABEL again.

4. Choose EXTENSION LINE. Draw extension lines for each end of the box. Use DELETE EXTENSION LINE to erase one of the lines, then restore it.

You are probably wondering why there are special draw and delete functions for extension lines when they look just like the lines drawn by DRAW LINE. The reason for this only becomes apparent when the drawing is output to a plotter. At that time the lines may be drawn according to the proper values for visible object lines, extension lines, etc. To that end a record of which line is which must be kept throughout the drafting process.

5. Choose DIMENSION LINE and draw a dimension line between the two extension lines. You will draw each end separately, specifying first the end of the shaft (tail) location and then the point of the arrow. Remember to leave enough room between the shaft ends for the dimension label. Erase the dimension line with DELETE DIMENSION LINE and then draw it in again.

6. Select CALCULATED DIMENSION LABEL. Use this to dimension the length of the box on the dimension line just established. Use a scale of 1 the first time. Now use DELETE LABEL to erase this dimension and use RESET SCALE. Use CALCULATED DIMENSION LABEL to dimension the length again at the new scale. Note that you could use KEYBOARD LABEL to put in whatever dimension you want. Repeat several times to get a feel for where the label must be placed. It is best to use EXACT PEN for this.

7. Now dimension the short side of the box by using EXTENSION LINE ⊥ TO LINE and DIMENSION LINE ⊥ TO EXTENSION LINE. The purpose of these options is to guarantee orthogonality. Erase and repeat several times to become familiar with their use.

■ NOTE The RESET SCALE FUNCTION is how you change the scale of the object drawing. Using it only affects the automatic dimension function CALCULATED DIMENSION LABEL. Otherwise you could just put in dimension labels according to the scale you wanted.

You have now been introduced to most of the drafting functions of NKCAD. To produce good drawings with it will take a little experience, but after 2 or 3 hours you should be good enough to consider producing a drawing for the plotter (if you have one). Remember, the plotter will produce perfect lines—unlike the raster display of your screen. However, before you spend a lot of effort on a drawing you should become familiar with the file management system. Otherwise you may lose a good drawing because you don't know how to handle it.

1. Practice saving and loading a file (drawing). Each drawing occupies a lot of storage, so don't expect to save many on the NKCAD disk, for example, because it is already almost full. After saving the file use CLEAR ALL START OVER before reloading your file to check the saving and loading process.

2. Do a printer dump. NKCAD has an APPLESOFT program called PR.DUMP. It is written for a Grappler interface card, but it would be easy to

modify for other cards such as Pkaso or Microbuffer.

3. Finally, try a plotter dump. Again, check for compatibility. You may have some rewriting to do, if you have a different plotter.

4. Once you are familiar with the file management system you can feel confident about investing a lot of time in a drawing. It would be a good idea to save major drawings to two different disks.

5. In the first menu an option called FMERGE is provided which will allow you to combine two drawing files. If you use this, make sure that you load the larger file first and that the two files have a common reference point.

REFERENCES

For those who wish to learn to program in BASIC on the Apple, Devon provides an efficient way to begin and Coan will provide a chance to further your skills. Computer graphics on the Apple is well covered by Myers, and Demel and Miller provide listings in both BASIC and FORTRAN for the IBM PC and the VAX, respectively. Disks are available for both of these texts (an Apple disk is available also for the Demel and Miller text). A more advanced text is that by Artwick. CAD is introduced in Zandi. More comprehensive treatments are given by Ryan (1984) and by Groover and Zimmer. A very good introduction to the way computers work is provided by Toong and Gupta. A recent review of computer software may be found in an issue of *Scientific American* devoted to the topic (vol. 251, no. 3, September 1984).

Artwick, Bruce A. *Applied Concepts in Microcomputer Graphics*. Englewood Cliffs, N.J.: Prentice-Hall, 1984.

Coan, James S. *Basic Apple Basic*. Rochelle, N.J.: Hayden Books, 1982.

Demel, John T., and Miller, Michael J. *Introduction to Computer Graphics*. Monterey, Calif.: Brooks/Cole, 1984.

Devon, Richard F. *The First Few Bytes*. Dubuque, Iowa: Kendall/Hunt, 1984.

Encarnacao, J., and Schlechtendahl, E. G. *Computer Aided Design*. Berlin: Springer-Verlag, 1983.

Gardan, Yvon, and Lucas, Michael. *Interactive Graphics in CAD*. London: Kogan Page, 1983.

Groover, Mikell P., and Zimmer, Emory W. *CAD/CAM Computer-Aided Design and Manufacturing*. Englewood Cliffs, N.J.: Prentice-Hall, 1984.

Myers, Roy E. *Microcomputer Graphics*. Reading, Mass.: Addison-Wesley, 1982.

Ryan, Daniel L. *Computer-Aided Graphics and Design*. New York: Marcel Dekker, 1979.

————*Principles of Automated Drafting*. New York: Marcel Dekker, 1984.

Toong, Hoo-min D., and Gupta, Amar. "Personal Computers," *Scientific American*, vol. 247, no. 6, pp. 86–107, December 1982.

Zandi, M. *Computer-Aided Design and Drafting: Concepts and Procedures*. Albany, New York: Delmar Publishers, 1985.

PROBLEMS

SECTION 2: COMPONENTS OF A CAD SYSTEM

1. What are the four major functions of a computer system?

2. What does "database" refer to?

3. What is the numeric system used by computers at their most basic level? Convert the decimal numbers 7, 23, and 35 to that system.

4. List and briefly describe three different input technologies.

5. What is the difference between raster-scan and stroke graphics?

6. What is virtual memory?

7. When a computer is described as 16-bit or 8-bit, what does this mean?

8. What is the difference between a daisy-wheel printer and a line printer?

SECTION 3: PROGRAMMING THE COMPUTER

9. Write a program that draws a picture of a corrugated surface. Hint: Use two sine curves with connecting lines.

10. Write a program that will compute the roots of $Y = AX^2 + BX + C$, where A, B and C are input by the user.

11. Write a program that will take as input the length of the three sides of a right triangle and give as output the three angles in degrees, minutes, and seconds.

12. Write a program that will find the maximum and minimum values of a set of data input by the user.

13. The factorial of 5 is $5 \times 4 \times 3 \times 2 \times 1$. Write a program that will calculate the factorial of any positive number input by the user.

14. Write a program that will draw a cube in isometric.

15. The equation for a parabola is $Y^2 = 4 \times A \times X$ where A is a constant for any given parabola. Write a program that will draw three parabolas by changing the value of A three times.

16. Modify the parabola program for Prob. 15 to draw the parabolas reflected as well.

SECTION 4: COMPUTER-AIDED DRAFTING

17. Describe two strengths and two weaknesses of CAD compared to traditional graphics.

18. What are some of the first choices you will be making when you log-on to a CAD system?

19. What is the snap function? What is a gravity field?

20. What is a zoom function? What is a window?

21. Provide two examples of transformation of a shape.

22. Suggest three ways that a CAD system might allow a user to define a circle.

23. Give two examples of mode choices when using CAD.

24. Provide two examples of function choices when using CAD.

SECTION 5: INTRODUCTION TO A CAD SYSTEM

For the problems below use NKCAD or some other CAD system. It is better if you type in the dimension labels to what they should be rather than let the computer calculate your screen dimensions, which will be only approximate.

25. Draw a front and a right side view for the object shown. Include labels and dimensions.

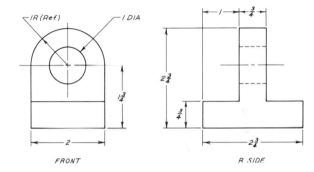

FRONT   R. SIDE

PROB. 25

26. Draw a front and a right side view for the object shown. Include labels and dimensions.

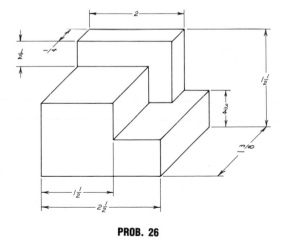

PROB. 26

27. For the object given in Prob. 26, draw an isometric pictorial. Omit dimensions and hidden lines.

28. Draw top and front views for the object shown. Include labels and dimensions.

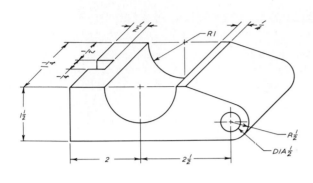

PROB. 28

29. Draw the front view and section AA for the object shown. Label, but do not dimension.

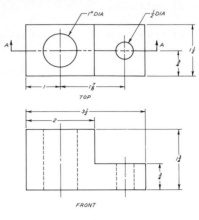

PROB. 29

30. Draw the front, top and right side views for the object given in Prob. 29. Label, but do not dimension.

31. Draw the top and front views of the object shown without labels and dimensions. Use the SCALE USING WINDOW function to draw the small details.

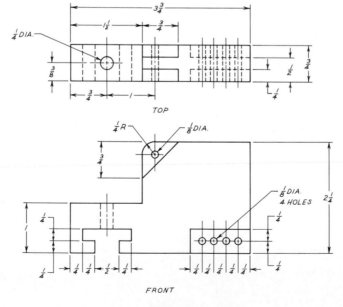

PROB. 31

32. Draw the front view and section BB for the object shown. Use metric dimensions as given. To do this DRAW DOTS and use the SCALE USING SCALE FACTOR (zoom) and set a zoom scale of 0.8. This will give you a screen grid of 5 (instead of 4) squares to the inch. Now use RESET SCALE and choose two grid points 5 squares apart and set the distance at 25. The CALCULATED DIMENSION LABEL will now give dimensions at 25 units per screen and plotter inch. Obviously, then, you will treat these units as millimeters since there are 25.4 millimeters to the inch. Your grid may be used as a guide as it is set at 5 millimeters per square. However, it will be better to type in the dimension labels to what they are supposed to be rather than let the CALCULATED DIMENSION LABEL produce exact accounts of your approximations.

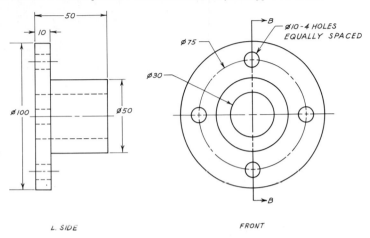

PROB. 32   METRIC.

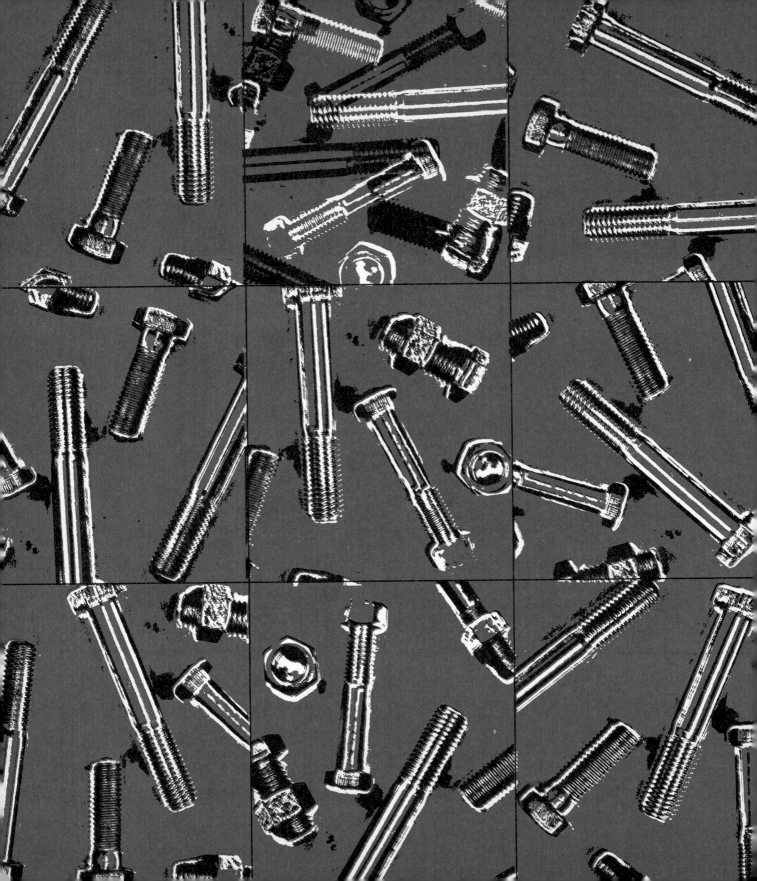

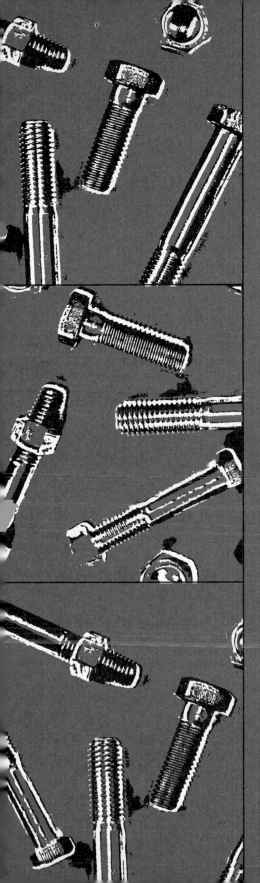

PART D

SPECIAL TOPICS

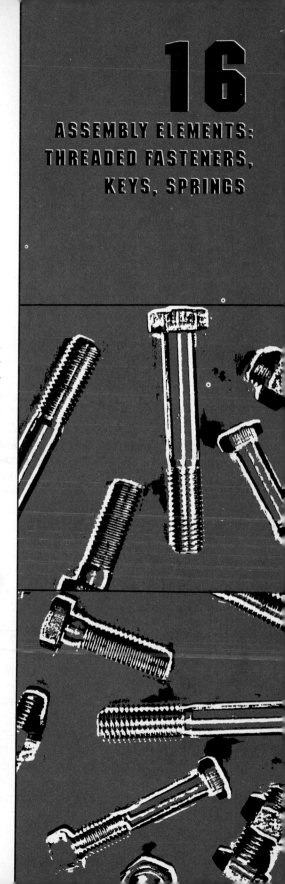

1. INTRODUCTION

This chapter is the first of three that deal with what can loosely be called "assembly elements." Chapters 16 and 17 discuss devices that hold parts together so they function as designed. The devices may be used to align one part to another, or may be used to transmit motion or force, as in a bolted drive-shaft flange. Welding and riveting, which are processes used for permanent assemblies, are covered in Chapter 18.

In this chapter threaded fasteners receive the most attention, since their use is so common. Keys and springs are covered in sufficient depth to acquaint the reader with their basic forms.

2. HISTORY OF THREADED FASTENERS

The earliest records of the screw are found in the writings of Archimedes (278 to 212 B.C.). Although specimens of ancient Greek and Roman screws are so rare as to indicate that they were seldom used, there are many from the later Middle Ages; and it is known that crude lathes and dies were used to cut threads in the latter period. Most early screws were made by hand, by forging the head, cutting the slot with a saw, and fashioning the screw with a file.

In America in colonial times wood screws were blunt on the ends; the gimlet point did not appear until 1846. Iron screws were made for each threaded hole. There was no interchanging of parts, and nuts had to be tied to their own bolts. In England, Sir Joseph Whitworth made the first attempt to set up a uniform standard in 1841. His system was generally adopted there but not in the United States.

The initial attempt to standardize screw threads in the United States came in 1864 with the adoption of a report prepared by a committee appointed by the Franklin Institute. The system, designed by William Sellers, came into general use and was known as the "Franklin Institute thread," the "Sellers thread," or the "United States thread." It fulfilled the need for a general-purpose thread at that

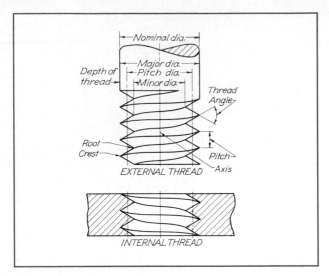

FIG. 1 Screw-thread terminology.

period; but with the coming of the automobile, the airplane, and other modern equipment, it became inadequate.

Through the efforts of the various engineering societies, the Bureau of Standards, and others, the National Screw Thread Commission was authorized by act of Congress in 1918 and inaugurated the present standards. The work has been carried on by ANSI (American National Standards Institute) and others. These organizations, working in cooperation with representatives of the British and Canadian governments and standards associations, developed an agreement covering a general-purpose thread that fulfilled the basic requirements for interchangeability of threaded products produced in the three countries. The "Declaration of Accord" establishing the Unified Screw Thread was signed in Washington, D.C., in 1948. Since then standards have been continually updated. The most recent standard for Unified Inch Screw Threads is ANSI B1.1–1982.

Metric thread has had its own separate development. Much effort has been given to reconciling the International Standards Organization (ISO) standards used in Europe with needs of American industry. After many meetings a compromise was reached between American needs and existing ISO standards. The resulting publication, ANSI B1.13–1983, provides metric screw threads to be used for general fasteners. The standard is in basic agreement with ISO screw thread standards and is now widely used in the United States.

3. SCREW THREAD TERMS

Figure 1 shows the primary terms useful in defining screw threads. These terms are:

1. *External thread (screw)*. A thread on the external surface of a cylinder.
2. *Internal thread (nut)*. A thread on the internal surface of a cylinder.
3. *Major diameter*. The largest diameter of a screw thread.
4. *Minor diameter*. The smallest diameter of a screw thread.
5. *Pitch diameter*. The diameter of an imaginary cylinder, the surface of which cuts the thread forms where the width of the thread and groove are equal.
6. *Crest*. The edge or surface that joins the sides of a thread and is farthest from the cylinder or cone from which the thread projects.
7. *Root*. The edge or surface that joins the sides of adjacent thread forms and coincides with the cylinder or cone from which the thread projects.
8. *Depth of thread*. The distance between crest and root measured normal to the axis.
9. *Pitch*. The distance between corresponding points on adjacent thread forms measured parallel to the axis.
10. *Lead*. The distance a threaded part moves axially, with respect to a fixed mating part, in one complete revolution. See Fig. 2.
11. *Threads per inch*. The reciprocal of the pitch and the value specified to govern the size of the thread form.
12. *Form*. The profile (cross section) of the thread. Figure 3 shows various forms.
13. *Right-hand thread*. A thread that when viewed axially winds in a clockwise and receding direction. Threads are always right-hand unless otherwise specified.

14. *Left-hand thread*. A thread that when viewed axially winds in a counterclockwise and receding direction. All left-hand threads are designated LH.

15. *Single thread*. A thread having the thread form produced on only one helix of the cylinder (Fig. 2). See multiple thread (below). On a single thread, the lead and pitch are equivalent. Threads are always single unless otherwise specified.

16. *Multiple thread*. A thread combination having the same form produced on two or more helices of the cylinder (Fig. 2). For a multiple thread, the lead is an integral multiple of the pitch; that is, on a *double thread*, lead is twice the pitch; on a *triple thread*, lead is three times the pitch, etc. A multiple thread permits a more rapid advance without a coarser (larger) thread form.

4. THREAD FORMS

"Thread form" describes the shape of the thread if one cut a threaded fastener in half along its axis and then looked at the thread configurations. Different thread forms have different uses and histories. Figure 3 shows a number of the more commonly known thread forms, only a few of which are widely used.

Metric thread is one thread form used worldwide in a broad range of general applications. Its use has increased in the United States because of increasing import-export trade in products such as automobiles and machine tools.

Another popular thread form is the unified (UN) form. It is heavily used in the United States; it is a combination of the older American national (N) and the British Whitworth. ANSI Standard B1.1–1982 includes a modified form of UN, called UNR, to supplement the UN form. Only external threads (shafts) may have a UNR form, which provides a rounded root for improved strength. Forms so far mentioned are known as V-shaped. They are excellent for assembly and adjustment purposes. A 60° stud is also a form of V thread, but it is less deep.

Shapes of the V form are not good for transmitting high loads. Stronger threads used for power transmission are square (rarely used now), Acme, stub Acme, but-

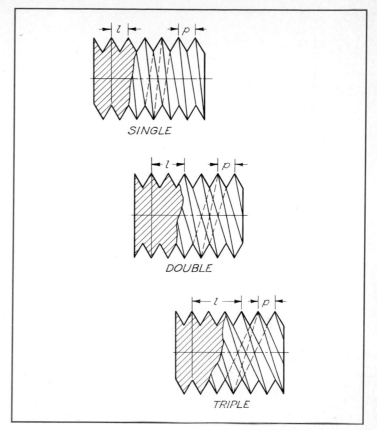

FIG. 2 Multiple threads. Note that, for a single thread, lead and pitch are identical values; for a double thread, lead is twice the pitch; and for a triple thread, lead is three times the pitch.

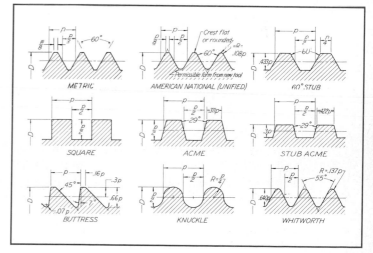

FIG. 3 Thread profiles.

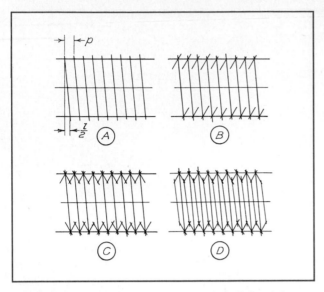

FIG. 4 Stages in construction for drawing a single V thread. Lay out pitch and half-lead and draw crest line *(A)*, start thread contour *(B)*, complete thread contour *(C)*, and draw root lines *(D)*.

tress, and B & S Worm (not illustrated). Knuckle thread is a unique form used on bottle caps, glass jars, and light-bulb bases.

5. SYMBOLS FOR STRAIGHT THREADS

A true representation of a thread is not needed on a working drawing. To draw a thread literally would be tedious and unnecessary. Symbols are used instead to show thread. Three types of symbols are in use: detailed, schematic, and simplified. The detailed symbol varies, depending on whether a V-shape, square, or Acme thread is shown. The schematic and simplified symbols are the same, regardless of thread form shown.

A. Detailed Symbols

■ **1. Detailed Symbol for UN/UNR Threads** In general, true pitch should be shown, although a small increase or decrease in pitch is permissible in order to have even units of measure when making the drawing. Thus seven threads per inch may be increased to eight, or four and one-half may be decreased to four. Remember that this is only a means to simplify the drawing—the actual threads per inch must be specified in the dimensioning.

To draw a thread you must know if it is external or internal; its form, major diameter, pitch, and multiplicity; and whether it is right- or left-hand.

Figure 4 illustrates the stages in drawing a UN/UNR thread. At A the diameter is laid out, and on it the pitch is measured on the upper line. This thread is a single thread; therefore, the pitch is equal to the lead, and the helix will advance $p/2 = l/2$ in 180°. This distance is laid off on the bottom diameter line, and the crest lines are drawn in. One side of the V form is drawn at *B*; it is completed at *C*. At *D*, the root lines are added.

Figure 5 shows in a larger illustration the detailed symbol used on both external and internal threads. The external thread does not have a chamfer as shown, but chamfers are commonly used to make engagement easier.

■ **2. Detailed Symbol for Square Threads** Figures 6 and 7 show the stages for constructing a square thread. The general process is similar to that for UN/UNR thread. Note that the thread in Fig. 6 is double and right-hand, whereas that in Fig. 7 is single and left-hand.

Observe in Fig. 7 that it is unnecessary to draw the threads the whole length of a long screw. If the thread is left-hand, the crest and root lines are slanted in the opposite direction from those shown in Figs. 4 to 6, as illustrated by the left-hand square thread in Fig. 7.

■ **3. Detailed Symbol for Acme Thread** Figure 8 relates the process for a single, right-hand thread. You will notice that the process is similar to those having to do with the detailed symbols already discussed.

B. Schematic Symbol The schematic symbol for thread is less realistic than the detailed symbols, but the schematic symbol is still very useful. Figure 9 shows the process for constructing an external schematic thread. Only the major diameter and length of thread need be known. The symbol is the same regardless of thread form depicted.

No attempt need be made to show the actual pitch of the threads or their depth by the spacing of lines in the symbol. Identical symbols may be used for several threads of the same diameter but of different pitch. Only in the larger sizes for which the symbols are used could the actual pitch and the true depth of thread be shown without creating a confusion of lines that would defeat the purpose of the symbol. The symbols should therefore

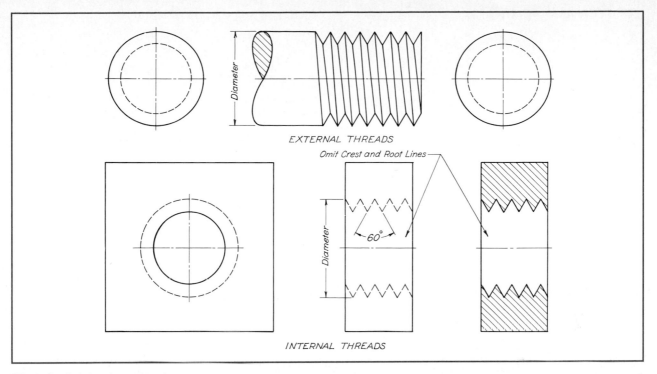

EXTERNAL THREADS

Omit Crest and Root Lines

$60°$

Diameter

INTERNAL THREADS

FIG. 5 Detailed thread representation.

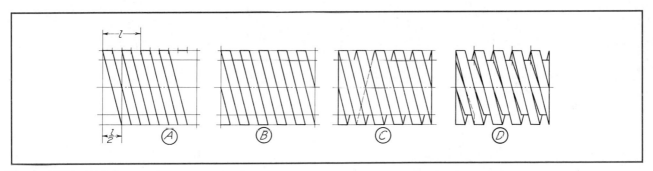

FIG. 6 Stages in drawing a double square thread. The lead is twice the pitch. Note that after spacing the lead and pitch (A), crest and root lines are drawn (B) and (C); then the base line of the thread is added (D).

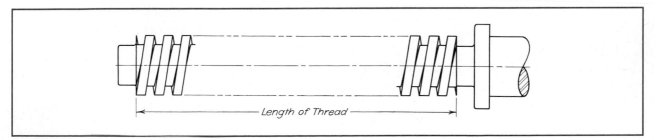

Length of Thread

FIG. 7 Thread representation on a long screw. The ''repeat'' lines save much drawing time.

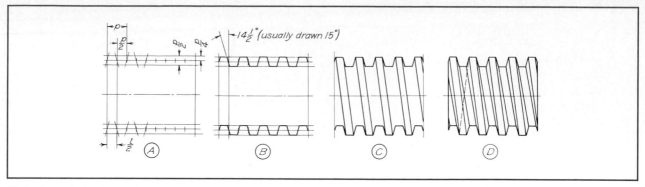

FIG. 8 Stages in drawing a single Acme thread. Because of the shape of the thread, the pitch diameter must first be located *(A)*; then the thread form is drawn *(B)*, and completed *(C)* and *(D)*.

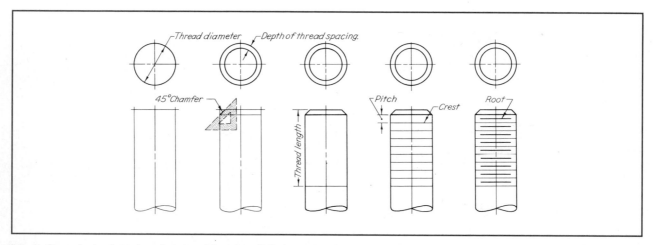

FIG. 9 Stages in drawing schematic external-thread symbols.

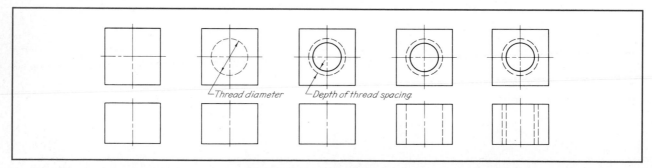

FIG. 10 Stages in drawing internal-thread symbols in external views. Schematic and simplified symbols are identical here.

be made so as to be read clearly on the drawing, without other considerations.

Internal threads using the schematic symbol are seen in Figs. 10 and 11. Again the lines representing depth of thread are not drawn to actual scale but are spaced to look better on the drawing.

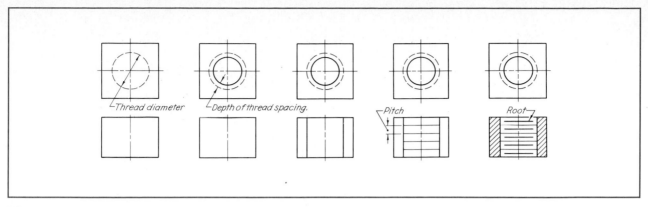

FIG. 11 Stages in drawing schematic internal-thread symbols in sectional views.

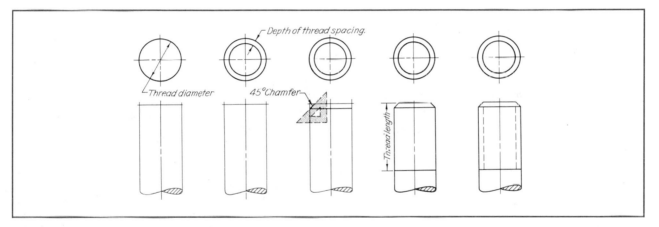

FIG. 12 Stages in drawing simplified external-thread symbols.

C. Simplified Symbol The simplified symbol is well named, for it is indeed a simple symbol. Its virtue lies in the speed with which it can be drawn, as seen in Fig. 12. The symbol is not particularly realistic and is normally used in cases where the dashed lines will not be confused with other features on the drawing. Internal thread is shown identically as that shown for schematic internal thread. Therefore, Fig. 10 serves for both schematic and simplified representations.

D. Threads in Section It is worth noting how the various threads are seen in section. To this point only Figs. 5

and 11 hint at this form of representation. Figures 13 and 14 illustrate well how schematic and simplified thread symbols appear when sectioned.

When pieces are screwed together, sectioning can increase the readability of the drawing, as in Fig. 15. Note that the smallest threaded piece (at the bottom) is left unsectioned to preserve its characteristic shape. Such treatment is subject to the judgment of the drafter.

E. Metric Thread Symbols It is common practice among drafters to draw metric threads in the same manner as UN/UNR threads. We will see in Sec. 7 that it is

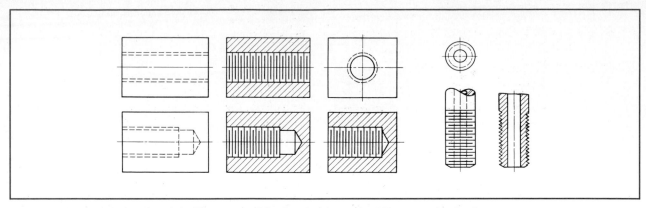

FIG. 13 ANSI schematic thread symbols. These are used for threads drawn under 1 in on assembly drawings.

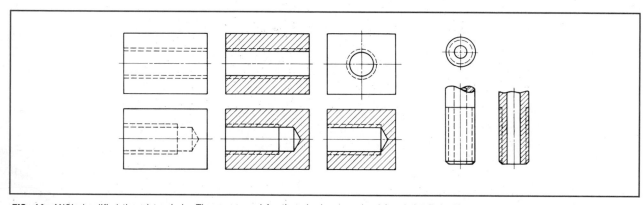

FIG. 14 ANSI simplified thread symbols. These are used for threads drawn under 1 in on detail drawings.

the *specification* of the thread that distinguishes between UN/UNR and metric types. The thread symbol, itself, is not sufficient to fully describe the thread.

6. SYMBOLS FOR TAPERED THREADS

All threads discussed so far have been *straight* threads; that is, there has been no change of major diameter as you move along the axis. Tapered threads are less common than straight threads, but tapered threads are used for assembly of pipes and are known as *pipe threads.* Three ANSI standards are available to fully describe pipe threads:

1. ANSI B2.1—1968, "American National Standards for Pipe Threads," covers six types of threads:

a. NPT: Taper pipe threads for general use
b. NPTR: Taper pipe threads for railing joints
c. ANPT: Aeronautical national form taper pipe threads
d. NPCS: Straight pipe threads in pipe couplings
e. NPSL: Straight pipe threads for loose-fitting mechanical joints with locknuts
f. NPSM: Straight pipe threads for free-fitting mechanical joints for fixtures

These threads usually required a sealant to prevent leakage. Note that pipe threads may be straight.

2. ANSI B1.20.3—1976 (R1982) (inch)

3. ANSI B1.20.4—1976 (metric)

The standards under 2 and 3 above cover eight different forms of what are known as *dryseal* pipe threads.

Such threads do not require a sealant, but will be leak-proof based on metal-to-metal contact between crest and root of the thread. The symbols for tapered pipe threads are seen in Fig. 16.

7. THREAD SPECIFICATIONS: UN/UNR, ACME

The thread symbols we have been discussing do a good job of representing thread forms, but rarely do symbols convey much sense of the *kind* of thread being shown. Drafters and designers must couple the thread symbol with the thread *specification*. Thread specifications will be given for the two most-used thread forms, UN/UNR and metric. Specifications are given in a note to accompany the thread symbol. Thread notes will be explained by using specific examples.

A. Specifications of UN/UNR Thread Figure 17 is an example of one possible thread note. The note conforms to ANSI B1.1—1982, "Unified Inch Screw Threads." Let us look at each part of the note to understand its meaning.

1. Length L: Length may be whatever a designer desires, but lengths usually come in even increments of an inch, dependent on major diameter. Length is independent of thread type. Length may be given as a dimension as shown, or as an addendum to the note itself, such as:

   ¾-16UNF-2A-LH, DOUBLE, 1¾ LG

   where 1¾ LG means 1¾ in long. A threaded hole (internal thread) would use DP for deep, rather than LG.

2. ¾: This is the major diameter of the thread, which may also be written as 0.750. Appendix 10 lists the various diameters available for UN/UNR thread.

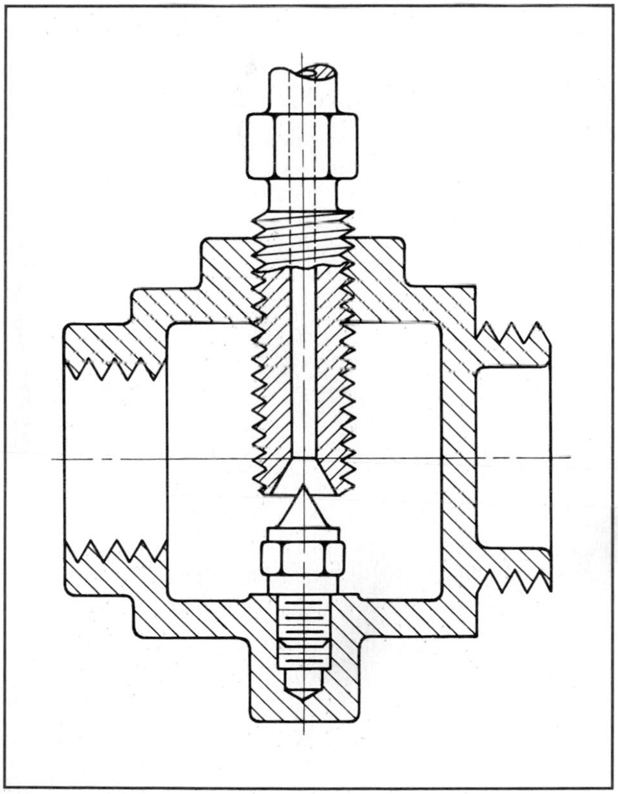

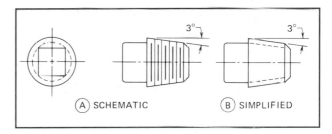

FIG. 16 Symbols for tapered thread.

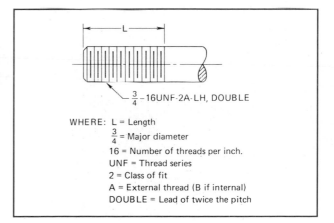

FIG. 17 Specification of a UNF thread.

FIG. 15 Threads in section, actual drawing size.

3. 16UNF: This is the number of threads per inch for a UNF (unified fine) thread. This combination is also seen in Appendix 10. The thread series is UNF, one of 11 available UN/UNR series threads in what is known as the Standard Series.

There is also available a Special Series, designated UNS/UNRS. This series contains all threads with diameter and pitch combinations not within the Standard Series. A discussion and listing of these special threads can be found in ANSI Standard B1.1—1982.

The coarse series, UNC/UNRC of the Standard Series, is used most for mass production of nuts and bolts. It is often used when materials are relatively weak (aluminum, plastic, etc.) because the larger threads, compared with other series, help prevent stripping of threads. Also, UNC/UNRC aids in rapid assembly, and is less prone to assembly problems from light corrosion or nicks.

The fine series, UNF/UNRF, is often used for higher-strength applications than is UN/UNRC because of the larger root diameter for a given size. Fine series also loosens less with vibration than coarse series, and provides greater precision for adjustment applications.

The extra-fine series, UNEF/UNREF, is used less often than the others but is ideal when extra precision of adjustment is needed, such as for retaining nuts for bearings.

The constant-pitch thread series is unique in that the pitch for a given series never changes regardless of the major diameter. Appendix 10 lists 8UN, 12UN, and 16UN versions of this series. The 8UN series is often used as a continuation of UNC for the larger sizes, 12UN for UNF, and 16UN for UNEF. Also existing in ANSI B1.1—1982, but less commonly used, are 4UN, 6UN, 20UN, 28UN, and 32UN series.

4. 2A: This class of fit is one of three (1A, 2A, 3A) that may be applied to external threads. The corresponding fits for internal threads would be 1B, 2B, 3B. The ANSI Standard B1.1—1982 lists exact limit dimensions for all classes of fit. *Class 1* is used where generous tolerances are desired to permit easy assembly. *Class 2* is the most common class of fit, being used for a wide range of applications. Precision is greater than for class 1. *Class 3* is used when precision of assembly or adjustment is the primary consideration.

5. LH: This stands for "left-hand," which means that the thread will tighten by going counterclockwise. This is the opposite of the more common right-hand thread. A left-hand thread is used where a right-hand thread might loosen under vibration or load, or where a particular motion is needed. For example, the standard bow compass has both right-hand and left-hand threads on the thumbwheel spindle to permit the needed motion. *Note:* When a thread is right-hand *no* notation is given in the thread specification. For example, a 1½-6UNC-1A thread is by implication right-hand.

6. Double: The factor of multiple thread was shown in Fig. 2. Double thread means that for one revolution of 360°, the lead equals twice the pitch. Most thread is single, that is, the lead equals the pitch. Triple thread is not common, being used only when very rapid advance is necessary (lead = 3P). *Note:* When a thread is single *no* notation is made in the thread specification. For example, a 0.375-16UNC-3B thread is by implication a single thread.

It also is possible to designate the thread gaging system desired. Thread gaging information can be added at the end of the thread note such as:

$$\text{7⁄16-20UNF-2A(23)}$$

where (23) relates to specific information available in ANSI B1.3—1979. Gaging information may be given instead by specifying it within documentation on the drawing.

All information within Sec. 7 of this chapter refers to threads that are not coated or plated. The effect of coating or plating on dimensions can be found in ANSI/ASME B1.1a—1984.

B. Specifications of Acme and Stub Acme Threads
These specialized types of threads are also specified by a note which closely follows that of UN/UNR thread. Notes are seen in Fig. 18. A detailed thread symbol has been used, but a schematic or simplified symbol could have been used also. Examples of other notes, with explanations, are:

1¾-4Acme-2G (general-purpose class 2 Acme; 1¾ in major diameter, 0.25 in pitch; single, right-hand)

1-5Acme-4C-LH (centralizing class 4 Acme; 1 in major diameter, 0.2 in pitch; single, left-hand)

2½-0.333p-0.666L-Acme-3G (general-purpose class 3 Acme; 2½ in major diameter, 0.333 in pitch, 0.666 in lead; double, right-hand)

¾-6Stub Acme (Stub Acme; ¾ in major diameter, 0.1667 in pitch; right-hand)

C. Tapped-Hole Specifications An internal thread in thick material can be tapped (threaded) only if a properly sized hole is first made. For every major diameter, there is a specific drill size to precede the tapping operation.

Appendix 11 lists the particular tap drill size for each thread size. Note that certain tap drills are given by fraction, others by number, and still others by letter. The conversions between number and letter sizes and the decimal-inch equivalents are given in Appendix 12.

It is common to use 75 percent of the theoretical depth of threads for tapped holes. This gives about 95 percent of the strength of a full thread and is easier to cut. A helpful guide uses an empirical formula based on fastener diameter and the material tapped. See Fig. 19 and Table 1.

8. THREAD SPECIFICATIONS: METRIC

Figure 20 shows a possible thread note that conforms to ANSI B.1.13M—1983, "Metric Screw Threads—M Profile." The designation "M Profile" covers general metric standards for a 60° symmetrical screw thread. There is also an MJ Profile (ANSI B1.21M—1978) used

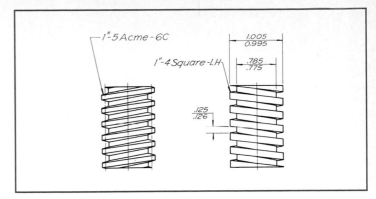

FIG. 18 Specifications of an Acme thread.

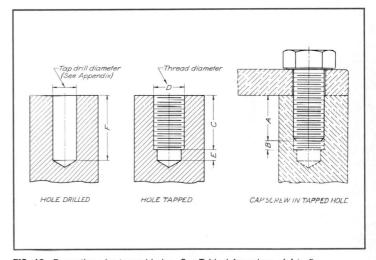

FIG. 19 Proportions for tapped holes. See Table 1 for values of A to F.

TABLE 1 DETAILED DEPTHS FOR DRILLING AND TAPPING HOLES IN COMMON MATERIALS

Material	Entrance length for cap screws, etc., A	Thread clearance at bottom of hole B	Thread depth C	Unthreaded portion at bottom of hole E	Depth of drilled hole F
Aluminum	2D	4/n	2D + 4/n	4/n	C + E
Cast iron	1½D	4/n	1½D + 4/n	4/n	C + E
Brass	1½D	4/n	1½D + 4/n	4/n	C + E
Bronze	1½D	4/n	1½D + 4/n	4/n	C + E
Steel	D	4/n	D + 4/n	4/n	C + E

A = entrance length for fastener.
B = thread clearance at bottom of hole.
C = total thread depth.
D = diameter of fastener.
E = unthreaded portion at bottom of hole.
F = depth of tap-drill hole.
n = threads per inch.

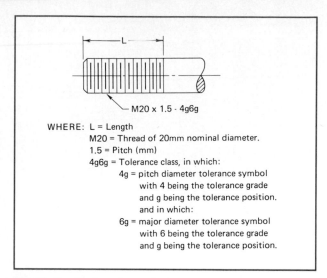

M20 × 1.5 - 4g6g

WHERE: L = Length
M20 = Thread of 20mm nominal diameter.
1.5 = Pitch (mm)
4g6g = Tolerance class, in which:
    4g = pitch diameter tolerance symbol
        with 4 being the tolerance grade
        and g being the tolerance position.
    and in which:
    6g = major diameter tolerance symbol
        with 6 being the tolerance grade
        and g being the tolerance position.

FIG. 20 Specification of a metric thread.

in the aeronautics industry. We will now look at each part of the thread note:

$$M20 \times 1.5\text{-}4g6g, \ 45 \ LG$$

1. Length L: As with UN/UNR thread, length is the choice of the designer. In this example, length is chosen to be 45 mm.

2. M20: This is the nominal diameter size or outside diameter. Appendix 13 lists all metric screw thread series conforming to ISO 261—1973. Appendix 14 lists selected tap drill sizes.

3. 1.5: This is the pitch (in millimeters) of the thread. Appendix 13 shows this to be a fine thread, whereas a pitch of 2.5 would be a coarse thread. *Note:* The screw thread is assumed to be right-hand, unless specifically noted to be left-hand. Therefore, our example is a right-hand thread. To convert to a left-hand thread, the note would become M20 × 1.5-4g6g-LH. The descriptive note so far, M20 × 1.5, is sometimes called a *basic* designation. It does not contain information on tolerance class. This is similar to the case with UN/UNR threads if one were to give the example of Fig. 17 as a ¾-16UNF thread, without mention of class of fit.

4. 4g6g: This is the tolerance class, which requires some explanation. *Tolerance class* is a combination of tolerance grade and tolerance position. It indicates the allowance and tolerance for the *pitch* and *major* diameters of *external* thread, and also the *pitch* and *minor* diameters of *internal* thread.

Tolerance grade is a set of tolerances the purpose of which is to give the same level of accuracy regardless of basic size, for a given tolerance grade number. Available grade numbers are as follows:

External threads:
    Pitch diameter   3, 4, 5, 6, 7, 8, 9
    Major diameter   4   6   8

Internal threads:
    Pitch diameter   4, 5, 6, 7, 8
    Minor diameter   4, 5, 6, 7, 8

Tolerances increase as the tolerance-grade numbers increase. A 4 would be a close tolerance, 6 a medium, and 8 a coarse tolerance. A tolerance grade of 6 is close to a class 2 of UN/UNR threads.

Tolerance position is the position of the tolerance zone relative to the basic size (for in-depth information see ANSI B1.13M—1983). It is expressed as a letter, which may be as follows:

External threads   e, g, h
Internal threads   G, H

Note the use of lowercase letters for external threads and uppercase letters for internal threads. The letter e permits a large allowance; g, G a small allowance; and h, H a zero allowance. Interestingly, only the letters g and H are used in ANSI B1.13M—1983. For other letters one must go to ISO 965/1.

Our example under discussion is M20 × 1.5-4g6g. Therefore, the 4g tells us that the pitch diameter has a close tolerance with a small allowance. The term 6g tells us that the major diameter has a medium tolerance and a small allowance.

5. Length of thread engagement: There are three lengths of thread engagement available: normal (N), short (S), and long (L). Most applications use normal lengths. Adjustments to tolerance grades are recommended as shown in the accompanying

table, if one switches from normal to short or long thread engagement.

TOLERANCE CLASS

Given normal thread engagement	Desire short thread engagement	Desire long thread engagement
6g*	5g6g	6e6g
4g6g	3g6g	4e6g
6h*	5h6h	6g6h
4h6h	3h6h	4g6h

Source: Adapted from ANSI B1.13M—1983.
*6g = 6g6g, 6h = 6h6h. Symbols need not be repeated if the two tolerance classes for a thread are the same.

9. THREADED FASTENERS

Most engineering products are composed of separate parts held together by some means of fastening. Threaded fasteners provide an advantage over permanent methods (e.g., welding, riveting) in that they are easily disassembled.

Threaded fasteners have descriptive names. Five types are the most common: the bolt, stud, cap screw, machine screw, and setscrew. These five are shown in Fig. 21 and will be discussed in order. Dimensions for drawing nearly all these fasteners will be found in the appendixes.

A. Bolts   A *bolt* (Fig. 21A), having an integral head on one end and a thread on the other end, is passed through clearance holes in two parts and draws them together by means of a nut screwed on the threaded end.

Two major groups of bolts have been standardized: round-head bolts and wrench-head bolts (sometimes called "machine bolts").

Round-head bolts are used as through fasteners with a nut, usually square. Eleven head types have standard proportions and include carriage bolts, step bolts, elevator bolts, and spline bolts. Several head types are intended for wood construction and have square sections, ribs, or fins under the head to prevent the bolts from turning. Appendix 17 shows the various head forms and gives nominal dimensions suitable for drawing purposes.

Wrench-head (machine) bolts have two standard head forms: square and hexagonal. Nuts to match the bolt-head form and grade are available, but any nut of correct

thread will fit a bolt. Machine bolts vary in grade from coarsely finished products resembling round-head bolts to a well-finished product matching a hexagon cap screw in appearance.

B. Studs   A *stud* (Fig. 21B) is a rod threaded on each end. The fastener passes through a clearance hole in one piece and screws permanently into a tapped hole in the

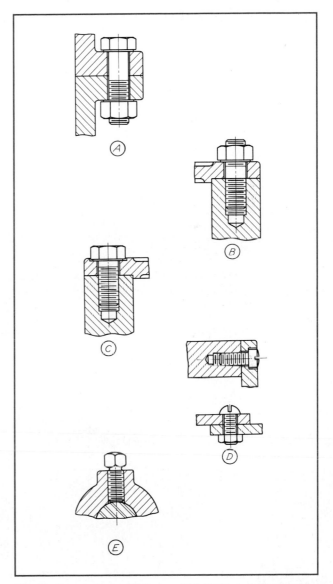

FIG. 21   Common types of fasteners. *(A)* bolt; *(B)* stud; *(C)* cap screw; *(D)* machine screws; *(E)* setscrew.

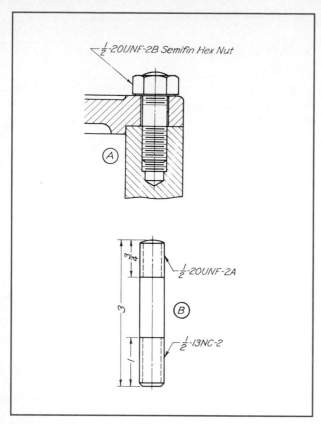

FIG. 22 Stud and nut. *(A)* is the assembly drawing and *(B)* is the detail drawing of the stud.

other. A nut then draws the parts together. The stud (Fig. 22) is used when through bolts are not suitable for parts that must be removed frequently, such as cylinder heads.

C Cap Screws A *cap screw* (Fig. 21C) passes through a clearance hole in one piece and screws into a tapped hole in the other. The head, an integral part of the screw, draws the parts together as the screw enters the tapped hole. Cap screws are used on machine tools and other products requiring close dimensions and finished appearances. They are well-finished products; for example, the heads of the slotted- and socket-head screws are machined, and all have chamfered points. The five types of heads shown in Fig. 23 are standard.

D. Machine Screws A *machine screw* (Fig. 21D) is a small fastener used with a nut to function in the same manner as a bolt; or without a nut, to function as a cap screw. Machine screws are small fasteners used principally in numbered diameter sizes. The ten standardized head shapes (Fig. 24), except for the hexagon, are available in slotted form as shown or with cross recesses (Fig. 25). The size of the recess varies with the size of the screw. The hexagon machine screw is not made with a cross recess but may be optionally slotted.

E. Setscrews A *setscrew* (Fig. 21E) screws into a tapped hole in an outer part, often a hub, and bears with

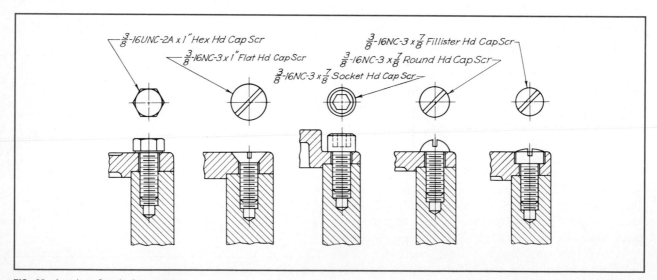

FIG. 23 American Standard cap screws.

its point against an inner part, usually a shaft. Setscrews are made of hardened steel and hold two parts in relative position by having the point set against the inner part. The American Standard square-head and headless screws are shown in Fig. 26. Types of points are shown in Fig. 27.

F. Nuts Nuts that accompany bolts and cap screws should be mentioned. These are available in two wrench-type styles: square and hexagonal. In addition to the plain form usually associated with bolts, several special-purpose styles are available. Jam, hexagonal slotted, and hexagonal castle nuts are shown in Fig. 28. Ma-

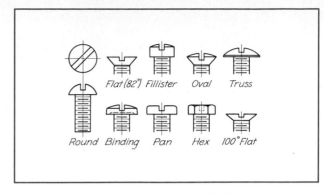

FIG. 24 American Standard machine-screw heads.

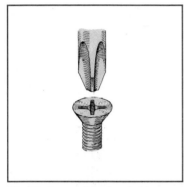

FIG. 25 Recessed head and driver.

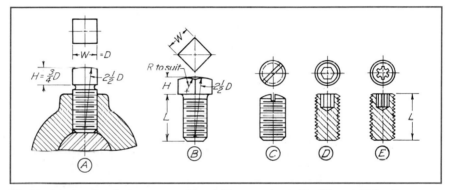

FIG. 26 American Standard setscrews. *(A)* and *(B)* square head; *(C)*, *(D)*, and *(E)* headless.

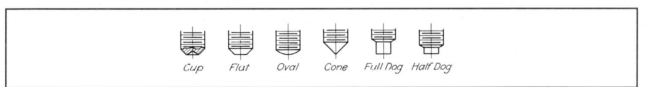

FIG. 27 American Standard setscrew points.

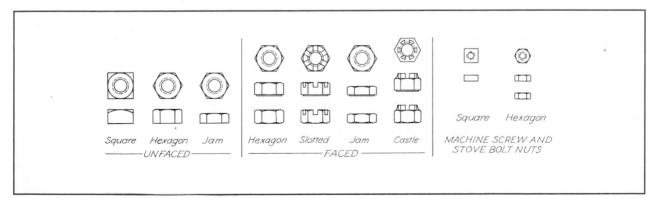

FIG. 28 American Standard nuts.

chine screw and stove-bolt nuts are made in small sizes only.

In the preceding discussion one should realize that the fastener type is independent of the thread form on a fastener. That is, a cap screw, for example, could come with a variety of thread forms, such as UNC, UNF, metric, etc.

10. FASTENER TERMS

1. *Nominal diameter.* The basic major diameter of the thread.

2. *Width across flats*, W. The distance separating parallel sides of the square or hexagonal head or nut, corresponding with the nominal size of the wrench.

3. *Tops of bolt heads and nuts.* The tops of heads and nuts are flat with a chamfer to remove the sharp corners. The angle of chamfer with the top surface is 25° for the square form and 30° for the hexagonal form; both are drawn at 30°. The diameter of the top circle is equal to the width across flats.

4. *Washer face.* The washer face is a circular boss turned or otherwise formed on the bearing surface of a bolt head or nut to make a smooth surface. The diameter is equal to the width across flats. A circular bearing surface can be obtained on a nut by chamfering the corners. The angle of chamfer with the bearing face is 30°; the diameter of the circle is equal to the width across flats.

5. *Fastener length.* The nominal length is the distance from the bearing surface to the point. For flat-head fasteners and for headless setscrews it is the overall length.

6. *Regular series.* Regular bolt heads and nuts are for general use. The dimensions and the resulting strengths are based on the theoretical analysis of the stresses and on results of numerous tests.

7. *Heavy series.* Bolt heads and nuts in this series are for use where greater bearing surface is necessary. Therefore, for the same nominal size, they are larger in overall dimensions than regular heads and nuts. They are used where a large clearance between the bolt and hole or a greater wrench-bearing surface is considered essential.

8. *Thick nuts.* These have the same dimensions as regular nuts, except that they are higher.

9. *Unfinished bolts and nuts* are not machined on any surface except the threads. The bearing surface is plain. Dimensional tolerances are as large as is practicable.

10. *Semifinished bolt heads and nuts.* These have a smooth bearing surface machined or formed at right angles to the axis (see Fig. 29). For bolt heads, this is a washer face; and for nuts, a washer face or a circular bearing surface produced by chamfering the corners. Dimensional tolerances of the fastener are otherwise the same as for the unfaced group.

11. *Finished bolts and nuts.* These differ from the

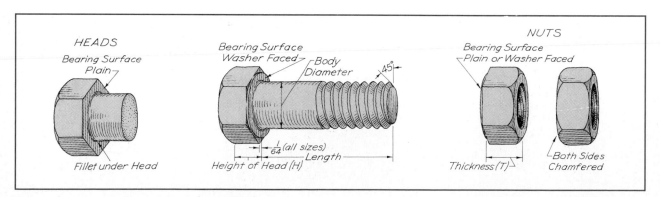

FIG. 29 Details of form for hexagonal bolts and nuts.

semifinished in two ways: The bearing surface may be washer-faced or produced by double chamfering (see Fig. 29). Dimensional tolerances are smaller than are those for the semifinished form. Finished hexagonal bolts have dimensional tolerances and artisanship similar to hexagonal cap screws.

11. STANDARD FASTENER SPECIFICATIONS

The following items may be included in a fastener specification in the order listed. Obviously, many of the items will not apply to some fastener types.

1. Thread specification (without length)
2. Fastener length
3. Fastener series (bolts only)
4. Class of finish (bolts only)
5. Material (if other than steel)
6. Head style or form
7. Point type (setscrews only)
8. Fastener group name

Sample specifications follow. Note carefully the machine-bolt notes.

½-13UNC-2A × 3
Reg Sq Bolt

¼-20UNC-2A × 4
Reg Semifin Hex Bolt

⅜-16UNC-2A × 3½
Hvy Semifin Hex Bolt

¾-16UNF-2A × 2½
Finished Hex Bolt (omit Reg)

⅝-11UNC-2A × 4¼
Hvy Finished Hex Bolt

#10-24NC-2 × 2
Brass Slotted Flat Hd Mach Scr

#6-32NC-2 × ½
Recess Fil Hd Mach Scr

¼-20UNC-2A × ¾
Sq Hd Cone Pt Setscrew

½-13UNC-3A × 2
Soc, Flt Pt Setscrew

Figure 23 illustrates cap-screw notes. Realize, also, that while the above examples are for UN/UNR types, a metric type such as, for example M10 × 1.5-4g6g × 50 Finished Hex Bolt, could just as well have been given.

12. DRAWING FASTENERS: SQUARE AND HEXAGONAL

Before drawing a fastener, you must know its *type, nominal diameter*, and *length*, if a bolt or screw. If you know the type and diameter, you can find the other dimensions in the appendixes.

Square and hexagonal heads and nuts are drawn "across corners" in all views showing the faces, unless a special reason exists for drawing them "across flats." Figure 30 shows stages in drawing square heads both across corners and across flats, and Fig. 31 shows the same for hexagonal heads. The principles apply equally to the drawing of nuts. The following information must be known: (1) type of head or nut, (2) weight—regular or heavy, (3) condition—unfinished or semifinished, and (4) nominal diameter. Using this information, obtain additional data W (width across flats), H (height of head), and T (thickness of nut) from the tables.

Figure 32 shows a regular semifinished hexagonal bolt and nut, drawn by the method of Fig. 31, showing the head across flats and the nut across corners. The length is selected from the bolt-length tables and the length of thread is determined from the bolt table (see appendix). Observe that the washer face shown in Fig. 31 will occur only with semifinished hexagonal fasteners and is sometimes omitted from the drawings of these.

Often the representation of fasteners may be approximate or even symbolic because the note specifications invariably accompany and exactly specify them. If an accurate drawing of the fastener is not essential, the W, H, and T dimensions of hexagonal and square heads and nuts may be obtained from the nominal diameter and the following formulas; the resulting values will be quite close to the actual dimensions. For the regular series, $W = 1\frac{1}{2}D$, $H = \frac{2}{3}D$, and $T = \frac{7}{8}D$; for the heavy series, $W = 1\frac{1}{2}D + \frac{1}{8}$ in, $H = \frac{3}{4}D$, and $T = D$.

A wide variety of *templates* to facilitate the drawing of fasteners is available.

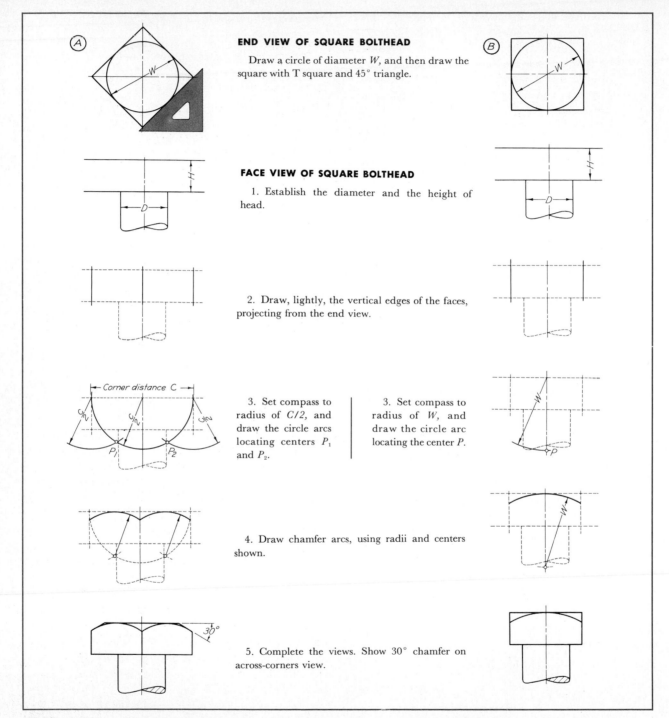

END VIEW OF SQUARE BOLTHEAD

Draw a circle of diameter W, and then draw the square with T square and 45° triangle.

FACE VIEW OF SQUARE BOLTHEAD

1. Establish the diameter and the height of head.

2. Draw, lightly, the vertical edges of the faces, projecting from the end view.

3. Set compass to radius of $C/2$, and draw the circle arcs locating centers P_1 and P_2.

3. Set compass to radius of W, and draw the circle arc locating the center P.

4. Draw chamfer arcs, using radii and centers shown.

5. Complete the views. Show 30° chamfer on across-corners view.

FIG. 30 *(A)* Stages in drawing a square head across corners. *(B)* Stages in drawing a square head across flats.

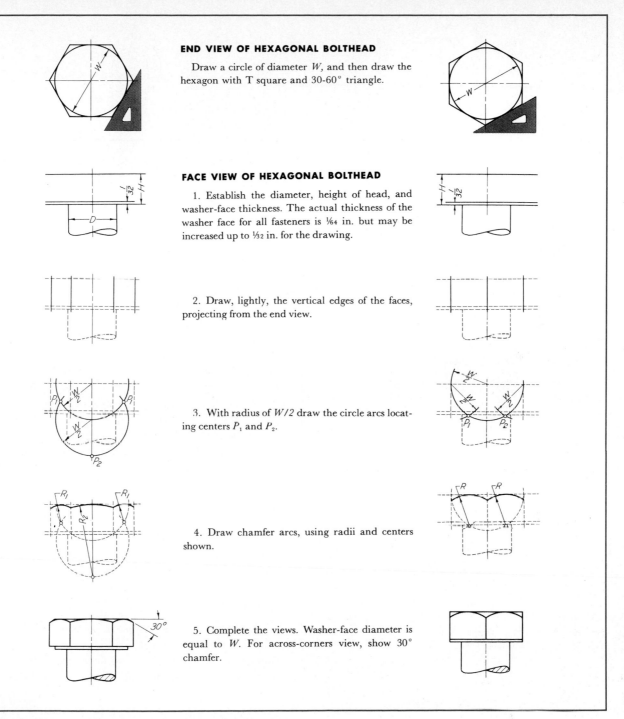

END VIEW OF HEXAGONAL BOLTHEAD

Draw a circle of diameter W, and then draw the hexagon with T square and 30-60° triangle.

FACE VIEW OF HEXAGONAL BOLTHEAD

1. Establish the diameter, height of head, and washer-face thickness. The actual thickness of the washer face for all fasteners is 1/64 in. but may be increased up to 1/32 in. for the drawing.

2. Draw, lightly, the vertical edges of the faces, projecting from the end view.

3. With radius of $W/2$ draw the circle arcs locating centers P_1 and P_2.

4. Draw chamfer arcs, using radii and centers shown.

5. Complete the views. Washer-face diameter is equal to W. For across-corners view, show 30° chamfer.

FIG. 31 *(A)* Stages in drawing a hexagonal head across corners. *(B)* Stages in drawing a hexagonal head across flats.

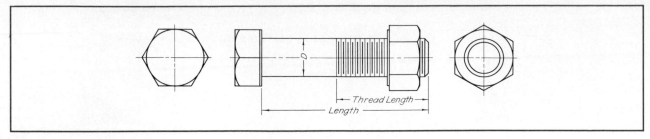

FIG. 32 American Standard regular hexagonal semifinished bolt and nut.

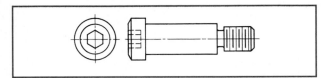

FIG. 33 American Standard shoulder screw. The body is accurately finished.

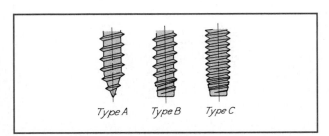

FIG. 34 American Standard tapping screw points. These are hardened and made to form their own thread in a drilled or punched hole.

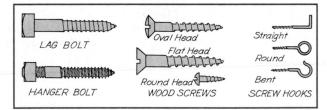

FIG. 35 Fasteners used in wood.

13. SPECIALIZED THREADED FASTENERS

Many specialized fasteners have been developed to answer particular needs. Several of the more common types will be discussed.

A. Shoulder Screws (Fig. 33) Shoulder screws are widely used for holding machine parts together and providing pivots, such as with cams, linkages, etc. They are also used with punch and die sets for attaching stripper plates and are then commonly called "stripper bolts."

B. Tapping Screws Tapping screws are hardened fasteners that form their own mating internal threads when driven into a hole of the proper size. For certain conditions and materials, these screws give a combination of speed and low production cost, which makes them preferred over others. Many special types are available. For drawing purposes, dimensions of machine-screw heads may be used. Sizes conform, in general, with machine-screw sizes. The ANSI provides three types of thread and point combinations (Fig. 34). The threads are 60° with flattened crest and root; types *A* and *B* are interrupted threads, and all fasteners are threaded to the head.

C. Fasteners for Wood Many forms of threaded fasteners for use in wood are available. Some are illustrated in Fig. 35. Threads are interrupted, 60° form, with a gimlet point. The ANSI has standardized lag bolts with square heads and also wood screws with flat, round, and oval heads. Lag bolt heads follow the same dimensions and, in general, the same nominal sizes as do regular square bolts.

D. Other Threaded Fasteners Figure 36 shows some of the many forms of threaded fasteners which we have not discussed in detail. Manufacturers' catalogs are a good source of current information on such fasteners.

14. OTHER SPECIALIZED FASTENERS

Not all fasteners are threaded. Many fasteners have no threads, and are ingenious in their simplicity and ease of use. We cannot cover all styles and variations, but a few of the specialized fasteners can be mentioned. Manufacturers' catalogs are a good source of information and should be consulted.

A. Push-On Fasteners These are spring-steel washerlike parts that are pushed onto an unthreaded stud, shaft, rod, wire, pin, or tube and make a permanent, quick, and inexpensive fastening by means of the gripping action of the specially designed hole in the part.

B. Rollpins This is a trade name for parts similar to standard straight or tapered pins but tubular in form with

a slotted and chamfered longitudinal opening. They are made of heat-treated spring steel and can be used to replace a grooved pin, key, rivet shaft, cotter pin, setscrew, clevis pin, hinge pin, dowel pin, bolt and nut, rivet, or taper pin.

C. Retaining Rings A retaining ring is a tempered spring-steel ring designed either as an external ring to fit into a groove on a shaft or as an internal ring to be fitted into a groove in a hub. The principal use of retaining rings is to locate or prevent axial movement of a shaft in a hub. They are made in a great variety of forms and sizes.

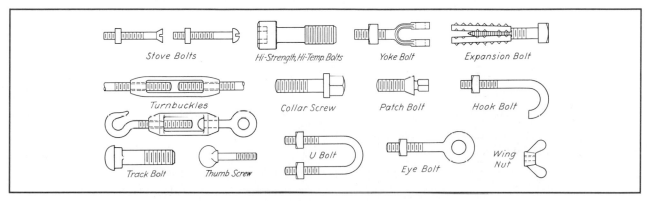

FIG. 36 Miscellaneous threaded fasteners.

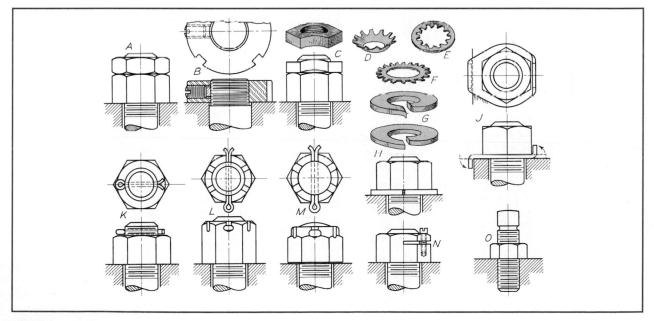

FIG. 37 Various locking devices.

15. LOCKNUTS AND LOCKING DEVICES

As seen in Fig. 37, many different locking devices are used to prevent nuts from working loose. A screw thread holds securely unless the parts are subject to impact and vibration, as in a railroad-track joint or an automobile engine. A common device is the *jam nut* shown at A. *Slotted nuts* (L) and *castle nuts* (M), to be held with a

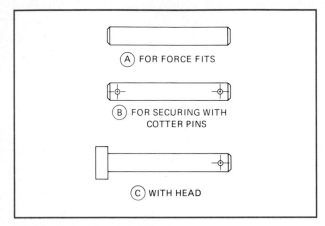

FIG. 38 Some styles of straight pins.

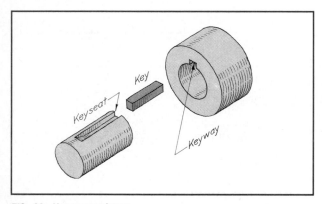

FIG. 39 Key nomenclature.

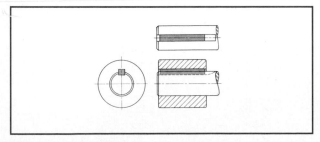

FIG. 40 Square (or flat) key.

cotter or wire, are commonly used in automotive and allied work.

At B is shown a *round nut* locked by means of a setscrew. A brass plug is placed under the setscrew to prevent damage to the thread. This is a common type of adjusting nut used in machine-tool practice. C is a *locknut*, in which the threads are deformed after cutting. Patented *spring washers*, such as are shown at D, E, and F, are common devices. Special patented nuts with plastic or fiber inserts or with distorted threads are in common use as locking devices.

The locking action of J, K, N, and O should be evident from the figure. At J a special tab is bent to secure the nut. At K, L, and M, a particular pin known as a *cotter pin* is inserted through both nut and bolt, and then spread as shown. Cotter pins come in a wide variety of sizes.

Also available are pins of constant diameter, used for force-fit assemblies, known as *straight pins*. It is possible to get straight pins with heads for certain uses. See Fig. 38.

16. WASHERS: PLAIN AND LOCK

Plain washers are commonly used in the assembly of nuts and bolts to provide a smooth surface for the nut or bolt to turn against. Specifications are given by inside diameter, outside diameter, thickness, and type, respectively. For example, a washer could be designated as $1.375 \times 2.500 \times 165$ type A plain washer. This would be used with 1¼-in-diameter fasteners.

Lock washers prevent a fastener from loosening due to vibration or stress. Spring washers have already been seen in Fig. 37, items D, E, F. These types are sometimes called *tooth lock washers*. The types of lock washers shown in Fig. 37, items G and H, are thicker and are known as *helical-spring lock washers*. Dimensions are given in Appendix 27 for both plain and lock washers.

17. KEYS

In making machine drawings there is frequent occasion for representing key fasteners, used to prevent the rotation of wheels, gears, etc., on their shafts. A key is a piece of metal (Fig. 39) placed so that part of it lies in a groove, called the "key seat," cut in a shaft. The key then extends somewhat above the shaft and fits into a "keyway" cut in a hub. After assembly the key is partly in the shaft and partly in the hub, locking the two together so that one cannot rotate without the other.

A. Key Types   The simplest key, geometrically, is the square key, placed *half* in the shaft and *half* in the hub (Fig. 40). A flat key is rectangular in cross section and is used in the same manner as the square key. The gibhead key (Fig. 41) is tapered on its upper surface and is driven in to form a very secure fastening. Both square and flat (parallel and tapered stock) keys have been standardized by the ANSI. Tables of standard sizes are given in Appendix 25A and B.

The Pratt and Whitney key (Fig. 42) is a variation on the square key. It is rectangular in cross section and has rounded ends. It is placed two-thirds in the shaft and one-third in the hub. The key is proportioned so that the key seat is square and the keyway is half as deep as it is wide. Sizes are given in Appendix 25E.

Perhaps the most common key is the Woodruff (Fig. 43). This key is a flat segmental disk with a flat (A) or round (B) bottom. The key seat is semicylindrical and cut to a depth so that *half* the *width* of the key extends above the shaft and into the hub. Tables of dimensions are given in Appendix 25C and D.

For very heavy duty, keys are not sufficiently strong, and splines (grooves) are cut in both shaft and hub, arranged so that they fit one within the other (Fig. 44). A and B are two forms of splines widely used instead of keys. B is the newer ANSI involute spline.

B. Specification of Keys   Keys are specified by note or number, depending upon the type.

Square and flat keys are specified by a note giving the width, height, and length; for example:

½ Square Key 2½ Lg

½ × ⅜ Flat Key 2½ Lg

Gib-head taper stock keys are specified by giving the same information, except for name, as that given for square or flat taper keys (see Appendix 25). For example:

¾ × ¾ × 2¼ Square Gib-head Taper Key

⅞ × ⅝ × 2½ Flat Gib-head Taper Key

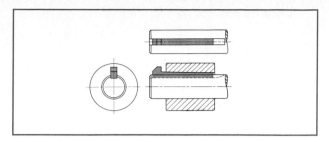

FIG. 41   Gib-head key. The head shape provides removal.

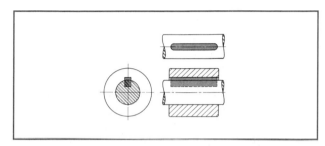

FIG. 42   Pratt and Whitney key. The ends of the key are rounded.

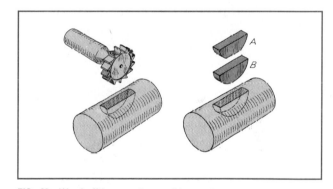

FIG. 43   Woodruff keys, cutter, and key seat.

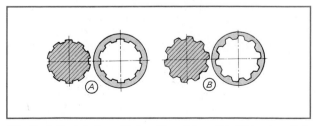

FIG. 44   Splined shafts and hubs. *(A)* straight-radial, *(B)* involute. Both provide strong resistance to torque forces.

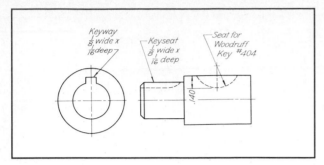

FIG. 45 Nominal dimensions of key seats and keyways (square and Woodruff).

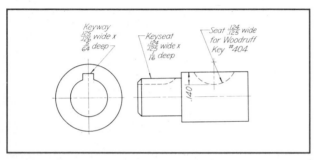

FIG. 46 Limit dimensions of key seats and keyways (square and Woodruff).

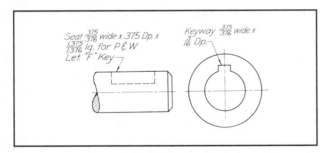

FIG. 47 Dimensioning of a Pratt and Whitney key seat and keyway.

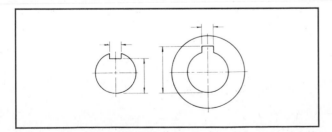

FIG. 48 Dimensions of keyway and key seat for interchangeable assembly.

Pratt and Whitney keys are specified by number or letter (see Appendix 25). For example:

Pratt and Whitney Key No. 6

Woodruff keys are specified by number (see Appendix 25).

Dimensions and specifications of other key types may be found in handbooks or manufacturers' catalogs.

C. Dimensioning for Keys The dimensioning of the *seat* and *way* for keys depends upon the purpose for which the drawing is intended. For unit production, when the keys are expected to be fitted by the machinist, nominal dimensions may be given, as in Fig. 45. For quantity production, the limits of width (and depth, if necessary) should be given as in Fig. 46. Pratt and Whitney key seats and keyways are dimensioned as in Fig. 47. Note that the *length* of the key seat is given to correspond with the specification of the key. If interchangeability is important and when careful gaging is necessary, the dimensions should be given as in Fig. 48 and the values expressed with limits.

18. SPRINGS

A *spring* can be defined as an elastic body designed to store energy when deflected. Springs are classified according to their geometric form: helical or flat.

A. Helical Springs These are further classified as (1) compression, (2) extension, or (3) torsion, according to the intended action. On working drawings, helical springs are drawn as a single-line convention, as in Fig. 49, or semiconventionally, as in Figs. 50 to 52, by laying out the diameter (D) and pitch (P) of coils and then drawing a construction circle for the wire size as the limiting positions and conventionalizing the helix with straight lines. Helical springs may be wound of round-, square-, or special-section wire.

Compression springs are wound with the coils separated so that the unit can be compressed, and the ends may be open or closed and may be left plain or ground, as shown in Fig. 50. The information that must be given for a compression spring is as follows:

1. Controlling diameter: (*a*) is outside, (*b*) is inside, (*c*) operates inside a tube, or (*d*) operates over a rod

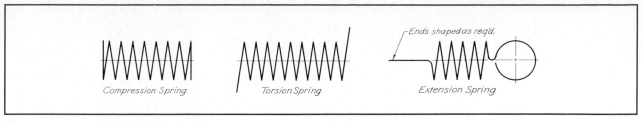

FIG. 49 Conventional representation of springs. The single-line treatment saves drawing time.

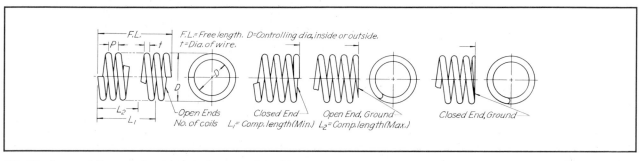

FIG. 50 Representation and dimensioning of compression springs.

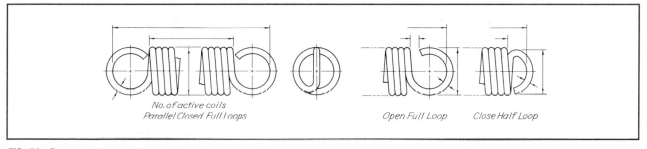

FIG. 51 Representation and dimensioning of extension springs.

2. Wire or bar size

3. Material (kind and grade)

4. Coils: (*a*) total number and (*b*) right- or left-hand type

5. Style of ends

6. Load at deflected length of —

7. Load rate between — inches and — inches

8. Maximum solid height

9. Minimum compressed height in use

Extension springs are wound with the loops in contact so that the unit can be extended, and the ends are usually made as a loop, as shown in Fig. 51. Special ends are sometimes required. The information that must be given for an extension spring is as follows:

1. Free length: (*a*) overall, (*b*) over coil, or (*c*) inside of hooks

2. Controlling diameter: (*a*) outside, (*b*) inside, or (*c*) inside a tube

3. Wire size

4. Material (kind and grade)

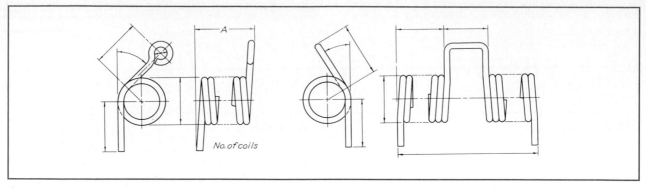

FIG. 52 Representation and dimensioning of torsion springs.

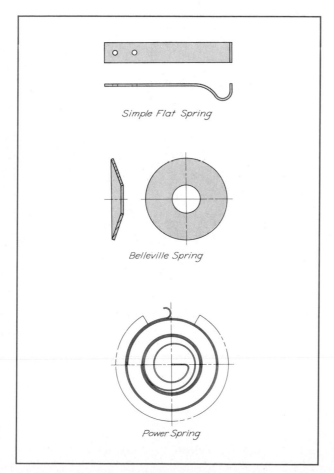

Simple Flat Spring

Belleville Spring

Power Spring

FIG. 53 Flat spring types. The variety is such that these require individual representation and specification.

5. Coils: (*a*) total number and (*b*) right- or left-hand type
6. Style of ends
7. Load at inside hooks
8. Load rate, pounds per 1-in deflection
9. Maximum extended length

Torsion springs are wound with closed or open coils, and the load is applied torsionally (at right angles to the spring axis). The ends may be shaped as hooks or as straight torsion arms (Fig. 52). The information that must be given for a torsion spring is as follows:

1. Free length (dimension A, Fig. 52)
2. Controlling diameter: (*a*) outside, (*b*) inside, (*c*) inside a hole, or (*d*) over a rod
3. Wire size
4. Material (kind and grade)
5. Coils: (*a*) total number and (*b*) right- or left-hand type
6. Torque, pounds at __ degrees of deflection
7. Maximum deflection (degrees from free position)
8. Style of ends

B. Flat Springs A *flat spring* can be defined as any spring made of flat or strip material. Flat springs (Fig. 53) are classified as (1) simple flat springs, formed so that the desired force will be applied when the spring is deflected in a direction opposite to the force: (2) power springs, made as a straight piece and then coiled inside an enclosing case; and (3) Belleville springs, stamped of thin material and shaped so as to store energy when deflected. The information that must be given for a flat spring is as follows:

1. Detailed shape and dimensions of the spring shown in a drawing

2. Material and heat treatment

3. Finish

PROBLEMS

GROUP 1. SCREW THREADS

1. Draw in section the following screw thread forms, 1-in pitch: American National, Acme, stub Acme, square.

2. Draw two views of a square-thread screw and a section of the nut separated; diameter, 2½ in; pitch, ¾ in; length of screw, 3 in.

3. Same as Prob. 2, but for V thread with ½-in pitch.

4. Draw screws 2 in in diameter and 3½ in long: single square thread, pitch ½ in; single V thread, pitch ¼ in; double V thread, pitch ½ in; left-hand double square thread, pitch ½ in.

5. Working space, 7 by 10½ in. Divide space as shown and in left space draw and label thread profiles, as follows: at A, sharp V, ½-in pitch; at B, American National (Unified), ½-in pitch; at C, square, 1-in pitch; at D, Acme, 1-in pitch. Show five threads each at A and B and three at C and D.

In right space, complete the views of the object by showing at E a 3½-4UNC-2B threaded hole in section. The thread runs out at a 3¹⁷⁄₃₂-in-diameter by ¼-in-wide thread relief. The lower line for the thread relief is shown. At F, show a 1¾-5UNC-3A thread. The thread is 1 in long and runs out at a neck that is 1⅜ in in diameter by ¼ in wide. The free end of the thread should be chamfered at 45°. At G, show the shank threaded ⅝-18UNF-3A by 1 in long; at H, show ⅜-16UNC-2B through tapped holes in section, four required. Completely specify all threads.

6. Working space 175 mm × 275 mm. Complete the views and show threaded features as follows: at A, an M12 × 1.25-6G through tapped hole; at B, an M16 × 2-4H6H by 20 mm deep tapped hole with tap drill 30 mm deep; at C, an M10 × 1.5-5H6H by 6 mm deep tapped hole with the tap drill through to the 25 mm hole; at D, an M42 × 4.5-6H8H threaded hole running out at the large cored hole. Completely specify all threads.

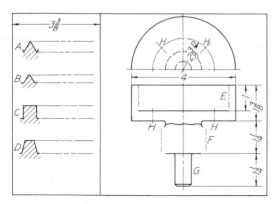

PROB. 5 Screw threads.

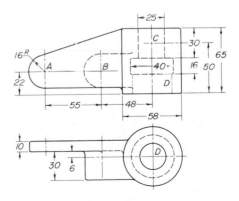

PROB. 6 METRIC.

7. Complete the offset support, showing threaded holes as follows: at *A*, 1⅜-6UNC-2B; at *B*, ½-13NC-2; at *C*, ⅝-18UNF-3B. *A* and *C* are through holes. *B* is a blind hole to receive a stud. Material is cast iron. Specify the threaded holes.

8. Complete the views of the aluminum rocker, showing threads as follows: on centerline *A-A*, M48 × 5-8H; on centerline *B-B*, an M12 × 1.75-6H tapped hole, 16 mm deep, with tap drill 22 mm deep; on centerline *C-C*, an M72 × 6-4g6g external thread; on centerline *D-D*, an M20 × 2.5-6g external thread.

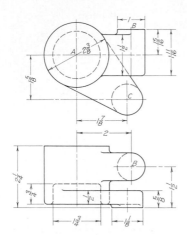

PROB. 7   Offset support.

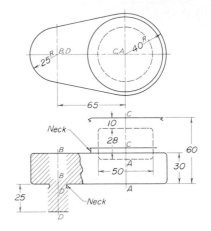

PROB. 8   METRIC.

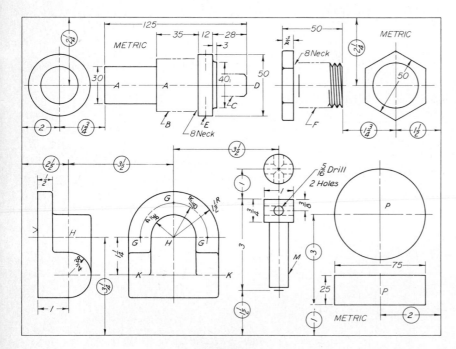

PROB. 9

9. Minimum working space, 10½ by 15 in. Complete the views of the objects, and show threads and other details as follows: *upper left*, on centerline *A-A*, show in section an M20 × 2.5 tapped hole, 40 mm deep, with tap-drill hole 65 mm deep. At *B*, show an M42 × 4 external thread in section. At *C*, show an M20 × 1.5 thread in section. From *D* on centerline *A-A*, show a 6-mm drilled hole extending to the tap-drill hole. At *E*, show six 6-mm drilled holes, 6 mm deep, equally spaced for spanner wrench. *Upper right*, at *F*, show an M36 × 2 thread. *Lower left*, at *G*, show (three) ⅜-16UNC-2B through holes. At *H*, show a ⅞-9UNC-2 tapped hole, 1⅛ in deep, with the tap-drill hole going through the piece. On centerline *K-K*, show a ⅞-14UNF-3B through tapped hole. *Lower center*, at *M*, show a ⅝-18UNF-3A thread 1¾ in long. *Lower right*, on centerline *P*, show an M48 × 5 hole in section.

Completely specify all threads.

GROUP 2. THREADED FASTENERS

10. Draw one view of a regular semifinished hexagonal bolt and nut, across corners; the diameter is 1 in; the length is 5 in.

11. Same as Prob. 10 for a heavy unfinished bolt and nut.

12. Same as Prob. 10 for a square bolt and nut.

13. Draw four ½- by 1½-in cap screws, each with a different kind of head. Specify each.

14. Show the pieces fastened together on centerline E-E with a 20-mm hexagonal-head cap screw and light lock washer. On centerline A-A show a 15 by 36 mm shoulder screw.

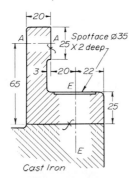

PROB. 14   METRIC.

15. Show pieces fastened together with a 18-mm square bolt and heavy square nut. Place nut at bottom.

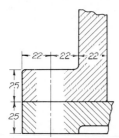

PROB. 15   METRIC.

16. Fasten pieces together with a ¾-in stud and regular semifinished hexagonal nut.

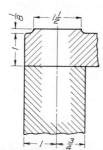

PROB. 16   Bracket and support.

17. Fasten pieces together with a ¾-in fillister-head cap screw.

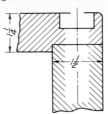

PROB. 17   Centered closure.

18. Draw one view of a round-head square-neck (carriage) bolt; diameter, 1 in; length, 5 in; with regular square nut.

19. Draw one view of a round-head rib-neck (carriage) bolt; diameter, ¾ in; length, 4 in; with regular square nut.

20. Draw two views of a shoulder screw; diameter, 20 mm; shoulder length, 130 mm.

21. Draw two views of a socket-head cap screw; diameter, 1¼ in; length, 6 in.

22. Draw the stuffing box and gland, showing the required fasteners. At A, show ½-in hexagonal-head cap screws (six required). At B, show ½-in studs and regular semifinished hexagonal nuts. Specify fasteners.

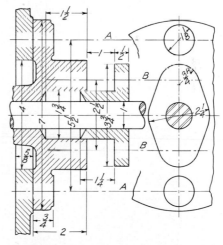

PROB. 22   Stuffing box and gland.

23. Draw the bearing plate, showing the required fasteners. (Sketch on page 530.) At C, show M12 regular semifinished hexagonal bolts and nuts (four required). At D, show M12 socket setscrew. At E, show M12 square-head setscrew. Setscrews are to have conc points. Specify fasteners.

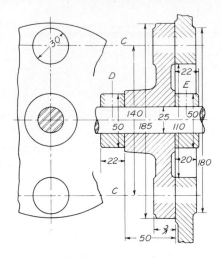

PROB. 23   METRIC.

24. Draw the ball-bearing head, showing the required fasteners. At A, show ½- by 1¾-in regular semifinished hexagonal bolts and nuts (six required), with heads to left and across flats. Note that this design prevents the heads from turning. At B, show ⁵⁄₁₆- by ¾-in fillister-head cap screws (four required). At C, show a ⅜-by ½-in slotted flat-point setscrew with fiber disk to protect threads of spindle. Specify fasteners.

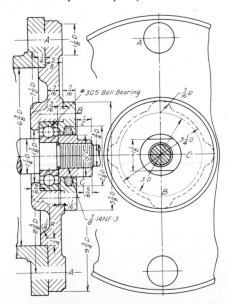

PROB. 24   Ball-bearing head.

GROUP 3. KEYS

Key sizes are given in the appendix.

25. Draw hub and shaft as shown, with a Woodruff key in position.

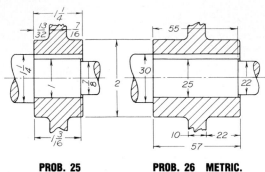

PROB. 25   **PROB. 26   METRIC.**

26. Draw hub and shaft as shown, with a square key 50 mm long in position. Select nearest inch equivalent and express as metric.

27. Draw hub and shaft as shown, with a gib-head key in position.

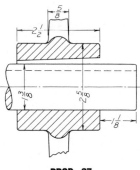

PROB. 27

28. Draw hub and shaft as shown, with a Pratt and Whitney key in position.

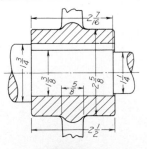

PROB. 28

GROUP 4. SPRINGS

29. Draw a compression spring as follows: inside diameter, ¾ in; wire size, ⅛ in diameter; coils 14, right-hand; squared and ground ends; free length, 3½ in.

30. Draw a compression spring as follows: outside diameter, 25 mm; wire size, 2.5 mm diameter; coils 12, left-hand; open ends, not ground; free length, 100 mm.

31. Draw an extension spring as follows: free length over coils, 2 in; outside diameter, 1⅛ in; wire size, ⅛ in diameter; coils 11, right-hand; the ends parallel, closed loops.

32. Draw an extension spring as follows: free length inside hooks, 2¾ in; inside diameter, ¾ in; wire size, ⅛ in diameter; coils 11, left-hand; the ends parallel, closed half loops.

33. Draw a torsion spring as follows: free length over coils, 20 mm; inside diameter, 28 mm; wire size, 3 mm diameter; coils 5, right-hand; ends straight and turned to follow radial lines to center of spring and extend 15 mm from outside diameter of spring.

34. A compression spring made of 26-gage steel wire with an outside diameter of 2 in has eight active coils wound to a pitch of ½ in. Determine the free length if:

    (a) The spring has open ends.
    (b) The spring has closed ends.
    (c) The spring has open ends ground.
    (d) The spring has closed ends ground.

35. If the above spring is compressed solid, determine its outside diameter.

35A. If the above spring has a pitch tolerance of ±0.015 in and an outside-diameter tolerance of ±0.015-in dia., determine the maximum outside diameter. (*Note:* The wire manufacturer's tolerance on the wire diameter is ±0.001 in.)

GROUP 5. DESIGN PROBLEMS

36. The shaft and bar shown are to be joined so that a light torque can be transmitted from one member to the other. The parts must be readily disassembled. Show at least three ways of solving this problem. In each case give the instructions and dimensions required to complete the design.

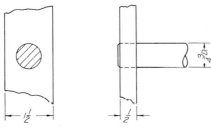

PROB. 36 Layout.

37. A shaft passing through a fixed panel is to have a plastic knob fastened to it. Its angular motion is to be limited to 45° of travel. Select appropriate standard fasteners, and make a drawing showing them installed. Give whatever instructions and dimensions are required to complete the design. (*Note:* This is a large production design and time and cost are important factors.)

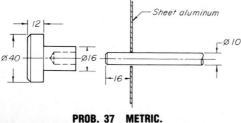

PROB. 37 METRIC.

38. The ½-in-diameter shaft shown must be able to move axially ¾ in in the frame without rotating. The block must be rigidly attached to this shaft. The two ⅜-in diameter shafts must be free to rotate in the block with no appreciable axial movement. Select appropriate standard fasteners to complete this design. Make necessary drawings showing any additional machining operations, and an assembly drawing. Give complete instructions and all necessary dimensions to complete this design.

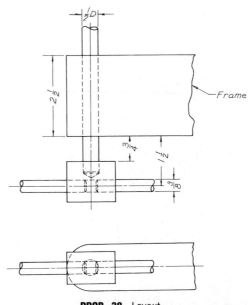

PROB. 38 Layout.

17

GEARS AND CAMS

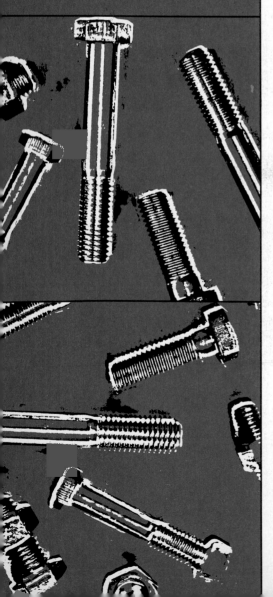

1. PURPOSE OF GEAR AND CAM STUDY

This chapter deals with two design components that are vital to the operation of many products in the field of mechanical design. Without gears and cams, a vast number of machines could not operate.

We are all familiar with the application of gears. For example, the automotive transmission is an intricate assembly of many gears, bearings, and shafts. There are also gears in the timing drive of an automotive engine and in the auto's differential drive.

Cams are also a common design component having widespread use. They can provide a prescribed motion in a specified part, such as the lift of a valve in an engine, or a particular contour of a part produced on a lathe.

In this chapter we will be concerned with the basic elements of gears and cams and how they may be drawn. This information is fundamental. Whether the actual drawings are manually drafted or done by computer-driven plotter does not affect a basic description of gears and cams. However, mass production of gears and cams with manual control is rare. Most such machining is now done by computer-controlled automatic tooling.

2. CLASSIFICATION OF GEARS

Gears are an adaptation of rolling cylinders and cones, designed to ensure positive motion. It is important to become familiar with the terms associated with gears, and with the basic proportions and formulas for calculation of parameters.

A. Basic Types There are numerous variations, but the basic forms (Fig. 1) are *spur gears*, for transmitting power from one shaft to a parallel shaft; *spur gear* and *rack*, for changing rotary motion to linear motion; *bevel gears*, for shafts whose axes intersect; and *worm gears*, for nonintersecting shafts at right angles to each other.

B. Gear Teeth The teeth of gears are projections designed to fit into the tooth spaces of the mating gear and contact mating teeth along a common line

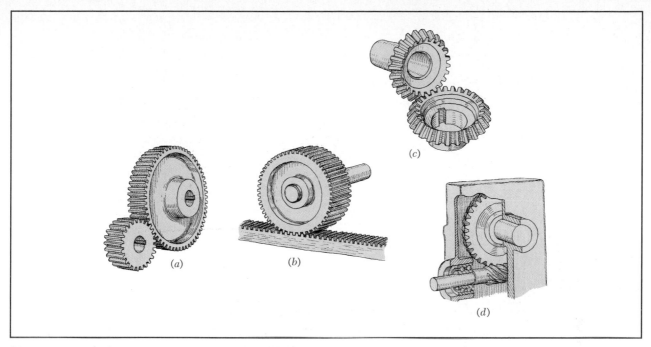

FIG. 1 Basic gear forms. *(a)* Spur gears; *(b)* spur gear and rack; *(c)* bevel gears; *(d)* worm and worm gear.

known as the "pressure line" (Fig. 2). The most common form for the tooth flank is the *involute,* and when it is made in this form the gears are known as "involute gears." The angle of the pressure line determines the particular involute the flank will have. The ANSI has standardized two pressure angles, 14½° and 20°. A composite 14½° tooth and a 20° stub tooth are also used.

3. SPUR GEARS

Within this section we will review letter symbols and formulas used with spur gears. The concept of gear ratio will be noted. We will deal also with spur-gear calculations and with the drawing of a spur gear and a rack (a derivative of a spur gear).

A. Letter Symbols and Formulas The names of the various portions of a spur gear and the teeth are given in Fig. 3. The standard letter symbols and formulas for calculation are as follows:

$$N = \text{number of teeth} = P_d \times D$$

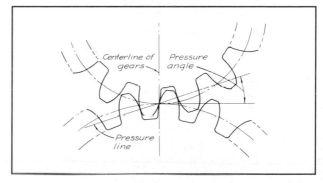

FIG. 2 Mating gear teeth. The tooth faces are involute surfaces that "roll" on each other.

P_d = diametral pitch = number of teeth on the gear for each inch of pitch diameter = N/D

D = pitch diameter = N/P_d

p = circular pitch = length of the arc of the pitch-diameter circle subtended by a tooth and a tooth space = $\pi D/N = \pi/P_d$

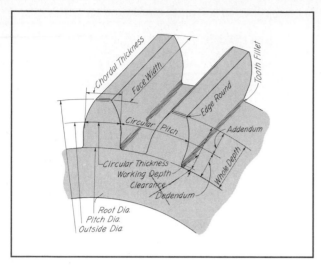

FIG. 3 Gear nomenclature.

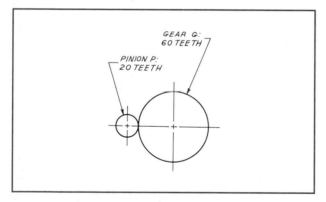

FIG. 4 Schematic of spur gears in mesh.

t = circular (tooth) thickness = length of the arc of the pitch-diameter circle subtended by a tooth = $p/2 = \pi D/2N = \pi/2P_d$

t_c = chordal thickness = length of the chord subtended by the circular thickness arc = $D \sin (90°/N)$

a = addendum = radial distance between the pitch-diameter circle and the top of a tooth = a constant/P_d = for standard 14½° or 20° involute teeth, $1/P_d$

b = dedendum = radial distance between the pitch-diameter circle and the bottom of a tooth

space = a constant/P_d = for standard 14½° or 20° involute teeth, $1.157/P_d$

c = clearance = radial distance between the top of a tooth and the bottom of a mating tooth space = a constant/P_d = for standard 14½° or 20° involute teeth, $0.157/P_d$

h_t = whole depth = radial distance between the top and bottom of a tooth = $a + b$ = for standard 14½° or 20° involute teeth, $2.157/P_d$

h_k = working depth = greatest depth a tooth of one gear extends into a tooth space of a mating gear = $2a$ = for standard 14½° or 20° involute teeth, $2/P_d$

D_O = outside diameter = diameter of the circle containing the top surfaces of the teeth = $D + 2a = (N + 2)/P_d$

D_R = root diameter = diameter of the circle containing the bottom surfaces of the tooth spaces = $D - 2b = (N - 2.314)/P_d$

F = face width = width of the tooth flank

f = tooth fillet = fillet joining the tooth flank and the bottom of the tooth space = $0.157/P_d$ max

r = edge round = radius of the circumferential edge of a gear tooth (to break the sharp corner)

n = revolutions per unit of time

B. Gear Ratios When two gears are in mesh, the smaller gear may be called the pinion P, leaving the larger gear to be called G. Gear ratio m may be defined in several ways that give an equivalent value of m.

Gear ratio m_G for the gear:

$$m_G = N_G/N_P = n_P/n_G = D_G/D_P$$

Gear ratio, m_p, for the pinion:

$$m_P = N_P/N_G = n_G/n_P = D_P/D_G$$

As an example, refer to Fig. 4. We note that the pinion has 20 teeth and the gear has 60 teeth. Then:

$$m_G = N_G/N_P = 60/20 = 3$$

and $$m_P = N_P/N_G = 20/60 = 1/3$$

Therefore, the gear ratio of the pinion is the inverse of that for the gear.

C. Calculations There are many different combinations producing various individual problems, but the following is typical and will illustrate the procedure.

Assume that the center distance and speeds are known.

Center distance = 7 in

Speed of gear 500 rpm, speed of pinion 1,500 rpm.

Then

$$m_G = n_P/n_G = 1,500/500 = 3/1 = D_G/D_P$$

Therefore

$$3D_P = D_G$$

From the center distance,

$$D_P + D_G = 14 \text{ in}$$

Substituting for D_G in terms of D_P:

$$D_P + 3D_P = 14$$

Solving,

$$D_P = 3\frac{1}{2} \text{ in}$$

Solving for $D_G = 3D_P$:

$$3 \times 3\frac{1}{2} = 10\frac{1}{2} \text{ in}$$

At this time, the diametral pitch will have to be assumed.

Assume $P_d = 5$. Then,

$$N_P = P_d \times D_P = 5 \times 3\frac{1}{2} = 17\frac{1}{2}$$

which is obviously impossible since a gear could not have a half tooth.

Assuming $P_d = 6$,

$$N_P = 6 \times 3\frac{1}{2} = 21$$
$$N_G = 6 \times 10\frac{1}{2} = 63$$

Now all the values such as addendum, dedendum, outside diameter, can be calculated from the equations in Sec. 3A. Note that if, in a given problem, the two pitch diameters are known, the center distance must be found; if one pitch diameter is known and the center distance is given, the second pitch diameter must be found, etc.

D. To Draw a Spur Gear (Fig. 5) To draw the teeth of a standard involute-toothed spur gear by an approximate circle-arc method, lay off the pitch circle, root circle, and outside circle. Start with the pitch point and divide the pitch circle into distances equal to the circular thickness. Through the pitch point draw a line of 75½° with the centerline for a 14½° involute tooth (for convenience, drafter use 75°), or 70° for a 20° tooth.

Draw the basic circle tangent to the pressure line. With compass set to a radius equal to one-fourth the radius of the pitch circle, describe arcs through the division points on the pitch circle, keeping the needle point on the base circle. Darken the arcs for the tops of the

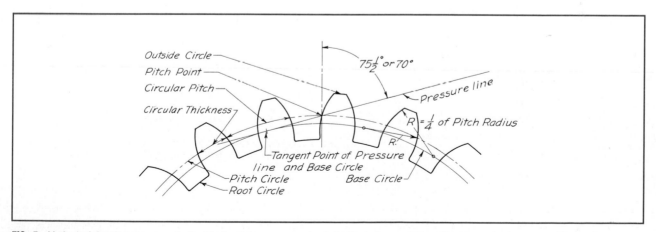

FIG. 5 Method of drawing spur gear teeth. The involutes are approximated with circle arcs. This method is used only on assembly drawings.

teeth and bottoms of the spaces and add the tooth fillets.

For 16 or fewer teeth, the radius value of one-fourth pitch radius *must be increased* to suit, in order to avoid the appearance of excessive undercut. For a small number of teeth, the radius may be as large as or equal to the pitch radius.

This method of drawing gear teeth is useful on display drawings, but on working drawings the tooth outlines are not drawn. Figure 6 illustrates the method of indicating the teeth and dimensioning a spur gear on a working drawing.

Gears must be precisely made in order to prevent noisy and faulty running characteristics. Thus, the dimensioning almost always includes surface roughness

and form-tolerance specifications as shown on Fig. 6. The tabular information given is for shop use in cutting and checking the teeth and as a record of tooth dimensions for checking with the mating gear.

E. To Draw a Rack (Fig. 7) A rack is, theoretically, a spur gear having an infinite pitch diameter. Therefore, compared with the mating gear, all circular dimensions become linear. To draw the teeth of a standard involute rack by an approximate method, draw the pitch line and lay off the addendum and dedendum distances. Divide the pitch line into spaces equal to the circular thickness of the mating gear. Through these points of division draw the tooth faces at 14½° (15° is used by drafter).

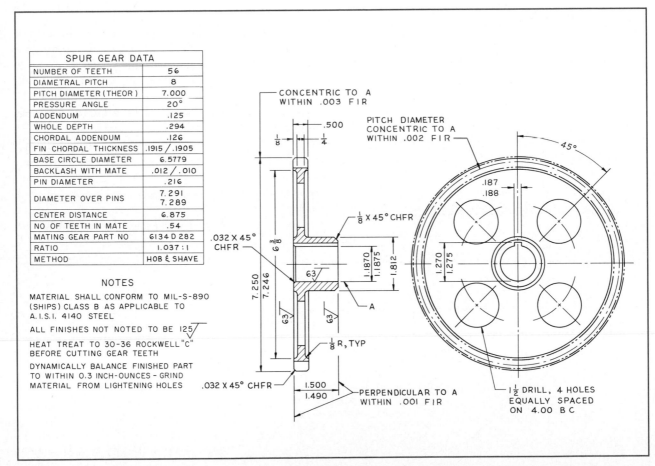

SPUR GEAR DATA	
NUMBER OF TEETH	56
DIAMETRAL PITCH	8
PITCH DIAMETER (THEOR)	7.000
PRESSURE ANGLE	20°
ADDENDUM	.125
WHOLE DEPTH	.294
CHORDAL ADDENDUM	.126
FIN CHORDAL THICKNESS	.1915 / .1905
BASE CIRCLE DIAMETER	6.5779
BACKLASH WITH MATE	.012 / .010
PIN DIAMETER	.216
DIAMETER OVER PINS	7.291 / 7.289
CENTER DISTANCE	6.875
NO OF TEETH IN MATE	.54
MATING GEAR PART NO	6134 D 282
RATIO	1.037 : 1
METHOD	HOB & SHAVE

NOTES

MATERIAL SHALL CONFORM TO MIL-S-890 (SHIPS) CLASS B AS APPLICABLE TO A.I.S.I. 4140 STEEL

ALL FINISHES NOT NOTED TO BE 125/

HEAT TREAT TO 30-36 ROCKWELL "C" BEFORE CUTTING GEAR TEETH

DYNAMICALLY BALANCE FINISHED PART TO WITHIN 0.3 INCH-OUNCES - GRIND MATERIAL FROM LIGHTENING HOLES

FIG. 6 Working drawing of a spur gear. Note that complete specifications include tooth details (tabular), surface roughness, form tolerances, and special notes.

Draw tops and bottoms and add the tooth fillets. For standard 20° full-depth or stub teeth, use 20° instead of 14½°. Specifications of rack teeth (to be given on a detail drawing) are axial (linear) pitch (equal to circular pitch of the mating gear), number of teeth, diametral pitch, whole depth.

4. BEVEL GEARS

For bevel gears the theoretical rolling surfaces become cones. As such, bevel gears are used to transmit a gear ratio through an angle, normally 90°. Applications are numerous for changing direction through gears. The motion involved with the rotation of a helicopter shaft is but one example.

A. Letter Symbols and Formulas The pitch diameter D of the gear, as shown in Fig. 8A, is the base diameter of the cone. The addendum and dedendum are calculated in the same way as for a spur gear and are measured on a cone, called the "back cone," whose elements are normal to the face-cone elements (Fig. 8A). The diametral pitch, circular pitch, etc., are the same as for a spur gear.

In addition to the letter symbols and calculations for spur gears, the dimensions shown in Fig. 8B, with formulas for calculating, are needed:

Γ = pitch-cone angle = angle between a pitch-cone element and the cone axis. tan Γ = $D_G/D_P = N_G/N_P$ (gear) = $D_P/D_G = N_P/N_G$ (pinion).

R_{PC} = pitch-cone radius = length of an element of the pitch cone. $R_{PC} = D/2 \sin \Gamma$.

α = addendum angle = angle between the pitch cone and the outside of a tooth. tan $\alpha = a/R_{PC}$.

δ = dedendum angle = angle between the pitch cone and the bottom of a tooth space. tan $\delta = b/R_{PC}$.

Γ_O = face angle = angle between a line *normal* to the gear axis and the outside (face) of the teeth. $\Gamma_O = 90° - (\Gamma + \alpha)$. (Note that this angle is the complement of the angle an element of the outside cone makes with the axis.)

Γ_R = root-cone angle = cutting angle = angle be-

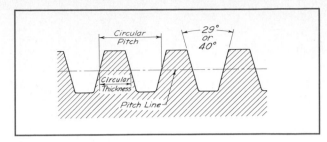

FIG. 7 Method of drawing an involute rack. Circular pitch and thickness are linear because the pitch diameter is infinite.

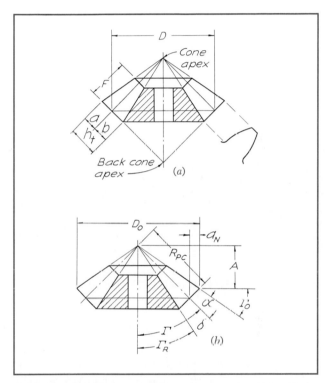

FIG. 8 Fundamental values and letter symbols for bevel-gear calculations.

tween the axis and an element of the root cone. $\Gamma_R = \Gamma - \delta$.

a_N = angular addendum = addendum distance measured *normal* to gear axis. $a_N = a \cos \Gamma$.

D_O = outside diameter = outside diameter at the base of the teeth. $D_O = D + 2a_N$.

A = apex distance = altitude of the outside-diameter cone. $A = (D_O/2) \tan \Gamma$.

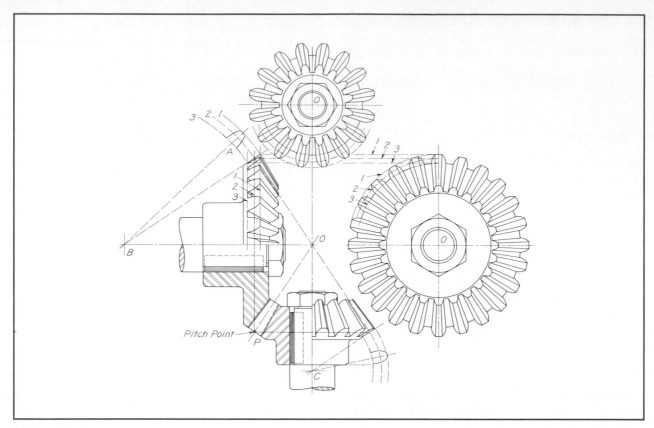

FIG. 9 Method of drawing the teeth of bevel gears. This is used only on assembly drawings.

B. To Draw a Bevel Gear (Fig. 9) To draw the teeth of an involute-toothed bevel gear by an approximate method (the "Tredgold method"), draw the centerlines, intersecting at O. Across the centerlines lay off the pitch diameters, and project them parallel to the centerlines until the projectors intersect at the pitch point P. From the pitch point, draw the pitch-circle diameters for each gear, and from their extremities, draw the "pitch cones" to the vertex or "cone center" O.

Lay off the addendum and dedendum distances for each gear on lines through the pitch points perpendicular to the cone elements. Extend one of these normals for each gear to intersect the axis, as at B and C, making the "back cones." With B as center, swing arcs 1, 2, and 3 for the top, pitch line, and bottom, respectively, of a developed tooth.

On a radial centerline AB, draw a tooth, by the method of Fig. 5. Start the plan view of the gear by projecting points 1, 2, and 3 across to its vertical centerline and drawing circles through the points. Lay off the radial centerlines for each tooth. With dividers take the circular thickness distances from A and transfer them to each tooth.

This will give three points on each side of each tooth through which a circle arc, found by trial, will pass, giving the foreshortened contour of the large end of the teeth in this view. From this point the drawing becomes a problem in projection drawing. In every view the lines converge at the cone center O; and by finding three points on the contour of each tooth, circle arcs can be found that will approximate the tooth shape.

This method is used for finished display drawings. Working drawings for cut bevel gears are drawn without tooth outlines and dimensioned as in Fig. 10, from calculations in Sec. 4A. Information on the drawing gives details for cutting and checking the teeth. The drawing is

dimensioned with limits where required. Surface roughness and form-tolerance specifications are sometimes
necessary.

5. WORM GEARS

Worm gears are used primarily to get great reductions in
relative speed and to obtain a large increase in effective
power. The worm is similar to a screw thread, and the
computing of pitch diameter, etc., is also similar. On a
section taken through the axis of the worm, the worm
gear and worm have the same relationship as a spur gear
and rack. Therefore, the tooth shape, addendum,
dedendum, etc., will be the same as for a spur gear and
rack.

   A. Letter Symbols and Formulas In addition to the
calculations necessary for a spur gear ar.d rack (addendum, dedendum, etc.), the following are needed:

- **WORM**

$$l = \text{lead of worm threads}$$
$$N_W = \text{multiplicity of worm threads}$$
$$p_{XW} = \text{axial pitch} = l/N_W$$
$$D_W = \text{pitch diameter of worm}$$
$$\lambda = \text{lead angle; } \tan \lambda = l/\pi D$$

- **WORM GEAR**

$$C = \text{center distance between worm and wheel}$$
$$D_G = \text{pitch diameter of worm gear}$$
$$D_{tp} = \text{pitch throat diameter} = \text{pitch diameter of wheel} = D_G$$
$$D_{to} = \text{outer throat diameter} = \text{pitch diameter plus twice the addendum} = D_G + 2a$$
$$D_{ti} = \text{inner throat diameter} = \text{pitch diameter minus twice the dedendum} = D_G - 2b$$

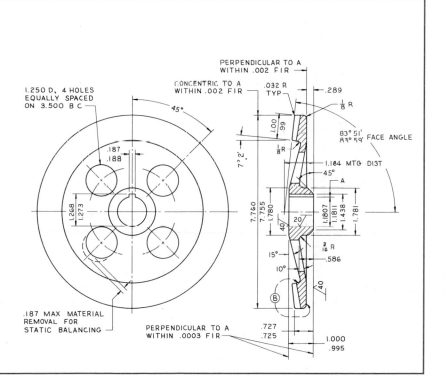

BEVEL GEAR DATA	
NUMBER OF TEETH	81
DIAMETRAL PITCH	10.4516
PITCH DIAMETER (THEOR)	7.750
PITCH ANGLE	82° 58'
ROOT ANGLE (BASIC)	80° 56'
PRESSURE ANGLE	20°
ADDENDUM (THEOR)	.042
WHOLE DEPTH (APPROX)	.181
CONE CENTER TO CROWN	.437
CONE DISTANCE	3.9044
BACKLASH WITH MATE	.010/.014
NO OF TEETH IN MATE	10
MATING GEAR PART NO	1620 D 10
DRIVING MEMBER	PINION

NOTES

(B) CARBURIZE AND HARDEN ENCLOSED
ZONE TO ROCKWELL "C" 58 MIN - FINISHED
CASE DEPTH .020-.025 DP - CORE AND
UNCARBURIZED AREAS TO BE ROCKWELL
"C" 27 MIN

FINISH UNLESS NOTED TO 100/ ALL OVER

GEAR AND PINION TO BE MATCHED AND
LAPPED AS A SET - ETCH GEAR WITH SET
NO, MOUNTING DIST, AND BACKLASH

MAGNETIC INSPECT PER MIL-I-6868

STATICALLY BALANCE TO WITHIN 0.06
INCH-OUNCES

PITCH DIAMETER OF GEAR CONCENTRIC
TO A WITHIN .002 FIR

FIG. 10 Working drawing of a bevel gear. Note that complete specifications include tooth details (tabular), surface roughness, form tolerances, and
special notes.

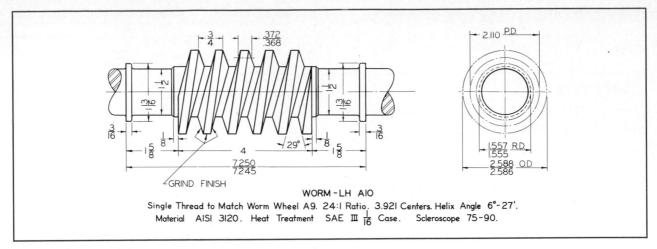

FIG. 11 Working drawing of a worm. This mates with the worm gear of Fig. 12.

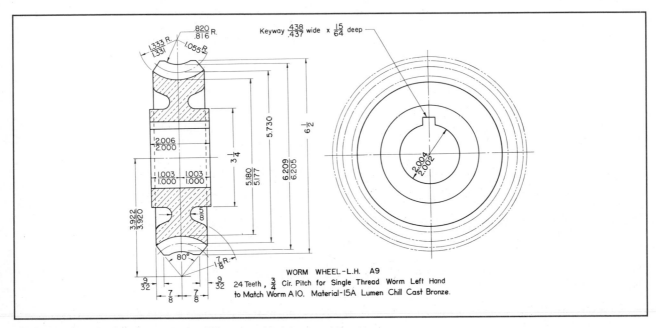

FIG. 12 Working drawing of a worm gear. This mates with the worm of Fig. 11.

R_{tp} = pitch throat radius = pitch radius of worm = $D_W/2$

R_{to} = outer throat radius = $(D_W/2) - a$

R_{ti} = inner throat radius = $(D_W/2) + b$

C_{Rt} = center of throat radius = $C = (D_G/2) + (D_W/2)$

B. To Draw a Worm Gear In assembly, a worm and worm gear are generally shown with the worm in section and the worm gear drawn as a conventional end view like the end view in Fig. 12. Detail drawings are made up as in Figs. 11 and 12. The calculated dimensions described in Sec. 5A are given as shown in the figures.

When tooth information is tabulated and given on

drawings, as in Figs. 6 and 10, it should include:

- FOR THE WORM WHEEL

1. Number of teeth
2. Addendum
3. Whole depth
4. Pitch diameter
5. Center distance between worm and wheel
6. Helix angle
7. Radius of throat
8. Face angle

- FOR THE WORM

1. Linear pitch
2. Addendum
3. Whole depth
4. Pitch diameter
5. Outside diameter
6. Root diameter
7. Helix angle
8. Minimum length of worm for complete action
9. Center distance between worm and wheel

6. CAMS

A cam is a machine element with surface or groove formed to produce special or irregular motion in a second part, called a "follower." The shape of the cam is dependent upon the motion that is required and the type of follower that is used. The type of cam is dictated by the required relationship of the parts and the motions of both.

A. Types of Cams   The direction of motion of the follower with respect to the cam axis determines two general types of cams: (1) radial or disk cams, in which the follower moves in a direction perpendicular to the cam axis; and (2) cylindrical or end cams, in which the follower moves in a direction parallel to the cam axis.

Figure 13 shows at (a) a *radial cam*, with a roller follower held against the cam by gravity or by a spring. As the cam revolves, the follower is raised and lowered. Followers are also made with pointed ends and with flat ends. At (b) is shown a *face cam*, with a roller follower at the end of an arm or link, the follower oscillating as the cam revolves. When the cam itself oscillates, the *toe* and *wiper* are used, as at (c). The toe, or follower, may also be made in the form of a swinging arm. A *yoke* or *positive-motion cam* is shown at (d), the enclosed follower making possible the application of force in either direction. The sum of the two distances from the center of the cam to the points of contact must always be equal to the distance between the follower surfaces. The cylindrical *groove cam* at (e) and the *end cam* at (f) both move the follower parallel to the cam axis, force being applied to the follower in both directions with the groove cam, and in only one direction with the end cam.

B. Kinds of Motions   Cams can be designed to move the follower with constant velocity, acceleration, or harmonic motion. In many cases combinations of these motions, together with surfaces arranged for sudden rise or fall, or to hold the follower stationary, make up the complete cam surface.

7. CAM DIAGRAMS

In studying the motion of the follower, a diagram showing the height of the follower for successive cam positions is useful and is frequently employed. The cam position is shown on the abscissa, the full 360° rotation of the cam being divided, generally, every 30° (intermediate points may be used if necessary). The follower positions are shown on the ordinate, divided into the same number of parts as the abscissa. These diagrams are generally made to actual size.

Constant velocity gives a uniform rise and fall and can be plotted, as at (a) in Fig. 14, by laying off the cam positions on the abscissa, measuring the total follower movement on the ordinate, and dividing it into the same number of parts as the abscissa. As the cam moves one unit of its rotation, the follower likewise moves one unit, producing the straight line of motion shown.

With constant acceleration, the distance traveled is proportional to the square of the time, or the total distance traveled is proportional to 1, 4, 9, 16, 25, etc.; and if the increments of follower distance are made proportional to 1, 3, 5, 7, etc., the curve can be plotted as shown at (b). Using a scale, divide the follower rise into the same number of parts as the abscissa, making the first

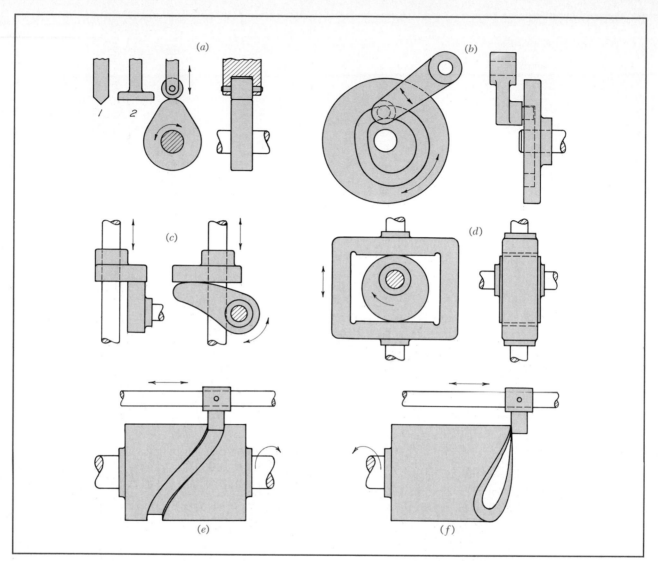

FIG. 13 Types of cams. *(a)* Radial; *(b)* face; *(c)* toe and wiper; *(d)* yoke; *(e)* cylindrical groove; *(f)* cylindrical end.

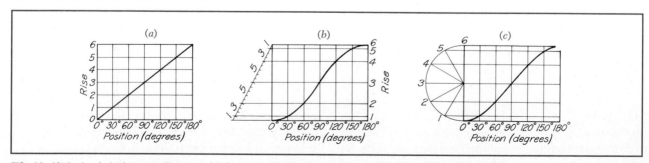

FIG. 14 Methods of plotting cam diagrams. *(a)* Constant velocity; *(b)* constant acceleration; *(c)* harmonic motion.

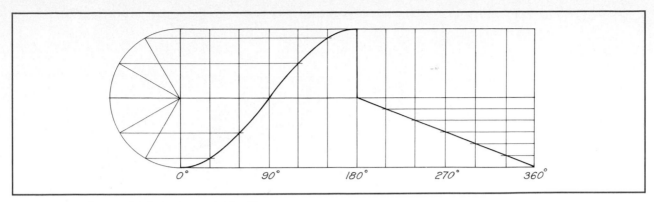

FIG. 15 A cam diagram. This is for the cam in Fig. 17.

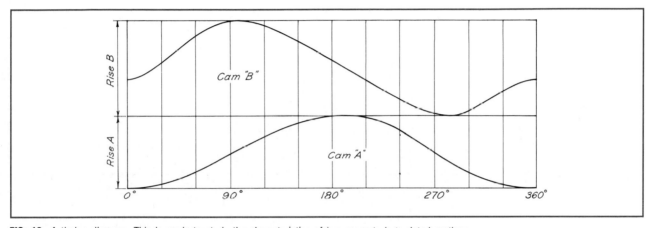

FIG. 16 A timing diagram. This is made to study the characteristics of two separate but related motions.

part one unit, the second three units, and so on. Plot points at the intersection of the coordinate lines, as shown. The curve at (b) accelerates and then decelerates to slow up the follower at the top of its rise.

Harmonic motion (sine curve) can be plotted as at (c) by measuring the rise and drawing a semicircle, dividing it into the same number of parts as the abscissa, and projecting the points on the semicircle as ordinate lines. Points are plotted at the intersection of the coordinate lines, as shown.

Figure 15 is the cam diagram for the cam in Fig. 17. The follower rises with harmonic motion in 180°, drops halfway down instantly, and then returns with uniform motion to the point of beginning.

8. TIMING DIAGRAMS

When two or more cams are used on the same machine and their functions are dependent on each other, the

"timing" and relative motions of each can be studied by means of a diagram showing each follower curve. The curves can be superimposed, but it is better to place one above the other as in Fig. 16.

9. DRAWING CAMS

Cam design is a rather specialized field and can involve advanced engineering practice. Our purpose here is simply to indicate how two basic types of cams may be drawn: the plate cam and the cylindrical cam.

A. To Draw a Plate Cam The same principle is involved in drawing all types of cams. The cam in Fig. 17, for which the diagram in Fig. 15 was made, will serve as an example. The point C is the center of the shaft, and A is the lowest and B the highest position of the center of the roller follower. Divide the rise into six parts harmonically proportional. Divide the semicircle ADE into as

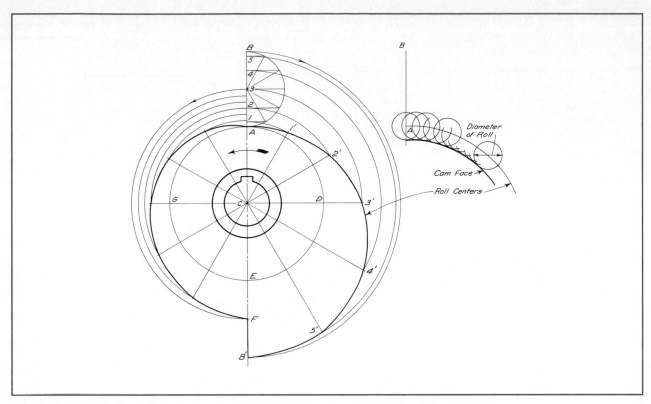

FIG. 17 Method of drawing a plate cam. Layout of harmonic motion and constant velocity are illustrated.

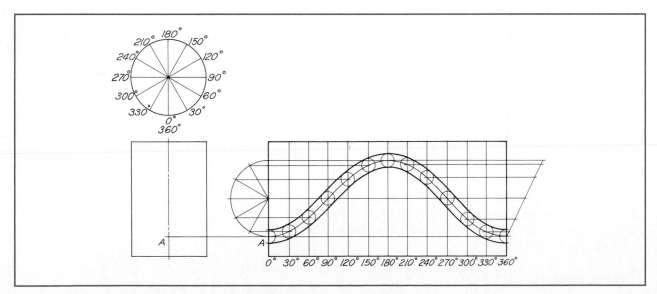

FIG. 18 Method of drawing a cylindrical cam. Development of the cylinder is employed.

many equal parts as there are spaces in the rise, and draw radial lines.

With C as center and radius C1, draw an arc intersecting the first radial line at 1'. In the same way locate points 2', 3', etc., and draw a smooth curve through them. If the cam is revolved in the direction of the arrow, it will raise the follower with the desired harmonic motion. Draw B'F equal to one-half AB. Divide A3 into six equal parts and the arc EGA into six equal parts. Then for equal angles the follower must fall equal distances. Circle arcs drawn as indicated will locate the required points on the cam outline.

This outline is for the center of the roller; allowance for the roller size can be made by drawing the roller in its successive positions and then drawing a tangent curve as shown in the auxiliary figure.

B. To Draw a Cylindrical Cam A drawing of a cylindrical cam differs somewhat from that of a plate cam, as in addition to the regular views it generally includes a developed view from which a template is made. Assume that the follower is to move upward 1½ in with harmonic motion in 180°, and then return with uniform acceleration. Top and front views of the cylinder are drawn (Fig. 18) and the development of the surface is laid out. Divide the surface as shown, also the top view to show the positions of points plotted. Divide the rise for

harmonic motion by drawing the semicircle and projecting the points (refer to Fig. 14C). Divide the return for acceleration as shown (refer to Fig. 14B). The curve thus obtained is for the center of the follower. Curves drawn tangent to circles representing positions of the follower will locate the working surfaces of the cam. The development made as described is the drawing used to make the cam.

C. Dimensioning a Cam Drawing In general, two practices are followed, depending upon the function of the cam. For cams where follower position is not critical, full-size layouts, as in Fig. 17, with necessary fractional or decimal dimensions suffice. Such cams may be made manually by profiling the surface by sawing, nibbling, and finishing, and by torch cutting, etc. If large quantities are to be produced, automatic tooling, computer-driven, would most likely be used.

For cams where the follower position and relative motion are critical, limit dimensioning from datum features is necessary. The dimensioning in Prob. 26, Chap. 3, is an example. Where the degree of precision requires, surface roughness and form tolerances are needed. Again, it is likely that a computer program would drive the machine tool so that repeatable accuracy would be available.

PROBLEMS

GROUP 1. SPUR GEARS

1. Make an assembly drawing of a pair of spur gears, from the following information: On an 11- by 17-in sheet locate centers for front view of gear B 4½ in from right border and 3½ in from bottom border. Gear A is to the left of gear B. Center distances between gears are 5.250 in. Gear A revolves at 300 rpm and has four spokes, elliptical in cross section, 1-in major and ½-in minor axes; inside flange diameter, 4⅜ in; hub, 2 in diameter, 1½ in long. Gear B revolves at 400 rpm and is disk-type with ½-in web; inside flange diameter, 3¼ in; hub, 2 in diameter, 1½ in long. Material is cast steel; face width, 1 in; diametral pitch, 4; shaft diameters, 1 in; ¼-in Woodruff keys. Draw front view and sectional top view.

2. The figure on the next page shows a gear box for a reducing mechanism using spur gears. Using the scale shown on the drawing, transfer the necessary distances to obtain the net available space for the gears and design gears as follows: center distance, 4 in; gear A revolves at 900 rpm, gear B at 1,500 rpm; diametral pitch, 4; face width, 1 in. Use standard 14½° involute teeth. Calculate the necessary values (shown in table in figure for Prob. 2). Draw the top view in section. Show four teeth on each gear on front view and complete the view with conventional lines.

3. Same as Prob. 2 but for A = 525 rpm, B = 675 rpm. Use 20° involute teeth.

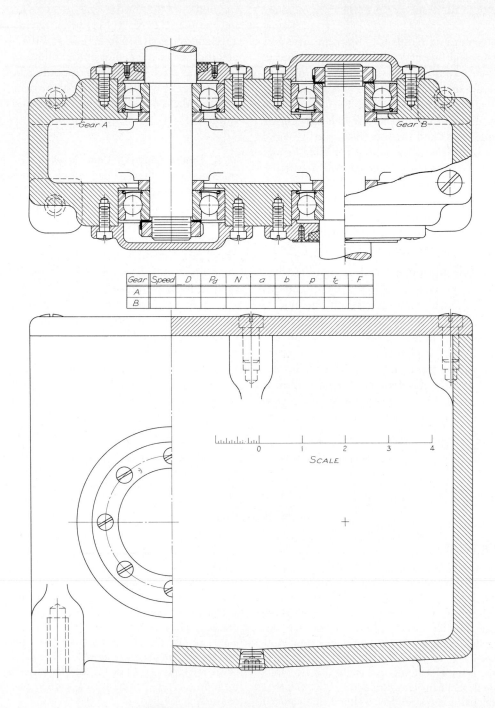

Gear	Speed	D	P_d	N	a	b	p	t_c	F
A									
B									

SCALE

PROB. 2 Spur-gear speed reducer.

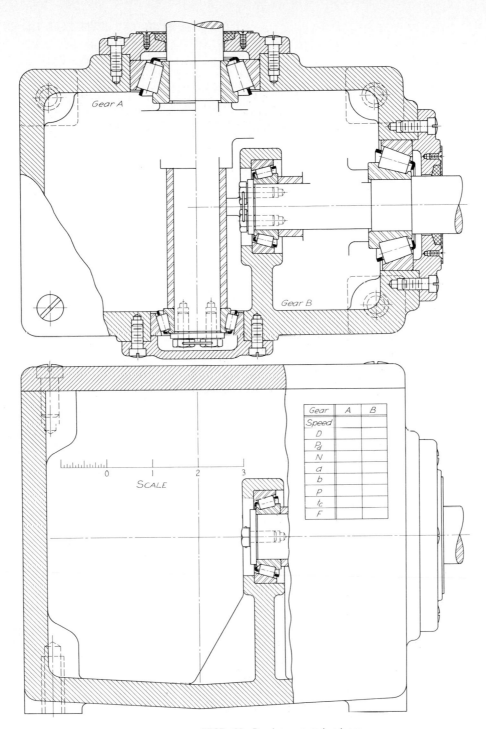

PROB. 10 Bevel-gear speed reducer.

4. Make working drawings for the gears in Prob. 2.

5. Make working drawings for the gears in Prob. 3.

6. A broken spur gear has been measured and the following information obtained: number of teeth, 33; outside diameter, 4⅜ in; width of face, 1 in; diameter of shaft, ⅞ in; length of hub, 1¼ in. Make a drawing of a gear blank with all dimensions and information necessary for making a new gear. The dimensions that are not specified here may be made to suit as the drawing is developed.

7. Make a drawing of a spur gear. The only information available is as follows: root diameter, 7.3372 in; outside diameter, 8.200 in; width of face, 1⅞ in; diameter of shaft, 1⅜ in; length of hub, 2 in.

GROUP 2. RACKS

8. Draw a spur gear and rack as follows: gear, 4 in pitch diameter, 20 teeth, face width, 25 mm; standard 14½° involute teeth. Rack is to move laterally 125 mm. Compute axial pitch, addendum, and dedendum for rack, and draw the rack and gear in assembly showing all teeth on rack and four teeth on gear.

9. Same as Prob. 8 but gear is 125 mm pitch diameter, has 25 teeth and 20 mm face width, with 20° standard involute teeth.

GROUP 3. BEVEL GEARS

10. The figure on the preceding page shows a gear box for a reducing mechanism using bevel gears. Using the scale shown on the drawing, transfer the necessary distances to obtain net available space for the gears and design gears as follows: gear A has 6 in pitch diameter, diametral pitch, 4; speed, 350 rpm; face width, 1 in. Gear B has a speed of 600 rpm. Use standard 14½° involute teeth. Compute the necessary values (shown in table in figure for Prob. 10). Draw top view in section and front view as an external view that shows all teeth.

11. Same as Prob. 10 but gear A has 5½ in pitch diameter and 22 teeth; speed, 750 rpm. Gear B has a speed of 1,100 rpm. Use standard 20° involute teeth. Face width is 1 in.

12. Make working drawings for the bevel gears in Prob. 10; refer to Sec. 4 to compute values needed.

13. Make working drawings for the bevel gears in Prob. 11; refer to Sec. 4 to compute values needed.

GROUP 4. SPUR- AND BEVEL-GEAR TRAINS

14. Make an assembly drawing of gear train, as follows: A and B are bevel gears, with ⅞ in face width and diametral pitch of 6. A is 3 in pitch diameter, revolves at 150 rpm. B revolves at 100 rpm. C and D are spur gears with diametral pitch of 8, 1 in face width. C engages D, which revolves at 40 rpm. All shafts are 1 in. Draw A in full section, B with lower half in section, C and D in full section, quarter-end view of gear B in space indicated, and end views of C and D.

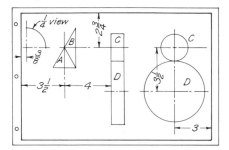

PROB. 14

15. A 3-in-diameter, 3-diametral-pitch bevel gear R on shaft AB running 1,120 rpm drives another bevel gear S on shaft AC at 840 rpm. On shaft AC centered at P, an 8-in-diameter, 4-diametral-pitch spur gear T drives a pinion U at 1,680 rpm. All shaft diameters are 1 in; face widths, 1 in. Hub diameters of R and S are 1¾ in. Gear C has four spokes, elliptical, ⅝ by 1 in; hub, 1⅞ in diameter; thickness of flange, ½ in. Draw gear R with upper half in section; S, T, and U in full section. Put quarter-end view of R in space indicated and end views of T and U on centerline MN.

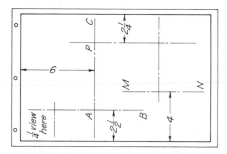

PROB. 15 Bevel-and-spur-gear train.

GROUP 5. WORM GEARS

16. Make an assembly drawing of a pair of worm gears, as follows: center distance, 6 in; pitch diameter of worm gear, 8½ in; face width of worm gear, 1⅞ in; pitch diameter of worm, 3½ in; standard 14½° involute worm teeth. Single-threaded worm, left-hand 4 in long, 1 in lead. Show appropriate spindle ends for worm and suitable hub for worm gear. Make assembly drawing as a half section through axis of worm.

17. Make detail drawings of the worm gear and worm of Prob. 16. Refer to Sec. 5 and Figs. 11 and 12 for calculation of necessary dimensions.

GROUP 6. CAMS

18. Make a drawing for a plate cam to satisfy the following conditions: On a vertical centerline, a point A is 20 mm above a point O, and a point B is 40 mm above A. With center at O, revolution clockwise, the follower starts at A and rises to B with uniform motion during one-third revolution, remains at rest one-third revolution, and drops with uniform motion the last one-third revolution to the starting point. Diameter of shaft, 20 mm; diameter of hub, 30 mm; thickness of plate, 12 mm; length of hub, 30 mm; diameter of roller, 12 mm.

19. Make a drawing for a face cam using the data of Prob. 18.

20. Make a drawing for a toe-and-wiper cam. The toe shaft is vertical, ¾ in in diameter. Starting at a point 1 in directly above center of wiper shaft, the toe is to move upward 2 in with simple harmonic motion, with 135° turn of the shaft. Wiper has 1¼-in-diameter hub; is 1¼ in long; has ¾-in-diameter shaft. Design toe to suit.

21. Make a drawing for a positive-motion cam. Starting at a point 1 in above center of cam shaft, upper follower surface is to move upward 1 in with simple harmonic motion in 180° turn of cam. Return is governed by necessary shape of cam. Follower is ½ in thick on ½-in vertical shaft. Cam is ½ in thick on ¾-in-diameter shaft; hub, 1¼ in diameter, 1¼ in long.

22. Make a drawing, with development, for a cylindrical cam. The 12 mm-diameter roller follower is to move 50 mm leftward with constant velocity in 180° turn of cylinder and return with simple harmonic motion. Cam axis is horizontal; cylinder, 100 mm diameter, 100 mm long on 25 mm shaft. Follower is pinned to 15 mm shaft 75 mm center to center from cylinder.

18

WELDING AND RIVETING

1. PURPOSE OF CHAPTER

This chapter will survey two useful methods for the permanent assembly of metal parts: welding and riveting. Each method has existed for many decades and each has been well refined in terms of approved procedures and standards to ensure a satisfactory assembly. For welding in particular, standards are extensive and must be used by the drafter if the user is to understand clearly what is desired. An excellent standard is AWS A2.41—1979, issued by the American Welding Society. Welding symbols within our study will conform to this standard.

2. WELDING PROCESSES

Welding processes may be divided into at least three classes: arc welding, gas welding, and resistance welding.

A. Arc Welding   Heat is generated by an electric arc that raises to the proper temperature the parts to be joined. In some case no filler material is used; the parts are simply fused together and become one. In other cases, a consumable electrode is used that provides a filler material. Arc welding can be used successfully on what would otherwise be difficult materials to join, such as stainless steels and nonferrous metals.

B. Gas Welding   This method involves the use of a combustible gas, such as a mixture of oxygen and acetylene. The burning gas raises the temperature of the parts to be joined. A welding rod provides filler material. Gas welding is a popular method for field work because it uses easily portable equipment and requires no electrical source.

Hot gases have a different use also. One may *cut* rather than *join* materials through the use of cutting torches. The hot gases from the torch cut through the material and follow a path desired by the person doing the cutting. Many people are familiar with scrap yards where worn-out equipment such as cars are cut up for the value of the scrap steel.

C. Resistance Welding Resistance welding is a heat-and-squeeze process. The parts that are to be welded, while being forced together by mechanical pressure, are raised to the temperature of fusion by the passage of a heavy electrical current through the junction.

Filler materials are not usually used. Lap and butt joints are the usual types involved. A form of resistance welding known as "spot welding" is extremely common and useful where sheet metal is to be joined at specified intervals along a seam. The robots recently given much publicity in the assembly of automobile bodies use resistance welding in many cases.

3. OTHER HEAT PROCESSES

Two other processes for joining metals using heat should be mentioned: *brazing* and *soldering*. Unlike welding, brazing does not melt the parent materials. Instead, a filler of a lower melting temperature than the materials to be joined is added in molten form around the parent materials. Capillary action causes the molten filler to flow between the parent materials, giving adhesion and sealing when cooled. Joints used are forms of lap and butt, also used in welding. A scarf joint (see Fig. 17) is also available.

Soldering is a method similar to brazing except that the filler material melts at a lower temperature than for brazing. Joints typically are not as strong as when brazed. However, joints are easier to make with the lower melting filler. Soldering is a universally used method for assembling many electrical circuits. It is also a popular method used in the home, in that one needs only a soldering iron or "gun" and a spool of solder containing flux.

4. TYPES OF WELDED JOINTS

There are five types of welded joints, as seen in Fig. 1. The types are named by the geometric relationship between the two parts. There are many types of welds that can be used to join the parts, as will be seen in the next section.

5. TYPES OF WELDS

A useful correlation of joint types and weld types is seen in Table 1. Examples of all weld types listed in Table 1

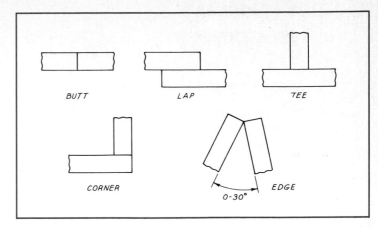

FIG. 1 Types of welded joints.

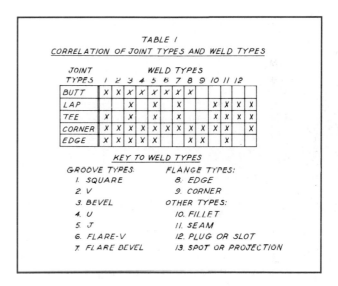

TABLE 1
CORRELATION OF JOINT TYPES AND WELD TYPES

JOINT TYPES	1	2	3	4	5	6	7	8	9	10	11	12	13
BUTT	X	X	X	X	X	X	X	X					
LAP			X		X		X			X	X	X	X
TEE	X		X		X		X			X	X	X	X
CORNER	X	X	X	X	X	X	X	X	X	X	X		X
EDGE	X	X	X	X	X			X	X		X		

KEY TO WELD TYPES

GROOVE TYPES:
1. SQUARE
2. V
3. BEVEL
4. U
5. J
6. FLARE-V
7. FLARE DEVEL

FLANGE TYPES:
8. EDGE
9. CORNER

OTHER TYPES:
10. FILLET
11. SEAM
12. PLUG OR SLOT
13. SPOT OR PROJECTION

are shown in Fig. 2, with the exception of a projection weld. A projection weld requires sheet metal with a small dimpled protruding area at each place a weld is to be made.

6. WELDING SYMBOLS

Weld types, sizes, and positions must be translated into symbols so users of drawings will know clearly all information about a particular weld. Fortunately, there are such symbols. The American Welding Society standard AWS A2.4—79 has compiled a listing of all necessary

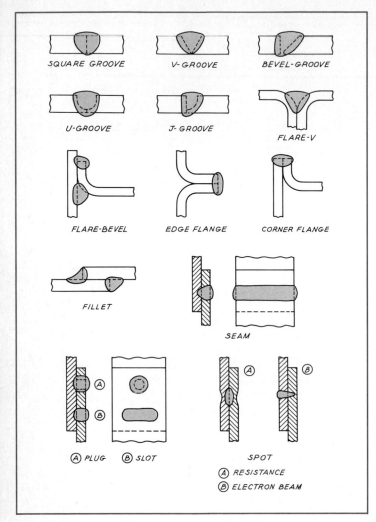

FIG. 2 Types of welds. These basic types may be intermixed on a single part if desired.

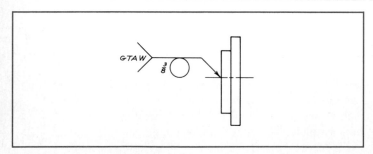

FIG. 3 Example of a welding symbol.

welding symbols. Figure 3 is an example of a welding symbol. What does it mean? It means we have an arrow-side gas tungsten-arc spot weld ⅜ inch in diameter. Perhaps this is not very enlightening at this point. We need to look at the general welding symbol and at particulars for the various types of welds. Then our understanding will increase.

A. General Welding Symbol Figure 4 gives the all-encompassing welding symbol. The general symbol serves for every case for every type of weld. Not every weld case would require all aspects of the general symbol. For example, a symbol could be as simple as that in Fig. 5, which is for an arrow-side bevel-groove weld.

B. Arrow Side vs. Other Side The converse to the term "arrow side," mentioned above, is "other side." The meaning of these terms is seen in Fig. 6. The general meaning is given in Fig. 6A, while Fig. 6B shows an arrow-side U-groove weld. Figure 6C gives an other-side U-groove weld. If the symbols appear on *both* sides of the reference line, the part is to be welded on both sides. See Fig. 10.

C. Specific Weld Symbols All available basic symbols are indicated in Figure 7. Symbols above the dashed lines are other-side symbols; symbols below the dashed lines are arrow-side symbols. The dashed lines correspond to the reference line of Fig. 4.

There are two symbols in Fig. 7 not discussed to this point, that for back welds and that for surfacing welds. Figure 8 gives the meaning of these two symbols. *Back welds* are used to fill in welds, as seen in Fig. 8A. *Surfacing welds* are used to build up a surface for the purpose of correcting dimensions or adding wear resistance.

Additional weld symbols are given in Fig. 9. These symbols are:

1. *Weld all around:* Means to weld completely around the particular joint indicated.
2. *Field weld:* Means welds are to be done on the construction site, not in the shop.
3. *Melt-thru:* Means the joint is to have 100 percent penetration in welds made from only one side.
4. *Backing or spacer material:* Means that a small extra piece has been added between the two parts to be joined to serve as a spacer around which the weld is made.

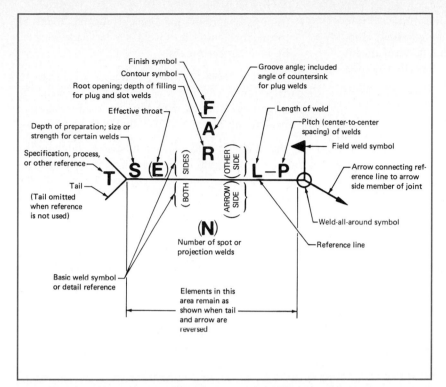

FIG. 4 This complete symbol includes all possible elements of the welding symbol. *(Courtesy American Welding Society.)*

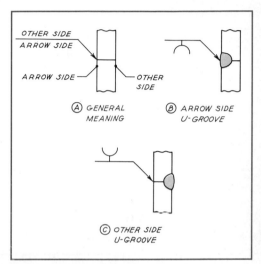

FIG. 5 Example of a simple welding symbol.

FIG. 6 Meaning of arrow side vs. other side.

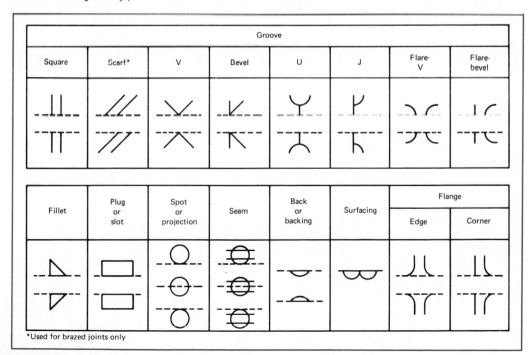

FIG. 7 Available basic symbols for welding. *(Courtesy American Welding Society.)*

*Used for brazed joints only

5. *Contours: Flush* means that a melt-thru weld is to be made flush with the surface by grinding, chipping, or other mechanical means. *Convex* means that a melt-thru weld is to be made to a convex (hump) shape by mechanical means. *Concave* means that a melt-thru weld is to be made concave by mechanical means.

D. Use of a Break in Arrow When a joint is to be prepared prior to welding (e.g., machine a bevel), the arrow may have a break. See Fig. 10 for an example. The arrow points to the member that is to be prepared.

E. Use of Metric Units The units used throughout the overall drawing should be used within the welding symbol. If the drawing is on the inch base, so must be the welding symbol units; if on the millimeter base, so the welding symbol. In no case, however, should dual units be used within the symbol.

F. Size of Welding Symbol Figure 11 gives suggested dimensions for the basic welding symbol. Weld symbols (fillet, V-groove, etc.) would of course be added. If the drawing were to be reduced in size by some reproduction means, the symbol could be proportionally larger.

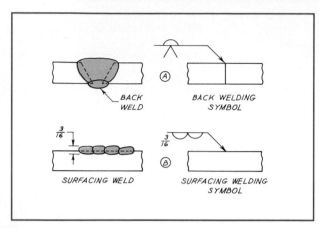

FIG. 8 Explanation of back and surfacing welds.

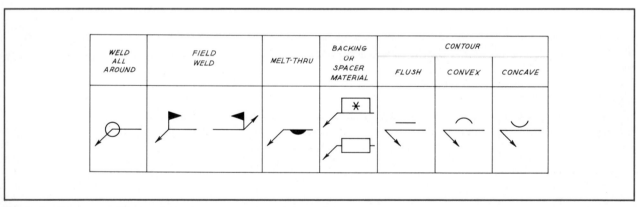

FIG. 9 Supplementary welding symbols. *(Courtesy American Welding Society.)*

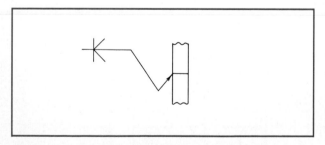

FIG. 10 Use of a break-in arrow. This joint is to have a bevel-groove weld on both sides after the bevels are first prepared.

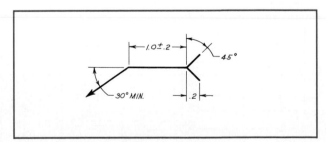

FIG. 11 Size of the basic welding symbol.

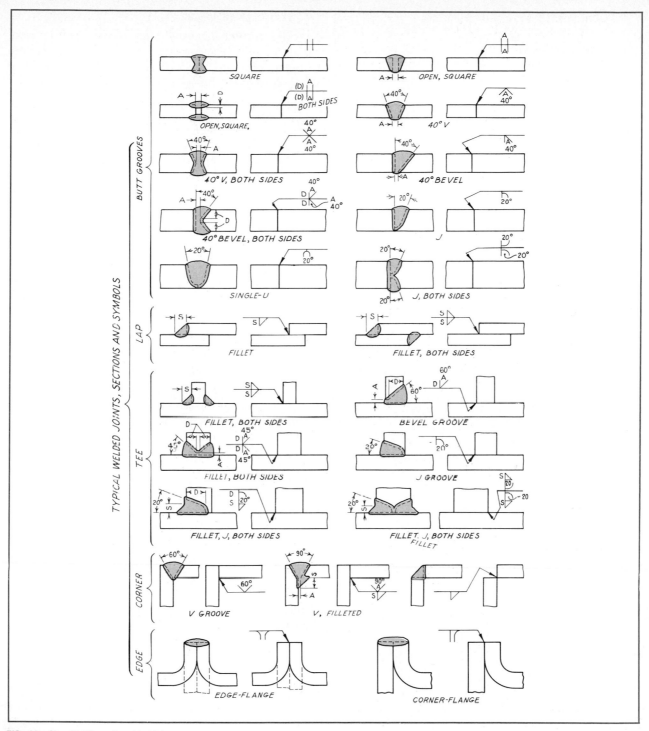

FIG. 12 Classification of welded joints.

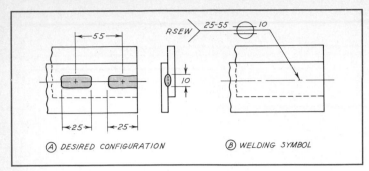

FIG. 13 Expressing an intermittent weld. Units are in millimeters.

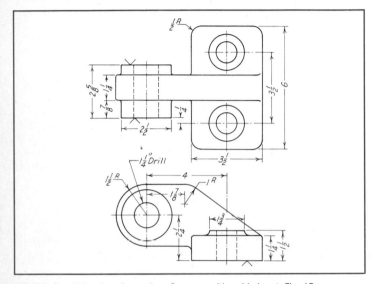

FIG. 14 Detail drawing of a casting. Compare with welded part, Fig. 15.

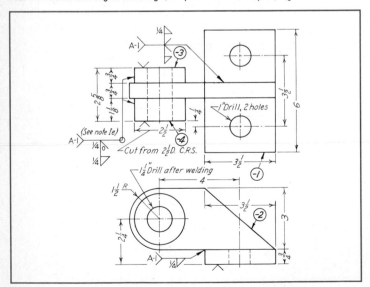

FIG. 15 Detail drawing of a welded part. Compare with cast part, Fig. 14.

7. EXAMPLES OF WELDS

Welds can come in a wide variety of types and combinations, as we have already seen. Refer to Fig. 12. There a number of welds are described both by illustration and by symbol. Note the opening or gap labeled A in a number of the butt welds. Gap A is known as a *root opening*.

Note also that some dimensions are in parentheses, others are not. The open, square butt weld (both sides) has (D) as part of the welding symbol. The letter D in parentheses indicates that D is the size of the *effective throat* of the weld. Sizes without parentheses are the *depth of preparation*.

It is possible to have combination welds where two weld symbols may appear. An example is the fillet, J-groove combination shown in Fig. 12. The weld symbols are merely stacked one above the other.

If one wishes to specify *intermittent* welds, the welding symbol can accommodate this need. For example, suppose the configuration of Fig. 13A is desired. The welding symbol of Fig. 13B gives the needed information. The numeral 10 is the size (mm), 25 the length, and 55 the pitch (distance from the center of one weld to the center of the next). Note that the symbol for a seam weld is symmetrical about the centerline. This means there is no arrow-side or other-side preference. The letters RSEW stand for resistance seam weld. See Appendix 43 for a complete listing of letters.

8. WELDING DRAWINGS

A welding drawing shows a unit or part made of several pieces of metal, with each welded joint described and specified. Figure 14 shows the detail drawing of a part made of cast iron, and Fig. 15 shows a part identical in function but made up by welding. Comparison of the two drawings shows the essential differences both in construction and in drawing technique. Note the absence of fillets and rounds in the welding drawing. Note also that all the pieces making up the welded part are dimensioned so that they can be cut easily from standard stock.

All joints between the individual pieces of the welded part must be shown, even though the joint would not appear as a line on the completed part. The lines marked A in Fig. 16 illustrate this principle. Each individual piece should be identified by number (Fig. 15).

When desired, general notes such as the following may be placed on a drawing to provide detailed information, and this information need not be repeated on the symbols.

Unless otherwise indicated, all fillet welds are 5/16 inch size.

Unless otherwise indicated, root openings for all groove welds are 3/16 inch.

9. BRAZED JOINTS

To indicate brazed joints one may use the conventional welding symbols. The only joint unique to brazing is the scarf joint. Figure 17 is an example of brazed joints. Note how the sizes are indicated within the symbol. Figure 17A shows a scarf joint, Fig. 17B a lap joint. The letters TB stand for torch brazing.

10. RIVETS

Rivets are used for making permanent fastenings, generally between pieces of sheet or rolled metal. They are round bars of steel with a head formed on one end and are often put in place red-hot so that a head can be formed on the other end by pressing or hammering. Rivet holes are punched, punched and reamed, or drilled larger than the diameter of the rivet, and the shank of the rivet is made just long enough to give sufficient metal to fill the hole completely and make the head.

Large rivets are used in structural-steel construction and in boiler and tank work. In structural work, only a few kinds of heads are normally needed: the button, high-button, and flattop countersunk heads (Fig. 18).

For boiler and tank work, the button, cone, round-top countersunk, and pan heads are used. Plates are connected by either lap or butt joints. Figure 19A is a single-riveted lap joint; (b) is a double-riveted lap joint; (c) is a single-strap butt joint; and (d) is a double-strap butt joint.

Large rivets are available in diameters of 1/2 to 1 3/4 in, by increments of even 1/8 in. The length needed is governed by the "grip," as shown in Fig. 19D, plus the length needed to form the head. Length of rivets for various grip distances may be found in the handbook "Steel Construction," published by the American Institute of Steel Construction.

The proportions for structural rivets are shown in Fig. 20. Note that the original lengths are given dashed lines. After assembly the rivets take the forms shown.

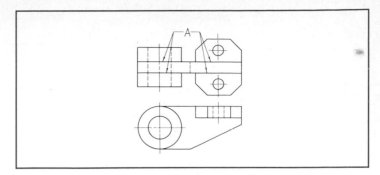

FIG. 16 Joint lines. These are shown on drawings of welded parts.

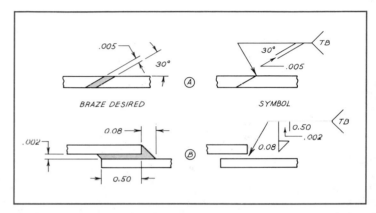

FIG. 17 Two different brazed joints. Case A, a scarf joint, is unique to brazing.

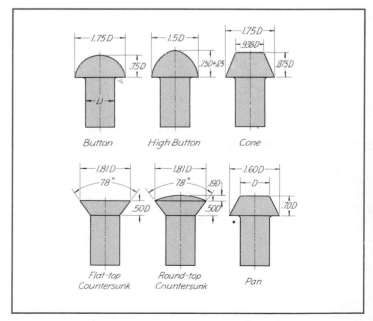

FIG. 18 Large rivet heads. These are used for structural steel.

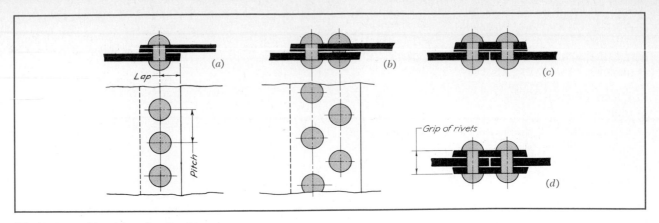

FIG. 19 Lap and butt joints. See Sec. 10.

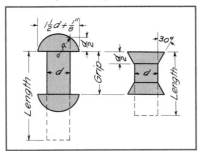

FIG. 20 Proportions of structural rivets.

Flat
A
A = 2.00 D
H = 0.33 D

Countersunk
A
90°
A = 1.850 D
H = 0.425 D

Button
A
H
A = 1.750 D
H = 0.750 D
r = 0.885 D

Pan
A
H
A = 1.720 D
H = 0.570 D
r_1 = 0.314 D
r_2 = 0.850 D
r_3 = 3.430 D

Truss or Wagon Box
A
H
A = 2.500 D
H = 0.330 D
r = 2.512 D

FIG. 21 American Standard small-rivet heads. These are used principally for thin material.

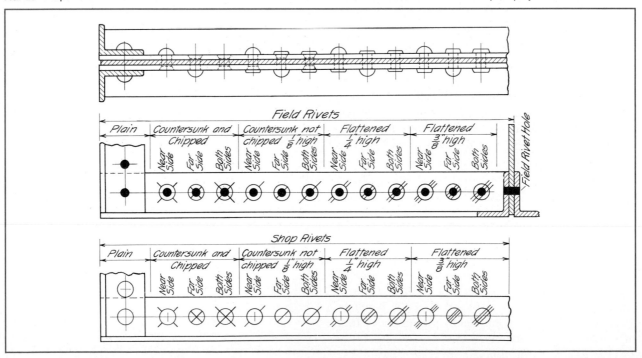

FIG. 22 American Standard rivet symbols.

Small rivets are used for fabricating light structural shapes and sheet metal. ANSI small-rivet heads are shown in Fig. 21. Small rivets are available in diameters of 3/32 to 7/16 in, by increments of even 1/32 in to 3/8 in diameter. Tinners', coopers', and belt rivets are used for fastening thin sheet metal, wood, leather, rubber, etc.

For large structures *rivet symbols* are used for representation because it is impossible to show the details of head form on a small-scale drawing. The ANSI symbols for indicating various kinds of heads are given in Fig. 22.

PROBLEMS

Note: Problems are on the inch base if dimensions involve fractions. Problems are on the millimeter base if dimensions involve whole numbers and/or decimals.

GROUP 1. WELDS

Study the welding symbols so that you will be able to write and read them without hesitation. Problems 1 and 2 give practice in reading, Probs. 3 and 4 in writing. Problems 5 to 9 give practice in use of the symbols on working drawings.

1 and 2. Make full-size cross-sectional sketches of the joints indicated. Dimension each sketch. Fractions are inch base. Whole numbers are millimeter base.

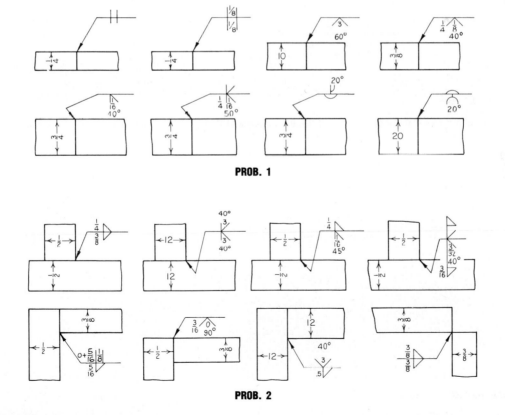

PROB. 1

PROB. 2

3 and 4. Sketch members and show welding symbol for each complete joint. Fractions are inch base. Whole numbers are millimeter base. Estimate weld size from the plate thickness.

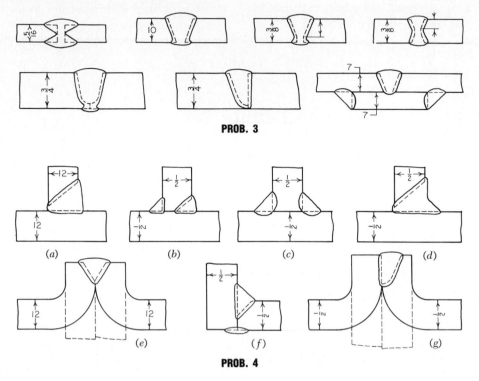

PROB. 3

(a)   *(b)*   *(c)*   *(d)*

(e)   *(f)*   *(g)*

PROB. 4

GROUP 2. WELDED PARTS

5 and 6. Make a complete welding drawing for each object. These problems are printed quarter-size. Draw full-size by scaling or transferring with dividers.

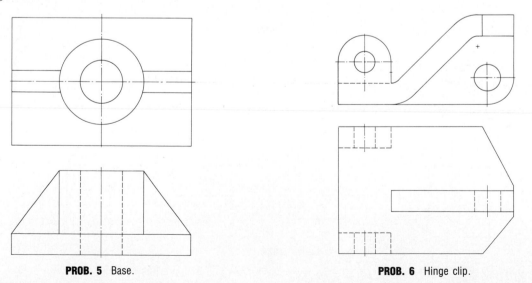

PROB. 5 Base.   **PROB. 6** Hinge clip.

7. Draw the views given, add welding symbols and dashed numbers for identification of the individual pieces, and complete the materials list.

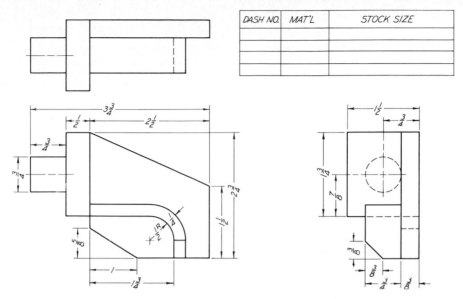

DASH NO.	MAT'L	STOCK SIZE

PROB. 7 Pivoted spacer.

8. Draw the views given and add welding symbols and dashed numbers. Make a materials list similar to the one used in Prob. 7.

9. Draw the views given and add welding symbols and dashed numbers. Make a materials list. Determine length of material for rim.

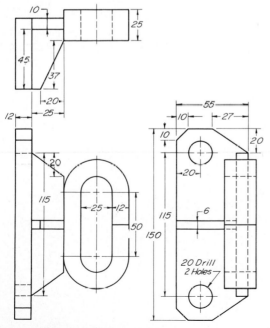

PROB. 8   METRIC. Belt-tightener bracket.

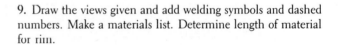

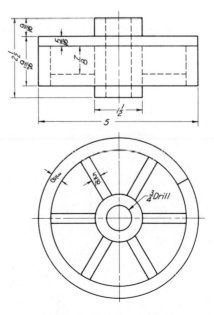

PROB. 9 Ribbed-disk wheel.

GROUP 3. RIVETS

10. Draw top view and section of a single-riveted butt joint 10⅝ in long. Pitch of rivets is 1¾ in. Use cone-head rivets.

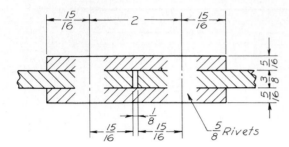

PROB. 10 Butt joint.

11. Draw a column section made of 15-in by 33.9-lb channels with cover plates as shown, using ⅞-in rivets (dimensions from the handbook of the American Institute of Steel Construction). Use button-head rivets on left side and flattop countersunk-head rivets on right side so that the outside surface is flush.

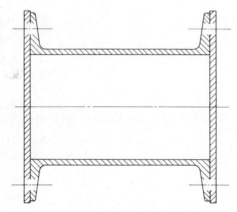

PROB. 11 Column.

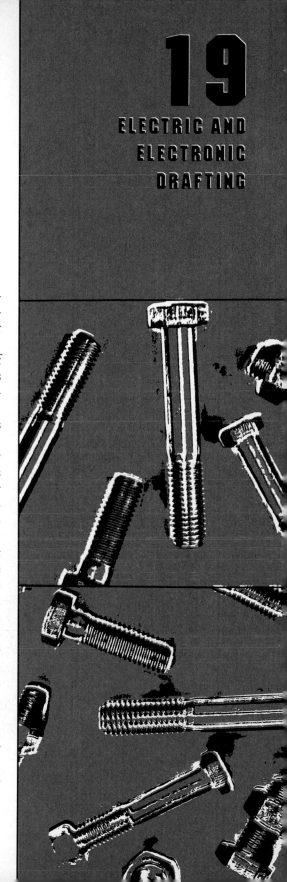

1. PURPOSE OF CHAPTER

A drafter who is involved in design work will at some point need to draw diagrams of electrical and electronic circuits. While this person need not be knowledgeable in the theory underlying what is to be drawn, an ability to use proper layout and symbols is vital.

Electric drafting deals with components related to the transmission and use of electrical power, whether for industry, business, or home. Typical components are generators, controls, transmission networks, lighting, and heating and cooling devices.

Electronic drafting, on the other hand, involves circuits of products such as radios, TVs, guidance systems, radar, computers, and a host of other examples. Power consumption or transmission rates within electronic circuits are typically far less than for products in the electrical area. A circuit breaker for a business involves energy levels far higher than a circuit breaker for a stereo amplifier, for example. However, electronic circuits have become very complex in many applications and involve high technology such as solid-state chips and miniaturization.

It is the intent of this chapter simply to introduce the reader to elementary yet fundamental practices in the drafting of electrical and electronic drawings. The standards used are well accepted ones: ANSI Y32.2—1975, "Graphic Symbols for Electrical and Electronic Diagrams"; USAS Y14.15—1973, "Electrical and Electronic Diagrams"; and ANSI Y14.15a—1971, "Interconnection Diagrams."

One must realize that electronics is one of the most rapidly evolving technologies of the modern age. A person working to any extent and depth in this field must keep current on drafting practices. A company's technical library, trade journals, and one's supervisors are good sources of information.

Realize, also, that in an area of drafting that uses repetitive symbols, as does electrical/electronic drafting, the role of computer-driven drafting is increasing rapidly. Much engineering thought is being expended on ways to make electrical/electronic drafting faster and more error-free by using computer assistance. Figure 1 gives a fairly simple example of a circuit drawn on a plotter driven by a computer.

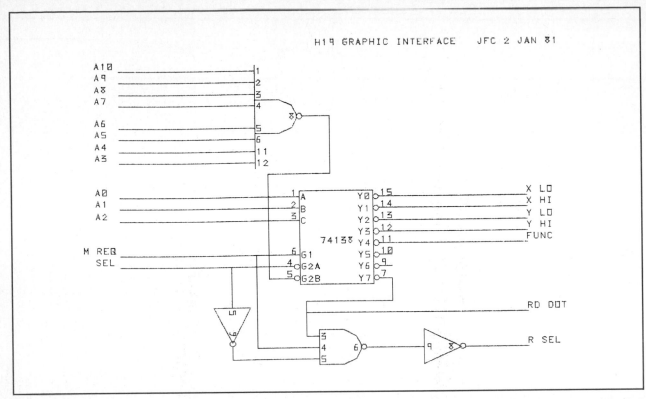

FIG. 1 Example of a simple schematic diagram drawn on a computer-driven plotter. (*Courtesy James Carras, The Pennsylvania State University.*)

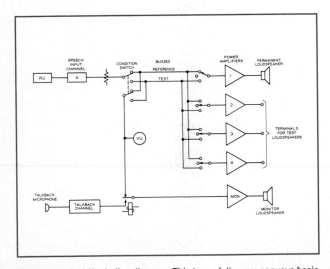

FIG. 2 A typical single-line diagram. This type of diagram conveys basic functional information but does not give detailed information. The one illustrated is for a complex audio system.

2. DIAGRAM TYPES

Four types of diagrams will be listed. These diagrams show arrangement of components and wiring connections. The definitions are quoted from USAS Y14.15—1973.

A. Single-Line or One-Line Diagram "A diagram which shows, by means of single lines and graphical symbols, the course of an electric circuit or system of circuits and the component devices or parts used therein." See Fig. 2.

B. Schematic or Elementary Diagram "A diagram which shows, by means of graphical symbols, the electrical connections and functions of a specific circuit arrangement. The schematic diagram facilitates tracing the circuit and its functions without regard to the actual physical size, shape, or location of the component device or parts." See Fig. 3.

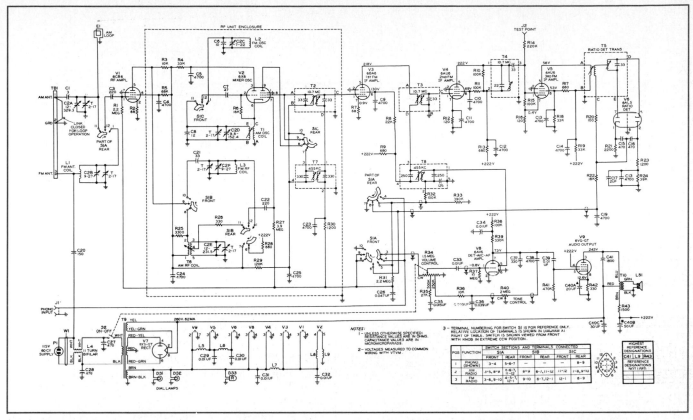

FIG. 3 A typical schematic diagram. This type of diagram shows all components by graphic symbols and gives details of all electrical connections.

C. Connection or Wiring Diagram "A diagram which shows the connections of an installation or its component devices or parts. It may cover internal or external connections, or both, and contains such detail as is needed to make or trace connections that are involved. The connection diagram usually shows general physical arrangement of the component devices or parts." See Fig. 4.

D. Interconnection Diagram "A form of connection or wiring diagram which shows only external connections between unit assemblies or equipment. The internal connections of the unit assemblies or equipment are usually omitted." See Fig. 5.

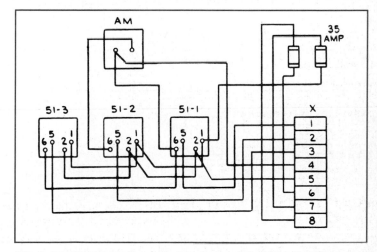

FIG. 4 Typical point-to-point connection diagram. *(USAS Y14.15—1973.)*

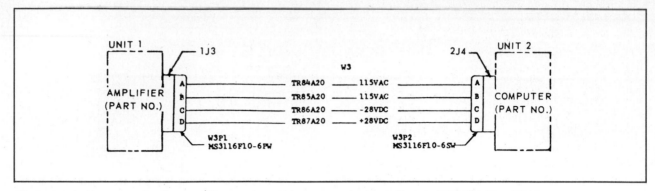

FIG. 5 A typical interconnection diagram. This example could be a communication application. *(ANSI Y14.15a—1971.)*

LINE APPLICATION	LINE THICKNESS
FOR GENERAL USE	MEDIUM
MECHANICAL-CONNECTING & SHIELDING LINE FUTURE CIRCUITS	MEDIUM
BRACKET-CONNECTING DASH LINE	MEDIUM
BRACKETS, LEADER LINES, ETC.	THIN
BOUNDARY OF MECHANICAL GROUPING	THIN
FOR EMPHASIS	THICK

USE OF THESE LINE THICK-NESSES OP-TIONAL

FIG. 6 Line conventions for electrical diagrams. Medium and thin lines are used for all lines normally needed on electrical diagrams. Thick lines are used only for emphasis.

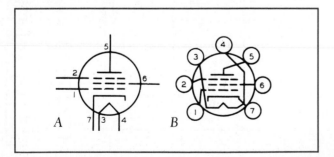

FIG. 7 Terminal identification for electron tube pins. *(a)* Arbitrary numbers are assigned; *(b)* physical location is shown when the tube is viewed from the bottom.

3. DRAFTING PRACTICES

We now will look at a few practices that make it as convenient as possible for a user to read electrical/electronic drawings.

A. Diagram Titles The title of a diagram is given to state the type of diagram and the system or component for which the diagram is made. Examples are: single-line diagram—audio circuit; schematic diagram—vertical oscillator.

B. Drawing Size and Format One should use sizes and formats given in ANSI Y14.1. Try to use smallest size feasible. Use cross-references when multiple sheet drawings are required.

C. Line Conventions and Lettering A line of medium thickness, as shown in Fig. 6, is recommended for general use on electrical diagrams. Thin lines are used for brackets and leaders. A heavy line is used to emphasize a special feature. However, the thickness of a line has no significance in meaning—the line thickness is chosen solely for readability of the diagram. When a drawing for reproduction is to be reduced in size, for example, for a manual, the line weights are made heavier to give good readability.

Lettering on electrical drawings and diagrams should be done in capital letters. For drawings made for direct reproduction (same size), the smallest letters should be not less than $\frac{1}{8}$ in high. Drawings reduced for manuals should be lettered in a size that will be not less than $\frac{1}{16}$ in high when reduced.

4. GRAPHIC SYMBOLS

Graphic symbols on electrical diagrams conform to "Graphical Symbols for Electrical Diagrams," ANSI-Y32.2, or other national-level standards if the symbol is not covered in the American Standard. A selection of symbols, adapted from ANSI-Y32.2, is given in the appendix. For special components, if no suitable symbol exists, an appropriate symbol may be made but must be explained in a note. Symbols that can be reversed or rotated without affecting their meaning may be so drawn to simplify the circuit layout. Symbols may be drawn to any convenient size that suits the diagram and produces good readability.

5. REPRESENTATION OF ELECTRICAL CONTACTS

Switch symbols (see appendix and ANSI-Y32.2) should be shown in the position of no applied operating force. For switches that may be in any of two or more positions with no operating force applied, or for switches actuated by mechanical or electrical means, a drawing note should identify the functional phase shown on the diagram. Relay contacts should be shown in the unenergized position.

6. NOTATION OF ELECTRON TUBE PINS

Tube-pin numbers are shown outside the tube envelope and directly above or at one side of the connecting line, as in Fig. 7A. An alternative method is shown at (B), where the circles correspond to approximate physical locations of the pins when the tube is viewed from the bottom. Starting with the pin adjacent to the tube-base key or similar point of reference, the circles are numbered consecutively in clockwise order.

7. LAYOUT OF DIAGRAMS

The best way to start an electrical diagram is to make a freehand sketch of the proposed arrangement. While doing this, place the symbols in the best position to eliminate long connecting lines and require as few crossovers as possible. In making the drawing use the sketch as a guide. The layout should prominently show the main features. Space the symbols and lines so that there is sufficient area for notes and reference information.

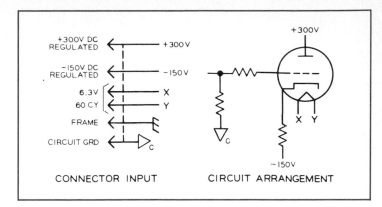

FIG. 8 Identification of interrupted lines. Left side of figure: a group of lines interrupted on the diagram. Right side of figure: single lines interrupted on the diagram.

However, avoid large blank spaces except when an area is needed for circuits or information to be added later. In general, electrical diagrams are drawn to follow the circuit, signal, or transmission path, from input to output, source to load, or in the order of functional sequence.

Connecting lines are drawn horizontally or vertically and with as few bends and crossovers as possible. No more than three lines should be shown connected at one point when an alternative arrangement is possible. Parallel lines are arranged in groups, preferably three to a group, with double space between groups. In grouping parallel lines, lines representing related functions are grouped together. Parallel lines should never be closer than $\frac{1}{16}$ in in final reduced size. See Fig. 4.

Interrupted lines are used as needed in a diagram, but each path must be identified and the destination shown. Letters, numbers, or abbreviations used for identification should be placed as close as possible to the point of interruption. Figure 8 shows two examples of interrupted paths, multiple paths on the left and single paths on the right.

Interrupted grouped lines are employed in a diagram as needed. The destination of the group is shown by bracketing and identifying as in Fig. 9, or by using a dashed line connecting the brackets as in Fig. 10. The dashed line is drawn so that it will not form a continuation of one of the lines of the interrupted group.

Abbreviations used on electrical diagrams should conform to "USA Standard Abbreviations for Use on

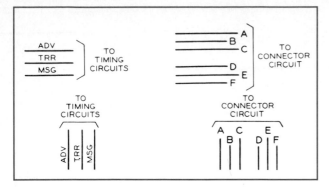

FIG. 9 Typical arrangement of line identifications and destinations.

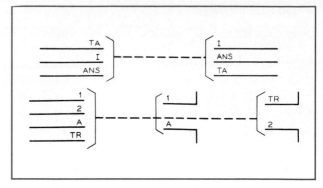

FIG. 10 Typical interrupted lines interconnected by dashed lines. The dashed line shows the interrupted paths that are to be connected. Individual line identifications indicate matching connections.

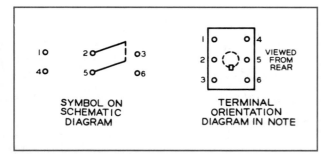

FIG. 11 Terminal identification for a toggle switch.

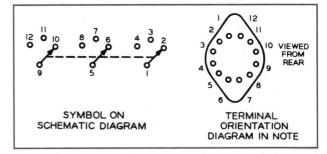

FIG. 12 Terminal identification for a rotary switch.

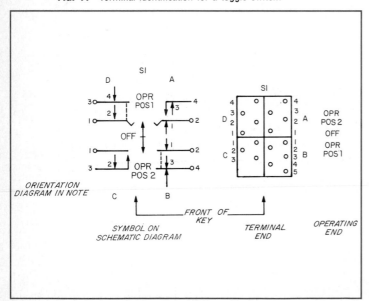

FIG. 13 Terminal identification for a lever switch.

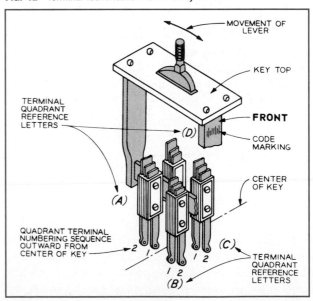

FIG. 14 Pictorial explanatory drawing of an electrical part. This shows the relationship and orientation of elements and gives the identification sequence to be used on a diagram.

Drawings," Z32.13 or other national-level standard if the abbreviation is not covered in the American Standard. If no suitable abbreviation exists, a special abbreviation may be used but must be explained by a note on the diagram.

8. IDENTIFICATION OF TERMINALS

Terminals exist on many parts, such as switches, relays, and transformers. Terminal identification should be added to graphic symbols to show actual markings on or near terminals of the part. We shall look at several examples: toggle, rotary, and lever-type switches and also terminal identification of adjustable resistors.

A. Toggle Switch. See Fig. 11.

B. Rotary Switch. See Fig. 12.

C. Lever Switch. See Fig. 13.

Figures 11, 12, and 13 show terminal identifications on the symbols of schematic diagrams and on orientation diagrams for the actual part. When terminal designations are arbitrarily assigned, the orientation diagram *must* be included in order to relate the designation to actual physical locations on the part. Pictorial drawings of a part, such as Fig. 14, are used when identification on a symbol and a simple terminal orientation note are insufficient. Pictorial drawings are often used in service manuals.

When terminals or leads are identified on the part by color code, letter, number, or geometric symbol, this identification is shown on or near the connecting line adjacent to the symbol. When no markings exist on a part, arbitrary number or letter designations may be assigned.

D. Adjustable Resistors If it is necessary to indicate direction of rotation, it is customary to refer to rotary motion as clockwise or counterclockwise when rotation is viewed from the knob or actuator end of the control. The preferred method of terminal identification is to designate with the letters CW the terminal adjacent to the movable contact when it is in the extreme clockwise position, as shown in Fig. 15A. Numbers may also be used with the resistor symbol in which No. 2 is assigned to the adjustable contact as shown at (B). Additional fixed taps are numbered sequentially, 4, 5, as at (C).

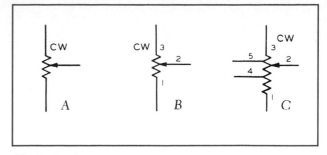

FIG. 15 Terminal identification. This example illustrates identifications on a variable resistor for *(a)* tap (CW) nearest variable contact in extreme clockwise position; *(b)* all taps (no. 2 assigned to variable contact); *(c)* additional fixed taps (4 and 5).

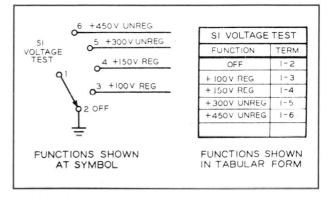

FIG. 16 Position and function identification for a rotary switch. The choice of method—diagram (left) or table (right)—is optional. These methods are recommended for simple switches.

9. SWITCH TERMINALS AND CIRCUIT FUNCTIONS

The relation of switch position to circuit function is shown on a schematic diagram by notation near the terminals as at the left in Fig. 16, or in tabular form as shown at the right. For simple toggle switches, the notations ON and OFF are usually sufficient.

Rotary switches that perform involved functions, as illustrated in Fig. 17, should be explained by using the tabular form shown. In tabular listings, dashes link the terminals that are connected.

10. IDENTIFICATION OF SEPARATED PARTS

It is sometimes necessary on a drawing to physically separate parts having several components. Portions of the

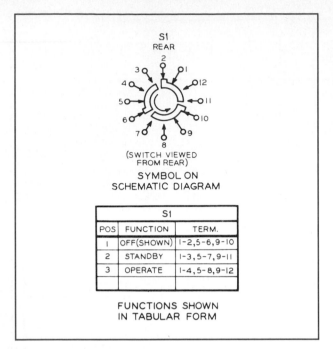

FIG. 17 Position and function identification for a rotary switch. Position and contact identification (only) are shown on the diagram proper. The tabular information then supplements by giving functional information. This method is recommended for complex switches.

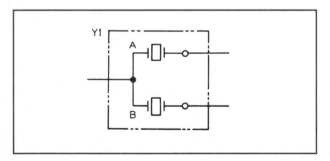

FIG. 18 Identification of parts by suffix letters. A and B are subdivisions of part Y1. Identifications are Y1A and Y1B.

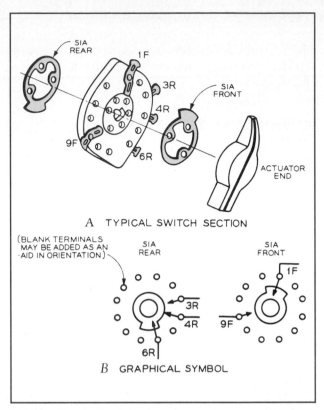

FIG. 19 Typical development of a graphic symbol (complex rotary switch). Portions of the part are oriented relative to each other for identification on the graphic symbols.

graphic symbol may be located at different locations on the schematic diagram.

Separated portions of parts can be identified by several means. We will look at a few examples. Consult USAS Y14.15—1973 for additional information.

A. Identification by Suffix Letter Subdivisions of parts are indicated by adding a suffix letter to the reference designation. For example, C1A and C1B will designate electrically separate sections of a dual capacitor C1.

Suffix letters are also used to identify subdivisions of a complete part when the individual parts are shown enclosed or associated in a unit, as shown in Fig. 18.

B. Use of Suffix Letters on Rotary Switches When parts or rotary switches are designated S1A, S1B, S1C, etc., the suffix letters start with A at the knob or actuator end and are assigned sequentially away from this position. Each section of the switch is shown viewed from the same end, as in Fig. 19A. When both sides of a rotary switch are used for separate functions, the front (actuator end) and rear symbols should be differentiated by modifying the reference designation, for example, S1A FRONT and S1A REAR as at (B).

C. Identification by the Term "Part of" When parts of connectors, terminal boards, or rotary-switch sections are functionally separated on the diagram, the words "part of" are used to precede the reference designation as in Fig. 20.

D. Identification by Individual Terminals When the separation of parts of connectors or terminal boards on the same drawing becomes extensive, the separated parts may be identified as individual terminals, as shown in Fig. 21.

11. REFERENCE DESIGNATIONS

These are combinations of letters and numbers that identify parts shown on single-line and schematic dia-grams and also on related drawings such as the assembly drawing and connection diagram.

The number portion of the designation follows the letter or letters without a space or hyphen and is of the same size as the letter(s), for example, S4, MG5, T6A, etc.

The assignment of numbers should start with the low-est numbers in the upper left-hand corner of the diagram and proceed consecutively from left to right throughout the drawing. When parts are eliminated as a result of a drawing revision, the remaining parts should not be re-numbered. As an aid in accounting for all reference des-ignations, a table, as in Fig. 22, may be included show-ing the highest reference designations and the reference designations not used.

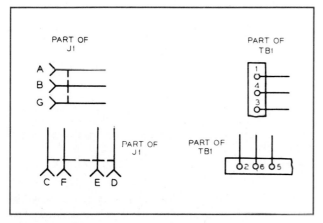

FIG. 20 Identification of parts by "part of" prefix. All "part of" portions (separated on the diagram) form a single unit on the actual connector or terminal board.

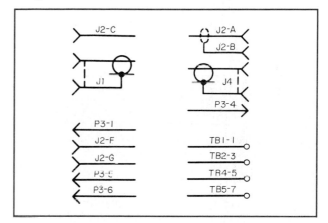

FIG. 21 Identification of parts as individual terminals.

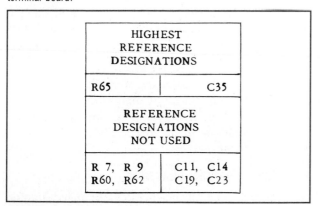

FIG. 22 Typical table indicating highest and omitted numerical reference designations.

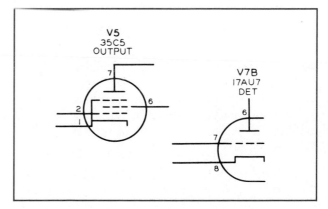

FIG. 23 Reference designations for electron tubes. The reference desig-nation is given on the top, the tube type number next, and then function.

		Symbol	
		Method	**Method**
Multiplier	**Prefix**	**1**	**2**
10^{12}	tera	T	T
10^9	giga	G	G
10^6 (1,000,000)	mega	M	M
10^3 (1000)	kilo	k	K
10^{-3} (.001)	milli	m	MILLI
10^{-6} (.000001)	micro	μ	U
10^{-9}	nano	n	N
10^{-12}	pico	p	P
10^{-15}	femto	f	F
10^{-18}	atto	a	A

FIG. 24 Recommended multipliers for basic units. *(USAS Y14.15—1973.)*

Range, ohms	**Express as**	**Example**
Less than 1000	ohms	.031 470
1000 to 99,999	ohms or kilohms	1800 15,853 10k 82k
100,000 to 999,999	kilohms or megohms	220k .22M
1,000,000 or more	megohms	3.3M

FIG. 25 Recommended manner to express resistance. *(USAS Y14.15—1973.)*

Range, picofarads	**Express as**	**Example**
Less than 10,000	picofarads	152.4pF 4700pF
10,000 or more	microfarads	.015μF 30μF

FIG. 26 Recommended manner to express capacitance. *(USAS Y14.15—1973.)*

Electron tubes are identified by tube type number in addition to the reference designation. The circuit function may also be given below the tube type number, as in Fig. 23.

12. NUMERICAL VALUES FOR RESISTANCE, CAPACITANCE, INDUCTANCE

It is good practice to express values of resistance, capacitance, and inductance using the fewest zeros. Figure 24 indicates how one may use the basic unit in combination with the multipliers. It should be noted that four-digit numbers should be shown without commas, such as 5200, not 5,200.

A. Resistance Values Values of resistance should be given as in Fig. 25.

B. Capacitance Values Values of capacitance should be given as in Fig. 26.

C. Inductance Values Values of inductance should be expressed in henries, millihenries, or microhenries. As examples, 5 H is proper, not .005 mH; also 3 mH should be given, not .003 H or 3000 μH.

D. Value Placement Numerical values of components are located as near as practical to the symbol. Preferred arrangements are shown in Fig. 27.

E. Notes It is quite proper to give a general note on a drawing to avoid repeating abbreviations of units of measurement. For example, one could say "Unless otherwise specified, resistance values are in ohms. Capacitance values are in microfarads."

13. SPECIAL DRAFTING PRACTICES

As in general drafting practices, there are techniques available to speed up drafting. These include electrical/electronic templates to aid in the quick drawing of symbols, photo drawings, and special uses of reproduction. Gummed overlays and press-ons can be used, as in Fig. 28. Many standard circuits, portions of circuits, and symbols are commercially available.

When the accuracy of an instrument drawing is not necessary, drafting time can be saved by making electrical diagrams freehand. Coordinate paper with divisions printed on the back promotes economy. The coordinate lines aid in keeping lines straight—particularly the long parallel connecting lines. The symbol sizes are not important but should be kept to proportions that look well and read easily on the diagram.

14. MODELS

When models of plants are used, the conduit is included with other details in a full representation, to scale, of all conductors. These are added to the model after machines, process piping, plumbing, special equipment, etc., are in place. The tagging of power and control lines on models is sometimes supplemented or replaced by a conduit schedule.

To speed up connection and tagging of wiring, or for service use, photo drawings are sometimes prepared. A photograph of the actual equipment is made, and the connections, instructions, or identifications are added to the screened positive.

15. DRAWINGS OF PRINTED CIRCUITS

Printed circuits can be obtained from very accurately drawn layouts. The method employs efficient laminated wiring on a wide variety of permanently bonded insulating plastics, from low-cost vinyls and polyethylenes to fluorocarbons. Printed circuits allow miniaturization and the elimination of circuit errors—advantages that

cannot be obtained by other methods. Once a pattern or suitable design is established, preparation of a black and white drawing can start. Scales for reduction, for example, 4 to 1, 3 to 1, or 2 to 1, are used. To ensure sufficient bonding area of the metal laminate during soldering operations, lines should not be less than $\frac{1}{32}$ inch in width when reduced. Line separation should never be closer than $\frac{1}{32}$ inch on the final circuit. Figure 29 illustrates the drawing of printed circuits. Computer-aided techniques are also available.

To produce the circuit, a sheet of copper foil is bonded to a plastic board. The copper foil is processed by printing photographically from the layout and then etching to leave the circuits as originally drawn. Holes are punched (or drilled) in the board (Fig. 30) and various components are inserted through them. The leads of the various components are cut and bent over the copper-foil wiring. The wiring side of the board is then dipped in molten solder to make all solder connections at once, and all copper-foil wiring becomes covered, thus increasing its ability to carry current. A coat of silicone resin varnish is applied to the wiring side of the board; this prevents short circuits caused by dust or moisture. The method is known as the etched-wiring method, and results in uniformity of wiring, compactness, and free-

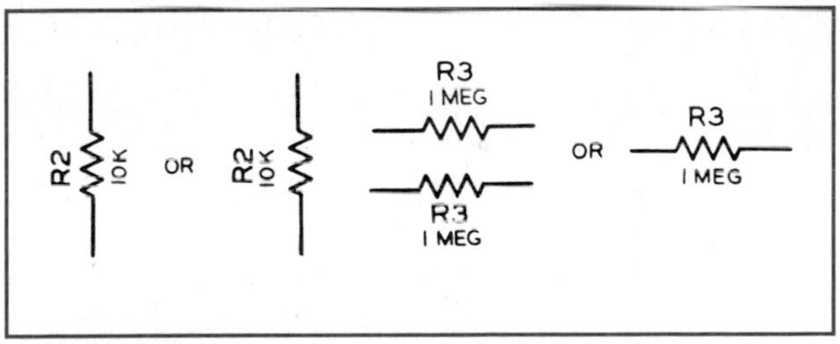

FIG. 27 Recommended placement of numerical values.

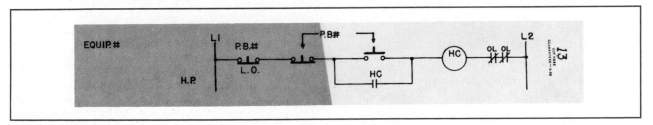

FIG. 28 A gummed overlay. The overlays are stripped from the backing paper and applied to the drawing. Many standard portions of circuits are available. *(Courtesy of Stanpat Co.)*

FIG. 29 Drawing a printed circuit. The drawing is carefully planned and executed for photographic reproduction on the circuit board.

FIG. 30 Punching holes in a circuit board. The holes are to receive component wires.

dom from wiring errors. It is used more widely than other methods, such as embossed wiring, stamped wiring, and pressed powdered wiring, because of its reliability, great flexibility, and low setup cost. Figure 31 is a photographic reproduction of a printed circuit board; the caption explains several details of printed circuitry.

16. LAYOUT OF INTEGRATED CIRCUITS

Integrated circuits were developed in the early 1970s when it became obvious that the transistors of that time could not be effectively miniaturized further. Transistors became smaller but the difficulty of interconnecting them remained. Therefore, the next step was to design complete logic units on one piece of silicon.

A number of components with all connections were assembled together on a single silicon chip of semiconductor material. The result was the *integrated circuit,* or IC. With time, producers of ICs compacted more and more circuitry onto a single chip, developing what are known as *very large scale integrated circuits,* or *microprocessors.* Microprocessors are an essential component of computers. More circuitry in less space has made possible increasingly more compact computers and other electronic equipment. Figure 32 gives an example of a very basic integrated circuit.

Design and drawing layout time are major variables in developing integrated circuits, due to their complexity. The traditional role of the drafter has been much modified to include considerable help from computers.

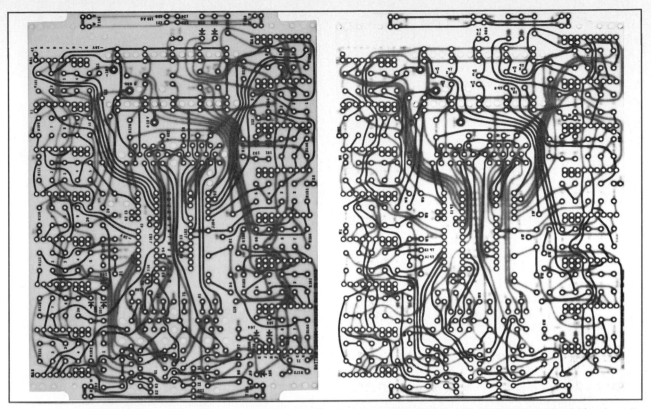

FIG. 31 A "double" printed circuit board. This type of circuit board has the circuit integrated and printed on both sides. The left illustration shows the resistor side of the board; the right illustration shows the side used for other components. The opposite side can be seen in both illustrations—the fainter, diffused lines. By printing both sides of the board, circuit crossovers are easily handled. Note that this complicated circuit can be produced in quantity without circuit errors.

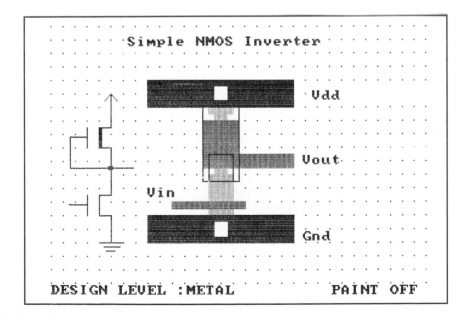

FIG. 32 An example of a five-level integrated circuit. The shading shows different levels for this N-channel metal oxide semiconductor (NMOS). The drawing was done on a dot matrix printer. (*Courtesy Electrical Engineering Department, The Pennsylvania State University.*)

For detailed information on layout of ICs, one must refer to specialized articles and texts available in a technical library. However, we will touch on three general methods for laying out IC drawings.

A. Manual Method Despite the name "manual method," the designer/drafter using this method actually has the aid of a digitizer and a computer graphics system. The designer sketches the locations where functionally related components are desired. Then, by the use of a digitizer, the locations and components are read into a computer, which drives a plotter, giving an accurate drawing. Corrections can be made efficiently by using the interactive computer-graphics terminal.

This method allows for the maximum use of a person, but the person is supported by the precision and ease of computer-plotted drawings that are modified interactively as needed. Industry has used this method widely in recent years.

B. Semiautomatic Method In this method software is developed to allow computer placement of standard parts and the subsequent routing of interconnections between the parts. Design time is several times faster than for the so-called manual method. Some manual effort is still required to refine the final stages of the design.

The silicon area required on the chip may be somewhat greater than when using the manual method. This may be seen as a disadvantage. However, the semiautomatic method has the advantage of using computer speed and accuracy with flexibility of design being retained by the designer. The interactive mode of the method allows the designer to repeat, correct, or modify any step of the layout. At the same time, the designer is not burdened with repetitive aspects of the layout, such as manually digitizing a large number of identical capacitors.

C. Automatic Method This method uses the newest technology in which computer algorithms, or subroutines, are automatically used after being programmed into the computer. The designer simply inputs to the computer the particular components needed and their basic geometry relative to the desired layout. The program software then dictates all geometric constraints and layout rules.

The computer operates in a trial-and-error procedure to find the best "fit" of components. This process can use a large amount of computer time, but no human time. The designer then may either accept or reject the layout plotted by the computer, but step-by-step interaction while the computer "thinks" is not a usual option.

If the layout needs changes, the designer may give a change of input variables and allow the computer to redesign the circuit. The eventual IC designed should perform as expected because all design rules and constraints have been programmed into the computer. Checking time is minimal.

D. Summary One can see that traditional drafting methods using manual instruments, templates, gum labels, etc., are insufficient within the technology of printed circuit layouts. Nevertheless, the requirements of recognizing and using graphic symbols and the rules of geometry still exist. A drafter can be trained to operate computer-driven design systems if he or she has a sound understanding of basic fundamentals of electronic drafting.

PROBLEMS

1. The drawing shows the arrangement of a circuit devised by an electrical engineer for testing the diode characteristics of a type-80 electronic tube. Draw the schematic diagram of this circuit. Include a voltmeter in parallel for measuring the voltage across the tube. It should be noted that the oscilloscope is a testing device that will trace a curve proportional to the voltages applied to the vertical and horizontal axes and that the electrical circuit will operate without this piece of equipment.

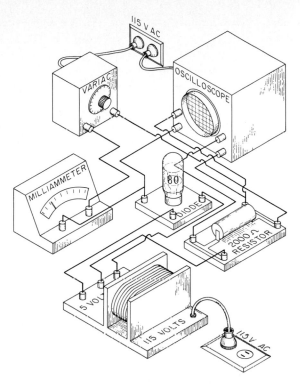

PROB. 1 A pictorial schematic diagram.

2. The drawing is a block diagram of the circuit for a simple regenerative radio. Draw the schematic diagram of this circuit on an 8½-by 11-in sheet of drawing paper.

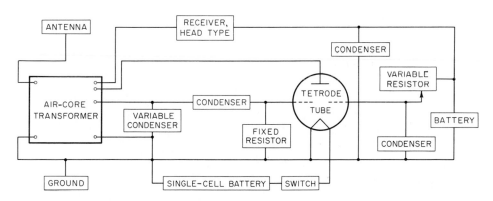

PROB. 2 Block diagram of a radio circuit.

3. Make a schematic drawing with proper symbols for the circuit shown. This is a simple circuit that could be built with ease.

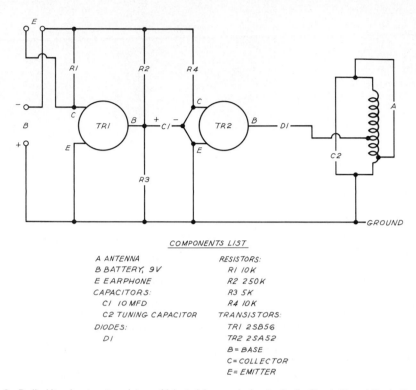

COMPONENTS LIST

A ANTENNA	RESISTORS:
B BATTERY, 9V	R1 10K
E EARPHONE	R2 250K
CAPACITORS:	R3 5K
C1 10 MFD	R4 10K
C2 TUNING CAPACITOR	TRANSISTORS:
DIODES:	TR1 2SB56
D1	TR2 2SA52
	B = BASE
	C = COLLECTOR
	E = EMITTER

PROB. 3 Radio kit using two transistors. (*Adapted from a design by Radio Shack Div. of Tandy Corp.*)

4. The block diagram of a simple electrical servo system is shown. In this system the voltage across the load is maintained almost constant when the load or the supply voltage is changed. For instance, if R_L or E would decrease, the voltage drop across R_1 would decrease. This decrease would in turn make the grid more positive and the current i_L would be increased back to (or near) the original value. Draw a schematic diagram of this circuit, making sure that the polarity is correct.

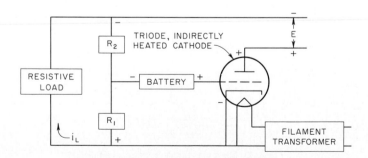

PROB. 4 Block diagram of a servo system.

5. The upper graph describes the flow of liquid raw materials necessary for a certain chemical process. These materials are forced to flow through a single solenoid-actuated automatic valve. In order to obtain the correct actuating mechanism, it is necessary to obtain the maximum dynamic requirements of the valve, that is, the maximum change in valve opening per unit of time. A manufacturer of automatic valves has supplied a curve of valve-flow rate capacity versus valve opening for the valve size required (lower graph). Draw the curve of valve opening versus time, and find the point on the curve where maximum change in valve opening occurs (the point where the slope of the curve is a maximum).

5A. Calculate the maximum change in valve opening per unit of time for Prob. 5. (Draw the slope of the curve and obtain $\dfrac{dy}{dx}\text{(max)} = \dfrac{\text{in}}{\text{sec}}$.) Show graphically how the solution was obtained.

5B. Draw a block diagram showing the combined mechanical and electrical system for Prob. 5. A pressure-operated metering switch actuates a relay that in turn actuates the valve solenoid.

5C. Draw a wiring diagram for the electrical system of Prob. 5B.

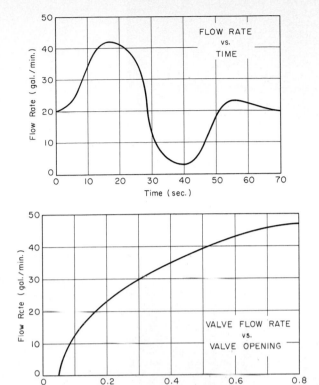

PROB. 5 Data for determination of value characteristics.

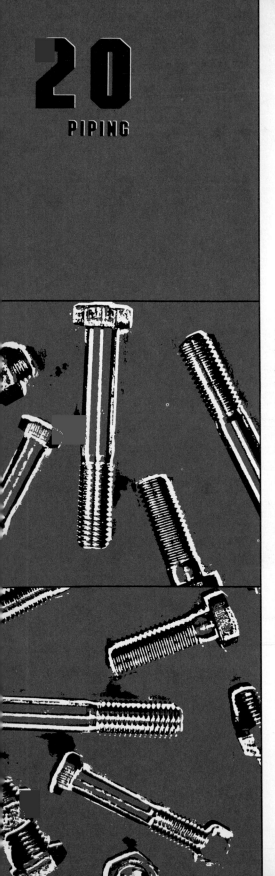

20
PIPING

1. FUNDAMENTAL CONSIDERATIONS

Piping systems are vital to modern society. There is scarcely a structure or industrial plant that does not contain piping systems. Some systems are quite complex, as in a petroleum distillation plant. Others may be more modest, say, for the water supply to a hotel. Some may be relatively simple, such as a system in a single home.

All systems share certain common elements. All would have the elements of the pipe or tubing itself, plus assorted valves and controls to maintain the proper flow and control of the liquids or gases involved. Fittings are used to join pipes together. Fittings are often screwed together, but they may be welded or attached via flanges and bolts.

The difference between piping and tubing is not particularly distinct. However, pipe dimensions for sizes at or below 12 inches contain actual outside dimensions greater than the nominal size. Tubes have outside diameters that are always identical to the nominal size. Also, tubing is more commonly bent to provide desired angles than is piping.

It is our purpose in this chapter to give a short, concise overview of the various pipes and tubings available, how they are attached to one another, and how one would represent graphically these components. Excellent references for drawing standards are available in the ANSI standards, and they will be referred to during the discussion.

2. TYPES OF PIPE

We will review three types of pipe from the viewpoint of materials: steel, other metals, and plastic.

A. Steel Pipe The outside diameter (OD) is set as the standard for pipe up through 12 inches. The hope of the early standards was that the resulting inside diameter would be close to the nominal size. Pipe of 14-inch OD and over has a nominal size the same as the OD.

Wall thickness has for years been designated as Standard (STD), Extra-Strong (XS), and Double Extra-Strong (XXS). Further designations known as schedule numbers were added as a convenience in ordering pipe. Theoretically, schedule numbers were to provide a system for pipe/wall combinations to yield numbers equal to $1000 \times P/S$, where P is pressure and S is the allowable stress. The schedule numbers turned out to be far off the theoretical ideal, but they were kept for the sake of convenience and are still used today.

A complete designation of steel pipe is usually given by nominal size and wall thickness. Wall thickness is selected by the designer so the pipe may properly resist the expected internal pressure. The designer may use a code such as ANSI B31, "Code for Pressure Piping." The selected wall thickness is matched then to the best fit in ANSI B36.10, an excellent standard for steel pipe.

For example, if a 0.68 wall thickness and an 8-inch nominal size were needed, an existing 8-inch nominal size, 0.719 wall thickness could be chosen from ANSI B36.10 as the closest safe thickness. The OD is 8.625; the Schedule Number happens to be 120. Alternatively, ANSI B36.10 permits a metric sizing, which in this example is 219.1 mm OD and 18.26 mm wall thickness. ANSI B36.10 covers sizes from ⅛ inch to 80 inches. Schedule Numbers range from 10 (thinnest wall) to 160 (thickest wall).

We have discussed briefly steel pipe and a useful corresponding standard ANSI B36.10. There exist at least 19 other standards of ANSI for special types of steel pipe, such as for stainless steel. One can easily see that piping design can be a complex, specialized aspect of engineering.

B. Cast-Iron Pipe We will not go into detail on this type of pipe. However, it is widely used in water and sewage systems, and has been since the 1800s. It corrodes only slowly and can be an excellent choice for low-pressure applications. One must be aware of its brittle nature relative to steel and not subject it to heavy loads.

C. Other Metal Pipe Brass, copper, stainless steel, and aluminum pipe have the same nominal diameters as iron pipe but some have thinner wall sections. There are two standard weights: regular and extra-strong. Commercial lengths are 12 ft, with longer lengths made to order.

Lead pipe and lead-lined pipe are used in chemical work. Cast-iron pipe is used for water and gas in underground mains and for drains in buildings.

Many other kinds of pipe are in more or less general use and are known by trade names, such as hydraulic pipe, merchant casing, API (American Petroleum Institute) pipe, etc. Details are given in manufacturers' catalogues.

Most small-line plumbing in homes, buildings, and industries, for hot- and cold-water installations, employ copper pipe with soldered-joint fittings, though plastic pipe is becoming more common.

D. Plastic Pipe Since plastic pipe does not corrode and has high resistance to a broad group of industrial chemicals, it is used extensively in place of metal pipe. Polyvinyl chloride, polyethylene, and styrene are basic plastic materials. Of these, polyvinyl chloride (PVC) is the most widely used. It does not support combustion, is nonmagnetic and nonsparking, imparts no odor or taste to contents, is light (½ the weight of aluminum), has low flow resistance, resists weathering, and is easily bent and fabricated by solvent cementing or, in heavy weights, by threading. Its chief limitations are higher cost (partially offset by lower installation cost), low temperature limit (150°F in continuous service), and low pressure limits. Also, it does not resist all solvents; and it requires more supports, and contracts and expands more (about five times) than steel.

Metal pipe lined with plastic has the advantage of combining the strength of metal with the chemical resistance of plastic. Seran rubber is also used for lining metal pipe.

3. TUBING

Seamless flexible metal tubing is used for conveying steam, gases, and liquids in all types of equipment such as locomotives, diesel engines, hydraulic presses, etc., where vibration is present, where outlets are not in alignment, and where there are moving parts.

Copper tubing is available in nominal diameters of ⅛ to 12 in and in four weights, known as classes K, L, M, and O. Class K is extra-heavy hard; class L is heavy hard; class M is standard hard; and class O is light hard. Boiler tubes in all sizes are designated by their outside diameters.

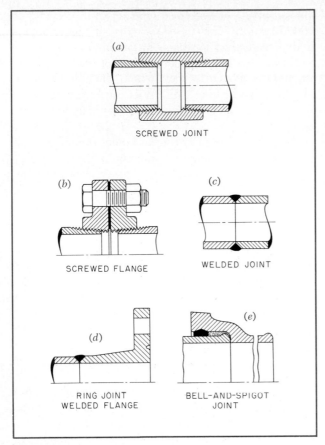

FIG. 1 Pipe joints. These are used for metal pipe.

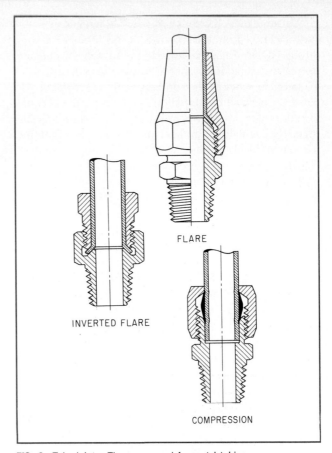

FIG. 2 Tube joints. These are used for metal tubing.

Tubing is made in a variety of materials—glass, steel, aluminum, copper, brass, aluminum bronze, asbestos, fiber, lead, and others.

4. PIPE JOINTS

Pipe is connected by methods dependent upon the material and the demands of service. Steel, brass, or bronze pipe is normally threaded and screwed into a coupling (or a fitting), as shown in Fig. 1 at (a). At (b) a screwed flange is illustrated; this joint is easily disassembled for cleaning or repair. At (c) a permanent welded joint is shown. When welded pipe must be disassembled periodically, ring joints (d) are used, where necessary, in the system; these joints are bolted together. Cast-iron pipe cannot be successfully welded or threaded; thus, a bell-and-spigot joint, caulked and leaded, as at (e) is used.

5. TUBE JOINTS

Tubing is commonly employed to connect small components for liquid or gas service. Figure 2 illustrates three common methods of connection. Both the *flare* and the *inverted flare* can be disassembled without serious injury to the joint, and can be used for fairly high pressures. The compression joint is used for lower pressures and when the joint is not expected to be taken apart and reassembled.

6. SPECIFICATION OF FITTINGS

Pipe fittings are used to connect pipe systems. Fittings are specified by giving the nominal pipe size, material, and name; for example, 1½″ Brass Tee. When a fitting connects more than one size of pipe, the size of the largest run opening is given first, followed by the size at the opposite end of the run. Figure 3 shows the order of specifying reducing fittings. The word "male" must follow the size of the opening if an external thread is wanted, for example,

$$2 \times 1 \text{ (male)} \times \tfrac{3}{4} \text{ M.I. Tee}$$

Where M.I. stands for malleable iron.

7. PIPE FITTINGS

Three common methods of joining pipes are by screwed, welded, and flanged joints. Screwed joints use taper pipe thread, which gives a tightly sealed joint. This method is very popular and will be discussed more below. Welded joints are formed by welding around the circumference of the pipe. This is a permanent method and one that has received publicity in recent years during construction of large-diameter systems, such as the Alaskan pipeline. Flanged joints use welded rather than screwed-on flanges. The flanges are then bolted together, making a connection which is excellent for high pressure, but

which can be disconnected when needed.

Screwed fittings of cast-iron or mallable iron are used often with steel pipe (Fig. 4). Brass and other alloys are employed for special purposes. Steel butt-welding fittings (Fig. 5) are used with steel pipe. Soldered-joint fittings

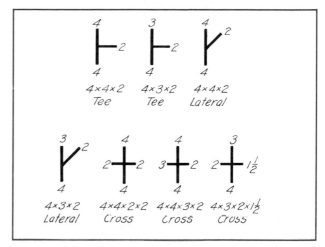

FIG. 3 Order of specifying the openings of reducing fittings.

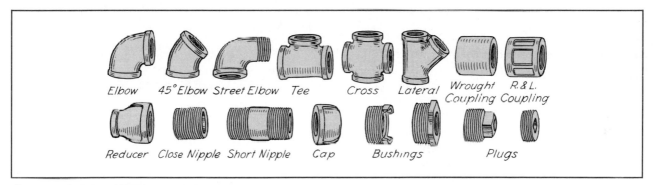

FIG. 4 Screwed fittings. See the appendix for detailed dimensions of malleable and cast-iron fittings.

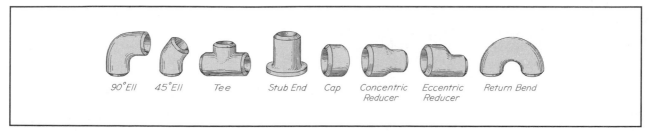

FIG. 5 Butt-welding fittings. See the appendix for detailed dimensions.

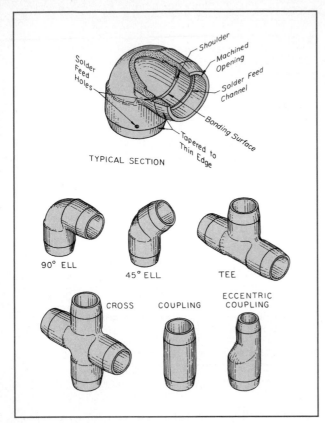

FIG. 6 Soldered-joint fittings. These are used principally for small-line plumbing. See appendix for detailed sizes.

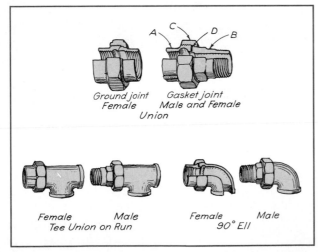

FIG. 7 Screwed unions and union fittings. These are used to "close" a pipe system.

(Fig. 6) are used with copper pipe. Cast-iron bell-and-spigot fittings are used with cast-iron pipe.

Elbows are used to change the direction of a pipeline either 90° or 45°. The *street elbow* has male threads on one end, thus eliminating one pipe joint if used at a fitting. *Tees* connect three pipes, and *crosses* connect four. *Laterals* are made with the third opening at 45° or 60° to the straight run.

Straight sections of pipe are made in 12- to 20-ft lengths and are connected by *couplings*. These are short cylinders, threaded on the inside. A right-hand coupling has right-hand threads at both ends. To close a system of piping, although a union is preferable, a *right-and-left coupling* is sometimes used.

A *reducer* is similar to a coupling but has the two ends threaded for different-sized pipe. Pipes are also connected by screwing them into cast-iron flanges and bolting the flanges together. Unless the pressures are very low, flanged fittings are recommended for all systems requiring pipe over 4 in in diameter.

Nipples are short pieces of pipe threaded on both ends. If the threaded portions meet, the fitting is a *close nipple*; if there is a short unthreaded portion, it is a *short nipple*. Long and extra-long nipples range in length up to 24 in.

A *cap* is used to close the end of a pipe. A *plug* is used to close an opening in a fitting. A *bushing* is used to reduce the size of an opening. *Unions* are used to close systems and to connect pipes that are to be disassembled occasionally. A screwed union (Fig. 7) is composed of three pieces, two of which, *A* and *B*, are screwed firmly on the ends of the pipes to be connected. The third piece *C* draws them together, the gasket *D* forming a tight joint. Unions are also made with ground joints or with special metallic joints instead of gaskets. Several forms of screwed unions and union fittings are shown in Fig. 7. Flange unions in a variety of forms are used for large sizes of pipe.

8. PIPE THREADS

When screwed fittings are used or when a connection must be made to a tapped hole, pipe is threaded on the ends for the purpose. ANSI provides two types of pipe thread: tapered and straight. The normal type employs a taper internal and taper external thread. This thread

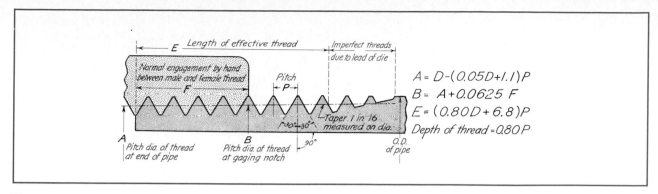

FIG. 8 American Standard taper pipe thread. "Wedging" action of the taper produces a tight seal.

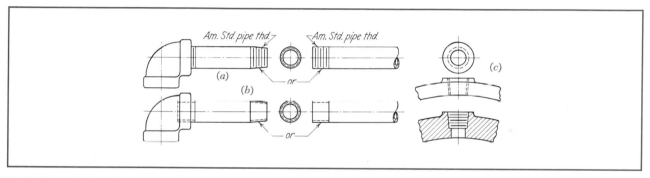

FIG. 9 Conventional methods of drawing pipe threads. *(a)* Regular method; *(b)* simplfied method; *(c)* hole tapped for pipe.

(originated in 1882 as the Briggs Standard) is illustrated in Fig. 8. The threads are cut on a taper of $\frac{1}{16}$ in per inch, measured on the diameter, thus fixing the distance a pipe enters a fitting and ensuring a tight joint. Taper threads are recommended by ANSI for all uses, with the exception of the following three types of joints: Type 1, free-fitting mechanical joints for fixtures; Type 2, loose-fitting mechanical joints with locknuts; Type 3, loose-fitting mechanical joints for hose couplings. For these joints, straight pipe threads can be used.

The number of threads per inch is the same in taper and straight pipe threads. Actual diameters vary for the different types of joints. When needed they can be obtained from ANSI standards. A common practice is to use a taper external thread with a straight internal thread, on the assumption that the materials are sufficiently ductile to allow the threads to adjust themselves to the taper thread. All pipe threads are assumed to be tapered unless otherwise specified.

Pipe threads are represented by the same conventional symbols as bolt threads. The taper is so slight that it shows only if exaggerated. It need not be indicated unless attention is being called to it, as in Fig. 9. In plan view *(c)*, the dashed circle should be the actual outside diameter of the pipe specified. The length of effective thread is $E = (0.80D + 6.8)P$ (Fig. 8).

Pipe threads are specified by giving the nominal pipe diameter, the number of threads per inch, and the standard letter symbol to denote the type of thread. The following ANSI symbols are used:

NPT = taper pipe thread
NPTF = taper pipe thread (dryseal)
NPSC = straight pipe thread in couplings
NPSI = intermediate internal straight pipe thread (dryseal)
NPSF = internal straight pipe thread (dryseal)
NPSM = straight pipe thread for mechanical joints

NPSL = straight pipe thread for locknuts and locknut pipe threads

NPSH = straight pipe thread for hose couplings and nipples

NPTR = taper pipe thread for railing fittings

Examples:

½-14NPT   2½-8NPTR

The specification for a tapped (pipe-thread) hole must include the tap drill size, for example:

⁵⁹⁄₆₄ Drill, ¾-14NPT

Dimensions of ANSI taper pipe threads (NPT) are given in the appendix. Dimensions of other pipe threads are given in ANSI B2.1 and in manufacturers' catalogs.

9. VALVES

Figure 10 shows a few types of valves used in piping. (*a*) is a gate valve, used for water and other liquids, as it allows a straight flow. (*b*) is a plug valve, opened and closed with a quarter turn; (*c*), a ball-check valve; and (*e*), a swing-check valve permitting flow in one direction. For heavy liquids the ball-check valve is preferred. (*d*) is a globe valve, used for throttling steam or fluids; (*f*) is a butterfly valve, opened and closed with a quarter turn, but not steam-tight, and used only as a check or damper.

Valves are specified by giving the nominal size, material, and type, for example:

1″ Iron Body Brass-mounted Globe Valve. (If a particular valve is required it is best to give, in addition, "Manufacturer's No.__ or equal.")

10. PIPING DRAWINGS

Two general systems are used: (1) *scale layout* and (2) *diagrammatic*. Scale layouts are used principally for large pipe, as in boiler and power-plant work where lengths are critical and especially when the pipe is not cut and fitted in the field. Also, smaller pipe can be thus detailed when the parts are cut and threaded and then shipped to the job. Figure 11 is an example of a scale layout. The

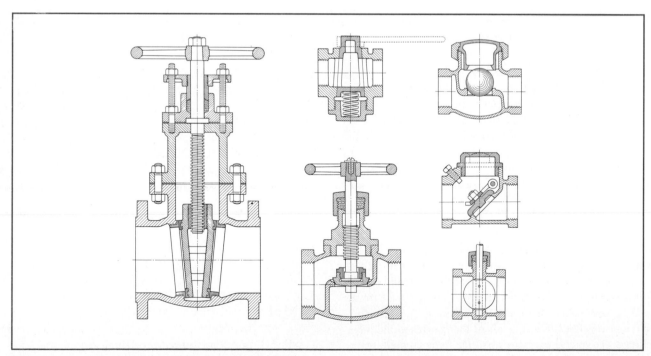

FIG. 10 Valves. The sectional drawings show their construction.

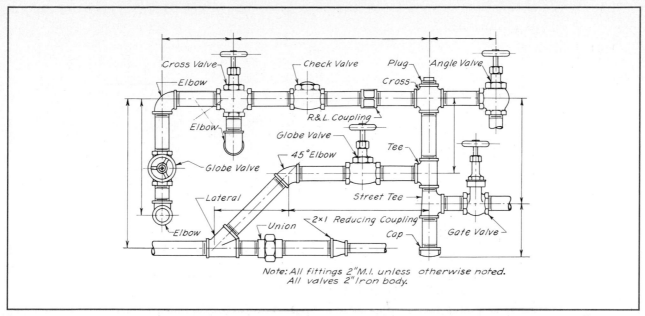

FIG. 11 Scale layout of piping. Pipe and fittings are drawn from dimensional specifications (see appendix).

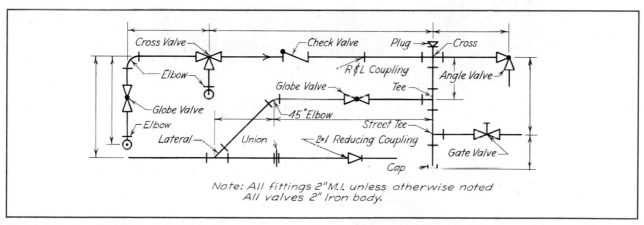

FIG. 12 Diagrammatic drawing of piping layout seen in Fig. 11. Fittings are indicated by standard symbols (see appendix).

fittings can be specified on the drawing, as in Fig. 11, or on a bill of material.

On small-scale drawings such as architectural plans, plant layouts, etc., or on sketches, the diagrammatic system is used. According to this system the fittings are shown by symbols (see appendix) and the runs of pipe are shown by a single line, regardless of the pipe diameters, as shown in Fig. 12. When lines carry different liquids

or different states of a liquid, they are identified by coded line symbols. The standard code for hot water, steam, cold water, etc., is given in the appendix. The single line should be made heavier than the other lines of the drawing.

The arrangement of views is generally in orthographic projection (Fig. 13A). Sometimes, however, it is clearer to rotate all the piping into one plane and make only one

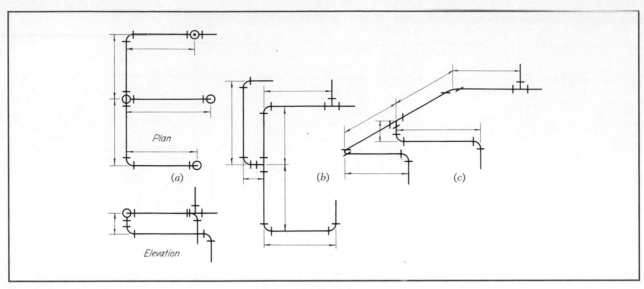

FIG. 13 Diagrammatic methods: *(a)* Orthographic; *(b)* developed; *(c)* pictorial.

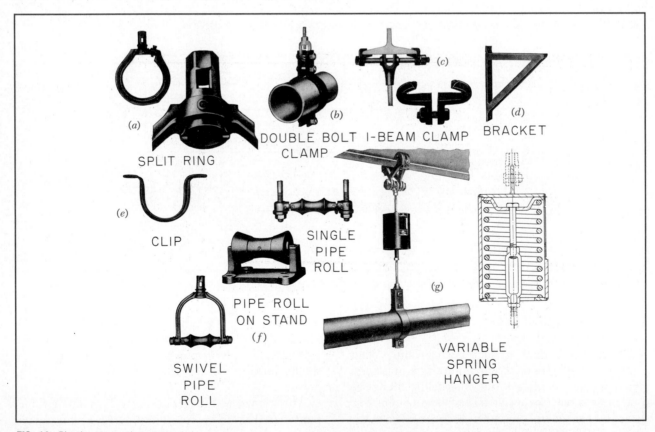

(a) SPLIT RING

(b) DOUBLE BOLT CLAMP

(c) I-BEAM CLAMP

(d) BRACKET

(e) CLIP

SINGLE PIPE ROLL

PIPE ROLL ON STAND

(f)

SWIVEL PIPE ROLL

(g)

VARIABLE SPRING HANGER

FIG. 14 Pipe hangers and supports.

"developed view" as at (B). Isometric and oblique diagrams, used alone or in conjunction with orthographic or developed drawings, are often employed in representing piping. The representation at (C) is drawn in oblique.

11. DIMENSIONING AND PIPE DRAWING

The dimensions on a piping drawing are principally *location* dimensions, all of which are made to centerlines, both in single-line diagrams and in scale layouts, as shown in Figs. 11 and 12. Valves and fittings are located by measurements to their centers, and the allowances for makeup are left to the pipe fitter. In designing a piping layout, take care to locate valves so that they are easily accessible and have ample clearance at the handwheels. The *sizes* of pipe should be specified by notes telling the nominal diameters, never by dimension lines on the drawing of the pipes. The fittings are specified by note, as described in Sec. 6. Very complete notes are an essential part of all piping drawings and sketches.

When it is necessary to dimension the actual length of a piece of pipe, the distance can be calculated by using the overall fitting dimensions and accounting for the entrance length of the pipe threads.

Dimensions for standard pipe and for various fittings are given in the appendix.

12. PIPE HANGERS AND SUPPORTS

Small pipe and light tubing in short runs can be supported by connections to various machines or fittings. Straps of various types are used to tie pipe to posts, columns, walls, ceilings, etc. Pipe hangers and supports are available for almost any size and type of installation. In accordance with ANSI B31.1, "Code for Pressure Piping," all piping systems require sway braces, guides, and supports. In Figure 14 a few commonly used hangers and supports are shown. A split ring (a) is used with a threaded rod attached to the building proper. The locking device prevents change of adjustment due to vibration and assures proper pitch of the line.

A double bolt clamp, shown at (b), is for service where it is desirable to have the yoke extend outside of pipe covering. The I-beam clamp, two styles shown at (c), is suitable for clamping on flanges ranging in width from 2 to 6½ inches. The steel bracket (d) can be bolted to a wall and pipes can be supported on, as well as hung from, the horizontal member. The strap or clip (e) can be used for small pipe; this type is used when pipe is to be flush with a ceiling or wall. Pipe rolls (f) are designed to support piping so that longitudinal movement resulting from expansion and contraction can take place. Three types are shown for varying support conditions. The variable spring hanger (g) can be obtained in several different sizes and arrangements.

PROBLEMS

GROUP 1. PIPE FITTINGS

1. In the upper left-hand corner of one sheet draw a 2-in tee (full size). Plug one outlet. In the second outlet place a 2- by 1½-in bushing; in the remaining outlet use a 2-in close nipple and on it screw a 2- by 1½-in reducing coupling. Lay out the remainder of the sheet so as to include the following 1½-in fittings; coupling, globe valve, R&L coupling, angle valve, 45° ell, 90° ell, 45° Y; cross, cap, three-part union, flange union.

Add extra pipe, nipples, and fittings so that the system will close at the reducing fitting first drawn.

2. Make a one-view drawing of a 1½-in globe valve. Use an 8½- by 11-in sheet. See appendix for proportions.

3. Same as Prob. 2 for a 1½-in angle globe valve.

4. Same as Prob. 2 for a 1¼-in gate valve.

GROUP 2. PIPING LAYOUTS TO SCALE

5. (*a*) Make a scale layout of the sodium hydroxide, NaOH + H_2O, lines of the hydrogen generator. Show a cross section of the headers, one cylinder of the pump, a portion of the sodium hydroxide tank, and a portion of the top of one generator. (*b*) Make a scale layout of the hydrogen (H_2) line of the hydrogen generator. Show a portion of each generator with the lines in place. Use 11- by 17-in paper. Scale, $3'' = 1'-0''$. Use standard pipe, malleable fittings, brass valves. Make a bill of material listing all pipe, valves, and fittings needed.

The operation of the unit is as follows: After the generators have been charged with ferrosilicon (Fe + Si), a mixture of sodium hydroxide (NaOH) and water (H_2O) is pumped into the generators. When the ferrosilicon, sodium hydroxide, and water are together in the generator, they react to form hydrogen (H_2) and a sludge (Fe, Na_2SiO_2, and H_2O). As the hydrogen gas forms, the internal pressure of the generator is built up. To keep the reaction going, additional water and sodium hydroxide solution has to be pumped into the generators against this pressure. As the hydrogen gas is formed, it is piped to a storage tank. The sludge is siphoned off periodically to prevent excessive accumulation that might retard the reaction.

A four-cylinder pump is used to provide a more uniform flow of fluid than would be obtained if a large single-cylinder pump were used. The four cylinders of the pump are connected to one intake line and to one exhaust line through headers (manifolds). Instead of building an intake and an exhaust valve into the cylinders of the pump, standard ball-type check valves are used and are connected as near as possible to the pump cylinders by means of standard pipe and fittings.

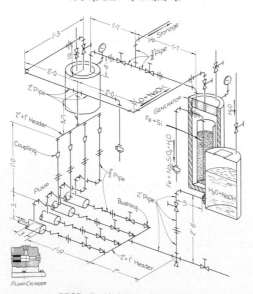

PROB. 5 Hydrogen generator.

6. Make a scale layout of the gas-burner installation. Specify fittings and give centerline dimensions. Use 11- by 17-in paper. Scale, $3'' = 1'-0''$.

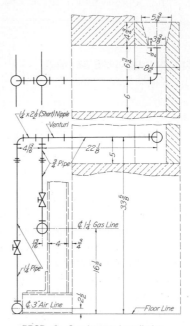

PROB. 6 Gas-burner installation.

7. Make a scale layout of a Grinnell industrial heating unit, closed return, gravity system. Use centerline distances and placement of fittings as shown in the diagram. Use 3-in supply main, 2-in pipe and fittings to unit, ¾-in pipe and fittings from unit to 2-in return main. Add all necessary notes and dimensions. Use an 11- by 17-in sheet. Scale, $3'' = 1'-0''$.

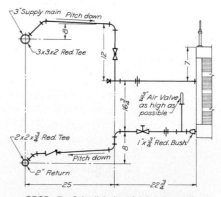

PROB. 7 Grinell industrial heating unit.

GROUP 3. PIPING LAYOUTS, DIAGRAMMATIC

8. A is a storage tank for supplying the mixing tanks B, C, and D and is located directly above them. The capacities of the mixers are in the ratios of 1, 2, and 3. Design (in one view) a piping system with sizes such that, neglecting frictional losses, the three tanks will fill in approximately the same time. So arrange the piping that any one of the tanks can be cut out or removed for repairs without disturbing the others. Use single-line conventional representation. Dimension to centerlines and specify the fittings.

9. The figure shows the arrangement of a set of mixing tanks. Make an isometric drawing of an overhead piping system to supply water to each tank. Water supply enters the building through a 2½-in main at point A, 3 ft below floor level. Place all pipe 10 ft above floor level, except riser from water main and drops to tanks, which are to end with globe valves 5 ft above floor level. Arrange the system to use as little pipe and as few fittings as possible. Neglecting frictional losses, sizes of pipe used should be such that they will deliver approximately an equal volume of water to each tank if all were being filled at the same time. The pipe size at the tank should not be less than ¾ in. Dimension and specify all pipe and fittings.

10. Make a drawing of the system of Prob. 9. Show the layout in a developed view. Dimension from center to center and specify all pipe and fittings.

11. Make a list of the pipe and fittings to be ordered for the system of Prob. 9. Arrange the list in a table, heading the columns as below.

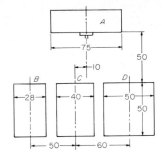

PROB. 8

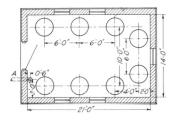

PROB. 9 Mixing tanks.

TABLE FOR PROBLEM 11

Size	Pipe lengths	Valves		Fittings		Material	Remarks (make, kinds of threads, etc.)
		Number	Kind	Number	Kind		

12. Make an oblique drawing of a system of piping to supply the tanks in Prob. 9. All piping except risers is to be in a trench 1 ft below floor level. Risers should not run higher than 6 ft above floor level. Other conditions as in Prob. 9.

13. The figure shows the outline of the right-hand half of a bank of eight heat-treating furnaces. X and Y are the lead-ins from the compressed-air and fuel mains. Draw the piping layout, using single-line representation, to distribute the air and fuel to the furnaces. The pipe sizes should be reduced proportionately as the oven leads are taken off. Each tail pipe should be removable without disturbing the other leads or closing down the other furnaces. Dimension the piping layout and make a bill of material for the pipe and fittings.

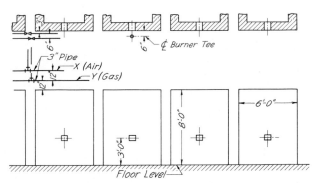

PROB. 13 Heat-treating furnaces.

14. The figure is the diagrammatic isometric layout of an industrial heater system. Using the basic layout as shown, assume positions for the preheating and reheating coils and make a complete, dimensioned, isometric diagrammatic layout.

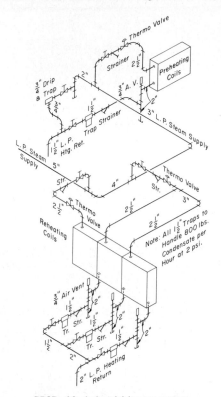

PROB. 14 Industrial heater system.

15. Make an orthographic diagrammatic layout for Prob. 14.

16. Make a developed diagrammatic layout for Prob. 14.

1. INTRODUCTION

The term "structures" covers a wide variety of projects. A structure can be a building, a dam, a bridge, a tower, or other large-size engineering endeavor. Materials used in structures are usually combinations of steel, concrete, wood, stone, and masonry.

Structural drawings include several types, but all are used in the design and implementation of particular plans. Drawing standards vary somewhat, depending on the material being used. That is, the standards for a steel framework for a building would be somewhat different than the standards used for a wooden framework. A drafter must pay careful attention to the drafting practices used in a particular construction technique.

This chapter will survey some of the basic concepts of drafting as applied to steel, wood, concrete, and masonry. The chief emphasis will be on steel structures.

2. GENERAL CLASSIFICATION OF STRUCTURAL DRAWINGS

We will look first at types of drawings needed regardless of the structural material used. Then we will look at drawings for specific materials. Virtually all structures would include the following types of drawings:

A. General Plan This will include a profile of the ground; location of the structure; elevations of ruling points in the structure; clearances; grades; direction of flow, high water, and low water (for a bridge); and all other data necessary for designing the substructure and superstructure.

B. Stress Diagram This will give the main dimensions of the structure, the loading, stresses in all members for the dead loads, live loads, wind loads, etc., itemized separately; total maximum stresses and minimum stresses; sizes of members; typical sections of all built members showing arrangement of material; and all information necessary for detailing the various parts of the structure.

C. Shop Drawings   Shop detail drawings should be made of all steel and ironwork, and detail drawings should be made of all timber, masonry, and concrete work.

D. Foundation or Masonry Plan   The foundation or masonry plan should contain detail drawings of all foundations, walls, piers, etc., that support the structure. The plans should show the loads on the foundations, the depth of footings, spacing of piles where used, the proportions for the concrete, the quality of masonry and mortar, the allowable bearing on the soil, and all data necessary for accurately locating and constructing the foundations.

E. Erection Diagram   The erection diagram should show the relative location of every part of the structure, shipping marks for the various members, all main dimensions, number of pieces in a member, packing of pins, size and grip of pins, and any special feature or information that may assist the erector in the field. The approximate weight of heavy pieces will materially assist the erector in designing his or her falsework and derricks.

F. Falsework Plans   For ordinary structures it is not common to prepare falsework plans in the office, this important detail being left to the erector in the field. For difficult or important work, erection plans should be worked out in the office and should show in detail all members and connections of the falsework and also give instructions for the successive steps in carrying out the work. Falsework plans are especially important for concrete and masonry arches and other concrete structures and for forms for all walls, piers, etc. Detail plans of travelers, derricks, etc., should also be furnished the erector.

In addition to the above types of drawings, certain lists are usually included in a complete set of plans. These lists are:

G. Bills of Material   Complete bills of material showing the different parts of the structure with their marks and shipping weights should be prepared. This is necessary to permit checking of shipping weights and shipment and arrival of materials.

H. Rivet List   The rivet list should show the dimensions and number of all field rivets, field bolts, spikes, etc., used in the erection of the structure.

I. List of Drawings   A list should be made showing the contents of all drawings belonging to the structure.

3. STRUCTURAL STEEL DRAFTING

We can cover only in an abbreviated way drafting for structural steel. An excellent reference for in-depth study is the *Manual of Steel Construction* published by the American Institute of Steel Construction, Inc., 400 North Michigan Avenue, Chicago, Illinois 60611.

There are two basic types of structural steel drawings: design drawings and shop drawings.

A. Design Drawings   These drawings are created within the engineering office for eventual use by the steel fabricator. Design drawings are all-encompassing in that not only are basic sizes and locations given for steel members, but also given are assumed loads, shear and moment diagrams, and axial forces. Types of high-strength bolts, if used, are indicated, as well as other assembly methods, such as welding and riveting. Types of beam and column connections are provided.

Included in design drawings are all sizes of members, cross sections of members (see Sec. 5), and locations of these members. Column centers are given along with overall dimensions. Unusual, unique, or complex aspects of the design are clarified by detail drawings as needed.

B. Shop Drawings   All details are accounted for in shop drawings. Exact dimensions are given for placement of bolts, welds, and rivets. Hole locations are given for all connections. Dimensions are typically given to the nearest $1/16$ inch.

Shop drawings distinguish between what work is to be done indoors within the shop and what work is left to be done on the site of the structure. Any sequence required for assembly of members is given also. A particular sequence could be indicated to minimize distortion from welding, for example.

4. TYPES OF STEEL FRAMING

It should be noted that actual design and shop drawings are greatly influenced by the type of framing desired by the designer. The drafter is not expected to know the theory of framing, but one should be aware of three general classes of framing.

A. Rigid Framing This type of framing is designed to have rigidity sufficient to essentially maintain the original design angles between intersecting members.

B. Simple Framing Framing of this type is unrestrained and free-ended. The ends of beams are connected for shear only and are free to rotate under gravity load.

C. Semirigid Framing Such framing is partially restrained and has a design performance intermediate between rigid and simple types of framing.

5. STRUCTURAL STEEL SHAPES

The drafter should be aware of basic cross-sectional shapes for construction steel. Steel structures are made up of "rolled shapes" fastened together with rivets, bolts, or welds. The function of structural-steel drawings is to show how these shapes are to be fabricated in the shop and then erected to form the various members of bridges, buildings, etc. Sections of the common shapes are shown in Fig. 1. Dimensions for detailing these and other less common shapes are shown in the *Manual of Steel Construction* of the American Institute of Steel Construction. Dimensions for detailing common shapes are given in the appendix.

Structural steel, as seen in Fig. 1, comes in various shapes. Shapes can be classified as:

A. Bars Bars may be round, square, or flat. Flat material is considered a bar if 6 inches or less in width.

B. Plates Plates are defined as flat sections having 8 or more inches of width.

C. Tubing Tubing by definition is hollow and closed and comes in round, square, or rectangular cross section.

D. Pipe Pipe usually has greater wall thickness than tubing, is round only, and is available in Standard, Extra-Strong, and Double Extra-Strong grades.

E. Shaped Steel When hot steel is rolled from billets it is possible to produce a variety of shapes. The most common shapes as seen in Fig. 1 are S (standard I-beam), W (wide flange), C (channel), and L (angle). There are variations on these basic shapes, as indicated in Fig. 1. Some of these shapes are made by cutting

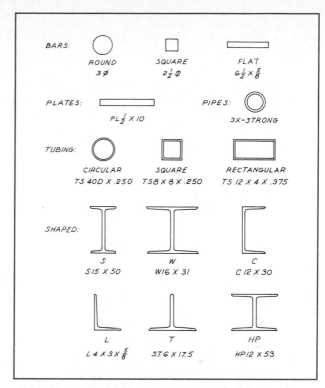

FIG. 1 Basic shapes for structural steel. Symbols and sizes are noted for the S, W, C, T, and HP shapes. Weights per foot are given.

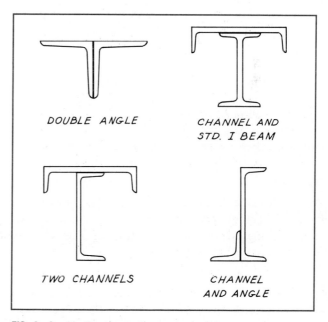

FIG. 2 Combination shapes for structural steel.

other shapes, for example, to make T shapes. Other shapes are HP (bearing pile), M (miscellaneous, cannot be classified S, W, or HP), and MC (cannot be classified M).

Note in Fig. 1 that example specifications of shapes are given with each shape. One example could be for the channel shape, C 12 × 30 × 20'-6", where 12 is the nominal depth and 30 is the weight in pounds per running foot. Overall length is 20'-6".

F. Combination Shapes   Figure 2 shows that by combining various shapes, others may be created. A designer is not limited to those shapes shown, but may develop other combinations to serve a need.

6. DETAILING PRACTICE FOR STRUCTURAL STEEL

Figure 3 is a typical shop drawing of a small roof truss, giving complete details. Such drawings are made about the stress-diagram lines (used in calculating the stresses and sizes of the members), which are then employed as the gravity lines of the members and form the skeleton, as illustrated separately to small scale in the box on the figure. The intersections of these lines are called "working points" and are the points from which all distances are figured. The length of each working line is computed accurately, and from it the intermediate dimensions are obtained. The design diagram is often put on the same sheet as the shop drawing, as in the drawing of the truss.

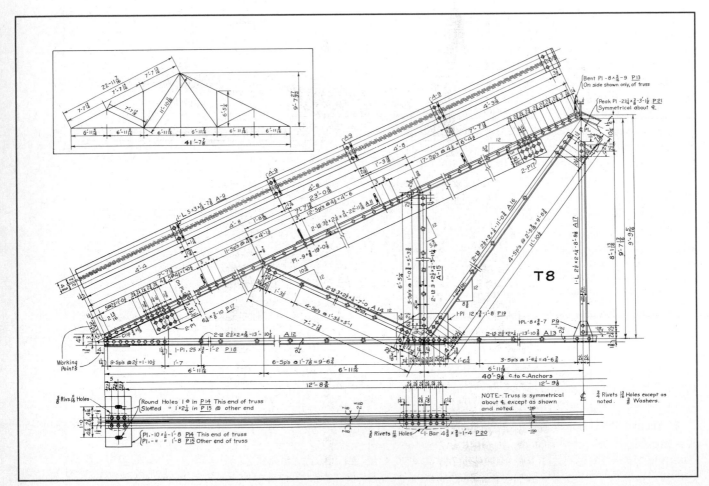

FIG. 3   A structural shop drawing. This is of a roof truss. The design diagram is included in the upper left corner.

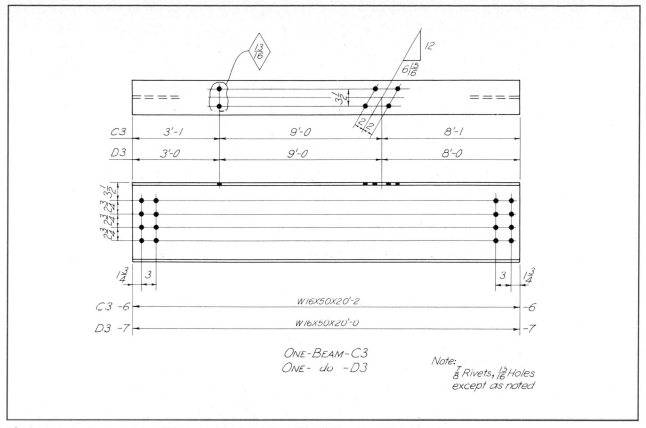

FIG. 4 A structural shop drawing. This is a detail of two beams, *C*3 and *D*3, which differ only in length.

Separate drawings made to a sufficiently large scale to carry complete information are called "shop detail drawings" (Fig. 4). When possible, the drawings of all members are shown in the same relative position that they will occupy in the completed structure: vertical, horizontal, or inclined. Long vertical or inclined members may be drawn in a horizontal position, a vertical member always having its lower end at the left, and an inclined member drawn in the direction it would fall. Except in plain building work, a diagram to small scale (showing by a heavy line the relative position of the member in the structure) should be drawn on every detail sheet.

In steel construction a member is composed of either a single rolled shape or a combination of two or more rolled shapes. Figure 4 is a shop detail drawing of a member made from a single rolled shape; Figs. 3, 5, and 6 are shop detail drawings of members made up of several rolled shapes. The figures illustrate detailing practice. Note in Fig. 3 that only one-half of the truss is shown. When so drawn it is always the left end, looking toward the side on which the principal connections are made.

In order to show the details clearly, the structural drafter often uses two scales in the same view, one for the centerlines or skeleton of the structure, showing the shape, and a larger one for the parts composing it. The scale used for the skeleton is determined by the size of the structure as compared with the sheet.

The scale of shop drawings ranges from $\frac{1}{4}'' = 1'$-0 to $1'' = 1'$-0. Often, for long members, the cross section is drawn to a larger scale than the length. Sometimes it is even advantageous to pay no attention to scaled length but to draw the member as though there were breaks in the length (but not shown on the drawing, as in Fig. 6)

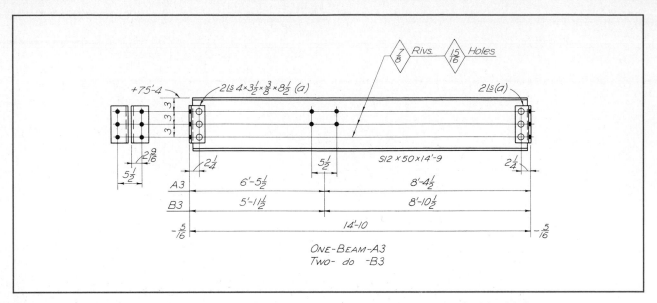

FIG. 5 A structural shop drawing. This is a detail of two beams, *A*3 and *B*3, which differ only in spacing of web holes.

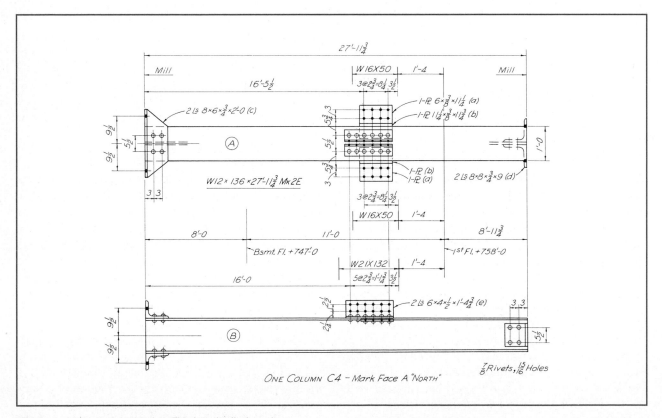

FIG. 6 A structural shop drawing. This is a detail of a column.

so that rivet spacings at the ends (and intermediately) can be drawn to the same scale as the cross section.

7. GUIDELINES FOR STRUCTURAL DRAFTING

We now offer guidelines that may assist the drafter as design and shop drawings are developed.

A. Dimensions Dimensions are placed above the dimension line, and the dimension lines are not broken but continuous. Length dimensions are expressed in feet and inches. All inch symbols are omitted unless there is the possibility of misunderstanding; thus 1 bolt should be 1″ bolt to distinguish between size and number. Inch symbols are omitted even though the dimensions are in feet and inches, and dimensions should be hyphenated as: 7′-0, 7′-0½, 7′-4. Plate widths and section depths of rolled shapes are given in inches. Dimensions are given to commercial sizes of materials.

Distances to centerlines of another connecting member are given, as in Figs. 4 and 5, by placing the distance, preceded by a minus sign (−), at the end of the length dimension. These distances to center are a great aid in checking the details with an assembly or layout drawing.

Inclinations of members and inclined centerlines, cuts, etc., are indicated by their tangents. As in Figs. 3 and 4, a small right triangle is drawn with its hypotenuse parallel to the line or on the line whose inclination is to be specified. The long leg of the triangle is always specified as 12.

B. Notes and Marks *Rolled shapes* are specified by abbreviated notes, as described in paragraph 5. The specification is given either with the length dimension as in Fig. 4 or near and parallel to the shape as in Fig. 5.

Erection marks are necessary in order to identify the member. These are indicated on the drawing by letters and numbers in the subtitle. The erection diagram then carries these marking numbers and identifies the position of the member. Problem 5 shows an example of an erection diagram with marking numbers.

The erection mark also identifies the member in the shop and serves as a shipping mark. The mark is composed of a letter and number. Capital letters are used, *B* for beam, *C* for column, *T* for truss, etc. The number gives the specific member in an assembly.

Assembly marks are used when the same shape is used in more than one place on a member. The member is completely specified *once* followed by the assembly mark (lowercase letters, to distinguish from erection marks). Then the complete specification is not repeated; only the assembly mark is given. As an example, see the angles "2∠ 4 × 3½ × ⅜ × 8½(a)" in Fig. 5.

General notes either accompany the general drawing or appear in the title of the shop drawing. Items included are painting, shipping instructions, etc.

C. Describing Rivets Rivets and holes are dimensioned in the view that shows them as circles. Open circles are used for shop rivets, but blackened circles are used for field rivets and bolts. Note this technique in Fig. 5.

Dimensions for a row of rivets are given in a single line as in Fig. 3. Gage dimensions, actually centerlines passing through holes, or rivets used for lineup of holes or rivets, are always given. The size of rivets and holes is given in a general note as in Fig. 4. Rivets are drawn to scale. Rivet symbols are used because it is impossible to show the details of head form on a small-scale drawing (see Chap. 18).

The length of rivets is not usually given on the drawing; the worker picks a rivet length long enough to go through the members and protrude far enough so that the head can be formed.

Common practice is to make rivet holes 1/16 in larger than the rivets (see Figs. 4 to 6).

If some of the rivets or holes for a member are of a different size, these are indicated by the size in a diamond, as in Figs. 4 and 5.

Rivet spacing is an important item in detailing, and there are a number of conditions that control the placement. Rivets spaced too close together or too close to a projecting part cannot be properly driven. Rivets placed too close to the edge or end of a member may weaken the member.

Rivets are spaced along "gage" lines, parallel to the axis of the member and at certain "pitch" distances along the gage line (Fig. 7). The minimum distance between rivets in any direction is three rivet diameters. The minimum edge distance (*e*, Fig. 8), from sheared edges of a member is 1½ rivet diameters. Distance to a rolled edge may be slightly less than 1½ diameters and is usually

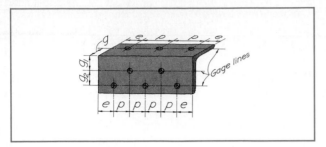

FIG. 7 Hole spacings. These are: gage (g), pitch (p), and edge (e) distances.

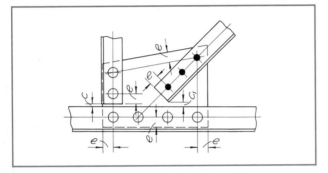

FIG. 8 Spacings on built-up members. These are edge (e) and clearance (c) distances.

controlled by the location of the gage line. Whenever possible the same edge distance is used on all members, and the distance is then given in a general note such as "Edge distance 1¼ except as noted."

Gage-line distances and minimum driving clearances for common rolled shapes are given in the appendix. Standard beam connections are also given in the appendix.

D. Other Considerations *Different members* may be detailed together on the same drawing when they differ only in length or spacing of rivets or holes, or when one member has special holes or an extra piece added. Figures 4 and 5 both show two beams detailed on the same drawing. Note that the different lengths are given with the erection mark at the end of the dimension line.

When one of two similar members has a hole or other feature on one member only, the special feature is indicated by drawing a freehand circle around the detail with a leader to a note, such as "On A-3 only"

Clearances between various members of a structure

are necessary so that there will be no interferences, because of manufacturing inaccuracies. More clearance should be allowed for field erection than for shop fabrication. Field clearance (C_1, Fig. 8) should be approximately ½ in, and shop clearance (C) should be about ¼ in. Note the clearances shown between the angles (fastened with a gusset plate) in Fig. 8.

Elevations, sections, and other views are placed in relation to each other by the rules of third-angle projection, except that when a view is given under a front view, as in Fig. 4, it is made as a section taken above the lower flange, looking down, instead of as a regular bottom view looking up. Large sections of materials are shown with uniform crosshatching. Small-scale sections are blacked-in solid, with white spaces left between adjacent parts.

Bent plates should be developed and the "stretchout" length of bent forged bars given. The length of a bent plate may be taken as the inside length of the bend plus half the thickness of the plate for each bend.

A *bill of material* always accompanies a structural drawing. This can be put on the drawing, but the best practice is to attach it as a separate "bill sheet," generally on 8½- by 11-in paper.

8. WOOD CONSTRUCTION

The representation of timber-framed structures involves no new principles but requires particular attention to details. Timber members are generally rectangular in section and are specified to nominal sizes in even inches, as 8″ × 12″. The general drawing must give center and other important distances accurately. Details drawn to larger scale give specific information as to separate parts. The particulars of joints, splices, methods of fastening, etc., are detailed.

Two scales are sometimes used to advantage on the general drawing, as is done in Fig. 9. Complete notes are an essential part of such drawings, especially when an attempt at dimensioning the smaller details results in confusion.

Joints in timber structures are fastened with nails or spikes, wood screws, bolts, or ring-shaped or flat *connectors,* similar in action to a dowel or key. Some common types are shown in Fig. 10. A split ring, shown at (a), is assembled in grooves in each piece and held together

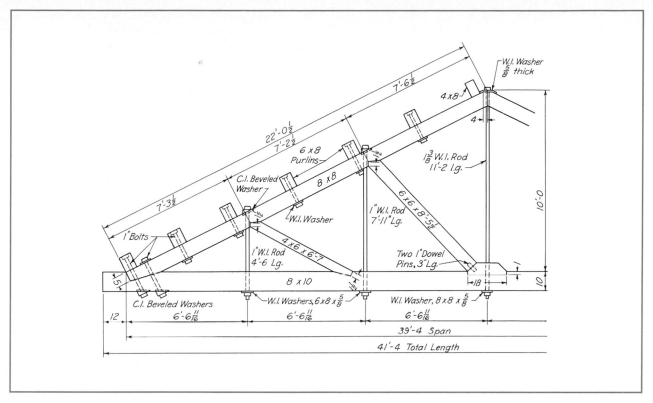

FIG. 9 A structural shop drawing. This is a detail of a timber truss.

with a bolt, as indicated at (*b*). The sharp projections on the alligator connector (*c*) are forced into the members by pressure. The clawplate connector (*d*) is used either in pairs, back to back, for timber-to-timber connections; or single for timber to metal. A typical assembly is shown at (*e*). The Kubler wood dowel connector (*f*) fits into a bored hole in each timber face, and a bolt holds the parts together.

The Forest Products Laboratory publication, "Wood Handbook," gives basic information on wood as a material of construction, including the connectors, and is available at small cost from the Superintendent of Documents, Washington, D.C.

9. CONCRETE CONSTRUCTION

Concrete can have a variable compressive strength, due mainly to different water-to-mix ratios and curing time. A typical compressive strength would be 5000 psi. How-

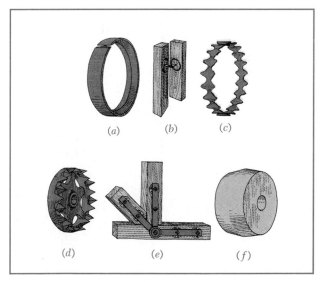

FIG. 10 Timber connectors. These are used for joints in wooden structures.

ever, plain concrete in *tensile* strength is relatively weak, about 500 psi.

To improve the tensile strength of concrete, steel reinforcing rods or wires can be added during construction. The steel serves to supplement the tensile strength while the concrete carries the compressive loads. Such concrete is called *reinforced concrete*. It is also possible to pretension steel which is embedded in the concrete, giving what is known as *prestressed concrete*.

The design and drafting involved in reinforced concrete is a highly specialized procedure. For example, it is difficult to show the shape and location of reinforcing bars in concrete without using a systematic scheme of symbols and markings. Assistance can be provided in publications of the American Concrete Institute.

Figure 11 shows a portion of the general structural plan for the first floor of a building. By means of the plan, the tabular data, the general notes, and the specifi-

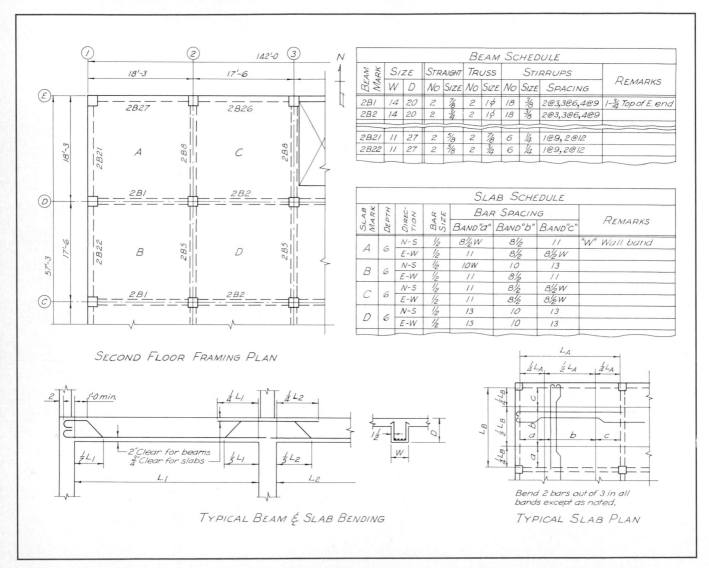

FIG. 11 A general reinforced concrete plan. The framing plan identifies beams and slabs by letter and number. Beam and slab schedules show the number, size, and spacing of reinforcing steel. Details show typical bending and location of reinforcing.

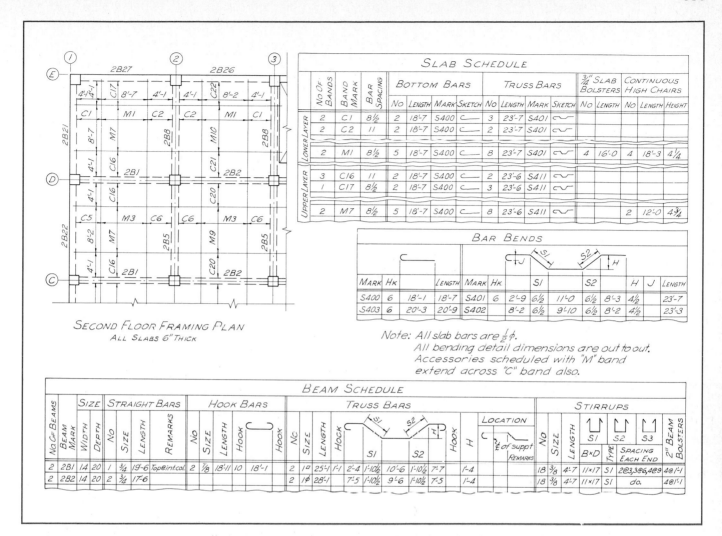

FIG. 12 A detail reinforced concrete plan. The framing plan identifies all beams and slabs by letter and number. The slab, beam, and bend schedules give all details for preparation and placement of the reinforcing steel.

cations, the engineer has completely specified the reinforced concrete floor of the building. Location of the reinforced beams is given on the plan; the size of the beam and reinforcing to be used are indicated in the beam schedule; and the typical beam sections give the basic information for bending the reinforcing bars. The type of floor slab used is shown on the plan and in the slab schedule by the letters A, B, etc. The slab schedule, in conjunction with the typical slab plan, indicates the direction and spacing of reinforcing.

General notes on the drawing or in the specifications

cover items such as maximum strength of concrete, grade of steel, and minimum cover.

The general drawing in Fig. 11, although completely specifying the reinforced concrete construction, does not give the exact details of how the bars are to be bent and placed. The contractor supplying the steel makes another drawing (Fig. 12), showing in detail the bending, spacing, etc., for all the steel.

Walls, columns, and other portions of reinforced structures are represented similarly to the beams and floor slabs illustrated.

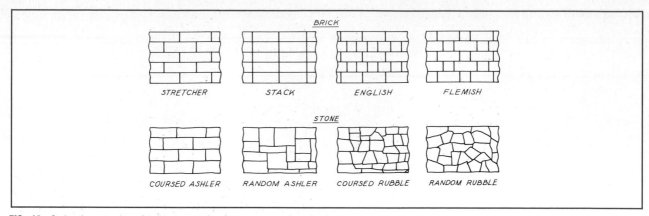

FIG. 13 Styles for coursing of brick and stone. Joints would include mortar.

10. MASONRY AND STONE CONSTRUCTION

Masonry is generally thought to apply to brick and tile construction. Since these are made-from-scratch products, a wide range of styles, colors, textures, and strengths are available. In most construction, the limiting factor in strength is that of the mortar holding together the various bricks or tiles. It is possible to reinforce masonry with steel rods to increase strength.

Stone is usually used where it can be readily seen; hence it has an ornamental and aesthetic value as well as structural. The use of stone in walls and paths is well known. Stone is seldom used for high structures, but is mainly limited to heights of two or three stories. Popular stone is granite, limestone, and to a limited degree, marble.

Both brick and stone can be laid in courses of various styles. A few of the styles are illustrated in Fig. 13.

PROBLEMS

1. Make shop drawing of item shown. Assume complete shop fabrication. Use ⅝-in rivets.

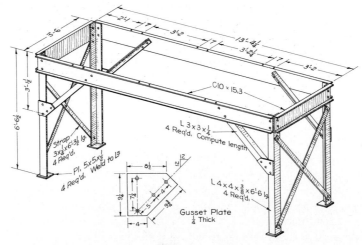

PROB. 1

1A. Make shop drawing of item shown in Prob. 1. Assume shop fabrication of all parts except long channels and 3- × 3- × ¼-in angle brace; use ⅝-in shop rivets, ⅝-in field bolts.

1B. Redesign Prob. 1A for welded construction.

2. Make a shop drawing of the column base. Assume complete assembly in the shop. Use ¾-in rivets. Open holes are for 1-in anchor bolts.

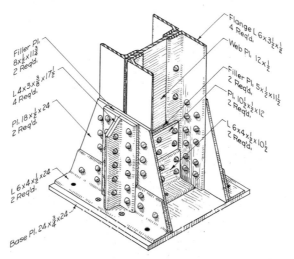

PROB. 2 Column base.

3. Make a shop drawing for the crane trolley. Assume fabrication as indicated in the figure. One end is shown; the other end is similar. Use ⅝-in rivets wherever possible.

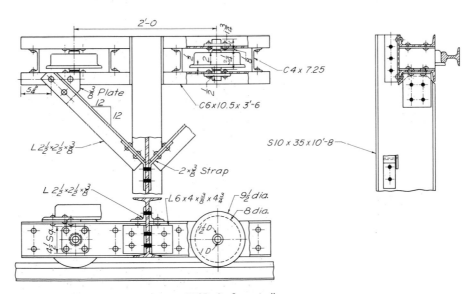

PROB. 3 Crane trolley.

3A. Make a bill of material of parts used in Prob. 3.

4. Make a shop drawing for the motor support. Assume shop fabrication as follows: (*a*) 8-in channels, 4- by 4- by ½-in angles, ½-in plate, and 6- by 4- by ⅜-in beam connectors (angles), two members required, one right-hand, one left-hand; (*b*) 21 by ¾ by 2'-4 plate; (*c*) 12-in I beam, two required with 4- by 4- by ⅜-in beam connectors temporarily bolted in place. Assume columns to be part of the existing structure. Use ¾-in rivets and ¾-in bolts.

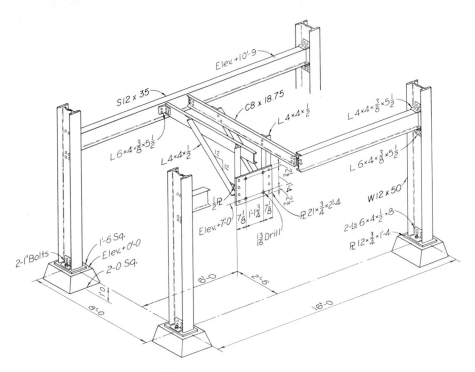

PROB. 4   Motor support.

4A. Make an erection drawing for the motor support. Assume columns to be part of the existing structure.

5. Make a shop drawing of roof truss *T*1 for the steel-frame mill building. This drawing is to be similar to the drawing shown in Fig. 3. For truss *T*1, use the same size members and the same roof pitch as is used for truss *T*8, Fig. 3. Detail only the left half of the truss to a scale suitable for 11- by 17-in paper.

5A. Prepare shop drawings for the following steel members used in fabricating the steel mill building: (*a*) bracket *M*1; (*b*) beams *B*1 and *B*3 (one detail for both); (*c*) beam *B*5 (assume ⅜-in plate to be shop-fabricated to column); and (*d*) beam *B*6. Assume shop and field fabrication indicated. Use ¾-in rivets. These details, along with the shipping list and general notes, may be drawn on 11- by 17-in paper to a scale of 1″ = 1'-0″.

5B. Prepare shop drawings for the following steel members used in fabricating the steel mill building: (*a*) bracket *M*1, and (*b*) column *C*4. Assume shop and field fabrication indicated. Use ¾-in rivets. Three faces of the column will have to be shown.

5C. Make a shop drawing of column *D*4. Assume shop and field fabrication as indicated. Use ¾-in rivets.

5D. Make a shop drawing of beam *B*7.

5E. Make a shop drawing of rafter *R*1.

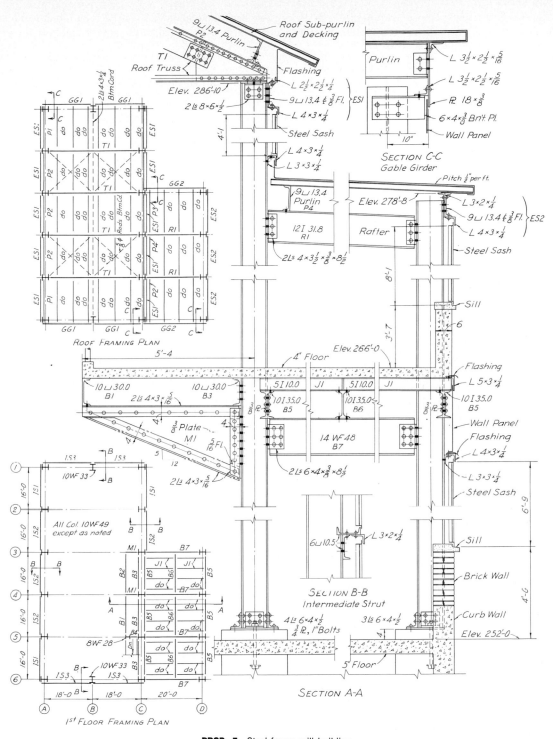

PROB. 5 Steel-frame mill building.

5F. Make a general structural plan, similar to Fig. 11, of the concrete floor at elevation 226'-0. Use ½-in round reinforcing bars spaced 9 in on centers, alternate bars are bent for the floor. In 6-in wall, use 3½-in round bars in bottom and ⅜-in round vertical bars at 18 in hooked 18 in into floor slab.

5G. Make a reinforcing detail drawing, similar to Fig. 11, for Prob. 5F.

Note that problem 5 uses the older forms of notation for sections in which I equals S section, ⊔ equals C section, and WF equals W section.

6. Make a detailed shop drawing of the structural members in a section of the pipe bridge. Do not include the bents.

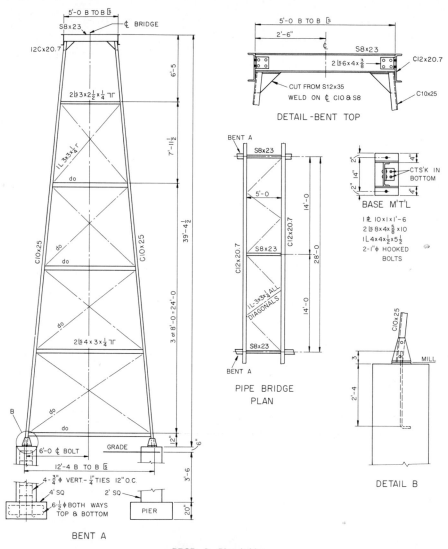

PROB. 6 Pipe bridge.

6A. Make a detailed drawing of the base section of bent A including the two bottom diagonal braces.

6B. Make a detailed shop drawing of the entire bent A.

6C. Make a detailed drawing of the pier support for bent A.

6D. Make a detail of a typical connection to the 10-in channel using the horizontal members as 2∠s 3 × 2½ × ¼ and each of the two diagonal members coming to this connection as 3 × 3 × ¼ angles.

6E. The pipe bridge carries 10-in, 7-in, 4-in, and 3-in insulated pipe lines suspended from the 8-in 23-lb S beams on each bent. Keep the pipes in the order indicated from left to right and show a detail of the strap supports and their connection to the bent.

7. The centerlines of a timber truss are shown. The joints are to be made using split-ring connectors—one at each joint—except at the junction of the short compression member and the rafter, where a 1- × 4-in wood scab plate nailed over the joint is all that is needed.

Complete the design for the necessary trusses to roof over the structure indicated. Allow for a 2-ft overhang all the way around. The end trusses must be made flush with the wall, and designed so that some form of sheeting to cover the gabled end can be nailed in place. The trusses are to be placed on 2-ft centers (approximately) and sheeted over with ½-in plywood. Plan for the transition trusses between the two wings to rest on the sheeting of the 50-ft wing. The design of these trusses can be varied to fit their decreasing length.

These trusses used as specified will carry approximately 35 lb per sq ft dead plus live load of the roof and 10 lb per sq ft dead load on the ceiling.

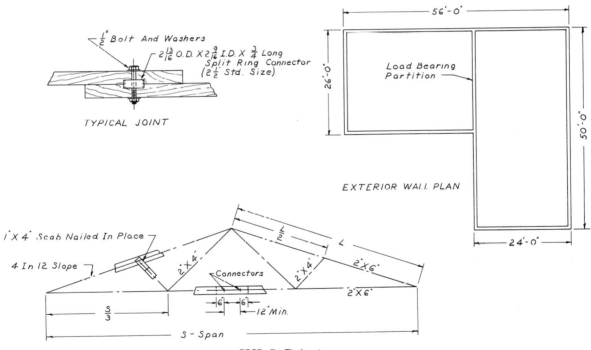

PROB. 7 Timber truss.

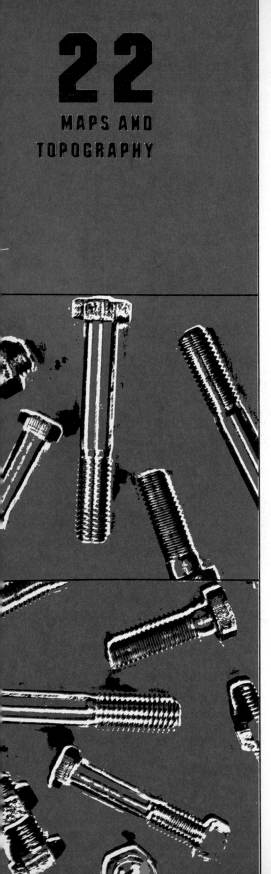

22

MAPS AND TOPOGRAPHY

1. MAPS AS ONE-VIEW DRAWINGS

So far in our consideration of drawing we have represented objects by drawing two or more views of them, with the exception of pictorials. In map drawing, however, it is possible to represent three features of the earth's surface. Maps can include height information by using the technique of contouring, as we shall see.

Persons using maps, such as technicians and engineers, should be familiar with map types and symbols used in map work. This chapter will give an overview of the classification and types of maps, with some examples. It is not our intent in this short chapter to make you an expert in the construction or use of maps, but to give you a working acquaintance with the various kinds of maps.

2. CLASSIFICATION

The content or information on maps is generally classified under three divisions:

1. The representation of boundaries, bearings, and distances, such as divisions between areas subject to different authority or ownership, either public or private; or lines indicating geometric measurements on land, on sea, and in the air. This division includes plats or land maps, farm surveys, city subdivisions, plats of mineral claims, and nautical and aeronautical charts.

2. The representation of real or material features of objects within the limits of the tract, showing their relative location or size and location, depending upon the purpose of the map. When relative location only is required, the scale may be small, and symbols may be employed to represent objects, such as houses, bridges, or even towns. When the size of the objects is an important consideration, the scale must be large and the map becomes a real orthographic top view.

3. The representation of the relative elevations of the surface of the ground. Maps with this feature are called "relief maps" or, if contours are used with elevations marked on them, "contour maps." Hydrographic maps show fathom-line depth curves.

Various combinations of these three divisions are required for different purposes. Maps are classified according to purpose, as follows:

1. *Geographical maps* include large areas and consequently must be to small scale. They show important towns and cities, streams and bodies of water, political boundaries, and relief.

2. *Topographic maps* are complete descriptions of certain areas and show to larger scale the geographical positions of the natural and constructed features. The relief is usually represented by contours.

3. *Engineering maps* are working maps for engineering projects and are designed for specific purposes to aid construction. They provide accurate horizontal and vertical control data and show objects on the site or along the right-of-way.

4. *Photogrammetric maps* represent features on the earth's surface from terrestrial and aerial photographs. These photographs are perspectives from which orthographic views are obtained by stereoscopic instruments. Ground-control stations are necessary to bring the photographs to a required datum.

   "Orthophoto maps," which are aerial photographs to which contours may be added, are a widely used type of photogrammetric map. A "planimetric map," which uses the basic information of orthophoto maps, but deletes unnecessary detail, may be produced from an orthophoto map.

5. *Cadastral maps* are very accurate control maps for cities and towns, made to large scale with all features drawn to size. They are used to control city development and operation, particularly taxation.

6. *Hydrographic maps* deal with information concerning bodies of water, such as shorelines, sounding depths, subaqueous contours, navigation aids, and water control.

7. *Nautical maps* or charts show aids to water navigation, such as buoys, beacons, lighthouses, lanes of traffic, sounding depths, shoals, and radio compass stations.

8. *Aeronautical maps* or charts give prominent landmarks of the terrain and accentuate the relief by layer tints, hachures, and 500- or 1,000-ft contours as aids to air navigation.

9. *Military maps* contain information of military importance in the area represented.

3. PLATS

A map plotted from a plane survey and having the third dimension omitted is called a "plat" or "land map." It is used in the description of any tract of land when it is not necessary to show relief, as in a farm survey or a city plat.

The plotting is done from field notes by (1) latitudes and departures, (2) bearings and distances, (3) azimuths and distances, (4) deflection angles and distances, or (5) rectangular coordinates. Or the plotting is done by the total latitude and departure from some fixed origin for each separate point; this method is necessary to distribute plotting errors over the entire survey. Angles are laid off from bearing or azimuth lines by plotting the tangent of the angle or the sine of half the angle, by sine-and-cosine method, or by an accurate protractor.

Simplicity is of first importance in the execution of this kind of drawing. Information should be clear, concise, and direct. The lettering should be done in single stroke, and the north point and border should be of the simplest character.

A. Plat of a Survey The plat of a survey should give clearly all the information necessary for the legal description of a parcel of land. It should contain:

1. Direction and length of each line
2. Acreage
3. Location and description of monuments found and set
4. Location of highways, streams, rights-of-way, and any appurtenances required
5. Official division lines within the tract
6. Names of owners of abutting property
7. Title, scale, date
8. North point with certification of horizontal control
9. Plat certification properly executed
10. Reference to state plane-coordinate system

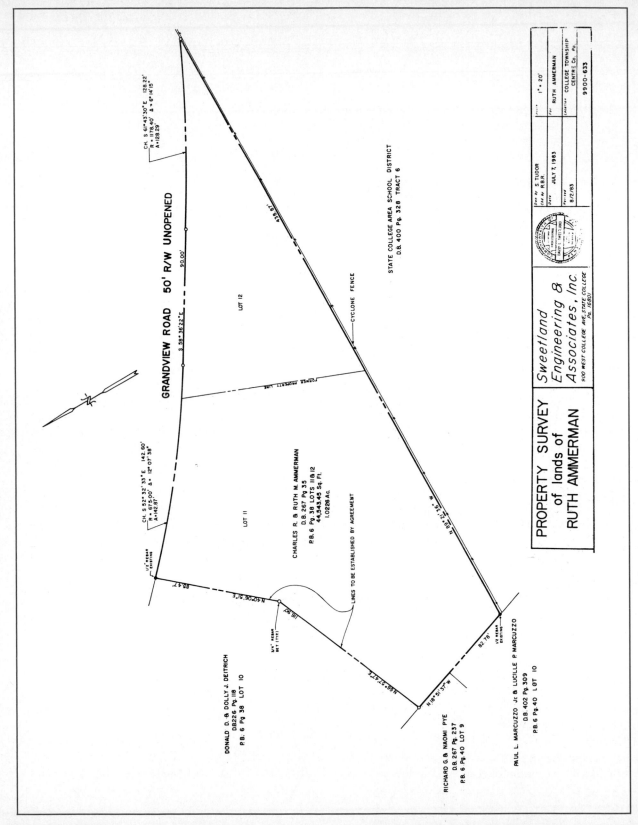

FIG. 1 Plot of a survey. A map that includes the information listed in Sec. 3A.

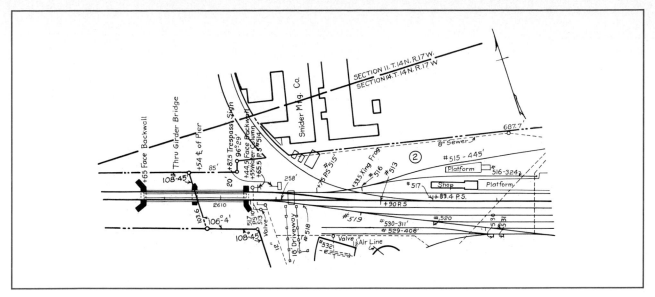

FIG. 2 Part of an industrial map. A railroad property map.

Figure 1 illustrates the general treatment of this kind of drawing. The size of the lettering used for the several features must be in proportion to their importance.

B. Industrial Plats Of the many kinds of plats used in industrial work only one is illustrated here, a portion of a railway situation or station map (Fig. 2). This might represent also a plant-valuation map, a type of plat often required. The information on such maps varies to meet the requirements of particular cases. In addition to the items in the preceding list, the map might include pipelines, fire hydrants, location and description of buildings, railroads and switch points, outdoor crane runways, etc.

C. Plats of Subdivisions The plats of subdivisions and allotments in towns are filed with the county recorder for record and must be complete in their information concerning the location and size of the various lots and parcels composing the subdivisions (Fig. 3). All monuments set should be shown and all directions and distances recorded, so that it will be possible to locate any lot with precision.

Figure 3 is a subdivision within a residential area and shows modern practice. As an interesting counterpoint, Fig. 4 shows a subdivision within a city. It is a subdivi-

sion that was done many years ago, perhaps in the 1930s, and represents a style very typical of its day. Such maps are still very much in existence and can be found as documents of record at courthouses and other governmental bodies.

D. Plat Showing a Sewer Plan It is essential that plats give information needed for a particular construction task. The placement of sewer lines within a subdivision is one such task. Figure 5 shows the general treatment involving sewer lines. Contour lines (see Fig. 9) are often included on sewer plans. Figure 5 is a small portion of the total sewer plan. The portion shown is for illustration purposes. To show the entire sewer plan would be impractical because to fit the page size of this text, detail would be so small as to be unreadable. Actual maps often cover several feet of length along any one side.

E. Plat Showing Horizontal Control Horizontal control is a system that provides points of reference for surveys. These accurately located controls can be thousands of feet apart, but serve to place all intervening points in context. Figure 6 shows a system done several decades ago for the city of Cleveland for geodetic and underground survey. Again, historical documentation such as this has value as subsequent maps are devised.

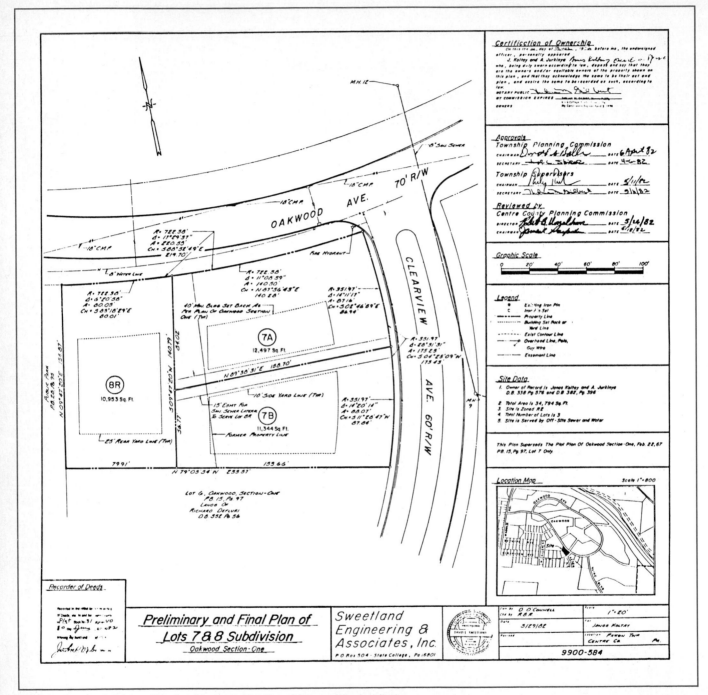

FIG. 3 Plot of a subdivision. Note that the size and location of lots are fully described.

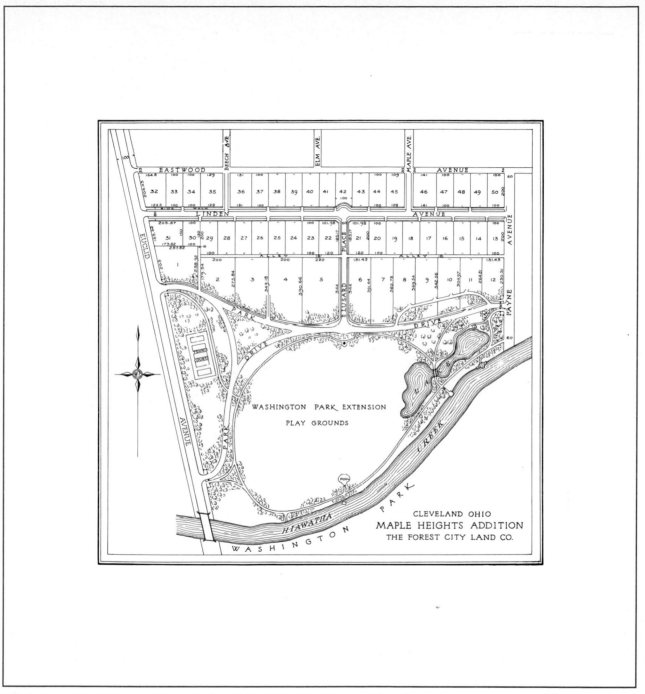

FIG. 4 A real estate display map. Shows the size of lots and streets but does not include the details given in a survey.

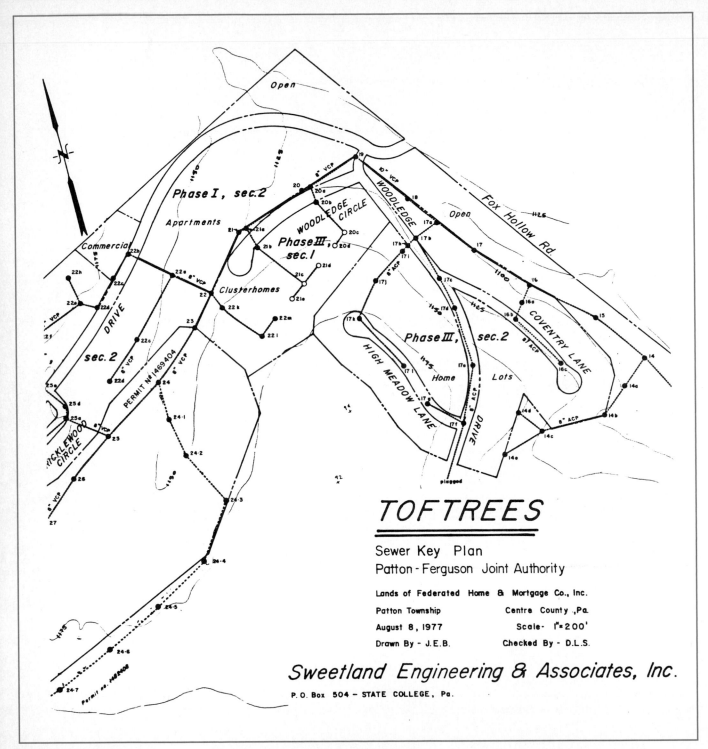

FIG. 5 Plot of a sewer plan. This portion of the map shows about one-fifth the actual total.

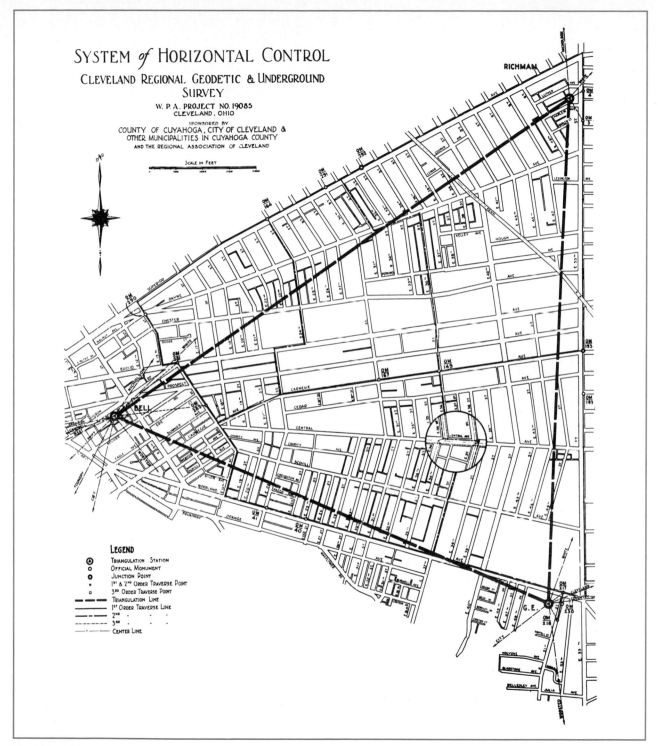

FIG. 6 Horizontal control. Locates accurate points of reference for surveys.

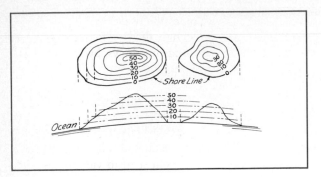

FIG. 7   Contours. These are theoretical lines of constant elevation.

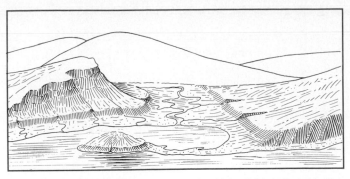

FIG. 8   A perspective view. A pictorial map in which relief is indicated by line shading. Maps of this type are used only for illustrative purposes.

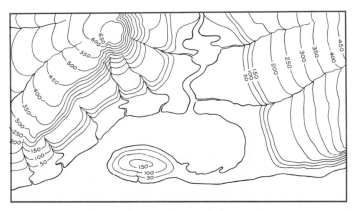

FIG. 9   A contour map. Map of the area shown in Fig. 8. The relief is read by noting the elevation, position, and spacing of the contours.

FIG. 10   A relief map. The area given in Fig. 8, in which the relief is shown by hachures. High areas are light; low areas are dark. This type of map shows only relative and approximate relief and is therefore not used for scientific purposes.

4. TOPOGRAPHIC DRAWING

Topographic maps are distinguished by the presence of contour lines running throughout the map. A "contour" is a theoretical line on the surface of the ground which at every point passes through the same elevation; thus the shoreline of a body of water represents a contour. If the water should rise 1 ft, the new shoreline would be another contour, with 1-ft "contour interval." A series of contours may be illustrated approximately, as in Fig. 7.

Figure 8 is a perspective view of a tract of land. Figure 9 is a contour map of the same area, and Fig. 10 shows the same surface with hill shading by hachures. Contours are drawn as fine full lines, with every fifth one of heavier weight and with the elevations in feet marked on them at intervals, usually with the sea level as datum.

Figure 11 is a topographic map of the site of a pro-

posed filtration plant and illustrates the use of the contour map as the necessary preliminary drawing for engineering projects. Often the same drawing shows, by lines of different character, both the existing contours and the required finished grades.

An interesting form of a topographic map is seen in Fig. 12. Called an orthophoto map, it is an aerial photo done by using special optical effects which correct for distortion that occurs at the edges of a photograph due to a nonvertical angle of viewing as compared to viewing directly downward (vertical). Contour lines are added as well as the lettering of important features (roads, malls, etc.).

When Fig. 12 is modified to the form shown in Fig. 13, the result is called a planimetric map. Superfluous detail of Fig. 12 has been deleted so that lots and other

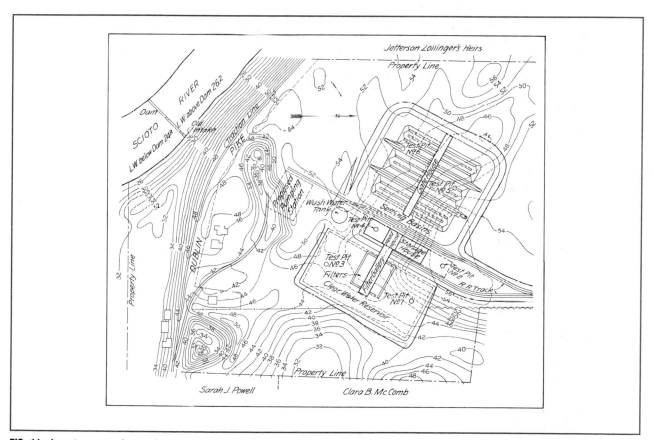

FIG. 11 A contour map of an engineering project. This type of map shows relief by contours and includes other physical features such as buildings, roadways, and streams.

FIG. 12 An orthophoto map. This is a portion of a much larger composite aerial photograph. *(Courtesy Centre Regional Planning Commission, State College, Pa.)*

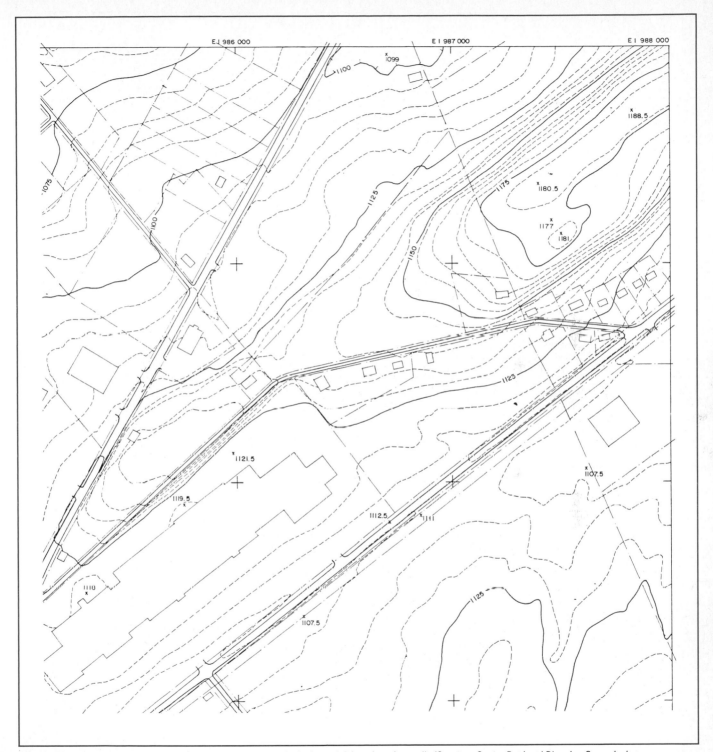

FIG. 13 A planimetric map derived from Fig. 12. The structure in the lower left is a shopping mall. *(Courtesy Centre Regional Planning Commission, State College, Pa.)*

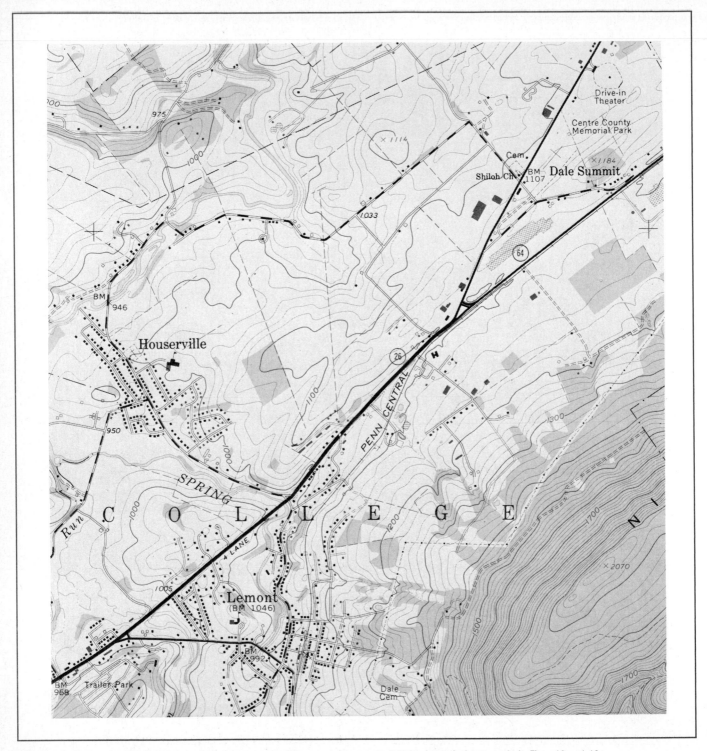

FIG. 14 A portion of a United States geological survey map. The area in the upper right is shown in larger scale in Figs. 12 and 13.

constructed features can be more readily seen. Also, note how much more distinct the contour lines in Fig. 13 are compared to those in Fig. 12.

Another popular form of topographic map is that provided by the U.S. Geological Survey, Washington, D.C. 20242. The quadrangle sheets issued by this agency are easily available for any area of the United States. Scale may vary depending upon whether the area is sparsely populated (e.g., Nevada), or heavily populated (e.g., Maryland).

A portion of a map from the U.S. Geological Survey is seen in Fig. 14. This area encompasses that seen in Figs. 12 and 13, though the smaller scale of 1:24000 allows more land area to be seen. Also, contour lines are coarser, being every 20 feet as opposed to the finer interval of every 5 feet in Figs. 12 and 13.

A last form of topographic map, drawn to a relatively large scale and showing many details, is known as a "landscape map." Though not illustrated here, such maps are required by architects and landscapers for use in planning buildings to fit the natural topographic features and in landscaping parks, playgrounds, and private estates. These are generally maps of small areas, and a scale of $1'' = 20'$ to $1'' = 50'$, depending upon the amount of detail, is used.

The contour interval varies from 6 in to 2 ft according to the ruggedness of the surface. The commonest interval is 1 ft. Landscape maps are often reproduced in black-line prints, upon which contours in different color are drawn to show the landscape treatment proposed. Natural features and culture are added in more detail than on ordinary topographic maps. Trees are designated as to size, species, and sometimes spread of branches and condition. It is often necessary to invent symbols suitable for a particular survey and to include a key or legend on the map. Roads, walks, streams, flower beds, houses, etc., should be plotted carefully to scale, so that measurements can be taken from them.

In summary, a complete topographic will contain:

1. The lines establishing the divisions of authority or ownership.
2. The geographical position of both natural and constructed features. It may also include information on the vegetation.
3. The relief, or indication of the relative elevations

and depressions. The relief, which is the third dimension, is represented in general by contours.

5. TOPOGRAPHIC SYMBOLS

The various symbols used in topographic drawing can be grouped under four heads:

1. Culture, or constructed features
2. Relief—relative elevations and depressions
3. Water features
4. Vegetation

Topographic symbols, used to represent characteristics on the earth's surface, are made, when possible, to resemble somewhat the features or objects represented as they would appear in plan or in elevation. No attempt is made here to give symbols for all the features that might occur in a map; indeed one may have to invent symbols for some particular conditions.

Features may be given by somewhat different symbols from one mapmaker to another. Efforts to totally standardize symbols have not been successful. However, symbols used by the United States Geological Survey are widely accepted.

Figure 15 illustrates a few of the conventional symbols used for constructed features. When the scale used is large, houses, bridges, roads, and even tree trunks can be plotted so that their principal dimensions can be scaled. The landscape architect is interested not only in the size of the trunk of a tree but also in the spread of its branches. A small-scale map can give by its symbols only the relative locations.

Symbols for aids to water navigation are shown in Fig. 16. The symbols of Fig. 17 are used to show relief; those of Fig. 18, water features; and those of Fig. 19, some of the common symbols for vegetation.

Keep in mind the purpose of the map and in some measure indicate the relative importance of the features, varying their prominence by the weight of lines used or by varying the scale of the symbol. For instance, in a map made for military maneuvers a cornfield might be an important feature; or in maps made to show the location of special features, such as fire hydrants, these objects would be plainly indicated. The map of an airport or a golf course would emphasize features significant for

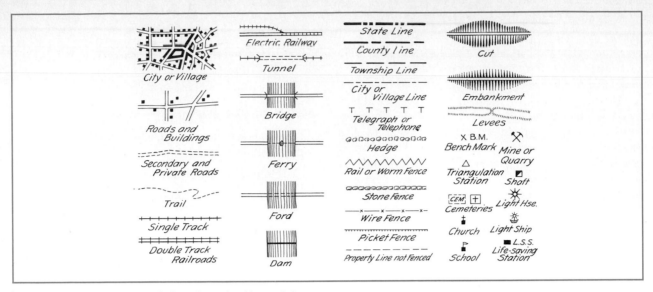

FIG. 15 Culture symbols, used to indicate the works of human beings.

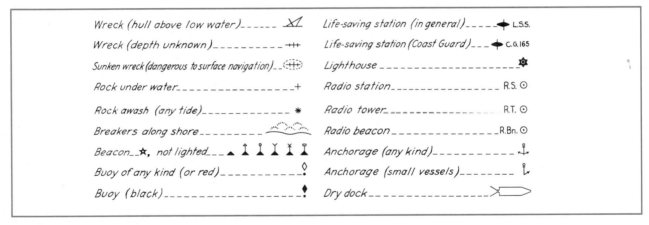

FIG. 16 Marine symbols, used on navigation maps for water craft.

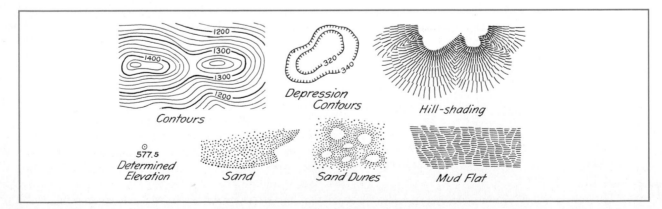

FIG. 17 Relief symbols. Although highly detailed, they may be used on contour and relief maps.

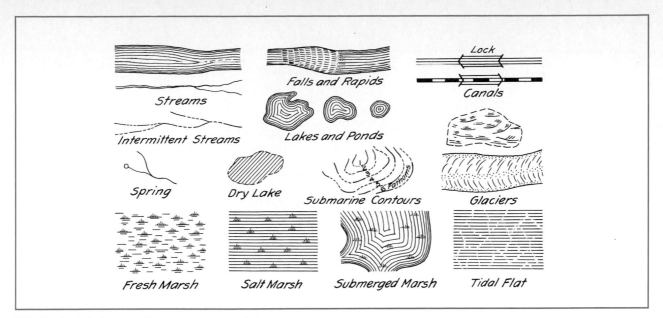

FIG. 18 Symbols for water features, used as needed on any map.

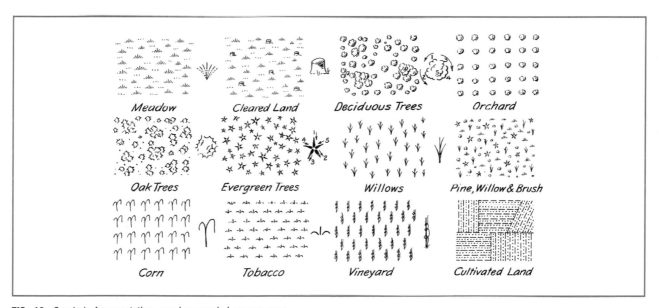

FIG. 19 Symbols for vegetation, used as needed on any map.

an airport or golf course. The principle calls for some originality in meeting various cases.

6. PROFILES

Perhaps no kind of drawing is used more by civil engineers than the ordinary "profile," which is simply a vertical section taken along a given line, either straight or curved. Such drawings are indispensable in problems of railroad construction, highway and street improvements, sewer construction, and many other problems where a study of the surface of the ground is required. Frequently engineers other than civil engineers are called upon to make profile drawings. Several different types of profile and cross-section paper are in use; they are described in the catalogs of the various firms dealing in drawing mate-

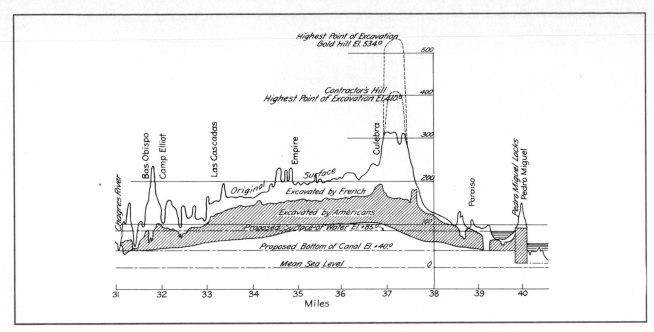

FIG. 20 A profile used to study surface variations. In this profile, the vertical scale is 50 times the horizontal scale.

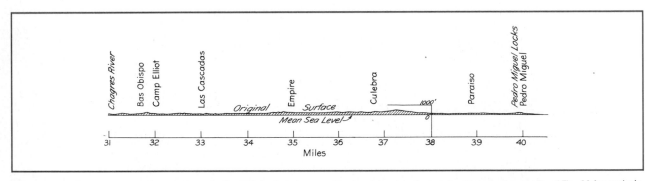

FIG. 21 A profile. The same surface as shown in Fig. 20 but plotted on equal horizontal and vertical scales. Note the superiority of Fig. 20 for analysis.

rials. One type of profile paper in common use is known as "Plate A" and has 4 divisions to the inch horizontally and 20 to the inch vertically. Other divisions in use are 4 × 30 to the inch and 5 × 25 to the inch. At intervals, both horizontally and vertically, heavier lines are made to facilitate reading.

Horizontal distances are plotted as abscissas and elevations as ordinates. Since the vertical distances represent elevations and are plotted to larger scale, a vertical exaggeration is obtained that is useful in studying profiles that are to be used for establishing grades. The vertical

exaggeration is sometimes confusing to the layperson or inexperienced engineer, but ordinarily a profile will fail in the purpose for which it was intended if the horizontal and vertical scales are the same. Furthermore, if the profile were not distorted in this way, it would be a long and unwieldy affair, perhaps even impossible to make. The difference between profiles with and without vertical exaggeration is shown in Figs. 20 and 21.

Figure 22 is a portion of a typical state highway alignment and profile sheet, plotted to a horizontal scale of $1'' = 100'$ and a vertical scale of $1'' = 10'$. The align-

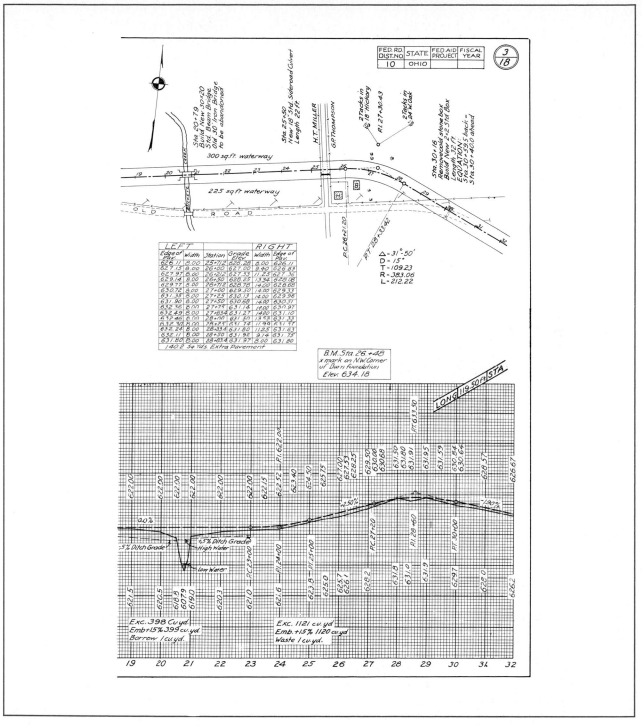

FIG. 22 Part of a state highway alignment and profile sheet. Gives fundamental information needed for construction of a highway.

ment and profile sheet is one of a set of drawings used in the construction of a highway for estimating cost and as a working drawing, by the contractor, during construction. Other drawings in the set would consist of a title sheet showing the location plan with detours provided, a sheet indicating conventional signs, a sheet giving an index to bound sheets, and a sheet with space reserved for declarations of approval and signatures of proper officials.

There would also be sheets of cross sections taken at each 100-ft station and all necessary intermediate stations to estimate earthwork for grading; working drawings for drainage structures; site plans for bridges; specifications for guardrails and other safety devices; standard or typical road sections for cut and fill and various other conditions; and finally, summary sheets for separate tables and quantities of materials for roadway, pavement, and structures.

7. COLORS AND LETTERING

Color can enhance the readability of maps. Common colors are black for cultural features, blue for water, green for vegetation, brown for contour lines, orange for roads, and black or purple for structures. The choice of color is quite wide, naturally, and may vary somewhat from one type of map to another. Technical pens are readily available for use. Also, felt-tip markers are increasingly accepted for coloring areas and where special effects are desired.

The style of lettering on a topographic map will depend upon the purpose for which the map is made. If it is for construction purposes, such as a contour map for the study of municipal problems, street grades, plants, or railroads, the single-stroke Gothic and Reinhardt is preferred. For a finished map, vertical Modern Roman letters should be used, capitals for important land features and lowercase for less important features, such as small towns and villages; for water features, inclined Roman and stump letters. The scale should always be drawn as well as stated.

8. TITLES

The design should be symmetrical, with the height of the letters proportioned to the relative importance of the line. A map title should contain as many of the following items as are necessary.

1. Kind—"Map of," etc.
2. Name
3. Location of tract
4. Purpose, if special features are represented
5. For whom made
6. Engineer in charge
7. Date (of survey)
8. Scale—stated and drawn, contour interval, datum
9. Authorities
10. Legend or key to symbols
11. North point, with certification of horizontal control
12. Certification, properly executed
13. Reference to state plane-coordinate system

PROBLEMS

In Probs. 1 and 2, contour maps, which can be obtained from local, county, state, or federal authorities, are needed.

GROUP 1. CONTOUR PROBLEMS

1. Obtain a contour map showing a river. Plot the watershed of the river. This is done by locating the series of highest points on each side of the river and drawing a watershed line connecting the points. Indicate the watershed area by crosshatching.

2. Obtain a contour map showing a river. At a point on the river where the banks are steep, or where there is a maximum rise above the river, locate a dam. Plot the lake formed by the dam. Represent the lake by crosshatching.

GROUP 2. PLATS OF SUBDIVISIONS

3. From Fig. 1, lay out the complete tract of land enclosed by the survey. Divide this tract into lots with (approximately) 70-foot frontage.

GROUP 3. PROFILES

4. Plot the profile of Fig. 20 to suitable scale on a 17- × 22-in sheet.

5. Plot the profile of a roadway similar to Fig. 22 but with these elevations:

STATION POINT AND ELEVATION											
Sta. 19	20	21	22	23	24	25	26	27	28	29	30
Elev. 511.0	511.0	511.0	511.0	511.2	511.5	513.5	516.0	518.5	520.5	520.9	519.8

Add a profile of ditch elevations deviating from the roadway as follows:

STATION POINT AND ELEVATION											
Sta. 19	20	21	22	23	24	25	26	27	28	29	30
Elev. −1.0	−1.5	−1.5	−1.6	−1.4	−1.0	−0.5	+1.0	+1.5	+2.5	+3.5	+4.5

APPENDIXES

CONTENTS

APPENDIX 1 GLOSSARY

PART A. SHOP TERMS

anneal (*v.*)   To soften a metal piece and remove internal stresses by heating to its critical temperature and allowing to cool very slowly.

arc-weld (*v.*)   To weld by electric-arc process.

bore (*v.*)   To enlarge a hole with a boring tool, as in a lathe or boring mill. Distinguished from *drill*.

boss (*n.*)   A projection of circular cross section, as on a casting or forging.

braze (*v.*)   To join by the use of hard solder.

broach (*v.*)   To finish the inside of a hole to a shape usually other than round. (*n.*) A tool with serrated edges pushed or pulled through a hole to enlarge it to a required shape.

buff (*v.*)   To polish with abrasive on a cloth wheel or other soft carrier.

burnish (*v.*)   To smooth or polish by a rolling or sliding tool under pressure.

bushing (*n.*)   A removable sleeve or liner for a bearing; also a guide for a tool in a jig or fixture.

carburize (*v.*)   To prepare a low-carbon steel for heat-treatment by packing in a box with carbonizing material, such as wood charcoal, and heating to about 2000°F for several hours, then allowing to cool slowly.

caseharden (*v.*)   To harden the surface of carburized steel by heating to critical temperature and quenching, as in an oil or lead bath.

castellate (*v.*)   To form into a shape resembling a castle battlement, as castellated nut. Often applied to a shaft with multiple integral keys milled on it.

chamfer (*v.*)   To bevel a sharp external edge. (*n.*) A beveled edge.

Boss

Bushing

Chamfer

Counterbore

Countersink

Fillet

Flange

Kerf

chase (*v.*)   To cut threads in a lathe, as distinguished from cutting threads with a die. (*n.*) A slot or groove.

chill (*v.*)   To harden the surface of cast iron by sudden cooling against a metal mold.

chip (*v.*)   To cut or clean with a chisel.

coin (*v.*)   To stamp and form a metal piece in one operation, usually with a surface design.

cold-work (*v.*)   To deform metal stock by hammering, forming, drawing, etc., while the metal is at ordinary room temperature.

color-harden (*v.*)   To caseharden to a very shallow depth, chiefly for appearance.

core (*v.*)   To form the hollow part of a casting, using a solid form made of sand, shaped in a core box, baked, and placed in the mold. After cooling, the core is easily broken up, leaving the casting hollow.

counterbore (*v.*)   To enlarge a hole to a given depth. (*n.*) 1. The cylindrical enlargement of the end of a drilled or bored hole. 2. A cutting tool for counterboring, having a piloted end the size of the drilled hole.

countersink (*v.*)   To form a depression to fit the conic head of a screw or the thickness of a plate so that the face will be level with the surface. (*n.*) A conic tool for countersinking.

crown (*n.*)   Angular or rounded contour, as on the face of a pulley.

die (*n.*)   1. One of a pair of hardened metal blocks for forming, impressing, or cutting out a desired shape. 2 (thread). A tool for cutting external threads. Opposite of *tap*.

die casting (*n.*)   A very accurate and smooth casting made by pouring a molten alloy (or composition, as Bakelite) usually under pressure into a metal mold or die. Distinguished from a casting made in sand.

die stamping (*n.*)   A piece, usually of sheet metal, formed or cut out by a die.

draw (*v.*)   1. To form by a distorting or stretching process. 2. To temper steel by gradual or intermittent quenching.

drill (*v.*)   To sink a hole with a drill, usually a twist drill. (*n.*) A pointed cutting tool rotated under pressure.

drop forging (*n.*)   A wrought piece formed hot between dies under a drop hammer, or by pressure.

face (*v.*)   To machine a flat surface perpendicular to the axis of rotation on a lathe. Distinguished from *turn*.

feather (*n.*)   A flat sliding key, usually fastened to the hub.

fettle (*v.*)   To remove fins and smooth the corners on unfired ceramic products.

file (*v.*)   To finish or trim with a file.

fillet (*n.*)   A rounded filling of the internal angle between two surfaces.

fin (*n.*)   A thin projecting rib. Also, excess ridge of material.

fit (*n.*)   The kind of contact between two machined surfaces. 1. *Drive, force,* or *press:* When the shaft is slightly larger than the hole and must be forced in with sledge or power press. 2. *Shrink:* When the shaft is slightly larger than the hole, the piece containing the hole is heated, thereby expanding the hole sufficiently to slip over the shaft. On cooling, the shaft will be seized firmly if the fit allowances have been correctly proportioned. 3. *Running* or *sliding:* When sufficient allowance has been made between sizes of shaft and hole to allow free running without seizing or heating. 4. *Wringing:* When the allowance is smaller than a running fit and the shaft will enter the hole by twisting it by hand.

flange (*n.*)   A projecting rim or edge for fastening or stiffening.

forge (*v.*)   To shape metal while hot and plastic by a hammering or forcing process either by hand or by machine.

galvanize (*v.*)   To treat with a bath of lead and zinc to prevent rusting.

graduate (*v.*)   To divide a scale or dial into regular spaces.

grind (*v.*)   To finish or polish a surface by means of an abrasive wheel.

harden (*v.*)   To heat hardenable steel above critical temperature and quench in bath.

hot-work (*v.*)   To deform metal stock by hammering, forming, drawing, etc., while the metal is heated to a plastic state.

kerf (*n.*)   The channel or groove cut by a saw or other tool.

key (*n.*) A small block or wedge inserted between shaft and hub to prevent circumferential movement.

keyway, key seat (*n.*) A groove or slot cut to fit a key. A key fits into a key seat and slides in a keyway.

knurl (*v.*) To roughen or indent a turned surface, as a knob or handle.

lap (*n.*) A piece of soft metal, wood, or leather charged with abrasive material, used for obtaining an accurate finish. (*v.*) To finish by lapping.

lug (*n.*) A projecting "ear," usually rectangular in cross section. Distinguished from *boss*.

malleable casting (*n.*) An ordinary casting toughened by annealing. Applicable to small castings with uniform metal thicknesses.

mill (*v.*) To machine with rotating toothed cutters on a milling machine.

neck (*v.*) To cut a groove around a shaft, usually near the end or at a change in diameter. (*n.*) A portion reduced in diameter between the ends of a shaft.

normalize (*v.*) To remove internal stresses by heating a metal piece to its critical temperature and allowing to cool very slowly.

pack-harden (*v.*) To carburize and case-harden.

pad (*n.*) A shallow projection. Distinguished from *boss* by shape or size.

peen (*v.*) To stretch, rivet, or clinch over by strokes with the peen of a hammer. (*n.*) The end of a hammer head opposite the face, as *ball peen*.

pickle (*v.*) To clean castings or forgings in a hot weak sulfuric acid bath.

plane (*v.*) To machine work on a planer having a fixed tool and reciprocating bed.

planish (*v.*) To finish sheet metal by hammering with polished-faced hammers.

plate (*v.*) The electrochemical coating of a metal piece with a different metal.

polish (*v.*) To make smooth or lustrous by friction with a very fine abrasive.

profile (*v.*) To machine an outline with a rotary cutter usually controlled by a master cam or die.

punch (*v.*) To perforate by pressing a non-rotating tool through the work.

ream (*v.*) To finish a drilled or punched hole very accurately with a rotating fluted tool of the required diameter.

relief (*n.*) The amount one plane surface of a piece is set below or above another plane, usually for clearance or for economy in machining.

rivet (*v.*) 1. To fasten with rivets. 2. To batter or upset the headless end of a pin used as a permanent fastening.

round (*n.*) A rounded exterior corner between two surfaces. Compare with *fillet*.

sandblast (*v.*) To clean castings or forgings by means of sand driven through a nozzle by compressed air.

shape (*v.*) To machine with a shaper, a machine tool differing from a planer in that the work is stationary and the tool reciprocating.

shear (*v.*) To cut off sheet or bar metal between two blades.

sherardize (*v.*) To galvanize with zinc by a dry heating process.

shim (*n.*) A thin spacer of sheet metal used for adjusting.

shoulder (*n.*) A plane surface on a shaft, normal to the axis and formed by a difference in diameter.

spin (*v.*) To shape sheet metal by forcing it against a form as it revolves.

spline (*n.*) A long keyway. Sometimes also a flat key.

spot-face (*v.*) To finish a round spot on a rough surface, usually around a drilled hole, to give a good seat to a screw or bolt-head, cut, usually 1/16 in deep, by a rotating milling cutter.

spot-weld (*v.*) To weld in spots by means of the heat of resistance to an electric current. Not applicable to sheet copper or brass.

steel casting (*n.*) Material used in machine construction. It is ordinary cast iron into which varying amounts of scrap steel have been added in the melting.

swage (*v.*) To shape metal by hammering or pressure with the aid of a form or anvil called a "swage block."

Keyway

Key and seat

Lug

Neck

Pad

Round

Spline

Spot-face

sweat (*v.*) To join metal pieces by clamping together with solder between and applying heat.

tack-weld (*v.*) To join at the edge by welding in short intermittent sections.

tap (*v.*) To cut threads in a hole with a rotating tool called a "tap," having threads on it and fluted to give cutting edges.

temper (*v.*) To change the physical characteristics of hardened steel by reheating to a temperature below the critical point and allowing to cool.

template, templet (*n.*) A flat pattern for laying out shapes, location of holes, etc.

trepan (*v.*) To cut an outside annular groove around a hole.

tumble (*v.*) To clean, smooth, or polish castings or forgings in a rotating barrel or drum by friction with each other, assisted by added mediums, as scraps, "jacks," balls, sawdust, etc.

turn (*v.*) To machine on a lathe. Distinguished from *face*.

undercut (*v.*) To cut, leaving an overhanging edge. (*n.*) A cut having inwardly sloping sides.

upset (*v.*) To forge a larger diameter or shoulder on a bar.

weld (*v.*) To join two pieces by heating them to the fusing point and pressing or hammering together.

PART B. STRUCTURAL TERMS

bar Square or round rod; also flat steel up to 6 in. in width.

batten plate A small plate used to hold two parts in their proper position when made up as one member.

batter A deviation from the vertical in upright members.

bay The distance between two trusses or transverse bents.

beam A horizontal member forming part of the frame of a building or structure.

bearing plate A steel plate, usually at the base of a column, used to distribute a load over a larger area.

bent A vertical framework usually consisting of a truss or beam supported at the ends on columns.

brace A diagonal member used to stiffen a framework.

buckle plate A flat plate with dished depression pressed into it to give transverse strength.

built-up member A member built from standard shapes to give one single stronger member.

camber Slight upward curve given to trusses and girders to avoid effect of sag.

cantilever A beam, girder, or truss overhanging one or both supports.

chord The principal member of a truss on either the top or bottom.

clearance Rivet driving clearance is distance from center of rivet to obstruction. Erection clearance is amount of space left between members for ease in assembly.

clevis U-shaped shackle for connecting a rod to a pin.

clip angle A small angle used for fastening various members together.

column A vertical compression member.

cope To cut out top or bottom of flanges and web so that one member will frame into another.

coping A projecting top course of concrete or stone.

counters Diagonal members in a truss to provide for reversal of shear due to live load.

cover plate A plate used in building up flanges, in a built-up member, to give greater strength and area or for protection.

crimp To offset the end of a stiffener to fit over the leg of an angle.

diagonals Diagonal members used for stiffening and wind bracing.

dowel An iron or wooden pin extending into, but not through, two timbers to connect them.

driftpin A tapered steel pin used to bring rivet holes fair in assembling steel work.

edge distance The distance from center of rivet to edge of plate or flange.

fabricate To cut, punch, and subassemble members in the shop.

Trepan

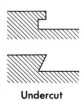

Undercut

fillers Either plate or ring fills used to take up space in riveting two members where a gusset is not used.

flange The projecting portion of a beam, channel, or column.

gage line The center line for rivet holes.

gin pole A guyed mast with block at the top for hoisting.

girder A horizontal member, either single or built up, acting as a principal beam.

girt A beam usually bolted to columns to support the side covering or serve as window lintels.

gusset plate A plate used to connect various members, such as in a truss.

hip The intersection between two sloping surfaces forming an exterior angle.

knee brace A corner brace used to prevent angular movement.

lacing or lattice bars Bars used diagonally to space and stiffen two parallel members, such as in a built-up column.

laterals Members used to prevent lateral deflection.

lintel A horizontal member used to carry a wall over an opening.

louvers Metal slats either movable or fixed, as in a monitor ventilator.

monitor ventilator A framework that carries fixed or movable louvers at the top of the roof.

panel The space between adjacent floor supports or purlins in a roof.

pitch Center distance between rivets parallel to axis of member. Also, for roofs, the ratio of rise to span.

plate Flat steel over 6 in. in width and ¼ in. or more in thickness.

purlins Horizontal members extending between trusses, used as beams for supporting the roof.

rafters Beams or truss members supporting the purlins.

sag ties Tie rods between purlins in the plane of the roof to carry the component of the roof load parallel to the roof.

separator Either a cast-iron spacer or wrought-iron pipe on a bolt for the purpose of holding members a fixed distance apart.

sheet Flat steel over 6 in. in width and less than ¼ in. in thickness.

shim A thin piece of wood or steel placed under a member to bring it to a desired elevation.

sleeve nut A long nut with right and left threads for connecting two rods to make an adjustable member.

span Distance between centers of supports of a truss, beam, or girder.

splice A longitudinal connection between the parts of a continuous member.

stiffener Angle, plate, or channel riveted to a member to prevent buckling.

stringer A longitudinal member used to support loads directly.

strut A compression member in a framework.

truss A rigid framework for carrying loads, formed in a series of triangles.

turnbuckle A coupling, threaded right and left or swiveled on one end, for adjustably connecting two rods.

valley The intersection between two sloping surfaces, forming a reentrant angle.

web The part of a channel, I beam, or girder between the flanges.

PART C. ARCHITECTURAL TERMS

apron The finished board placed against the wall surface, immediately below a window stool.

ashlar Thin, squared, and dressed stone facing of a masonry wall.

backing The inner portion of a wall; that which is used behind the facing.

batten A strip of wood used for nailing across two other pieces of wood to hold them together and cover a crack.

batter boards Boards set up at the corners of a proposed building from which the lines marking off the walls are stretched.

bearing wall A wall that supports loads other than its own weight.

bond The joining together of building materials to ensure solidity.

bridging The braces or system of bracing used between joists or other structural members to stiffen them and to distribute the load.

centering A substructure of temporary nature, usually of timber or planks, on which a masonry arch or vault is built.

coffer An ornamental panel deeply recessed, usually in a dome or portico ceiling.

corbel A bracket formed on a wall by building out successive courses of masonry.

curtain wall A wall that carries no building load other than its own weight.

fenestration The arrangement and proportioning of window and door openings.

flashing The sheet metal built into the joints of a wall, or covering the valleys, ridges, and hips of a roof for the purpose of preventing leakage.

footing A course or series of courses projecting at the base of a wall for the purpose of distributing the load from above over a greater area, thereby preventing excessive settlement.

furring The application of thin wood, metal, or other building material to a wall, beam, ceiling, or the like to level a surface for lathing, boarding, etc., or to make an air space within a wall.

glazing The act of furnishing or fitting with glass.

ground Strips of wood, flush with the plastering, to which moldings, etc., are attached. Grounds are usually installed first and the plastering floated flush with them.

grout A thin mortar used for filling up spaces where heavier mortar will not penetrate.

head The horizontal piece forming the top of a wall opening, as a door or window.

hip The intersection of two roof surfaces, which form on the plan an external angle.

jamb The vertical piece forming the side of a wall opening.

lintel The horizontal structural member that supports the wall over an opening.

millwork The finish woodwork, machined and in some cases partly assembled at the mill.

miter To match together, as two pieces of molding, on a line bisecting the angle of junction.

mullion A vertical division of a window opening.

muntin The thin members that separate the individual lights of glass in a window frame.

party wall A division wall common to two adjacent pieces of property.

plate A horizontal member that carries other structural members; usually the top timber of a wall that carries the roof trusses or rafters directly.

rail A horizontal piece in a frame or paneling.

return The continuation in a different direction, most often at right angles, of the face of a building or any member, as a colonnade or molding; applied to the shorter in contradistinction to the longer.

reveal The side of a wall opening; the whole thickness of the wall; the jamb.

riser The upright piece of a step, from tread to tread.

saddle A small double-sloping roof to carry water away from behind chimneys, etc.

scratch coat The first coat in plastering, roughened by scratching or scoring so that the next coat will firmly adhere to it.

screeds A strip of plaster of the thickness proposed for the work, applied to the wall at intervals of 4 or 5 ft, to serve as guides.

shoring A prop, as a timber, placed against the side of a structure; a prop placed beneath anything, as a beam, to prevent sinking or sagging.

sill The horizontal piece, as a timber, which forms the lowest member of a frame.

sleepers The timbers laid on a firm foundation to carry and secure the superstructure.

soffit The underside of subordinate parts and members of buildings, such as staircases, beams, arches, etc.

stile A vertical piece in a frame or paneling.

stool The narrow shelf fitted on the inside of a window against the actual sill.

threshold The stone, wood, or metal piece that lies directly under a door.

trap A water seal in a sewage system to prevent sewer gas from entering the building.

tread The upper horizontal piece of a step, on which the foot is placed.

valley The intersection of two roof surfaces which form, on the plan, a reentrant angle.

PART D. WELDING TERMS

air-acetylene welding A gas-welding process in which coalescence is produced by heating with a gas flame or flames obtained from the combustion of acetylene with air, without the application of pressure and with or without the use of filler metal.

arc cutting A group of cutting processes in which the severing of metals is effected by melting with the heat of an arc between an electrode and the base metal.

arc welding A group of welding processes in which coalescence is produced by heating with an electric arc or arcs, with or without the application of pressure and with or without the use of filler metal.

As-welded The condition of weld metal, welded joints, and weldments after welding prior to any subsequent thermal or mechanical treatment.

atomic-hydrogen welding An arc-welding process in which coalescence is produced by heating with an electric arc maintained between two metal electrodes in an atmosphere of hydrogen. Shielding is obtained from the hydrogen. Pressure may or may not be used, and filler metal may or may not be used.

automatic welding Welding with equipment which performs the entire welding operation without constant observation and adjustment of the controls by an operator. The equipment may or may not perform the loading and unloading of the work.

axis of a weld A line through the length of a weld, perpendicular to the cross section at its center of gravity.

backing Material (metal, weld metal, asbestos, carbon, granular flux, etc.) backing up the joint during welding to facilitate obtaining a sound weld at the root.

bevel A type of edge preparation.

braze A weld in which coalescence is produced by heating to suitable temperatures above 800°F and by using a nonferrous filler metal having a melting point below that of the base metals. The filler metal is distributed between the closely fitted surfaces of the joint by capillary attraction.

butt joint A joint between two members lying approximately in the same plane.

coalesce To unite or merge into a single body or mass. Fusion.

die welding A forge-welding process in which coalescence is produced by heating in a furnace and by applying pressure by means of dies.

edge joint A joint between the edges of two or more parallel or nearly parallel members.

filler metal Metal to be added in making a weld.

fillet weld A weld of approximately triangular cross section joining two surfaces approximately at right angles to each other in a lap joint, tee joint, or corner joint.

forge welding A group of welding processes in which coalescence is produced by heating in a forge or other furnace and by applying pressure or blows.

fusion The melting together of filler metal and base metal, or of base metal only, which results in coalescence.

gas welding A group of welding processes in which coalescence is produced by heating with a gas flame or flames, with or without the application of pressure, and with or without the use of filler metal.

groove weld A weld made in the groove between two members to be joined. The standard types of groove welds are: square-groove weld; single-V-groove weld; single-bevel-groove weld; single-U-groove weld; single-J-groove weld; double-V-groove weld; double-bevel-groove weld; double-U-groove weld; double-J-groove weld.

hammer welding A forge-welding process in which coalescence is produced by heating in a forge or other furnace and by applying pressure by means of hammer blows.

intermittent welding Welding in which the

continuity is broken by recurring unwelded spaces.

joint penetration The minimum depth a groove weld extends from its face into a joint, exclusive of reinforcement.

kerf The space from which metal has been removed by a cutting process.

lap joint A joint between two overlapping members.

oxy-acetylene welding A gas-welding process in which coalescence is produced by heating with a gas flame or flames obtained from a combustion of acetylene with oxygen, with or without the application of pressure and with or without the use of filler metal.

peening The mechanical working of metals by means of hammer blows.

plug weld A circular weld made by either arc or gas welding through one member of a lap or tee joint joining that member to the other. The weld may or may not be made through a hole in the first member.

pressure welding Any welding process or method in which pressure is used to complete the weld.

projection welding A resistance-welding process in which coalescence is produced by the heat obtained from resistance to the flow of electric current through the work parts held together under pressure by electrodes.

root opening The separation between the members to be joined, at the root of the joint.

seam weld A weld consisting of a series of overlapping spot welds, made by seam welding or spot welding.

slot weld A weld made in an elongated hole in one member of a lap or tee joint joining that member to the portion of the surface of the other member which is exposed through the hole.

spot welding A resistance-welding process in which coalescence is produced by the heat obtained from resistance to the flow of electric current through the work parts, which are held together under pressure by electrodes.

tack weld A weld made to hold parts of a weldment in proper alignment until the final welds are made.

tee joint A joint between two members located approximately at right angles to each other in the form of a T.

upset welding A resistance-welding process in which coalescence is produced, simultaneously over the entire area of abutting surfaces or progressively along a joint, by the heat obtained from resistance to the flow of electric current through the area of contact of those surfaces. Pressure is applied before heating is started and is maintained throughout the heating period.

APPENDIX 2 SI UNIT PREFIXES

Multiplication factor		Prefix	Symbol	Pronunciation (USA)*	Term (USA)	Term (other countries)
1 000 000 000 000 000 000	= 10^{18}	exa	E	as in Texas	one quintillion†	one trillion
1 000 000 000 000 000	= 10^{15}	peta	P	as in petal	one quadrillion†	one thousand billion
1 000 000 000 000	= 10^{12}	tera	T	as in terrace	one trillion†	one billion
1 000 000 000	= 10^{9}	giga	G	jig' a (a as in about)	one billion†	one millard
1 000 000	= 10^{6}	mega	M	as in megaphone	one million	
1 000	= 10^{3}	kilo	k	as in kilowatt	one thousand	
100	= 10^{2}	hecto	h	heck'toe	one hundred	
10	= 10	deka	da	deck' a (a as in about)	ten	
0.1	= 10^{-1}	deci	d	as in decimal	one tenth	
0.01	= 10^{-2}	centi	c	as in sentiment	one hundredth	
0.001	= 10^{-3}	milli	m	as in military	one thousandth	
0.000 001	= 10^{-6}	micro	µ	as in microphone	one millionth	
0.000 000 001	= 10^{-9}	nano	n	nan' oh (nan as in Nancy)	one billionth†	one milliardth
0.000 000 000 001	= 10^{-12}	pico	p	peek' oh	one trillionth†	one billionth
0.000 000 000 000 001	= 10^{-15}	femto	f	fem' toe (fem as in feminine)	one quadrillionth†	one thousand billionth
0.000 000 000 000 000 001	= 10^{-18}	atto	a	as in anatomy	one quintillionth†	one trillionth

Abstracted from material by American National Metric Council, 1977.

*The first syllable of every prefix is accented to ensure that the prefix will retain its identity

†These terms should be avoided in technical writing because the denominations above one million and below one millionth are different in most countries, as indicated in the last column. Instead, use the prefixes or 10 raised to an integral power

APPENDIX 3   COMMON SI UNITS

Quantity	Some common units	Symbol	Equivalent	Symbol
Length	kilometer	km		
	meter	m		
	centimeter	cm		
	millimeter	mm		
	micrometer	μm		
Area	square kilometer	km^2		
	square hectometer	hm^2	hectare	ha
	square meter	m^2		
	square centimeter	cm^2		
	square millimeter	mm^2		
Volume	cubic meter	m^3		
	cubic decimeter	dm^3	liter	L
	cubic centimeter	cm^3	milliliter	mL
Plane angle	degree	°		
Speed or velocity	meter per second	m/s		
	kilometer per hour	km/h		
Acceleration	meter per second squared	m/s^2		
Frequency	megahertz	MHz		
	kilohertz	kHz		
	hertz	Hz		
Rotational frequency	revolution per second	r/s		
	revolution per minute	r/min		
Mass	megagram	Mg	metric ton	t
	kilogram	kg		
	gram	g		
	milligram	mg		
Density	kilogram per cubic meter	kg/m^3	gram per liter	g/L
Force	kilonewton	kN		
	newton	N		
Moment of force	newton-meter	N·m		
Pressure (or vacuum)	kilopascal	kPa		
Stress	megapascal	MPa		
Viscosity (dynamic)	millipascal second	mPa·s		
Viscosity (kinematic)	square millimeter per second	mm^2/s		
Energy, work, or quantity of heat	joule	J		
	kilowatthour	kW·h	kilowatthour	kWh
Power, or heat flow rate	kilowatt	kW		
	watt	W		
Temperature, or temperature interval	kelvin	K		
	degree Celsius	°C		
Electric current	ampere	A		

Quantity	Some common units	Symbol	Equivalent	Symbol
Quantity of electricity	coulomb	C		
	ampere-hour	A·h		Ah
Electromotive force	volt	V		
Electric resistance	ohm	Ω		
Luminous intensity	candela	cd		
Luminous flux	lumen	lm		
Illuminance	lux	lx		
Sound level	decibel	dB		

Abstracted from material by American National Metric Council, 1977.

APPENDIX 4 DECIMAL INCH-MILLIMETER CONVERSIONS

Decimal Equivalents of Inch Fractions

Fraction	Equiv.	Fraction	Equiv.	Fraction	Equiv.	Fraction	Equiv.
1/64	0.015625	17/64	0.265625	33/64	0.515625	49/64	0.765625
1/32	0.03125	9/32	0.28125	17/32	0.53125	25/32	0.78125
3/64	0.046875	19/64	0.296875	35/64	0.546875	51/64	0.796875
1/16	0.0625	5/16	0.3125	9/16	0.5625	13/16	0.8125
5/64	0.078125	21/64	0.328125	37/64	0.578125	53/64	0.828125
3/32	0.09375	11/32	0.34375	19/32	0.59375	27/32	0.84375
7/64	0.109375	23/64	0.359375	39/64	0.609375	55/64	0.859375
1/8	0.1250	3/8	0.3750	5/8	0.6250	7/8	0.8750
9/64	0.140625	25/64	0.390625	41/64	0.640625	57/64	0.890625
5/32	0.15625	13/32	0.40625	21/32	0.65625	29/32	0.90625
11/64	0.171875	27/64	0.421875	43/64	0.671875	59/64	0.921875
3/16	0.1875	7/16	0.4375	11/16	0.6875	15/16	0.9375
13/64	0.203125	29/64	0.453125	45/64	0.703125	61/64	0.953125
7/32	0.21875	15/32	0.46875	23/32	0.71875	31/32	0.96875
15/64	0.234375	31/64	0.484375	47/64	0.734375	63/64	0.984375
1/4	0.2500	1/2	0.5000	3/4	0.7500	1	1.0000

Metric Equivalents

Mm	In.*	Mm	In.	In.	Mm †	In.	Mm
1 = 0.0394		17 = 0.6693		1/32 = 0.794		17/32 = 13.494	
2 = 0.0787		18 = 0.7087		1/16 = 1.588		9/16 = 14.288	
3 = 0.1181		19 = 0.7480		3/32 = 2.381		19/32 = 15.081	
4 = 0.1575		20 = 0.7874		1/8 = 3.175		5/8 = 15.875	
5 = 0.1969		21 = 0.8268		5/32 = 3.969		21/32 = 16.669	
6 = 0.2362		22 = 0.8662		3/16 = 4.762		11/16 = 17.462	
7 = 0.2756		23 = 0.9055		7/32 = 5.556		23/32 = 18.256	
8 = 0.3150		24 = 0.9449		1/4 = 6.350		3/4 = 19.050	
9 = 0.3543		25 = 0.9843		9/32 = 7.144		25/32 = 19.844	
10 = 0.3937		26 = 1.0236		5/16 = 7.938		13/16 = 20.638	
11 = 0.4331		27 = 1.0630		11/32 = 8.731		27/32 = 21.431	
12 = 0.4724		28 = 1.1024		3/8 = 9.525		7/8 = 22.225	
13 = 0.5118		29 = 1.1418		13/32 = 10.319		29/32 = 23.019	
14 = 0.5512		30 = 1.1811		7/16 = 11.112		15/16 = 23.812	
15 = 0.5906		31 = 1.2205		15/32 = 11.906		31/32 = 24.606	
16 = 0.6299		32 = 1.2599		1/2 = 12.700		1 = 25.400	

*Rounded to fourth decimal place.
†Rounded to third decimal place.

Angle	Sine		Cosine		Tangent		Cotangent		Angle
	Nat.	Log.	Nat.	Log.	Nat.	Log.	Nat.	Log.	
0° 00′	.0000	∞	1.0000	0.0000	.0000	∞	∞	∞	90° 00′
10	.0029	7.4637	1.0000	0000	.0029	7.4637	343.77	2.5363	50
20	.0058	7648	1.0000	0000	.0058	7648	171.89	2352	40
30	.0087	9408	1.0000	0000	.0087	9409	114.59	0591	30
40	.0116	8.0658	.9999	0000	.0116	8.0658	85.940	1.9342	20
50	.0145	1627	.9999	0000	.0145	1627	68.750	8373	10
1° 00′	.0175	8.2419	.9998	9.9999	.0175	8.2419	57.290	1.7581	89° 00′
10	.0204	3088	.9998	9999	.0204	3089	49.104	6911	50
20	.0233	3668	.9997	9999	.0233	3669	42.964	6331	40
30	.0262	4179	.9997	9999	.0262	4181	38.188	5819	30
40	.0291	4637	.9996	9998	.0291	4638	34.368	5362	20
50	.0320	5050	.9995	9998	.0320	5053	31.242	4947	10
2° 00′	.0349	8.5428	.9994	9.9997	.0349	8.5431	28.636	1.4569	88° 00′
10	.0378	5776	.9993	9997	.0378	5779	26.432	4221	50
20	.0407	6097	.9992	9996	.0407	6101	24.542	3899	40
30	.0436	6397	.9990	9996	.0437	6401	22.904	3599	30
40	.0465	6677	.9989	9995	.0466	6682	21.470	3318	20
50	.0494	6940	.9988	9995	.0495	6945	20.206	3055	10
3° 00′	.0523	8.7188	.9986	9.9994	.0524	8.7194	19.081	1.2806	87° 00′
10	.0552	7423	.9985	9993	.0553	7429	18.075	2571	50
20	.0581	7645	.9983	9993	.0582	7652	17.169	2348	40
30	.0610	7857	.9981	9992	.0612	7865	16.350	2135	30
40	.0640	8059	.9980	9991	.0641	8067	15.605	1933	20
50	.0669	8251	.9978	9990	.0670	8261	14.924	1739	10
4° 00′	.0698	8.8436	.9976	9.9989	.0699	8.8446	14.301	1.1554	86° 00′
10	.0727	8613	.9974	9989	.0729	8624	13.727	1376	50
20	.0756	8783	.9971	9988	.0758	8795	13.197	1205	40
30	.0785	8946	.9969	9987	.0787	8960	12.706	1040	30
40	.0814	9104	.9967	9986	.0816	9118	12.251	0882	20
50	.0843	9256	.9964	9985	.0846	9272	11.826	0728	10
5° 00′	.0872	8.9403	.9962	9.9983	.0875	8.9420	11.430	1.0580	85° 00′
10	.0901	9545	.9959	9982	.0904	9563	11.059	0437	50
20	.0929	9682	.9957	9981	.0934	9701	10.712	0299	40
30	.0958	9816	.9954	9980	.0963	9836	10.385	0164	30
40	.0987	9945	.9951	9979	.0992	9966	10.078	0034	20
50	.1016	9.0070	.9948	9977	.1022	9.0093	9.7882	0.9907	10
6° 00′	.1045	9.0192	.9945	9.9976	.1051	9.0216	9.5144	0.9784	84° 00′
10	.1074	0311	.9942	9975	.1080	0336	9.2553	9664	50
20	.1103	0426	.9939	9973	.1110	0453	9.0098	9547	40
30	.1132	0539	.9936	9972	.1139	0567	8.7769	9433	30
40	.1161	0648	.9932	9971	.1169	0678	8.5555	9322	20
50	.1190	0755	.9929	9969	.1198	0786	8.3450	9214	10
7° 00′	.1219	9.0859	.9925	9.9968	.1228	9.0891	8.1443	0.9109	83° 00′
10	.1248	0961	.9922	9966	.1257	0995	7.9530	9005	50
20	.1276	1060	.9918	9964	.1287	1096	7.7704	8904	40
	Nat.	Log.	Nat.	Log.	Nat.	Log.	Nat.	Log.	
Angle	Cosine		Sine		Cotangent		Tangent		Angle

Angle	Sine		Cosine		Tangent		Cotangent		Angle
	Nat.	Log.	Nat.	Log.	Nat.	Log.	Nat.	Log.	
30	.1305	1157	.9914	9963	.1317	1194	7.5958	8806	30
40	.1334	1252	.9911	9961	.1346	1291	7.4287	8709	20
50	.1363	1345	.9907	9959	.1376	1385	7.2687	8615	10
8° 00′	.1392	9.1436	.9903	9.9958	.1405	9.1478	7.1154	0.8522	82° 00′
10	.1421	1525	.9899	9956	.1435	1569	6.9682	8431	50
20	.1449	1612	.9894	9954	.1465	1658	6.8269	8342	40
30	.1478	1697	.9890	9952	.1495	1745	6.6912	8255	30
40	.1507	1781	.9886	9950	.1524	1831	6.5606	8169	20
50	.1536	1863	.9881	9948	.1554	1915	6.4348	8085	10
9° 00′	.1564	9.1943	.9877	9.9946	.1584	9.1997	6.3138	0.8003	81° 00′
10	.1593	2022	.9872	9944	.1614	2078	6.1970	7922	50
20	.1622	2100	.9868	9942	.1644	2158	6.0844	7842	40
30	.1650	2176	.9863	9940	.1673	2236	5.9758	7764	30
40	.1679	2251	.9858	9938	.1703	2313	5.8708	7687	20
50	.1708	2324	.9853	9936	.1733	2389	5.7694	7611	10
10° 00′	.1736	9.2397	.9848	9.9934	.1763	9.2463	5.6713	0.7537	80° 00′
10	.1765	2468	.9843	9931	.1793	2536	5.5764	7464	50
20	.1794	2538	.9838	9929	.1823	2609	5.4845	7391	40
30	.1822	2606	.9833	9927	.1853	2680	5.3955	7320	30
40	.1851	2674	.9827	9924	.1883	2750	5.3093	7250	20
50	.1880	2740	.9822	9922	.1914	2819	5.2257	7181	10
11° 00′	.1908	9.2806	.9816	9.9919	.1944	9.2887	5.1446	0.7113	79° 00′
10	.1937	2870	.9811	9917	.1974	2953	5.0658	7047	50
20	.1965	2934	.9805	9914	.2004	3020	4.9894	6980	40
30	.1994	2997	.9799	9912	.2035	3085	4.9152	6915	30
40	.2022	3058	.9793	9909	.2065	3149	4.8430	6851	20
50	.2051	3119	.9787	9907	.2095	3212	4.7729	6788	10
12° 00′	.2079	9.3179	.9781	9.9904	.2126	9.3275	4.7046	0.6725	78° 00′
10	.2108	3238	.9775	9901	.2156	3336	4.6382	6664	50
20	.2136	3296	.9769	9899	.2186	3397	4.5736	6603	40
30	.2164	3353	.9763	9896	.2217	3458	4.5107	6542	30
40	.2193	3410	.9757	9893	.2247	3517	4.4494	6483	20
50	.2221	3466	.9750	9890	.2278	3576	4.3897	6424	10
13° 00′	.2250	9.3521	.9744	9.9887	.2309	9.3634	4.3315	0.6366	77° 00′
10	.2278	3575	.9737	9884	.2339	3691	4.2747	6309	50
20	.2306	3629	.9730	9881	.2370	3748	4.2193	6252	40
30	.2334	3682	.9724	9878	.2401	3804	4.1653	6196	30
40	.2363	3734	.9717	9875	.2432	3859	4.1126	6141	20
50	.2391	3786	.9710	9872	.2462	3914	4.0611	6086	10
14° 00′	.2419	9.3837	.9703	9.9869	.2493	9.3968	4.0108	0.6032	76° 00′
10	.2447	3887	.9696	9866	.2524	4021	3.9617	5979	50
20	.2476	3937	.9689	9863	.2555	4074	3.9136	5926	40
30	.2504	3986	.9681	9859	.2586	4127	3.8667	5873	30
40	.2532	4035	.9674	9856	.2617	4178	3.8208	5822	20
50	.2560	4083	.9667	9853	.2648	4230	3.7760	5770	10
	Nat.	Log.	Nat.	Log.	Nat.	Log.	Nat.	Log.	
Angle	Cosine		Sine		Cotangent		Tangent		Angle

Angle	Sine		Cosine		Tangent		Cotangent		Angle
	Nat.	Log.	Nat.	Log.	Nat.	Log.	Nat.	Log.	
15° 00′	.2588	9.4130	.9659	9.9849	.2679	9.4281	3.7321	0.5719	75° 00′
10	.2616	4177	.9652	9846	.2711	4331	3.6891	5669	50
20	.2644	4223	.9644	9843	.2742	4381	3.6470	5619	40
30	.2672	4269	.9636	9839	.2773	4430	3.6059	5570	30
40	.2700	4314	.9628	9836	.2805	4479	3.5656	5521	20
50	.2728	4359	.9621	9832	.2836	4527	3.5261	5473	10
16° 00′	.2756	9.4403	.9613	9.9828	.2867	9.4575	3.4874	0.5425	74° 00′
10	.2784	4447	.9605	9825	.2899	4622	3.4495	5378	50
20	.2812	4491	.9596	9821	.2931	4669	3.4124	5331	40
30	.2840	4533	.9588	9817	.2962	4716	3.3759	5284	30
40	.2868	4576	.9580	9814	.2994	4762	3.3402	5238	20
50	.2896	4618	.9572	9810	.3026	4808	3.3052	5192	10
17° 00′	.2924	9.4659	.9563	9.9806	.3057	9.4853	3.2709	0.5147	73° 00′
10	.2952	4700	.9555	9802	.3089	4898	3.2371	5102	50
20	.2979	4741	.9546	9798	.3121	4943	3.2041	5057	40
30	.3007	4781	.9537	9794	.3153	4987	3.1716	5013	30
40	.3035	4821	.9528	9790	.3185	5031	3.1397	4969	20
50	.3062	4861	.9520	9786	.3217	5075	3.1084	4925	10
18° 00′	.3090	9.4900	.9511	9.9782	.3249	9.5118	3.0777	0.4882	72° 00′
10	.3118	4939	.9502	9778	.3281	5161	3.0475	4839	50
20	.3145	4977	.9492	9774	.3314	5203	3.0178	4797	40
30	.3173	5015	.9483	9770	.3346	5245	2.9887	4755	30
40	.3201	5052	.9474	9765	.3378	5287	2.9600	4713	20
50	.3228	5090	.9465	9761	.3411	5329	2.9319	4671	10
19° 00′	.3256	9.5126	.9455	9.9757	.3443	9.5370	2.9042	0.4630	71° 00′
10	.3283	5163	.9446	9752	.3476	5411	2.8770	4589	50
20	.3311	5199	.9436	9748	.3508	5451	2.8502	4549	40
30	.3338	5235	.9426	9743	.3541	5491	2.8239	4509	30
40	.3365	5270	.9417	9739	.3574	5531	2.7980	4469	20
50	.3393	5306	.9407	9734	.3607	5571	2.7725	4429	10
20° 00′	.3420	9.5341	.9397	9.9730	.3640	9.5611	2.7475	0.4389	70° 00′
10	.3448	5375	.9387	9725	.3673	5650	2.7228	4350	50
20	.3475	5409	.9377	9721	.3706	5689	2.6985	4311	40
30	.3502	5443	.9367	9716	.3739	5727	2.6746	4273	30
40	.3529	5477	.9356	9711	.3772	5766	2.6511	4234	20
50	.3557	5510	.9346	9706	.3805	5804	2.6279	4196	10
21° 00′	.3584	9.5543	.9336	9.9702	.3839	9.5842	2.6051	0.4158	69° 00′
10	.3611	5576	.9325	9697	.3872	5879	2.5826	4121	50
20	.3638	5609	.9315	9692	.3906	5917	2.5605	4083	40
30	.3665	5641	.9304	9687	.3939	5954	2.5386	4046	30
40	.3692	5673	.9293	9682	.3973	5991	2.5172	4009	20
50	.3719	5704	.9283	9677	.4006	6028	2.4960	3972	10
22° 00′	.3746	9.5736	.9272	9.9672	.4040	9.6064	2.4751	0.3936	68° 00′
10	.3773	5767	.9261	9667	.4074	6100	2.4545	3900	50
20	.3800	5798	.9250	9661	.4108	6136	2.4342	3864	40
	Nat.	Log.	Nat.	Log.	Nat.	Log.	Nat.	Log.	
Angle	Cosine		Sine		Cotangent		Tangent		Angle

Angle	Sine		Cosine		Tangent		Cotangent		Angle
	Nat.	Log.	Nat.	Log.	Nat.	Log.	Nat.	Log.	
30	.3827	5828	.9239	9656	.4142	6172	2.4142	3828	30
40	.3854	5859	.9228	9651	.4176	6208	2.3945	3792	20
50	.3881	5889	.9216	9646	.4210	6243	2.3750	3757	10
23° 00′	.3907	9.5919	.9205	9.9640	4245	9.6279	2.3559	0.3721	67° 00′
10	.3934	5948	.9194	9635	.4279	6314	2.3369	3686	50
20	.3961	5978	.9182	9629	.4314	6348	2.3183	3652	40
30	.3987	6007	.9171	9624	.4348	6383	2.2998	3617	30
40	.4014	6036	.9159	9618	.4383	6417	2.2817	3583	20
50	.4041	6065	.9147	9613	.4417	6452	2.2637	3548	10
24° 00′	.4067	9.6093	.9135	9.9607	.4452	9.6486	2.2460	0.3514	66° 00′
10	.4094	6121	.9124	9602	.4487	6520	2.2286	3480	50
20	.4120	6149	.9112	9596	.4522	6553	2.2113	3447	40
30	.4147	6177	.9100	9590	.4557	6587	2.1943	3413	30
40	.4173	6205	.9088	9584	.4592	6620	2.1775	3380	20
50	.4200	6232	.9075	9579	.4628	6654	2.1609	3346	10
25° 00′	.4226	9.6259	.9063	9.9573	.4663	9.6687	2.1445	0.3313	65° 00′
10	.4253	6286	.9051	9567	.4699	6720	2.1283	3280	50
20	.4279	6313	.9038	9561	.4734	6752	2.1123	3248	40
30	.4305	6340	.9026	9555	.4770	6785	2.0965	3215	30
40	.4331	6366	.9013	9549	.4806	6817	2.0809	3183	20
50	.4358	6392	.9001	9543	.4841	6850	2.0655	3150	10
26° 00′	.4384	9.6418	.8988	9.9537	.4877	9.6882	2.0503	0.3118	64° 00′
10	.4410	6444	.8975	9530	.4913	6914	2.0353	3086	50
20	.4436	6470	.8962	9524	.4950	6946	2.0204	3054	40
30	.4462	6495	.8949	9518	.4986	6977	2.0057	3023	30
40	.4488	6521	.8936	9512	.5022	7009	1.9912	2991	20
50	.4514	6546	.8923	9505	.5059	7040	1.9768	2960	10
27° 00′	.4540	9.6570	.8910	9.9499	.5095	9.7072	1.9626	0.2928	63° 00′
10	.4566	6595	.8897	9492	.5132	7103	1.9486	2897	50
20	.4592	6620	.8884	9486	.5169	7134	1.9347	2866	40
30	.4617	6644	.8870	9479	.5206	7165	1.9210	2835	30
40	.4643	6668	.8857	9473	.5243	7196	1.9074	2804	20
50	.4669	6692	.8843	9466	.5280	7226	1.8940	2774	10
28° 00′	.4695	9.6716	.8829	9.9459	.5317	9.7257	1.8807	0.2743	62° 00′
10	.4720	6740	.8816	9453	.5354	7287	1.8676	2713	50
20	.4746	6763	.8802	9446	.5392	7317	1.8546	2683	40
30	.4772	6787	.8788	9439	.5430	7348	1.8418	2652	30
40	.4797	6810	.8774	9432	.5467	7378	1.8291	2622	20
50	.4823	6833	.8760	9425	.5505	7408	1.8165	2592	10
29° 00′	.4848	9.6856	.8746	9.9418	.5543	9.7438	1.8040	0.2562	61° 00′
10	.4874	6878	.8732	9411	.5581	7467	1.7917	2533	50
20	.4899	6901	.8718	9404	.5619	7497	1.7796	2503	40
30	.4924	6923	.8704	9397	.5658	7526	1.7675	2474	30
40	.4950	6946	.8689	9390	.5696	7556	1.7556	2444	20
50	.4975	6968	.8675	9383	.5735	7585	1.7437	2415	10
	Nat.	Log.	Nat.	Log.	Nat.	Log.	Nat.	Log.	
Angle	Cosine		Sine		Cotangent		Tangent		Angle

Angle	Sine		Cosine		Tangent		Cotangent		Angle
	Nat.	Log.	Nat.	Log.	Nat.	Log.	Nat.	Log.	
30° 00′	.5000	9.6990	.8660	9.9375	.5774	9.7614	1.7321	0.2386	60° 00′
10	.5025	7012	.8646	9368	.5812	7644	1.7205	2356	50
20	.5050	7033	.8631	9361	.5851	7673	1.7090	2327	40
30	.5075	7055	.8616	9353	.5890	7701	1.6977	2299	30
40	.5100	7076	.8601	9346	.5930	7730	1.6864	2270	20
50	.5125	7097	.8587	9338	.5969	7759	1.6753	2241	10
31° 00′	.5150	9.7118	.8572	9.9331	.6009	9.7788	1.6643	0.2212	59° 00′
10	.5175	7139	.8557	9323	.6048	7816	1.6534	2184	50
20	.5200	7160	.8542	9315	.6088	7845	1.6426	2155	40
30	.5225	7181	.8526	9308	.6128	7873	1.6319	2127	30
40	.5250	7201	.8511	9300	.6168	7902	1.6212	2098	20
50	.5275	7222	.8496	9292	.6208	7930	1.6107	2070	10
32° 00′	.5299	9.7242	.8480	9.9284	.6249	9.7958	1.6003	0.2042	58° 00′
10	.5324	7262	.8465	9276	.6289	7986	1.5900	2014	50
20	.5348	7282	.8450	9268	.6330	8014	1.5798	1986	40
30	.5373	7302	.8434	9260	.6371	8042	1.5697	1958	30
40	.5398	7322	.8418	9252	.6412	8070	1.5597	1930	20
50	.5422	7342	.8403	9244	.6453	8097	1.5497	1903	10
33° 00′	.5446	9.7361	.8387	9.9236	.6494	9.8125	1.5399	0.1875	57° 00′
10	.5471	7380	.8371	9228	.6536	8153	1.5301	1847	50
20	.5495	7400	.8355	9219	.6577	8180	1.5204	1820	40
30	.5519	7419	.8339	9211	.6619	8208	1.5108	1792	30
40	.5544	7438	.8323	9203	.6661	8235	1.5013	1765	20
50	.5568	7457	.8307	9194	.6703	8263	1.4919	1737	10
34° 00′	.5592	9.7476	.8290	9.9186	.6745	9.8290	1.4826	0.1710	56° 00′
10	.5616	7494	.8274	9177	.6787	8317	1.4733	1683	50
20	.5640	7513	.8258	9169	.6830	8344	1.4641	1656	40
30	.5664	7531	.8241	9160	.6873	8371	1.4550	1629	30
40	.5688	7550	.8225	9151	.6916	8398	1.4460	1602	20
50	.5712	7568	.8208	9142	.6959	8425	1.4370	1575	10
35° 00′	.5736	9.7586	.8192	9.9134	.7002	9.8452	1.4281	0.1548	55° 00′
10	.5760	7604	.8175	9125	.7046	8479	1.4193	1521	50
20	.5783	7622	.8158	9116	.7089	8506	1.4106	1494	40
30	.5807	7640	.8141	9107	.7133	8533	1.4019	1467	30
40	.5831	7657	.8124	9098	.7177	8559	1.3934	1441	20
50	.5854	7675	.8107	9089	.7221	8586	1.3848	1414	10
36° 00′	.5878	9.7692	.8090	9.9080	.7265	9.8613	1.3764	0.1387	54° 00′
10	.5901	7710	.8073	9070	.7310	8639	1.3680	1361	50
20	.5925	7727	.8056	9061	.7355	8666	1.3597	1334	40
30	.5948	7744	.8039	9052	.7400	8692	1.3514	1308	30
40	.5972	7761	.8021	9042	.7445	8718	1.3432	1282	20
50	.5995	7778	.8004	9033	.7490	8745	1.3351	1255	10
37° 00′	.6018	9.7795	.7986	9.9023	.7536	9.8771	1.3270	0.1229	53° 00′
10	.6041	7811	.7969	9014	.7581	8797	1.3190	1203	50
20	.6065	7828	.7951	9004	.7627	8824	1.3111	1176	40
	Nat.	Log.	Nat.	Log.	Nat.	Log.	Nat.	Log.	
Angle	Cosine		Sine		Cotangent		Tangent		Angle

Angle	Sine		Cosine		Tangent		Cotangent		Angle
	Nat.	Log.	Nat.	Log.	Nat.	Log.	Nat.	Log.	
30	.6088	7844	.7934	8995	.7673	8850	1.3032	1150	30
40	.6111	7861	.7916	8985	.7720	8876	1.2954	1124	20
50	.6134	7877	.7898	8975	.7766	8902	1.2876	1098	10
38° 00'	.6157	9.7893	.7880	9.8965	.7813	9.8928	1.2799	0.1072	52° 00'
10	.6180	7910	.7862	8955	.7860	8954	1.2723	1046	50
20	.6202	7926	.7844	8945	.7907	8980	1.2647	1020	40
30	.6225	7941	.7826	8935	.7954	9006	1.2572	0994	30
40	.6248	7957	.7808	8925	.8002	9032	1.2497	0968	20
50	.6271	7973	.7790	8915	.8050	9058	1.2423	0942	10
39° 00'	.6293	9.7989	.7771	9.8905	.8098	9.9084	1.2349	0.0916	51° 00'
10	.6316	8004	.7753	8895	.8146	9110	1.2276	0890	50
20	.6338	8020	.7735	8884	.8195	9135	1.2203	0865	40
30	.6361	8035	.7716	8874	.8243	9161	1.2131	0839	30
40	.6383	8050	.7698	8864	.8292	9187	1.2059	0813	20
50	.6406	8066	.7679	8853	.8342	9212	1.1988	0788	10
40° 00'	.6428	9.8081	.7660	9.8843	.8391	9.9238	1.1918	0.0762	50° 00'
10	.6450	8096	.7642	8832	.8441	9264	1.1847	0736	50
20	.6472	8111	.7623	8821	.8491	9289	1.1778	0711	40
30	.6494	8125	.7604	8810	.8541	9315	1.1708	0685	30
40	.6517	8140	.7585	8800	.8591	9341	1.1640	0659	20
50	.6539	8155	.7566	8789	.8642	9366	1.1571	0634	10
41° 00'	.6561	9.8169	.7547	9.8778	.8693	9.9392	1.1504	0.0608	49° 00'
10	.6583	8184	.7528	8767	.8744	9417	1.1436	0583	50
20	.6604	8198	.7509	8756	.8796	9443	1.1369	0557	40
30	.6626	8213	.7490	8745	.8847	9468	1.1303	0532	30
40	.6648	8227	.7470	8733	.8899	9494	1.1237	0506	20
50	.6670	8241	.7451	8722	.8952	9519	1.1171	0481	10
42° 00'	.6691	9.8255	.7431	9.8711	.9004	9.9544	1.1106	0.0456	48° 00'
10	.6713	8269	.7412	8699	.9057	9570	1.1041	0430	50
20	.6734	8283	.7392	8688	.9110	9595	1.0977	0405	40
30	.6756	8297	.7373	8676	.9163	9621	1.0913	0379	30
40	.6777	8311	.7353	8665	.9217	9646	1.0850	0354	20
50	.6799	8324	.7333	8653	.9271	9671	1.0786	0329	10
43° 00'	.6820	9.8338	.7314	9.8641	.9325	9.9697	1.0724	0.0303	47° 00'
10	.6841	8351	.7294	8629	.9380	9722	1.0661	0278	50
20	.6862	8365	.7274	8618	.9435	9747	1.0599	0253	40
30	.6884	8378	.7254	8606	.9490	9772	1.0538	0228	30
40	.6905	8391	.7234	8594	.9545	9798	1.0477	0202	20
50	.6926	8405	.7214	8582	.9601	9823	1.0416	0177	10
44° 00'	.6947	9.8418	.7193	9.8569	.9657	9.9848	1.0355	0.0152	46° 00'
10	.6967	8431	.7173	8557	.9713	9874	1.0295	0126	50
20	.6988	8444	.7153	8545	.9770	9899	1.0235	0101	40
30	.7009	8457	.7133	8532	.9827	9924	1.0176	0076	30
40	.7030	8469	.7112	8520	.9884	9949	1.0117	0051	20
50	.7050	8482	.7092	8507	.9942	9975	1.0058	0025	10
45° 00'	.7071	9.8495	.7071	9.8495	1.0000	0.0000	1.0000	0.0000	45° 00'
	Nat.	Log.	Nat.	Log.	Nat.	Log.	Nat.	Log	
Angle	Cosine		Sine		Cotangent		Tangent		Angle

LENGTH OF CHORD FOR CIRCLE ARCS OF 1-IN. RADIUS

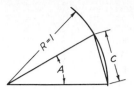

°	0′	10′	20′	30′	40′	50′
0	0.0000	0.0029	0.0058	0.0087	0.0116	0.0145
1	0.0175	0.0204	0.0233	0.0262	0.0291	0.0320
2	0.0349	0.0378	0.0407	0.0436	0.0465	0.0494
3	0.0524	0.0553	0.0582	0.0611	0.0640	0.0669
4	0.0698	0.0727	0.0756	0.0785	0.0814	0.0843
5	0.0872	0.0901	0.0931	0.0960	0.0989	0.1018
6	0.1047	0.1076	0.1105	0.1134	0.1163	0.1192
7	0.1221	0.1250	0.1279	0.1308	0.1337	0.1366
8	0.1395	0.1424	0.1453	0.1482	0.1511	0.1540
9	0.1569	0.1598	0.1627	0.1656	0.1685	0.1714
10	0.1743	0.1772	0.1801	0.1830	0.1859	0.1888
11	0.1917	0.1946	0.1975	0.2004	0.2033	0.2062
12	0.2091	0.2119	0.2148	0.2177	0.2206	0.2235
13	0.2264	0.2293	0.2322	0.2351	0.2380	0.2409
14	0.2437	0.2466	0.2495	0.2524	0.2553	0.2582
15	0.2611	0.2639	0.2668	0.2697	0.2726	0.2755
16	0.2783	0.2812	0.2841	0.2870	0.2899	0.2927
17	0.2956	0.2985	0.3014	0.3042	0.3071	0.3100
18	0.3129	0.3157	0.3186	0.3215	0.3244	0.3272
19	0.3301	0.3330	0.3358	0.3387	0.3416	0.3444
20	0.3473	0.3502	0.3530	0.3559	0.3587	0.3616
21	0.3645	0.3673	0.3702	0.3730	0.3759	0.3788
22	0.3816	0.3845	0.3873	0.3902	0.3930	0.3959
23	0.3987	0.4016	0.4044	0.4073	0.4101	0.4130
24	0.4158	0.4187	0.4215	0.4244	0.4272	0.4300
25	0.4329	0.4357	0.4386	0.4414	0.4442	0.4471
26	0.4499	0.4527	0.4556	0.4584	0.4612	0.4641
27	0.4669	0.4697	0.4725	0.4754	0.4782	0.4810
28	0.4838	0.4867	0.4895	0.4923	0.4951	0.4979
29	0.5008	0.5036	0.5064	0.5092	0.5120	0.5148
30	0.5176	0.5204	0.5233	0.5251	0.5289	0.5317
31	0.5345	0.5373	0.5401	0.5429	0.5457	0.5485
32	0.5513	0.5541	0.5569	0.5597	0.5625	0.5652
33	0.5680	0.5708	0.5736	0.5764	0.5792	0.5820
34	0.5847	0.5875	0.5903	0.5931	0.5959	0.5986
35	0.6014	0.6042	0.6070	0.6097	0.6125	0.6153
36	0.6180	0.6208	0.6236	0.6263	0.6291	0.6319
37	0.6346	0.6374	0.6401	0.6429	0.6456	0.6484
38	0.6511	0.6539	0.6566	0.6594	0.6621	0.6649
39	0.6676	0.6704	0.6731	0.6758	0.6786	0.6813
40	0.6840	0.6868	0.6895	0.6922	0.6950	0.6977
41	0.7004	0.7031	0.7059	0.7086	0.7113	0.7140
42	0.7167	0.7195	0.7222	0.7249	0.7276	0.7303
43	0.7330	0.7357	0.7384	0.7411	0.7438	0.7465
44	0.7492	0.7519	0.7546	0.7573	0.7600	0.7627
45 *	0.7654	0.7681	0.7707	0.7734	0.7761	0.7788

* For angles between 45° and 90°, draw 90° angle and lay off complement from 90° line.

Given	Multiply by	To obtain
Acceleration by gravity	980.665	centimeters per second
Acceleration by gravity	32.16	feet per second
Acres	0.4047	hectares
Acres	4.356×10^4	square feet
Acre	1.56×10^{-3}	square miles
Angstrom	10^{-10}	meters
Ares	119.6	square yards
Atmospheres (atm)	29.921	inches of mercury
Atmospheres	33.934	feet of water
Atmospheres	1.033228	kilograms per square centimeter
Atmospheres	14.6959	pounds per square inch
Atmospheres	1.013×10^3	kilopascals
British thermal units (BTU)	0.252	calories
British thermal units	778.0	foot-pounds
BTU per minute	1.758×10^{-2}	kilowatt
BTU per minute	2.357×10^{-2}	horsepower
BTU per pound	2.325	joules per gram
Calories (cal)	3.9682	British thermal units
Calories	3.088×10^3	foot-pounds
Carats	3.086	grains
Carats	200.0	milligrams
Centiares	1.0	square meters
Celsius, degrees (°C)	$\dfrac{9}{5}\,°C + 32$	Fahrenheit, degrees
Centigrams (cg)	0.1543	grains
Centiliters	3.38×10^{-2}	ounces, fluid
Centimeters (cm)	3.28×10^{-2}	feet
Centimeters	0.3937	inches
Circle (angular)	360.0	degrees
Circumference of the earth at the equator	21,600.0	miles, nautical
Circumference of the earth at the equator	24,874.5	miles, statute
Cord (cd), of wood, (4 × 4 × 8)	128.0	cubic feet
Cubic centimeters	6.102×10^{-2}	cubic inches
Cubic centimeters	1×10^{-3}	liters
Cubic feet	1.728×10^3	cubic inches
Cubic feet	2.83×10^{-2}	cubic meters
Cubic feet	7.4805	gallons, U.S.
Cubic feet	28.3163	liters
Cubic feet of water at 39.1 degrees Fahrenheit (°F)	28.3156	kilograms
Cubic feet of water at 39.1 degrees Fahrenheit	62.4245	pounds
Cubic inches	16.3872	cubic centimeters
Cubic inches	4.32×10^{-3}	gallons, U.S.
Cubic inches	1.64×10^{-2}	liters
Cubic inches	1.488×10^{-2}	quarts, dry
Cubic inches	1.73×10^{-2}	quarts, liquid

Given	Multiply by	To obtain
Cubic meters	35.3133	cubic feet
Cubic meters	1.3079	cubic yards
Decigrams	1.5432	grains
Decaliters	10.0	liters
Degrees (deg or °)	60.0	minutes
Degrees (arc)	1.75×10^{-2}	radians
Degrees (at the equator)	60.0	miles, nautical
Degrees (at the equator)	69.168	miles, statute
Drams (dr), apothecaries	60.0	grains
Drams, apothecaries	3.543	grams
Drams, avoirdupois	27.344	grains
Drams, avoirdupois	1.772	grams
Drams, avoirdupois	6.25×10^{-2}	ounces, avoirdupois
Drams, fluid	0.2256	cubic inches
Drams, fluid	3.6966	milliliters
Dynes	10^{-5}	newtons
Ergs	10^{-7}	joules
Ergs	1.0	dyne-centimeters
Fahrenheit	$\dfrac{5(°F - 32)}{9}$	Celsius, degrees
Fathoms	6.0	feet
Fathoms	1.8288	meters
Feet	12.0	inches
Feet	0.3048	meters
Feet	1.89×10^{-4}	miles
Feet	1.654×10^{-4}	miles, nautical
Feet	6.061×10^{-2}	rods
Feet of water at 62 degrees Fahrenheit	304.442	kilograms per square meter
Feet of water at 62 degrees Fahrenheit	62.355	pounds per square foot
Feet per second (fps)	0.5921	knots
Feet per second	0.6816	miles per hour
Foot-pounds (ft-lb)	1.29×10^{-3}	British thermal units
Foot-pounds	3.2×10^{-4}	calories
Foot-pounds-force	1.356	joules
Foot-pounds-force	0.13835	meter kilograms
Foot-pounds per minute	3×10^{-6}	horsepower
Furlongs	660.0	feet
Furlongs	201.17	meters
Gallons, imperial	1.2009	gallons, U.S.
Gallons, imperial	4.54607	liters
Gallons, U.S.	0.1337	cubic feet
Gallons, U.S.	3.8×10^{-3}	cubic meters
Gallons, U.S.	0.8327	gallons, imperial
Gallons, U.S.	3.7878	liters
Gallons, U.S.	128.0	ounces, U.S. fluid
Gallons, U.S., water	8.5	pounds
Grains	3.66×10^{-2}	drams, avoirdupois
Grains	6.48×10^{-2}	grams
Grains	2.29×10^{-3}	ounces, avoirdupois

Given	Multiply by	To obtain
Grains	2.08×10^{-3}	ounces, troy and apothecaries'
Grams	15.4475	grains
Grams	3.53×10^{-2}	ounces, avoirdupois
Grams per cubic centimeter	62.4	pounds per cubic foot
Grams per cubic centimeter	3.613×10^{-2}	pounds per cubic inch
Gross	12.0	dozen
Hectares (ha)	100.0	ares
Hectograms	100.0	grams
Hectograms	3.5274	ounces, avoirdupois
Hectometers	100.0	meters
Horsepower (hp)	76.042	kilogram-meters per second
Horsepower	550.0	foot-pounds per second
Horsepower	1.0139	metric horsepower
Horsepower	746.0	watts per minute
Horsepower, metric	0.9862	horsepower
Horsepower, metric	542.5	foot-pounds per second
Horsepower, metric	75.0	kilogram meters per second
Inches (in.)	2.5400	centimeters
Inches	2.54×10^{-2}	meters
Inches	1×10^{3}	mils
Inches of mercury	1.1341	feet of water
Inches of mercury	34.542	grams per square centimeter
Inches of mercury	0.49115	pounds per square inch
Inches of water	2.537	grams per square centimeter
Inches of water	5.1052	pounds per square foot
Joules	10^{7}	ergs
Joules	9.4845×10^{-4}	British thermal units
Joules	0.7376	foot-pounds-force
Joules	2.778×10^{-7}	kilowatt-hours
Joules	0.102	meter-kilograms
Kilocycles	1×10^{3}	cycles per second
Kilogram-meters (kg-m)	7.2330	pound-feet
Kilogram-meters per second	1.305×10^{-2}	horsepower
Kilograms	2.2046	pounds, avoirdupois
Kilograms	1×10^{-3}	tons, metric
Kilograms per cubic meter (kg per cu m or kg/m^3)	6.243×10^{-2}	pounds per cubic foot
Kilograms per meter	0.6721	pounds per foot
Kilograms per square meter	1.42×10^{-3}	pounds per square inch
Kilometers	0.5396	miles, nautical
Kilometers	0.6214	miles, statute
Kilometers per hour	0.5396	knots
Kilometers per hour	0.62138	miles per hour
Kilowatts (kw)	4.426×10^{-2}	foot-pounds per minute
Kilowatts	3.6×10^{6}	joules per hour
Kilowatt-hours (kwhr)	3.413×10^{3}	British thermal units per hour
Kilowatt-hours	1.3414	horsepower hours
Kilowatt-hours	3.671×10^{5}	meter-kilograms
Knots	1.6889	feet per second

Given	Multiply by	To obtain
Knots	1.8532	kilometers per hour
Knots	1.1516	miles per hour
Knots	1.0	nautical miles per hour
Leagues, land	4.83	kilometers
Leagues, marine	5.56	kilometers
Links	1×10^{-2}	chains
Links	0.66	feet
Links	4.0×10^{-2}	rods
Liters	61.02398	cubic inches
Liters	0.2199	gallons, imperial
Liters	0.2641	gallons, U.S.
Liters	1.0567	quarts, liquid
Long tons	2.24×10^3	pounds, avoirdupois
Meter-kilograms	9.297×10^{-3}	British-thermal units
Meter-kilograms	7.233	foot-pounds-force
Meter-kilograms	9.807	joules
Meter-kilograms	2.724×10^6	kilowatt-hours
Meters	39.370	inches
Meters	5.41×10^{-4}	miles, nautical
Meters	6.22×10^{-4}	miles, U.S.
Meter-kilograms (m-kg)	7.2330	foot-pounds
Meters per second	1.9425	knots
Meters per second	2.2369	miles per hour
Microns (μ)	3.9×10^{-5}	inches
Microns	1×10^{-6}	meters
Miles, nautical	1.85325	kilometers
Miles, nautical	1.1516	miles, statute
Miles, statute	5.28×10^3	feet
Miles, statute	8.0	furlongs
Miles, statute	1.60935×10^3	meters
Miles per hour (mph or mi/h)	1.4667	feet per second
Miles per hour	1.6093	kilometers per hour
Miles per hour	0.8684	knots
Miles per hour	0.4470	meters per second
Milliliters	3.38×10^{-2}	ounces, fluid
Millimeters (mm)	3.3937×10^{-2}	inches
Millimeters	1×10^3	microns
Mils	1×10^{-3}	inches
Minutes (min)	60.0	seconds
Newtons	10^5	dynes
Newtons	1	joule per meter
Newtons	0.2248	pounds-force
Ounces (oz), apothecaries'	8.0	drams, apothecaries'
Ounces, avoirdupois	16.0	drams, avoirdupois
Ounces, avoirdupois	437.5	grains
Ounces, avoirdupois	28.3495	grams
Ounces, avoirdupois	1.0971	ounces, troy and apothecaries'
Ounces, avoirdupois	6.25×10^{2}	pounds, avoirdupois
Ounces, troy and apothecaries'	480.0	grains

Given	Multiply by	To obtain
Ounces, troy and apothecaries'	31.10348	grams
Ounces, U.S. fluid	1.805	cubic inches
Ounces, U.S. fluid	8.0	drams, fluid
Ounces, U.S. fluid	7.81×10^{-3}	gallons, U.S.
Ounces, U.S. fluid	2.96×10^{-2}	liters
Pascal	1.0	newton per square meter
Pecks (pk)	0.25	bushel, U.S.
Pecks	537.61	cubic inches
Pecks	8.8096	liters
Pints, dry	33.60	cubic inches
Pints, dry	0.5506	liters
Pints, dry	0.5	quarts, dry
Pints, liquid	28.875	cubic inches
Pints, liquid	0.4732	liters
Poundals	3.113×10^{-2}	pounds, avoirdupois
Pounds, avoirdupois	7×10^3	grains
Pounds, avoirdupois	453.5924	grams
Pounds, avoirdupois	16.0	ounces, avoirdupois
Pounds, avoirdupois	4.448	newtons
Pounds, avoirdupois	3.11×10^{-2}	slugs
Pound-feet (lb-ft)	0.1383	kilogram-meters
Pounds per cubic foot (lb per cu ft)	1.602×10^{-2}	grams per cubic centimeter
Pounds per cubic foot	16.0184	kilograms per cubic meter
Pounds per square foot (psf)	0.1922	inches of water
Pounds per square foot	4.8824	kilograms per square meter
Pounds per square inch (psi)	6.80×10^{-2}	atmospheres
Pounds per square inch	2.0360	inches of mercury
Pounds per square inch	703.0669	kilograms per square meter
Quadrants	90.0	degrees
Quarts, dry, U.S.	2.0	pints, dry
Quarts, liquid	0.94636	liters
Quintals	1×10^5	grams
Quintals	220.46	pounds, avoirdupois
Radians	57.2958	degrees, arc
Radians	0.1591	revolutions
Radians per second	9.4460	revolutions per minute
Ream, printing paper	500.0	sheets
Revolutions	6.2832	radians
Rods	0.25	chains
Rods	16.5	feet
Rods	40.0	furlongs
Rods	25.0	links
Rods	5.029	meters
Score	20.0	units
Seconds	1.667×10^{-2}	minutes
Slugs	32.1740	pounds
Square centimeters	0.1550	square inches
Square chains	0.1	acres
Square chains	16.0	square rods

Given	Multiply by	To obtain
Square feet (sq ft or ft^2)	2.2988×10^{-5}	acres
Square feet	9.29×10^{-2}	square meters
Square inches	6.4516	square centimeters
Square kilometers (sq km or km^2)	100.0	hectares
Square kilometers	0.3861	square miles
Square meters	10.7639	square feet
Square miles	2.590	square kilometers
Square miles	640.0	acres
Tons, long	2.24×10^3	pounds, avoirdupois
Tons, metric	1×10^3	kilograms
Tons, metric	2.2046×10^3	pounds, avoirdupois
Tons, register	100.0	cubic feet
Tons, short	907.18	kilograms
Tons, short	2×10^3	pounds, avoirdupois
Watts (W)	1×10^7	ergs per second
Yards	3.0	feet
Yards	4.545×10^{-3}	furlongs

APPENDIX 8A RUNNING AND SLIDING FITS

Limits are in thousandths of an inch.

Limits for hole and shaft are applied algebraically to the basic size to obtain the limits of size for the parts.

Data in bold face are in accordance with ABC agreements.

Symbols H5, g5, etc., are Hole and Shaft designations used in ABC System

Nominal Size Range Inches		Class RC 1			Class RC 2			Class RC 3			Class RC 4		
		Limits of Clearance	Standard Limits		Limits of Clearance	Standard Limits		Limits of Clearance	Standard Limits		Limits of Clearance	Standard Limits	
Over	To		Hole H5	Shaft g4		Hole H6	Shaft g5		Hole H7	Shaft f6		Hole H8	Shaft f7
0	− 0.12	0.1 / 0.45	+ 0.2 / 0	− 0.1 / − 0.25	0.1 / 0.55	+ 0.25 / 0	− 0.1 / − 0.3	0.3 / 0.95	+ 0.4 / 0	− 0.3 / − 0.55	0.3 / 1.3	+ 0.6 / 0	− 0.3 / − 0.7
0.12	− 0.24	0.15 / 0.5	+ 0.2 / 0	− 0.15 / − 0.3	0.15 / 0.65	+ 0.3 / 0	− 0.15 / − 0.35	0.4 / 1.2	+ 0.5 / 0	− 0.4 / − 0.7	0.4 / 1.6	+ 0.7 / 0	− 0.4 / − 0.9
0.24	− 0.40	0.2 / 0.6	+ 0.25 / 0	− 0.2 / − 0.35	0.2 / 0.85	+ 0.4 / 0	− 0.2 / − 0.45	0.5 / 1.5	+ 0.6 / 0	− 0.5 / − 0.9	0.5 / 2.0	+ 0.9 / 0	− 0.5 / − 1.1
0.40	− 0.71	0.25 / 0.75	+ 0.3 / 0	− 0.25 / − 0.45	0.25 / 0.95	+ 0.4 / 0	− 0.25 / − 0.55	0.6 / 1.7	+ 0.7 / 0	− 0.6 / − 1.0	0.6 / 2.3	+ 1.0 / 0	− 0.6 / − 1.3
0.71	− 1.19	0.3 / 0.95	+ 0.4 / 0	− 0.3 / − 0.55	0.3 / 1.2	+ 0.5 / 0	− 0.3 / − 0.7	0.8 / 2.1	+ 0.8 / 0	− 0.8 / − 1.3	0.8 / 2.8	+ 1.2 / 0	− 0.8 / − 1.6
1.19	− 1.97	0.4 / 1.1	+ 0.4 / 0	− 0.4 / − 0.7	0.4 / 1.4	+ 0.6 / 0	− 0.4 / − 0.8	1.0 / 2.6	+ 1.0 / 0	− 1.0 / − 1.6	1.0 / 3.6	+ 1.6 / 0	− 1.0 / − 2.0
1.97	− 3.15	0.4 / 1.2	+ 0.5 / 0	− 0.4 / − 0.7	0.4 / 1.6	+ 0.7 / 0	− 0.4 / − 0.9	1.2 / 3.1	+ 1.2 / 0	− 1.2 / − 1.9	1.2 / 4.2	+ 1.8 / 0	− 1.2 / − 2.4
3.15	− 4.73	0.5 / 1.5	+ 0.6 / 0	− 0.5 / − 0.9	0.5 / 2.0	+ 0.9 / 0	− 0.5 / − 1.1	1.4 / 3.7	+ 1.4 / 0	− 1.4 / − 2.3	1.4 / 5.0	+ 2.2 / 0	− 1.4 / − 2.8
4.73	− 7.09	0.6 / 1.8	+ 0.7 / 0	− 0.6 / − 1.1	0.6 / 2.3	+ 1.0 / 0	− 0.6 / − 1.3	1.6 / 4.2	+ 1.6 / 0	− 1.6 / − 2.6	1.6 / 5.7	+ 2.5 / 0	− 1.6 / − 3.2
7.09	− 9.85	0.6 / 2.0	+ 0.8 / 0	− 0.6 / − 1.2	0.6 / 2.6	+ 1.2 / 0	− 0.6 / − 1.4	2.0 / 5.0	+ 1.8 / 0	− 2.0 / − 3.2	2.0 / 6.6	+ 2.8 / 0	− 2.0 / − 3.8
9.85	−12.41	0.8 / 2.3	+ 0.9 / 0	− 0.8 / − 1.4	0.7 / 2.8	+ 1.2 / 0	− 0.7 / − 1.6	2.5 / 5.7	+ 2.0 / 0	− 2.5 / − 3.7	2.2 / 7.2	+ 3.0 / 0	− 2.2 / − 4.2
12.41	−15.75	1.0 / 2.7	+ 1.0 / 0	− 1.0 / − 1.7	0.7 / 3.1	+ 1.4 / 0	− 0.7 / − 1.7	3.0 / 6.6	+ 2.2 / 0	− 3.0 / − 4.4	2.5 / 8.2	+ 3.5 / 0	− 2.5 / − 4.7
15.75	−19.69	1.2 / 3.0	+ 1.0 / 0	− 1.2 / − 2.0	0.8 / 3.4	+ 1.6 / 0	− 0.8 / − 1.8	4.0 / 8.1	+ 2.5 / 0	− 4.0 / − 5.6	2.8 / 9.3	+ 4.0 / 0	− 2.8 / − 5.3
19.69	−30.09	1.6 / 3.7	+ 1.2 / 0	− 1.6 / − 2.5	1.6 / 4.8	+ 2.0 / 0	− 1.6 / − 2.8	5.0 / 10.0	+ 3.0 / 0	− 5.0 / − 7.0	5.0 / 13.0	+ 5.0 / 0	− 5.0 / − 8.0
30.09	−41.49	2.0 / 4.6	+ 1.6 / 0	− 2.0 / − 3.0	2.0 / 6.1	+ 2.5 / 0	− 2.0 / − 3.6	6.0 / 12.5	+ 4.0 / 0	− 6.0 / − 8.5	6.0 / 16.0	+ 6.0 / 0	− 6.0 / −10.0
41.49	−56.19	2.5 / 5.7	+ 2.0 / 0	− 2.5 / − 3.7	2.5 / 7.5	+ 3.0 / 0	− 2.5 / − 4.5	8.0 / 16.0	+ 5.0 / 0	− 8.0 / −11.0	8.0 / 21.0	+ 8.0 / 0	− 8.0 / −13.0
56.19	−76.39	3.0 / 7.1	+ 2.5 / 0	− 3.0 / − 4.6	3.0 / 9.5	+ 4.0 / 0	− 3.0 / − 5.5	10.0 / 20.0	+ 6.0 / 0	−10.0 / −14.0	10.0 / 26.0	+10.0 / 0	−10.0 / −16.0
76.39	−100.9	4.0 / 9.0	+ 3.0 / 0	− 4.0 / − 6.0	4.0 / 12.0	+ 5.0 / 0	− 4.0 / − 7.0	12.0 / 25.0	+ 8.0 / 0	−12.0 / −17.0	12.0 / 32.0	+12.0 / 0	−12.0 / −20.0
100.9	−131.9	5.0 / 11.5	+ 4.0 / 0	− 5.0 / − 7.5	5.0 / 15.0	+ 6.0 / 0	− 5.0 / − 9.0	16.0 / 32.0	+10.0 / 0	−16.0 / −22.0	16.0 / 42.0	+16.0 / 0	−16.0 / −26.0
131.9	−171.9	6.0 / 14.0	+ 5.0 / 0	− 6.0 / − 9.0	6.0 / 19.0	+ 8.0 / 0	− 6.0 / −11.0	18.0 / 38.0	+12.0 / 0	−18.0 / −26.0	18.0 / 50.0	+20.0 / 0	−18.0 / −30.0
171.9	−200	8.0 / 18.0	+ 6.0 / 0	− 8.0 / −12.0	8.0 / 22.0	+10.0 / 0	− 8.0 / −12.0	22.0 / 48.0	+16.0 / 0	−22.0 / −32.0	22.0 / 63.0	+25.0 / 0	−22.0 / −38.0

RUNNING AND SLIDING FITS

Limits are in thousandths of an inch.

Limits for hole and shaft are applied algebraically to the basic size to obtain the limits of size for the parts

Data in bold face are in accordance with ABC agreements

Symbols H8, e7, etc., are Hole and Shaft designations used in ABC System

Class RC 5			Class RC 6			Class RC 7			Class RC 8			Class RC 9			Nominal Size Range Inches	
Limits of Clearance	Hole H8	Shaft e7	Limits of Clearance	Hole H9	Shaft e8	Limits of Clearance	Hole H9	Shaft d8	Limits of Clearance	Hole H10	Shaft c9	Limits of Clearance	Hole H11	Shaft	Over	To
0.6 1.6	+0.6 −0	−0.6 −1.0	0.6 2.2	+1.0 −0	−0.6 −1.2	1.0 2.6	+1.0 0	−1.0 −1.6	2.5 5.1	+1.6 0	−2.5 −3.5	4.0 8.1	+2.5 0	−4.0 −5.6	0	0.12
0.8 2.0	+0.7 −0	−0.8 −1.3	0.8 2.7	+1.2 −0	−0.8 −1.5	1.2 3.1	+1.2 0	−1.2 −1.9	2.8 5.8	+1.8 0	−2.8 −4.0	4.5 9.0	+3.0 0	−4.5 −6.0	0.12	0.24
1.0 2.5	+0.9 −0	−1.0 −1.6	1.0 3.3	+1.4 −0	−1.0 −1.9	1.6 3.9	+1.4 0	−1.6 −2.5	3.0 6.6	+2.2 0	−3.0 −4.4	5.0 10.7	+3.5 0	−5.0 −7.2	0.24	0.40
1.2 2.9	+1.0 −0	−1.2 −1.9	1.2 3.8	+1.6 −0	−1.2 −2.2	2.0 4.6	+1.6 0	−2.0 −3.0	3.5 7.9	+2.8 0	−3.5 −5.1	6.0 12.8	+4.0 −0	−6.0 −8.8	0.40	0.71
1.6 3.6	+1.2 −0	−1.6 −2.4	1.6 4.8	+2.0 −0	−1.6 −2.8	2.5 5.7	+2.0 0	−2.5 −3.7	4.5 10.0	+3.5 0	−4.5 −6.5	7.0 15.5	+5.0 0	−7.0 −10.5	0.71	1.19
2.0 4.6	+1.6 −0	−2.0 −3.0	2.0 6.1	+2.5 −0	−2.0 −3.6	3.0 7.1	+2.5 0	−3.0 −4.6	5.0 11.5	+4.0 0	−5.0 −7.5	8.0 18.0	+6.0 0	−8.0 −12.0	1.19	1.97
2.5 5.5	+1.8 −0	−2.5 −3.7	2.5 7.3	+3.0 −0	−2.5 −4.3	4.0 8.8	+3.0 0	−4.0 −5.8	6.0 13.5	+4.5 0	−6.0 −9.0	9.0 20.5	+7.0 0	−9.0 −13.5	1.97	3.15
3.0 6.6	+2.2 −0	−3.0 −4.4	3.0 8.7	+3.5 −0	−3.0 −5.2	5.0 10.7	+3.5 0	−5.0 −7.2	7.0 15.5	+5.0 0	−7.0 −10.5	10.0 24.0	+9.0 0	−10.0 −15.0	3.15	4.73
3.5 7.6	+2.5 −0	−3.5 −5.1	3.5 10.0	+4.0 −0	−3.5 −6.0	6.0 12.5	+4.0 0	−6.0 −8.5	8.0 18.0	+6.0 0	−8.0 −12.0	12.0 28.0	+10.0 0	−12.0 −18.0	4.73	7.09
4.0 8.6	+2.8 −0	−4.0 −5.8	4.0 11.3	+4.5 0	−4.0 −6.8	7.0 14.3	+4.5 0	−7.0 −9.8	10.0 21.5	+7.0 0	−10.0 −14.5	15.0 34.0	+12.0 0	−15.0 −22.0	7.09	9.85
5.0 10.0	+3.0 0	−5.0 −7.0	5.0 13.0	+5.0 0	−5.0 −8.0	8.0 16.0	+5.0 0	−8.0 −11.0	12.0 25.0	+8.0 0	−12.0 −17.0	18.0 38.0	+12.0 0	−18.0 −26.0	9.85	12.41
6.0 11.7	+3.5 0	−6.0 −8.2	6.0 15.5	+6.0 0	−6.0 −9.5	10.0 19.5	+6.0 0	−10.0 −13.5	14.0 29.0	+9.0 0	−14.0 −20.0	22.0 45.0	+14.0 0	−22.0 −31.0	12.41	15.75
8.0 14.5	+4.0 0	−8.0 −10.5	8.0 18.0	+6.0 0	−8.0 −12.0	12.0 22.0	+6.0 0	−12.0 −16.0	16.0 32.0	+10.0 0	−16.0 −22.0	25.0 51.0	+16.0 0	−25.0 −35.0	15.75	19.69
10.0 18.0	+5.0 0	−10.0 −13.0	10.0 23.0	+8.0 0	−10.0 −15.0	16.0 29.0	+8.0 0	−16.0 −21.0	20.0 40.0	+12.0 0	−20.0 −28.0	30.0 62.0	+20.0 0	−30.0 −42.0	19.69	30.09
12.0 22.0	+6.0 0	−12.0 −16.0	12.0 28.0	+10.0 0	−12.0 −18.0	20.0 36.0	+10.0 0	−20.0 −26.0	25.0 51.0	+16.0 0	−25.0 −35.0	40.0 81.0	+25.0 0	−40.0 −56.0	30.09	41.49
16.0 29.0	+8.0 0	−16.0 −21.0	16.0 36.0	+12.0 0	−16.0 −24.0	25.0 45.0	+12.0 0	−25.0 −33.0	30.0 62.0	+20.0 0	−30.0 −42.0	50.0 100	+30.0 0	−50.0 −70.0	41.49	56.19
20.0 36.0	+10.0 0	−20.0 −26.0	20.0 46.0	+16.0 0	−20.0 −30.0	30.0 56.0	+16.0 0	−30.0 −40.0	40.0 81.0	+25.0 0	−40.0 −56.0	60.0 125	+40.0 0	−60.0 −85.0	56.19	76.39
25.0 45.0	+12.0 0	−25.0 −33.0	25.0 57.0	+20.0 0	−25.0 −37.0	40.0 72.0	+20.0 0	−40.0 −52.0	50.0 100	+30.0 0	−50.0 −70.0	80.0 160	+50.0 0	−80.0 −110	76.39	100.9
30.0 56.0	+16.0 0	−30.0 −40.0	30.0 71.0	+25.0 0	−30.0 −46.0	50.0 91.0	+25.0 0	−50.0 −66.0	60.0 125	+40.0 0	−60.0 −85.0	100 200	+60.0 0	−100 −140	100.9	131.9
35.0 67.0	+20.0 0	−35.0 −47.0	35.0 85.0	+30.0 0	−35.0 −55.0	60.0 110.0	+30.0 0	−60.0 −80.0	80.0 160	+50.0 0	−80.0 −110	130 260	+80.0 0	−130 −180	131.9	171.9
45.0 86.0	+25.0 0	−45.0 −61.0	45.0 110.0	+40.0 0	−45.0 −70.0	80.0 145.0	+40.0 0	−80.0 −105.0	100 200	+60.0 0	−100 −140	150 310	+100 0	−150 −210	171.9	200

APPENDIX 8B Locational clearance fits

Limits are in thousandths of an inch.

Limits for hole and shaft are applied algebraically to the basic size to obtain the limits of size for the parts.

Data in bold face are in accordance with ABC agreements.

Symbols H6, h5, etc., are Hole and Shaft designations used in ABC System

Nominal Size Range Inches Over	To	Class LC 1 Limits of Clearance	Hole H6	Shaft h5	Class LC 2 Limits of Clearance	Hole H7	Shaft h6	Class LC 3 Limits of Clearance	Hole H8	Shaft h7	Class LC 4 Limits of Clearance	Hole H10	Shaft h9	Class LC 5 Limits of Clearance	Hole H7	Shaft g6
0 —	0.12	0 / 0.45	+0.25 / -0	+0 / -0.2	0 / 0.65	+0.4 / -0	+0 / -0.25	0 / 1	+0.6 / -0	+0 / -0.4	0 / 2.6	+1.6 / -0	+0 / -1.0	0.1 / 0.75	+0.4 / -0	-0.1 / -0.35
0.12—	0.24	0 / 0.5	+0.3 / -0	+0 / -0.2	0 / 0.8	+0.5 / -0	+0 / -0.3	0 / 1.2	+0.7 / -0	+0 / -0.5	0 / 3.0	+1.8 / -0	+0 / -1.2	0.15 / 0.95	+0.5 / -0	-0.15 / -0.45
0.24—	0.40	0 / 0.65	+0.4 / -0	+0 / -0.25	0 / 1.0	+0.6 / -0	+0 / -0.4	0 / 1.5	+0.9 / -0	+0 / -0.6	0 / 3.6	+2.2 / -0	+0 / -1.4	0.2 / 1.2	+0.6 / -0	-0.2 / -0.6
0.40—	0.71	0 / 0.7	+0.4 / -0	+0 / -0.3	0 / 1.1	+0.7 / -0	+0 / -0.4	0 / 1.7	+1.0 / -0	+0 / -0.7	0 / 4.4	+2.8 / -0	+0 / -1.6	0.25 / 1.35	+0.7 / -0	-0.25 / -0.65
0.71—	1.19	0 / 0.9	+0.5 / -0	+0 / -0.4	0 / 1.3	+0.8 / -0	+0 / -0.5	0 / 2	+1.2 / -0	+0 / -0.8	0 / 5.5	+3.5 / -0	+0 / -2.0	0.3 / 1.6	+0.8 / -0	-0.3 / -0.8
1.19—	1.97	0 / 1.0	+0.6 / -0	+0 / -0.4	0 / 1.6	+1.0 / -0	+0 / -0.6	0 / 2.6	+1.6 / -0	+0 / -1	0 / 6.5	+4.0 / -0	+0 / -2.5	0.4 / 2.0	+1.0 / -0	-0.4 / -1.0
1.97—	3.15	0 / 1.2	+0.7 / -0	+0 / -0.5	0 / 1.9	+1.2 / -0	+0 / -0.7	0 / 3	+1.8 / -0	+0 / -1.2	0 / 7.5	+4.5 / -0	+0 / -3	0.4 / 2.3	+1.2 / -0	-0.4 / -1.1
3.15—	4.73	0 / 1.5	+0.9 / -0	+0 / -0.6	0 / 2.3	+1.4 / -0	+0 / -0.9	0 / 3.6	+2.2 / -0	+0 / -1.4	0 / 8.5	+5.0 / -0	+0 / -3.5	0.5 / 2.8	+1.4 / -0	-0.5 / -1.4
4.73—	7.09	0 / 1.7	+1.0 / -0	+0 / -0.7	0 / 2.6	+1.6 / -0	+0 / -1.0	0 / 4.1	+2.5 / -0	+0 / -1.6	0 / 10	+6.0 / -0	+0 / -4	0.6 / 3.2	+1.6 / -0	-0.6 / -1.6
7.09—	9.85	0 / 2.0	+1.2 / -0	+0 / -0.8	0 / 3.0	+1.8 / -0	+0 / -1.2	0 / 4.6	+2.8 / -0	+0 / -1.8	0 / 11.5	+7.0 / -0	+0 / -4.5	0.6 / 3.6	+1.8 / -0	-0.6 / -1.8
9.85—	12.41	0 / 2.1	+1.2 / -0	+0 / -0.9	0 / 3.2	+2.0 / -0	+0 / -1.2	0 / 5	+3.0 / -0	+0 / -2.0	0 / 13	+8.0 / -0	+0 / -5	0.7 / 3.9	+2.0 / -0	-0.7 / -1.9
12.41—	15.75	0 / 2.4	+1.4 / -0	+0 / -1.0	0 / 3.6	+2.2 / -0	+0 / -1.4	0 / 5.7	+3.5 / -0	+0 / -2.2	0 / 15	+9.0 / -0	+0 / -6	0.7 / 4.3	+2.2 / -0	-0.7 / -2.1
15.75—	19.69	0 / 2.6	+1.6 / -0	+0 / -1.0	0 / 4.1	+2.5 / -0	+0 / -1.6	0 / 6.5	+4 / -0	+0 / -2.5	0 / 16	+10.0 / -0	+0 / -6	0.8 / 4.9	+2.5 / -0	-0.8 / -2.4
19.69—	30.09	0 / 3.2	+2.0 / -0	+0 / -1.2	0 / 5.0	+3 / -0	+0 / -2	0 / 8	+5 / -0	+0 / -3	0 / 20	+12.0 / -0	+0 / -8	0.9 / 5.9	+3.0 / -0	-0.9 / -2.9
30.09—	41.49	0 / 4.1	+2.5 / -0	+0 / -1.6	0 / 6.5	+4 / -0	+0 / -2.5	0 / 10	+6 / -0	+0 / -4	0 / 26	+16.0 / -0	+0 / -10	1.0 / 7.5	+4.0 / -0	-1.0 / -3.5
41.49—	56.19	0 / 5.0	+3.0 / -0	+0 / -2.0	0 / 8.0	+5 / -0	+0 / -3	0 / 13	+8 / -0	+0 / -5	0 / 32	+20.0 / -0	+0 / -12	1.2 / 9.2	+5.0 / -0	-1.2 / -4.2
56.19—	76.39	0 / 6.5	+4.0 / -0	+0 / -2.5	0 / 10	+6 / -0	+0 / -4	0 / 16	+10 / -0	+0 / -6	0 / 41	+25.0 / -0	+0 / -16	1.2 / 11.2	+6.0 / -0	-1.2 / -5.2
76.39—	100.9	0 / 8.0	+5.0 / -0	+0 / -3.0	0 / 13	+8 / -0	+0 / -5	0 / 20	+12 / -0	+0 / -8	0 / 50	+30.0 / -0	+0 / -20	1.4 / 14.4	+8.0 / -0	-1.4 / -6.4
100.9 —	131.9	0 / 10.0	+6.0 / -0	+0 / -4.0	0 / 16	+10 / -0	+0 / -6	0 / 26	+16 / -0	+0 / -10	0 / 65	+40.0 / -0	+0 / -25	1.6 / 17.6	+10.0 / -0	-1.6 / -7.6
131.9 —	171.9	0 / 13.0	+8.0 / -0	+0 / -5.0	0 / 20	+12 / -0	+0 / -8	0 / 32	+20 / -0	+0 / -12	0 / 80	+50.0 / -0	+0 / -30	1.8 / 21.8	+12.0 / -0	-1.8 / -9.8
171.9 —	200	0 / 16.0	+10.0 / -0	+0 / -6.0	0 / 26	+16 / -0	+0 / -10	0 / 41	+25 / -0	+0 / -16	0 / 100	+60.0 / -0	+0 / -40	1.8 / 27.8	+16.0 / -0	-1.8 / -11.8

LOCATIONAL CLEARANCE FITS

Limits are in thousandths of an inch.

Limits for hole and shaft are applied algebraically to the basic size to obtain the limits of size for the parts.

Data in bold face are in accordance with ABC agreements.
Symbols H9, f8, etc., are Hole and Shaft designations used in ABC System

Class LC 6 Limits of Clearance	Class LC 6 Hole H9	Class LC 6 Shaft f8	Class LC 7 Limits of Clearance	Class LC 7 Hole H10	Class LC 7 Shaft e9	Class LC 8 Limits of Clearance	Class LC 8 Hole H10	Class LC 8 Shaft d9	Class LC 9 Limits of Clearance	Class LC 9 Hole H11	Class LC 9 Shaft c10	Class LC 10 Limits of Clearance	Class LC 10 Hole H12	Class LC 10 Shaft	Class LC 11 Limits of Clearance	Class LC 11 Hole H13	Class LC 11 Shaft	Nominal Size Range Inches Over — To
0.3 / 1.9	+1.0 / 0	−0.3 / −0.9	0.6 / 3.2	+1.6 / 0	−0.6 / −1.6	1.0 / 3.6	+1.6 / −0	−1.0 / −2.0	2.5 / 6.6	+2.5 / −0	−2.5 / −4.1	4 / 12	+4 / −0	−4 / −8	5 / 17	+6 / −0	−5 / −11	0 — 0.12
0.4 / 2.3	+1.2 / 0	−0.4 / −1.1	0.8 / 3.8	+1.8 / 0	−0.8 / −2.0	1.2 / 4.2	+1.8 / −0	−1.2 / −2.4	2.8 / 7.6	+3.0 / −0	−2.8 / −4.6	4.5 / 14.5	+5 / −0	−4.5 / −9.5	6 / 20	+7 / −0	−6 / −13	0.12 — 0.24
0.5 / 2.8	+1.4 / 0	−0.5 / −1.4	1.0 / 4.6	+2.2 / 0	−1.0 / −2.4	1.6 / 5.2	+2.2 / −0	−1.6 / −3.0	3.0 / 8.7	+3.5 / −0	−3.0 / −5.2	5 / 17	+6 / −0	−5 / −11	7 / 25	+9 / −0	−7 / −16	0.24 — 0.40
0.6 / 3.2	+1.6 / 0	−0.6 / −1.6	1.2 / 5.6	+2.8 / 0	−1.2 / −2.8	2.0 / 6.4	+2.8 / −0	−2.0 / −3.6	3.5 / 10.3	+4.0 / −0	−3.5 / −6.3	6 / 20	+7 / −0	−6 / −13	8 / 28	+10 / −0	−8 / −18	0.40 — 0.71
0.8 / 4.0	+2.0 / 0	−0.8 / −2.0	1.6 / 7.1	+3.5 / 0	−1.6 / −3.6	2.5 / 8.0	+3.5 / −0	−2.5 / −4.5	4.5 / 13.0	+5.0 / −0	−4.5 / −8.0	7 / 23	+8 / −0	−7 / −15	10 / 34	+12 / −0	−10 / −22	0.71 — 1.19
1.0 / 5.1	+2.5 / 0	−1.0 / −2.6	2.0 / 8.5	+4.0 / 0	−2.0 / −4.5	3.0 / 9.5	+4.0 / −0	−3.0 / −5.5	5 / 15	+6 / −0	−5 / 9	8 / 28	+10 / −0	−8 / −18	12 / 44	+16 / −0	−12 / −28	1.19 — 1.97
1.2 / 6.0	+3.0 / 0	−1.2 / −3.0	2.5 / 10.0	+4.5 / 0	−2.5 / −5.5	4.0 / 11.5	+4.5 / −0	−4.0 / −7.0	6 / 17.5	+7 / −0	−6 / −10.5	10 / 34	+12 / −0	−10 / −22	14 / 50	+18 / −0	−14 / −32	1.97 — 3.15
1.4 / 7.1	+3.5 / 0	−1.4 / −3.6	3.0 / 11.5	+5.0 / 0	−3.0 / −6.5	5.0 / 13.5	+5.0 / −0	−5.0 / −8.5	7 / 21	+9 / −0	−7 / −12	11 / 39	+14 / −0	−11 / −25	16 / 60	+22 / −0	−16 / −38	3.15 — 4.73
1.6 / 8.1	+4.0 / 0	−1.6 / −4.1	3.5 / 13.5	+6.0 / 0	−3.5 / −7.5	6 / 16	+6 / −0	−6 / −10	8 / 24	+10 / −0	−8 / −14	12 / 44	+16 / −0	−12 / −28	18 / 68	+25 / −0	−18 / −43	4.73 — 7.09
2.0 / 9.3	+4.5 / 0	−2.0 / −4.8	4.0 / 15.5	+7.0 / 0	−4.0 / −8.5	7 / 18.5	+7 / −0	−7 / −11.5	10 / 29	+12 / −0	−10 / −17	16 / 52	+18 / −0	−16 / −34	22 / 78	+28 / −0	−22 / −50	7.09 — 9.85
2.2 / 10.2	+5.0 / 0	−2.2 / −5.2	4.5 / 17.5	+8.0 / 0	−4.5 / −9.5	7 / 20	+8 / −0	−7 / −12	12 / 32	+12 / −0	−12 / −20	20 / 60	+20 / −0	−20 / −40	28 / 88	+30 / −0	−28 / −58	9.85 — 12.41
2.5 / 12.0	+6.0 / 0	−2.5 / −6.0	5.0 / 20.0	+9.0 / 0	−5 / −11	8 / 23	+9 / −0	−8 / −14	14 / 37	+14 / −0	−14 / −23	22 / 66	+22 / −0	−22 / −44	30 / 100	+35 / −0	−30 / −65	12.41 — 15.75
2.8 / 12.8	+6.0 / 0	−2.8 / −6.8	5.0 / 21.0	+10.0 / 0	−5 / −11	9 / 25	+10 / −0	−9 / −15	16 / 42	+16 / −0	−16 / −26	25 / 75	+25 / −0	−25 / −50	35 / 115	+40 / −0	−35 / −75	15.75 — 19.69
3.0 / 16.0	+8.0 / 0	−3.0 / −8.0	6.0 / 26.0	+12.0 / −0	−6 / −14	10 / 30	+12 / −0	−10 / −18	18 / 50	+20 / −0	−18 / −30	28 / 88	+30 / −0	−28 / −58	40 / 140	+50 / −0	−40 / −90	19.69 — 30.09
3.5 / 19.5	+10.0 / 0	−3.5 / −9.5	7.0 / 33.0	+16.0 / −0	−7 / −17	12 / 38	+16 / −0	−12 / −22	20 / 61	+25 / −0	−20 / −36	30 / 110	+40 / −0	−30 / −70	45 / 165	+60 / −0	−45 / −105	30.09 — 41.49
4.0 / 24.0	+12.0 / 0	−4.0 / −12.0	8.0 / 40.0	+20.0 / −0	−8 / −20	14 / 46	+20 / −0	−14 / −26	25 / 75	+30 / −0	−25 / −45	40 / 140	+50 / −0	−40 / −90	60 / 220	+80 / −0	−60 / −140	41.49 — 56.19
4.5 / 30.5	+16.0 / 0	−4.5 / −14.5	9.0 / 50.0	+25.0 / −0	−9 / −25	16 / 57	+25 / −0	−16 / −32	30 / 95	+40 / −0	−30 / −55	50 / 170	+60 / −0	−50 / 110	70 / 270	+100 / −0	−70 / −170	56.19 — 76.39
5.0 / 37.0	+20.0 / 0	−5 / −17	10.0 / 60.0	+30.0 / −0	−10 / −30	18 / 68	+30 / −0	−18 / −38	35 / 115	+50 / −0	−35 / −65	50 / 210	+80 / −0	−50 / −130	80 / 330	+125 / −0	−80 / −205	76.39 — 100.9
6.0 / 47.0	+25.0 / 0	−6 / −22	12.0 / 67.0	+40.0 / −0	−12 / −27	20 / 85	+40 / −0	−20 / −45	40 / 140	+60 / −0	−40 / −80	60 / 260	+100 / −0	−60 / −160	90 / 410	+160 / −0	−90 / −250	100.9 — 131.9
7.0 / 57.0	+30.0 / 0	−7 / −27	14.0 / 94.0	+50.0 / −0	−14 / −44	25 / 105	+50 / −0	−25 / −55	50 / 180	+80 / −0	−50 / −100	80 / 330	+125 / −0	−80 / −205	100 / 500	+200 / −0	−100 / −300	131.9 — 171.9
7.0 / 72.0	+40.0 / 0	−7 / −32	14.0 / 114.0	+60.0 / −0	−14 / −54	25 / 125	+60 / −0	−25 / −65	50 / 210	+100 / −0	−50 / −110	90 / 410	+160 / −0	−90 / −250	125 / 625	+250 / −0	−125 / −375	171.9 — 200

APPENDIX 8C Locational transitional fits

Limits are in thousandths of an inch.

Limits for hole and shaft are applied algebraically to the basic size to obtain the limits of size for the mating parts.

Data in bold face are in accordance with ABC agreements.

"Fit" represents the maximum interference (minus values) and the maximum clearance (plus values).

Symbols H7, js6, etc., are Hole and Shaft designations used in ABC System

Nominal Size Range Inches Over	To	Class LT 1 Fit	LT 1 Hole H7	LT 1 Shaft js6	Class LT 2 Fit	LT 2 Hole H8	LT 2 Shaft js7	Class LT 3 Fit	LT 3 Hole H7	LT 3 Shaft k6	Class LT 4 Fit	LT 4 Hole H8	LT 4 Shaft k7	Class LT 5 Fit	LT 5 Hole H7	LT 5 Shaft n6	Class LT 6 Fit	LT 6 Hole H7	LT 6 Shaft n7
0 –	0.12	−0.10 / +0.50	+0.4 / −0	+0.10 / −0.10	−0.2 / +0.8	+0.6 / −0	+0.2 / −0.2							−0.5 / +0.15	+0.4 / −0	+0.5 / +0.25	−0.65 / +0.15	+0.4 / −0	−0.65 / +0.25
0.12 –	0.24	−0.15 / +0.65	+0.5 / −0	+0.15 / −0.15	−0.25 / +0.95	+0.7 / −0	+0.25 / −0.25							−0.6 / +0.2	+0.5 / −0	+0.6 / +0.3	−0.8 / +0.2	+0.5 / −0	+0.8 / +0.3
0.24 –	0.40	−0.2 / +0.8	+0.6 / −0	+0.2 / −0.2	−0.3 / +1.2	+0.9 / −0	+0.3 / −0.3	−0.5 / +0.5	+0.6 / −0	+0.5 / +0.1	−0.7 / +0.8	+0.9 / −0	+0.7 / +0.1	−0.8 / +0.2	+0.6 / −0	+0.8 / +0.4	−1.0 / +0.2	+0.6 / −0	+1.0 / +0.4
0.40 –	0.71	−0.2 / +0.9	+0.7 / −0	+0.2 / −0.2	−0.35 / +1.35	+1.0 / −0	+0.35 / −0.35	−0.5 / +0.6	+0.7 / −0	+0.5 / +0.1	−0.8 / +0.9	+1.0 / −0	+0.8 / +0.1	−0.9 / +0.2	+0.7 / −0	+0.9 / +0.5	−1.2 / +0.2	+0.7 / −0	+1.2 / +0.5
0.71 –	1.19	−0.25 / +1.05	+0.8 / −0	+0.25 / −0.25	−0.4 / +1.6	+1.2 / −0	+0.4 / −0.4	−0.6 / +0.7	+0.8 / −0	+0.6 / +0.1	−0.9 / +1.1	+1.2 / −0	+0.9 / +0.1	−1.1 / +0.2	+0.8 / −0	+1.1 / +0.6	−1.4 / +0.2	+0.8 / −0	+1.4 / +0.6
1.19 –	1.97	−0.3 / +1.3	+1.0 / −0	+0.3 / −0.3	−0.5 / +2.1	+1.6 / −0	+0.5 / −0.5	−0.7 / +0.9	+1.0 / −0	+0.7 / +0.1	−1.1 / +1.5	+1.6 / −0	+1.1 / +0.1	−1.3 / +0.3	+1.0 / −0	+1.3 / +0.7	−1.7 / +0.3	+1.0 / −0	+1.7 / +0.7
1.97 –	3.15	−0.3 / +1.5	+1.2 / −0	+0.3 / −0.3	−0.6 / +2.4	+1.8 / −0	+0.6 / −0.6	−0.8 / +1.1	+1.2 / −0	+0.8 / +0.1	−1.3 / +1.7	+1.8 / −0	+1.3 / +0.1	−1.5 / +0.4	+1.2 / −0	+1.5 / +0.8	−2.0 / +0.4	+1.2 / −0	+2.0 / +0.8
3.15 –	4.73	−0.4 / +1.8	+1.4 / −0	+0.4 / −0.4	−0.7 / +2.9	+2.2 / −0	+0.7 / −0.7	−1.0 / +1.3	+1.4 / −0	+1.0 / +0.1	−1.5 / +2.1	+2.2 / −0	+1.5 / +0.1	−1.9 / +0.4	+1.4 / −0	+1.9 / +1.0	−2.4 / +0.4	+1.4 / −0	+2.4 / +1.0
4.73 –	7.09	−0.5 / +2.1	+1.6 / −0	+0.5 / −0.5	−0.8 / +3.3	+2.5 / −0	+0.8 / −0.8	−1.1 / +1.5	+1.6 / −0	+1.1 / +0.1	−1.7 / +2.4	+2.5 / −0	+1.7 / +0.1	−2.2 / +0.4	+1.6 / −0	+2.2 / +1.2	−2.8 / +0.4	+1.6 / −0	+2.8 / +1.2
7.09 –	9.85	−0.6 / +2.4	+1.8 / −0	+0.6 / −0.6	−0.9 / +3.7	+2.8 / −0	+0.9 / −0.9	−1.4 / +1.6	+1.8 / −0	+1.4 / +0.2	−2.0 / +2.6	+2.8 / −0	+2.0 / +0.2	−2.6 / +0.4	+1.8 / −0	+2.6 / +1.4	−3.2 / +0.4	+1.8 / −0	+3.2 / +1.4
9.85 –	12.41	−0.6 / +2.6	+2.0 / −0	+0.6 / −0.6	−1.0 / +4.0	+3.0 / −0	+1.0 / −1.0	−1.4 / +1.8	+2.0 / −0	+1.4 / +0.2	−2.2 / +2.8	+3.0 / −0	+2.2 / +0.2	−2.6 / +0.6	+2.0 / −0	+2.6 / +1.4	−3.4 / +0.6	+2.0 / −0	+3.4 / +1.4
12.41 –	15.75	−0.7 / +2.9	+2.2 / −0	+0.7 / −0.7	−1.0 / +4.5	+3.5 / −0	+1.0 / −1.0	−1.6 / +2.0	+2.2 / −0	+1.6 / +0.2	−2.4 / +3.3	+3.5 / −0	+2.4 / +0.2	−3.0 / +0.6	+2.2 / −0	+3.0 / +1.6	−3.8 / +0.6	+2.2 / −0	+3.8 / +1.6
15.75 –	19.69	−0.8 / +3.3	+2.5 / −0	+0.8 / −0.8	−1.2 / +5.2	+4.0 / −0	+1.2 / −1.2	−1.8 / +2.3	+2.5 / −0	+1.8 / +0.2	−2.7 / +3.8	+4.0 / −0	+2.7 / +0.2	−3.4 / +0.7	+2.5 / −0	+3.4 / +1.8	−4.3 / +0.7	+2.5 / −0	+4.3 / +1.8

APPENDIX 8D Locational interference fits

Limits are in thousandths of an inch.
Limits for hole and shaft are applied algebraically to the
basic size to obtain the limits of size for the parts.

Data in bold face are in accordance with ABC agreements,
Symbols H7, p6, etc., are Hole and Shaft designations
used in ABC System

Nominal Size Range Inches Over — To	Class LN 1			Class LN 2			Class LN 3		
	Limits of Interference	Standard Limits		Limits of Interference	Standard Limits		Limits of Interference	Standard Limits	
		Hole H6	Shaft n5		Hole H7	Shaft p6		Hole H7	Shaft r6
0 — 0.12	0 / 0.45	+ 0.25 / − 0	+0.45 / +0.25	0 / 0.65	+ 0.4 / − 0	+ 0.65 / + 0.4	0.1 / 0.75	+ 0.4 / − 0	+ 0.75 / + 0.5
0.12 — 0.24	0 / 0.5	+ 0.3 / − 0	+0.5 / +0.3	0 / 0.8	+ 0.5 / − 0	+ 0.8 / + 0.5	0.1 / 0.9	+ 0.5 / 0	+ 0.9 / + 0.6
0.24 — 0.40	0 / 0.65	+ 0.4 / − 0	+0.65 / +0.4	0 / 1.0	+ 0.6 / − 0	+ 1.0 / + 0.6	0.2 / 1.2	+ 0.6 / − 0	+ 1.2 / + 0.8
0.40 — 0.71	0 / 0.8	+ 0.4 / − 0	+0.8 / +0.4	0 / 1.1	+ 0.7 / − 0	+ 1.1 / + 0.7	0.3 / 1.4	+ 0.7 / − 0	+ 1.4 / + 1.0
0.71 — 1.19	0 / 1.0	+ 0.5 / − 0	+1.0 / +0.5	0 / 1.3	+ 0.8 / − 0	+ 1.3 / + 0.8	0.4 / 1.7	+ 0.8 / − 0	+ 1.7 / + 1.2
1.19 — 1.97	0 / 1.1	+ 0.6 / − 0	+1.1 / +0.6	0 / 1.6	+ 1.0 / − 0	+ 1.6 / + 1.0	0.4 / 2.0	+ 1.0 / − 0	+ 2.0 / + 1.4
1.97 — 3.15	0.1 / 1.3	+ 0.7 / − 0	+1.3 / +0.8	0.2 / 2.1	+ 1.2 / − 0	+ 2.1 / + 1.4	0.4 / 2.3	+ 1.2 / − 0	+ 2.3 / + 1.6
3.15 — 4.73	0.1 / 1.6	+ 0.9 / − 0	+1.6 / +1.0	0.2 / 2.5	+ 1.4 / − 0	+ 2.5 / + 1.6	0.6 / 2.9	+ 1.4 / − 0	+ 2.9 / + 2.0
4.73 — 7.09	0.2 / 1.9	+ 1.0 / − 0	+1.9 / +1.2	0.2 / 2.8	+ 1.6 / − 0	+ 2.8 / + 1.8	0.9 / 3.5	+ 1.6 / − 0	+ 3.5 / + 2.5
7.09 — 9.85	0.2 / 2.2	+ 1.2 / − 0	+2.2 / +1.4	0.2 / 3.2	+ 1.8 / − 0	+ 3.2 / + 2.0	1.2 / 4.2	+ 1.8 / − 0	+ 4.2 / + 3.0
9.85 — 12.41	0.2 / 2.3	+ 1.2 / − 0	+2.3 / +1.4	0.2 / 3.4	+ 2.0 / − 0	+ 3.4 / + 2.2	1.5 / 4.7	+ 2.0 / − 0	+ 4.7 / + 3.5
12.41 — 15.75	0.2 / 2.6	+ 1.4 / − 0	+2.6 / +1.6	0.3 / 3.9	+ 2.2 / − 0	+ 3.9 / + 2.5	2.3 / 5.9	+ 2.2 / − 0	+ 5.9 / + 4.5
15.75 — 19.69	0.2 / 2.8	+ 1.6 / − 0	+2.8 / +1.8	0.3 / 4.4	+ 2.5 / − 0	+ 4.4 / + 2.8	2.5 / 6.6	+ 2.5 / − 0	+ 6.6 / + 5.0
19.69 — 30.09		+ 2.0 / − 0		0.5 / 5.5	+ 3 / − 0	+ 5.5 / + 3.5	4 / 9	+ 3 / − 0	+ 9 / + 7
30.09 — 41.49		+ 2.5 / − 0		0.5 / 7.0	+ 4 / − 0	+ 7.0 / + 4.5	5 / 11.5	+ 4 / − 0	+11.5 / + 9
41.49 — 56.19		+ 3.0 / − 0		1 / 9	+ 5 / − 0	+ 9 / + 6	7 / 15	+ 5 / − 0	+15 / +12
56.19 — 76.39		+ 4.0 / − 0		1 / 11	+ 6 / − 0	+11 / + 7	10 / 20	+ 6 / − 0	+20 / +16
76.39 — 100.9		+ 5.0 / − 0		1 / 14	+ 8 / − 0	+14 / + 9	12 / 25	+ 8 / − 0	+25 / +20
100.9 — 131.9		+ 6.0 / − 0		2 / 18	+10 / − 0	+18 / +12	15 / 31	+10 / − 0	+31 / +25
131.9 — 171.9		+ 8.0 / − 0		4 / 24	+12 / − 0	+24 / +16	18 / 38	+12 / − 0	+38 / +30
171.9 — 200		+10.0 / − 0		4 / 30	+16 / − 0	+30 / +20	24 / 50	+16 / − 0	+50 / +40

661

APPENDIX 8E Force and shrink fits

Limits are in thousandths of an inch.

Limits for hole and shaft are applied algebraically to the basic size to obtain the limits of size for the parts.

Data in bold face are in accordance with ABC agreements.

Symbols H7, s6, etc., are Hole and Shaft designations used in ABC System

Nominal Size Range Inches Over	To	Class FN 1 Limits of Interference	Hole H6	Shaft	Class FN 2 Limits of Interference	Hole H7	Shaft s6	Class FN 3 Limits of Interference	Hole H7	Shaft t6	Class FN 4 Limits of Interference	Hole H7	Shaft u6	Class FN 5 Limits of Interference	Hole H8	Shaft x7
0	0.12	0.05	+0.25	+0.5	0.2	+0.4	+0.85				0.3	+0.4	+0.95	0.3	+0.6	+1.3
		0.5	−0	+0.3	0.85	−0	+0.6				0.95	−0	+0.7	1.3	−0	+0.9
0.12	0.24	0.1	+0.3	+0.6	0.2	+0.5	+1.0				0.4	+0.5	+1.2	0.5	+0.7	+1.7
		0.6	−0	+0.4	1.0	−0	+0.7				1.2	−0	+0.9	1.7	−0	+1.2
0.24	0.40	0.1	+0.4	+0.75	0.4	+0.6	+1.4				0.6	+0.6	+1.6	0.5	+0.9	+2.0
		0.75	−0	+0.5	1.4	−0	+1.0				1.6	−0	+1.2	2.0	−0	+1.4
0.40	0.56	0.1	+0.4	+0.8	0.5	+0.7	+1.6				0.7	+0.7	+1.8	0.6	+1.0	+2.3
		0.8	−0	+0.5	1.6	−0	+1.2				1.8	−0	+1.4	2.3	−0	+1.6
0.56	0.71	0.2	+0.4	+0.9	0.5	+0.7	+1.6				0.7	+0.7	+1.8	0.8	+1.0	+2.5
		0.9	−0	+0.6	1.6	−0	+1.2				1.8	−0	+1.4	2.5	−0	+1.8
0.71	0.95	0.2	+0.5	+1.1	0.6	+0.8	+1.9				0.8	+0.8	+2.1	1.0	+1.2	+3.0
		1.1	−0	+0.7	1.9	−0	+1.4				2.1	−0	+1.6	3.0	−0	+2.2
0.95	1.19	0.3	+0.5	+1.2	0.6	+0.8	+1.9	0.8	+0.8	+2.1	1.0	+0.8	+2.3	1.3	+1.2	+3.3
		1.2	−0	+0.8	1.9	−0	+1.4	2.1	−0	+1.6	2.3	−0	+1.8	3.3	−0	+2.5
1.19	1.58	0.3	+0.6	+1.3	0.8	+1.0	+2.4	1.0	+1.0	+2.6	1.5	+1.0	+3.1	1.4	+1.6	+4.0
		1.3	−0	+0.9	2.4	−0	+1.8	2.6	−0	+2.0	3.1	−0	+2.5	4.0	−0	+3.0
1.58	1.97	0.4	+0.6	+1.4	0.8	+1.0	+2.4	1.2	+1.0	+2.8	1.8	+1.0	+3.4	2.4	+1.6	+5.0
		1.4	−0	+1.0	2.4	−0	+1.8	2.8	−0	+2.2	3.4	−0	+2.8	5.0	−0	+4.0
1.97	2.56	0.6	+0.7	+1.8	0.8	+1.2	+2.7	1.3	+1.2	+3.2	2.3	+1.2	+4.2	3.2	+1.8	+6.2
		1.8	−0	+1.3	2.7	−0	+2.0	3.2	−0	+2.5	4.2	−0	+3.5	6.2	−0	+5.0
2.56	3.15	0.7	+0.7	+1.9	1.0	+1.2	+2.9	1.8	+1.2	+3.7	2.8	+1.2	+4.7	4.2	+1.8	+7.2
		1.9	−0	+1.4	2.9	−0	+2.2	3.7	−0	+3.0	4.7	−0	+4.0	7.2	−0	+6.0
3.15	3.94	0.9	+0.9	+2.4	1.4	+1.4	+3.7	2.1	+1.4	+4.4	3.6	+1.4	+5.9	4.8	+2.2	+8.4
		2.4	−0	+1.8	3.7	−0	+2.8	4.4	−0	+3.5	5.9	−0	+5.0	8.4	−0	+7.0
3.94	4.73	1.1	+0.9	+2.6	1.6	+1.4	+3.9	2.6	+1.4	+4.9	4.6	+1.4	+6.9	5.8	+2.2	+9.4
		2.6	−0	+2.0	3.9	−0	+3.0	4.9	−0	+4.0	6.9	−0	+6.0	9.4	−0	+8.0
4.73	5.52	1.2	+1.0	+2.9	1.9	+1.6	+4.5	3.4	+1.6	+6.0	5.4	+1.6	+8.0	7.5	+2.5	+11.6
		2.9	−0	+2.2	4.5	−0	+3.5	6.0	−0	+5.0	8.0	−0	+7.0	11.6	−0	+10.0
5.52	6.30	1.5	+1.0	+3.2	2.4	+1.6	+5.0	3.4	+1.6	+6.0	5.4	+1.6	+8.0	9.5	+2.5	+13.6
		3.2	−0	+2.5	5.0	−0	+4.0	6.0	−0	+5.0	8.0	−0	+7.0	13.6	−0	+12.0
6.30	7.09	1.8	+1.0	+3.5	2.9	+1.6	+5.5	4.4	+1.6	+7.0	6.4	+1.6	+9.0	9.5	+2.5	+13.6
		3.5	−0	+2.8	5.5	−0	+4.5	7.0	−0	+6.0	9.0	−0	+8.0	13.6	−0	+12.0
7.09	7.88	1.8	+1.2	+3.8	3.2	+1.8	+6.2	5.2	+1.8	+8.2	7.2	+1.8	+10.2	11.2	+2.8	+15.8
		3.8	−0	+3.0	6.2	−0	+5.0	8.2	−0	+7.0	10.2	−0	+9.0	15.8	−0	+14.0
7.88	8.86	2.3	+1.2	+4.3	3.2	+1.8	+6.2	5.2	+1.8	+8.2	8.2	+1.8	+11.2	13.2	+2.8	+17.8
		4.3	−0	+3.5	6.2	−0	+5.0	8.2	−0	+7.0	11.2	−0	+10.0	17.8	−0	+16.0
8.86	9.85	2.3	+1.2	+4.3	4.2	+1.8	+7.2	6.2	+1.8	+9.2	10.2	+1.8	+13.2	13.2	+2.8	+17.8
		4.3	−0	+3.5	7.2	−0	+6.0	9.2	−0	+8.0	13.2	−0	+12.0	17.8	−0	+16.0
9.85	11.03	2.8	+1.2	+4.9	4.0	+2.0	+7.2	7.0	+2.0	+10.2	10.0	+2.0	+13.2	15.0	+3.0	+20.0
		4.9	−0	+4.0	7.2	−0	+6.0	10.2	−0	+9.0	13.2	−0	+12.0	20.0	−0	+18.0
11.03	12.41	2.8	+1.2	+4.9	5.0	+2.0	+8.2	7.0	+2.0	+10.2	12.0	+2.0	+15.2	17.0	+3.0	+22.0
		4.9	−0	+4.0	8.2	−0	+7.0	10.2	−0	+9.0	15.2	−0	+14.0	22.0	−0	+20.0
12.41	13.98	3.1	+1.4	+5.5	5.8	+2.2	+9.4	7.8	+2.2	+11.4	13.8	+2.2	+17.4	18.5	+3.5	+24.2
		5.5	−0	+4.5	9.4	−0	+8.0	11.4	−0	+10.0	17.4	−0	+16.0	24.2	+0	+22.0
13.98	15.75	3.6	+1.4	+6.1	5.8	+2.2	+9.4	9.8	+2.2	+13.4	15.8	+2.2	+19.4	21.5	+3.5	+27.2
		6.1	−0	+5.0	9.4	−0	+8.0	13.4	−0	+12.0	19.4	−0	+18.0	27.2	−0	+25.0
15.75	17.72	4.4	+1.6	+7.0	6.5	+2.5	+10.6	9.5	+2.5	+13.6	17.5	+2.5	+21.6	24.0	+4.0	+30.5
		7.0	−0	+6.0	10.6	−0	+9.0	13.6	−0	+12.0	21.6	−0	+20.0	30.5	−0	+28.0
17.72	19.69	4.4	+1.6	+7.0	7.5	+2.5	+11.6	11.5	+2.5	+15.6	19.5	+2.5	+23.6	26.0	+4.0	+32.5
		7.0	−0	+6.0	11.6	−0	+10.0	15.6	−0	+14.0	23.6	−0	+22.0	32.5	−0	+30.0

FORCE AND SHRINK FITS

Limits are in thousandths of an inch.

Limits for hole and shaft are applied algebraically to the basic size to obtain the limits of size for the parts.

Data in bold face are in accordance with ABC agreements.

Symbols H7, s6, etc., are Hole and Shaft designations used in ABC System

Nominal Size Range Inches Over — To	Class FN 1 Limits of Interference	Standard Limits Hole H6	Shaft	Class FN 2 Limits of Interference	Standard Limits Hole H7	Shaft s6	Class FN 3 Limits of Interference	Standard Limits Hole H7	Shaft t6	Class FN 4 Limits of Interference	Standard Limits Hole H7	Shaft u6	Class FN 5 Limits of Interference	Standard Limits Hole H8	Shaft x7
19.69 — 24.34	6.0	+ 2.0	+ 9.2	9.0	+ 3.0	+ 14.0	15.0	+ 3.0	+ 20.0	22.0	+ 3.0	+ 27.0	30.0	+ 5.0	+ 38.0
	9.2	− 0	+ 8.0	14.0	− 0	+ 12.0	20.0	− 0	+ 18.0	27.0	− 0	+ 25.0	38.0	− 0	+ 35.0
24.34 — 30.09	7.0	+ 2.0	+10.2	11.0	+ 3.0	+ 16.0	17.0	+ 3.0	+ 22.0	27.0	+ 3.0	+ 32.0	35.0	+ 5.0	+ 43.0
	10.2	− 0	+ 9.0	16.0	− 0	+ 14.0	22.0	− 0	+ 20.0	32.0	− 0	+ 30.0	43.0	− 0	+ 40.0
30.09 — 35.47	7.5	+ 2.5	+11.6	14.0	+ 4.0	+ 20.5	21.0	+ 4.0	+ 27.5	31.0	+ 4.0	+ 37.5	44.0	+ 6.0	+ 54.0
	11.6	− 0	+10.0	20.5	− 0	+ 18.0	27.5	− 0	+ 25.0	37.5	− 0	+ 35.0	54.0	− 0	+ 50.0
35.47 — 41.49	9.5	+ 2.5	+13.6	16.0	+ 4.0	+ 22.5	24.0	+ 4.0	+ 30.5	36.0	+ 4.0	+ 43.5	54.0	+ 6.0	+ 64.0
	13.6	− 0	+12.0	22.5	− 0	+ 20.0	30.5	− 0	+ 28.0	43.5	− 0	+ 40.0	64.0	− 0	+ 60.0
41.49 — 48.28	11.0	+ 3.0	+16.0	17.0	+ 5.0	+ 25.0	30.0	+ 5.0	+ 38.0	45.0	+ 5.0	+ 53.0	62.0	+ 8.0	+ 75.0
	16.0	− 0	+14.0	25.0	− 0	+ 22.0	38.0	− 0	+ 35.0	53.0	− 0	+ 50.0	75.0	− 0	+ 70.0
48.28 — 56.19	13.0	+ 3.0	+18.0	20.0	+ 5.0	+ 28.0	35.0	+ 5.0	+ 43.0	55.0	+ 5.0	+ 63.0	72.0	+ 8.0	+ 85.0
	18.0	− 0	+16.0	28.0	− 0	+ 25.0	43.0	− 0	+ 40.0	63.0	− 0	+ 60.0	85.0	− 0	+ 80.0
56.19 — 65.54	14.0	+ 4.0	+20.5	24.0	+ 6.0	+ 34.0	39.0	+ 6.0	+ 49.0	64.0	+ 6.0	+ 74.0	90.0	+10.0	+106
	20.5	− 0	+18.0	34.0	− 0	+ 30.0	49.0	− 0	+ 45.0	74.0	− 0	+ 70.0	106	− 0	+100
65.54 — 76.39	18.0	+ 4.0	+24.5	29.0	+ 6.0	+ 39.0	44.0	+ 6.0	+ 54.0	74.0	+ 6.0	+ 84.0	110	+10.0	+126
	24.5	− 0	+22.0	39.0	− 0	+ 35.0	54.0	− 0	+ 50.0	84.0	− 0	+ 80.0	126	− 0	+120
76.39 — 87.79	20.0	+ 5.0	+28.0	32.0	+ 8.0	+ 45.0	52.0	+ 8.0	+ 65.0	82.0	+ 8.0	+ 95.0	128	+12.0	+148
	28.0	− 0	+25.0	45.0	− 0	+ 40.0	65.0	− 0	+ 60.0	95.0	− 0	+ 90.0	148	− 0	+140
87.79 — 100.9	23.0	+ 5.0	+31.0	37.0	+ 8.0	+ 50.0	62.0	+ 8.0	+ 75.0	92.0	+ 8.0	+105	148	+12.0	+168
	31.0	− 0	+28.0	50.0	− 0	+ 45.0	75.0	− 0	+ 70.0	105	− 0	+100	168	− 0	+160
100.9 — 115.3	24.0	+ 6.0	+34.0	40.0	+10.0	+ 56.0	70.0	+10.0	+ 86.0	110	+10.0	+126	164	+16.0	+190
	34.0	− 0	+30.0	56.0	− 0	+ 50.0	86.0	− 0	+ 80.0	126	− 0	+120	190	− 0	+180
115.3 — 131.9	29.0	+ 6.0	+39.0	50.0	+10.0	+ 66.0	80.0	+10.0	+ 96.0	130	+10.0	+146	184	+16.0	+210
	39.0	− 0	+35.0	66.0	− 0	+ 60.0	96.0	− 0	+ 90.0	146	− 0	+140	210	− 0	+200
131.9 — 152.2	37.0	+ 8.0	+50.0	58.0	+12.0	+ 78.0	88.0	+12.0	+108	148	+12.0	+168	200	+20.0	+232
	50.0	− 0	+45.0	78.0	− 0	+ 70.0	108	− 0	+100	168	− 0	+160	232	− 0	+220
152.2 — 171.9	42.0	+ 8.0	+55.0	68.0	+12.0	+ 88.0	108	+12.0	+128	168	+12.0	+188	230	+20.0	+262
	55.0	− 0	+50.0	88.0	− 0	+ 80.0	128	− 0	+120	188	− 0	+180	262	− 0	+250
171.9 — 200	50.0	+10.0	+66.0	74.0	+16.0	+100	124	+16.0	+150	184	+16.0	+210	275	+ 25	+316
	66.0	− 0	+60.0	100	− 0	+ 90	150	− 0	+140	210	− 0	+200	316	− 0	+300

APPENDIX 9 — AMERICAN NATIONAL STANDARD PREFERRED METRIC LIMITS AND FITS (ANSI B4.2—1978)

Dimensions in mm.

PREFERRED HOLE BASIS CLEARANCE FITS

BASIC SIZE		LOOSE RUNNING Hole H11	Shaft c11	Fit	FREE RUNNING Hole H9	Shaft d9	Fit	CLOSE RUNNING Hole H8	Shaft f7	Fit	SLIDING Hole H7	Shaft g6	Fit	LOCATIONAL CLEARANCE Hole H7	Shaft h6	Fit
1	MAX	1.060	0.940	0.180	1.025	0.980	0.070	1.014	0.994	0.030	1.010	0.998	0.018	1.010	1.000	0.016
	MIN	1.000	0.880	0.060	1.000	0.955	0.020	1.000	0.984	0.006	1.000	0.992	0.002	1.000	0.994	0.000
1.2	MAX	1.260	1.140	0.180	1.225	1.180	0.070	1.214	1.194	0.030	1.210	1.198	0.018	1.210	1.200	0.016
	MIN	1.200	1.080	0.060	1.200	1.155	0.020	1.200	1.184	0.006	1.200	1.192	0.002	1.200	1.194	0.000
1.6	MAX	1.660	1.540	0.180	1.625	1.580	0.070	1.614	1.594	0.030	1.610	1.598	0.018	1.610	1.600	0.016
	MIN	1.600	1.480	0.060	1.600	1.555	0.020	1.600	1.584	0.006	1.600	1.592	0.002	1.600	1.594	0.000
2	MAX	2.060	1.940	0.180	2.025	1.980	0.070	2.014	1.994	0.030	2.010	1.998	0.018	2.010	2.000	0.016
	MIN	2.000	1.880	0.060	2.000	1.955	0.020	2.000	1.984	0.006	2.000	1.992	0.002	2.000	1.994	0.000
2.5	MAX	2.560	2.440	0.180	2.525	2.480	0.070	2.514	2.494	0.030	2.510	2.498	0.018	2.510	2.500	0.016
	MIN	2.500	2.380	0.060	2.500	2.455	0.020	2.500	2.484	0.006	2.500	2.492	0.002	2.500	2.494	0.000
3	MAX	3.060	2.940	0.180	3.025	2.980	0.070	3.014	2.994	0.030	3.010	2.998	0.018	3.010	3.000	0.016
	MIN	3.000	2.880	0.060	3.000	2.955	0.020	3.000	2.984	0.006	3.000	2.992	0.002	3.000	2.994	0.000
4	MAX	4.075	3.930	0.220	4.030	3.970	0.090	4.018	3.990	0.040	4.012	3.996	0.024	4.012	4.000	0.020
	MIN	4.000	3.855	0.070	4.000	3.940	0.030	4.000	3.978	0.010	4.000	3.988	0.004	4.000	3.992	0.000
5	MAX	5.075	4.930	0.220	5.030	4.970	0.090	5.018	4.990	0.040	5.012	4.996	0.024	5.012	5.000	0.020
	MIN	5.000	4.855	0.070	5.000	4.940	0.030	5.000	4.978	0.010	5.000	4.988	0.004	5.000	4.992	0.000
6	MAX	6.075	5.930	0.220	6.030	5.970	0.090	6.018	5.990	0.040	6.012	5.996	0.024	6.012	6.000	0.020
	MIN	6.000	5.855	0.070	6.000	5.940	0.030	6.000	5.978	0.010	6.000	5.988	0.004	6.000	5.992	0.000
8	MAX	8.090	7.920	0.260	8.036	7.960	0.112	8.022	7.987	0.050	8.015	7.995	0.029	8.015	8.000	0.024
	MIN	8.000	7.830	0.080	8.000	7.924	0.040	8.000	7.972	0.013	8.000	7.986	0.005	8.000	7.991	0.000
10	MAX	10.090	9.920	0.260	10.036	9.960	0.112	10.022	9.987	0.050	10.015	9.995	0.029	10.015	10.000	0.024
	MIN	10.000	9.830	0.080	10.000	9.924	0.040	10.000	9.972	0.013	10.000	9.986	0.005	10.000	9.991	0.000
12	MAX	12.110	11.905	0.315	12.043	11.950	0.136	12.027	11.984	0.061	12.018	11.994	0.035	12.018	12.000	0.029
	MIN	12.000	11.795	0.095	12.000	11.907	0.050	12.000	11.966	0.016	12.000	11.983	0.006	12.000	11.989	0.000
16	MAX	16.110	15.905	0.315	16.043	15.950	0.136	16.027	15.984	0.061	16.018	15.994	0.035	16.018	16.000	0.029
	MIN	16.000	15.795	0.095	16.000	15.907	0.050	16.000	15.966	0.016	16.000	15.983	0.006	16.000	15.989	0.000
20	MAX	20.130	19.890	0.370	20.052	19.935	0.169	20.033	19.980	0.074	20.021	19.993	0.041	20.021	20.000	0.034
	MIN	20.000	19.760	0.110	20.000	19.883	0.065	20.000	19.959	0.020	20.000	19.980	0.007	20.000	19.987	0.000
25	MAX	25.130	24.890	0.370	25.052	24.935	0.169	25.033	24.980	0.074	25.021	24.993	0.041	25.021	25.000	0.034
	MIN	25.000	24.760	0.110	25.000	24.883	0.065	25.000	24.959	0.020	25.000	24.980	0.007	25.000	24.987	0.000
30	MAX	30.130	29.890	0.370	30.052	29.935	0.169	30.033	29.980	0.074	30.021	29.993	0.041	30.021	30.000	0.034
	MIN	30.000	29.760	0.110	30.000	29.883	0.065	30.000	29.959	0.020	30.000	29.980	0.007	30.000	29.987	0.000

APPENDIX 9A Preferred hole basis clearance fits (Cont.)

BASIC SIZE		LOOSE RUNNING			FREE RUNNING			CLOSE RUNNING			SLIDING			LOCATIONAL CLEARANCE		
		Hole H11	Shaft c11	Fit	Hole H9	Shaft d9	Fit	Hole H8	Shaft f7	Fit	Hole H7	Shaft g6	Fit	Hole H7	Shaft h6	Fit
40	MAX	40.160	39.880	0.440	40.062	39.920	0.204	40.039	39.975	0.089	40.025	39.991	0.050	40.025	40.000	0.041
	MIN	40.000	39.720	0.120	40.000	39.858	0.080	40.000	39.950	0.025	40.000	39.975	0.009	40.000	39.984	0.000
50	MAX	50.160	49.870	0.450	50.062	49.920	0.204	50.039	49.975	0.089	50.025	49.991	0.050	50.025	50.000	0.041
	MIN	50.000	49.710	0.130	50.000	49.858	0.080	50.000	49.950	0.025	50.000	49.975	0.009	50.000	49.984	0.000
60	MAX	60.190	59.860	0.520	60.074	59.900	0.248	60.046	59.970	0.106	60.030	59.990	0.059	60.030	60.000	0.049
	MIN	60.000	59.670	0.140	60.000	59.826	0.100	60.000	59.940	0.030	60.000	59.971	0.010	60.000	59.981	0.000
80	MAX	80.190	79.850	0.530	80.074	79.900	0.248	80.046	79.970	0.106	80.030	79.990	0.059	80.030	80.000	0.049
	MIN	80.000	79.660	0.150	80.000	79.826	0.100	80.000	79.940	0.030	80.000	79.971	0.010	80.000	79.981	0.000
100	MAX	100.220	99.830	0.610	100.087	99.880	0.294	100.054	99.964	0.125	100.035	99.988	0.069	100.035	100.000	0.057
	MIN	100.000	99.610	0.170	100.000	99.793	0.120	100.000	99.929	0.036	100.000	99.966	0.012	100.000	99.978	0.000
120	MAX	120.220	119.820	0.620	120.087	119.880	0.294	120.054	119.964	0.125	120.035	119.988	0.069	120.035	120.000	0.057
	MIN	120.000	119.600	0.180	120.000	119.793	0.120	120.000	119.929	0.036	120.000	119.966	0.012	120.000	119.978	0.000
160	MAX	160.250	159.790	0.710	160.100	159.855	0.345	160.063	159.957	0.146	160.040	159.986	0.079	160.040	160.000	0.065
	MIN	160.000	159.540	0.210	160.000	159.755	0.145	160.000	159.917	0.043	160.000	159.961	0.014	160.000	159.975	0.000
200	MAX	200.290	199.760	0.820	200.115	199.830	0.400	200.072	199.950	0.168	200.046	199.985	0.090	200.046	200.000	0.075
	MIN	200.000	199.470	0.240	200.000	199.715	0.170	200.000	199.904	0.050	200.000	199.956	0.015	200.000	199.971	0.000
250	MAX	250.290	249.720	0.860	250.115	249.830	0.400	250.072	249.950	0.168	250.046	249.985	0.090	250.046	250.000	0.075
	MIN	250.000	249.430	0.280	250.000	249.715	0.170	250.000	249.904	0.050	250.000	249.956	0.015	250.000	249.971	0.000
300	MAX	300.320	299.670	0.970	300.130	299.810	0.450	300.081	299.944	0.189	300.052	299.983	0.101	300.052	300.000	0.084
	MIN	300.000	299.350	0.330	300.000	299.680	0.190	300.000	299.892	0.056	300.000	299.951	0.017	300.000	299.968	0.000
400	MAX	400.360	399.600	1.120	400.140	399.790	0.490	400.089	399.938	0.208	400.057	399.982	0.111	400.057	400.000	0.093
	MIN	400.000	399.240	0.400	400.000	399.650	0.210	400.000	399.881	0.062	400.000	399.946	0.018	400.000	399.964	0.000
500	MAX	500.400	499.520	1.280	500.155	499.770	0.540	500.097	499.932	0.228	500.063	499.980	0.123	500.063	500.000	0.103
	MIN	500.000	499.120	0.480	500.000	499.615	0.230	500.000	499.869	0.068	500.000	499.940	0.020	500.000	499.960	0.000

APPENDIX 9B   Preferred hole basis transition and interference fits

BASIC SIZE		LOCATIONAL TRANSN. Hole H7	Shaft k6	Fit	LOCATIONAL TRANSN. Hole H7	Shaft n6	Fit	LOCATIONAL INTERF. Hole H7	Shaft p6	Fit	MEDIUM DRIVE Hole H7	Shaft s6	Fit	FORCE Hole H7	Shaft u6	Fit
1	MAX	1.010	1.006	0.010	1.010	1.010	0.006	1.010	1.012	0.004	1.010	1.020	-0.004	1.010	1.024	-0.008
	MIN	1.000	1.000	-0.006	1.000	1.004	-0.010	1.000	1.006	-0.012	1.000	1.014	-0.020	1.000	1.018	-0.024
1.2	MAX	1.210	1.206	0.010	1.210	1.210	0.006	1.210	1.212	0.004	1.210	1.220	-0.004	1.210	1.224	-0.008
	MIN	1.200	1.200	-0.006	1.200	1.204	-0.010	1.200	1.206	-0.012	1.200	1.214	-0.020	1.200	1.218	-0.024
1.6	MAX	1.610	1.606	0.010	1.610	1.610	0.006	1.610	1.612	0.004	1.610	1.620	-0.004	1.610	1.624	-0.008
	MIN	1.600	1.600	-0.006	1.600	1.604	-0.010	1.600	1.606	-0.012	1.600	1.614	-0.020	1.600	1.618	-0.024
2	MAX	2.010	2.006	0.010	2.010	2.010	0.006	2.010	2.012	0.004	2.010	2.020	-0.004	2.010	2.024	-0.008
	MIN	2.000	2.000	-0.006	2.000	2.004	-0.010	2.000	2.006	-0.012	2.000	2.014	-0.020	2.000	2.018	-0.024
2.5	MAX	2.510	2.506	0.010	2.510	2.510	0.006	2.510	2.512	0.004	2.510	2.520	-0.004	2.510	2.524	-0.008
	MIN	2.500	2.500	-0.006	2.500	2.504	-0.010	2.500	2.506	-0.012	2.500	2.514	-0.020	2.500	2.518	-0.024
3	MAX	3.010	3.006	0.010	3.010	3.010	0.006	3.010	3.012	0.004	3.010	3.020	-0.004	3.010	3.024	-0.008
	MIN	3.000	3.000	-0.006	3.000	3.004	-0.010	3.000	3.006	-0.012	3.000	3.014	-0.020	3.000	3.018	-0.024
4	MAX	4.012	4.009	0.011	4.012	4.016	0.004	4.012	4.020	0.000	4.012	4.027	-0.007	4.012	4.031	-0.011
	MIN	4.000	4.001	-0.009	4.000	4.008	-0.016	4.000	4.012	-0.020	4.000	4.019	-0.027	4.000	4.023	-0.031
5	MAX	5.012	5.009	0.011	5.012	5.016	0.004	5.012	5.020	0.000	5.012	5.027	-0.007	5.012	5.031	-0.011
	MIN	5.000	5.001	-0.009	5.000	5.008	-0.016	5.000	5.012	-0.020	5.000	5.019	-0.027	5.000	5.023	-0.031
6	MAX	6.012	6.009	0.011	6.012	6.016	0.004	6.012	6.020	0.000	6.012	6.027	-0.007	6.012	6.031	-0.011
	MIN	6.000	6.001	-0.009	6.000	6.008	-0.016	6.000	6.012	-0.020	6.000	6.019	-0.027	6.000	6.023	-0.031
8	MAX	8.015	8.010	0.014	8.015	8.019	0.005	8.015	8.024	0.000	8.015	8.032	-0.008	8.015	8.037	-0.013
	MIN	8.000	8.001	-0.010	8.000	8.010	-0.019	8.000	8.015	-0.024	8.000	8.023	-0.032	8.000	8.028	-0.037
10	MAX	10.015	10.010	0.014	10.015	10.019	0.005	10.015	10.024	0.000	10.015	10.032	-0.008	10.015	10.037	-0.013
	MIN	10.000	10.001	-0.010	10.000	10.010	-0.019	10.000	10.015	-0.024	10.000	10.023	-0.032	10.000	10.028	-0.037
12	MAX	12.018	12.012	0.017	12.018	12.023	0.006	12.018	12.029	0.000	12.018	12.039	-0.010	12.018	12.044	-0.015
	MIN	12.000	12.001	-0.012	12.000	12.012	-0.023	12.000	12.018	-0.029	12.000	12.028	-0.039	12.000	12.033	-0.044
16	MAX	16.018	16.012	0.017	16.018	16.023	0.006	16.018	16.029	0.000	16.018	16.039	-0.010	16.018	16.044	-0.015
	MIN	16.000	16.001	-0.012	16.000	16.012	-0.023	16.000	16.018	-0.029	16.000	16.028	-0.039	16.000	16.033	-0.044
20	MAX	20.021	20.015	0.019	20.021	20.028	0.006	20.021	20.035	-0.001	20.021	20.048	-0.014	20.021	20.054	-0.020
	MIN	20.000	20.002	-0.015	20.000	20.015	-0.028	20.000	20.022	-0.035	20.000	20.035	-0.048	20.000	20.041	-0.054
25	MAX	25.021	25.015	0.019	25.021	25.028	0.006	25.021	25.035	-0.001	25.021	25.048	-0.014	25.021	25.061	-0.027
	MIN	25.000	25.002	-0.015	25.000	25.015	-0.028	25.000	25.022	-0.035	25.000	25.035	-0.048	25.000	25.048	-0.061
30	MAX	30.021	30.015	0.019	30.021	30.028	0.006	30.021	30.035	-0.001	30.021	30.048	-0.014	30.021	30.061	-0.027
	MIN	30.000	30.002	-0.015	30.000	30.015	-0.028	30.000	30.022	-0.035	30.000	30.035	-0.048	30.000	30.048	-0.061

APPENDIX 9B (Cont.)

PREFERRED HOLE BASIS TRANSITION AND INTERFERENCE FITS

Dimensions in mm.

BASIC SIZE		LOCATIONAL TRANSN. Hole H7	Shaft k6	Fit	LOCATIONAL TRANSN. Hole H7	Shaft n6	Fit	LOCATIONAL INTERF. Hole H7	Shaft p6	Fit	MEDIUM DRIVE Hole H7	Shaft s6	Fit	FORCE Hole H7	Shaft u6	Fit
40	MAX	40.025	40.018	0.023	40.025	40.033	0.008	40.025	40.042	-0.001	40.025	40.059	-0.018	40.025	40.076	-0.035
	MIN	40.000	40.002	-0.018	40.000	40.017	-0.033	40.000	40.026	-0.042	40.000	40.043	-0.059	40.000	40.060	-0.076
50	MAX	50.025	50.018	0.023	50.025	50.033	0.008	50.025	50.042	-0.001	50.025	50.059	-0.018	50.025	50.086	-0.045
	MIN	50.000	50.002	-0.018	50.000	50.017	-0.033	50.000	50.026	-0.042	50.000	50.043	-0.059	50.000	50.070	-0.086
60	MAX	60.030	60.021	0.028	60.030	60.039	0.010	60.030	60.051	-0.002	60.030	60.072	-0.023	60.030	60.106	-0.057
	MIN	60.000	60.002	-0.021	60.000	60.020	-0.039	60.000	60.032	-0.051	60.000	60.053	-0.072	60.000	60.087	-0.106
80	MAX	80.030	80.021	0.028	80.030	80.039	0.010	80.030	80.051	-0.002	80.030	80.078	-0.029	80.030	80.121	-0.072
	MIN	80.000	80.002	-0.021	80.000	80.020	-0.039	80.000	80.032	-0.051	80.000	80.059	-0.078	80.000	80.102	-0.121
100	MAX	100.035	100.025	0.032	100.035	100.045	0.012	100.035	100.059	-0.002	100.035	100.093	-0.036	100.035	100.146	-0.089
	MIN	100.000	100.003	-0.025	100.000	100.023	-0.045	100.000	100.037	-0.059	100.000	100.071	-0.093	100.000	100.124	-0.146
120	MAX	120.035	120.025	0.032	120.035	120.045	0.012	120.035	120.059	-0.002	120.035	120.101	-0.044	120.035	120.166	-0.109
	MIN	120.000	120.003	-0.025	120.000	120.023	-0.045	120.000	120.037	-0.059	120.000	120.079	-0.101	120.000	120.144	-0.166
160	MAX	160.040	160.028	0.037	160.040	160.052	0.013	160.040	160.068	-0.003	160.040	160.125	-0.060	160.040	160.215	-0.150
	MIN	160.000	160.003	-0.028	160.000	160.027	-0.052	160.000	160.043	-0.068	160.000	160.100	-0.125	160.000	160.190	-0.215
200	MAX	200.046	200.033	0.042	200.046	200.060	0.015	200.046	200.079	-0.004	200.046	200.151	-0.076	200.046	200.265	-0.190
	MIN	200.000	200.004	-0.033	200.000	200.031	-0.060	200.000	200.050	-0.079	200.000	200.122	-0.151	200.000	200.236	-0.265
250	MAX	250.046	250.033	0.042	250.046	250.060	0.015	250.046	250.079	-0.004	250.046	250.169	-0.094	250.046	250.313	-0.238
	MIN	250.000	250.004	-0.033	250.000	250.031	-0.060	250.000	250.050	-0.079	250.000	250.140	-0.169	250.000	250.284	-0.313
300	MAX	300.052	300.036	0.048	300.052	300.066	0.018	300.052	300.088	-0.004	300.052	300.202	-0.118	300.052	300.382	-0.298
	MIN	300.000	300.004	-0.036	300.000	300.034	-0.066	300.000	300.056	-0.088	300.000	300.170	-0.202	300.000	300.350	-0.382
400	MAX	400.057	400.040	0.053	400.057	400.073	0.020	400.057	400.098	-0.005	400.057	400.244	-0.151	400.057	400.471	-0.378
	MIN	400.000	400.004	-0.040	400.000	400.037	-0.073	400.000	400.062	-0.098	400.000	400.208	-0.244	400.000	400.435	-0.471
500	MAX	500.063	500.045	0.058	500.063	500.080	0.023	500.063	500.108	-0.005	500.063	500.292	-0.189	500.063	500.580	-0.477
	MIN	500.000	500.005	-0.045	500.000	500.040	-0.080	500.000	500.068	-0.108	500.000	500.252	-0.292	500.000	500.540	-0.580

APPENDIX 9C Preferred shaft basis clearance fits

Dimensions in mm.

BASIC SIZE		LOOSE RUNNING Hole C11	Shaft h11	Fit	FREE RUNNING Hole D9	Shaft h9	Fit	CLOSE RUNNING Hole F8	Shaft h7	Fit	SLIDING Hole G7	Shaft h6	Fit	LOCATIONAL CLEARANCE Hole H7	Shaft h6	Fit
1	MAX	1.120	1.000	0.180	1.045	1.000	0.070	1.020	1.000	0.030	1.012	1.000	0.018	1.010	1.000	0.016
	MIN	1.060	0.940	0.060	1.020	0.975	0.020	1.006	0.990	0.006	1.002	0.994	0.002	1.000	0.994	0.000
1.2	MAX	1.320	1.200	0.180	1.245	1.200	0.070	1.220	1.200	0.030	1.212	1.200	0.018	1.210	1.200	0.016
	MIN	1.260	1.140	0.060	1.220	1.175	0.020	1.206	1.190	0.006	1.202	1.194	0.002	1.200	1.194	0.000
1.6	MAX	1.720	1.600	0.180	1.645	1.600	0.070	1.620	1.600	0.030	1.612	1.600	0.018	1.610	1.600	0.016
	MIN	1.660	1.540	0.060	1.620	1.575	0.020	1.606	1.590	0.006	1.602	1.594	0.002	1.600	1.594	0.000
2	MAX	2.120	2.000	0.180	2.045	2.000	0.070	2.020	2.000	0.030	2.012	2.000	0.018	2.010	2.000	0.016
	MIN	2.060	1.940	0.060	2.020	1.975	0.020	2.006	1.990	0.006	2.002	1.994	0.002	2.000	1.994	0.000
2.5	MAX	2.620	2.500	0.180	2.545	2.500	0.070	2.520	2.500	0.030	2.512	2.500	0.018	2.510	2.500	0.016
	MIN	2.560	2.440	0.060	2.520	2.475	0.020	2.506	2.490	0.006	2.502	2.494	0.002	2.500	2.494	0.000
3	MAX	3.120	3.000	0.180	3.045	3.000	0.070	3.020	3.000	0.030	3.012	3.000	0.018	3.010	3.000	0.016
	MIN	3.060	2.940	0.060	3.020	2.975	0.020	3.006	2.990	0.006	3.002	2.994	0.002	3.000	2.994	0.000
4	MAX	4.145	4.000	0.220	4.060	4.000	0.090	4.028	4.000	0.040	4.016	4.000	0.024	4.012	4.000	0.020
	MIN	4.070	3.925	0.070	4.030	3.970	0.030	4.010	3.988	0.010	4.004	3.992	0.004	4.000	3.992	0.000
5	MAX	5.145	5.000	0.220	5.060	5.000	0.090	5.028	5.000	0.040	5.016	5.000	0.024	5.012	5.000	0.020
	MIN	5.070	4.925	0.070	5.030	4.970	0.030	5.010	4.988	0.010	5.004	4.992	0.004	5.000	4.992	0.000
6	MAX	6.145	6.000	0.220	6.060	6.000	0.090	6.028	6.000	0.040	6.016	6.000	0.024	6.012	6.000	0.020
	MIN	6.070	5.925	0.070	6.030	5.970	0.030	6.010	5.988	0.010	6.004	5.992	0.004	6.000	5.992	0.000
8	MAX	8.170	8.000	0.260	8.076	8.000	0.112	8.035	8.000	0.050	8.020	8.000	0.029	8.015	8.000	0.024
	MIN	8.080	7.910	0.080	8.040	7.964	0.040	8.013	7.985	0.013	8.005	7.991	0.005	8.000	7.991	0.000
10	MAX	10.170	10.000	0.260	10.076	10.000	0.112	10.035	10.000	0.050	10.020	10.000	0.029	10.015	10.000	0.024
	MIN	10.080	9.910	0.080	10.040	9.964	0.040	10.013	9.985	0.013	10.005	9.991	0.005	10.000	9.991	0.000
12	MAX	12.205	12.000	0.315	12.093	12.000	0.136	12.043	12.000	0.061	12.024	12.000	0.035	12.018	12.000	0.029
	MIN	12.095	11.890	0.095	12.050	11.957	0.050	12.016	11.982	0.016	12.006	11.989	0.006	12.000	11.989	0.000
16	MAX	16.205	16.000	0.315	16.093	16.000	0.136	16.043	16.000	0.061	16.024	16.000	0.035	16.018	16.000	0.029
	MIN	16.095	15.890	0.095	16.050	15.957	0.050	16.016	15.982	0.016	16.006	15.989	0.006	16.000	15.989	0.000
20	MAX	20.240	20.000	0.370	20.117	20.000	0.169	20.053	20.000	0.074	20.028	20.000	0.041	20.021	20.000	0.034
	MIN	20.110	19.870	0.110	20.065	19.948	0.065	20.020	19.979	0.020	20.007	19.987	0.007	20.000	19.987	0.000
25	MAX	25.240	25.000	0.370	25.117	25.000	0.169	25.053	25.000	0.074	25.028	25.000	0.041	25.021	25.000	0.034
	MIN	25.110	24.870	0.110	25.065	24.948	0.065	25.020	24.979	0.020	25.007	24.987	0.007	25.000	24.987	0.000
30	MAX	30.240	30.000	0.370	30.117	30.000	0.169	30.053	30.000	0.074	30.028	30.000	0.041	30.021	30.000	0.034
	MIN	30.110	29.870	0.110	30.065	29.948	0.065	30.020	29.979	0.020	30.007	29.987	0.007	30.000	29.987	0.000

APPENDIX 9C (Cont.)

PREFERRED SHAFT BASIS CLEARANCE FITS

Dimensions in mm.

BASIC SIZE		LOOSE RUNNING Hole C11	Shaft h11	Fit	FREE RUNNING Hole D9	Shaft h9	Fit	CLOSE RUNNING Hole F8	Shaft h7	Fit	SLIDING Hole G7	Shaft h6	Fit	LOCATIONAL CLEARANCE Hole H7	Shaft h6	Fit
40	MAX	40.280	40.000	0.440	40.142	40.000	0.204	40.064	40.000	0.089	40.034	40.000	0.050	40.025	40.000	0.041
	MIN	40.120	39.840	0.120	40.080	39.938	0.080	40.025	39.975	0.025	40.009	39.984	0.009	40.000	39.984	0.000
50	MAX	50.290	50.000	0.450	50.142	50.000	0.204	50.064	50.000	0.089	50.034	50.000	0.050	50.025	50.000	0.041
	MIN	50.130	49.840	0.130	50.080	49.938	0.080	50.025	49.975	0.025	50.009	49.984	0.009	50.000	49.984	0.000
60	MAX	60.330	60.000	0.520	60.174	60.000	0.248	60.076	60.000	0.106	60.040	60.000	0.059	60.030	60.000	0.049
	MIN	60.140	59.810	0.140	60.100	59.926	0.100	60.030	59.970	0.030	60.010	59.981	0.010	60.000	59.981	0.000
80	MAX	80.340	80.000	0.530	80.174	80.000	0.248	80.076	80.000	0.106	80.040	80.000	0.059	80.030	80.000	0.049
	MIN	80.150	79.810	0.150	80.100	79.926	0.100	80.030	79.970	0.030	80.010	79.981	0.010	80.000	79.981	0.000
100	MAX	100.390	100.000	0.610	100.207	100.000	0.294	100.090	100.000	0.125	100.047	100.000	0.069	100.035	100.000	0.057
	MIN	100.170	99.780	0.170	100.120	99.913	0.120	100.036	99.965	0.036	100.012	99.978	0.012	100.000	99.978	0.000
120	MAX	120.400	120.000	0.620	120.207	120.000	0.294	120.090	120.000	0.125	120.047	120.000	0.069	120.035	120.000	0.057
	MIN	120.180	119.780	0.180	120.120	119.913	0.120	120.036	119.965	0.036	120.012	119.978	0.012	120.000	119.978	0.000
160	MAX	160.460	160.000	0.710	160.245	160.000	0.345	160.106	160.000	0.146	160.054	160.000	0.079	160.040	160.000	0.065
	MIN	160.210	159.750	0.210	160.145	159.900	0.145	160.043	159.960	0.043	160.014	159.975	0.014	160.000	159.975	0.000
200	MAX	200.530	200.000	0.820	200.285	200.000	0.400	200.122	200.000	0.168	200.061	200.000	0.090	200.046	200.000	0.075
	MIN	200.240	199.710	0.240	200.170	199.885	0.170	200.050	199.954	0.050	200.015	199.971	0.015	200.000	199.971	0.000
250	MAX	250.570	250.000	0.860	250.285	250.000	0.400	250.122	250.000	0.168	250.061	250.000	0.090	250.046	250.000	0.075
	MIN	250.280	249.710	0.280	250.170	249.885	0.170	250.050	249.954	0.050	250.015	249.971	0.015	250.000	249.971	0.000
300	MAX	300.650	300.000	0.970	300.320	300.000	0.450	300.137	300.000	0.189	300.069	300.000	0.101	300.052	300.000	0.084
	MIN	300.330	299.680	0.330	300.190	299.870	0.190	300.056	299.948	0.056	300.017	299.968	0.017	300.000	299.968	0.000
400	MAX	400.760	400.000	1.120	400.350	400.000	0.490	400.151	400.000	0.208	400.075	400.000	0.111	400.057	400.000	0.093
	MIN	400.400	399.640	0.400	400.210	399.860	0.210	400.062	399.943	0.062	400.018	399.964	0.018	400.000	399.964	0.000
500	MAX	500.880	500.000	1.280	500.385	500.000	0.540	500.165	500.000	0.228	500.083	500.000	0.123	500.063	500.000	0.103
	MIN	500.480	499.600	0.480	500.230	499.845	0.230	500.068	499.937	0.068	500.020	499.960	0.020	500.000	499.960	0.000

APPENDIX 9D Preferred shaft basis transition and interference fits

Dimensions in mm.

BASIC SIZE		LOCATIONAL TRANSN. Hole K7	Shaft h6	Fit	LOCATIONAL TRANSN. Hole N7	Shaft h6	Fit	LOCATIONAL INTERF. Hole P7	Shaft h6	Fit	MEDIUM DRIVE Hole S7	Shaft h6	Fit	FORCE Hole U7	Shaft h6	Fit
1	MAX	1.000	1.000	0.006	0.996	1.000	0.002	0.994	1.000	0.000	0.986	1.000	−0.008	0.982	1.000	−0.012
	MIN	0.990	0.994	−0.010	0.986	0.994	−0.014	0.984	0.994	−0.016	0.976	0.994	−0.024	0.972	0.994	−0.028
1.2	MAX	1.200	1.200	0.006	1.196	1.200	0.002	1.194	1.200	0.000	1.186	1.200	−0.008	1.182	1.200	−0.012
	MIN	1.190	1.194	−0.010	1.186	1.194	−0.014	1.184	1.194	−0.016	1.176	1.194	−0.024	1.172	1.194	−0.028
1.6	MAX	1.600	1.600	0.006	1.596	1.600	0.002	1.594	1.600	0.000	1.586	1.600	−0.008	1.582	1.600	−0.012
	MIN	1.590	1.594	−0.010	1.586	1.594	−0.014	1.584	1.594	−0.016	1.576	1.594	−0.024	1.572	1.594	−0.028
2	MAX	2.000	2.000	0.006	1.996	2.000	0.002	1.994	2.000	0.000	1.986	2.000	−0.008	1.982	2.000	−0.012
	MIN	1.990	1.994	−0.010	1.986	1.994	−0.014	1.984	1.994	−0.016	1.976	1.994	−0.024	1.972	1.994	−0.028
2.5	MAX	2.500	2.500	0.006	2.496	2.500	0.002	2.494	2.500	0.000	2.486	2.500	−0.008	2.482	2.500	−0.012
	MIN	2.490	2.494	−0.010	2.486	2.494	−0.014	2.484	2.494	−0.016	2.476	2.494	−0.024	2.472	2.494	−0.028
3	MAX	3.000	3.000	0.006	2.996	3.000	0.002	2.994	3.000	0.000	2.986	3.000	−0.008	2.982	3.000	−0.012
	MIN	2.990	2.994	−0.010	2.986	2.994	−0.014	2.984	2.994	−0.016	2.976	2.994	−0.024	2.972	2.994	−0.028
4	MAX	4.003	4.000	0.011	3.996	4.000	0.004	3.992	4.000	0.000	3.985	4.000	−0.007	3.981	4.000	−0.011
	MIN	3.991	3.992	−0.009	3.984	3.992	−0.016	3.980	3.992	−0.020	3.973	3.992	−0.027	3.969	3.992	−0.031
5	MAX	5.003	5.000	0.011	4.996	5.000	0.004	4.992	5.000	0.000	4.985	5.000	−0.007	4.981	5.000	−0.011
	MIN	4.991	4.992	−0.009	4.984	4.992	−0.016	4.980	4.992	−0.020	4.973	4.992	−0.027	4.969	4.992	−0.031
6	MAX	6.003	6.000	0.011	5.996	6.000	0.004	5.992	6.000	0.000	5.985	6.000	−0.007	5.981	6.000	−0.011
	MIN	5.991	5.992	−0.009	5.984	5.992	−0.016	5.980	5.992	−0.020	5.973	5.992	−0.027	5.969	5.992	−0.031
8	MAX	8.005	8.000	0.014	7.996	8.000	0.005	7.991	8.000	0.000	7.983	8.000	−0.008	7.978	8.000	−0.013
	MIN	7.990	7.991	−0.010	7.981	7.991	−0.019	7.976	7.991	−0.024	7.968	7.991	−0.032	7.963	7.991	−0.037
10	MAX	10.005	10.000	0.014	9.996	10.000	0.005	9.991	10.000	0.000	9.983	10.000	−0.008	9.978	10.000	−0.013
	MIN	9.990	9.991	−0.010	9.981	9.991	−0.019	9.976	9.991	−0.024	9.968	9.991	−0.032	9.963	9.991	−0.037
12	MAX	12.006	12.000	0.017	11.995	12.000	0.006	11.989	12.000	0.000	11.979	12.000	−0.010	11.974	12.000	−0.015
	MIN	11.988	11.989	−0.012	11.977	11.989	−0.023	11.971	11.989	−0.029	11.961	11.989	−0.039	11.956	11.989	−0.044
16	MAX	16.006	16.000	0.017	15.995	16.000	0.006	15.989	16.000	0.000	15.979	16.000	−0.010	15.974	16.000	−0.015
	MIN	15.988	15.989	−0.012	15.977	15.989	−0.023	15.971	15.989	−0.029	15.961	15.989	−0.039	15.956	15.989	−0.044
20	MAX	20.006	20.000	0.019	19.993	20.000	0.006	19.986	20.000	0.000	19.973	20.000	−0.014	19.967	20.000	−0.020
	MIN	19.985	19.987	−0.015	19.972	19.987	−0.028	19.965	19.987	−0.035	19.952	19.987	−0.048	19.946	19.987	−0.054
25	MAX	25.006	25.000	0.019	24.993	25.000	0.006	24.986	25.000	0.000	24.973	25.000	−0.014	24.960	25.000	−0.027
	MIN	24.985	24.987	−0.015	24.972	24.987	−0.028	24.965	24.987	−0.035	24.952	24.987	−0.048	24.939	24.987	−0.061
30	MAX	30.006	30.000	0.019	29.993	30.000	0.006	29.986	30.000	0.000	29.973	30.000	−0.014	29.960	30.000	−0.027
	MIN	29.985	29.987	−0.015	29.972	29.987	−0.028	29.965	29.987	−0.035	29.952	29.987	−0.048	29.939	29.987	−0.061

APPENDIX 9D (Cont.)

PREFERRED SHAFT BASIS TRANSITION AND INTERFERENCE FITS

Dimensions in mm.

BASIC SIZE		LOCATIONAL TRANSN. Hole K7	Shaft h6	Fit	LOCATIONAL TRANSN. Hole N7	Shaft h6	Fit	LOCATIONAL INTERF. Hole P7	Shaft h6	Fit	MEDIUM DRIVE Hole S7	Shaft h6	Fit	FORCE Hole U7	Shaft h6	Fit
40	MAX	40.007	40.000	0.023	39.992	40.000	0.008	39.983	40.000	-0.001	39.966	40.000	-0.018	39.949	40.000	-0.035
	MIN	39.982	39.984	-0.018	39.967	39.984	-0.033	39.958	39.984	-0.042	39.941	39.984	-0.059	39.924	39.984	-0.076
50	MAX	50.007	50.000	0.023	49.992	50.000	0.008	49.983	50.000	-0.001	49.966	50.000	-0.018	49.939	50.000	-0.045
	MIN	49.982	49.984	-0.018	49.967	49.984	-0.033	49.958	49.984	-0.042	49.941	49.984	-0.059	49.914	49.984	-0.086
60	MAX	60.009	60.000	0.028	59.991	60.000	0.010	59.979	60.000	-0.002	59.958	60.000	-0.023	59.924	60.000	-0.057
	MIN	59.979	59.981	-0.021	59.961	59.981	-0.039	59.949	59.981	-0.051	59.928	59.981	-0.072	59.894	59.981	-0.106
80	MAX	80.009	80.000	0.028	79.991	80.000	0.010	79.979	80.000	-0.002	79.952	80.000	-0.029	79.909	80.000	-0.072
	MIN	79.979	79.981	-0.021	79.961	79.981	-0.039	79.949	79.981	-0.051	79.922	79.981	-0.078	79.879	79.981	-0.121
100	MAX	100.010	100.000	0.032	99.990	100.000	0.012	99.976	100.000	-0.002	99.942	100.000	-0.036	99.889	100.000	-0.089
	MIN	99.975	99.978	-0.025	99.955	99.978	-0.045	99.941	99.978	-0.059	99.907	99.978	-0.093	99.854	99.978	-0.146
120	MAX	120.010	120.000	0.032	119.990	120.000	0.012	119.976	120.000	-0.002	119.934	120.000	-0.044	119.869	120.000	-0.109
	MIN	119.975	119.978	-0.025	119.955	119.978	-0.045	119.941	119.978	-0.059	119.899	119.978	-0.101	119.834	119.978	-0.166
160	MAX	160.012	160.000	0.037	159.988	160.000	0.013	159.972	160.000	-0.003	159.915	160.000	-0.060	159.825	160.000	-0.150
	MIN	159.972	159.975	-0.028	159.948	159.975	-0.052	159.932	159.975	-0.068	159.875	159.975	-0.125	159.785	159.975	-0.215
200	MAX	200.013	200.000	0.042	199.986	200.000	0.015	199.967	200.000	-0.004	199.895	200.000	-0.076	199.781	200.000	-0.190
	MIN	199.967	199.971	-0.033	199.940	199.971	-0.060	199.921	199.971	-0.079	199.849	199.971	-0.151	199.735	199.971	-0.265
250	MAX	250.013	250.000	0.042	249.986	250.000	0.015	249.967	250.000	-0.004	249.877	250.000	-0.094	249.733	250.000	-0.238
	MIN	249.967	249.971	-0.033	249.940	249.971	-0.060	249.921	249.971	-0.079	249.831	249.971	-0.169	249.687	249.971	-0.313
300	MAX	300.016	300.000	0.048	299.986	300.000	0.018	299.964	300.000	-0.004	299.850	300.000	-0.118	299.670	300.000	-0.298
	MIN	299.964	299.968	-0.036	299.934	299.968	-0.066	299.912	299.968	-0.088	299.798	299.968	-0.202	299.618	299.968	-0.382
400	MAX	400.017	400.000	0.053	399.984	400.000	0.020	399.959	400.000	-0.005	399.813	400.000	-0.151	399.586	400.000	-0.378
	MIN	399.960	399.964	-0.040	399.927	399.964	-0.073	399.902	399.964	-0.098	399.756	399.964	-0.244	399.529	399.964	-0.471
500	MAX	500.018	500.000	0.058	499.983	500.000	0.023	499.955	500.003	-0.005	499.771	500.000	-0.189	499.483	500.000	-0.477
	MIN	499.955	499.960	-0.045	499.920	499.960	-0.080	499.892	499.960	-0.108	499.708	499.960	-0.292	499.420	499.960	-0.580

APPENDIX 10 AMERICAN NATIONAL STANDARD UNIFIED INCH SCREW THREADS (UN AND UNR THREAD FORM) (ANSI B1.1—1982)

SIZES Primary	SIZES Secondary	BASIC MAJOR DIAMETER	Coarse UNC	Fine UNF	Extra fine UNEF	4UN	6UN	8UN	12UN	16UN	20UN	28UN	32UN	SIZES
0		0.0600	—	80	—	—	—	—	—	—	—	—	—	0
	1	0.0730	64	72	—	—	—	—	—	—	—	—	—	1
2		0.0860	56	64	—	—	—	—	—	—	—	—	—	2
	3	0.0990	48	56	—	—	—	—	—	—	—	—	—	3
4		0.1120	40	48	—	—	—	—	—	—	—	—	—	4
5		0.1250	40	44	—	—	—	—	—	—	—	—	—	5
6		0.1380	32	40	—	—	—	—	—	—	—	—	UNC	6
8		0.1640	32	36	—	—	—	—	—	—	—	—	UNC	8
10		0.1900	24	32	—	—	—	—	—	—	—	—	UNF	10
	12	0.2160	24	28	32	—	—	—	—	—	—	UNF	UNEF	12
1/4		0.2500	20	28	32	—	—	—	—	—	UNC	UNF	UNEF	1/4
5/16		0.3125	18	24	32	—	—	—	—	—	20	28	UNEF	5/16
3/8		0.3750	16	24	32	—	—	—	—	UNC	20	28	UNEF	3/8
7/16		0.4375	14	20	28	—	—	—	—	16	UNF	UNEF	32	7/16
1/2		0.5000	13	20	28	—	—	—	—	16	UNF	UNEF	32	1/2
9/16		0.5625	12	18	24	—	—	—	UNC	16	20	28	32	9/16
5/8		0.6250	11	18	24	—	—	—	12	16	20	28	32	5/8
	11/16	0.6875	—	—	24	—	—	—	12	16	20	28	32	11/16
3/4		0.7500	10	16	20	—	—	—	12	UNF	UNEF	28	32	3/4
	13/16	0.8125	—	—	20	—	—	—	12	16	UNEF	28	32	13/16
7/8		0.8750	9	14	20	—	—	—	12	16	UNEF	28	32	7/8
	15/16	0.9375	—	—	20	—	—	—	12	16	UNEF	28	32	15/16
1		1.0000	8	12	20	—	—	UNC	UNF	16	UNEF	28	32	1
	1 1/16	1.0625	—	—	18	—	—	8	12	16	20	28	—	1 1/16
1 1/8		1.1250	7	12	18	—	—	8	UNF	16	20	28	—	1 1/8
	1 3/16	1.1875	—	—	18	—	—	8	12	16	20	28	—	1 3/16
1 1/4		1.2500	7	12	18	—	—	8	UNF	16	20	28	—	1 1/4
	1 5/16	1.3125	—	—	18	—	—	8	12	16	20	28	—	1 5/16
1 3/8		1.3750	6	12	18	—	UNC	8	UNF	16	20	28	—	1 3/8
	1 7/16	1.4375	—	—	18	—	6	8	12	16	20	28	—	1 7/16
1 1/2		1.5000	6	12	18	—	UNC	8	UNF	16	20	28	—	1 1/2
	1 9/16	1.5625	—	—	18	—	6	8	12	16	20	—	—	1 9/16
1 5/8		1.6250	—	—	18	—	6	8	12	16	20	—	—	1 5/8
	1 11/16	1.6875	—	—	18	—	6	8	12	16	20	—	—	1 11/16
1 3/4		1.7500	5	—	—	—	6	8	12	16	20	—	—	1 3/4
	1 13/16	1.8125	—	—	—	—	6	8	12	16	20	—	—	1 13/16
1 7/8		1.8750	—	—	—	—	6	8	12	16	20	—	—	1 7/8
	1 15/16	1.9375	—	—	—	—	6	8	12	16	20	—	—	1 15/16
2		2.0000	4 1/2	—	—	—	6	8	12	16	20	—	—	2
	2 1/8	2.1250	—	—	—	—	6	8	12	16	20	—	—	2 1/8
2 1/4		2.2500	4 1/2	—	—	—	6	8	12	16	20	—	—	2 1/4
	2 3/8	2.3750	—	—	—	—	6	8	12	16	20	—	—	2 3/8
2 1/2		2.5000	4	—	—	UNC	6	8	12	16	20	—	—	2 1/2
	2 5/8	2.6250	—	—	—	4	6	8	12	16	20	—	—	2 5/8
2 3/4		2.7500	4	—	—	UNC	6	8	12	16	20	—	—	2 3/4
	2 7/8	2.8750	—	—	—	4	6	8	12	16	20	—	—	2 7/8
3		3.0000	4	—	—	UNC	6	8	12	16	20	—	—	3
	3 1/8	3.1250	—	—	—	4	6	8	12	16	—	—	—	3 1/8
3 1/4		3.2500	4	—	—	UNC	6	8	12	16	—	—	—	3 1/4
	3 3/8	3.3750	—	—	—	4	6	8	12	16	—	—	—	3 3/8
3 1/2		3.5000	4	—	—	UNC	6	8	12	16	—	—	—	3 1/2
	3 5/8	3.6250	—	—	—	4	6	8	12	16	—	—	—	3 5/8
3 3/4		3.7500	4	—	—	UNC	6	8	12	16	—	—	—	3 3/4
	3 7/8	3.8750	—	—	—	4	6	8	12	16	—	—	—	3 7/8
4		4.0000	4	—	—	UNC	6	8	12	16	—	—	—	4
	4 1/8	4.1250	—	—	—	4	6	8	12	16	—	—	—	4 1/8
4 1/4		4.2500	—	—	—	4	6	8	12	16	—	—	—	4 1/4
	4 3/8	4.3750	—	—	—	4	6	8	12	16	—	—	—	4 3/8
4 1/2		4.5000	—	—	—	4	6	8	12	16	—	—	—	4 1/2
	4 5/8	4.6250	—	—	—	4	6	8	12	16	—	—	—	4 5/8
4 3/4		4.7500	—	—	—	4	6	8	12	16	—	—	—	4 3/4
	4 7/8	4.8750	—	—	—	4	6	8	12	16	—	—	—	4 7/8
5		5.0000	—	—	—	4	6	8	12	16	—	—	—	5
	5 1/8	5.1250	—	—	—	4	6	8	12	16	—	—	—	5 1/8
5 1/4		5.2500	—	—	—	4	6	8	12	16	—	—	—	5 1/4
	5 3/8	5.3750	—	—	—	4	6	8	12	16	—	—	—	5 3/8
5 1/2		5.5000	—	—	—	4	6	8	12	16	—	—	—	5 1/2
	5 5/8	5.6250	—	—	—	4	6	8	12	16	—	—	—	5 5/8
5 3/4		5.7500	—	—	—	4	6	8	12	16	—	—	—	5 3/4
	5 7/8	5.8750	—	—	—	4	6	8	12	16	—	—	—	5 7/8
6		6.0000	—	—	—	4	6	8	12	16	—	—	—	6

*Series designation shown indicates the UN thread form, however, the UNR thread form may be specified by substituting UNR in place of UN in all designations for external use only.

APPENDIX 11 TAP DRILL SIZES FOR SPECIFIC UNIFIED INCH SCREW THREADS

Threads per inch for coarse, fine, extra-fine, 8-thread, 12-thread, and 16-thread series [b] [tap-drill sizes for approximately 75 per cent depth of thread (not American Standard)]

Nominal size (basic major diam.)	Coarse-thd. series UNC and NC[c] in classes 1A, 1B, 2A, 2B, 3A, 3B, 2, 3		Fine-thd. series UNF and NF[c] in classes 1A, 1B, 2A, 2B, 3A, 3B, 2, 3		Extra-fine thd. series UNEF and NEF[d] in classes 2A, 2B, 2, 3		8-thd. series 8N[c] in classes 2A, 2B, 2, 3		12-thd. series 12UN and 12N[d] in classes 2A, 2B, 2, 3		16-thd. series 16UN and 16N[d] in classes 2A, 2B, 2, 3	
	Thd./in.	Tap drill	Thd./in.	Tap drill	Thd./in.	Tap drill	Thd./in.	Tap drill	Thd./in.	Tap drill	Thd./in.	Tap drill
0(0.060)	....	.....	80	$\frac{3}{64}$								
1(0.073)	64	No. 53	72	No. 53								
2(0.086)	56	No. 50	64	No. 50								
3(0.099)	48	No. 47	56	No. 45								
4(0.112)	40	No. 43	48	No. 42								
5(0.125)	40	No. 38	44	No. 37								
6(0.138)	32	No. 36	40	No. 33								
8(0.164)	32	No. 29	36	No. 29								
10(0.190)	24	No. 25	32	No. 21								
12(0.216)	24	No. 16	28	No. 14	32	No. 13						
$\frac{1}{4}$	20	No. 7	28	No. 3	32	$\frac{7}{32}$						
$\frac{5}{16}$	18	Let. F	24	Let. I	32	$\frac{9}{32}$						
$\frac{3}{8}$	16	$\frac{5}{16}$	24	Let. Q	32	$\frac{11}{32}$						
$\frac{7}{16}$	14	Let. U	20	$\frac{25}{64}$	28	$\frac{13}{32}$						
$\frac{1}{2}$	13	$\frac{27}{64}$	20	$\frac{29}{64}$	28	$\frac{15}{32}$	..	...	12	$\frac{27}{64}$		
$\frac{9}{16}$	12	$\frac{31}{64}$	18	$\frac{33}{64}$	24	$\frac{33}{64}$	..	...	12	$\frac{31}{64}$		
$\frac{5}{8}$	11	$\frac{17}{32}$	18	$\frac{37}{64}$	24	$\frac{37}{64}$	..	...	12	$\frac{35}{64}$		
$\frac{11}{16}$	....	.....	..		24	$\frac{41}{64}$	..	...	12	$\frac{39}{64}$		
$\frac{3}{4}$	10	$\frac{21}{32}$	16	$\frac{11}{16}$	20	$\frac{45}{64}$	..	...	12	$\frac{43}{64}$	16	$\frac{11}{16}$
$\frac{13}{16}$	....	.....	..	.....	20	$\frac{49}{64}$	..	...	12	$\frac{47}{64}$	16	$\frac{3}{4}$
$\frac{7}{8}$	9	$\frac{49}{64}$	14	$\frac{13}{16}$	20	$\frac{53}{64}$	..	...	12	$\frac{51}{64}$	16	$\frac{13}{16}$
$\frac{15}{16}$	....	.....	..	.....	20	$\frac{57}{64}$	..	...	12	$\frac{55}{64}$	16	$\frac{7}{8}$
1			14	$\frac{15}{16}$	..	.....	8	$\frac{7}{8}$				
1	8	$\frac{7}{8}$	12	$\frac{59}{64}$	20	$\frac{61}{64}$	..	...	12	$\frac{59}{64}$	16	$\frac{15}{16}$
1 $\frac{1}{16}$	....	.....	..	.....	18	1	..	...	12	$\frac{63}{64}$	16	1
1 $\frac{1}{8}$	7	$\frac{63}{64}$	12	1 $\frac{3}{64}$	18	1 $\frac{5}{64}$	8	1	12	1 $\frac{3}{64}$	16	1 $\frac{1}{16}$
1 $\frac{3}{16}$	....	.....	..	.....	18	1 $\frac{9}{64}$	..	...	12	1 $\frac{7}{64}$	16	1 $\frac{1}{8}$
1 $\frac{1}{4}$	7	1 $\frac{7}{64}$	12	1 $\frac{11}{64}$	18	1 $\frac{3}{16}$	8	1 $\frac{1}{8}$	12	1 $\frac{11}{64}$	16	1 $\frac{3}{16}$
1 $\frac{5}{16}$	....	.....	..	.....	18	1 $\frac{17}{64}$	..	...	12	1 $\frac{15}{64}$	16	1 $\frac{1}{4}$
1 $\frac{3}{8}$	6	1 $\frac{7}{32}$	12	1 $\frac{19}{64}$	18	1 $\frac{5}{16}$	8	1 $\frac{1}{4}$	12	1 $\frac{19}{64}$	16	1 $\frac{5}{16}$
1 $\frac{7}{16}$	....	.....	..	.....	18	1 $\frac{3}{8}$	..	...	12	1 $\frac{23}{64}$	16	1 $\frac{3}{8}$
1 $\frac{1}{2}$	6	1 $\frac{11}{32}$	12	1 $\frac{27}{64}$	18	1 $\frac{7}{16}$	8	1 $\frac{3}{8}$	12	1 $\frac{27}{64}$	16	1 $\frac{7}{16}$
1 $\frac{9}{16}$	....	.....	..	.....	18	1 $\frac{1}{2}$	..	...	..	.....	16	1 $\frac{1}{2}$
1 $\frac{5}{8}$	....	.....	..	.....	18	1 $\frac{9}{16}$	8	1 $\frac{1}{2}$	12	1 $\frac{35}{64}$	16	1 $\frac{9}{16}$
1 $\frac{11}{16}$	....	.....	..	.....	18	1 $\frac{5}{8}$	..	...	..	.....	16	1 $\frac{5}{8}$

Tap drill sizes are for approximately 75 percent depth of thread.

American Standard Unified and American Thread Series (*Cont.*)

Nominal size (basic major diam.)	Coarse-thd. series UNC and NC[c] in classes 1A, 1B, 2A, 2B, 3A, 3B, 2, 3		Fine-thd. series UNF and NF[c] in classes 1A, 1B, 2A, 2B, 3A, 3B, 2, 3		Extra-fine thd. series UNEF and NEF[d] in classes 2A, 2B, 2, 3		8-thd. series 8N[c] in classes 2A, 2B, 2, 3		12-thd. series 12UN and 12N[d] in classes 2A, 2B, 2, 3		16-thd. series 16UN and 16N[d] in classes 2A, 2B, 2, 3	
	Thd. /in.	Tap drill	Thd. /in.	Tap drill	Thd. /in.	Tap drill	Thd. /in.	Tap drill	Thd. /in.	Tap drill	Thd. /in.	Tap drill
1 3/4	5	1 9/16	..	.....	16	1 11/66	8[e]	1 5/8	12	1 43/64	16	1 11/16
1 13/16	....	.....	..	.....	..	.....	..	...	..	.....	16	1 3/4
1 7/8	....	.....	..	.....	..	.....	8	1 3/4	12	1 51/64	16	1 13/16
1 15/16	....	.....	..	.....	..	.....	..	...	..	.....	16	1 7/8
2	4 1/2	1 25/32	..	.....	16	1 15/16	8[e]	1 7/8	12	1 59/64	16	1 15/16
2 1/16	....	.....	..	.....	..	.....	..	...	..	.....	16	2
2 1/8	....	.....	..	.....	..	.....	8	2	12	2 3/64	16	2 1/16
2 3/16	....	.....	..	.....	..	.....	..	...	..	.....	16	2 1/8
2 1/4	4 1/2	2 1/32	..	.....	..	.....	8[e]	2 1/8	12	2 11/64	16	2 3/16
2 5/16	....	.....					..	.....	..	.....	16	2 1/4
2 3/8	....						..	...	12	2 19/64	16	2 5/16
2 7/16											16	2 3/8
2 1/2	4	2 1/4	..	.....	..	.....	8[e]	2 3/8	12	2 27/64	16	2 7/16
2 5/8							..	...	12	2 35/64	16	2 9/16
2 3/4	4	2 1/2	..	.....	..	.....	8[e]	2 5/8	12	2 43/64	16	2 11/16
2 7/8							..	...	12	2 51/64	16	2 13/16
3	4	2 3/4	..	.....	..	.....	8[e]	2 7/8	12	2 59/64	16	2 15/16
3 1/8	....						..	...	12	3 3/64	16	3 1/16
3 1/4	4	3	..	.....	..	.....	8[e]	3 1/8	12	3 11/64	16	3 3/16
3 3/8	....						..	...	12	3 19/64	16	3 5/16
3 1/2	4	3 1/4	..	.....	..	.....	8[e]	3 3/8	12	3 27/64	16	3 7/16
3 5/8	....						..	...	12	3 35/64	16	3 9/16
3 3/4	4	3 1/2	..	.....	..	.....	8[e]	3 5/8	12	3 43/64	16	3 11/16
3 7/8	....						..	...	12	3 51/64	16	3 13/16
4	4	3 3/4	..	.....	..	.....	8[e]	3 7/8	12	3 59/64	16	3 15/16
4 1/4	....	.....	..	.....	..	.....	8[e]	4 1/8	12	4 11/64	16	4 3/16
4 1/2	....	.....	..	.....	..	.....	8[e]	4 3/8	12	4 27/64	16	4 7/16
4 3/4	....	.....	..	.....	..	.....	8[e]	4 5/8	12	4 43/64	16	4 11/16
5	....	.....	..	.....	..	.....	8[e]	4 7/8	12	4 59/64	16	4 15/16
5 1/4	....	.....	..	.....	..	.....	8[e]	5 1/8	12	5 11/64	16	5 3/16
5 1/2	....	.....	..	.....	..	.....	8[e]	5 3/8	12	5 27/64	16	5 7/16
5 3/4	....	.....	..	.....	..	.....	8[e]	5 5/8	12	5 43/64	16	5 11/16
6	....	.....	..	.....	..	.....	8[e]	5 7/8	12	5 59/64	16	5 15/16

No.	Size	No.	Size	No.	Size	Letter	Size
80	0.0135	53	0.0595	26	0.1470	A	0.2340
79	0.0145	52	0.0635	25	0.1495	B	0.2380
78	0.0160	51	0.0670	24	0.1520	C	0.2420
77	0.0180	50	0.0700	23	0.1540	D	0.2460
76	0.0200	49	0.0730	22	0.1570	E	0.2500
75	0.0210	48	0.0760	21	0.1590	F	0.2570
74	0.0225	47	0.0785	20	0.1610	G	0.2610
73	0.0240	46	0.0810	19	0.1660	H	0.2660
72	0.0250	45	0.0820	18	0.1695	I	0.2720
71	0.0260	44	0.0860	17	0.1730	J	0.2770
70	0.0280	43	0.0890	16	0.1770	K	0.2810
69	0.0292	42	0.0935	15	0.1800	L	0.2900
68	0.0310	41	0.0960	14	0.1820	M	0.2950
67	0.0320	40	0.0980	13	0.1850	N	0.3020
66	0.0330	39	0.0995	12	0.1890	O	0.3160
65	0.0350	38	0.1015	11	0.1910	P	0.3230
64	0.0360	37	0.1040	10	0.1935	Q	0.3320
63	0.0370	36	0.1065	9	0.1960	R	0.3390
62	0.0380	35	0.1100	8	0.1990	S	0.3480
61	0.0390	34	0.1110	7	0.2010	T	0.3580
60	0.0400	33	0.1130	6	0.2040	U	0.3680
59	0.0410	32	0.1160	5	0.2055	V	0.3770
58	0.0420	31	0.1200	4	0.2090	W	0.3860
57	0.0430	30	0.1285	3	0.2130	X	0.3970
56	0.0465	29	0.1360	2	0.2210	Y	0.4040
55	0.0520	28	0.1405	1	0.2280	Z	0.4130
54	0.0550	27	0.1440				

APPENDIX 13 METRIC SCREW THREAD SERIES ISO 261 Diameter/Pitch Combinations (ANSI B1.13M—1983)

The standard M profile diameter/pitch combinations are shown in boldface.

Col. 1 1st Choice	Col. 2 2nd Choice	Col. 3 3rd Choice	Coarse	3	2	1.5	1.25	1	0.75	0.5	0.35	0.25	0.2
1.6			**0.35**	...	...	...	...	...	...	...	...	...	0.2
	1.8		0.35	...	...	...	...	...	...	...	...	...	0.2
2			**0.4**	...	...	...	...	...	...	...	...	0.25	...
	2.2		0.45	...	...	...	...	...	...	...	...	0.25	...
2.5			**0.45**	...	...	...	...	...	...	...	0.35	...	...
3			**0.5**	...	...	...	...	...	...	...	0.35	...	...
	3.5		**0.6**	...	...	...	...	...	...	...	0.35	...	...
4			**0.7**	...	...	...	...	...	...	0.5	...	...	...
	4.5		0.75	...	...	...	...	...	...	0.5	...	...	...
5			**0.8**	...	...	...	...	...	...	0.5	...	...	...
		5.5	...	...	...	...	...	...	...	0.5	...	...	...
6			**1**	...	...	...	...	...	0.75	...	...	...	...
		7	1	...	...	...	...	...	0.75	...	...	...	...
8			**1.25**	...	...	...	...	1	0.75	...	...	...	...
		9	1.25	...	...	...	...	1	0.75	...	...	...	...
10			**1.5**	...	...	...	**1.25**	1	**0.75**	...	...	...	...
		11	1.5	...	...	...	...	1	0.75	...	...	...	...
12			**1.75**	...	...	**1.5**[d]	**1.25**	1	...	...	...	...	...
	14		**2**	...	...	**1.5**	**1.25**[a]	1	...	...	...	...	...
		15	...	...	...	1.5	...	1	...	...	...	...	...
16			**2**	...	...	**1.5**	...	1	...	...	...	...	...
		17	...	...	...	1.5	...	1	...	...	...	...	...
	18		2.5	...	2	**1.5**	...	1	...	...	...	...	...
20			**2.5**	...	2	**1.5**	...	1	...	...	...	...	...
	22		2.5[c]	...	2	**1.5**	...	1	...	...	...	...	...
24			**3**	...	**2**	1.5	...	1	...	...	...	...	...
		25	...	...	2	**1.5**	...	1	...	...	...	...	...
		26	...	...	...	1.5	...	...	...	...	...	...	...
	27		3[c]	...	**2**	1.5	...	1	...	...	...	...	...
		28	...	...	2	1.5	...	1	...	...	...	...	...
30			**3.5**	(3)[e]	**2**	1.5	...	1	...	...	...	...	...
		32	...	...	2	1.5	...	...	...	...	...	...	...
	33		3.5	(3)[e]	**2**	1.5	...	...	...	...	...	...	...
		35[b]	...	...	...	**1.5**	...	...	...	...	...	...	...
36			**4**	3	**2**	1.5	...	...	...	...	...	...	...
		38	...	...	...	1.5	...	...	...	...	...	...	...
	39		4	3	**2**	1.5	...	...	...	...	...	...	...

(cont'd.)

(a) Only for spark plugs for engines.
(b) Only for nuts for bearings.
(c) Only for high strength structural steel bolts.

(d) Only for wheel studs and nuts.
(e) Pitches shown in brackets are to be avoided as far as possible.

Appendix 13 (Cont.)
ISO 261 Diameter/Pitch Combinations

Nominal Diameters			Pitches					
Col. 1 1st Choice	Col. 2 2nd Choice	Col. 3 3rd Choice	Coarse	Fine				
				6	4	3	2	1.5
		40	...	...	...	3	2	1.5
42			4.5	...	4	3	2	1.5
		45	4.5	...	4	3	2	1.5
48			5	...	4	3	2	1.5
		50	...	...	...	3	2	1.5
	52		5	...	4	3	2	1.5
		55	...	...	4	3	2	1.5
56			5.5	...	4	3	2	1.5
		58	...	...	4	3	2	1.5
	60		5.5	...	4	3	2	1.5
		62	...	...	4	3	2	1.5
64			6	...	4	3	2	1.5
		65	...	...	4	3	2	1.5
	68		6	...	4	3	2	1.5
		70	...	6	4	3	2	1.5
72			6	6	4	3	2	1.5
		75	...	...	4	3	2	1.5
	76		...	6	4	3	2	1.5
		78	...	...	...	...	2	...
80			6	6	4	3	2	1.5
		82	...	...	...	...	2	...
	85		...	6	4	3	2	...
90			6	6	4	3	2	...
	95		...	6	4	3	2	...
100			6	6	4	3	2	...
	105		...	6	4	3	2	...
110			...	6	4	3	2	...
	115		...	6	4	3	2	...
	120		...	6	4	3	2	...
125			...	6	4	3	2	...
	130		...	6	4	3	2	...
		135	...	6	4	3	2	...
140			...	6	4	3	2	...
		145	...	6	4	3	2	...
	150		...	6	4	3	2	...
		155	...	6	4	3	...	...
160			...	6	4	3	...	...
		165	...	6	4	3	...	...
	170		...	6	4	3	...	...
		175	...	6	4	3	...	...
180			...	6	4	3	...	...
		185	...	6	4	3	...	...
	190		...	6	4	3	...	...
		195	...	6	4	3	...	...
200			...	6	4	3	...	...

ISO 261 Diameter/Pitch Combinations

Nominal Diameters			Pitches					
Col. 1 1st Choice	Col. 2 2nd Choice	Col. 3 3rd Choice	Coarse	Fine				
				6	4	3	2	1.5
		205	...	6	4	3	...	...
	210		...	6	4	3	...	...
		215	...	6	4	3	...	...
220			...	6	4	3	...	...
		225	...	6	4	3	...	...
		230	...	6	4	3	...	...
		235	...	6	4	3	...	...
	240		...	6	4	3	...	...
		245	...	6	4	3	...	...
250			...	6	4	3	...	...
		255	...	6	4	...	...	...
	260		...	6	4	...	...	...
		265	...	6	4	...	...	...
		270	...	6	4	...	...	...
		275	...	6	4	...	...	...
280			...	6	4	...	...	...
		285	...	6	4	...	...	...
		290	...	6	4	...	...	...
		295	...	6	4	...	...	...
	300		...	6	4	...	...	...

APPENDIX 14 TAP DRILL SIZES FOR SPECIFIC METRIC SCREW THREADS

Nominal Dia. mm.	Equiv. Dia. inches	Coarse		Fine	
		Pitch mm	Tap drill mm	Pitch mm	Tap drill mm
1	0.0394	0.25	0.75		
1.6	0.0630	0.35	1.25		
2	0.0787	0.40	1.60		
2.5	0.0984	0.45	2.05		
3	0.1181	0.5	2.50		
3.5	0.1378	0.6	2.90		
4	0.1575	0.7	3.30		
5	0.1969	0.8	4.20		
6	0.2362	1	5.00		
6.3	0.2488	1	5.30		
8	0.3150	1.25	6.80	1	7.00
10	0.3937	1.5	8.50	1.25	8.75
12	0.4724	1.75	10.20	1.25	10.75
14	0.5512	2	12.00	1.5	12.50
16	0.6299	2	14.00	1.5	14.50
18	0.7087	2.5	15.50	2	16.00
20	0.7874	2.5	17.50	1.5	18.50
22	0.8661	2.5	19.50	1.5	20.50
24	0.9449	3	21.00	2	22.00
27	1.0630	3	24.00	2	25.00
30	1.1811	3.5	26.50	2	28.00
33	1.2992	3.5	29.50	2	31.00
36	1.4173	4	32.00	3	33.00
42	1.6535	4.5	37.50	3	39.00
48	1.8898	5	43.00	3	45.00
56	2.2047	5.5	50.50		
64	2.5197	6	58.00		

Based on Approx. 75% thread.

APPENDIX 15 ACME AND STUB ACME THREADS (ANSI B1.5—1977)

Nominal (major) diam.	Threads/in.	Nominal (major) diam.	Threads/in.	Nominal (major) diam.	Threads/in.	Nominal (major) diam.	Threads/in.
¼	16	¾	6	1½	4	3	2
5/16	14	⅞	6	1¾	4	3½	2
⅜	12	1	5	2	4	4	2
7/16	12	1⅛	5	2¼	3	4½	2
½	10	1¼	5	2½	3	5	2
⅝	8	1⅜	4	2¾	3		

Appendix 16A American National Standard square bolts (ANSI B18.2.1—1981)

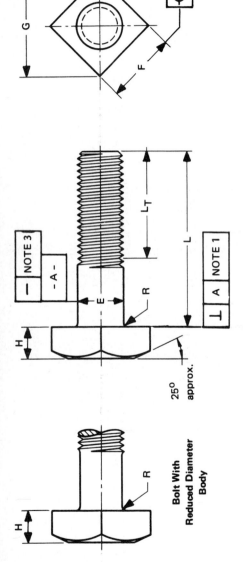

Bolt With Reduced Diameter Body

25° approx.

DIMENSIONS OF SQUARE BOLTS

Nominal Size or Basic Product Dia		E Body Dia Max	F Width Across Flats			G Width Across Corners			H Height				R Radius of Fillet			L_T Thread Length For Bolt Lengths	
			Basic	Max	Min	Max	Min	Basic	Max	Min		Max	Min	6 in. and shorter Basic	over 6 in. Basic
1/4	0.2500	0.260	3/8	0.375	0.362	0.530	0.498	11/64	0.188	0.156		0.03	0.01	0.750	1.000
5/16	0.3125	0.324	1/2	0.500	0.484	0.707	0.665	13/64	0.220	0.186		0.03	0.01	0.875	1.125
3/8	0.3750	0.388	9/16	0.562	0.544	0.795	0.747	1/4	0.268	0.232		0.03	0.01	1.000	1.250
7/16	0.4375	0.452	5/8	0.625	0.603	0.884	0.828	19/64	0.316	0.278		0.03	0.01	1.125	1.375
1/2	0.5000	0.515	3/4	0.750	0.725	1.061	0.995	21/64	0.348	0.308		0.03	0.01	1.250	1.500
5/8	0.6250	0.642	15/16	0.938	0.906	1.326	1.244	27/64	0.444	0.400		0.06	0.02	1.500	1.750
3/4	0.7500	0.768	1 1/8	1.125	1.088	1.591	1.494	1/2	0.524	0.476		0.06	0.02	1.750	2.000
7/8	0.8750	0.895	1 5/16	1.312	1.269	1.856	1.742	19/32	0.620	0.568		0.06	0.02	2.000	2.250
1	1.0000	1.022	1 1/2	1.500	1.450	2.121	1.991	21/32	0.684	0.628		0.09	0.03	2.250	2.500
1 1/8	1.1250	1.149	1 11/16	1.688	1.631	2.386	2.239	3/4	0.780	0.720		0.09	0.03	2.500	2.750
1 1/4	1.2500	1.277	1 7/8	1.875	1.812	2.652	2.489	27/32	0.876	0.812		0.09	0.03	2.750	3.000
1 3/8	1.3750	1.404	2 1/16	2.062	1.994	2.917	2.738	29/32	0.940	0.872		0.09	0.03	3.000	3.250
1 1/2	1.5000	1.531	2 1/4	2.250	2.175	3.182	2.986	1	1.036	0.964		0.09	0.03	3.250	3.500

[1]Bearing surface. A die seam across the bearing surface is permissible. Bearing surface shall be perpendicular to the axis of the body within a tolerance of 3° for 1-in size and smaller, and 2° for sizes larger than 1 in. Angularity measurement shall be taken at a location to avoid interference from a die seam.

[2]True position of head. The axis of the head shall be located at true position with respect to the axis of the body (determined over a distance under the head equal to one diameter) within a tolerance zone having a diameter equivalent to 6 percent of the maximum width across flats, regardless of feature size.

[3]Straightness. Shanks of bolts shall be straight within the following limits: for bolts with nominal lengths to and including 12 in, the maximum camber shall be 0.006 in per inch of bolt length. Bolts with nominal lengths over 12 in to and including 24 in, the maximum camber shall be 0.008 in per

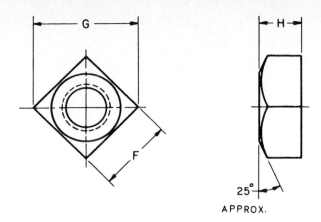

Dimensions of Square Nuts

Nominal Size or Basic Major Dia of Thread		F Width Across Flats			G Width Across Corners		H Thickness		
		Basic	Max	Min	Max	Min	Basic	Max	Min
1/4	0.2500	7/16	0.438	0.425	0.619	0.584	7/32	0.235	0.203
5/16	0.3125	9/16	0.562	0.547	0.795	0.751	17/64	0.283	0.249
3/8	0.3750	5/8	0.625	0.606	0.884	0.832	21/64	0.346	0.310
7/16	0.4375	3/4	0.750	0.728	1.061	1.000	3/8	0.394	0.356
1/2	0.5000	13/16	0.812	0.788	1.149	1.082	7/16	0.458	0.418
5/8	0.6250	1	1.000	0.969	1.414	1.330	35/64	0.569	0.525
3/4	0.7500	1 1/8	1.125	1.088	1.591	1.494	21/32	0.680	0.632
7/8	0.8750	1 5/16	1.312	1.269	1.856	1.742	49/64	0.792	0.740
1	1.0000	1 1/2	1.500	1.450	2.121	1.991	7/8	0.903	0.847
1 1/8	1.1250	1 11/16	1.688	1.631	2.386	2.239	1	1.030	0.970
1 1/4	1.2500	1 7/8	1.875	1.812	2.652	2.489	1 3/32	1.126	1.062
1 3/8	1.3750	2 1/16	2.062	1.994	2.917	2.738	1 13/64	1.237	1.169
1 1/2	1.5000	2 1/4	2.250	2.175	3.182	2.986	1 5/16	1.348	1.276

Appendix 16C American National Standard hex bolts (ANSI B18.2.1—1981)

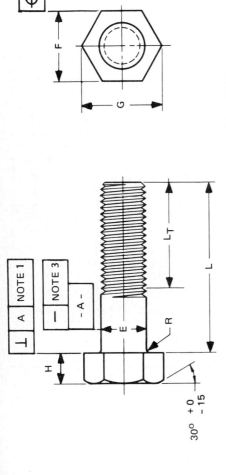

$30° \begin{array}{l} +0 \\ -15 \end{array}$

DIMENSIONS OF HEX BOLTS

Nominal Size or Basic Product Dia		E Body Dia	F Width Across Flats			G Width Across Corners		H Height			R Radius of Fillet		L_T Thread Length For Bolt Lengths	
													6 in. and shorter	over 6 in.
		Max	Basic	Max	Min	Max	Min	Basic	Max	Min	Max	Min	Basic	Basic
1/4	0.2500	0.260	7/16	0.438	0.425	0.505	0.484	11/64	0.188	0.150	0.03	0.01	0.750	1.000
5/16	0.3125	0.324	1/2	0.500	0.484	0.577	0.552	7/32	0.235	0.195	0.03	0.01	0.875	1.125
3/8	0.3750	0.388	9/16	0.562	0.544	0.650	0.620	1/4	0.268	0.226	0.03	0.01	1.000	1.250
7/16	0.4375	0.452	5/8	0.625	0.603	0.722	0.687	19/64	0.316	0.272	0.03	0.01	1.125	1.375
1/2	0.5000	0.515	3/4	0.750	0.725	0.866	0.826	11/32	0.364	0.302	0.03	0.01	1.250	1.500
5/8	0.6250	0.642	15/16	0.928	0.906	1.083	1.033	27/64	0.444	0.378	0.06	0.02	1.500	1.750
3/4	0.7500	0.768	1 1/8	1.125	1.088	1.299	1.240	1/2	0.524	0.455	0.06	0.02	1.750	2.000
7/8	0.8750	0.895	1 5/16	1.312	1.269	1.516	1.447	37/64	0.604	0.531	0.06	0.02	2.000	2.250
1	1.0000	1.022	1 1/2	1.500	1.450	1.732	1.653	43/64	0.700	0.591	0.09	0.03	2.250	2.500
1 1/8	1.1250	1.149	1 11/16	1.688	1.631	1.949	1.859	3/4	0.780	0.658	0.09	0.03	2.500	2.750
1 1/4	1.2500	1.277	1 7/8	1.875	1.812	2.165	2.066	27/32	0.876	0.749	0.09	0.03	2.750	3.000
1 3/8	1.3750	1.404	2 1/16	2.062	1.994	2.382	2.273	29/32	0.940	0.810	0.09	0.03	3.000	3.250
1 1/2	1.5000	1.531	2 1/4	2.250	2.175	2.598	2.480	1	1.036	0.902	0.09	0.03	3.250	3.500
1 3/4	1.7500	1.785	2 5/8	2.625	2.538	3.031	2.893	1 5/32	1.196	1.054	0.12	0.04	3.750	4.000
2	2.0000	2.039	3	3.000	2.900	3.464	3.306	1 11/32	1.388	1.175	0.12	0.04	4.250	4.500
2 1/4	2.2500	2.305	3 3/8	3.375	3.262	3.897	3.719	1 1/2	1.548	1.327	0.19	0.06	4.750	5.000
2 1/2	2.5000	2.559	3 3/4	3.750	3.625	4.330	4.133	1 21/32	1.708	1.479	0.19	0.06	5.250	5.500
2 3/4	2.7500	2.827	4 1/8	4.125	3.988	4.763	4.546	1 13/16	1.869	1.632	0.19	0.06	5.750	6.000
3	3.0000	3.081	4 1/2	4.500	4.350	5.196	4.959	2	2.060	1.815	0.19	0.06	6.250	6.500
3 1/4	3.2500	3.335	4 7/8	4.875	4.712	5.629	5.372	2 3/16	2.251	1.936	0.19	0.06	6.750	7.000
3 1/2	3.5000	3.589	5 1/4	5.250	5.075	6.062	5.786	2 5/16	2.380	2.057	0.19	0.06	7.250	7.500
3 3/4	3.7500	3.858	5 5/8	5.625	5.437	6.495	6.198	2 1/2	2.572	2.241	0.19	0.06	7.750	8.000
4	4.0000	4.111	6	6.000	5.800	6.928	6.612	2 11/16	2.764	2.424	0.19	0.06	8.250	8.500

Note: For notes 1, 2, 3, see Appendix 16A.

APPENDIX 16D American National Standard hex nuts (ANSI B18.2.2—1972)

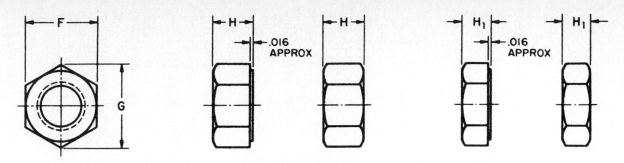

Dimensions of Hex Nuts and Hex Jam Nuts

Nominal Size or Basic Major Dia of Thread		F Width Across Flats			G Width Across Corners		H Thickness Hex Nuts			H₁ Thickness Hex Jam Nuts		
		Basic	Max	Min	Max	Min	Basic	Max	Min	Basic	Max	Min
1/4	0.2500	7/16	0.438	0.428	0.505	0.488	7/32	0.226	0.212	5/32	0.163	0.150
5/16	0.3125	1/2	0.500	0.489	0.577	0.557	17/64	0.273	0.258	3/16	0.195	0.180
3/8	0.3750	9/16	0.562	0.551	0.650	0.628	21/64	0.337	0.320	7/32	0.227	0.210
7/16	0.4375	11/16	0.688	0.675	0.794	0.768	3/8	0.385	0.365	1/4	0.260	0.240
1/2	0.5000	3/4	0.750	0.736	0.866	0.840	7/16	0.448	0.427	5/16	0.323	0.302
9/16	0.5625	7/8	0.875	0.861	1.010	0.982	31/64	0.496	0.473	5/16	0.324	0.301
5/8	0.6250	15/16	0.938	0.922	1.083	1.051	35/64	0.559	0.535	3/8	0.387	0.363
3/4	0.7500	1 1/8	1.125	1.088	1.299	1.240	41/64	0.665	0.617	27/64	0.446	0.398
7/8	0.8750	1 5/16	1.312	1.269	1.516	1.447	3/4	0.776	0.724	31/64	0.510	0.458
1	1.0000	1 1/2	1.500	1.450	1.732	1.653	55/64	0.887	0.831	35/64	0.575	0.519
1 1/8	1.1250	1 11/16	1.688	1.631	1.949	1.859	31/32	0.999	0.939	39/64	0.639	0.579
1 1/4	1.2500	1 7/8	1.875	1.812	2.165	2.066	1 1/16	1.094	1.030	23/32	0.751	0.687
1 3/8	1.3750	2 1/16	2.062	1.994	2.382	2.273	1 11/64	1.206	1.138	25/32	0.815	0.747
1 1/2	1.5000	2 1/4	2.250	2.175	2.598	2.480	1 9/32	1.317	1.245	27/32	0.880	0.808

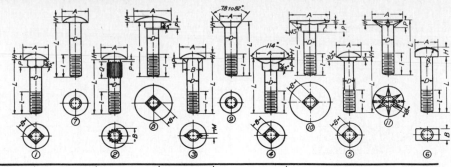

American Standard Round-head Bolts[a]

Proportions for drawing purposes

Fastener name		Nominal diam.[b] (basic major diam.)	A	H	P or M	B	
Carriage bolts	Round-head square-neck bolt[c]	No. 10, $\frac{1}{4}''$ to $\frac{1}{2}''$ by ($\frac{1}{16}''$), $\frac{5}{8}''$ to $1''$ by ($\frac{1}{8}''$)	$2D + \frac{1}{16}$	$\frac{D}{2}$	$\frac{D}{2}$	D	
	Round-head ribbed-neck bolt[d]	No. 10, $\frac{1}{4}''$ to $\frac{1}{2}''$ by ($\frac{1}{16}''$), $\frac{5}{8}''$ and $\frac{3}{4}''$	$2D + \frac{1}{16}$	$\frac{D}{2}$	$\frac{1}{16}$	$D + \frac{1}{16}$	$Q \begin{cases} \frac{3}{16}'' \text{ for } L = \frac{7}{8}'' \text{ or less} \\ \frac{5}{16}'' \text{ for } L = 1'' \text{ and } 1\frac{1}{8}'' \\ \frac{1}{2}'' \text{ for } L = 1\frac{1}{4}'' \text{ or more} \end{cases}$
	Round-head fin.-neck bolt[c]	No. 10, $\frac{1}{4}''$ to $\frac{1}{2}''$ by ($\frac{1}{16}''$)	$2D + \frac{1}{16}$	$\frac{D}{2}$	$\frac{3}{8}D$	$1\frac{1}{2}D + \frac{1}{16}$	
	114° counter-sunk square-neck bolt[c]	No. 10, $\frac{1}{4}''$ to $\frac{1}{2}''$ by ($\frac{1}{16}''$), $\frac{5}{8}''$ and $\frac{3}{4}''$	$2D + \frac{1}{8}$	$\frac{1}{32}$	$D + \frac{1}{32}$	D	
Round-head short square-neck bolt[e]		$\frac{1}{4}''$ to $\frac{1}{2}''$ by ($\frac{1}{16}''$), $\frac{5}{8}''$ and $\frac{3}{4}''$	$2D + \frac{1}{16}$	$\frac{D}{2}$	$\frac{D}{4} + \frac{1}{32}$	D	
T-head bolt[d]		$\frac{1}{4}''$ to $\frac{1}{2}''$ by ($\frac{1}{16}''$), $\frac{5}{8}''$ to $1''$ by ($\frac{1}{8}''$)	$2D$	$\frac{7}{8}D$	$1\frac{5}{8}D$	D	
Round-head bolt[c] (buttonhead bolt)		No. 10, $\frac{1}{4}''$ to $\frac{1}{2}''$ by ($\frac{1}{16}''$), $\frac{5}{8}''$ to $1''$ by ($\frac{1}{8}''$)	$2D + \frac{1}{16}$	$\frac{D}{2}$			
Step bolt[c]		No. 10, $\frac{1}{4}''$ to $\frac{1}{2}''$ by ($\frac{1}{16}''$)	$3D + \frac{1}{16}$	$\frac{D}{2}$	$\frac{D}{2}$	D	
Countersunk bolt[c] (may be slotted if so specified)		$\frac{1}{4}''$ to $\frac{1}{2}''$ by ($\frac{1}{16}''$), $\frac{5}{8}''$ to $1\frac{1}{2}''$ by ($\frac{1}{8}''$)	$1.8D$	Obtain by projection			
Elevator bolt, flat head, counter-sunk[c]		No. 10, $\frac{1}{4}''$ to $\frac{1}{2}''$ by ($\frac{1}{16}''$)	$2\frac{1}{2}D + \frac{5}{16}$	$\frac{D}{3}$	$\frac{D}{2} + \frac{1}{16}$	D	Angle $C = 16D + 5°$ (approx.)
Elevator bolt, ribbed head[d] (slotted or un-slotted as specified)		$\frac{1}{4}''$, $\frac{5}{16}''$, $\frac{3}{8}''$	$2D + \frac{1}{16}$	$\frac{D}{2} - \frac{1}{32}$	$\frac{D}{2} + \frac{3}{64}$	$\frac{D}{8} + \frac{1}{16}$	

[a] The proportions in this table are in some instances approximate and are intended for drawing purposes only. For exact dimensions see ASA B18.5— 1959, from which this table was compiled. Dimensions are in inches.

[b] Fractions in parentheses show diameter increments, *e.g.*, $\frac{1}{4}$ in. to $\frac{1}{2}$ in. by ($\frac{1}{16}$ in.) includes the diameters $\frac{1}{4}$ in., $\frac{5}{16}$ in., $\frac{3}{8}$ in., $\frac{7}{16}$ in., and $\frac{1}{2}$ in

Threads are coarse series, class 2A.

Minimum thread length l: $2D + \frac{1}{4}$ in. for bolts 6 in. or less in length; $2D + \frac{1}{2}$ in. for bolts over 6 in. in length.

For bolt length increments see table page 720.

[c] Full-size body bolts furnished unless undersize body is specified.

[d] Only full-size body bolts furnished.

[e] Undersize body bolts furnished unless full-size body is specified.

Bolt-length Increments*

Bolt diameter		1/4	5/16	3/8	7/16	1/2	5/8	3/4	7/8	1
Length increments	1/4	3/4–3	3/4–4	3/4–6	1–3	1–6	1–6	1–6	1–4½	
	1/2	3–4	4–5	6–9	3–6	6–13	6–10	6–15	4½–6	3–6
	1	4–5	....	9–12	6–8	13–24	10–22	15–24	6–20	6–12
	2	....	....	.....	...	.....	22–30	24–30	20–30	12–30

* Compiled from manufacturers' catalogues.

Example. 1/4-in. bolt lengths increase by 1/4-in. increments from 3/4- to 3-in. length. 1/2-in. bolt lengths increase by 1/2-in. increments from 6- to 13-in. length. 1-in. bolt lengths increase by 2-in. increments from 12- to 30-in. length.

APPENDIX 19 AMERICAN STANDARD CAP SCREWS[a]—SOCKET[b] AND SLOTTED HEADS[c]

For hexagon-head screws, see page A64.

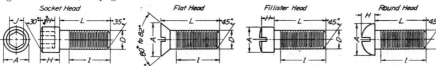

Nominal diam.	Socket head[d]			Flat head[e]	Fillister head[e]		Round head[e]	
	A	H	J	A	A	H	A	H
0	0.096	0.060	0.050					
1	0.118	0.073	0.050					
2	0.140	0.086	1/16					
3	0.161	0.099	5/64					
4	0.183	0.112	5/64					
5	0.205	0.125	3/32					
6	0.226	0.138	3/32					
8	0.270	0.164	1/8					
10	5/16	0.190	5/32					
12	11/32	0.216	5/32					
1/4	3/8	1/4	3/16	1/2	3/8	11/64	7/16	3/16
5/16	7/16	5/16	7/32	5/8	7/16	13/64	9/16	15/64
3/8	9/16	3/8	5/16	3/4	9/16	1/4	5/8	17/64
7/16	5/8	7/16	5/16	13/16	5/8	19/64	3/4	5/16
1/2	3/4	1/2	3/8	7/8	3/4	21/64	13/16	11/32
9/16	13/16	9/16	3/8	1	13/16	3/8	15/16	13/32
5/8	7/8	5/8	1/2	1 1/8	7/8	27/64	1	7/16
3/4	1	3/4	9/16	1 3/8	1	1/2	1 1/4	17/32
7/8	1 1/8	7/8	9/16	1 5/8	1 1/8	19/32		
1	1 5/16	1	5/8	1 7/8	1 5/16	21/32		
1 1/8	1 1/2	1 1/8	3/4					
1 1/4	1 3/4	1 1/4	3/4					
1 3/8	1 7/8	1 3/8	3/4					
1 1/2	2	1 1/2	1					

[a] Dimensions in inches.
[b] See ANSI B.18.3—1983 for complete specifications.
[c] See ANSI B.18.6.2—1982 for complete specifications.
[d] Thread coarse or fine, Approximate thread length *l*: coarse thread, $2D + 1/2$ in.; fine thread, $1\frac{1}{2}D + 1/2$ in.
[e] Thread coarse, fine, or 8-pitch, class 2A. Thread length *l*: $2D + 1/4$ in.

Slot proportions vary with size of screw; draw to look well. All body-length increments for screw lengths 1/4 in. to 1 in. = 1/8 in., for screw lengths 1 in. to 8 in. = 1/4 in.

APPENDIX 20 AMERICAN STANDARD MACHINE SCREWS*

Heads may be slotted or recessed

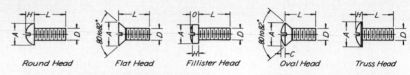

Round Head Flat Head Fillister Head Oval Head Truss Head

Nominal diam.	Round head		Flat head	Fillister head			Oval head		Truss head	
	A	H	A	A	H	O	A	C	A	H
0	0.113	0.053	0.119	0.096	0.045	0.059	0.119	0.021		
1	0.138	0.061	0.146	0.118	0.053	0.071	0.146	0.025	0.194	0.053
2	0.162	0.069	0.172	0.140	0.062	0.083	0.172	0.029	0.226	0.061
3	0.187	0.078	0.199	0.161	0.070	0.095	0.199	0.033	0.257	0.069
4	0.211	0.086	0.225	0.183	0.079	0.107	0.225	0.037	0.289	0.078
5	0.236	0.095	0.252	0.205	0.088	0.120	0.252	0.041	0.321	0.086
6	0.260	0.103	0.279	0.226	0.096	0.132	0.279	0.045	0.352	0.094
8	0.309	0.120	0.332	0.270	0.113	0.156	0.332	0.052	0.384	0.102
10	0.359	0.137	0.385	0.313	0.130	0.180	0.385	0.060	0.448	0.118
12	0.408	0.153	0.438	0.357	0.148	0.205	0.438	0.068	0.511	0.134
1/4	0.472	0.175	0.507	0.414	0.170	0.237	0.507	0.079	0.573	0.150
5/16	0.590	0.216	0.635	0.518	0.211	0.295	0.635	0.099	0.698	0.183
3/8	0.708	0.256	0.762	0.622	0.253	0.355	0.762	0.117	0.823	0.215
7/16	0.750	0.328	0.812	0.625	0.265	0.368	0.812	0.122	0.948	0.248
1/2	0.813	0.355	0.875	0.750	0.297	0.412	0.875	0.131	1.073	0.280
9/16	0.938	0.410	1.000	0.812	0.336	0.466	1.000	0.150	1.198	0.312
5/8	1.000	0.438	1.125	0.875	0.375	0.521	1.125	0.169	1.323	0.345
3/4	1.250	0.547	1.375	1.000	0.441	0.612	1.375	0.206	1.573	0.410

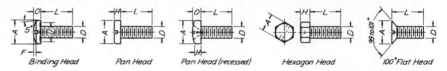

Binding Head Pan Head Pan Head (recessed) Hexagon Head 100° Flat Head

Nominal diam.	Binding head				Pan head			Hexagon head		100° flat head
	A	O	F	U	A	H	O	A	H	A
2	0.181	0.046	0.018	0.141	0.167	0.053	0.062	0.125	0.050	
3	0.208	0.054	0.022	0.162	0.193	0.060	0.071	0.187	0.055	
4	0.235	0.063	0.025	0.184	0.219	0.068	0.080	0.187	0.060	0.225
5	0.263	0.071	0.029	0.205	0.245	0.075	0.089	0.187	0.070	
6	0.290	0.080	0.032	0.226	0.270	0.082	0.097	0.250	0.080	0.279
8	0.344	0.097	0.039	0.269	0.322	0.096	0.115	0.250	0.110	0.332
10	0.399	0.114	0.045	0.312	0.373	0.110	0.133	0.312	0.120	0.385
12	0.454	0.130	0.052	0.354	0.425	0.125	0.151	0.312	0.155	
1/4	0.513	0.153	0.061	0.410	0.492	0.144	0.175	0.375	0.190	0.507
5/16	0.641	0.193	0.077	0.513	0.615	0.178	0.218	0.500	0.230	0.635
3/8	0.769	0.234	0.094	0.615	0.740	0.212	0.261	0.562	0.295	0.762

*ANSI B18.6.3—1962. Dimensions given are maximum values, all in inches.

Thread length: screws 2 in. long or less, thread entire length; screws over 2 in. long, thread length $l = 1\frac{3}{4}$ in. Threads are coarse or fine series, class 2. Heads may be slotted or recessed as specified, excepting hexagon form, which is plain or may be slotted if so specified. Slot and recess proportions vary with size of fastener; draw to look well.

See ANSI B18.6.3—1972 (R1977) for complete specifications.

APPENDIX 21 AMERICAN STANDARD MACHINE-SCREW* AND STOVE-BOLT† NUTS‡

Nominal size	0	1	2	3	4	5	6	8	10	12	¼	5/16	⅜
"W"	5/32	5/32	3/16	3/16	¼	5/16	5/16	11/32	⅜	7/16	7/16	9/16	⅝
"T"	3/64	3/64	1/16	1/16	3/32	7/64	7/64	⅛	⅛	5/32	3/16	7/32	¼

*Machine-screw nuts are hexagonal and square.
†Stove-bolt nuts are square.
‡ANSI B18.6.3—1962. Dimensions are in inches.
Thread is coarse series for square nuts and coarse or fine series for hexagon nuts; class 2B.
See ANSI B18.6.3—1972 (R1977) for complete specifications.

APPENDIX 22 AMERICAN STANDARD HEXAGON SOCKET,* SLOTTED HEAD-LESS,† AND SQUARE-HEAD‡ SETSCREWS

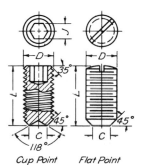

Cup Point   Flat Point

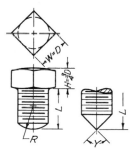

Oval Point   Cone Point

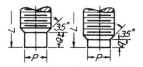

Full Dog Point   Half Dog Point

(All six point types are available in all three head types)

Diam. D	Cup and flat-point diam. C	Oval-point radius R	Cone-point angle Y 118° for these lengths and shorter	90° for these lengths and longer	Full and half dog points Diam. P	Length Full Q	Length Half q	Socket width J
5	1/16	3/32	⅛	3/16	0.083	0.06	0.03	1/16
6	0.069	7/64	⅛	3/16	0.092	0.07	0.03	1/16
8	5/64	⅛	3/16	¼	0.109	0.08	0.04	5/64
10	3/32	9/64	3/16	¼	0.127	0.09	0.04	3/32
12	7/64	5/32	3/16	¼	0.144	0.11	0.06	3/32
¼	⅛	3/16	¼	5/16	5/32	⅛	1/16	⅛
5/16	11/64	15/64	5/16	⅜	13/64	5/32	5/64	5/32
⅜	13/64	9/32	⅜	7/16	¼	3/16	3/32	3/16
7/16	15/64	21/64	7/16	½	19/64	7/32	7/64	7/32
½	9/32	⅜	½	9/16	11/32	¼	⅛	¼
9/16	5/16	27/64	9/16	⅝	25/64	9/32	9/64	¼
⅝	23/64	15/32	⅝	¾	15/32	5/16	5/32	5/16
¾	7/16	9/16	¾	⅞	9/16	⅜	3/16	⅜
⅞	33/64	21/32	⅞	1	21/32	7/16	7/32	½
1	19/32	¾	1	1⅛	¾	½	¼	9/16
1⅛	43/64	27/32	1⅛	1¼	27/32	9/16	9/32	9/16
1¼	¾	15/16	1¼	1½	15/16	⅝	5/16	⅝
1⅜	53/64	1 1/32	1⅜	1⅝	1 1/32	11/16	11/32	⅝
1½	29/32	1⅛	1½	1¾	1⅛	¾	⅜	¾
1¾	1 1/16	1 5/16	1¾	2	1 5/16	⅞	7/16	1
2	1 7/32	1½	2	2¼	1½	1	½	1

*ANSI B18.3—1982. Dimensions are in inches. Threads coarse or fine, class 3A. Length increments: ⅛ in. to ½ in. by (1/16 in.); ½ in. to 1 in. by (⅛ in.); 1 in. to 2 in. by (¼ in.); 2 in. to 6 in. by (½ in.).
†ANSI B 18.6.2—1983. Threads coarse or fine, class 2A. Slotted headless screws standardized in sizes No. 0 to ¾ in. only. Slot proportions vary with diameter. Draw to look well.
‡ANSI B18.6.2—1983. Threads coarse, fine, or 8-pitch, class 2A. Square-head setscrews standardized in sizes No. 10 to 1½ in. only.
§Note that full dog point is not available in hex socket.

APPENDIX 23 AMERICAN STANDARD SOCKET-HEAD SHOULDER SCREWS[a]

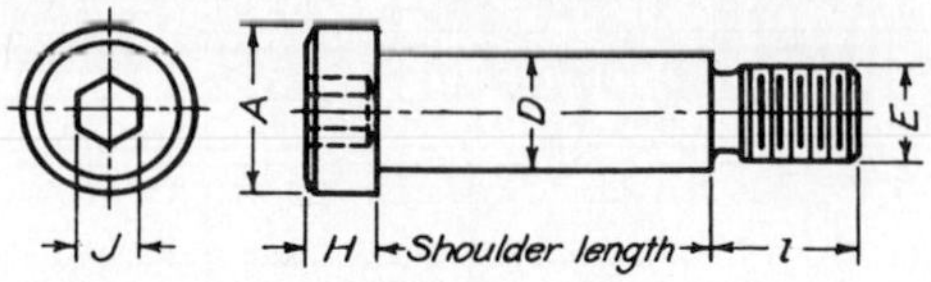

Shoulder diameter D			Head[b]			Thread		Shoulder lengths[d]
Nominal	Max.	Min.	Diam. A	Height H	Hexagon[c] J	Specification E	Length l	
1/4	0.2480	0.2460	3/8	3/16	1/8	10-24NC-3	3/8	3/4–2 1/2
5/16	0.3105	0.3085	7/16	7/32	5/32	1/4-20NC-3	7/16	1 –3
3/8	0.3730	0.3710	9/16	1/4	3/16	5/16-18NC-3	1/2	1 –4
1/2	0.4980	0.4960	3/4	5/16	1/4	3/8-16NC-3	5/8	1 1/4–5
5/8	0.6230	0.6210	7/8	3/8	5/16	1/2-13NC-3	3/4	1 1/2–6
3/4	0.7480	0.7460	1	1/2	3/8	5/8-11NC-3	7/8	1 1/2–8
1	0.9980	0.9960	1 5/16	5/8	1/2	3/4-10NC-3	1	1 1/2–8
1 1/4	1.2480	1.2460	1 3/4	3/4	5/8	7/8-9NC-3	1 1/8	1 1/2–8

[a]ANSI B18.3—1976 Dimensions are in inches.
[b]Head chamfer is 30° to 45°.
[c]Socket depth = 2/3 H approx.
[d]Shoulder-length increments: shoulder lengths from 1/4 in. to 3/4 in., 1/8-in. intervals; shoulder lengths from 3/4 in. to 5 in., 1/4-in. intervals; shoulder lengths from 5 in. to 8 in., 1/2-in. intervals.

APPENDIX 24 AMERICAN STANDARD WOOD SCREWS*

Nominal size	Basic diam. of screw D	No. of threads per in.†	Slot width‡ J (all heads)	Round head§		Flat head	Oval head	
				A	H	A	A	H
0	0.060	32	0.023	0.113	0.053	0.119	0.119	0.035
1	0.073	28	0.026	0.138	0.061	0.146	0.146	0.043
2	0.086	26	0.031	0.162	0.069	0.172	0.172	0.051
3	0.099	24	0.035	0.187	0.078	0.199	0.199	0.059
4	0.112	22	0.039	0.211	0.086	0.225	0.225	0.067
5	0.125	20	0.043	0.236	0.095	0.252	0.252	0.075
6	0.138	18	0.048	0.260	0.103	0.279	0.279	0.083
7	0.151	16	0.048	0.285	0.111	0.305	0.305	0.091
8	0.164	15	0.054	0.309	0.120	0.332	0.332	0.100
9	0.177	14	0.054	0.334	0.128	0.358	0.358	0.108
10	0.190	13	0.060	0.359	0.137	0.385	0.385	0.116
12	0.216	11	0.067	0.408	0.153	0.438	0.438	0.132
14	2.242	10	0.075	0.457	0.170	0.507	0.507	0.153
16	0.268	9	0.075	0.506	0.187	0.544	0.544	0.164
18	0.294	8	0.084	0.555	0.204	0.635	0.635	0.191
20	0.320	8	0.084	0.604	0.220	0.650	0.650	0.196
24	0.372	7	0.094	0.702	0.254	0.762	0.762	0.230

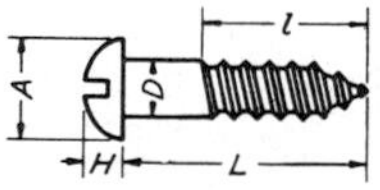

Round Head

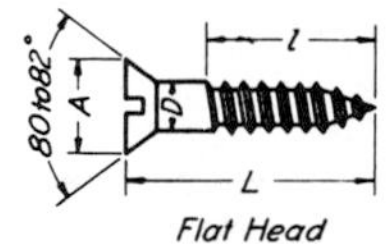

Flat Head

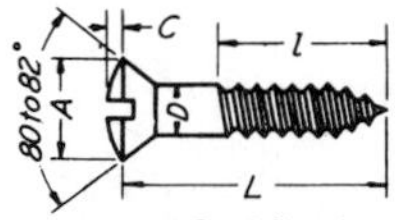

Oval Head

*ANSI B18.6.1—1981. Dimensions given are maximum values, all in inches. Heads may be slotted or recessed as specified.
†Thread length l = 2/3 L.
‡Slot depths and recesses vary with type and size of screw; draw to look well.
§It is recommended that round heads be replaced by pan heads in new applications.

APPENDIX 25 KEYS AND KEYSEATS

Appendix 25A

Widths and Heights of Standard Square- and Flat-stock Keys with Corresponding Shaft Diameters*

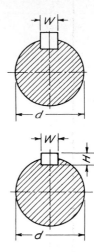

Shaft diam. d (inclusive)	Square-stock keys W	Flat-stock keys, $W \times H$	Shaft diam. d (inclusive)	Square-stock keys W	Flat-stock keys, $W \times H$
½ – 9⁄16	⅛	⅛ × 3⁄32	2⅞–3¼	¾	¾ × ½
⅝ – ⅞	3⁄16	3⁄16 × ⅛	3⅜–3¾	⅞	⅞ × ⅝
15⁄16–1¼	¼	¼ × 3⁄16	3⅞–4½	1	1 × ¾
1 5⁄16–1⅜	5⁄16	5⁄16 × ¼			
1 7⁄16–1¾	⅜	⅜ × ¼	4¾–5½	1¼	1¼ × ⅞
1 13⁄16–2¼	½	½ × ⅜	5¾–6	1½	1½ × 1
2 5⁄16–2¾	⅝	⅝ × 7⁄16			

* Compiled from manufacturers' catalogues.

Appendix 25B   Gib-head dimensions (USAS B17.1—1967)

Gib Head Nominal Dimensions

Nominal key size width, W	Square			Rectangular		
	H	d	B	H	A	B
⅛	⅛	¼	¼	3⁄32	3⁄16	⅛
3⁄16	3⁄16	5⁄16	5⁄16	⅛	¼	¼
¼	¼	7⁄16	⅜	3⁄16	5⁄16	5⁄16
5⁄16	5⁄16	½	⁷⁄16	¼	7⁄16	⅜
⅜	⅜	⅝	½	¼	7⁄16	⅜
½	½	⅞	⅝	⅜	⅝	½
⅝	⅝	1	¾	7⁄16	¾	9⁄16
¾	¾	1-¼	⅞	½	⅞	⅝
⅞	⅞	1-⅜	1	⅝	1	¾
1	1	1-⅝	1-⅛	¾	1-¼	⅞
1-¼	1-¼	2	1-7⁄16	⅞	1-⅜	1
1-½	1-½	2-⅜	1-¾	1	1-⅝	1-⅛
1-¾	1-¾	2-¾	2	1-½	2-⅜	1-¾
2	2	3-½	2-¼	1-½	2-⅜	1-¾
2-½	2-½	4	3	1-¾	2-¾	2
3	3	5	3-½	2	3-½	2-¼
3-½	3-½	6	4	2-½	4	3

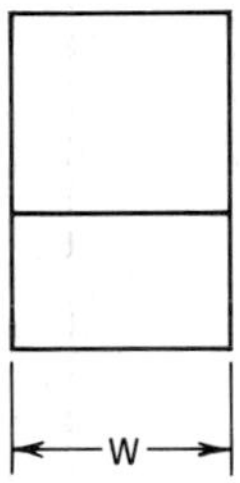

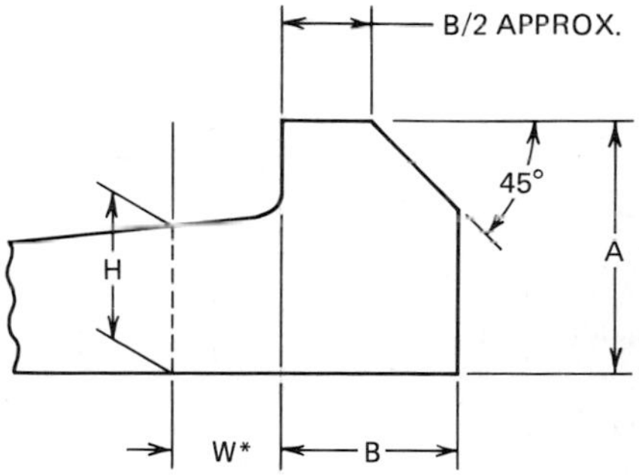

*For locating position of dimension H.
For larger sizes the following relationships are suggested as guides for establishing A and B.

$$A = 1.8 H \qquad B = 1.2 H$$

All dimensions given in inches. COURTESY USAS B17.1—1967

Woodruff-key Dimensions

Key* No.	Nominal size $A \times B$	Max. width of key A	Max. diam. of key B	Max. height of key		Distance below center E
				C	D	
204	$\frac{1}{16} \times \frac{1}{2}$	0.0635	0.500	0.203	0.194	$\frac{3}{64}$
304	$\frac{3}{32} \times \frac{1}{2}$	0.0948	0.500	0.203	0.194	$\frac{3}{64}$
305	$\frac{3}{32} \times \frac{5}{8}$	0.0948	0.625	0.250	0.240	$\frac{1}{16}$
404	$\frac{1}{8} \times \frac{1}{2}$	0.1260	0.500	0.203	0.194	$\frac{3}{64}$
405	$\frac{1}{8} \times \frac{5}{8}$	0.1260	0.625	0.250	0.240	$\frac{1}{16}$
406	$\frac{1}{8} \times \frac{3}{4}$	0.1260	0.750	0.313	0.303	$\frac{1}{16}$
505	$\frac{5}{32} \times \frac{5}{8}$	0.1573	0.625	0.250	0.240	$\frac{1}{16}$
506	$\frac{5}{32} \times \frac{3}{4}$	0.1573	0.750	0.313	0.303	$\frac{1}{16}$
507	$\frac{5}{32} \times \frac{7}{8}$	0.1573	0.875	0.375	0.365	$\frac{1}{16}$
606	$\frac{3}{16} \times \frac{3}{4}$	0.1885	0.750	0.313	0.303	$\frac{1}{16}$
607	$\frac{3}{16} \times \frac{7}{8}$	0.1885	0.875	0.375	0.365	$\frac{1}{16}$
608	$\frac{3}{16} \times 1$	0.1885	1.000	0.438	0.428	$\frac{1}{16}$
609	$\frac{3}{16} \times 1\frac{1}{8}$	0.1885	1.125	0.484	0.475	$\frac{5}{64}$
807	$\frac{1}{4} \times \frac{7}{8}$	0.2510	0.875	0.375	0.365	$\frac{1}{16}$
808	$\frac{1}{4} \times 1$	0.2510	1.000	0.438	0.428	$\frac{1}{16}$
809	$\frac{1}{4} \times 1\frac{1}{8}$	0.2510	1.125	0.484	0.475	$\frac{5}{64}$
810	$\frac{1}{4} \times 1\frac{1}{4}$	0.2510	1.250	0.547	0.537	$\frac{5}{64}$
811	$\frac{1}{4} \times 1\frac{3}{8}$	0.2510	1.375	0.594	0.584	$\frac{3}{32}$
812	$\frac{1}{4} \times 1\frac{1}{2}$	0.2510	1.500	0.641	0.631	$\frac{7}{64}$
1008	$\frac{5}{16} \times 1$	0.3135	1.000	0.438	0.428	$\frac{1}{16}$
1009	$\frac{5}{16} \times 1\frac{1}{8}$	0.3135	1.125	0.484	0.475	$\frac{5}{64}$
1010	$\frac{5}{16} \times 1\frac{1}{4}$	0.3135	1.250	0.547	0.537	$\frac{5}{64}$
1011	$\frac{5}{16} \times 1\frac{3}{8}$	0.3135	1.375	0.594	0.584	$\frac{3}{32}$
1012	$\frac{5}{16} \times 1\frac{1}{2}$	0.3135	1.500	0.641	0.631	$\frac{7}{64}$
1210	$\frac{3}{8} \times 1\frac{1}{4}$	0.3760	1.250	0.547	0.537	$\frac{5}{64}$
1211	$\frac{3}{8} \times 1\frac{3}{8}$	0.3760	1.375	0.594	0.584	$\frac{3}{32}$
1212	$\frac{3}{8} \times 1\frac{1}{2}$	0.3760	1.500	0.641	0.631	$\frac{7}{64}$

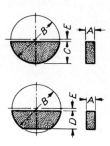

*Dimensions in inches. Key numbers indicate the nominal key dimensions. The last two digits give the nominal diameter B in eighths of an inch, and the digits preceeding the last two give the nominal width A in thirty-seconds of an inch. Thus 204 indicates a key $\frac{2}{32}$ by $\frac{1}{8}$, or $\frac{1}{16}$ by $\frac{1}{2}$ in.
See USAS B17.2—1967 for complete specifications.

Woodruff-key-seat Dimensions

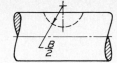

Key* No.	Nominal size	Key slot			
		Width W		Depth H	
		Max.	Min.	Max.	Min.
204	1/16 × 1/2	0.0630	0.0615	0.1718	0.1668
304	3/32 × 1/2	0.0943	0.0928	0.1561	0.1511
305	3/32 × 5/8	0.0943	0 0928	0.2031	0.1981
404	1/8 × 1/2	0.1255	0.1240	0.1405	0.1355
405	1/8 × 5/8	0.1255	0.1240	0.1875	0.1825
406	1/8 × 3/4	0.1255	0.1240	0.2505	0.2455
505	5/32 × 5/8	0.1568	0.1553	0.1719	0.1669
506	5/32 × 3/4	0.1568	0.1553	0.2349	0.2299
507	5/32 × 7/8	0.1568	0.1553	0.2969	0.2919
606	3/16 × 3/4	0.1880	0.1863	0.2193	0.2143
607	3/16 × 7/8	0.1880	0.1863	0.2813	0.2763
608	3/16 × 1	0.1880	0.1863	0.3443	0.3393
609	3/16 × 1 1/8	0.1880	0.1863	0.3903	0.3853
807	1/4 × 7/8	0.2505	0.2487	0.2500	0.2450
808	1/4 × 1	0.2505	0.2487	0.3130	0.3080
809	1/4 × 1 1/8	0.2505	0.2487	0.3590	0.3540
810	1/4 × 1 1/4	0.2505	0.2487	0.4220	0.4170
811	1/4 × 1 3/8	0.2505	0.2487	0.4690	0.4640
812	1/4 × 1 1/2	0.2505	0.2487	0.5160	0.5110
1008	5/16 × 1	0.3130	0.3111	0.2818	0.2768
1009	5/16 × 1 1/8	0.3130	0.3111	0.3278	0.3228
1010	5/16 × 1 1/4	0.3130	0.3111	0.3908	0.3858
1011	5/16 × 1 3/8	0.3130	0.3111	0.4378	0.4328
1012	5/16 × 1 1/2	0.3130	0.3111	0.4848	0.4798
1210	3/8 × 1 1/4	0.3755	0.3735	0.3595	0.3545
1211	3/8 × 1 3/8	0.3755	0.3735	0.4060	0.4015
1212	3/8 × 1 1/2	0.3755	0.3735	0.4535	0.4485

* Dimensions in inches. Key numbers indicate the nominal key dimensions. The last two digits give the nominal diameter B in eighths of an inch, and the digits preceding the last two give the nominal width A in thirty-seconds of an inch. Thus 204 indicates a key 2/32 by 4/8, or 1/16 by 1/2 in. See USAS B17.2 for complete specifications.

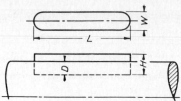

Dimensions of Pratt and Whitney Keys

Key No.	L	W	H	D	Key No.	L	W	H	D
1	1/2	1/16	3/32	1/16	22	1 3/8	1/4	3/8	1/4
2	1/2	3/32	9/64	3/32	23	1 3/8	5/16	15/32	5/16
3	1/2	1/8	3/16	1/8	F	1 3/8	3/8	9/16	3/8
4	5/8	3/32	9/64	3/32	24	1 1/2	1/4	3/8	1/4
5	5/8	1/8	3/16	1/8	25	1 1/2	5/16	15/32	5/16
6	5/8	5/32	15/64	5/32	G	1 1/2	3/8	9/16	3/8
7	3/4	1/8	3/16	1/8	51	1 3/4	1/4	3/8	1/4
8	3/4	5/32	15/64	5/32	52	1 3/4	5/16	15/32	5/16
9	3/4	3/16	9/32	3/16	53	1 3/4	3/8	9/16	3/8
10	7/8	5/32	15/64	5/32	26	2	3/16	9/32	3/16
11	7/8	3/16	9/32	3/16	27	2	1/4	3/8	1/4
12	7/8	7/32	21/64	7/32	28	2	5/16	15/32	5/16
A	7/8	1/4	3/8	1/4	29	2	3/8	9/16	3/8
13	1	3/16	9/32	3/16	54	2 1/4	1/4	3/8	1/4
14	1	7/32	21/64	7/32	55	2 1/4	5/16	15/16	5/16
15	1	1/4	3/8	1/4	56	2 1/4	3/8	9/16	3/8
B	1	5/16	15/32	5/16	57	2 1/4	7/16	21/32	7/16
16	1 1/8	3/16	9/32	3/16	58	2 1/2	5/16	15/32	5/16
17	1 1/8	7/32	21/64	7/32	59	2 1/2	3/8	9/16	3/8
18	1 1/8	1/4	3/8	1/4	60	2 1/2	7/16	21/32	7/16
C	1 1/8	5/16	15/32	5/16	61	2 1/2	1/2	3/4	1/2
19	1 1/4	3/16	9/32	3/16	30	3	3/8	9/16	3/8
20	1 1/4	7/32	21/64	7/32	31	3	7/16	21/32	7/16
21	1 1/4	1/4	3/8	1/4	32	3	1/2	3/4	1/2
D	1 1/4	5/16	15/32	5/16	33	3	9/16	27/32	9/16
E	1 1/4	3/8	9/16	3/8	34	3	5/8	15/16	5/8

Dimensions in inches. Key is 2/3 in shaft; 1/3 in hub. Keys are 0.001 in. oversize in width to ensure proper fitting in keyway. Keyway size: width $= W$; depth $= H - D$. Length L should never be less than $2W$.

Tapers. Taper means the difference in diameter or width in 1 ft of length; see figure below. *Taper pins*, much used for fastening cylindrical parts and for doweling, have a standard taper of ¼ in. per ft.

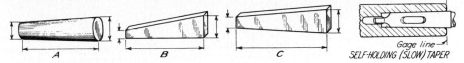

Machine tapers. The ANSI Standard for self-holding (slow) machine tapers is designed to replace the various former standards. The table below shows its derivation. Detailed dimensions and tolerances for taper tool shanks and taper sockets will be found in ANSI B5.10—1963.

Dimensions of Taper Pins

Taper ¼ in. per ft

Size No.	Diam., large end	Drill size for reamer	Max. length
000000	0.072	53	⅝
00000	0.092	47	⅝
0000	0.108	42	¾
000	0.125	37	¾
00	0.147	31	1
0	0.156	28	1
1	0.172	25	1¼
2	0.193	19	1½
3	0.219	12	1¾
4	0.250	3	2
5	0.289	¼	2¼
6	0.341	⁹⁄₃₂	3¼
7	0.409	¹¹⁄₃₂	3¾
8	0.492	¹³⁄₃₂	4½
9	0 591	³¹⁄₆₄	5¼
10	0.706	¹⁹⁄₃₂	6
11	0.857	²³⁄₃₂	7¼
12	1.013	⁵⁵⁄₆₄	8¾
13	1.233	1 ¹⁄₆₄	10¾

All dimensions in inches.

ANSI Standard Machine Tapers,* Self-holding (Slow) Taper Series

Basic dimensions

Origin of series	No. of taper	Taper /ft	Diam. at gage line	Means of driving and holding
Brown and Sharpe taper series		0.239	0.500	0.239
		0.299	0.500	0.299
		0.375	0.500	0.375
Morse taper series	1	0.600	0.475	Tongue drive with shank held in by friction / Tongue drive with shank held in by key / Key drive with shank held in by key / Key drive with shank held in by drawbolt
	2	0.600	0.700	
	3	0.602	0.938	
	4	0.623	1.231	
	4½	0.623	1.500	
	5	0.630	1.748	
¾-in./ft taper series	200	0.750	2.000	
	250	0.750	2.500	
	300	0.750	3.000	
	350	0.750	3.500	
	400	0.750	4.000	
	500	0.750	5.000	
	600	0.750	6.000	
	800	0.750	8.000	
	1,000	0.750	10.000	
	1,200	0.750	12.000	

* ANSI B5.10—All dimensions in inches.

APPENDIX 27A Plain washers (ANSI B18.22.1—1965, R1981)

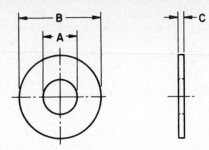

Dimensions of Preferred Sizes of Type A Plain Washers**

Nominal Washer Size***			Inside Diameter A			Outside Diameter B			Thickness C		
			Basic	Tolerance		Basic	Tolerance		Basic	Max	Min
				Plus	Minus		Plus	Minus			
—	—		0.078	0.000	0.005	0.188	0.000	0.005	0.020	0.025	0.016
—	—		0.094	0.000	0.005	0.250	0.000	0.005	0.020	0.025	0.016
—	—		0.125	0.008	0.005	0.312	0.008	0.005	0.032	0.040	0.025
No. 6	0.138		0.156	0.008	0.005	0.375	0.015	0.005	0.049	0.065	0.036
No. 8	0.164		0.188	0.008	0.005	0.438	0.015	0.005	0.049	0.065	0.036
No. 10	0.190		0.219	0.008	0.005	0.500	0.015	0.005	0.049	0.065	0.036
3/16	0.188		0.250	0.015	0.005	0.562	0.015	0.005	0.049	0.065	0.036
No. 12	0.216		0.250	0.015	0.005	0.562	0.015	0.005	0.065	0.080	0.051
1/4	0.250	N	0.281	0.015	0.005	0.625	0.015	0.005	0.065	0.080	0.051
1/4	0.250	W	0.312	0.015	0.005	0.734*	0.015	0.007	0.065	0.080	0.051
5/16	0.312	N	0.344	0.015	0.005	0.688	0.015	0.007	0.065	0.080	0.051
5/16	0.312	W	0.375	0.015	0.005	0.875	0.030	0.007	0.083	0.104	0.064
3/8	0.375	N	0.406	0.015	0.005	0.812	0.015	0.007	0.065	0.080	0.051
3/8	0.375	W	0.438	0.015	0.005	1.000	0.030	0.007	0.083	0.104	0.064
7/16	0.438	N	0.469	0.015	0.005	0.922	0.015	0.007	0.065	0.080	0.051
7/16	0.438	W	0.500	0.015	0.005	1.250	0.030	0.007	0.083	0.104	0.064
1/2	0.500	N	0.531	0.015	0.005	1.062	0.030	0.007	0.095	0.121	0.074
1/2	0.500	W	0.562	0.015	0.005	1.375	0.030	0.007	0.109	0.132	0.086
9/16	0.562	N	0.594	0.015	0.005	1.156*	0.030	0.007	0.095	0.121	0.074
9/16	0.562	W	0.625	0.015	0.005	1.469*	0.030	0.007	0.109	0.132	0.086
5/8	0.625	N	0.656	0.030	0.007	1.312	0.030	0.007	0.095	0.121	0.074
5/8	0.625	W	0.688	0.030	0.007	1.750	0.030	0.007	0.134	0.160	0.108
3/4	0.750	N	0.812	0.030	0.007	1.469	0.030	0.007	0.134	0.160	0.108
3/4	0.750	W	0.812	0.030	0.007	2.000	0.030	0.007	0.148	0.177	0.122
7/8	0.875	N	0.938	0.030	0.007	1.750	0.030	0.007	0.134	0.160	0.108
7/8	0.875	W	0.938	0.030	0.007	2.250	0.030	0.007	0.165	0.192	0.136
1	1.000	N	1.062	0.030	0.007	2.000	0.030	0.007	0.134	0.160	0.108
1	1.000	W	1.062	0.030	0.007	2.500	0.030	0.007	0.165	0.192	0.136
1 1/8	1.125	N	1.250	0.030	0.007	2.250	0.030	0.007	0.134	0.160	0.108
1 1/8	1.125	W	1.250	0.030	0.007	2.750	0.030	0.007	0.165	0.192	0.136
1 1/4	1.250	N	1.375	0.030	0.007	2.500	0.030	0.007	0.165	0.192	0.136
1 1/4	1.250	W	1.375	0.030	0.007	3.000	0.030	0.007	0.165	0.192	0.136
1 3/8	1.375	N	1.500	0.030	0.007	2.750	0.030	0.007	0.165	0.192	0.136
1 3/8	1.375	W	1.500	0.045	0.010	3.250	0.045	0.010	0.180	0.213	0.153
1 1/2	1.500	N	1.625	0.030	0.007	3.000	0.030	0.007	0.165	0.192	0.136
1 1/2	1.500	W	1.625	0.045	0.010	3.500	0.045	0.010	0.180	0.213	0.153
1 5/8	1.625		1.750	0.045	0.010	3.750	0.045	0.010	0.180	0.213	0.153
1 3/4	1.750		1.875	0.045	0.010	4.000	0.045	0.010	0.180	0.213	0.153
1 7/8	1.875		2.000	0.045	0.010	4.250	0.045	0.010	0.180	0.213	0.153
2	2.000		2.125	0.045	0.010	4.500	0.045	0.010	0.180	0.213	0.153
2 1/4	2.250		2.375	0.045	0.010	4.750	0.045	0.010	0.220	0.248	0.193
2 1/2	2.500		2.625	0.045	0.010	5.000	0.045	0.010	0.238	0.280	0.210
2 3/4	2.750		2.875	0.065	0.010	5.250	0.065	0.010	0.259	0.310	0.228
3	3.000		3.125	0.065	0.010	5.500	0.065	0.010	0.284	0.327	0.249

*The 0.734 in., 1.156 in., and 1.469 in. outside diameters avoid washers which could be used in coin operated devices.
**Preferred sizes are for the most part from series previously designated "Standard Plate" and "SAE." Where common sizes existed in the two series, the SAE size is designated "N" (narrow) and the Standard Plate "W" (wide). These sizes as well as all other sizes of Type A Plain Washers are to be ordered by ID, OD, and thickness dimensions.
***Nominal washer sizes are intended for use with comparable nominal screw or bolt sizes.

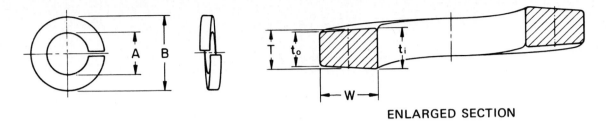

ENLARGED SECTION

Dimensions of Regular Helical Spring Lock Washers[1]

Nominal Washer Size		A		B	T	W
		Inside Diameter		Outside Diameter	Mean Section Thickness $\left(\dfrac{t_i + t_o}{2}\right)$	Section Width
		Max	Min	Max[2]	Min	Min
No. 2	0.086	0.094	0.088	0.172	0.020	0.035
No. 3	0.099	0.107	0.101	0.195	0.025	0.040
No. 4	0.112	0.120	0.114	0.209	0.025	0.040
No. 5	0.125	0.133	0.127	0.236	0.031	0.047
No. 6	0.138	0.148	0.141	0.250	0.031	0.047
No. 8	0.164	0.174	0.167	0.293	0.040	0.055
No. 10	0.190	0.200	0.193	0.334	0.047	0.062
No. 12	0.216	0.227	0.220	0.377	0.056	0.070
¼	0.250	0.262	0.254	0.489	0.062	0.109
5⁄16	0.312	0.326	0.317	0.586	0.078	0.125
⅜	0.375	0.390	0.380	0.683	0.094	0.141
7⁄16	0.438	0.455	0.443	0.779	0.109	0.156
½	0.500	0.518	0.506	0.873	0.125	0.171
9⁄16	0.562	0.582	0.570	0.971	0.141	0.188
⅝	0.625	0.650	0.635	1.079	0.156	0.203
11⁄16	0.688	0.713	0.698	1.176	0.172	0.219
¾	0.750	0.775	0.760	1.271	0.188	0.234
13⁄16	0.812	0.843	0.824	1.367	0.203	0.250
⅞	0.875	0.905	0.887	1.464	0.219	0.266
15⁄16	0.938	0.970	0.950	1.560	0.234	0.281
1	1.000	1.042	1.017	1.661	0.250	0.297
1 1⁄16	1.062	1.107	1.080	1.756	0.266	0.312
1 ⅛	1.125	1.172	1.144	1.853	0.281	0.328
1 3⁄16	1.188	1.237	1.208	1.950	0.297	0.344
1 ¼	1.250	1.302	1.271	2.045	0.312	0.359
1 5⁄16	1.312	1.366	1.334	2.141	0.328	0.375
1 ⅜	1.375	1.432	1.398	2.239	0.344	0.391
1 7⁄16	1.438	1.497	1.462	2.334	0.359	0.406
1 ½	1.500	1.561	1.525	2.430	0.375	0.422

[1] Formerly designated Medium Helical Spring Lock Washers.
[2] The maximum outside diameters specified allow for the commercial tolerances on cold drawn wire.

APPENDIX 28 COTTER PINS

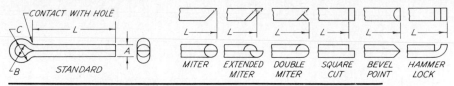

Pin diameter A			Eye diameter, min.		Recommended hole diam., drill size
Nominal	Max.	Min.	Inside B	Outside C	
$\frac{1}{32}$ (0.031)	0.032	0.028	$\frac{1}{32}$	$\frac{1}{16}$	$\frac{3}{64}$ (0.0469)
$\frac{3}{64}$ (0.047)	0.048	0.044	$\frac{3}{64}$	$\frac{3}{32}$	$\frac{1}{16}$ (0.0625)
$\frac{1}{16}$ (0.062)	0.060	0.056	$\frac{1}{16}$	$\frac{1}{8}$	$\frac{5}{64}$ (0.0781)
$\frac{5}{64}$ (0.078)	0.076	0.072	$\frac{5}{64}$	$\frac{5}{32}$	$\frac{3}{32}$ (0.0937)
$\frac{3}{32}$ (0.094)	0.090	0.086	$\frac{3}{32}$	$\frac{3}{16}$	$\frac{7}{64}$ (0.1094)
$\frac{1}{8}$ (0.125)	0.120	0.116	$\frac{1}{8}$	$\frac{1}{4}$	$\frac{9}{64}$ (0.1406)
$\frac{5}{32}$ (0.156)	0.150	0.146	$\frac{5}{32}$	$\frac{5}{16}$	$\frac{11}{64}$ (0.1719)
$\frac{3}{16}$ (0.188)	0.176	0.172	$\frac{3}{16}$	$\frac{3}{8}$	$\frac{13}{64}$ (0.2031)
$\frac{7}{32}$ (0.219)	0.207	0.202	$\frac{7}{32}$	$\frac{7}{16}$	$\frac{15}{64}$ (0.2344)
$\frac{1}{4}$ (0.250)	0.225	0.220	$\frac{1}{4}$	$\frac{1}{2}$	$\frac{17}{64}$ (0.2656)
$\frac{5}{16}$ (0.312)	0.280	0.275	$\frac{5}{16}$	$\frac{5}{8}$	$\frac{5}{16}$ (0.3125)
$\frac{3}{8}$ (0.375)	0.335	0.329	$\frac{3}{8}$	$\frac{3}{4}$	$\frac{3}{8}$ (0.3750)
$\frac{1}{2}$ (0.500)	0.473	0.467	$\frac{1}{2}$	1	$\frac{1}{2}$ (0.5000)

*Preferred points per ANSI B18.8.1—1972.
See this standard for complete specifications.

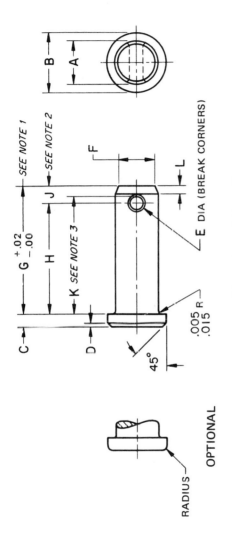

OPTIONAL

RADIUS

.005 R / .015 R

E DIA (BREAK CORNERS)

Dimensions of Clevis Pins

Nominal Size[1] or Basic Pin Diameter	A Shank Diameter Max	A Shank Diameter Min	B Head Diameter Max	B Head Diameter Min	C Head Height Max	C Head Height Min	D Head Chamfer ±0.01	E Hole Diameter Max	E Hole Diameter Min	F Point Diameter Max	F Point Diameter Min	G[2] Pin Length Basic	H Head to Center of Hole Max	H Head to Center of Hole Min	J[3] End to Center Ref Basic	K[4] Head to Edge of Hole Ref Max	K[4] Head to Edge of Hole Ref Min	L Point Length Max	L Point Length Min	Recommended Cotter Pin Nominal Size	
3/16 0.188	0.186	0.181	0.32	0.30	0.07	0.05	0.02	0.088	0.073	0.15	0.14	0.58	0.504	0.484	0.09	0.548	0.520	0.055	0.035	1/16	0.062
1/4 0.250	0.248	0.243	0.38	0.36	0.10	0.08	0.03	0.088	0.073	0.21	0.20	0.77	0.692	0.672	0.09	0.736	0.708	0.055	0.035	1/16	0.062
5/16 0.312	0.311	0.306	0.44	0.42	0.10	0.08	0.03	0.119	0.104	0.26	0.25	0.94	0.832	0.812	0.12	0.892	0.864	0.071	0.049	3/32	0.093
3/8 0.375	0.373	0.368	0.51	0.49	0.13	0.11	0.03	0.119	0.104	0.33	0.32	1.06	0.958	0.938	0.12	1.018	0.990	0.071	0.049	3/32	0.093
7/16 0.438	0.436	0.431	0.57	0.55	0.16	0.14	0.04	0.119	0.104	0.39	0.38	1.19	1.082	1.062	0.12	1.142	1.114	0.071	0.049	3/32	0.093
1/2 0.500	0.496	0.491	0.63	0.61	0.16	0.14	0.04	0.151	0.136	0.44	0.43	1.36	1.223	1.203	0.15	1.298	1.271	0.089	0.063	1/8	0.125
5/8 0.625	0.621	0.616	0.82	0.80	0.21	0.19	0.06	0.151	0.136	0.56	0.55	1.61	1.473	1.453	0.15	1.548	1.521	0.089	0.063	1/8	0.125
3/4 0.750	0.746	0.741	0.94	0.92	0.26	0.24	0.07	0.182	0.167	0.68	0.67	1.91	1.739	1.719	0.18	1.830	1.802	0.110	0.076	5/32	0.156
7/8 0.875	0.871	0.866	1.04	1.02	0.32	0.30	0.09	0.182	0.167	0.80	0.79	2.16	1.989	1.969	0.18	2.080	2.052	0.110	0.076	5/32	0.156
1 1.000	0.996	0.991	1.19	1.17	0.35	0.33	0.10	0.182	0.167	0.93	0.92	2.41	2.239	2.219	0.18	2.330	2.302	0.110	0.076	5/32	0.156

1 Where specifying nominal size in decimals, zeros preceding decimal shall be omitted.
2 Lengths tabulated are intended for use with standard clevises, without spacers. When required, it is recommended that other pin lengths be limited wherever possible to nominal lengths in 0.06 in. increments.
3 Basic "J" dimension (distance from centerline of hole to end of pin) is specified for calculating hole location from underside of head on pins of lengths not tabulated.
4 Reference dimension provided for convenience in design layout and is not subject to inspection.

697

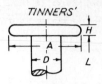

TINNERS'

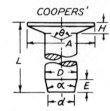

COOPERS'

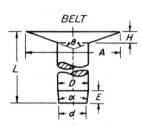

BELT

Tinner's			Cooper's			Belt		
	D	L		D	L		D	L
Size No.†	Diam. body	Length	Size No.	Diam. body	Length	Size No.‡	Diam. body	Length
8 oz	0.089	0.16	1 lb	0.109	0.219	7	0.180	
12	0.105	0.19	1½	0.127	0.256	8	0.165	
1 lb	0.111	0.20	2	0.141	0.292	9	0.148	From ⅜ to ¾ by ⅛″ increments
1½	0.130	0.23	2½	0.148	0.325	10	0.134	
2	0.144	0.27	3	0.156	0.358	11	0.120	
2½	0.148	0.28	4	0.165	0.392	12	0.109	
3	0.160	0.31	6	0.203	0.466	13	0.095	
4	0.176	0.34	8	0.238	0.571			
6	0.203	0.39	10	0.250	0.606			
8	0.224	0.44	12	0.259	0.608			
10	0.238	0.47	14	0.271	0.643			
12	0.259	0.50	16	0.281	0.677			
14	0.284	0.52						
16	0.300	0.53						

Approx. proportions:

Belt:

$A = 2.8 \times D, d = 0.9 \times D$

$E = 0.4 \times D, H = 0.3 \times D$

Cooper's:

$A = 2.25 \times D,\ d = 0.90 \times D$

$E = 0.40 \times D, H = 0.30 \times D$

Included $\angle\theta = 144°$

$\angle\alpha = 18°$

Tinner's:

Approx. proportions:

$A = 2.25 \times D$

$H = 0.30 \times D$

*Adapted from ANSI B18.1.1—1983. All dimensions given in inches.

†Size numbers refer to the approximate weight of 1,000 rivets.

‡Size number refers to the Stubs iron-wire gage number of the stock used in the body of the rivet.

APPENDIX 31 USA STANDARD PIPE THREADS

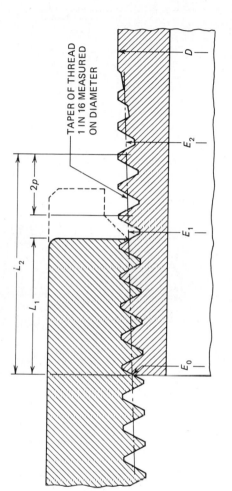

TAPER OF THREAD 1 IN 16 MEASURED ON DIAMETER

Basic Dimensions of USA (American) Standard Taper Pipe Thread, NPT[1]

Nominal Pipe Size	Outside Diameter of Pipe, D	Threads per inch, n	Pitch of Thread, P	Pitch Diameter at beginning of External Thread, E_0	Handtight Engagement — Length, L_1 In.	Handtight Engagement — Length, L_1 Thds.	Dia, E_1	Effective Thread, External — Length, L_2 In.	Effective Thread, External — Length, L_2 Thds.	Dia, E_2
1	2	3	4	5	6	7	8	9	10	11
1/16	0.3125	27	0.03704	0.27118	0.160	4.32	0.28118	0.2611	7.05	0.28750
1/8	0.405	27	0.03704	0.36351	0.1615	4.36	0.37360	0.2639	7.12	0.38000
1/4	0.540	18	0.05556	0.47739	0.2278	4.10	0.49163	0.4018	7.23	0.50250
3/8	0.675	18	0.05556	0.61201	0.240	4.32	0.62701	0.4078	7.34	0.63750
1/2	0.840	14	0.07143	0.75843	0.320	4.48	0.77843	0.5337	7.47	0.79179
3/4	1.050	14	0.07143	0.96768	0.339	4.75	0.98887	0.5457	7.64	1.00179
1	1.315	11.5	0.08696	1.21363	0.400	4.60	1.23863	0.6828	7.85	1.25630
1 1/4	1.660	11.5	0.08696	1.55713	0.420	4.83	1.58338	0.7068	8.13	1.60130
1 1/2	1.900	11.5	0.08696	1.79609	0.420	4.83	1.82234	0.7235	8.32	1.84130
2	2.375	11.5	0.08696	2.26902	0.436	5.01	2.29627	0.7565	8.70	2.31630
2 1/2	2.875	8	0.12500	2.71953	0.682	5.46	2.76216	1.1375	9.10	2.79062
3	3.500	8	0.12500	3.34062	0.766	6.13	3.38850	1.2000	9.60	3.41562
3 1/2	4.000	8	0.12500	3.83750	0.821	6.57	3.88881	1.2500	10.00	3.91562
4	4.500	8	0.12500	4.33438	0.844	6.75	4.38712	1.3000	10.40	4.41562
5	5.563	8	0.12500	5.39073	0.937	7.50	5.44929	1.4063	11.25	5.47862
6	6.625	8	0.12500	6.44609	0.958	7.66	6.50597	1.5125	12.10	6.54062
8	8.625	8	0.12500	8.43359	1.063	8.50	8.50003	1.7125	13.70	8.54062
10	10.750	8	0.12500	10.54531	1.210	9.68	10.62094	1.9250	15.40	10.66562
12	12.750	8	0.12500	12.53281	1.360	10.88	12.61781	2.1250	17.00	12.66562
14 OD	14.000	8	0.12500	13.77500	1.562	12.50	13.87262	2.2500	18.00	13.91562
16 OD	16.000	8	0.12500	15.76250	1.812	14.50	15.87575	2.4500	19.60	15.91562
18 OD	18.000	8	0.12500	17.75000	2.000	16.00	17.87500	2.6500	21.20	17.91562
20 OD	20.000	8	0.12500	19.73750	2.125	17.00	19.87031	2.8500	22.80	19.91562
24 OD	24.000	8	0.12500	23.71250	2.375	19.00	23.86094	3.2500	26.00	23.91562

[1] Dimensions are in inches. The four or five decimal places for dimensions imply a greater degree of precision than usual, but these dimensions are the basis of gage-dimension calculations.
[2] For example, ⅜ NPT or 0.375 NPT.

ABSTRACTED FROM USAS B2.1—1968.

CLASS 250 FLANGED FITTINGS

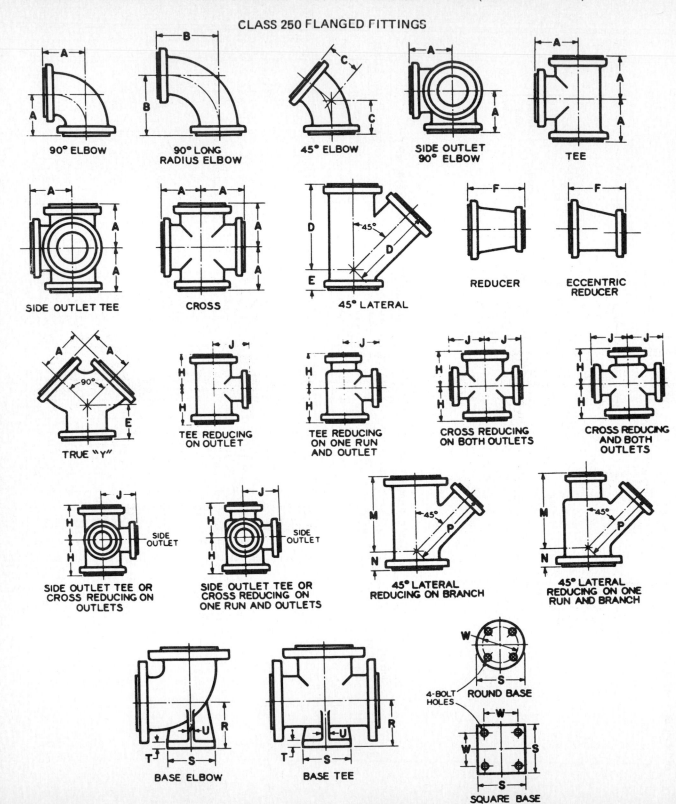

90° ELBOW

90° LONG RADIUS ELBOW

45° ELBOW

SIDE OUTLET 90° ELBOW

TEE

SIDE OUTLET TEE

CROSS

45° LATERAL

REDUCER

ECCENTRIC REDUCER

TRUE "Y"

TEE REDUCING ON OUTLET

TEE REDUCING ON ONE RUN AND OUTLET

CROSS REDUCING ON BOTH OUTLETS

CROSS REDUCING AND BOTH OUTLETS

SIDE OUTLET TEE OR CROSS REDUCING ON OUTLETS

SIDE OUTLET TEE OR CROSS REDUCING ON ONE RUN AND OUTLETS

45° LATERAL REDUCING ON BRANCH

45° LATERAL REDUCING ON ONE RUN AND BRANCH

BASE ELBOW

BASE TEE

4-BOLT HOLES ROUND BASE

SQUARE BASE

Dimensions of Class 250 Cast Iron Flanged Fittings

Dimensions in Inches

Nominal Pipe Size	Flanges		General Fittings			Straight Fittings						Reducing (Short Body)		
												Tees, Crosses		Center to Face Outlet or Side Outlet J
	Dia. of Flange	Thickness of Flange Q (Min)	Dia. of Raised Face	Inside Dia. of Fittings	Body Wall Thickness	Center to Face 90 Deg Elbow Tees, Crosses and True "Y" A	Center to Face 90 Deg Long Radius Elbow B	Center to Face 45 Deg Elbow C	Center to Face Lateral D	Short Center to Face True "Y" and Lateral E	Face to Face Reducer F	NPS Size of Outlet and Smaller	Center to Face Run H	
1	4.88	0.69	2.69	1.00	0.44	4.00	5.00	2.00	6.50	2.00	–			
1¼	5.25	0.75	3.06	1.25	0.44	4.25	5.50	2.50	7.25	2.25	–			
1½	6.12	0.81	3.56	1.50	0.44	4.50	6.00	2.75	8.50	2.50	–			
2	6.50	0.88	4.19	2.00	0.44	5.00	6.50	3.00	9.00	2.50	5.0			
2½	7.50	1.00	4.94	2.50	0.50	5.50	7.00	3.50	10.50	2.50	5.5			
3	8.25	1.12	5.69	3.00	0.56	6.00	7.75	3.50	11.00	3.00	6.0			
3½	9.00	1.19	6.31	3.50	0.56	6.50	8.50	4.00	12.50	3.00	6.5			
4	10.00	1.25	6.94	4.00	0.62	7.00	9.00	4.50	13.50	3.00	7.0			
5	11.00	1.38	8.31	5.00	0.69	8.00	10.25	5.00	15.00	3.50	8.0			
6	12.50	1.44	9.69	6.00	0.75	8.50	11.50	5.50	17.50	4.00	9.0			
8	15.00	1.62	11.94	8.00	0.81	10.00	14.00	6.00	20.50	5.00	11.0			
10	17.50	1.88	14.06	10.00	0.94	11.50	16.50	7.00	24.00	5.50	12.0			
12	20.50	2.00	16.44	12.00	1.00	13.00	19.00	8.00	27.50	6.00	14.0	12	14.0	17.0
14	23.00	2.12	18.94	13.25	1.12	15.00	21.50	8.50	31.00	6.50	16.0	14	15.5	18.5
16	25.50	2.25	21.06	15.25	1.25	16.50	24.00	9.50	34.50	7.50	18.0	16	17.0	21.5
18	28.00	2.38	23.31	17.00	1.38	18.00	26.50	10.00	37.50	8.00	19.0			
20	30.50	2.50	25.56	19.00	1.50	19.50	29.00	10.50	40.50	8.50	20.0	20	20.5	25.5
24	36.00	2.75	30.31	23.00	1.62	22.50	34.00	12.00	47.50	10.00	24.0			
30	43.00	3.00	37.19	29.00	2.00	27.50	41.50	15.00	–	–	30.0			

All reducing tees and crosses, sizes 16 in. and smaller, shall have same center to face dimensions as straight size fittings, corresponding to the size of the largest opening.

See ANSI B16.1—1975 for information on additional classes: 25, 125, and 800.

CAST IRON THREADED FITTINGS— CLASS 250

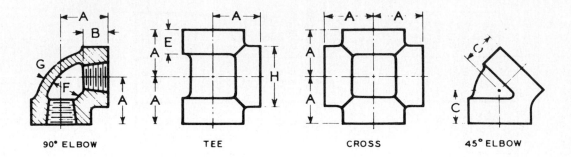

90° ELBOW   TEE   CROSS   45° ELBOW

Dimensions of 90 and 45 Deg. Elbows, Tees, and Crosses
(Straight Sizes)

Nominal Pipe Size	Center-to-End, Elbows, Tees, and Crosses A	Center-to-End, 45 Deg. Elbows C	Length of Thread, Min. B	Width of Band, Min. E	Inside Diameter of Fitting F		Metal Thick-ness G	Outside Diameter of Band, Min. H
					Max.	Min.		
¼	0.94	0.81	0.43	0.49	0.58	0.54	0.18	1.17
⅜	1.06	0.88	0.47	0.55	0.72	0.67	0.18	1.36
½	1.25	1.00	0.57	0.60	0.90	0.84	0.20	1.59
¾	1.44	1.13	0.64	0.68	1.11	1.05	0.23	1.88
1	1.63	1.31	0.75	0.76	1.38	1.31	0.28	2.24
1¼	1.94	1.50	0.84	0.88	1.73	1.66	0.33	2.73
1½	2.13	1.69	0.87	0.97	1.97	1.90	0.35	3.07
2	2.50	2.00	1.00	1.12	2.44	2.37	0.39	3.74
2½	2.94	2.25	1.17	1.30	2.97	2.87	0.43	4.60
3	3.38	2.50	1.23	1.40	3.60	3.50	0.48	5.36
3½	3.75	2.63	1.28	1.49	4.10	4.00	0.52	5.98
4	4.13	2.81	1.33	1.57	4.60	4.50	0.56	6.61
5	4.88	3.19	1.43	1.74	5.66	5.56	0.66	7.92
6	5.63	3.50	1.53	1.91	6.72	6.62	0.74	9.24
8	7.00	4.31	1.72	2.24	8.72	8.62	0.90	11.73
10	8.63	5.19	1.93	2.58	10.85	10.75	1.08	14.37
12	10.00	6.00	2.13	2.91	12.85	12.75	1.24	16.84

All dimensions given in inches.
The Class 250 standard for threaded fittings covers only the straight sizes of 90 and 45 deg. elbows, tees, and crosses.

APPENDIX 34 PIPE BUSHINGS*

Dimensions of outside-head, inside-head, and face bushings in inches

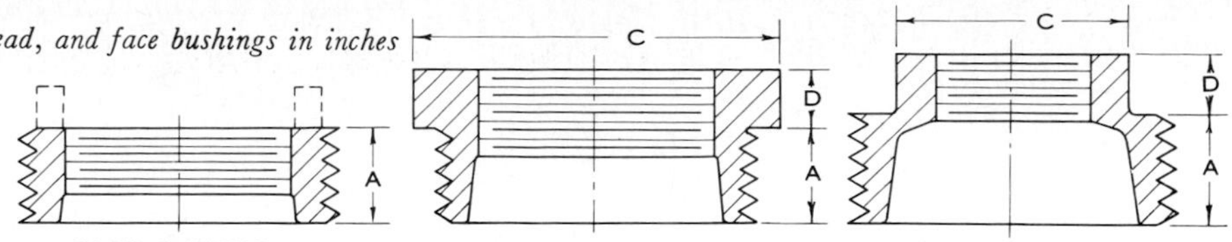

FACE BUSHING   OUTSIDE HEAD   INSIDE HEAD

Size	Length of external thread,† min., A	Height of head, min., D	Width of head,‡ min., C Outside	Width of head,‡ min., C Inside	Size	Length of external thread,‡ min., A	Height of head, min., D	Width of head,† min., C Outside	Width of head,† min., C Inside
¼ × ⅛	0.44	0.14	0.64		1½ × ¼	0.83	0.37	....	1.12
⅜ × ¼	0.48	0.16	0.68		2 × 1½	0.88	0.34	2.48	
⅜ × ⅛	0.48	0.16	0.68		2 × 1¼	0.88	0.34	2.48	
½ × ⅜	0.56	0.19	0.87		2 × 1	0.88	0.41	....	1.95
½ × ¼	0.56	0.19	0.87		2 × ¾	0.88	0.41	....	1.63
½ × ⅛	0.56	0.19	0.87		2 × ½	0.88	0.41	....	1.34
¾ × ½	0.63	0.22	1.15		2 × ⅜	0.88	0.41	....	1.12
¾ × ⅜	0.63	0.22	1.15		2 × ¼	0.88	0.41	....	1.12
¾ × ¼	0.63	0.22	1.15		2½ × 2	1.07	0.37	2.98	
¾ × ⅛	0.63	0.22	1.15		2½ × 1½	1.07	0.44	2.68	
1 × ¾	0.75	0.25	1.42		2½ × 1¼	1.07	0.44	....	2.39
1 × ½	0.75	0.25	1.42		2½ × 1	1.07	0.44	....	1.95
1 × ⅜	0.75	0.30	....	1.12	2½ × ¾	1.07	0.44	....	1.63
1 × ¼	0.75	0.30	....	1.12	2½ × ½	1.07	0.44	....	1.34
1 × ⅛	0.75	0.30	....	1.12	3 × 2½	1.13	0.40	3.86	
1¼ × 1	0.80	0.28	1.76		3 × 2	1.13	0.48	3.28	
1¼ × ¾	0.80	0.28	1.76		3 × 1½	1.13	0.48	....	2.68
1¼ × ½	0.80	0.34	....	1.34	3 × 1¼	1.13	0.48	....	2.39
1¼ × ⅜	0.80	0.34	....	1.12	3 × 1	1.13	0.48	....	1.95
1¼ × ¼	0.80	0.34	....	1.12	3 × ¾	1.13	0.48	....	1.63
1½ × 1¼	0.83	0.31	2.00		3 × ½	1.13	0.48	....	1.34
1½ × 1	0.83	0.31	2.00						
1½ × ¾	0.83	0.37	....	1.63					
1½ × ½	0.83	0.37	....	1.34					
1½ × ⅜	0.83	0.37	....	1.12					

*Adapted from ANSI B16.14—1983.

†In the case of outside-head bushings, length A includes provisions for imperfect threads.

‡Heads of bushings shall be hexagonal or octagonal, except that on the larger sizes of outside-head bushings the heads may be made round with lugs instead of hexagonal or octagonal.

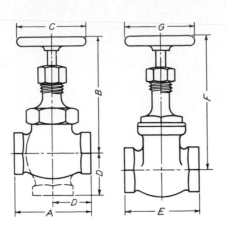

Size	A (globe only)	B (open)	C	D (angle only)	E	F (open)	G
⅛	2	4	1¾	1			
¼	2	4	1¾	1	1⅞	5⅛	1¾
⅜	2¼	4½	2	1⅛	2	5⅛	1¾
½	2¾	5¼	2½	1¼	2⅛	5½	2
¾	3³⁄₁₆	6	2¾	1½	2⅜	6⅝	2½
1	3¾	6¾	3	1¾	2⅞	7⅞	2¾
1¼	4¼	7¼	3⅝	2	3¼	9½	3
1½	4¾	8¼	4	2¼	3½	10⅞	3⅝
2	5¾	9½	4¾	2¾	3⅞	13⅛	4
2½	6¾	11	6	3¼	4½	15⅜	4¾
3	8	12¼	7	3¾	5	17⅞	5⅜

* Dimensions in inches and compiled from manufacturers' catalogues for drawing purposes.

Size	Length		Size	Length		Size	Length		Size	Length	
	Close	Short		Close	Short		Close	Short		Close	Short
⅛	¾	1½	½	1⅛	1½	1¼	1⅝	2½	2½	2½	3
¼	⅞	1½	¾	1⅜	2	1½	1¾	2½	3	2⅝	3
⅜	1	1½	1	1½	2	2	2	2½			

* Compiled from manufacturers' catalogues. Dimensions in inches.

Long-nipple lengths: from short-nipple lengths to 6 in. in ½ in. increments; from 6 in. nipple lengths to 12 in. in 1 in. increments; from 12 in. nipple lengths to 24 in. in 2 in. increments.

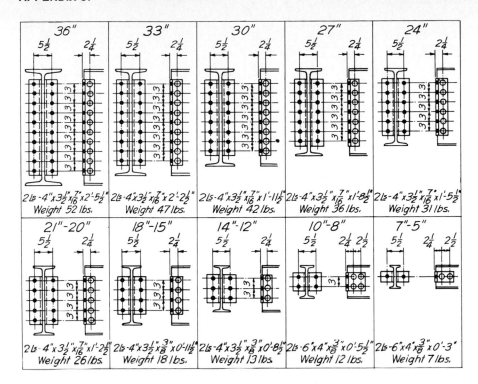

36"	33"	30"	27"	24"
2 ls - 4" x 3½" x 7/16" x 2'-5½" Weight 52 lbs.	2 ls - 4" x 3½" x 7/16" x 2'-2½" Weight 47 lbs.	2 ls - 4" x 3½" x 7/16" x 1'-11½" Weight 42 lbs.	2 ls - 4" x 3½" x 7/16" x 1'-8½" Weight 36 lbs.	2 ls - 4" x 3½" x 7/16" x 1'-5½" Weight 31 lbs.

21"-20"	18"-15"	14"-12"	10"-8"	7"-5"
2 ls - 4" x 3½" x 7/16" x 1'-2½" Weight 26 lbs.	2 ls - 4" x 3½" x 3/8" x 0'-11½" Weight 18 lbs.	2 ls - 4" x 3½" x 3/8" x 0'-8½" Weight 13 lbs.	2 ls - 6" x 4" x 3/8" x 0'-5½" Weight 12 lbs.	2 ls - 6" x 4" x 3/8" x 0'-3" Weight 7 lbs.

APPENDIX 38 **DRIVING CLEARANCES FOR RIVETING**

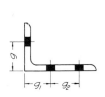

				Diam. of rivet					
	$\frac{1}{2}$	$\frac{5}{8}$	$\frac{3}{4}$	$\frac{7}{8}$	1	$1\frac{1}{8}$	$1\frac{1}{4}$	$1\frac{3}{8}$	$1\frac{1}{2}$
D	$1\frac{3}{4}$	2	$2\frac{1}{4}$	$2\frac{1}{2}$	$2\frac{3}{4}$	3	$3\frac{1}{4}$	$3\frac{1}{2}$	$3\frac{3}{4}$
C	1	$1\frac{1}{8}$	$1\frac{1}{4}$	$1\frac{3}{8}$	$1\frac{1}{2}$	$1\frac{5}{8}$	$1\frac{3}{4}$	$1\frac{7}{8}$	2

Dimensions in inches.

APPENDIX 39 **GAGE AND MAXIMUM RIVET SIZE FOR ANGLES**

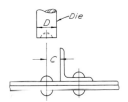

Leg	8	7	6	5	4	$3\frac{1}{2}$	3	$2\frac{1}{2}$	2	$1\frac{3}{4}$	$1\frac{1}{2}$	$1\frac{3}{8}$	$1\frac{1}{4}$	1
g	$4\frac{1}{2}$	4	$3\frac{1}{2}$	3	$2\frac{1}{2}$	2	$1\frac{3}{4}$	$1\frac{3}{8}$	$1\frac{1}{8}$	1	$\frac{7}{8}$	$\frac{7}{8}$	$\frac{3}{4}$	$\frac{5}{8}$
g_1	3	$2\frac{1}{2}$	$2\frac{1}{4}$	2										
g_2	3	3	$2\frac{1}{2}$	$1\frac{3}{4}$										
Max. rivet	$1\frac{1}{8}$	$1\frac{1}{8}$	1	1	$\frac{7}{8}$	$\frac{7}{8}$	$\frac{7}{8}$	$\frac{3}{4}$	$\frac{5}{8}$	$\frac{1}{2}$	$\frac{3}{8}$	$\frac{3}{8}$	$\frac{3}{8}$	$\frac{1}{4}$

Dimensions in inches.

APPENDIX 40 SELECTED STRUCTURAL SHAPES*

Dimensions for detailing

Name	Depth of section, in.	Weight per foot, lb	Flange		Web		Distances				Grip, in.	Max. flange rivet, in.	Usual gage g, in.	Clearance m
			Width, in.	Mean thickness, in.	Thickness, in.	Half thickness, in.	T, in.	k, in.	g_1, in.	c, in.				
Channels	18	58.0	$4\frac{1}{4}$	$\frac{5}{8}$	$\frac{11}{16}$	$\frac{3}{8}$	$15\frac{3}{8}$	$1\ \frac{5}{16}$	$2\frac{3}{4}$	$\frac{3}{4}$	$\frac{5}{8}$	1	$2\frac{1}{2}$	
	15	40.0	$3\frac{1}{2}$	$\frac{5}{8}$	$\frac{9}{16}$	$\frac{1}{4}$	$12\frac{3}{8}$	$1\ \frac{5}{16}$	$2\frac{3}{4}$	$\frac{5}{8}$	$\frac{5}{8}$	1	2	
	12	30.0	$3\frac{1}{8}$	$\frac{1}{2}$	$\frac{1}{2}$	$\frac{1}{4}$	$9\frac{7}{8}$	$1\ \frac{1}{16}$	$2\frac{1}{2}$	$\frac{9}{16}$	$\frac{1}{2}$	$\frac{7}{8}$	$1\frac{3}{4}$	
	10	15.3	$2\frac{5}{8}$	$\frac{7}{16}$	$\frac{1}{4}$	$\frac{1}{8}$	$8\frac{1}{8}$	$\frac{15}{16}$	$2\frac{1}{2}$	$\frac{5}{16}$	$\frac{7}{16}$	$\frac{3}{4}$	$1\frac{1}{2}$	
	9	13.4	$2\frac{3}{8}$	$\frac{7}{16}$	$\frac{1}{4}$	$\frac{1}{8}$	$7\frac{1}{4}$	$\frac{7}{8}$	$2\frac{1}{2}$	$\frac{5}{16}$	$\frac{3}{8}$	$\frac{3}{4}$	$1\frac{3}{8}$	
	8	18.75	$2\frac{1}{2}$	$\frac{3}{8}$	$\frac{1}{2}$	$\frac{1}{4}$	$6\frac{3}{8}$	$\frac{13}{16}$	$2\frac{1}{4}$	$\frac{9}{16}$	$\frac{3}{8}$	$\frac{3}{4}$	$1\frac{1}{2}$	
	7	12.25	$2\frac{1}{4}$	$\frac{3}{8}$	$\frac{5}{16}$	$\frac{3}{16}$	$5\frac{3}{8}$	$\frac{13}{16}$	2	$\frac{3}{8}$	$\frac{3}{8}$	$\frac{5}{8}$	$1\frac{1}{4}$	
	6	10.5	2	$\frac{3}{8}$	$\frac{5}{16}$	$\frac{3}{16}$	$4\frac{1}{2}$	$\frac{3}{4}$	2	$\frac{3}{8}$	$\frac{3}{8}$	$\frac{5}{8}$	$1\frac{1}{8}$	
	5	9.0	$1\frac{7}{8}$	$\frac{5}{16}$	$\frac{5}{16}$	$\frac{3}{16}$	$3\frac{5}{8}$	$\frac{11}{16}$	2	$\frac{3}{8}$	$\frac{5}{16}$	$\frac{1}{2}$	$1\frac{1}{8}$	
	4	7.25	$1\frac{3}{4}$	$\frac{5}{16}$	$\frac{5}{16}$	$\frac{3}{16}$	$2\frac{3}{4}$	$\frac{5}{8}$	2	$\frac{3}{8}$	$\frac{5}{16}$	$\frac{1}{2}$	1	
	3	6.0	$1\frac{5}{8}$	$\frac{1}{4}$	$\frac{3}{8}$	$\frac{3}{16}$	$1\frac{3}{4}$	$\frac{5}{8}$	...	$\frac{7}{16}$	$\frac{5}{16}$	$\frac{1}{2}$	$\frac{7}{8}$	
WF shapes	21†($21\frac{1}{4}$)	127	13	1	$\frac{9}{16}$	$\frac{5}{16}$	$17\frac{3}{4}$	$1\ \frac{3}{4}$	3	$\frac{3}{8}$	. . . .		$5\frac{1}{2}$	25
	16 ($16\frac{3}{8}$)	96	$11\frac{1}{2}$	$\frac{7}{8}$	$\frac{9}{16}$	$\frac{5}{16}$	$13\frac{1}{8}$	$1\ \frac{5}{8}$	$2\frac{3}{4}$	$\frac{3}{8}$	. . . .		$5\frac{1}{2}$	20
	14 ($14\frac{1}{8}$)	84	12	$\frac{3}{4}$	$\frac{7}{16}$	$\frac{1}{4}$	$11\frac{3}{8}$	$1\ \frac{3}{8}$	$2\frac{3}{4}$	$\frac{5}{16}$	. . . .		$5\frac{1}{2}$	$18\frac{5}{8}$
	14 ($13\frac{3}{4}$)	48	8	$\frac{9}{16}$	$\frac{3}{8}$	$\frac{3}{16}$	$11\frac{3}{8}$	$1\ \frac{3}{16}$	$2\frac{1}{2}$	$\frac{1}{4}$	. . . .		$5\frac{1}{2}$	16
	12 ($12\frac{1}{4}$)	50	$8\frac{1}{8}$	$\frac{5}{8}$	$\frac{3}{8}$	$\frac{3}{16}$	$9\frac{3}{4}$	$1\ \frac{1}{4}$	$2\frac{1}{2}$	$\frac{1}{4}$	. . . .		$5\frac{1}{2}$	$14\frac{5}{8}$
	10 (10)	49	10	$\frac{9}{16}$	$\frac{3}{8}$	$\frac{3}{16}$	$7\frac{7}{8}$	$1\ \frac{1}{16}$	$2\frac{1}{2}$	$\frac{1}{4}$	. . . .		$5\frac{1}{2}$	$14\frac{1}{8}$
	10 ($9\frac{3}{4}$)	33	8	$\frac{7}{16}$	$\frac{5}{16}$	$\frac{3}{16}$	$7\frac{7}{8}$	$\frac{15}{16}$	$2\frac{1}{4}$	$\frac{1}{4}$	. . . .		$5\frac{1}{2}$	$12\frac{5}{8}$
	8 (8)	28	$6\frac{1}{2}$	$\frac{7}{16}$	$\frac{5}{16}$	$\frac{1}{8}$	$6\frac{3}{8}$	$\frac{13}{16}$	$2\frac{1}{4}$	$\frac{3}{16}$	. . . .		$3\frac{1}{2}$	$10\frac{1}{2}$
Beams	24	120.0	8	$1\ \frac{1}{8}$	$\frac{13}{16}$	$\frac{7}{16}$	$20\frac{1}{8}$	$1\frac{15}{16}$	$3\frac{1}{4}$	$\frac{1}{2}$	$1\ \frac{1}{8}$	1	4	
	20	85.0	7	$\frac{15}{16}$	$\frac{11}{16}$	$\frac{5}{16}$	$16\frac{1}{2}$	$1\ \frac{3}{4}$	$3\frac{1}{4}$	$\frac{3}{8}$	$\frac{7}{8}$	1	4	
	18	70.0	$6\frac{1}{4}$	$\frac{11}{16}$	$\frac{3}{4}$	$\frac{3}{8}$	$15\frac{1}{4}$	$1\ \frac{3}{8}$	$2\frac{3}{4}$	$\frac{7}{16}$	$\frac{11}{16}$	$\frac{7}{8}$	$3\frac{1}{2}$	
	15	50.0	$5\frac{5}{8}$	$\frac{5}{8}$	$\frac{9}{16}$	$\frac{5}{16}$	$12\frac{1}{2}$	$1\ \frac{1}{4}$	$2\frac{3}{4}$	$\frac{3}{8}$	$\frac{9}{16}$	$\frac{3}{4}$	$3\frac{1}{2}$	
	12	31.8	5	$\frac{9}{16}$	$\frac{3}{8}$	$\frac{3}{16}$	$9\frac{3}{4}$	$1\ \frac{1}{8}$	$2\frac{1}{2}$	$\frac{1}{4}$	$\frac{1}{2}$	$\frac{3}{4}$	3	
	10	35.0	5	$\frac{1}{2}$	$\frac{5}{8}$	$\frac{5}{16}$	8	1	$2\frac{1}{2}$	$\frac{3}{8}$	$\frac{1}{2}$	$\frac{3}{4}$	$2\frac{3}{4}$	
	8	23.0	$4\frac{1}{8}$	$\frac{7}{16}$	$\frac{7}{16}$	$\frac{1}{4}$	$6\frac{1}{4}$	$\frac{7}{8}$	$2\frac{1}{4}$	$\frac{5}{16}$	$\frac{7}{16}$	$\frac{3}{4}$	$2\frac{1}{4}$	
	7	20.0	$3\frac{7}{8}$	$\frac{3}{8}$	$\frac{7}{16}$	$\frac{1}{4}$	$5\frac{3}{8}$	$\frac{13}{16}$	2	$\frac{5}{16}$	$\frac{3}{8}$	$\frac{5}{8}$	$2\frac{1}{4}$	
	6	17.25	$3\frac{5}{8}$	$\frac{3}{8}$	$\frac{1}{2}$	$\frac{1}{4}$	$4\frac{1}{2}$	$\frac{3}{4}$	2	$\frac{5}{16}$	$\frac{3}{8}$	$\frac{5}{8}$	2	
	5	10.0	3	$\frac{5}{16}$	$\frac{1}{4}$	$\frac{1}{8}$	$3\frac{5}{8}$	$\frac{11}{16}$	2	$\frac{3}{16}$	$\frac{5}{16}$	$\frac{1}{2}$	$1\frac{3}{4}$	
	4	9.5	$2\frac{3}{4}$	$\frac{5}{16}$	$\frac{5}{16}$	$\frac{3}{16}$	$2\frac{3}{4}$	$\frac{5}{8}$	2	$\frac{1}{4}$	$\frac{5}{16}$	$\frac{1}{2}$	$1\frac{1}{2}$	
	3	7.5	$2\frac{1}{2}$	$\frac{1}{4}$	$\frac{3}{8}$	$\frac{3}{16}$	$1\frac{7}{8}$	$\frac{9}{16}$	...	$\frac{1}{4}$	$\frac{1}{4}$	$\frac{3}{8}$	$1\frac{1}{2}$	

CHANNEL

WF SHAPE

BEAM

*These values are approximate. For exact sizes and other sizes, check the Manual of Steel Construction, American Institute of Steel Construction, Inc.

†Nominal depth; ( ) indicates actual depth.

Dimensions in decimal parts of an inch

No. of gage	American or Brown and Sharpe[a]	Washburn & Moen or American Steel & Wire Co.[b]	Birmingham or Stubs iron wire[c]	Music wire[d]	Imperial wire gage[e]	U.S. Std. for plate[f]
0000000	. . . . .	0.4900	. . . .	. . . . .	0.5000	0.5000
000000	0.5800	0.4615	. . . . .	0.004	0.4640	0.4688
00000	0.5165	0.4305	0.500	0.005	0.4320	0.4375
0000	0.4600	0.3938	0.454	0.006	0.4000	0.4063
000	0.4096	0.3625	0.425	0.007	0.3720	0.3750
00	0.3648	0.3310	0.380	0.008	0.3480	0.3438
0	0.3249	0.3065	0.340	0.009	0.3240	0.3125
1	0.2893	0.2830	0.300	0.010	0.3000	0.2813
2	0.2576	0.2625	0.284	0.011	0.2760	0.2656
3	0.2294	0.2437	0.259	0.012	0.2520	0.2500
4	0.2043	0.2253	0.238	0.013	0.2320	0.2344
5	0.1819	0.2070	0.220	0.014	0.2120	0.2188
6	0.1620	0.1920	0.203	0.016	0.1920	0.2031
7	0 1443	0.1770	0.180	0.018	0.1760	0.1875
8	0.1285	0.1620	0.165	0.020	0.1600	0.1719
9	0.1144	0.1483	0.148	0.022	0.1440	0.1563
10	0.1019	0.1350	0.134	0.024	0.1280	0.1406
11	0.0907	0.1205	0.120	0.026	0.1160	0.1250
12	0.0808	0.1055	0.109	0.029	0.1040	0.1094
13	0.0720	0.0915	0.095	0.031	0.0920	0.0938
14	0.0641	0.0800	0.083	0.033	0.0800	0.0781
15	0.0571	0.0720	0.072	0.035	0.0720	0.0703
16	0.0508	0.0625	0.065	0.037	0.0640	0.0625
17	0.0453	0.0540	0.058	0.039	0.0560	0.0563
18	0.0403	0.0475	0.049	0.041	0.0480	0.0500
19	0.0359	0.0410	0.042	0.043	0.0400	0.0438
20	0.0320	0.0348	0.035	0.045	0.0360	0.0375
21	0.0285	0.0317	0.032	0.047	0.0320	0.0344
22	0.0253	0.0286	0.028	0.049	0.0280	0.0313
23	0.0226	0.0258	0.025	0.051	0.0240	0.0281
24	0.0201	0.0230	0.022	0.055	0.0220	0.0250
25	0.0179	0.0204	0.020	0.059	0.0200	0.0219
26	0.0159	0.0181	0.018	0.063	0.0180	0.0188
27	0.0142	0.0173	0.016	0.067	0.0164	0.0172
28	0.0126	0.0162	0.014	0.071	0.0148	0.0156
29	0.0113	0.0150	0.013	0.075	0.0136	0.0141
30	0.0100	0.0140	0.012	0.080	0.0124	0.0125
31	0.0089	0.0132	0.010	0.085	0.0116	0.0109
32	0.0080	0.0128	0.009	0.090	0.0108	0.0102
33	0.0071	0.0118	0.008	0.095	0.0100	0.0094
34	0.0063	0.0104	0.007	0.100	0.0092	0.0086
35	0.0056	0.0095	0.005	0.106	0.0084	0.0078
36	0.0050	0.0090	0.004	0.112	0.0076	0.0070
37	0.0045	0.0085	. . . . .	0.118	0.0068	0.0066
38	0.0040	0.0080	. . . . .	0.124	0.0060	0.0063
39	0.0035	0.0075	. . . . .	0.130	0.0052	
40	0.0031	0.0070	. . . . .	0.138	0.0048	

[a] Recognized standard in the United States for wire and sheet metal of copper and other metals except steel and iron.
[b] Recognized standard for steel and iron wire. Called the "U.S. steel wire gage."
[c] Formerly much used, now nearly obsolete.
[d] American Steel & Wire Co.'s music wire gage. Recommended by U.S. Bureau of Standards.
[e] Official British Standard.
[f] Legalized U.S. Standard for iron and steel plate, although plate is now always specified by its thickness in decimals of an inch.

Preferred thicknesses for uncoated thin flat metals (under 0.250 in.): ANSI B32.1—1952 (R 1977) gives recommended sizes for sheets. See B32.3—1977 for preferred metric sizes for flat metal products.

APPENDIX 42   AMERICAN STANDARD GRAPHIC SYMBOLS FOR PIPING AND HEATING*

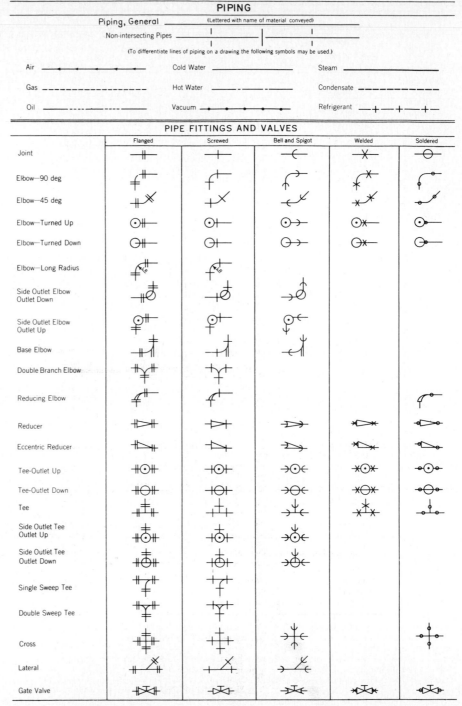

PIPING

Piping, General _____ (Lettered with name of material conveyed)

Non-intersecting Pipes

(To differentiate lines of piping on a drawing the following symbols may be used.)

Air	Cold Water	Steam
Gas	Hot Water	Condensate
Oil	Vacuum	Refrigerant

PIPE FITTINGS AND VALVES

	Flanged	Screwed	Bell and Spigot	Welded	Soldered
Joint					
Elbow—90 deg					
Elbow—45 deg					
Elbow—Turned Up					
Elbow—Turned Down					
Elbow—Long Radius					
Side Outlet Elbow Outlet Down					
Side Outlet Elbow Outlet Up					
Base Elbow					
Double Branch Elbow					
Reducing Elbow					
Reducer					
Eccentric Reducer					
Tee-Outlet Up					
Tee-Outlet Down					
Tee					
Side Outlet Tee Outlet Up					
Side Outlet Tee Outlet Down					
Single Sweep Tee					
Double Sweep Tee					
Cross					
Lateral					
Gate Valve					

*Z32.2.3—1949 (reaffirmed 1953).

PIPING

	Flanged	Screwed	Bell and Spigot	Welded	Soldered
Globe Valve					
Angle Globe Valve					
Angle Gate Valve					
Check Valve					
Angle Check Valve					
Stop Cock					
Safety Valve					
Quick Opening Valve					
Float Operating Valve					
Motor Operated Gate Valve					
Motor Operated Globe Valve					
Expansion Joint Flanged					
Reducing Flange					
Union	(See Joint)				
Sleeve					
Bushing					

HEATING AND VENTILATING

Lock and Shield Valve	Tube Radiator	(Plan) (Elev.)	Exhaust Duct, Section
Reducing Valve	Wall Radiator	(Plan) (Elev.)	Butterfly Damper — (Plan or Elev.) (Elev. or Plan)
Diaphragm Valve	Pipe Coil	(Plan) (Elev.)	Deflecting Damper Rectangular Pipe
Thermostat	Indirect Radiator	(Plan) (Elev.)	Vanes
Radiator Trap (Plan) (Elev.)	Supply Duct, Section		Air Supply Outlet
			Exhaust Inlet

HEAT-POWER APPARATUS

Flue Gas Reheater (Intermediate Superheater)	Steam Turbine	Automatic By-pass Valve
Steam Generator (Boiler)	Condensing Turbine	Automatic Valve Operated by Governor
Live Steam Superheater	Open Tank	Pumps Air Service Boiler Feed Condensate Circulating Water Reciprocating
Feed Heater With Air Outlet	Closed Tank	
Surface Condenser	Automatic Reducing Valve	Dynamic Pump (Air Ejector)

Welding and allied processes	Letter designation	Welding and allied processes	Letter designation
adhesive bonding	ABD	resistance welding	RW
arc welding	AW	flash welding	FW
atomic hydrogen welding	AHW	high frequency resistance welding	HFRW
bare metal arc welding	BMAW	percussion welding	PEW
carbon arc welding	CAW	projection welding	RPW
electrogas welding*	EGW	resistance seam welding	RSEW
flux cored arc welding	FCAW	resistance spot welding	RSW
gas metal arc welding	GMAW	upset welding	UW
gas tungsten arc welding	GTAW	soldering	S
plasma arc welding	PAW	dip soldering	DS
shielded metal arc welding	SMAW	furnace soldering	FS
stud arc welding	SW	induction soldering	IS
submerged arc welding	SAW	infrared soldering	IRS
arc welding process variations		iron soldering	INS
gas carbon arc welding	CAW-G	resistance soldering	RS
gas metal arc welding—		torch soldering	TS
pulsed arc	GMAW-P	wave soldering	WS
gas metal arc welding—		solid state welding	SSW
short circuiting arc	GMAW-S	cold welding	CW
gas tungsten arc welding—		diffusion welding	DFW
pulsed arc	GTAW-P	explosion welding	EXW
series submerged arc welding	SAW-S	forge welding	FOW
shielded carbon arc welding	CAW-S	friction welding	FRW
twin carbon arc welding	CAW-T	hot pressure welding	HPW
brazing	B	roll welding	ROW
arc brazing	AB	ultrasonic welding	USW
block brazing	BB	thermal cutting	TC
diffusion brazing	DFB	arc cutting	AC
dip brazing	DB	air carbon arc cutting	AAC
flow brazing	FLB	carbon arc cutting	CAC
furnace brazing	FB	gas metal arc cutting	GMAC
induction brazing	IB	gas tungsten arc cutting	GTAC
infrared brazing	IRB	metal arc cutting	MAC
resistance brazing	RB	plasma arc cutting	PAC
torch brazing	TB	shielded metal arc cutting	SMAC
twin carbon arc brazing	TCAB	electron beam cutting	EBC
other welding processes		laser beam cutting	LBC
electron beam welding	EBW	oxygen cutting	OC
electroslag welding	ESW	chemical flux cutting	FOC
flow welding	FLOW	metal powder cutting	POC
induction welding	IW	oxyfuel gas cutting	OFC
laser beam welding	LBW	oxyacetylene cutting	OFC-A
thermit welding	TW	oxyhydrogen cutting	OFC-H
oxyfuel gas welding	OFW	oxynatural gas cutting	OFC-N
air acetylene welding	AAW	oxypropane cutting	OFC-P
oxyacetylene welding	OAW	oxygen arc cutting	AOC
oxyhydrogen welding	OHW	oxygen lance cutting	LOC
pressure gas welding	PGW	thermal spraying	THSP
		electric arc spraying	EASP
		flame spraying	FLSP
		plasma spraying	PSP

*In the 1976 edition of AWS A2.4, electrogas welding was designated FCAW-EG and GMAW-EG depending on the manner of application.

1. Qualifying Symbols

1.1 Adjustability
Variability

1.2 Special-Property Indicators

t°   x   τ   ∫

1.3 Radiation Indicators

1.4 Physical State Recognition Symbols

1.5 Test-Point Recognition Symbol

1.6 Polarity Markings

+   —

1.7 Direction of Flow of Power, Signal, or Information

1.8 Kind of Current

1.9 Connection Symbols

1.10 Envelope
Enclosure

1.11 Shield
Shielding

1.12 Special Connector or Cable Indicator

1.13 Electret

2. Fundamental Items

2.1 Resistor

2.2 Capacitor

2.3 Antenna

2.4 Attenuator

2.5 Battery

2.6 Delay Function
Delay Line
Slow-Wave Structure

2.7 Oscillator
Generalized Alternating-Current Source

2.8 Permanent Magnet

2.9 Pickup
Head

2.10 Piezoelectric Crystal Unit

2.11 Primary Detector
Measuring Transducer

2.12 Squib, Electrical

2.13 Thermocouple

2.14 Thermal Element
Thermomechanical Transducer

2.15 Spark gap
Igniter gap

2.16 Continuous Loop Fire Detector
(temperature sensor)

2.17 Ignitor Plug

3. Transmission Path

3.1 Transmission Path
Conductor
Cable
Wiring

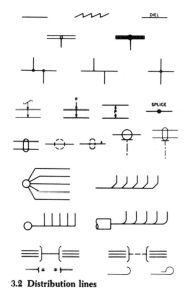

3.2 Distribution lines
Transmission lines

F   S   T   V

3.3 Alternative or Conditioned Wiring

3.4 Associated or Future

3.5 Intentional Isolation of Direct-Current Path in Coaxial or Waveguide Applications

3.6 Waveguide

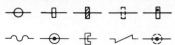

3.7 Strip-Type Transmission Line

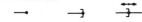

3.8 Termination

3.9 Circuit Return

3.10 Pressure-Tight Bulkhead Cable Gland
Cable Sealing End

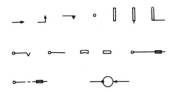

4. Contacts, Switches, Contactors, and Relays

4.1 Switching Function

4.2 Electrical Contact

4.3 Basic Contact Assemblies

4.4 Magnetic Blowout Coil

4.5 Operating Coil
Relay Coil

4.6 Switch

4.7 Pushbutton, Momentary or Spring-Return

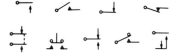

4.8 Two-Circuit, Maintained or Not Spring-Return

4.9 Nonlocking Switch, Momentary or Spring-Return

4.10 Locking Switch

4.11 Combination Locking and Non-locking Switch

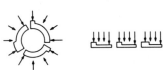

4.12 Key-Type Switch
Lever Switch

4.13 Selector or Multiposition Switch

4.14 Limit Switch
Sensitive Switch

4.15 Safety Interlock

4.16 Switches with Time-Delay Feature

4.17 Flow-Actuated Switch

4.18 Liquid-Level-Actuated Switch

4.19 Pressure- or Vacuum-Actuated Switch

4.20 Temperature-Actuated Switch

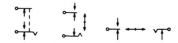

4.21 Thermostat

4.22 Flasher
Self-interrupting switch

4.23 Foot-Operated Switch
Foot Switch

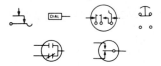

4.24 Switch Operated by Shaft Rotation and Responsive to Speed or Direction

4.25 Switches with Specific Features

4.26 Telegraph Key

4.27 Governor
Speed Regulator

4.28 Vibrator
Interrupter

4.29 Contactor

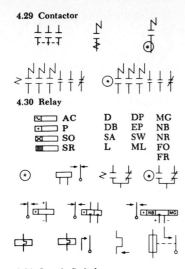

4.30 Relay

AC	D	DP	MG
P	DB	EP	NB
SO	SA	SW	NR
SR	L	ML	FO
			FR

4.31 Inertia Switch

4.32 Mercury Switch

4.33 Aneroid Capsule

5. Terminals and Connectors

5.1 Terminals

5.2 Cable Termination

5.3 Connector
Disconnecting Device

5.4 Connectors of the Type Commonly Used for Power-Supply Purposes

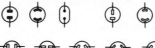

5.5 Test Blocks

5.6 Coaxial Connector

5.7 Waveguide Flanges
Waveguide junction

6. Transformers, Inductors, and Windings

6.1 Core

6.2 Inductor
Winding
Reactor
Radio frequency coil
Telephone retardation coil

6.3 Transductor

6.4 Transformer
Telephone induction coil
Telephone repeating coil

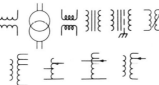

6.5 Linear Coupler

7. Electron Tubes and Related Devices

7.1 Electron Tube

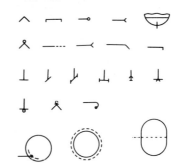

7.2 General Notes

7.3 Typical Applications

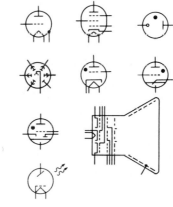

7.4 Solion
Ion-Diffusion Device

7.5 Coulomb Accumulator
Electrochemical Step-
Function Device

7.6 Conductivity cell

713

7.7 Nuclear-Radiation Detector
Ionization Chamber
Proportional Counter Tube
Geiger-Müller Counter Tube

8. Semiconductor Devices

8.1 Semiconductor Device
Transistor
Diode

8.2 Element Symbols

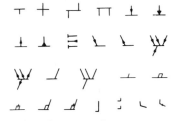

8.3 Special Property Indicators

8.4 Rules for Drawing Style 1 Symbols

8.5 Typical Applications: Two-Terminal Devices

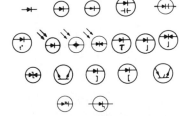

8.6 Typical Applications: Three- (or More) Terminal Devices

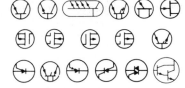

8.7 Photosensitive Cell

8.8 Semiconductor Thermocouple

8.9 Hall Element
Hall Generator

8.10 Photon-coupled isolator

8.11 Solid-state-thyratron

9. Circuit Protectors

9.1 Fuse

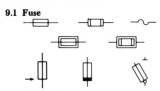

9.2 Current Arrester

9.3 Lightning Arrester
Arrester
Gap

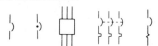

9.4 Circuit Breaker

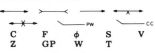

9.5 Protective Relay

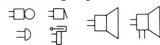

C   F   φ   S   V
Z   GP   W   T

10. Acoustic Devices

10.1 Audible-Signaling Device

10.2 Microphone

10.3 Handset
Operator's Set

10.4 Telephone Receiver
Earphone
Hearing-Aid Receivers

11. Lamps and Visual-Signaling Devices

11.1 Lamp

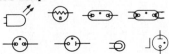

11.2 Visual-Signaling Device

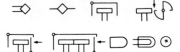

12. Readout Devices

12.1 Meter
Instrument

A	DB	I	OP	RF	VA
AH	DBM	INT	OSCG	SY	VAR
C	DM	μA	PH	TLM	VARH
CMA	DTR	UA	PI	t°	VI
CMC	MA	PF	THC	VU	
CMV	G	NM	RD	TT	W
CRO	GD	OHM	REC	V	WH

12.2 Electromagnetically Operated
Counter
Message Register

13. Rotating Machinery

13.1 Rotating Machine

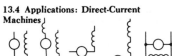

13.2 Field, Generator or Motor

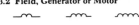

13.3 Winding Connection Symbols

13.4 Applications: Direct-Current Machines

13.5 Applications: Alternating-Current Machines

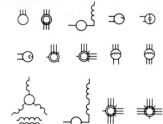

13.6 Applications: Alternating-Current Machines with Direct-Current Field Excitation

13.7 Applications: Alternating- and Direct-Current Composite

13.8 Synchro

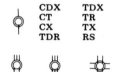

CDX	TDX
CT	TR
CX	TX
TDR	RS

14. Mechanical Functions

**14.1 Mechanical Connection
Mechanical Interlock**

14.2 Mechanical Motion

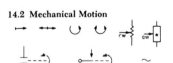

**14.3 Clutch
Brake**

14.4 Manual Control

15. Commonly Used in Connection with VHF, UHF, SHF Circuits

15.1 Discontinuity

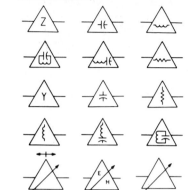

15.2 Coupling

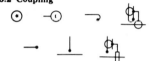

15.3 Directional Coupler

**15.4 Hybrid
Directionally Selective
Transmission Devices**

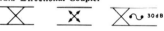

15.5 Mode Transducer

15.6 Mode Suppression

15.7 Rotary Joint

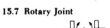

15.8 Non-reciprocal devices

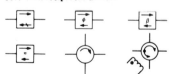

**15.9 Resonator
Tuned Cavity**

15.10 Resonator (Cavity Type) Tube

15.11 Magnetron

15.12 Velocity-Modulation (Velocity-Variation) Tube

15.13 Transmit-Receive (TR) Tube

15.14 Traveling-Wave-Tube

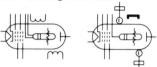

15.15 Balun

15.16 Filter

15.17 Phase shifter

15.18 Ferrite bead rings

15.19 Line stretcher

16. Composite Assemblies

**16.1 Circuit assembly
Circuit subassembly
Circuit element**

EQ	FL-BP	RC	TPR
FAX	FL-HP	RU	TTY
FL	FL-LP	DIAL	CLK
FL-BE	PS	TEL	IND
ST-INV			

16.2 Amplifier

BDG	EXP	PRE
BST	LIM	PWR
CMP	MON	TRQ
DC	PGM	

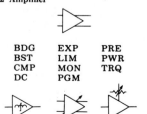

16.3 Rectifier

16.4 Repeater

16.5 Network

16.6 Phase Shifter
Phase-Changing Network

16.7 Chopper

16.8 Diode-type ring demodulator
Diode-type ring modulator

16.9 Gyro
Gyroscope
Gyrocompass

16.10 Position Indicator

16.11 Position Transmitter

16.12 Fire Extinguisher Actuator
Head

17. Analog Functions

17.1 Operational Amplifier

17.2 Summing Amplifier

17.3 Integrator

17.4 Electronic Multiplier

17.5 Electronic Divider

17.6 Electronic Function Generator

17.7 Generalized Integrator

17.8 Positional Servo-mechanism

17.9 Function Potentiometer

18. Digital Logic Functions

18.1 Digital Logic Functions
(See cross references)

19. Special Purpose Maintenance
Diagrams

19.1 Data flow code signals

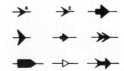

19.2 Functional Circuits

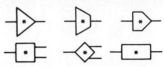

20. System Diagrams, Maps and
Charts

20.1 Radio station

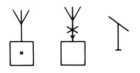

20.2 Space station

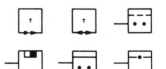

20.3 Exchange equipment

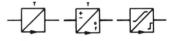

20.4 Telegraph repeater

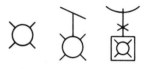

20.5 Telegraph equipment

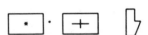

20.6 Telephone set

21. System Diagrams, Maps and
Charts

21.1 Generating station

21.2 Hydroelectric generating station

21.3 Thermoelectric generating station

21.4 Prime mover

21.5 Substation

22. Class Designation Letters

A	DS	J	PU	TP
AR	E	K	Q	TR
AT	EQ	L	R	U
B	F	LS	RE	V
BT	FL	M	RT	VR
C	G	MG	RV	W
CB	H	MK	S	WT
CP	HP	MP	SQ	X
CR	HR	MT	SR	Y
D	HS	N	T	Z
DC	HT	P	TB	
DL	HY	PS	TC	

APPENDIX 45 WEIGHTS OF MATERIALS

Metals	lb/cu in.	Wood	lb/cu in.
Aluminum alloy, cast	0.099	Ash	0.024
Aluminum, cast	0.094	Balsa	0.0058
Aluminum, wrought	0.097	Cedar	0.017
Babbitt metal	0.267	Cork	0.009
Brass, cast or rolled	0.303–0.313	Hickory	0.0295
Brass, drawn	0.323	Maple	0.025
Bronze, aluminum cast	0.277	Oak (white)	0.028
Bronze, phosphor	0.315–0.321	Pine (white)	0.015
Chromium	0.256	Pine (yellow)	0.025
Copper, cast	0.311	Poplar	0.018
Copper, rolled, drawn or wire	0.322	Walnut (black)	0.023
Dowmetal *A*	0.065	**Miscellaneous materials**	**lb/ cu ft**
Duralumin	0.101	Asbestos	175
Gold	0.697	Bakelite	79.5
Iron, cast	0.260	Brick, common	112
Iron, wrought	0.283	Brick, fire	144
Lead	0.411	Celluloid	86.4
Magnesium	0.063	Earth, packed	100
Mercury	0.491	Fiber	89.9
Monel metal	0.323	Glass	163
Silver	0.379	Gravel	109
Steel, cast or rolled	0.274–0.281	Limestone	163
Steel, tool	0.272	Plexiglass	74.3
Tin	0.263	Sandstone	144
Zinc	0.258	Water	62.4

Abstracted from the SAE Aerospace-Automotive Drawing Standards (Section Z-1).

Word	Abbreviation	Word	Abbreviation
Abbreviate	ABBR	American Wire Gage	AWG
Absolute	ABS	Ammeter	AM
Accelerate	ACCEL	Amount	AMT
Acceleration due to gravity	G	Ampere	AMP
Access panel	AP	Ampere hour	AMP HR
Accessory	ACCESS.	Amplifier	AMPL
Actual	ACT.	Anneal	ANL
Adapter	ADPT	Antenna	ANT.
Addendum	ADD.	Apparatus	APP
Adjust	ADJ	Approved	APPD
Advance	ADV	Approximate	APPROX
After	AFT.	Arc weld	ARC/W
Aggregate	AGGR	Area	A
Aileron	AIL	Armature	ARM.
Air-break switch	ABS	Arrange	ARR.
Air-circuit breaker	ACB	Arrester	ARR.
Aircraft	ACFT	Asbestos	ASB
Airplane	APL	Assemble	ASSEM
Airtight	AT	Assembly	ASSY
Alarm	ALM	Atomic	AT
Allowance	ALLOW	Attach	ATT
Alloy	ALY	Audio-frequency	AF
Alteration	ALT	Automatic	AUTO
Alternate	ALT	Auto-transformer	AUTO TR
Alternating current	AC	Auxiliary	AUX
Alternator	ALT	Average	AVG
Altitude	ALT		
Aluminum	AL	Babbitt	BAB
American Standard	AMER STD	Back to back	B to B

Word	Abbreviation	Word	Abbreviation
Baffle	BAF	Cast (used with other materials)	C
Balance	BAL	Cast iron	CI
Ball bearing	BB	Cast-iron pipe	CIP
Base line	BL	Cast steel	CS
Base plate	BP	Casting	CSTG
Battery	BAT.	Castle nut	CAS NUT
Bearing	BRG	Cement	CEM
Bent	BT	Center	CTR
Between	BET.	Center line	CL
Between centers	BC	Center to center	C to C
Between perpendiculars	BP	Centering	CTR
Bevel	BEV	Centigrade	C
Bill of material	B/M	Centigram	CG
Birmingham Wire Gage	BWG	Centiliter	CL
Blank	BLK	Centimeter	CM
Block	BLK	Centrifugal	CENT.
Blueprint	BP	Centrifugal force	CF
Bolt circle	BC	Ceramic	CER
Bottom	BOT	Chain	CH
Bottom chord	BC	Chamfer	CHAM
Brake	BK	Change	CHG
Brass	BRS	Change notice	CN
Brazing	BRZG	Change order	CO
Break	BRK	Channel	CHAN
Breaker	BKR	Check	CHK
Brinnell hardness	BH	Check valve	CV
British Standard	BR STD	Chemical	CHEM
British thermal units	BTU	Chord	CHD
Broach	BRO	Chrome molybdenum	CR MOLY
Bronze	BRZ	Chromium plate	CR PL
Brown & Sharp	B&S	Chrome vanadium	CR VAN
Brush	BR	Circle	CIR
Burnish	BNH	Circuit	CKT
Bushing	BUSH.	Circular	CIR
Bypass	BYP	Circular pitch	CP
		Circulate	CIRC
Cadmium plate	CD PL	Circumference	CIRC
Calculate	CALC	Clamp	CLP
Calibrate	CAL	Class	CL
Calking	CLKG	Clear	CLR
Capacitor	CAP	Clearance	CL
Capacity	CAP	Clockwise	CW
Cap screw	CAP. SCR	Closing	CL
Carburize	CARB	Clutch	CL
Caseharden	CH	Coated	CTD
Casing	CSG	Coaxial	COAX

Word	Abbreviation	Word	Abbreviation
Coefficient	COEF	Counterbore	CBORE
Cold drawn	CD	Counterdrill	CDRILL
Cold-drawn steel	CDS	Counterpunch	CPUNCH
Cold rolled	CR	Countersink	CSK
Cold-rolled steel	CRS	Countersink other side	CSK-O
Column	COL	Coupling	CPLG
Combination	COMB.	Cover	COV
Combustion	COMB	Crank	CRK
Communication	COMM	Cross connection	XCONN
Commutator	COMM	Cross section	XSECT
Complete	COMPL	Cubic	CU
Composite	CX	Current	CUR
Composition	COMP	Cyanide	CYN
Compressor	COMPR	Cycle	CY
Concentric	CONC	Cycles per minute	CPM
Concrete	CONC	Cycles per second	CPS
Condition	COND	Cylinder	CYL
Conduct	COND		
Conductor	COND	Decibel	DB
Conduit	CND	Decimal	DEC
Connect	CONN	Dedendum	DED
Constant	CONST	Deep drawn	DD
Contact	CONT	Deflect	DEFL
Container	CNTR	Degree	(°) DEG
Continue	CONT	Density	D
Continuous wave	CW	Describe	DESCR
Contract	CONT	Design	DSGN
Contractor	CONTR	Designation	DESIG
Control	CONT	Detail	DET
Control relay	CR	Detector	DET
Control switch	CS	Detonator	DET
Controller	CONT	Develop	DEV
Convert	CONV	Diagonal	DIAG
Conveyor	CNVR	Diagram	DIAG
Cooled	CLD	Diameter	DIA
Copper oxide	CUO	Diametral pitch	DP
Copper plate	COP. PL	Diaphragm	DIAPH
Cord	CD	Differential	DIFF
Correct	CORR	Dimension	DIM.
Corrosion resistant	CRE	Diode	DIO
Corrosion-resistant steel	CRES	Direct current	DC
Corrugate	CORR	Directional	DIR
Cotter	COT	Discharge	DISCH
Counter	CTR	Disconnect	DISC.
Counterclockwise	CCW	Distance	DIST
Counterbalance	CBAL	Distribute	DISTR

Word	Abbreviation	Word	Abbreviation
Ditto	DO.	Fahrenheit	F
Double	DBL	Fairing	FAIR.
Dovetail	DVTL	Farad	F
Dowel	DWL	Far side	FS
Down	DN	Feed	FD
Drafting	DFTG	Feeder	FDR
Draftsman	DFTSMN	Feet	(′) FT
Drain	DR	Feet per minute	FPM
Drawing	DWG	Feet per second	FPS
Drawing list	DL	Female	FEM
Drill	DR	Fiber	FBR
Drill rod	DR	Field	FLD
Drive	DR	Figure	FIG.
Drive fit	DF	Filament	FIL
Drop	D	Fillet	FIL
Drop forge	DF	Filling	FILL.
Duplex	DX	Fillister	FIL
Duplicate	DUP	Filter	FLT
Dynamic	DYN	Finish	FIN.
Dynamo	DYN	Finish all over	FAO
		Fireproof	FPRF
Each	EA	Fitting	FTG
Eccentric	ECC	Fixture	FIX.
Effective	EFF	Flange	FLG
Electric	ELEC	Flashing	FL
Elevation	EL	Flat	F
Enclose	ENCL	Flat head	FH
End to end	E to E	Flexible	FLEX.
Envelope	ENV	Float	FLT
Equal	EQ	Floor	FL
Equation	EQ	Fluid	FL
Equipment	EQUIP.	Fluorescent	FLUOR
Equivalent	EQUIV	Flush	FL
Estimate	EST	Focus	FOC
Evaporate	EVAP	Foot	(′) FT
Excavate	EXC	Force	F
Exhaust	EXH	Forging	FORG
Expand	EXP	Forward	FWD
Exterior	EXT	Foundation	FDN
External	EXT	Foundry	FDRY
Extra heavy	X HVY	Fractional	FRAC
Extra strong	X STR	Frame	FR
Extrude	EXTR	Freezing point	FP
		Frequency	FREQ
Fabricate	FAB	Frequency, high	HF
Face to face	F to F	Frequency, low	LF

Word	Abbreviation	Word	Abbreviation
Frequency, medium	MF	Harden	HDN
Frequency modulation	FM	Hardware	HDW
Frequency, super high	SHF	Head	HD
Frequency, ultra high	UHF	Headless	HDLS
Frequency, very high	VHF	Heat	HT
Frequency, very low	VLF	Heat treat	HT TR
Friction horsepower	FHP	Heater	HTR
From below	FR BEL	Heavy	HVY
Front	FR	Height	HGT
Fuel	F	Henry	H
Furnish	FURN	Hexagon	HEX
Fusible	FSBL	High	H
Fusion point	FNP	High frequency	HF
		High point	H PT
Gage or Gauge	GA	High pressure	HP
Gallon	GAL	High speed	HS
Galvanize	GALV	High-speed steel	HSS
Galvanized iron	GI	High tension	HT
Galvanized steel	GS	High voltage	HV
Galvanized steel wire rope	GSWR	Highway	HWY
Gas	G	Holder	HLR
Gasket	GSKT	Hollow	HOL
Gasoline	GASO	Horizontal	HOR
General	GEN	Horsepower	HP
Glaze	GL	Hot rolled	HR
Government	GOVT	Hot-rolled steel	HRS
Government furnished equipment	GFE	Hour	HR
Governor	GOV	Hydraulic	HYD
Grade	GR		
Graduation	GRAD	Identify	IDENT
Gram	G	Ignition	IGN
Graphic	GRAPH.	Illuminate	ILLUM
Graphite	GPH	Illustrate	ILLUS
Grating	GRTG	Impact	IMP
Gravity	G	Impedance	IMP.
Grid	G	Inch	(") IN.
Grind	GRD	Inches per second	IPS
Groove	GRV	Include	INCL
Ground	GRD	Increase	INCR
		Indicate	IND
Half hard	½H	Inductance or induction	IND
Half round	½RD	Industrial	IND
Handle	HDL	Information	INFO
Hanger	HGR	Injection	INJ
Hard	H	Inlet	IN
Hard-drawn	HD	Inspect	INSP

Word	Abbreviation	Word	Abbreviation
Install	INSTL	Lacquer	LAQ
Instantaneous	INST	Laminate	LAM
Instruct	INST	Lateral	LAT
Instrument	INST	Lead-coated metal	LCM
Insulate	INS	Lead covered	LC
Interchangeable	INTCHG	Leading edge	LE
Interior	INT	Left	L
Interlock	INTLK	Left hand	LH
Intermediate	INTER	Length	LG
Intermittent	INTMT	Length over all	LOA
Internal	INT	Letter	LTR
Interrupt	INTER	Light	LT
Interrupted continuous wave	ICW	Limit	LIM
Interruptions per minute	IPM	Line	L
Interruptions per second	IPS	Linear	LIN
Intersect	INT	Link	LK
Inverse	INV	Liquid	LIQ
Invert	INV	Liter	L
Iron	I	Locate	LOC
Iron-pipe size	IPS	Long	LG
Irregular	IRREG	Longitude	LONG.
Issue	ISS	Low explosive	LE
		Low frequency	LF
Jack	J	Low pressure	LP
Job order	JO	Low tension	LT
Joint	JT	Low voltage	LV
Junction	JCT	Low speed	LS
		Low torque	LT
Kelvin	K	Lubricate	LUB
Key	K	Lubricating oil	LO
Keyseat	KST	Lumen	L
Keyway	KWY	Lumens per watt	LPW
Kilo	K		
Kilocycle	KC	Machine	MACH
Kilocycles per second	KC	Magnet	MAG
Kilogram	KG	Main	MN
Kiloliter	KL	Male and female	M&F
Kilometer	KM	Malleable	MALL
Kilovolt	KV	Malleable iron	MI
Kilovolt-ampere	KVA	Manual	MAN.
Kilovolt-ampere hour	KVAH	Manufacture	MFR
Kilowatt	KW	Manufactured	MFD
Kilowatt-hour	KWH	Manufacturing	MFG
Kip (1000 lb)	K	Material	MATL
Knots	KN	Material list	ML
		Maximum	MAX
Laboratory	LAB		

Word	Abbreviation	Word	Abbreviation
Maximum working pressure	MWP	Mounted	MTD
Mean effective pressure	MEP	Mounting	MTG
Mechanical	MECH	Multiple	MULT
Mechanism	MECH	Multiple contact	MC
Medium	MED		
Mega	M	National	NATL
Megacycles	MC	Natural	NAT
Megawatt	MW	Near face	NF
Megohm	MEG	Near side	NS
Melting point	MP	Negative	NEG
Metal	MET.	Network	NET
Meter (instrument or measure		Neutral	NEUT
of length)	M	Nickel-silver	NI-SIL
Micro	μ or U	Nipple	NIP.
Microampere	μA or UA	Nominal	NOM
Microfarad	μF or UF	Normal	NOR
Microhenry	μH or UH	Normally closed	NC
Micro-inch-root-mean square	μ-IN-RMS or	Normally open	NO
	U-IN-RMS	Not to scale	NTS
Micrometer	MIC	Number	NO.
Micron	μ or U		
Microvolt	μV or UV	Obsolete	OBS
Microwatt	μW or UW	Octagon	OCT
Miles	MI	Ohm	Ω
Miles per gallon	MPG	Oil-circuit breaker	OCB
Miles per hour	MPH	Oil insulated	OI
Milli	M	Oil switch	OS
Milliampere	MA.	On center	OC
Milligram	MG	One pole	1 P
Millihenry	MH	Opening	OPNG
Millimeter	MM	Operate	OPR
Milliseconds	MS	Opposite	OPP
Millivolt	MV	Optical	OPT
Milliwatt	MW	Ordnance	ORD
Minimum	MIN	Orifice	ORF
Minute	(') MIN	Original	ORIG
Miscellaneous	MISC	Oscillate	OSC
Mixture	MIX.	Ounce	OZ
Model	MOD	Out to Out	O to O
Modify	MOD	Outlet	OUT.
Modulated continuous wave	MCW	Output	OUT.
Modulator	MOD	Outside diameter	OD
Molecular weight	MOL WT	Outside face	OF
Monument	MON	Outside radius	OR
Morse taper	MOR T	Over-all	OA
Motor	MOT	Overhead	OVHD

Word	Abbreviation	Word	Abbreviation
Overload	OVLD	Pounds per cubic foot	PCF
Overvoltage	OVV	Pounds per square foot	PSF
Oxidized	OXD	Pounds per square inch	PSI
		Pounds per square inch absolute	PSIA
Pack	PK	Power	PWR
Packing	PKG	Power amplifier	PA
Painted	PTD	Power directional relay	PDR
Pair	PR	Power factor	PF
Panel	PNL	Preamplifier	PREAMP
Parallel	PAR.	Precast	PRCST
Part	PT	Prefabricated	PREFAB
Pattern	PATT	Preferred	PFD
Perforate	PERF	Premolded	PRMLD
Permanent	PERM	Prepare	PREP
Permanent magnet	PM	Press	PRS
Perpendicular	PERP	Pressure	PRESS.
Phase	PH	Pressure angle	PA.
Phosphor bronze	PH BRZ	Primary	PRI
Photograph	PHOTO	Process	PROC
Physical	PHYS	Production	PROD
Piece	PC	Profile	PF
Piece mark	PC MK	Project	PROJ
Pierce	PRC	Punch	PCH
Pipe Tap	PT	Purchase	PUR
Pitch	P	Push-pull	P-P
Pitch circle	PC		
Pitch diameter	PD	Quadrant	QUAD
Plastic	PLSTC	Quality	QUAL
Plate	PL	Quantity	QTY
Plotting	PLOT.	Quart	QT
Pneumatic	PNEU	Quarter	QTR
Point	PT	Quarter hard	¼ H
Point of compound curve	PCC	Quarter round	¼ RD
Point of curve	PC	Quartz	QTZ
Point of intersection	PI		
Point of reverse curve	PRC	Radial	RAD
Point of switch	PS	Radio frequency	RF
Point of tangent	PT	Radius	R
Polar	POL	Reactive	REAC
Pole	P	Reactive kilovolt-ampere	KVAR
Polish	POL	Reactive volt-ampere	VAR
Port	P	Reactive voltmeter	RVM
Position	POS	Reactor	REAC
Positive	POS	Ream	RM
Potential	POT.	Reassemble	REASM
Pound	LB	Received	RECD

Word	Abbreviation	Word	Abbreviation
Receiver	REC	Saddle	SDL
Receptacle	RECP	Safe working pressure	SWP
Recriprocate	RECIP	Safety	SAF
Recirculate	RECIRC	Sand blast	SD BL
Reclosing	RECL	Saturate	SAT.
Record	REC	Schedule	SCH
Rectangle	RECT	Schematic	SCHEM
Rectifier	RECT	Scleroscope hardness	SH
Reduce	RED.	Screen	SCRN
Reference	REF	Screw	SCR
Reference line	REF L	Second	SEC
Regulator	REG	Section	SECT
Reinforce	REINF	Segment	SEG
Relay	REL	Select	SEL
Release	REL	Semifinished	SF
Relief	REL	Semifixed	SFXD
Remove	REM	Semisteel	SS
Repair	REP	Separate	SEP
Replace	REPL	Sequence	SEQ
Reproduce	REPRO	Serial	SER
Require	REQ	Series	SER
Required	REQD	Serrate	SERR
Resistance	RES	Service	SERV
Resistor	RES	Set screw	SS
Retainer	RET.	Shaft	SFT
Retard	RET.	Shield	SHLD
Return	RET.	Shipment	SHPT
Reverse	REV	Shop order	SO
Revise	REV	Short wave	SW
Revolution	REV	Shunt	SH
Revolutions per minute	RPM	Side	S
Revolutions per second	RPS	Signal	SIG
Rheostat	RHEO	Sink	SK
Right	R	Sketch	SK
Right hand	RH	Sleeve	SLV
Ring	R	Slide	SL
Rivet	RIV	Slotted	SLOT.
Rockwell hardness	RH	Small	SM
Roller bearing	RB	Smoke	SMK
Root diameter	RD	Smokeless	SMKLS
Root mean square	RMS	Socket	SOC
Rotary	ROT.	Soft	S
Rotate	ROT.	Solder	SLD
Rough	RGH	Solenoid	SOL
Round	RD	Sound	SND
Rubber	RUB.	South	S

Word	Abbreviation	Word	Abbreviation
Space	SP	Tangent	TAN.
Spare	SP	Taper	TPR
Speaker	SPKR	Technical	TECH
Special	SPL	Tee	T
Specific	SP	Teeth per inch	TPI
Specific gravity	SP GR	Television	TV
Specific heat	SP HT	Temperature	TEMP
Specification	SPEC	Template	TEMP
Speed	SP	Tensile strength	TS
Spherical	SPHER	Tension	TENS.
Spindle	SPDL	Terminal	TERM.
Split phase	SP PH	Terminal board	TB
Spot-faced	SF	That is	IE
Spring	SPG	Theoretical	THEO
Square	SQ	Thermal	THRM
Stabilize	STAB	Thermostat	THERMO
Stainless	STN	Thick	THK
Standard	STD	Thousand	M
Static pressure	SP	Thread	THD
Station	STA	Throttle	THROT
Stationary	STA	Through	THRU
Steel	STL	Time	T
Stiffener	STIFF.	Time delay	TD
Stock	STK	Time-delay closing	TDC
Storage	STG	Time-delay opening	TDO
Straight	STR	Tinned	TD
Strip	STR	Tobin bronze	TOB BRZ
Structural	STR	Toggle	TGL
Substitute	SUB	Tolerance	TOL
Suction	SUCT	Tongue and groove	T&G
Summary	SUM.	Tool steel	TS
Supervise	SUPV	Tooth	T
Supply	SUP	Total	TOT
Surface	SUR	Total indicator reading	TIR
Survey	SURV	Trace	TR
Switch	SW	Tracer	TCR
Symbol	SYM	Transfer	TRANS
Symmetrical	SYM	Transformer	TRANS
Synchronous	SYN	Transmission	XMSN
Synthetic	SYN	Transmitter	XMTR
System	SYS	Transmitting	XMTG
		Transportation	TRANS
Tabulate	TAB.	Transverse	TRANSV
Tachometer	TACH	Trimmer	TRIM.
Tandem	TDM	Triode	TRI

Word	Abbreviation	Word	Abbreviation
True air speed	TAS	Voltmeter	VM
Truss	T	Volts per mil	VPM
Tubing	TUB	Volume	VOL
Tuned radio frequency	TRF		
Turbine	TURB	Washer	WASH
Typical	TYP	Water	W
		Water line	WL
Ultimate	ULT	Watertight	WT
Ultra-high frequency	UHF	Watt	W
Under voltage	UV	Watt-hour	WHR
Unit	U	Watt-hour meter	WHM
United States Gage	USG	Wattmeter	WM
United States Standard	USS	Weight	WT
Universal	UNIV	West	W
		Wet bulb	WB
Vacuum	VAC	Width	W
Vacuum tube	VT	Wind	WD
Valve	V	Winding	WDG
Vapor proof	VAP PRF	Wire	W
Variable	VAR	With	W/
Variable-frequency oscillator	VFO	With equipment and spare parts	W/E&SP
Velocity	V	Without	W/O
Ventilate	VENT.	Without equipment and spare parts	W/O E&SP
Versed sine	VERS	Wood	WD
Versus	VS	Woodruff	WDF
Vertical	VERT	Working point	WP
Very-high frequency	VHF	Working pressure	WP
Very-low frequency	VLF	Wrought	WRT
Video-frequency	VDF	Wrought iron	WI
Vibrate	VIB		
Viscosity	VISC	Yard	YD
Vitreous	VIT	Year	YR
Voice frequency	VF	Yield point	YP
Volt	V	Yield strength	YS
Volt-ampere	VA		

Abbreviations for Colors

Amber	AMB	Green	GRN
Black	BLK	Orange	ORN
Blue	BLU	White	WHT
Brown	BRN	Yellow	YEL

Partial List of Chemical Symbols

Word	Abbreviation	Word	Abbreviation
Aluminum	Al	Molybdenum	Mo
Antimony (stibium)	Sb	Neon	Ne
Barium	Ba	Nickel	Ni
Beryllium	Be	Nitrogen	N
Bismuth	Bi	Oxygen	O
Boron	B	Phosphorus	P
Bromine	Br	Platinum	Pt
Cadmium	Cd	Potassium (kalium)	K
Calcium	Ca	Radium	Ra
Carbon	C	Rhodium	Rh
Chlorine	Cl	Ruthenium	Ru
Chromium	Cr	Selenium	Se
Cobalt	Co	Silicon	Si
Copper	Cu	Silver (argentum)	Ag
Fluorine	F	Sodium (natrium)	Na
Gold (aurium)	Au	Strontium	Sr
Helium	He	Sulfur	S
Hydrogen	H	Tantalum	Ta
Indium	In	Tellurium	Te
Iodine	I	Thallium	Tl
Iridium	Ir	Tin (stannum)	Sn
Iron (ferrum)	Fe	Titanium	Ti
Lead (plumbum)	Pb	Tungsten (wolframium)	W
Lithium	Li	Uranium	U
Magnesium	Mg	Vanadium	V
Manganese	Mn	Zinc	Zn
Mercury (hydrargyrum)	Hg	Zirconium	Zr

INDEX